T0145374

Lecture Notes in Computer Science　14073

Founding Editors

Gerhard Goos
Juris Hartmanis

Editorial Board Members

The series Lecture Notes in Computer Science (LNCS), including its subseries Lecture Notes in Artificial Intelligence (LNAI) and Lecture Notes in Bioinformatics (LNBI), has established itself as a medium for the publication of new developments in computer science and information technology research, teaching, and education.

LNCS enjoys close cooperation with the computer science R & D community, the series counts many renowned academics among its volume editors and paper authors, and collaborates with prestigious societies. Its mission is to serve this international community by providing an invaluable service, mainly focused on the publication of conference and workshop proceedings and postproceedings. LNCS commenced publication in 1973.

Jiří Mikyška · Clélia de Mulatier ·
Maciej Paszynski · Valeria V. Krzhizhanovskaya ·
Jack J. Dongarra · Peter M. A. Sloot
Editors

Computational Science – ICCS 2023

23rd International Conference
Prague, Czech Republic, July 3–5, 2023
Proceedings, Part I

 Springer

Editors
Jiří Mikyška
Czech Technical University in Prague
Prague, Czech Republic

Clélia de Mulatier
University of Amsterdam
Amsterdam, The Netherlands

Maciej Paszynski
AGH University of Science and Technology
Krakow, Poland

Valeria V. Krzhizhanovskaya
University of Amsterdam
Amsterdam, The Netherlands

Jack J. Dongarra
University of Tennessee at Knoxville
Knoxville, TN, USA

Peter M. A. Sloot
University of Amsterdam
Amsterdam, The Netherlands

ISSN 0302-9743   ISSN 1611-3349 (electronic)
Lecture Notes in Computer Science
ISBN 978-3-031-35994-1   ISBN 978-3-031-35995-8 (eBook)
https://doi.org/10.1007/978-3-031-35995-8

This Springer imprint is published by the registered company Springer Nature Switzerland AG
The registered company address is: Gewerbestrasse 11, 6330 Cham, Switzerland

Preface

Welcome to the 23rd annual International Conference on Computational Science (ICCS - https://www.iccs-meeting.org/iccs2023/), held on July 3–5, 2023 at the Czech Technical University in Prague, Czechia.

In keeping with the new normal of our times, ICCS featured both in-person and online sessions. Although the challenges of such a hybrid format are manifold, we have always tried our best to keep the ICCS community as dynamic, creative, and productive as possible. We are proud to present the proceedings you are reading as a result.

ICCS 2023 was jointly organized by the Czech Technical University in Prague, the University of Amsterdam, NTU Singapore, and the University of Tennessee.

Standing on the Vltava River, Prague is central Europe's political, cultural, and economic hub.

The Czech Technical University in Prague (CTU) is one of Europe's largest and oldest technical universities and the highest-rated in the group of Czech technical universities. CTU offers 350 accredited study programs, 100 of which are taught in a foreign language. Close to 19,000 students are studying at CTU in 2022/2023. The Faculty of Nuclear Sciences and Physical Engineering (FNSPE), located along the river bank in Prague's beautiful Old Town (Staré Mesto) and host to ICCS 2023, is the only one in Czechia to offer studies in a broad range of fields related to Nuclear Physics and Engineering. The Faculty operates both fission (VR-1) and fusion (GOLEM Tokamak) reactors and hosts several cutting-edge research projects, collaborating with a number of international research centers (CERN, ITER, BNL-STAR, ELI).

The International Conference on Computational Science is an annual conference that brings together researchers and scientists from mathematics and computer science as basic computing disciplines, as well as researchers from various application areas who are pioneering computational methods in sciences such as physics, chemistry, life sciences, engineering, arts, and humanitarian fields, to discuss problems and solutions in the area, identify new issues, and shape future directions for research.

Since its inception in 2001, ICCS has attracted increasingly higher-quality attendees and papers, and this year is not an exception, with over 300 participants. The proceedings series have become a primary intellectual resource for computational science researchers, defining and advancing the state of the art in this field.

The theme for 2023, "**Computation at the Cutting Edge of Science**", highlights the role of Computational Science in assisting multidisciplinary research. This conference was a unique event focusing on recent developments in scalable scientific algorithms; advanced software tools; computational grids; advanced numerical methods; and novel application areas. These innovative novel models, algorithms, and tools drive new science through efficient application in physical systems, computational and systems biology, environmental systems, finance, and others.

ICCS is well known for its excellent lineup of keynote speakers. The keynotes for 2023 were:

- **Helen Brooks**, United Kingdom Atomic Energy Authority (UKAEA), UK
- **Jack Dongarra**, University of Tennessee, USA
- **Derek Groen**, Brunel University London, UK
- **Anders Dam Jensen**, European High Performance Computing Joint Undertaking (EuroHPC JU), Luxembourg
- **Jakub Šístek**, Institute of Mathematics of the Czech Academy of Sciences & Czech Technical University in Prague, Czechia

This year we had 531 submissions (176 to the main track and 355 to the thematic tracks). In the main track, 54 full papers were accepted (30.7%); in the thematic tracks, 134 full papers (37.7%). A higher acceptance rate in the thematic tracks is explained by the nature of these, where track organizers personally invite many experts in a particular field to participate in their sessions. Each submission received at least 2 single-blind reviews (2.9 reviews per paper on average).

ICCS relies strongly on our thematic track organizers' vital contributions to attract high-quality papers in many subject areas. We would like to thank all committee members from the main and thematic tracks for their contribution to ensuring a high standard for the accepted papers. We would also like to thank *Springer, Elsevier,* and *Intellegibilis* for their support. Finally, we appreciate all the local organizing committee members for their hard work in preparing for this conference.

We are proud to note that ICCS is an A-rank conference in the CORE classification.

We hope you enjoyed the conference, whether virtually or in person.

July 2023

Jiří Mikyška
Clélia de Mulatier
Maciej Paszynski
Valeria V. Krzhizhanovskaya
Jack J. Dongarra
Peter M. A. Sloot

Organization

The Conference Chairs

General Chair

Valeria Krzhizhanovskaya   University of Amsterdam, The Netherlands

Main Track Chair

Clélia de Mulatier   University of Amsterdam, The Netherlands

Thematic Tracks Chair

Maciej Paszynski   AGH University of Science and Technology, Poland

Scientific Chairs

Peter M. A. Sloot   University of Amsterdam, The Netherlands | Complexity Institute NTU, Singapore

Jack Dongarra   University of Tennessee, USA

Local Organizing Committee

LOC Chair

Jiří Mikyška   Czech Technical University in Prague, Czechia

LOC Members

Pavel Eichler   Czech Technical University in Prague, Czechia
Radek Fučík   Czech Technical University in Prague, Czechia
Jakub Klinkovský   Czech Technical University in Prague, Czechia
Tomáš Oberhuber   Czech Technical University in Prague, Czechia
Pavel Strachota   Czech Technical University in Prague, Czechia

Thematic Tracks and Organizers

Advances in High-Performance Computational Earth Sciences: Applications and Frameworks – IHPCES

Takashi Shimokawabe, Kohei Fujita, Dominik Bartuschat

Artificial Intelligence and High-Performance Computing for Advanced Simulations – AIHPC4AS

Maciej Paszynski, Robert Schaefer, Victor Calo, David Pardo, Quanling Deng

Biomedical and Bioinformatics Challenges for Computer Science – BBC

Mario Cannataro, Giuseppe Agapito, Mauro Castelli, Riccardo Dondi, Rodrigo Weber dos Santos, Italo Zoppis

Computational Collective Intelligence – CCI

Marcin Maleszka, Ngoc Thanh Nguyen

Computational Diplomacy and Policy – CoDiP

Michael Lees, Brian Castellani, Bastien Chopard

Computational Health – CompHealth

Sergey Kovalchuk, Georgiy Bobashev, Anastasia Angelopoulou, Jude Hemanth

Computational Modelling of Cellular Mechanics – CMCM

Gabor Zavodszky, Igor Pivkin

Computational Optimization, Modelling, and Simulation – COMS

Xin-She Yang, Slawomir Koziel, Leifur Leifsson

Computational Social Complexity – CSCx

Vítor V. Vasconcelos. Debraj Roy, Elisabeth Krüger, Flávio Pinheiro, Alexander J. Stewart, Victoria Garibay, Andreia Sofia Teixeira, Yan Leng, Gabor Zavodszky

Computer Graphics, Image Processing, and Artificial Intelligence – CGIPAI

Andres Iglesias, Lihua You, Akemi Galvez-Tomida

Machine Learning and Data Assimilation for Dynamical Systems – MLDADS

Rossella Arcucci, Cesar Quilodran-Casas

MeshFree Methods and Radial Basis Functions in Computational Sciences – MESHFREE

Vaclav Skala, Samsul Ariffin Abdul Karim

Multiscale Modelling and Simulation – MMS

Derek Groen, Diana Suleimenova

Network Models and Analysis: From Foundations to Complex Systems – NMA

Marianna Milano, Pietro Cinaglia, Giuseppe Agapito

Quantum Computing – QCW

Katarzyna Rycerz, Marian Bubak

Simulations of Flow and Transport: Modeling, Algorithms, and Computation – SOFTMAC

Shuyu Sun, Jingfa Li, James Liu

Smart Systems: Bringing Together Computer Vision, Sensor Networks and Machine Learning – SmartSys

Pedro Cardoso, Roberto Lam, Jânio Monteiro, João Rodrigues

Solving Problems with Uncertainties – SPU

Vassil Alexandrov, Aneta Karaivanova

Teaching Computational Science – WTCS

Angela Shiflet, Nia Alexandrov

Reviewers

Zeeshan Abbas
Samsul Ariffin Abdul Karim
Tesfamariam Mulugeta Abuhay
Giuseppe Agapito
Elisabete Alberdi
Vassil Alexandrov
Nia Alexandrov
Alexander Alexeev
Nuno Alpalhão
Julen Alvarez-Aramberri
Domingos Alves
Sergey Alyaev
Anastasia Anagnostou
Anastasia Angelopoulou
Fabio Anselmi
Hideo Aochi
Rossella Arcucci
Konstantinos Asteriou
Emanouil Atanassov
Costin Badica
Daniel Balouek-Thomert
Krzysztof Banaś
Dariusz Barbucha
Luca Barillaro
João Barroso
Dominik Bartuschat
Pouria Behnodfaur
Jörn Behrens
Adrian Bekasiewicz
Gebrail Bekdas
Mehmet Belen
Stefano Beretta
Benjamin Berkels
Daniel Berrar
Piotr Biskupski
Georgiy Bobashev
Tomasz Boiński
Alessandra Bonfanti

Carlos Bordons
Bartosz Bosak
Lorella Bottino
Roland Bouffanais
Lars Braubach
Marian Bubak
Jérémy Buisson
Aleksander Byrski
Cristiano Cabrita
Xing Cai
Barbara Calabrese
Nurullah Çalık
Victor Calo
Jesús Cámara
Almudena Campuzano
Cristian Candia
Mario Cannataro
Pedro Cardoso
Eddy Caron
Alberto Carrassi
Alfonso Carriazo
Stefano Casarin
Manuel Castañón-Puga
Brian Castellani
Mauro Castelli
Nicholas Chancellor
Ehtzaz Chaudhry
Théophile Chaumont-Frelet
Thierry Chaussalet
Sibo Cheng
Siew Ann Cheong
Lock-Yue Chew
Su-Fong Chien
Marta Chinnici
Bastien Chopard
Svetlana Chuprina
Ivan Cimrak
Pietro Cinaglia

Noélia Correia
Adriano Cortes
Ana Cortes
Anna Cortes
Enrique Costa-Montenegro
David Coster
Carlos Cotta
Peter Coveney
Daan Crommelin
Attila Csikasz-Nagy
Javier Cuenca
António Cunha
Luigi D'Alfonso
Alberto d'Onofrio
Lisandro Dalcin
Ming Dao
Bhaskar Dasgupta
Clélia de Mulatier
Pasquale Deluca
Yusuf Demiroglu
Quanling Deng
Eric Dignum
Abhijnan Dikshit
Tiziana Di Matteo
Jacek Długopolski
Anh Khoa Doan
Sagar Dolas
Riccardo Dondi
Rafal Drezewski
Hans du Buf
Vitor Duarte
Rob E. Loke
Amir Ebrahimi Fard
Wouter Edeling
Nadaniela Egidi
Kareem Elsafty
Nahid Emad
Christian Engelmann
August Ernstsson
Roberto R. Expósito
Fangxin Fang
Giuseppe Fedele
Antonino Fiannaca
Christos Filelis-Papadopoulos
Piotr Frąckiewicz

Alberto Freitas
Ruy Freitas Reis
Zhuojia Fu
Kohei Fujita
Takeshi Fukaya
Wlodzimierz Funika
Takashi Furumura
Ernst Fusch
Marco Gallieri
Teresa Galvão Dias
Akemi Galvez-Tomida
Luis Garcia-Castillo
Bartłomiej Gardas
Victoria Garibay
Frédéric Gava
Piotr Gawron
Bernhard Geiger
Alex Gerbessiotis
Josephin Giacomini
Konstantinos Giannoutakis
Alfonso Gijón
Nigel Gilbert
Adam Glos
Alexandrino Gonçalves
Jorge González-Domínguez
Yuriy Gorbachev
Pawel Gorecki
Markus Götz
Michael Gowanlock
George Gravvanis
Derek Groen
Lutz Gross
Tobias Guggemos
Serge Guillas
Xiaohu Guo
Manish Gupta
Piotr Gurgul
Zulfiqar Habib
Yue Hao
Habibollah Haron
Mohammad Khatim Hasan
Ali Hashemian
Claire Heaney
Alexander Heinecke
Jude Hemanth

Marcin Hernes
Bogumila Hnatkowska
Maximilian Höb
Rolf Hoffmann
Tzung-Pei Hong
Muhammad Hussain
Dosam Hwang
Mauro Iacono
Andres Iglesias
Mirjana Ivanovic
Alireza Jahani
Peter Janků
Jiri Jaros
Agnieszka Jastrzebska
Piotr Jedrzejowicz
Gordan Jezic
Zhong Jin
Cedric John
David Johnson
Eleda Johnson
Guido Juckeland
Gokberk Kabacaoglu
Piotr Kalita
Aneta Karaivanova
Takahiro Katagiri
Mari Kawakatsu
Christoph Kessler
Faheem Khan
Camilo Khatchikian
Petr Knobloch
Harald Koestler
Ivana Kolingerova
Georgy Kopanitsa
Pavankumar Koratikere
Sotiris Kotsiantis
Sergey Kovalchuk
Slawomir Koziel
Dariusz Król
Elisabeth Krüger
Valeria Krzhizhanovskaya
Sebastian Kuckuk
Eileen Kuehn
Michael Kuhn
Tomasz Kulpa
Julian Martin Kunkel

Krzysztof Kurowski
Marcin Kuta
Roberto Lam
Rubin Landau
Johannes Langguth
Marco Lapegna
Ilaria Lazzaro
Paola Lecca
Michael Lees
Leifur Leifsson
Kenneth Leiter
Yan Leng
Florin Leon
Vasiliy Leonenko
Jean-Hugues Lestang
Xuejin Li
Qian Li
Siyi Li
Jingfa Li
Che Liu
Zhao Liu
James Liu
Marcellino Livia
Marcelo Lobosco
Doina Logafatu
Chu Kiong Loo
Marcin Łoś
Carlos Loucera
Stephane Louise
Frederic Loulergue
Thomas Ludwig
George Lykotrafitis
Lukasz Madej
Luca Magri
Peyman Mahouti
Marcin Maleszka
Alexander Malyshev
Tomas Margalef
Osni Marques
Stefano Marrone
Maria Chiara Martinis
Jaime A. Martins
Paula Martins
Pawel Matuszyk
Valerie Maxville

Pedro Medeiros
Wen Mei
Wagner Meira Jr.
Roderick Melnik
Pedro Mendes Guerreiro
Yan Meng
Isaak Mengesha
Ivan Merelli
Tomasz Michalak
Lyudmila Mihaylova
Marianna Milano
Jaroslaw Miszczak
Dhruv Mittal
Miguel Molina-Solana
Fernando Monteiro
Jânio Monteiro
Andrew Moore
Anabela Moreira Bernardino
Eugénia Moreira Bernardino
Peter Mueller
Khan Muhammad
Daichi Mukunoki
Judit Munoz-Matute
Hiromichi Nagao
Kengo Nakajima
Grzegorz J. Nalepa
I. Michael Navon
Vittorio Nespeca
Philipp Neumann
James Nevin
Ngoc-Thanh Nguyen
Nancy Nichols
Marcin Niemiec
Sinan Melih Nigdeli
Hitoshi Nishizawa
Algirdas Noreika
Manuel Núñez
Joe O'Connor
Frederike Oetker
Lidia Ogiela
Ángel Javier Omella
Kenji Ono
Eneko Osaba
Rongjiang Pan
Nikela Papadopoulou

Marcin Paprzycki
David Pardo
Anna Paszynska
Maciej Paszynski
Łukasz Pawela
Giulia Pederzani
Ebo Peerbooms
Alberto Pérez de Alba Ortíz
Sara Perez-Carabaza
Dana Petcu
Serge Petiton
Beata Petrovski
Toby Phillips
Frank Phillipson
Eugenio Piasini
Juan C. Pichel
Anna Pietrenko-Dabrowska
Gustavo Pilatti
Flávio Pinheiro
Armando Pinho
Catalina Pino Muñoz
Pietro Pinoli
Yuri Pirola
Igor Pivkin
Robert Platt
Dirk Pleiter
Marcin Płodzień
Cristina Portales
Simon Portegies Zwart
Roland Potthast
Małgorzata Przybyła-Kasperek
Ela Pustulka-Hunt
Vladimir Puzyrev
Ubaid Qadri
Rick Quax
Cesar Quilodran-Casas
Issam Rais
Andrianirina Rakotoharisoa
Célia Ramos
Vishwas H. V. S. Rao
Robin Richardson
Heike Riel
Sophie Robert
João Rodrigues
Daniel Rodriguez

Marcin Rogowski
Sergio Rojas
Diego Romano
Albert Romkes
Debraj Roy
Adam Rycerz
Katarzyna Rycerz
Mahdi Saeedipour
Arindam Saha
Ozlem Salehi
Alberto Sanchez
Ayşin Sancı
Gabriele Santin
Vinicius Santos Silva
Allah Bux Sargano
Azali Saudi
Ileana Scarpino
Robert Schaefer
Ulf D. Schiller
Bertil Schmidt
Martin Schreiber
Gabriela Schütz
Jan Šembera
Paulina Sepúlveda-Salas
Ovidiu Serban
Franciszek Seredynski
Marzia Settino
Mostafa Shahriari
Vivek Sheraton
Angela Shiflet
Takashi Shimokawabe
Alexander Shukhman
Marcin Sieniek
Joaquim Silva
Mateusz Sitko
Haozhen Situ
Leszek Siwik
Vaclav Skala
Renata Słota
Oskar Slowik
Grażyna Ślusarczyk
Sucha Smanchat
Alexander Smirnovsky
Maciej Smołka
Thiago Sobral

Isabel Sofia
Piotr Sowiński
Christian Spieker
Michał Staniszewski
Robert Staszewski
Alexander J. Stewart
Magdalena Stobinska
Tomasz Stopa
Achim Streit
Barbara Strug
Dante Suarez
Patricia Suarez
Diana Suleimenova
Shuyu Sun
Martin Swain
Edward Szczerbicki
Tadeusz Szuba
Ryszard Tadeusiewicz
Daisuke Takahashi
Osamu Tatebe
Carlos Tavares Calafate
Andrey Tchernykh
Andreia Sofia Teixeira
Kasim Terzic
Jannis Teunissen
Sue Thorne
Ed Threlfall
Alfredo Tirado-Ramos
Pawel Topa
Paolo Trunfio
Hassan Ugail
Carlos Uriarte
Rosarina Vallelunga
Eirik Valseth
Tom van den Bosch
Ana Varbanescu
Vítor V. Vasconcelos
Alexandra Vatyan
Patrick Vega
Francesc Verdugo
Gytis Vilutis
Jackel Chew Vui Lung
Shuangbu Wang
Jianwu Wang
Peng Wang

Contents – Part I

ICCS 2023 Main Track Full Papers

Improving the Resiliency of Decentralized Crowdsourced Blockchain Oracles

Adrian Fuertes Blanco[1], Zeshun Shi[1,2], Debraj Roy[1],
and Zhiming Zhao[1(✉)]

[1] Informatics Institute, University of Amsterdam, Amsterdam, The Netherlands
adrifuertes@me.com, {d.roy,z.zhao}@uva.nl
[2] Cyber Security Group, Delft University of Technology, Delft, The Netherlands
z.shi-2@tudelft.nl

Abstract. The emergence of blockchain technologies has created the possibility of transforming business processes in the form of immutable agreements called smart contracts. Smart contracts suffer from a major limitation; they cannot authenticate the trustworthiness of real-world data sources, creating the need for intermediaries called oracles. Oracles are trusted entities that connect on-chain systems with off-chain data, allowing smart contracts to operate on real-world inputs in a trustworthy manner. A popular oracle protocol is a crowdsourced oracle, where unrelated individuals attest to facts through voting mechanisms in smart contracts. Crowdsourced oracles have unique challenges: the trustworthiness and correctness of outcomes cannot be explicitly verified. These problems are aggravated by inherent vulnerabilities to attacks, such as Sybil attacks. To address this weakness, this paper proposes a reputation-based mechanism, where oracles are given a reputation value depending on the implied correctness of their actions over time. This reputation score is used to eliminate malicious agents from the participant pool. Additionally, two reputation-based voting mechanisms are proposed. The effectiveness of the proposed mechanism is evaluated using an agent-based simulation of a crowdsourced oracle platform, where a pool of oracles performs evaluate Boolean queries.

Keywords: Blockchain · Reputation-based consensus · Decentralized oracle · Voting

1 Introduction

Blockchain technologies can enhance business processes by overcoming the requirement of trust in contracts through the use of smart contracts. Smart contracts are programs stored on the blockchain; each invocation that modifies the state of the smart contract will be recorded on the blockchain, making the transaction immutable and visible to all authorized parties [4]. While smart contracts can bring trustworthiness and transparency to business processes, they have one major limitation: they cannot directly authenticate external APIs and

© The Author(s), under exclusive license to Springer Nature Switzerland AG 2023
J. Mikyška et al. (Eds.): ICCS 2023, LNCS 14073, pp. 3–17, 2023.
https://doi.org/10.1007/978-3-031-35995-8_1

make use of real-world data. Trusted entities, called oracles, act as an interface between the real-world and smart contracts [12].

There exist different types of oracles, depending on the process at hand. Software oracles can extract data from applications (i.e. web services), hardware oracles can extract data from the physical world (i.e. IoT devices) and crowdsourced oracles can leverage the power of human judgment to make decisions (i.e. data labeling) [6]. Oracles usually utilize multiple independent data sources, and the source quality and integrity is a core cause of concern in smart contracts [2]. To make the process sustainable and to incentivize participation, oracles usually charges a fee for each transaction. The remuneration of oracle tasks can then be modeled as an economic problem in a market, with oracles exchanging their labor for a fee. While the market structure creates an incentive for efficient participation, it also creates an incentive for malicious interference. The potential for malicious interference depends on the protocol used and the oracle type. In this paper, we considered the most popular permissionless, voting-based crowdsourced oracles that operate binary tasks (queries with a True or False answer).

The crowdsourcing process involves requesters posting tasks to be completed by workers. Tasks can be unrelated and workers can choose or be assigned tasks. In decentralized oracles, requesters and workers are anonymous and independent. To ensure correctness, multiple workers are assigned to a task and the majority vote is selected as the final answer. Workers who vote with the majority are rewarded, while others who don't might lose their deposit. Voting systems can improve truth-telling and oracle accuracy but create vulnerabilities, such as manipulation of majority results by malicious actors [16]. These vulnerabilities can be addressed through incentive structures that make malicious behavior unprofitable or through trust management by estimating the trustworthiness of individuals and designing voting and participation rules accordingly.

In this paper, three reputation-based mechanisms are introduced to improve the resiliency of crowdsourced oracles by controlling the participation of agents and altering the voting mechanism based on reputation. A reputation system is first designed to reward participants who vote truthfully and eliminate participants who act maliciously. The simple majority voting system is replaced by a reputation-based weighted voting system, which is designed to limit the influence of malicious agents in the event of attacks. Two variations of this system are further proposed to tackle Sybil and Camouflage Attacks, respectively. Additionally, an agent-based simulation of a crowdsourced oracle platform is introduced. The simulation is used to validate the aforementioned strategies, providing insight into the resiliency of the system under various attacks. All of the reputation-based mechanisms are tested under Simple Attacks, Camouflage Attacks, and Sybil Attacks. Lastly, a novel "certifier" agent is defined as a bounded-rational, profit-maximizing player, which utilizes past voting outcomes and innate accuracy in a probabilistic decision-making model. This design improves upon existing methods by relaxing rationality assumptions and incorporating stochasticity into the decision-making process, characteristic of crowdsourcing systems. This agent is used in the agent-based simulation in an honest certifier role.

The rest of the paper is organized as follows: Sect. 2 discusses related work on the resilience of crowdsourced oracles; Sect. 3 presents the reputation-based model and details the simulation components; Sect. 4 displays the simulation settings and experiment results; Sect. 5 discusses the obtained results and Sect. 6 concludes the paper.

2 Related Work

The existing frameworks for decentralized crowdsourced oracles can be divided on whether they tackle vulnerabilities through incentive design or reputation management. Frameworks that focus on incentive design, such as [1,12] and [15], propose guidelines on reward quantities [15] or variations on the majority voting system [1,12] designed to make malicious behavior unprofitable for rogue agents. These frameworks argue their incentive systems induce truth-telling by proving honesty is optimal in a Nash Equilibrium. The optimality proof requires the assumption that all participants are fully rational and have a sufficiently large accuracy, which might not hold for all agents or tasks in reality [3].

These vulnerabilities still leave room for a reputation-based system to control unwanted behavior. To our best knowledge, no explicit reputation-based control system has been proposed for permissionless crowdsourced oracles. Though, decentralized crowdsourcing platforms [9,10] outline reputation systems to optimize worker [10] and requester [9] behavior, their systems (like traditional crowdsourcing) require the verification of output by a third party. Output verification is useful in a crowdsourcing context, where output quality is easily verifiable, but it is not feasible to implement in oracle models, where the underlying oracle output is the closest thing to ground truth itself. This creates the need for a reputation-based system that is not based on external verification and can be supported by the oracle's output.

Reputation-based voting is a popular form of achieving consensus in governance systems such as [13]. Reputation is used both as a way to deter unwanted behavior and a way to reward active participants. As users gain relative reputation, their influence increases, since reputation acts as weight in a weighted voting system. While [12,15] and [1] suggest the use of a reputation-based system might be a beneficial addition to their incentive systems, no explicit implementation details are suggested.

A major shortcoming in all the aforementioned protocols is the lack of practical evidence. While incentive optimization techniques might suggest optimal parameters, experimental evidence suggests human behavior might not align with theoretical predictions [3]. The implementation of blockchain-based systems can be very costly [8], a problem accentuated by the immutability of smart contracts. This immutability is particularly problematic in incentive design and reputation management, where parameter choice can have a significant impact on system performance. This results in a higher potential cost of failure, on top of the high implementation cost.

A solution to high smart contract development costs is using simulation systems like Agent-Based Model (ABM). They study the properties of complex systems and have been used for crowdsourcing, voting, and smart contract manipulation. Simulation allows developers to test parameters, strategies, and threats before deployment, reducing cost and uncertainty [7].

3   Proposed Model

This section explains the components of the proposed model, starting with a system overview. The design of the agents is covered afterward, followed by a description of reputation voting and control models. Lastly, experiment evaluation metrics are discussed.

3.1   System Overview

There are two roles in a decentralized crowdsourced oracle: requesters and certifiers. Within the oracle platform, requesters submit queries (q) with an associated budget to them. Certifiers can then choose to engage in these queries to potentially earn a reward. We assume each query q has an underlying true value of True (T) or False (F). The objective of the oracle is to reach this true value via a voting mechanism performed by the certifiers. Since the true value is not known, certifiers are rewarded based on the value estimated by the oracle. To incentivize honesty, certifiers submit a participation fee that they are liable to lose if their behavior does not align with the oracle's output. The oracles considered in this paper are permissionless, meaning no approval is required to join the system. To join the pool, certifiers do have to register with their wallet address, which gives them a starting reputation and the ability to participate in games.

We assume the queries are of a monitoring nature, meaning a query is constantly being evaluated by requesters and it's bound to change at any moment. As noted by [14,15], an example of this would be monitoring the compliance of a Service Level Agreement (SLA) for cloud computing services, where the oracle is used to attest for the up-time of the service (certifying whether a violation in the SLA has occurred or not). These tasks are characterized by a significantly higher likelihood of the true value being $False$ (no violation has occurred), meaning true $Trues$ are rare and not guaranteed. For this reason, it is assumed that the reward associated with voting $False$ is lower than the reward associated with voting $True$, as agents are then incentivized to exert more effort when a $True$ outcome is observed, avoiding a lazy equilibrium in which every agent reports $False$ for a guaranteed income with minimum effort.

In this context, malicious agents are incentivized to influence the oracle towards a $False$ outcome, since it is likely to go unnoticed and it guarantees a low but steady reward. Adversarial agents can influence requesters by either consistently or overwhelmingly reporting $False$. If these attacks are successful, honest certifiers (who are assumed to be profit-maximizing agents) will be incentivized to also submit *False* reports, as the expected profit of doing so would be higher than that of true reporting. We consider three different threat models which will be described in detail in Sect. 3.4. Besides, the reputation mechanism is designed to limit malicious agent influence in two ways: by excluding agents with a bad reputation and by limiting their voting power. The iteration of the voting game is shown in Fig. 1, and the implementation details are discussed in Sect. 3.3.

Fig. 1. Diagram showcasing a single iteration of the voting game.

3.2 Agents

Certifiers. Certifiers are assumed to be honest and profit-maximizing players. Each player i is assumed to have an innate accuracy value $\gamma_i \in [0,1]$. Intuitively this value represents the ability of the player to assess the query at hand, and therefore the probability they will be correct about their assessment. Players use their accuracy to form a private assessment of the query. Additionally, each player has a fixed memory value m_i ($m_i \geq 0$). Their memory represents their ability to look at past votes, either through the public transactions stored on the blockchain or through the memory of past games. An overview of the decision-making process behind agent's votes is outlined in Fig. 2.

Each agent balances both their innate accuracy and their memory of past votes to create a profit expectation of voting truthfully (submitting their private opinion) or not (submitting the opposite of their private opinion). The balance between these two parameters is parametrized by a recency bias $RB \in [0,1]$, which acts as a weight between the profit estimate from private opinion and the profit estimate from the available voting history. Formally, a certifier A_i with accuracy γ_i, memory m_i and recency bias RB_i submits a vote $v_i \in [F,T]$. A_i will submit T if $E[\pi(v_i = T)] \geqslant E[\pi(v_i = F)]$, otherwise they will submit F.

The expected profit of submitting T is calculated as:

$$E[\pi(v_i = T)]$$

$$= Pr(MV = T \mid v_i = T) \cdot \pi(MV_{v_i=MV} = T)$$

$$+(1 - Pr(MV = T \mid v_i = T)) \cdot \pi(MV_{v_i \neq MV} = F)$$

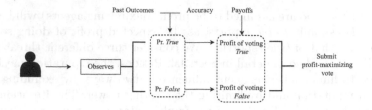

Fig. 2. Overview of the decision-making process of an honest certifier.

where $MV \in [T, F]$ represents the majority vote, or the vote chosen by the majority voting process.

If $v_i = T$, the probability of a consensus being formed is:

$$Pr(MV = T \mid v_i = t)$$

$$= RB_i \cdot Pr(MV = r \mid a_i = r, memory = m_i) + (1 - RB_i) \cdot \gamma_i$$

Where $Pr(MV = FT \mid v_i = T, memory = m_i)$ measures the expected likelihood of the consensus being formed around F, given the votes available to the players' memory. This is modeled using a Beta-Binomial distribution as such:

$$Pr(MV = T \mid v_i = T, memory = m_i)$$

$$= 1 - BinomialCDF(p, k, n)$$

The p parameter follows a $Beta(\alpha, \beta)$ with $\alpha = \sum_{i=1}^{n_{memory}} 1 \cdot [v_i = True]$ and $\beta = \sum_{i=1}^{n_{memory}} 1 \cdot [v_i = False]$. Then, $k = n_{players} \cdot MajorityThreshold - 1$, with the majority threshold being the percentage of votes required for consensus in a game (50% for the simple majority). Lastly, n is the number of players in a game.

The profit estimation is identical for $E[\pi(v_i = F)]$, though the payoffs change and the accuracy value is reversed since it is assumed to refer to the accuracy of assessing a true $True$ signal.

For the purpose of the simulation, we assume all agents share the same memory and recency bias. These values are set at 10 for memory, such that agents have access to the votes cast in the last 10 games, and 0.3 for the recency bias, such that agents weigh their experience 30% while making a profit estimation. The accuracy is sampled from a Beta distribution with $\alpha = 10$ and $\beta = 2.75$, which creates a visual center around 0.85 accuracy, as observed from Fig. 3. This ensures the majority of agents have relatively high accuracy, with some exceptions. Lastly, we assume a simple majority is necessary to achieve consensus.

This formulation ensures that both an agent's accuracy as well as their experience are considered in the decision-making process. Additionally, the modeling of accuracy creates more realistic agents, as real participants are not always 100% accurate and are liable to make mistakes.

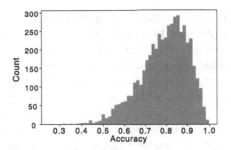

Fig. 3. Histogram including 5000 sampled values from a $Beta(\alpha, \beta)$ distribution with $alpha = 10$ and $beta = 2.75$

Malicious. Malicious agents are assumed to have 100% accuracy and they always follow their assigned voting strategy regardless of expected profits. The voting strategies of malicious agents are outlined in Sect. 3.4.

3.3 Reputation

For each agent, their reputation value R_i, $R_i \geq 0$ represents their trustworthiness based on past voting outcomes. All certifiers start with a reputation value R_s, $R_s > 0$. As players participate in games, their reputation can increase by R_+ if they vote according to the consensus, or decrease by R_- if they vote against the consensus. R_s, R_+ and R_- are model parameters.

Participation Control. As a way to control malicious agents, agents will be permanently excluded from the system if their reputation value reaches zero. This can prevent agents from engaging in a Simple Attack, where agents continuously perform adversarial behavior. Since the behavior is constant and individual, the attacker is likely to vote against the consensus enough times for them to be expelled from the pool. This holds true as long as the percentage of the malicious agent is less than 50%, such that the oracle is likely to produce the correct result on average.

This does not prevent agents from engaging in Camouflage Attacks or Sybil Attacks. Agents involved in Camouflage Attacks pretend to be honest until they achieve a high reputation status when they start attacking the system. Given their high reputation, if enough malicious agents are in the system, it's possible they will influence the oracle's opinion before their reputation goes to zero. In Sybil Attacks, an attacker impersonates a large number of players in order to influence game outcomes. If the attacker floods the pool with enough agents, such that over 51% of the players in a game are malicious, the oracle will always vote in accordance with the malicious agents, making it such that their reputation never reaches zero.

Voting Systems. Reputation-based, weighted voting systems can mitigate Camouflage and Sybil Attacks by redistributing voting power and minimizing the voting influence of malicious agents. In a simple voting system, every vote has the same weight, which results in every voter having the same voting power. In a system with 10 voters, gaining control over 4 votes can already give an attacker decision power over the system outcome. By weighting votes by a user's reputation value, higher voting power is assigned to voters with a proven history of truthful behavior. This mitigates Sybil Attacks, since all newly created player accounts will have a relatively low reputation, assuming more experienced players are participating in the game. This would increase the necessary amount of players an attacker has to control, which given the random player selection and a sufficiently large user base, would make this attacks impossible.

The weight of each candidates vote, x_i, is calculated as follows:

$$x_i = \frac{R_i}{\sum_j^n R_j}, i \neq j$$

The total votes for each outcome can be calculated as follows:

$$V_{True} = \sum_{i=1}^{n} x_i \cdot v_i [v_i = True]$$

$$V_{False} = \sum_{i=1}^{n} x_i \cdot v_i [v_i = False]$$

Then, if $V_{True} \geq V_{False}$ the outcome will be decided as $True$. Otherwise, the outcome will be decided as $False$.

While this offers an improvement upon simple voting, weighted voting suffers from seniority-induced advantages which can disincentive participation. This is known as the Matthew Effect, which dictates that those who begin with advantage accumulate more advantage over time and those who begin with disadvantage become more disadvantaged over time [5]. The earlier an honest certifier joins the platform, the more opportunities they will have to gain a reputation. This leads to the earliest certifiers having disproportionate voting power over newcomers, which also increases the potential negative influence they can have over the system if they act maliciously. For example, a player who has played 1010 games, can have 1000 times more voting power than a certifier who has played 10 games. Despite players not having access to others' reputation scores, this could easily disincentive new players from joining the certifier pool. Stratified voting can solve Matthew Effect by weighting votes relative to the game participants, rather than to the whole certifier pool. This can be achieved by dividing the pool into k different partitions, where agents in each partition are given the same voting weight. Agents are assigned to partitions depending on their global reputation value, but since the divisions are made relative to the participants in the game, an agent with extensive experience can't have disproportionate voting power, since their voting power will be the same as the second most reputed player in the game. The number of partitions k is a model parameter and it can be set depending on the number of participants. The relative vote weight assigned to each partition, w_k, is also a model parameter. The combination of k and w_k can affect the ultimate voting power of each agent, though finding the optimal combination is out of the scope of this paper. For the purpose of this simulation, a linear weight scale will be tested.

To assign agents to a partition, participants are sorted based on their reputation score and assigned to partition 1 to k, or the closest possible number. The number of partitions, k, acts as an upper bound, since it cannot be guaranteed that agents will have different reputation values with random sortition.

Each agent's weight would then be calculated as follows:

$$x_i = w_k, A_i \in k$$

$$w_k = \sigma(k)$$

Where the w_k is the weight of the partition A_i belongs to and σ defines the weight scale depending on the partition number. For a linear partition $w_k = k$.

3.4   Threat Models

Three different threat models are considered in this paper: Simple Attacks, Camouflage Attacks, and Sybil Attacks. Each reputation-based control mechanism will be assessed under all three threat models.

Simple Attacks. In a Simple Attack, a malicious agent exists passively in the agent pool, voting *False* whenever it enters a game. In this attack, a percentage of the agents in the pool, $p_a \in [0, 1]$, are malicious, and as $p_a \to 0.5$ their influence surpasses that of honest agents, inducing the oracle to always output *False*.

Camouflage Attacks. In the Camouflage Attack, a malicious agent masquerades as an honest agent for n_c turns or until it has accrued a high enough reputation value, R_t. When $R_i \geq R_t$ the agent starts attacking the system by always voting *False*. Since malicious agents are assumed to be 100% accurate $n_c = R_t$ in this model. n_c and R_t are model parameters.

Sybil Attacks. In a Sybil Attack, an attacker forges multiple accounts, flooding the agent pool with malicious agents that always vote *False*. The number of malicious agents that enter the pool is parameterized as $n_f = \alpha n$, where n is the total number of agents in the pool. Since in this model game participants are randomly selected, α should be larger than one (under simple voting) to perform a successful attack. The attack lasts for $i_a \in [1, \inf]$ turns.

3.5   Rewards

Since the game is assumed to be monitoring in nature, the reward for a *True* outcome, π_{TT}, is m times larger than that of a *False* outcome, π_{FF}. We follow the payment structure outlined in [15] $\pi_{FF} = 1$, and an agent loses the full amount of their deposit if they vote *False* when the oracle outputs *True*, making $\pi_{FT} = -1$. An agent faces no loss when the oracle outputs *True* and they vote *False*, making $\pi_{TF} = 0$. We set $m = 2$, instead of the proposed $m = 10$, which increases the influence of attacks while keeping the system accurate without threats. The payoff matrix is outlined in Table 1.

Table 1. Payoff matrix relating an agent's payoff relative to the oracle output

Player i	Consensus	
	True	False
True	2	−1
False	0	1

3.6   Evaluation

Ground Truth Model. To model accuracy, the underlying truth distribution needs to be defined. The underlying truth is modeled as a sign square wave, which oscillates between *True* (-1) and *False* (1) in phases. The formulation is as follows:

$$S(t) = \text{sgn}(\sin 2\pi f t)$$

With the square wave following a duty cycle as $duty = \frac{sin(m\pi t+1)}{2}$. m has been fixed to 38.8 and f has been fixed to 58 with $t = 500$. This results in an average of 40% *True* periods, such that the majority of agents are able to experience both outcomes, while the True outcome remains less frequent. A sample of this signal for 500 periods is shown in Fig. 4.

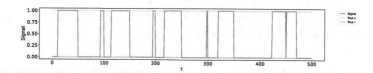

Fig. 4. Model of a ground truth signal using a square sign wave over 500 periods. The signal oscillates between 1 and −1, indicating a *False* and *True* actual signal.

Accuracy. The primary measure of success for a single experiment in a simulation is its accuracy in the final period. The accuracy is measured as the percentage of iterations in which the oracle predicted the ground truth. Formally:

$$accuracy = \frac{\sum_{i=1}^{n} 1 \cdot [o_i = t_i]}{n}$$

Where n refers to the total number of iterations, o_i refers to the oracles output in iteration i and t_i refers to the true signal in iteration i.

4   Experiments and Simulation

This section outlines the details of each experiment performed in this paper followed by the results.

4.1   Simulation Settings

All simulation experiments share the following parameters:

- 15 voters are randomly selected for each game.
- Each player follows the payoffs aligned in Sect. 3.5.
- Each simulation follows the true signal outlined in Sect. 3.6.
- The voting power distribution is calculated over the last 5 games played by each agent.

The number of game participants is chosen to be odd which prevents ties during voting. Additionally, it is sufficiently large such that an average player will experience enough games to gain a meaningful reputation (or be banned) and have ample voting opportunities, which could influence their voting choices.

Each of the proposed mechanisms is tested using a Simple Attack, a Camouflage Attack, and a Sybil Attack; all compared against a baseline with no interventions. The Simple Attack is simulated with $p_a \in [0.05, 0.50]$ in 0.025 intervals. The Camouflage Attack is simulated with $n_c = 5$ and $p_a \in [0.05, 0.50]$ in 0.025 intervals. Lastly, the Sybil Attack is simulated with $\alpha = 3$ and $i_a \in [0, 65]$ in 5 game intervals. Each Sybil Attack simulation includes 5 attacks. The timing of attacks is random, with at least 50 games between attacks, such that the system is under attack at most 50% of the time.

Each attack is tested with increasing p_a (Simple and Camouflage) or i_a (Sybil) to test the resilience of the system, as an increased presence of malicious agents impedes the oracle from achieving high accuracy.

Each agent i starts with 1 reputation point, $R_s = 1$, with $R_+ = 1$ and $R_- = 1$. On average, an agent can accumulate 15 reputation points, and they can last a minimum of 1 turn in the agent pool, under participation control.

Due to the stochastic nature of agent decision-making and agent accuracy sampling, each experiment consists of multiple iterations, with the results being presented as a sample average. Specifically, each experiment consists of 100 iterations, with 500 consecutive games per iteration. In each iteration, a pool of 500 players is generated.

The simulations are built in Python, using the ABM framework MESA [11].

4.2   Participation Control

In the participation control experiments, agents are permanently banned from the agent pool when their reputation reaches 0, with the intention of spelling malicious agents when they vote maliciously. The experiment results for each attack tested are displayed in Fig. 5.

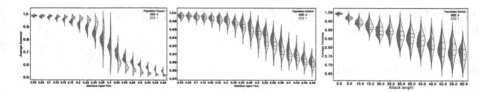

Fig. 5. Participation control simulation results with subfigures displaying the Simple, Camouflage, and Sybil attacks respectively. The violin plots show the density of average accuracy values from control and treatment simulations, each with 100 observations.

Simple Attack. The percentage of malicious agents can significantly affect the accuracy of the system under a Simple Attack, with average accuracy values decreasing to 50% as p_a approaches 0.5. The participation control system helps mitigate this decline under lower threat levels, resulting in higher average accuracy up to $p_a \sim 0.25$. When $p_a \in [0.25, 0.3]$, the accuracy under the treatment becomes uniformly distributed - meaning sampled accuracy is as often very high as it is very low. In the remaining p_a percentages, the population-controlled simulations have significantly lower accuracy than the control. **Camouflage Attack.** The reputation control strategy does not improve resiliency when compared to the control, with the average treatment accuracy being lower than the control. The lower accuracy is caused by mistake-making honest being eliminated, which then increases the relative percentage of malicious agents once the camouflage period is over. **Sybil Attack.** The reputation control mechanism does not improve resiliency against the simulated Sybil Attacks.

4.3   Weighted Voting

In the weighted voting experiments, simple majority voting is replaced with weighted voting, with reputation values acting as weights. By giving more reputable agents (which are likely to be honest certifiers) higher voting power, the oracle is expected to produce more accurate results under threat. The experiment results for each attack tested are shown in Fig. 6.

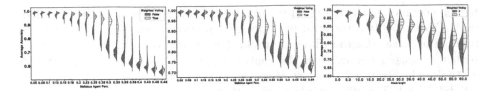

Fig. 6. Weighted voting simulation results with subfigures displaying the Simple, Camouflage, and Sybil attacks respectively. The violin plots show the density of average accuracy values from control and treatment simulations, each with 100 observations.

Simple Attack. The weighted voting strategy increases resiliency compared to the control, being able to maintain a higher average accuracy under larger p_a values. The strategy results in significantly increased accuracy until $p_a \sim 0.35$, and maintains a relatively higher accuracy for the remaining p_a. **Camouflage Attack.** The weighted voting strategy consistently achieves higher average accuracy than the control, demonstrating it is effective in increasing resiliency against a Camouflage Attack. **Sybil Attack.** A weighted voting system can improve resiliency against Sybil Attacks. Although simulations using this system show increased average accuracy under all attack lengths, the differences are lower than under other attacks.

4.4   Stratified Voting

In the stratified voting experiment, agents are assigned to k partitions, which dictate their relative voting power. Each game will consist of up to 5 partitions ($k = 5$) with linear voting power differences between partitions. By assigning linear weights between partitions, individual agents have less voting power when compared to weighted voting, which is intended to limit the voting power of more experienced agents. The experiment results for each attack tested are shown in Fig. 7.

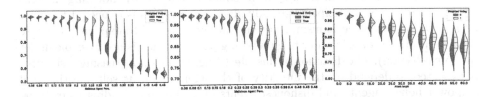

Fig. 7. Stratified voting simulation results with subfigures displaying the Simple, Camouflage, and Sybil attacks respectively. The violin plots show the density of average accuracy values from control and treatment simulations, each with 100 observations

Simple Attack. The stratified voting strategy increases resiliency when compared to the control. Simulations using stratified voting result in significantly higher accuracy until $p_a \sim 0.3$, maintaining an average accuracy higher than 90%. From $p_a \sim 0.3$, the stratified voting simulations result in higher or equal average accuracy. **Camouflage Attack.** The stratified voting strategy shows higher average accuracy than the control under a Camouflage Attack. **Sybil Attack.** The Stratified voting strategy can increase the oracles' resiliency under a Sybil Attack. The efficacy is lower than that showcased by the weighted voting strategy, particularly for higher attack lengths.

5   Discussion

Results from all experiments show that an increase in malicious agents leads to a decline in oracle accuracy across all threat models, revealing the vulnerability of the crowdsourced oracle without threat prevention. Participation control improves resiliency for lower p_a levels under Simple Attack, however it worsens accuracy under Camouflage and Simple Attacks due to malicious agents exploiting the control mechanism. Alternative voting mechanisms improve resiliency against Simple Attacks, but still produce low accuracy as p_a approaches 0.5. Simple weighted voting is the only mechanism to significantly improve resiliency under Sybil Attacks, due to the importance of players with large reputation values. The high accuracy seen under Camouflage Attack suggests that 500 game periods are insufficient to observe Matthew Effect. The efficacy of Weighted and stratified voting systems remains to be tested over longer simulation cycles.

6   Conclusion

In this paper, we propose three reputation-based methods to improve the resiliency of crowdsourced oracles. We leverage simulation techniques to test the validity of the proposed methods under multiple threat models. A bounded-rational agent model was also introduced, enabling the simulation to capture oracles dynamics over time. Finally, we discuss the strengths and weaknesses of the mechanism proposed, highlighting the mechanism behind them based on the simulation results.

In future work, we plan to expand the simulation scope by modeling the self-selection process of participants to individual games. This is expected to affect all reputation-based mechanism, as agents are incentivized to explicitly increase their reputation scores in order to gain voting power (which is not possible under random selection). The direction of this effect (increasing or decreasing resilience) is expected to depend on the integrity of the agent pool, as self-selection will empower honest agents and dishonest agents alike, possibly leading to extreme results in either direction. We also aim to implement different weighting functions for the stratified voting method, as results suggest that relative voting power differences are a key aspect of the improvement in resilience. Lastly, we plan to conduct empirical experiments on crowdsourced oracle systems. During this experiment, we aim to identify the voting and selection pattern of certifiers, which would serve to inform the selection and decision-making process of the model. These experiments would also serve as validation of the simulation model.

Acknowledgements. This research is funded by the European Union's Horizon 2020 research and innovation program under grant agreements 825134 (ARTICONF project), 862409 (BlueCloud project) and 824068 (ENVRI-FAIR project). The work is also partially supported by LifeWatch ERIC.

References

1. Adler, J., Berryhill, R., Veneris, A., Poulos, Z., Veira, N., Kastania, A.: Astraea: a decentralized blockchain oracle. arXiv:1808.00528 [cs] (2018). arXiv: 1808.00528
2. Caldarelli, G., Ellul, J.: The blockchain oracle problem in decentralized finance-a multivocal approach. Appl. Sci. **11**(1616), 7572 (2021)
3. Camerer, C.F.: Behavioral Game Theory: Experiments In Strategic Interaction. Princeton University Press, Princeton (2003)
4. Chen, Y., Meng, L., Zhou, H., Xue, G.: A blockchain-based medical data sharing mechanism with attribute-based access control and privacy protection. Wirel. Commun. Mob. Comput. **2021**, 1–12 (2021)
5. Dannefer, D.: Cumulative advantage/disadvantage and the life course: cross-fertilizing age and social science theory. J. Gerontol. Series B **58**(6), S327–S337 (2003)
6. Di Ciccio, C., Meroni, G., Plebani, P.: On the adoption of blockchain for business process monitoring. Softw. Syst. Model. **21**(3), 915–937 (2022)
7. Janssen, M.A., Ostrom, E.: Empirically based, agent-based models. Ecol. Soc. **11**(2) (2006)
8. Khan, S.N., Loukil, F., Ghedira-Guegan, C., Benkhelifa, E., Bani-Hani, A.: Blockchain smart contracts: applications, challenges, and future trends. Peer-to-Peer Networking Appl. **14**(5), 2901–2925 (2021)
9. Li, C., Qu, X., Guo, Y.: Tfcrowd: a blockchain-based crowdsourcing framework with enhanced trustworthiness and fairness. EURASIP J. Wirel. Commun. Nctw. **2021**(1), 168 (2021)
10. LI, M., et al.: CrowdBC: a blockchain-based decentralized framework for crowdsourcing. IEEE Trans. Parallel Distrib. Syst. **30**(6), 1251–1266 (2019)
11. Masad, D., Kazil, J.: MESA: an agent-based modeling framework. In: 14th PYTHON in Science Conference, pp. 51–58 (2015)
12. Nelaturu, K., et al.: On public crowdsource-based mechanisms for a decentralized blockchain oracle. IEEE Trans. Eng. Manage. **67**(4), 1444–1458 (2020)
13. Orange: https://www.orangeprotocol.io/
14. Shi, Z., Farshidi, S., Zhou, H., Zhao, Z.: An auction and witness enhanced trustworthy SLA model for decentralized cloud marketplaces. In: Proceedings of the Conference on Information Technology for Social Good, pp. 109–114 (2021)
15. Zhou, H., Ouyang, X., Ren, Z., Su, J., de Laat, C., Zhao, Z.: A blockchain based witness model for trustworthy cloud service level agreement enforcement. In: IEEE INFOCOM 2019 - IEEE Conference on Computer Communications, pp. 1567–1575 (2019)
16. Zhuang, Q., Liu, Y., Chen, L., Ai, Z.: Proof of reputation: a reputation-based consensus protocol for blockchain based systems. In: Proceedings of the 2019 International Electronics Communication Conference, pp. 131–138. ACM, Okinawa Japan (2019)

Characterization of Pedestrian Contact Interaction Trajectories

Jaeyoung Kwak[1(✉)], Michael H. Lees[2], and Wentong Cai[1]

[1] Nanyang Technological University, Singapore 639798, Singapore
{jaeyoung.kwak,aswtcai}@ntu.edu.sg
[2] University of Amsterdam, 1098XH Amsterdam, The Netherlands
m.h.lees@uva.nl

Abstract. A spreading process can be observed when a particular behavior, substance, or disease spreads through a population over time in social and biological systems. It is widely believed that contact interactions among individual entities play an essential role in the spreading process. Although the contact interactions are often influenced by geometrical conditions, little attention has been paid to understand their effects especially on contact duration among pedestrians. To examine how the pedestrian flow setups affect contact duration distribution, we have analyzed trajectories of pedestrians in contact interactions collected from pedestrian flow experiments of uni-, bi- and multi-directional setups. Based on standardized maximal distance, we have classified types of motions observed in the contact interactions. We have found that almost all motion in the unidirectional flow setup can be characterized as subdiffusive motion, suggesting that the empirically measured contact duration tends to be longer than one estimated by ballistic motion assumption. However, Brownian motion is more frequently observed from other flow setups, indicating that the contact duration estimated by ballistic motion assumption shows good agreement with the empirically measured one. Furthermore, when the difference in relative speed distributions between the experimental data and ballistic motion assumption is larger, more subdiffusive motions are observed. This study also has practical implications. For instance, it highlights that geometrical conditions yielding smaller difference in the relative speed distributions are preferred when diseases can be transmitted through face-to-face interactions.

Keywords: Pedestrian flow · Contact interaction · Brownian motion · Subdiffusive motion · Contact duration

1 Introduction

Modeling contact interactions among individual entities is essential to understand spreading processes in social and biological systems, such as information diffusion in human populations [1,2] and transmission of infectious disease in animal and human groups [3,4]. For the spreading processes in social and biological systems, one can observe a contact interaction when two individual entities

© The Author(s), under exclusive license to Springer Nature Switzerland AG 2023
J. Mikyška et al. (Eds.): ICCS 2023, LNCS 14073, pp. 18–32, 2023.
https://doi.org/10.1007/978-3-031-35995-8_2

are within a close distance, so they can exchange substance and information or transmit disease from one to the other one. In previous studies, macroscopic patterns of contact interactions are often estimated based on simple random walking behaviors including ballistic motion. For example, Rast [5] simulated continuous-space-time random walks based on ballistic motion of non-interacting random walkers. Although such random walk models have widely applied to estimate contact duration for human contact networks, little work has been done to study the influence of pedestrian flow geometrical conditions on the distribution of contact duration.

To examine how the geometrical conditions of pedestrian flow affect the contact duration distribution, we perform trajectory analysis for the experimental dataset collected from a series of experiments performed for various pedestrian flow setups. The trajectory analysis of moving organisms, including proteins in living cells, animals in nature, and humans, has been a popular research topic in various fields such as biophysics [6–8], movement ecology [9–11], and epidemiology [12,13]. Single particle tracking (SPT) analysis, a popular trajectory analysis approach frequently applied in biophysics and its neighboring disciplines, characterizes the movement dynamics of individual entities based on observed trajectories [7,14]. According to SPT analysis, one can identify different types of diffusion, for instance, directed diffusion in which individuals move in a clear path and confined diffusion in which individuals tend to move around the initial position. The most common method for identifying diffusion types is based on the mean-squared displacement (MSD), which reflects the deviation of an individual's position with respect to the initial position after time lag [14,15]. Motion types can be identified based on the diffusion exponent. MSD has been widely applied for various trajectory analysis studies in biophysics [16,17]. For pedestrian flow trajectory analysis, Murakami et al. [18,19] analyzed experimental data of bidirectional pedestrian flow and reported diffusive motion in individual movements perpendicular to the flow direction. They suggested that uncertainty in predicting neighbors' future motion contributes to the appearance of diffusive motion in pedestrian flow.

Previous studies have demonstrated usefulness of SPT analysis in examining movement of individuals. However, SPT analysis does not explicitly consider relative motions among individuals in contact, suggesting that analyzing the relative motions can reveal patterns that might not be noticeable from the SPT analysis approach. For example, if two nearby individuals are walking in parallel directions together with a similar speed, one might be able to see various shapes of relative motion trajectories although the individual trajectories are nearly straight lines. For contact interaction analysis, the analysis of relative motion trajectories can be utilized to predict the length of contact duration and identify contact interaction characteristics such as when the interacting individuals change walking direction significantly. Regarding the spreading processes, understanding of relative motion trajectory can be applied to identify optimal geometrical conditions that can minimize contract duration when diseases can be transmitted through face-to-face interactions.

Although MSD is simple to apply, MSD has limitations. Refs. [20–22] pointed out that MSD might not be suitable for short trajectories to extract meaningful information. Additionally, due to its power-law form, the estimation of the diffusion exponent in MSD is prone to estimation errors [23,24]. As an alternative to MSD, various approaches have been proposed, including the statistical test approach [20,21,25,26] and machine learning approach [27].

In this work, we analyze trajectories collected from different experiment setups including uni-, bi-, and multi-directional flow for pedestrian contact interactions. Rather than using individual trajectories, we analyze the relative motion of interactions to understand *pedestrian contact interactions*. We identify different types of motion observed in pedestrian contact interactions based on a statistical test procedure. For the statistical test procedure, we measured a standardized value of largest distance traveled by an individual from their starting point during the contact interaction. Our results demonstrate that examining the interactions in this way can provide important insights regarding contact duration, and hence help estimate transmission risk in different pedestrian flow conditions.

The remainder of this paper is organized as follows. Section 2 describes the datasets including pedestrian flow experiment setups and some descriptive statistics. The statistical test procedure and trajectory classification results are presented in Sect. 3. We discuss the findings of our analysis in Sect. 4.

2   Datasets

Figure 1 shows the sketches of various experiment setup: uni-directional flow, bi-directional flow, 2-way crossing flow, 3-way crossing flow, and 4-way crossing flow. In the uni-directional flow setup, pedestrians were walking to the right in a straight corridor of 5 m wide and 18 m long. In a bi-directional flow, two groups of pedestrians were entering a straight corridor of 4 m wide and 10 m long through 4 m wide entrance and then walking opposite directions. They left the corridor through the open passage once they reached the other side of the corridor. In 2-way, 3-way, and 4-way crossing flows, different groups of pedestrians were entering the corridor through 4 m wide entrance and walked 5 m before and after passing through an intersection (4 m by 4 m rectangle in 2-way crossing and 4-way crossing flows, and 4 m wide equilateral triangle in 3-way crossing flow). Similar to the setup of bi-directional flow, pedestrian groups left the corridor through the open passage after they reached the end of corridors. A more detailed description of the experiment setups can be found in Refs. [28–30].

From the experimental data, we extracted pairs of individuals in contact and their relative motion trajectories. We considered a pair of individuals is in contact when the two individuals are within a contact radius. The contact radius r_c would depend on the form of transmission in question. In this paper, we assume a 2 m radius based on previous studies [12,31–33]. For the analysis, we considered trajectories with at least 10 data points based on literature [21,25]. Table 1 shows the basic statistics of representative scenarios including the number of

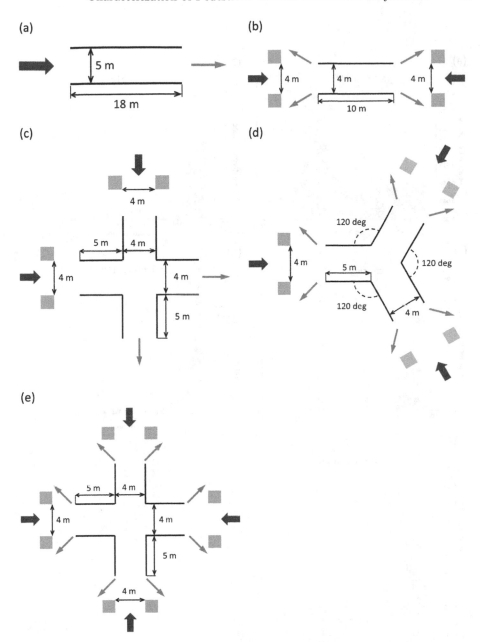

Fig. 1. Schematic representation of experiment setups: (a) unidirectional flow (scenario name: uni-05), (b) bidirectional flow (scenario name: bi-b01), (c) 2-way crossing flow (scenario name: crossing-90-d08), (d) 3-way crossing flow (scenario name: crossing-120-b01), and (e) 4-way crossing flow crossing flow (scenario name: crossing-90-a10). Here, blue thick arrows show the walking direction of incoming pedestrians entering the corridors and red thin arrows indicates the walking direction of outgoing pedestrians leaving the corridors. A pair of gray rectangles is placed to set up an entrance of pedestrian group entering the corridor. (Color figure online)

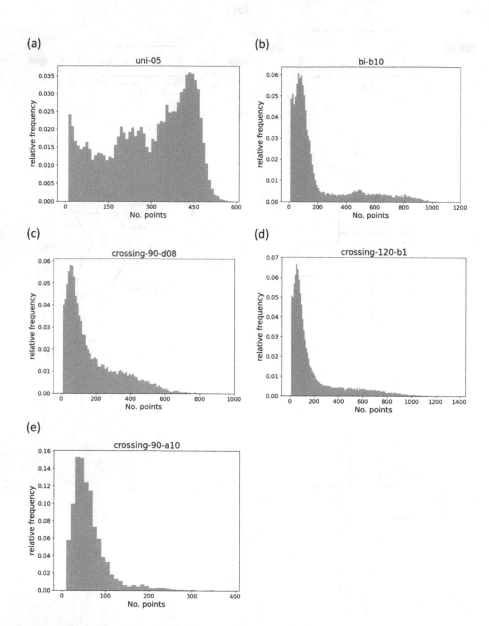

Fig. 2. Histogram of the number of data points per contact interaction trajectory (a) unidirectional flow (scenario name: uni-05), (b) bidirectional flow (scenario name: bi-b01), (c) 2-way crossing flow (scenario name: crossing-90-d08), (d) 3-way crossing flow (scenario name: crossing-120-b01), and (e) 4-way crossing flow crossing flow (scenario name: crossing-90-a10).

Table 1. Basic descriptive statistics of representative scenarios.

Setup	Scenario name	N	Period (s)	No. contacts
Uni-directional	uni-05	905	157.68	25390
Bi-directional	bi-b10	736	324.56	57126
2-way crossing	crossing-90-d08	592	147.88	41648
3-way crossing	crossing-120-b01	769	215.63	93156
4-way crossing	crossing-90-a10	324	94.08	9270

individuals N, experiment period, and the number of contacts. The number of contacts was given as the number of interacting individual pairs. Figure 2 presents histograms of trajectory length which is given in terms of the number of data points per contact interaction trajectory.

3   Data Analysis

Based on previous studies [20–22, 25], we evaluate a standardized value of maximal distance T_n for a trajectory containing n data points of position:

$$T_n = \frac{D_n}{\sqrt{(t_n - t_0)\hat{\sigma}^2}}, \tag{1}$$

where D_n is the maximal distance traveled from the initial position during the contact interaction, t_0 and t_n are the start and end time of contact interaction, and $\hat{\sigma}$ is the consistent estimator of the standard deviation of D_n. The maximal distance D_n is defined as

$$D_n = \max_{i=1,2,..,n} \| X(t_i) - X(t_0) \|. \tag{2}$$

Here, $X(t_i)$ denotes the position at time instance i and $X(t_0)$ for the position at the start of contact interaction. The consistent estimator $\hat{\sigma}$ is given as

$$\hat{\sigma}^2 = \frac{1}{2n\Delta t} \sum_{j=1}^{n} \| X(t_j) - X(t_{j-1}) \|^2, \tag{3}$$

where Δt is the time step size.

We can characterize different pedestrian contact interactions based on the value of T_n. A small T_n indicates that the individuals stay close to their initial position during the contact interaction, implying subdiffusive motion in Fig. 3(a). On the other hand, when individuals travel far away from their initial position during the contact interaction, one can observe large T_n, hinting at the possibility of Brownian motion. As can be seen from Fig. 3(b), individual j is entering and leaving the contact circle without changing walking direction significantly. It should be noted that subdiffusive motion in contact interaction trajectories

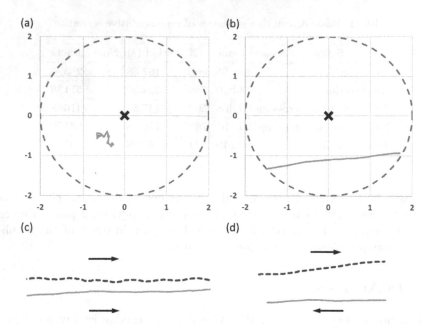

Fig. 3. Representative types of pedestrian contact interaction trajectories characterized based on the value of T_n. Small T_n implies subdiffusive motion while large T_n suggests Brownian motion: (a) Subdiffusive motion with $T_n = 0.525$ (individual id 25 and id 46 in unidirectional flow scenario uni-05) and (b) Brownian motion with $T_n = 2.5$ (individual id 65 and id 99 in 4-way crossing scenario crossing-90-a10). The position of focal individuals (id 25 in (a) and id 65 in (b)) is indicated at $(0, 0)$ by a black cross symbol $\times$. Blue dashed circles show contact range of a focal individual ($r_c = 2$ m). The relative motion of pedestrians interacting with the focal individuals (id 46 in (a) and id 99 in (b)) denoted by red solid lines. In the lower panels, blue dashed lines indicate the ground truth trajectories of focal individuals (id 25 in (c) and id 65 in (d)), and red solid lines for pedestrians interacting with the focal individuals (id 46 in (c) and id 99 in (d)). Arrows are guide for the eyes, indicating the walking direction of individuals. (Color figure online)

does not necessarily suggest that ground truth trajectories display subdiffusive motions. In the case of subdiffusive motion (see Fig. 3(c)), the selected individuals move in parallel along a straight line, showing directed motions.

We classify different motion types in line with the statistical test procedure presented in previous studies [20, 21, 25]. We set Brownian motion as the null hypothesis H_0 and subdiffusive motion as the alternative hypothesis H_1. In the context of the disease spreading processes, quantifying the motion types is useful to determine whether the contact interaction is brief or long-lived, suggesting risk of virus exposure in human face-to-face interactions. In this study, subdiffusive motion is characterized by small T_n, implying that the risk level is high, while Brownian motion yields high T_n, indicating that the risk level is lower. We define a critical region based on the knowledge of T_n distribution under the hypothesis H_0

$$q_\alpha \leq T_n, \tag{4}$$

where q_α is the quantile of T_n distribution, indicating that T_n lies in the critical region with the probability $1 - \alpha$. We use $q_\alpha = 0.785$ for $\alpha = 2.5\%$ according to Refs. [20,21,25].

Table 2. Summary of motion type classification results.

Setup	Scenario name	No. contacts	Subdiffusive motion	Brownian motion
Uni-directional	uni-05	25390	20848 (82.11%)	4542 (17.89%)
Bi-directional	bi-b10	57126	14465 (25.32%)	42661 (74.68%)
2-way crossing	crossing-90-d08	41648	12540 (30.11%)	29108 (69.89%)
3-way crossing	crossing-120-b01	93156	14492 (15.56%)	78664 (84.44%)
4-way crossing	crossing-90-a10	9270	532 (5.74%)	8738 (94.26%)

Figure 4 shows histograms of the standardized maximal distance T_n for representative experiment scenarios. The summary of motion type classification results can be found from Table 2. We can observe that Brownian motion is more frequently observed in most of the experiment scenarios, the exception being the unidirectional setup (scenario name uni-05). In contrast, almost all the motion in the unidirectional setup are categorized as subdiffusive motion, hinting at the possibility that the actual contact duration is much longer than one estimated based on the ballistic motion assumption. Similar to previous studies [1,2,5], we estimated the contact duration of ballistic motion assumption $t_{c,b}$ as

$$t_{c,b} = \frac{2r_c \left| \cos \theta_i - \cos \theta_j \right|}{v_0 (1 - \cos \theta_{ij})}, \tag{5}$$

where $r_c = 2$ m is the contact radius and v_0 is the initial value of relative speed between individuals i and j measured at the beginning of contact interaction. The heading of individuals i and j are denoted by θ_i and θ_j respectively, and the contact angle between the individuals is given as $\theta_{ij} = |\theta_i - \theta_j|$. Similar to v_0, the heading of individuals and the contact angle between them (i.e., θ_i, θ_j, and θ_{ij}) are measured at the beginning of contact interaction. Figure 5 illustrates the histograms of contact duration that measured from the experimental datasets t_c and the estimated contact duration under the ballistic motion assumption $t_{c,b}$. In the case of the unidirectional setup, one can see a striking difference between the distributions of t_c and that of $t_{c,b}$. In other setups, the distribution of $t_{c,b}$ shows good agreement with that of t_c.

To examine the reasons for the discrepancies in contact duration distributions in Fig. 5, we compared the distributions of relative speed of pedestrians in contact interactions measured from experimental data and those generated based on the ballistic motion assumption, see Fig. 6. Note that the average value of relative speed during contact interaction is used for the distribution of experimental data

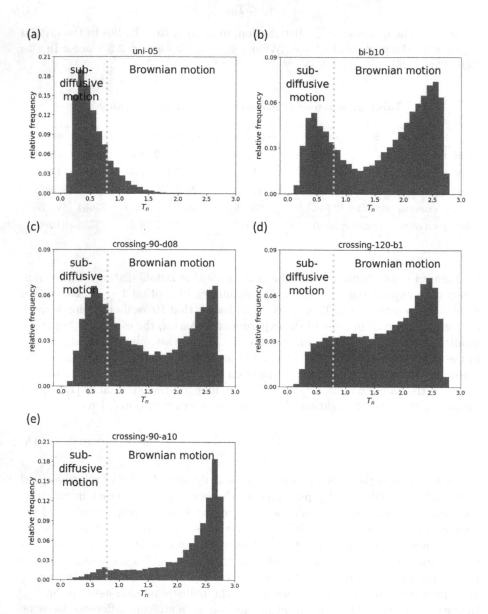

Fig. 4. Histogram of the standardized maximal distance T_n in trajectories (a) unidirectional flow (scenario name: uni-05), (b) bidirectional flow (scenario name: bi-b01), (c) 2-way crossing flow (scenario name: crossing-90-d08), (d) 3-way crossing flow (scenario name: crossing-120-b01), and (e) 4-way crossing flow crossing flow (scenario name: crossing-90-a10). The red dashed vertical lines indicate $\alpha = 2.5\%$ quantile of T_n distribution suggested in the previous studies [20,21]. (Color figure online)

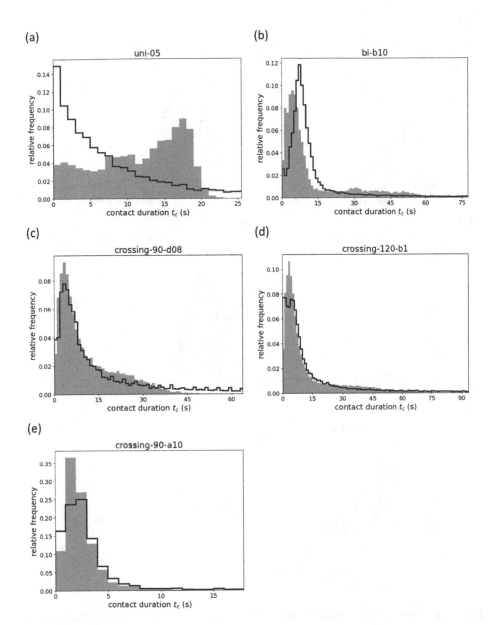

Fig. 5. Histogram of the contact duration. Light blue areas indicate the histogram of actual contact duration measured from the presented experimental datasets, i.e., t_c. Orange lines show the histogram of the contact duration estimated based on ballistic motion assumption, i.e., $t_{c,b}$ (see Eq. 5). (a) unidirectional flow (scenario name: uni-05), (b) bidirectional flow (scenario name: bi-b01), (c) 2-way crossing flow (scenario name: crossing-90-d08), (d) 3-way crossing flow (scenario name: crossing-120-b01), and (e) 4-way crossing flow crossing flow (scenario name: crossing-90-a10). (Color figure online)

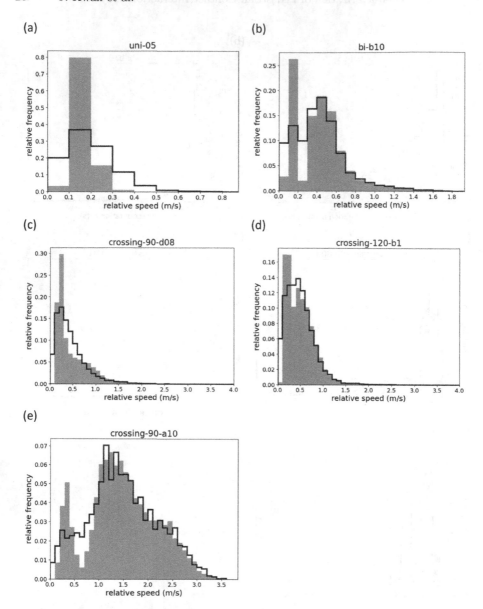

Fig. 6. Histogram of the relative speed of pedestrians in contact interactions. Light blue areas indicate the histogram of actual relative speed measured from the presented experimental datasets. Orange lines are for the relative speed of contact interaction estimated based on ballistic motion assumption (v_0 in Eq. 5). Note that the average value of relative speed during contact interaction is used for the distribution of experimental data and the initial value of relative speed v_0 measured at the beginning of contact interaction is used for the distribution of ballistic motion assumption. (a) unidirectional flow (scenario name: uni-05), (b) bidirectional flow (scenario name: bi-b01), (c) 2-way crossing flow (scenario name: crossing-90-d08), (d) 3-way crossing flow (scenario name: crossing-120-b01), and (e) 4-way crossing flow crossing flow (scenario name: crossing-90-a10). (Color figure online)

and the initial value of relative speed v_0 measured at the beginning of contact interaction is used for the distribution of ballistic motion assumption. As can be seen from Fig. 6, difference in the relative frequency distribution of relative speed is significant for the case of the unidirectional flow setup, but it is less significant for other experiment setups.

Table 3. Wasserstein distance metric between relative speed distributions presented in Fig. 6.

Setup	Scenario name	% Subdiffusive motion	Wasserstein distance
Uni-directional	uni-05	82.11%	0.0742
Bi-directional	bi-b10	25.32%	0.0145
2-way crossing	crossing-90-d08	30.11%	0.0155
3-way crossing	crossing-120-b01	15.56%	0.0065
4-way crossing	crossing-90-a10	5.74%	0.0036

Analogous to Refs. [34, 35], we quantify the difference in the relative frequency distributions shown in Fig. 6 by means of the Wasserstein distance. We use the 1-Wasserstein distance W_1 which is given as

$$W_1 = \sum_k |F(k) - F_b(k)| . \tag{6}$$

Here, $F(k)$ and $F_b(k)$ are the cumulative distribution of relative speed relative frequency from the experimental data and the one based on the ballistic motion assumption, respectively. We compute the cumulative distribution as $F(k) = \sum_{k' \leq k} f(k')$, where $f(k')$ is the the relative frequency of relative speed measured for histogram bin k'. Table 3 presents the Wasserstein distance measured for the presented experiment scenarios. The results show a general tendency that a higher proportion of subdiffusive motion are observed for larger values of Wasserstein distance. That is, the difference in relative speed distributions of experimental data and the ballistic motion assumption contributes considerably to the discrepancies in contact duration distributions.

Our analysis results suggest that random walk models based on ballistic motion have limitations in accounting for the influence of pedestrian flow geometrical conditions on contact duration distributions. In the case of unidirectional flow setup, the relative speed distribution of ballistic motion assumption shows a notable difference with that of experimental data. This results in an unrealistic contact duration distribution. Furthermore, geometrical conditions yielding larger difference in the relative speed distributions tend to generate more subdiffusive motions, suggesting higher risk of disease spreading. Thus, it is desirable to have smaller difference in the relative speed distributions especially for lower relative speed.

4   Conclusion

To examine the influence of pedestrian flow geometrical conditions on the contact duration distribution, we have analyzed pedestrian contact interaction trajectories of uni-, bi- and multi-directional flow setups in experimental data [28–30]. We have classified types of motions observed in the contact interactions based on standardized maximal distance T_n. In the unidirectional flow setup, most contact interaction trajectories have small T_n values. That is, the individuals stay close to their initial position during the contact interactions, thus subdiffusive motion is frequently observed. In contrast, other experiment setups yield higher T_n values. This indicates that individuals travel far away from their initial positions during the contact interactions, so Brownian motion is more frequently observed. It is noted that random walk models based on ballistic motion might not be able to generate realistic contact duration distributions depending on the geometrical conditions, especially for the case of unidirectional flow setup. This study also highlights that geometrical conditions yielding smaller difference in the relative speed distributions are preferred when diseases can be transmitted through face-to-face interactions.

A few selected experimental scenarios have been analyzed to study the fundamental role of pedestrian flow setups (e.g., uni-, bi-, and multi-directional flow) in the distribution of pedestrian motion types and contact duration. To generalize the findings of this study, the presented analysis should be further performed with larger number of scenarios and different layouts of pedestrian facilities. Another interesting extension of the presented study can be planned in line with machine learning algorithms, for instance, developing a prediction model and performing a feature importance analysis to identify factors influencing on the motion types and duration of contact interactions [22, 27, 36–38].

Acknowledgements. This research is supported by Ministry of Education (MOE) Singapore under its Academic Research Fund Tier 1 Program Grant No. RG12/21 MoE Tier 1.

References

1. Samar, P.M., Wicker, S.B.: Link dynamics and protocol design in a multihop mobile environment. IEEE Trans. Mob. Comput. **5**, 1156–1172 (2006)
2. Wu, Y.T., Liao, W., Tsao, C.L., Lin, T.N.: Impact of node mobility on link duration in multihop mobile networks. IEEE Trans. Veh. Technol. **58**, 2435–2442 (2008)
3. Hu, H., Nigmatulina, K., Eckhoff, P.: The scaling of contact rates with population density for the infectious disease models. Math. Biosci. **244**, 125–134 (2013)
4. Manlove, K., et al.: Defining an epidemiological landscape that connects movement ecology to pathogen transmission and pace-of-life. Ecol. Lett. **25**, 1760–1782 (2022)
5. Rast, M.P.: Contact statistics in populations of noninteracting random walkers in two dimensions. Phys. Rev. E **105**, 014103 (2022)
6. Saxton, M.J., Jacobson, K.: Single-particle tracking: applications to membrane dynamics. Annu. Rev. Biophys. Biomol. Struct. **26**, 373–399 (1997)

7. Manzo, C., Garcia-Parajo, M.F.: A review of progress in single particle tracking: from methods to biophysical insights. Rep. Prog. Phys. **78**, 124601 (2015)
8. Shen, H., et al.: Single particle tracking: from theory to biophysical applications. Chem. Rev. **117**, 7331–7376 (2017)
9. Benhamou, S.: How many animals really do the Lévy walk? Ecology **88**, 1962–1969 (2007)
10. Edelhoff, H., Signer, J., Balkenhol, N.: Path segmentation for beginners: an overview of current methods for detecting changes in animal movement patterns. Mov. Ecol. **4**, 21 (2016)
11. Getz, W.M., Saltz, D.: A framework for generating and analyzing movement paths on ecological landscapes. Proc. Natl. Acad. Sci. **105**, 19066–19071 (2008)
12. Rutten, P., Lees, M.H., Klous, S., Heesterbeek, H., Sloot, P.: Modelling the dynamic relationship between spread of infection and observed crowd movement patterns at large scale events. Sci. Rep. **12**, 14825 (2022)
13. Wilber, M.Q., et al.: A model for leveraging animal movement to understand spatio-temporal disease dynamics. Ecol. Lett. **25**, 1290–1304 (2022)
14. Qian, H., Sheetz, M.P., Elson, E.L.: Single particle tracking. Analysis of diffusion and flow in two-dimensional systems. Biophys. J. **60**, 910–921 (1991)
15. Michalet, X.: Mean square displacement analysis of single-particle trajectories with localization error: Brownian motion in an isotropic medium. Phys. Rev. E **82**, 041914 (2010)
16. Goulian, M., Simon, S.M.: Tracking single proteins within cells. Biophys. J. **79**, 2188–2198 (2000)
17. Hubicka, K., Janczura, J.: Time-dependent classification of protein diffusion types: a statistical detection of mean-squared-displacement exponent transitions. Phys. Rev. E **1010**, 022107 (2020)
18. Murakami, H., Feliciani, C., Nishinari, K.: Lévy walk process in self-organization of pedestrian crowds. J. R. Soc. Interface **16**, 20180939 (2019)
19. Murakami, H., Feliciani, C., Nishiyama, Y., Nishinari, K.: Mutual anticipation can contribute to self-organization in human crowds. Sci. Adv. **7**, eabe7758 (2021)
20. Briane, V., Vimond, M., Kervrann, C.: An adaptive statistical test to detect non Brownian diffusion from particle trajectories. In: 2016 IEEE 13th International Symposium on Biomedical Imaging (ISBI), pp. 972–975, IEEE, Prague (2016)
21. Briane, V., Kervrann, C., Vimond, M.: Statistical analysis of particle trajectories in living cells. Phys. Rev. E **97**, 062121 (2018)
22. Janczura, J., Kowalek, P., Loch-Olszewska, H., Szwabiński, J., Weron, A.: Classification of particle trajectories in living cells: machine learning versus statistical testing hypothesis for fractional anomalous diffusion. Phys. Rev. E **102**, 032402 (2020)
23. Kepten, E., Weron, A., Sikora, G., Burnecki, K., Garini, Y.: Guidelines for the fitting of anomalous diffusion mean square displacement graphs from single particle tracking experiments. PLoS ONE **10**, e0117722 (2015)
24. Burnecki, K., Kepten, E., Garini, Y., Sikora, G., Weron, A.: Estimating the anomalous diffusion exponent for single particle tracking data with measurement errors- An alternative approach. Sci. Rep. **10**, 11306 (2015)
25. Weron, A., Janczura, J., Boryczka, E., Sungkaworn, T., Calebiro, D.: Statistical testing approach for fractional anomalous diffusion classification. Phys. Rev. E **99**, 042149 (2019)
26. Janczura, J., Burnecki, K., Muszkieta, M., Stanislavsky, A., Weron, A.: Classification of random trajectories based on the fractional Lévy stable motion. Chaos, Solitons Fractals **154**, 111606 (2022)

27. Kowalek, P., Loch-Olszewska, H., Łaszczuk, Ł, Opała, J., Szwabiński, J.: Boosting the performance of anomalous diffusion classifiers with the proper choice of features. J. Phys. A: Math. Theor. **55**, 244005 (2022)
28. Holl, S.: Methoden für die Bemessung der Leistungsfähigkeit multidirektional genutzter Fußverkehrsanlagen. Bergische Universität, Wuppertal (2016)
29. Cao, S., Seyfried, A., Zhang, J., Holl, S., Song, W.: Fundamental diagrams for multidirectional pedestrian flows. J. Stat. Mech: Theory Exp. **2017**, 033404 (2017)
30. Data archive of experimental data from studies about pedestrian dynamics. https://ped.fz-juelich.de/da/doku.php. Accessed 1 Mar 2023
31. Han, E., et al.: Lessons learnt from easing COVID-19 restrictions: an analysis of countries and regions in Asia Pacific and Europe. Lancet **396**, 1525–1534 (2020)
32. Ronchi, E., Lovreglio, R.: EXPOSED: An occupant exposure model for confined spaces to retrofit crowd models during a pandemic. Saf. Sci. **130**, 104834 (2020)
33. Garcia, W., Mendez, S., Fray, B., Nicolas, A.: Model-based assessment of the risks of viral transmission in non-confined crowds. Saf. Sci. **144**, 105453 (2021)
34. Chkhaidze, K., et al.: Spatially constrained tumour growth affects the patterns of clonal selection and neutral drift in cancer genomic data. PLoS Comput. Biol. **15**, e1007243 (2019)
35. Flam-Shepherd, D., Zhu, K., Aspuru-Guzik, A.: Language models can learn complex molecular distributions. Nat. Commun. **13**, 3293 (2022)
36. Kowalek, P., Loch-Olszewska, H., Szwabiński, J.: Classification of diffusion modes in single-particle tracking data: feature-based versus deep-learning approach. Phys. Rev. E **100**, 032410 (2019)
37. Wagner, T., Kroll, A., Haramagatti, C.R., Lipinski, H.G., Wiemann, M.: Classification and segmentation of nanoparticle diffusion trajectories in cellular micro environments. PLoS ONE **12**, e0170165 (2017)
38. Pinholt, H.D., Bohr, S.S.R., Iversen, J.F., Boomsma, W., Hatzakis, N.S.: Single-particle diffusional fingerprinting: a machine-learning framework for quantitative analysis of heterogeneous diffusion. Proc. Natl. Acad. Sci. **118**, e2104624118 (2021)

Siamese Autoencoder-Based Approach for Missing Data Imputation

Ricardo Cardoso Pereira[1]([⊠]), Pedro Henriques Abreu[1],
and Pedro Pereira Rodrigues[2]

[1] Centre for Informatics and Systems of the University of Coimbra,
Department of Informatics Engineering, University of Coimbra, 3030-290 Coimbra,
Portugal
{rdpereira,pha}@dei.uc.pt
[2] Center for Health Technology and Services Research,
Faculty of Medicine (MEDCIDS), University of Porto, 4200-319 Porto, Portugal
pprodrigues@med.up.pt

Abstract. Missing data is an issue that can negatively impact any task performed with the available data and it is often found in real-world domains such as healthcare. One of the most common strategies to address this issue is to perform imputation, where the missing values are replaced by estimates. Several approaches based on statistics and machine learning techniques have been proposed for this purpose, including deep learning architectures such as generative adversarial networks and autoencoders. In this work, we propose a novel siamese neural network suitable for missing data imputation, which we call Siamese Autoencoder-based Approach for Imputation (SAEI). Besides having a deep autoencoder architecture, SAEI also has a custom loss function and triplet mining strategy that are tailored for the missing data issue. The proposed SAEI approach is compared to seven state-of-the-art imputation methods in an experimental setup that comprises 14 heterogeneous datasets of the healthcare domain injected with Missing Not At Random values at a rate between 10% and 60%. The results show that SAEI significantly outperforms all the remaining imputation methods for all experimented settings, achieving an average improvement of 35%.

Keywords: Missing Data · Imputation · Siamese Autoencoder · Missing Not At Random

1 Introduction

Missing data can be described by the absence of values in the instances of a dataset. It is one of the most common data issues, affecting most real-word domains. The existence of missing values has a deep impact on the conclusions that can be drawn from the data [10]. Within machine learning, the majority of the classification and regression models cannot cope with missing data, or suffer a decrease in their performance. In domains such as healthcare, missing values can compromise the results to the point where the study becomes unfeasible [12].

© The Author(s), under exclusive license to Springer Nature Switzerland AG 2023
J. Mikyška et al. (Eds.): ICCS 2023, LNCS 14073, pp. 33–46, 2023.
https://doi.org/10.1007/978-3-031-35995-8_3

However, different types of missing values may have different impacts. Missing data can be classified according to three different mechanisms which are related to the missingness causes [8,15]:

- Missing Completely At Random (MCAR), which is described by Eq. 1, where $P(x)$ is the probability function, R is the binary matrix of the missing values in Y, Y is a combination of the observed values Y_{obs} and the missing values Y_{mis}, and ψ are the parameters of the missing data model. In this mechanism the missingness causes are purely random, and therefore only depended on ψ. An example would be if someone simply forgot to fill in a field in a questionnaire.

$$P(R = 0|Y_{obs}, Y_{mis}, \psi) = P(R = 0|\psi) \tag{1}$$

- Missing At Random (MAR), which is described by Eq. 2 and the missingness nature is related to observed data in one or more features, therefore depending on both Y_{obs} and ψ. For example, people may not have results for a specific medical exam because they are too young to be doing such an exam.

$$P(R = 0|Y_{obs}, Y_{mis}, \psi) = P(R = 0|Y_{obs}, \psi) \tag{2}$$

- Missing Not At Random (MNAR), where the missingness causes are unknown since the missing values are related with themselves or to external data not available. As a consequence, the missingness depends on Y_{obs}, Y_{mis} and ψ. For example, people that drink a lot of alcohol may be apprehensive about answering how many drinks they have per day.

A solution often used to addressed the missing data issue is imputation, which consists in the replacing the missing values by estimates [19]. There are several statistical and machine learning-based methods to perform imputation, and each one may be more suitable for a specific missing mechanism [10]. However, most approaches are tailored for MCAR and MAR, while MNAR still is the less addressed mechanism with few solutions, mostly because of its relation with unknown data. The proposal of new approaches that perform well under the MNAR mechanism is an important open challenge when considering that most real-world contexts suffer from this mechanism, including critical domains such as healthcare [13].

Recently, several deep learning architectures have been explored to perform imputation assuming the different mechanisms. The state-of-the-art architectures in this scope are generative adversarial networks (GAN) [21], denoising autoencoders (DAE) [5,20] and variational autoencoders (VAE) [11,13]. In this work, we explore the use of siamese networks [6] to perform missing data imputation. We propose a new model called Siamese Autoencoder-based Approach for Imputation (SAEI), which extends and adapts the vanilla siamese network for the imputation task by adding a deep autoencoder architecture, a custom loss function and a custom triplet mining strategy. To the best of our knowledge, this is the first time siamese networks have been used for this purpose. We compared our SAEI model with a baseline of seven state-of-the-art imputation methods in a experimental setup encompassing 14 datasets of the healthcare domain injected with MNAR values at different missingness levels (10% to 60% missing rates). The achieved results proved that our SAEI model significantly outperformed the remaining baseline methods in all experimented settings, achieving an average improvement of 35%.

The remainder of the paper is organized in the following way: Sect. 2 presents related work on the missing data field; Sect. 3 describes with detail the proposed SAEI model; Sect. 4 presents the experimental setup used to validate the effectiveness of our SAEI model; Sect. 5 analyzes and discusses the results obtained from the experiments; and Sect. 6 presents our conclusions and the future work.

2 Related Work

As previously stated, the missing data problem cannot be neglected, otherwise it will have a major impact in any tasks performed with the data. An approach often used to deal with this issue is to remove the missing values, which is called case deletion. Two different deletion approaches can be applied [10]: listwise deletion, where all instances containing at least one missing value are eliminated, which is a very simple procedure but can lead to the lost of a considerable amount of information, being for that reason only recommended in big data scenarios or when the instances with missing values are less than 5% of the data; and pairwise deletion, where only the instances containing missing values for the features of interest are deleted, which suffers from the same problems of listwise deletion but reduces the loss of information. To apply any deletion strategies, the missing mechanism must be considered since deleting data may remove relations between instances and features, and for that reason this approach should only be applied with missing values under MCAR [10].

When deleting instances is not a suitable option, imputation tends to be the preferred strategy [19]. The key idea is to generate plausible new estimates to replace the missing values. Among the existent methods, the statistical-based ones are frequently used [8]. A common approach is to use the mean or mode of the feature containing missing values for numeric and nominal variables, respectively [10]. Another strategy is to use a regression to model one or more dependent variables (the ones containing missing values) using as independent variables the remaining features of the dataset [10]. The fitting process must only consider the instances that are complete for the independent features, and the type of regression (e.g., linear or non-linear) must be chosen taking in consideration the nature of the data. Both latter approaches are said to be single imputation strategies, since only one value is used for the imputation of each missing value. For this reason, the imputation results may be biased since the uncertainty of the generated values is not accounted for. To address this issue multiple imputation strategies can be applied. The key idea is to perform the imputation M times, generating different but complete results each time (different imputation methods may be used). The M complete datasets are then analyzed and combined, being the result of this combination the final dataset

[16]. The state-of-the-art method that uses this approach is the Multiple Imputation by Chained Equations (MICE) [4]. It creates a series of regressions where each one is modeled for a different variable with missing values, meaning that each feature is modeled conditionally upon the other features. The process is repeated throughout several iterations until the parameters of the regressions converge to stable values [1].

Other approaches based on statistical and algebra concepts often used are matrix completion methods, such as the SoftImpute. The goal is to perform the imputation by finding a low-rank matrix that is an approximation of the original one. The process usually relies on matrix factorization, where a matrix with (m, n) dimensions is decomposed into two matrices with dimensions (m, k) and (k, n), where k is the rank and must be smaller than m and n. The idea is to find the latent features that better describe the available values. The low-rank approximation matrix can finally be obtained by multiplying the resulting matrices from the decomposition process [18].

The imputation task can also be performed by using machine learning models. In theory, any algorithm that can be trained to predict new values is suitable for imputation [8]. Generating new estimates for numerical features may be seen as a regression problem, or a classification problem for nominal features.

An algorithm often used for this purpose is the k-nearest neighbors (KNN) [2,7]. It tries to find the k most similar instances to the one that contains missing values using only the features that are complete. To calculate this similarity a distance function is used, and it must be chosen taking in consideration the data type: the euclidean distance is suitable for numeric data but not for nominal. For this latter case the data can be transformed through one-hot encoding or a different distance function must be used (e.g., hamming distance works with nominal data) [3]. When $k > 1$ the values of the neighbors' instances must be combined to produce the new value. For numerical data a common approach used is the simple or weighted mean, and for categorical values a vote of majority may be applied.

Neural networks are also often used for imputation, particularly autoencoders and generative adversarial networks (GAN). Autoencoders are a type of neural network that learns a representation of the data from the input layer and tries to reproduce it at the output layer. Among its variants, the denoising autoencoder (DAE) is the one more often used for missing data imputation because it is designed to recover noisy data, which can exist due to data corruption via some additive mechanism or by the introduction of missing values [5]. However, the variational autoencoder (VAE) has recently been used for the same purpose. This architecture learns the multidimensional parameters of a Gaussian distribution and generates the estimates to replace the missing values by sampling from that distribution [13]. Regarding GANs, these are trained through an adversarial process and are based on a generative model that learns the data distribution and a discriminative model that outputs the probability of a sample being from the training data rather than the generative model [9]. GANs are directly applied to the missing data imputation task through the generative adversarial imputation nets (GAIN) [21], where the generator performs the imputation and the discriminator tries to distinguish between the original and the imputed data.

A type of neural network that has not yet been used for missing data imputation is the siamese network. This architecture is based on two or more subnetworks that share the same architecture and parameters. The goal is to output embedding representations in a way that similar concepts would have a similar latent space embedding [6,17]. In this work, we propose an extension to the siamese network tailored for missing data imputation, which is, to the best of our knowledge, the first time that such architecture is being used for this purpose. Since this network uses a triplet loss function, which is based on distances between anchor, positive and negative samples, it can learn from reduced quantities of data (especially when compared to other neural network-based approaches). This quality makes this type of network feasible to be used with both lower and higher missing rates. Furthermore, the number of complete samples (i.e., without containing missing values) needed to train the model can be rather small when compared to other network-based imputation models. We leverage these characteristics of the siamese networks and propose an adapted method for missing data imputation with a novel deep autoencoder architecture, custom loss function and custom triplet mining strategy.

3   Siamese Autoencoder-Based Approach for Imputation

The Siamese Autoencoder-based Approach for Imputation (SAEI) is an extension of a vanilla siamese network, and it is tailored for missing data imputation by comprising three adaptations:

- A deep autoencoder architecture that allows the network to reproduce the input data at the output layer in an unsupervised fashion;
- A custom loss function that includes both the distance-based triplet loss and the reconstruction error of the autoencoder component of the network;
- A custom triplet mining strategy that was designed specifically for the missing data issue by creating hard triplets based on the existing missing values.

These three adaptations are independently described in the next subsections.

3.1   Deep Autoencoder Architecture

The architecture of our SAEI model is roughly inspired in the well-know ZFNet [22], although it presents several changes motivated by it being aimed at tabular data. Figure 1 and Fig. 2 depict graphical descriptions of the encoder and decoder networks, respectively. The encoder network is composed of two one-dimensional convolutional layers with 16 filters, ReLU as the activation function, and kernel sizes of five and three. Such layers are followed by max-pooling layers with two strides, which are then followed by two regularization layers that perform batch normalization and dropout at a rate of 25%. Moreover, a residual connection is also used two skip the second convolutional layer. Finally, the output

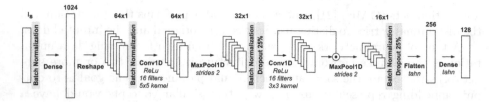

Fig. 1. SAEI encoder architecture. I_s represents the input shape. The encoder network is composed of two one-dimensional convolutional layers with 16 filters followed by max-pooling layers with two strides. Regularization is performed through batch normalization and dropout at a rate of 25%. A residual connection is also used two skip the second convolutional layer. The latent output is obtained from a dense layer with 128 units.

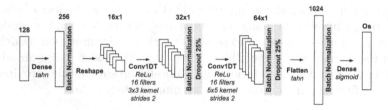

Fig. 2. SAEI decoder architecture. O_s represents the output shape. The decoder network is symmetric to the encoder. Therefore, it has the same layers and regularization but in reserve order. The residual connection is not applied by the decoder.

from the last pooling layer is flattened and passed through a hyperbolic tangent activation function, and the latent output is obtained from a final dense layer with 128 units and the latter activation function. The decoder network is symmetric to the encoder, presenting the same architecture but in reserve order (without the residual connection). To perform the deconvolution operation, the one-dimensional convolutional and the max pooling layers are replaced by one-dimensional transposed convolutional layers with two strides. The output dense layer uses the sigmoid activation function so that the data can be normalized within $[0, 1]$.

Convolutional layers operate based on the spatial positioning of the data. In other words, the position where each value is placed is relevant for the feature extraction process. Tabular data does not present this behavior because the features' positions are irrelevant. To surpass this limitation, the SAEI model feeds the input data to a dense layer with 1024 units and without activation function. The purpose of this layer is to learn an abstract representation of the input data where the spatial relation between the new abstract features is meaningful. Therefore, our SAEI model delegates the task of learning the spatial structure of the features to the network. The first convolutional layer is then fed with the output of the mentioned dense layer.

3.2 Custom Loss Function

One of the original loss functions proposed to train a siamese network was the triplet loss. It tries to minimize the distance between an anchor and a positive sample which represent the same concept, while maximizing the distance between that same anchor and a negative sample of a different concept. The formulation is presented in Eq. 3, where $f(x)$ is a function of the embedding representation, N is the number of triples composed by an anchor (x_i^a), a positive sample (x_i^p) and a negative one (x_i^n) [17]. Furthermore, α is a margin used to ensure a minimum distance between positive and negative samples.

$$TL_i = \sum_i^N \left[\|f(x_i^a) - f(x_i^p)\|_2^2 - \|f(x_i^a) - f(x_i^n)\|_2^2 + \alpha \right] \tag{3}$$

Using the triplet loss function is ideal for a vanilla siamese network since the goal is for the model to distinguish between positive and negative concepts in comparison to the anchor, while outputting embedding representations that incorporate such differences. However, our SAEI model relies on an autoencoder architecture that outputs a reconstruction of the input data based on the anchor latent representation. Therefore, this reconstruction component must also be reflected in the loss function of the model. Equation 4 presents our custom loss function, where the triplet loss (TL_i) from Eq. 3 is added to the mean squared error between the anchor and positive sample, similarly to what would happen in a denoising autoencoder (i.e., the anchor is the corrupted version of the ground truth represented by the positive sample).

$$TL_i + \sum_i^N (\hat{x}_i^a - x_i^p)^2 \tag{4}$$

3.3 Custom Triplet Mining

The triplet selection is a key step for successfully training a siamese network. To ensure that the network is able to learn how to distinguish between the positive and negative concepts, it is imperative to select triplets that the network is unable to differentiate before being trained. These triplets are composed by the so-called hard positives and hard negatives, which are samples that violate the constraint of the triplet loss, presented in Eq. 5 [17]. If the network is trained with easy triples where the triplet loss constraint is already satisfied before training, it would not gain the capacity of distinguishing between positive and negative concepts.

$$\|f(x_i^a) - f(x_i^p)\|_2^2 + \alpha < \|f(x_i^a) - f(x_i^n)\|_2^2 \tag{5}$$

When extending this type of network and its respective training procedure to address missing data imputation, the definitions of anchor, negative and positive samples must be redefined within the missing data scope. Our SAEI model proposes the following definitions:

- Anchors are the samples containing missing values, which are pre-imputed with the mean of the available values in each feature (or the mode in categorical features). This pre-imputation step is required since neural networks are unable to be trained with data containing missing values.
- Positives are the original samples without containing missing values (i.e., the ground truth of the anchor samples). This implies that we must have a portion of complete data to train the SAEI model, but such assumption is common to all deep learning-based imputation methods. Furthermore, the anchor samples are created by artificially generating missing values in the positive samples, according to a pre-established missing data mechanism.
- Negatives are the same as anchor samples but with the missing values being replaced by Gaussian noise sampled according to Eq. 6, where D represents the entire dataset and λ represents the variance of the noise domain.

$$\mathcal{N}\left(\frac{\max(D) - \min(D)}{2}, \lambda\right)$$ (6)

The rational behind the mean value is to use the midpoint within the domain of the data, therefore keeping the Gaussian noise centered on that domain. Such strategy works specially well when the data is normalized within a specific range (e.g., $[0, 1]$). Moreover, the λ parameter acts as a control variable to define how hard or easy should the triplets be: a low λ value leads to Gaussian noise mostly or completely contained within the data domain, therefore leading to harder triplets since the negative sample is likely to be partially overlapped with the positive one; on the other hand, a high λ value increases the noise domain and creates a negative sample that is more dissimilar to the positive, therefore creating easier triplets where the negative and positive samples are immediately distinguishable by the model. As a consequence, this parameter should be defined aiming to generate hard triplets while considering for the data domain.

4   Experimental Setup

The quality of the imputation results achieved by the SAEI model was assessed by comparing it with a baseline of seven state-of-the-art imputation methods introduced in Sect. 2: kNN with $k = 5$, Mean/Mode, MICE with 100 iterations, SoftImpute with 100 iterations, DAE, VAE and GAIN. The DAE and VAE were defined according to the following architecture and hyperparameters: a hidden layer with half of the input dimension and ReLU as the activation function (although VAE has two additional layers for the Gaussian distribution parameters), Adam as the optimizer with a learning rate of 0.001, Sigmoid as the activation function of the output layer (forcing the data to be normalized within $[0, 1]$), batches of 64 samples, 200 training epochs, a dropout rate of 10% for regularization, Mean Squared Error as the reconstruction loss, and early stop and learning rate reduction by 80% if the validation loss does not improve over 100 epochs. The VAE loss function also includes the Kullback-Leibler divergence for regularization. Furthermore, both DAE and VAE require pre-imputation, which is performed with the mean/mode of the features. The GAIN method was used with the hyperparameters proposed by its authors. Our SAEI model follows the architecture described in Sect. 3.1 while all the remaining hyperparameters are the same used by the DAE and VAE. Moreover, the Gaussian noise of the negative instances was sampled from $\mathcal{N}(0.5, 0.05^{0.5})$, and the margin of the triplet loss function was set to $\alpha = 0.2$ as the original authors proposed. With the exception of GAIN, the described hyperparameters of all methods were defined through a grid search process. This is a standard procedure that aims to achieve hyperparameters that conform to common use cases. Moreover, the DAE and VAE were implemented with the Keras library, the GAIN implementation was obtained from its original authors[1], and the remaining methods were obtained and used from the Scikit-learn library. Our SAEI model was also coded with the Keras library and is available on GitHub[2].

The experiments were conducted over 14 datasets of the healthcare domain which cover different types of clinical data that was collected for several pathologies. We chose to cover this medical domain since it often suffers from the missing data issue, particularly missing values under the MNAR mechanism, which creates deep challenges to any subsequent analysis or task performed with the data [12]. In this health context, each instance usually represents a group of values collected for a patient, and each value belongs to a feature that could be a measurement, an exam result, or any other medical input. The missing values may appear at any feature and/or instance. All the 14 datasets are public and available at the UC Irvine Machine Learning[3] and Kaggle[4] repositories, and they present heterogeneous characteristics as seen in Table 1.

The datasets were normalized within [0, 1] and split into train and test sets with 70% and 30% of the instances, respectively. The scaler learns the minimum and maximum from the train set and transforms both sets. This strategy ensures the test data is not biased with information from the training data. However, in the presence of high missing rates (usually above 50%), the test set may contain values unseen by the scaler, which leads to the normalization boundaries being slightly extended from the expected [0, 1] domain. Also, the autoencoder-based methods use 20% of the train set as the validation set. Furthermore, the categorical nominal features were transformed through the one-hot encoding procedure so that they can be supported by all imputation methods. Finally, in order to be able to calculate the imputation error of the experimented methods, the datasets must be complete and the test set must be injected with artificially generated missing values so that we can compare the estimated values with the original ones (i.e., ground truth). We choose to generate MNAR values since it is the hardest mechanism to address and the one more often found in real-world contexts such as healthcare [12,13]. Our generation

---

[1] https://github.com/jsyoon0823/GAIN.
[2] https://github.com/ricardodcpereira/SAEI.
[3] https://archive.ics.uci.edu/ml.
[4] https://www.kaggle.com/datasets.

Table 1. Datasets characteristics.

Dataset	# Instances	# Features	
		Categorical	Continuous
diabetic-retinopathy	1151	4	16
ecoli	336	1	7
ctg-2c	2126	1	21
new-thyroid-N-vs-HH	215	1	5
kala-azar	68	2	5
immunotherapy	90	3	5
saheart	462	2	8
bc-coimbra	116	1	9
cleveland-0-vs-4	173	1	13
newthyroid-v1	185	1	5
biomed	194	1	5
cryotherapy	90	3	4
thyroid-3-vs-2	703	1	21
pima	768	1	8

strategy is based on removing the smaller values of several features at once (upon a certain missing rate) in a multivariate procedure. Therefore, the missing rate is defined for the entire dataset and the imputation is performed on all features simultaneously. The obtained results were evaluated with the Mean Absolute Error (MAE) between the estimates and the ground truth values. Regarding the missing rate, we considered four levels of missingness: 10%, 20%, 40% and 60%.

To avoid the impact of stochastic behaviors in the results, the experiment was executed 30 independent times, with the train/test split being randomly performed in every run. The MAE results of the experiment are the average of the 30 runs. Moreover, each run was executed in a computer with the following specifications: Windows 11, CPU AMD Ryzen 5600X, 16 GB RAM, and GPU NVIDIA GeForce GTX 1060 6GB. The time complexity of each imputation method was not measured, but all methods were computed in a feasible amount of time.

5   Results

The imputation results obtained from the experimental setup are displayed with detail in Table 2. The MAE results (average and standard deviation of the 30 independent runs) are individually displayed for each dataset and missing rate. Furthermore, the overall results for each imputation method and per missing rate (i.e., considering all datasets) are displayed in Fig. 3.

In an overall analysis, our SAEI model clearly outperforms all the remaining imputation methods, achieving smaller MAE values for every dataset and

Table 2. Experimental results per dataset and missing rate (average ± standard deviation of MAE). The best results are bolded and highlighted.

Dataset	MR (%)	kNN	Mean/Mode	MICE	SoftImpute	DAE	VAE	GAIN	SAEI
diabetic-retinopathy	10	0.133 ± 0.02	0.194 ± 0.01	0.119 ± 0.02	0.165 ± 0.02	0.152 ± 0.01	0.178 ± 0.01	0.192 ± 0.02	**0.109 ± 0.01**
	20	0.157 ± 0.02	0.213 ± 0.02	0.128 ± 0.02	0.185 ± 0.03	0.164 ± 0.02	0.175 ± 0.01	0.216 ± 0.03	**0.100 ± 0.01**
	40	0.207 ± 0.03	0.252 ± 0.03	0.179 ± 0.03	0.236 ± 0.03	0.199 ± 0.02	0.163 ± 0.01	0.264 ± 0.04	**0.131 ± 0.01**
	60	0.324 ± 0.05	0.358 ± 0.05	0.292 ± 0.05	0.354 ± 0.04	0.298 ± 0.05	0.242 ± 0.04	0.367 ± 0.06	**0.231 ± 0.04**
ecoli	10	0.213 ± 0.03	0.304 ± 0.04	0.229 ± 0.03	0.286 ± 0.04	0.265 ± 0.03	0.328 ± 0.03	0.337 ± 0.06	**0.173 ± 0.03**
	20	0.224 ± 0.04	0.309 ± 0.04	0.236 ± 0.04	0.297 ± 0.04	0.273 ± 0.03	0.306 ± 0.03	0.327 ± 0.06	**0.151 ± 0.03**
	40	0.363 ± 0.08	0.413 ± 0.07	0.363 ± 0.08	0.367 ± 0.05	0.347 ± 0.07	0.352 ± 0.05	0.456 ± 0.08	**0.229 ± 0.06**
	60	0.750 ± 0.21	0.759 ± 0.20	0.730 ± 0.20	0.703 ± 0.18	0.674 ± 0.21	0.672 ± 0.20	0.800 ± 0.22	**0.589 ± 0.20**
ctg-2c	10	0.151 ± 0.02	0.288 ± 0.03	0.137 ± 0.02	0.155 ± 0.02	0.188 ± 0.02	0.255 ± 0.02	0.261 ± 0.02	**0.121 ± 0.02**
	20	0.191 ± 0.02	0.296 ± 0.02	0.161 ± 0.01	0.167 ± 0.01	0.201 ± 0.02	0.227 ± 0.01	0.271 ± 0.02	**0.116 ± 0.01**
	40	0.318 ± 0.03	0.362 ± 0.03	0.286 ± 0.04	0.233 ± 0.02	0.293 ± 0.04	0.205 ± 0.02	0.347 ± 0.03	**0.191 ± 0.03**
	60	0.498 ± 0.05	0.526 ± 0.05	0.488 ± 0.05	0.399 ± 0.05	0.508 ± 0.11	0.356 ± 0.06	0.511 ± 0.06	**0.354 ± 0.06**
new-thyroid-N-vs-HH	10	0.264 ± 0.07	0.270 ± 0.05	0.266 ± 0.08	0.258 ± 0.05	0.272 ± 0.06	0.413 ± 0.04	0.361 ± 0.08	**0.124 ± 0.04**
	20	0.273 ± 0.04	0.264 ± 0.04	0.256 ± 0.05	0.227 ± 0.04	0.238 ± 0.04	0.408 ± 0.04	0.369 ± 0.08	**0.143 ± 0.03**
	40	0.269 ± 0.04	0.296 ± 0.04	0.269 ± 0.04	0.204 ± 0.03	0.247 ± 0.04	0.427 ± 0.04	0.392 ± 0.09	**0.156 ± 0.03**
	60	0.355 ± 0.10	0.432 ± 0.09	0.391 ± 0.09	0.257 ± 0.07	0.374 ± 0.11	0.502 ± 0.07	0.532 ± 0.11	**0.239 ± 0.12**
kala-azar	10	0.317 ± 0.07	0.358 ± 0.07	0.288 ± 0.06	0.300 ± 0.07	0.356 ± 0.06	0.488 ± 0.09	0.398 ± 0.08	**0.217 ± 0.00**
	20	0.409 ± 0.05	0.401 ± 0.05	0.369 ± 0.04	0.357 ± 0.05	0.404 ± 0.07	0.508 ± 0.05	0.455 ± 0.07	**0.232 ± 0.05**
	40	0.468 ± 0.07	0.472 ± 0.07	0.468 ± 0.08	0.408 ± 0.04	0.441 ± 0.09	0.589 ± 0.07	0.540 ± 0.12	**0.277 ± 0.05**
	60	0.951 ± 0.35	0.952 ± 0.35	0.953 ± 0.35	0.805 ± 0.31	0.908 ± 0.35	1.025 ± 0.34	0.983 ± 0.35	**0.796 ± 0.36**
immunotherapy	10	0.346 ± 0.08	0.384 ± 0.08	0.341 ± 0.09	0.362 ± 0.08	0.348 ± 0.08	0.433 ± 0.09	0.388 ± 0.09	**0.285 ± 0.08**
	20	0.370 ± 0.07	0.372 ± 0.07	0.349 ± 0.07	0.384 ± 0.06	0.337 ± 0.06	0.447 ± 0.06	0.386 ± 0.07	**0.268 ± 0.06**
	40	0.463 ± 0.08	0.463 ± 0.08	0.460 ± 0.09	0.469 ± 0.07	0.396 ± 0.08	0.524 ± 0.07	0.498 ± 0.08	**0.345 ± 0.08**
	60	1.026 ± 0.38	1.028 ± 0.38	1.015 ± 0.36	0.958 ± 0.36	0.929 ± 0.36	1.084 ± 0.36	1.027 ± 0.38	**0.871 ± 0.36**
saheart	10	0.289 ± 0.03	0.367 ± 0.03	0.295 ± 0.03	0.245 ± 0.03	0.315 ± 0.03	0.344 ± 0.03	0.387 ± 0.05	**0.231 ± 0.03**
	20	0.335 ± 0.04	0.389 ± 0.04	0.336 ± 0.03	0.268 ± 0.03	0.336 ± 0.04	0.332 ± 0.02	0.410 ± 0.05	**0.216 ± 0.03**
	40	0.464 ± 0.07	0.486 ± 0.06	0.464 ± 0.07	0.366 ± 0.05	0.442 ± 0.07	0.361 ± 0.04	0.531 ± 0.07	**0.296 ± 0.06**
	60	0.876 ± 0.28	0.884 ± 0.28	0.869 ± 0.28	0.732 ± 0.27	0.854 ± 0.29	0.729 ± 0.29	0.934 ± 0.29	**0.677 ± 0.26**
bc-coimbra	10	0.232 ± 0.04	0.300 ± 0.05	0.232 ± 0.04	0.209 ± 0.03	0.328 ± 0.04	0.464 ± 0.04	0.326 ± 0.06	**0.177 ± 0.05**
	20	0.277 ± 0.03	0.317 ± 0.03	0.269 ± 0.03	0.226 ± 0.02	0.319 ± 0.06	0.482 ± 0.02	0.357 ± 0.05	**0.175 ± 0.03**
	40	0.391 ± 0.06	0.411 ± 0.06	0.390 ± 0.07	0.293 ± 0.05	0.363 ± 0.08	0.559 ± 0.05	0.427 ± 0.08	**0.221 ± 0.05**
	60	0.661 ± 0.15	0.691 ± 0.16	0.656 ± 0.14	0.499 ± 0.13	0.675 ± 0.14	0.799 ± 0.14	0.697 ± 0.17	**0.459 ± 0.15**
cleveland-0-vs-4	10	0.340 ± 0.04	0.381 ± 0.04	0.346 ± 0.04	0.273 ± 0.04	0.332 ± 0.04	0.406 ± 0.04	0.347 ± 0.05	**0.255 ± 0.05**
	20	0.358 ± 0.04	0.401 ± 0.04	0.364 ± 0.03	0.280 ± 0.03	0.305 ± 0.03	0.399 ± 0.02	0.383 ± 0.04	**0.224 ± 0.03**
	40	0.526 ± 0.08	0.550 ± 0.08	0.526 ± 0.08	0.412 ± 0.04	0.434 ± 0.11	0.499 ± 0.09	0.524 ± 0.09	**0.303 ± 0.09**
	60	0.917 ± 0.17	0.931 ± 0.16	0.919 ± 0.17	0.761 ± 0.17	0.884 ± 0.18	0.922 ± 0.19	0.920 ± 0.17	**0.733 ± 0.22**
newthyroid-v1	10	0.250 ± 0.05	0.303 ± 0.05	0.240 ± 0.06	0.271 ± 0.06	0.205 ± 0.06	0.408 ± 0.04	0.368 ± 0.00	**0.157 ± 0.04**
	20	0.268 ± 0.05	0.300 ± 0.04	0.257 ± 0.05	0.302 ± 0.05	0.254 ± 0.05	0.409 ± 0.05	0.396 ± 0.11	**0.155 ± 0.03**
	40	0.299 ± 0.05	0.327 ± 0.05	0.311 ± 0.05	0.335 ± 0.05	0.240 ± 0.04	0.434 ± 0.04	0.449 ± 0.13	**0.161 ± 0.03**
	60	0.444 ± 0.15	0.494 ± 0.15	0.468 ± 0.15	0.437 ± 0.10	0.351 ± 0.12	0.554 ± 0.12	0.571 ± 0.19	**0.290 ± 0.12**
biomed	10	0.210 ± 0.05	0.288 ± 0.04	0.208 ± 0.05	0.277 ± 0.07	0.232 ± 0.04	0.409 ± 0.04	0.379 ± 0.10	**0.159 ± 0.03**
	20	0.228 ± 0.04	0.299 ± 0.03	0.211 ± 0.04	0.255 ± 0.04	0.217 ± 0.04	0.418 ± 0.03	0.381 ± 0.10	**0.149 ± 0.02**
	40	0.283 ± 0.04	0.343 ± 0.04	0.261 ± 0.04	0.274 ± 0.04	0.257 ± 0.05	0.448 ± 0.04	0.424 ± 0.08	**0.179 ± 0.03**
	60	0.474 ± 0.11	0.513 ± 0.11	0.448 ± 0.11	0.383 ± 0.10	0.399 ± 0.14	0.573 ± 0.12	0.572 ± 0.14	**0.335 ± 0.14**
cryotherapy	10	0.302 ± 0.08	0.371 ± 0.10	0.317 ± 0.09	0.295 ± 0.09	0.354 ± 0.10	0.453 ± 0.08	0.412 ± 0.11	**0.267 ± 0.09**
	20	0.347 ± 0.09	0.415 ± 0.07	0.339 ± 0.08	0.361 ± 0.07	0.401 ± 0.07	0.465 ± 0.08	0.425 ± 0.10	**0.284 ± 0.06**
	40	0.497 ± 0.10	0.521 ± 0.10	0.477 ± 0.10	0.455 ± 0.09	0.464 ± 0.09	0.573 ± 0.10	0.547 ± 0.11	**0.390 ± 0.10**
	60	1.373 ± 1.05	1.396 ± 1.05	1.371 ± 1.05	1.252 ± 1.02	1.352 ± 1.07	1.419 ± 1.06	1.392 ± 1.06	**1.218 ± 1.03**
thyroid-3-vs-2	10	0.094 ± 0.01	0.099 ± 0.01	0.085 ± 0.01	0.059 ± 0.01	0.068 ± 0.01	0.095 ± 0.01	0.111 ± 0.03	**0.045 ± 0.01**
	20	0.106 ± 0.02	0.109 ± 0.02	0.099 ± 0.02	0.062 ± 0.01	0.072 ± 0.02	0.092 ± 0.01	0.119 ± 0.03	**0.041 ± 0.01**
	40	0.137 ± 0.02	0.135 ± 0.02	0.138 ± 0.02	0.069 ± 0.01	0.093 ± 0.02	0.086 ± 0.01	0.145 ± 0.03	**0.051 ± 0.01**
	60	0.226 ± 0.04	0.224 ± 0.04	0.222 ± 0.04	0.112 ± 0.04	0.170 ± 0.06	0.119 ± 0.04	0.234 ± 0.05	**0.110 ± 0.04**
pima	10	0.266 ± 0.03	0.338 ± 0.03	0.273 ± 0.02	0.239 ± 0.02	0.297 ± 0.03	0.306 ± 0.03	0.384 ± 0.07	**0.207 ± 0.03**
	20	0.275 ± 0.02	0.336 ± 0.03	0.279 ± 0.02	0.234 ± 0.02	0.290 ± 0.03	0.269 ± 0.02	0.368 ± 0.06	**0.168 ± 0.02**
	40	0.360 ± 0.03	0.391 ± 0.04	0.354 ± 0.04	0.281 ± 0.03	0.341 ± 0.04	0.261 ± 0.03	0.452 ± 0.08	**0.199 ± 0.03**
	60	0.569 ± 0.12	0.575 ± 0.12	0.558 ± 0.12	0.435 ± 0.10	0.526 ± 0.13	0.406 ± 0.09	0.628 ± 0.14	**0.392 ± 0.13**

missing rate considered, as seen in Table 2. In fact, the average improvement of the SAEI model in comparison to the remaining baseline methods is 35%, peaking at the 20% missing rate with an improvement of 42%. For the lower missing rate (10%) the improvement is 36%. For the higher missing rates (40% and 60%), the improvement rates are 39% and 24%, respectively. Such results

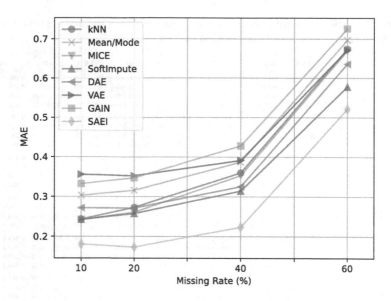

Fig. 3. Overall Mean Absolute Error of all imputation methods per missing rate. Our SAEI model significantly outperforms the remaining methods in all settings.

show the consistency of our SAEI model with different levels of missingness and with data presenting different characteristics.

In order to validate if the obtained results were statistically significant, we applied the Three-Way ANOVA on ranks test with a significance level of 5%. We considered as factors the dataset, the missing rate and the imputation method, while the MAE was set as the dependent variable. The normality assumptions were not met, so the data was transformed into rankings using the Ordered Quantile normalization [14]. Such assumptions were also ensured for the data subgroups. The obtained p-values show that the results are statistically significant for all factors, with $p < 0.001$. Additionally, to validate if our SAEI model outperformed the remaining methods with a statistical significance of 5%, the post-hoc Tukey's HSD test was applied to this factor. The obtained p-values show that the SAEI model significantly outperformed all the baseline of imputation methods, with $p < 0.001$ in all scenarios.

6 Conclusions

In this work we propose a new model for missing data imputation called Siamese Autoencoder-based Approach for Imputation (SAEI), which is an extension of the vanilla siamese networks adapted for this imputation task. The model incorporates three main adaptations: a deep autoencoder architecture that is tailored for missing data reconstruction, a custom loss function that encompasses both the triplet and reconstruction losses, and a triplet mining strategy tailored for the missing data issue that is capable of generating hard triplets that are meaningful for the training procedure. To the best of our knowledge, this is the first time a siamese architecture is being used and adapted for missing data imputation. We compared our SAEI model with seven state-of-the-art imputation methods from both statistical and machine learning backgrounds, in an experimental setup that used 14 datasets of the healthcare domain injected with MNAR values in a multivariate fashion at 10%, 20%, 40% and 60% missing rates. Our SAEI model significantly outperformed all the baseline methods used for comparison in all the datasets and missing rates, with a statistical significance of 5%, achieving an average improvement of 35%.

In the future we want to extend this work to test other missing data mechanisms, and we want to explore automatic approaches to evolve our deep autoencoder architecture so it can be even further optimized.

Acknowledgements. This work is supported in part by the FCT - Foundation for Science and Technology, I.P., Research Grant SFRH/BD/149018/2019. This work is also funded by the FCT - Foundation for Science and Technology, I.P./MCTES through national funds (PIDDAC), within the scope of CISUC R&D Unit - UIDB/00326/2020 or project code UIDP/00326/2020.

References

1. Azur, M.J., Stuart, E.A., Frangakis, C., Leaf, P.J.: Multiple imputation by chained equations: what is it and how does it work? Int. J. Methods Psychiatr. Res. **20**(1), 40–49 (2011)
2. Batista, G., Monard, M.: Experimental comparison of k-nearest neighbor and mean or mode imputation methods with the internal strategies used by c4. 5 and CN2 to treat missing data. Univ. Sao Paulo **34** (2003)
3. Batista, G.E., Monard, M.C., et al.: A study of k-nearest neighbour as an imputation method. HIS **87**(251–260), 48 (2002)
4. Buuren, S.v., Groothuis-Oudshoorn, K.: MICE: multivariate imputation by chained equations in R. J. Stat. Softw. 1–68 (2010)
5. Charte, D., Charte, F., García, S., del Jesus, M.J., Herrera, F.: A practical tutorial on autoencoders for nonlinear feature fusion: taxonomy, models, software and guidelines. Inf. Fusion **44**, 78–96 (2018)
6. Chicco, D.: Siamese neural networks: an overview. Artif. Neural Netw. 73–94 (2021)
7. García-Laencina, P.J., Abreu, P.H., Abreu, M.H., Afonoso, N.: Missing data imputation on the 5-year survival prediction of breast cancer patients with unknown discrete values. Comput. Biol. Med. **59**, 125–133 (2015)
8. García-Laencina, P.J., Sancho-Gómez, J.L., Figueiras-Vidal, A.R.: Pattern classification with missing data: a review. Neural Comput. Appl. **19**(2), 263–282 (2010)
9. Goodfellow, I., et al.: Generative adversarial networks. Commun. ACM **63**(11), 139–144 (2020)
10. Little, R.J., Rubin, D.B.: Statistical Analysis with Missing Data, vol. 793. Wiley, Hoboken (2019)
11. McCoy, J.T., Kroon, S., Auret, L.: Variational autoencoders for missing data imputation with application to a simulated milling circuit. IFAC-PapersOnLine **51**(21), 141–146 (2018)

12. Peek, N., Rodrigues, P.P.: Three controversies in health data science. Int. J. Data Sci. Anal. **6**(3), 261–269 (2018)
13. Pereira, R.C., Abreu, P.H., Rodrigues, P.P.: Partial multiple imputation with variational autoencoders: tackling not at randomness in healthcare data. IEEE J. Biomed. Health Inform. **26**(8), 4218–4227 (2022)
14. Peterson, R.A., Cavanaugh, J.E.: Ordered quantile normalization: a semiparametric transformation built for the cross-validation era. J. Appl. Stat. **47**, 1–16 (2019)
15. Rubin, D.B.: Inference and missing data. Biometrika **63**(3), 581–592 (1976)
16. Rubin, D.B.: Multiple Imputation for Nonresponse in Surveys, vol. 81. Wiley, Hoboken (2004)
17. Schroff, F., Kalenichenko, D., Philbin, J.: FaceNet: a unified embedding for face recognition and clustering. In: Proceedings of the IEEE Conference on Computer Vision and Pattern Recognition, pp. 815–823 (2015)
18. Udell, M., Horn, C., Zadeh, R., Boyd, S., et al.: Generalized low rank models. Found. Trends Mach. Learn. **9**(1), 1–118 (2016)
19. Van Buuren, S.: Flexible Imputation of Missing Data. Chapman and Hall/CRC (2018)
20. Vincent, P., Larochelle, H., Bengio, Y., Manzagol, P.A.: Extracting and composing robust features with denoising autoencoders. In: Proceedings of the 25th International Conference on Machine learning, pp. 1096–1103 (2008)
21. Yoon, J., Jordon, J., Schaar, M.: GAIN: missing data imputation using generative adversarial nets. In: International Conference on Machine Learning, pp. 5689–5698. PMLR (2018)
22. Zeiler, M.D., Fergus, R.: Visualizing and understanding convolutional networks. In: Fleet, D., Pajdla, T., Schiele, B., Tuytelaars, T. (eds.) ECCV 2014. LNCS, vol. 8689, pp. 818–833. Springer, Cham (2014). https://doi.org/10.1007/978-3-319-10590-1_53

An Intelligent Transportation System for Tsunamis Combining CEP, CPN and Fuzzy Logic

Gregorio Díaz[1](✉)(ID), Hermenegilda Macià[1](ID), Enrique Brazález[1](ID), Juan Boubeta-Puig[2](ID), M. Carmen Ruiz[1](ID), and Valentín Valero[1](ID)

[1] School of Computer Science, Universidad de Castilla-La Mancha, Albacete, Spain
{Gregorio.Diaz,Hermenegilda.Macia,Enrique.Brazalez,
MCarmen.Ruiz,Valentin.Valero}@uclm.es
[2] Department of Computer Science and Engineering, University of Cadiz,
Puerto Real, Cádiz, Spain
juan.boubeta@uca.es

Abstract. Tsunamis and earthquakes have a great impact in human lives, infrastructures and economy. Although preventing tsunamis from occurring is impossible, minimizing their negative effects is in our hands. The aim of the Intelligent Transportation System (ITS) proposed in this paper is to provide safer routes for emergency and rescue vehicles. This system must consider the information regarding the tsunami alert system and the road state combined with the vehicle performance. Complex Event Processing (CEP) technology allows us to gather and process the information provided by authorities to establish the alert level. A Fuzzy Inference System (FIS) can be used to consider the uncertain regarding the road-status related concepts, such as, flood, objects and alert levels, and to assist authorities to determine whether roads are accessible. The information obtained through these technologies can then be used in a Colored Petri Net (CPN) model in order to obtain safer routes. This proposal has been applied to the Spanish city of Cádiz, due to its population density and its location in a small peninsula close to an active tectonic rift.

Keywords: Intelligent Transportation System · Tsunami · Complex Event Processing · Fuzzy Logic · Colored Petri Nets

1 Introduction

Sadly, earthquakes and its consequences such as tsunamies and aftershocks have a deadly impact in our societies. Recently, on early and mid February 2023, a series of earthquakes and aftershocks with magnitudes reaching 7.5 and 7.8 (Mw) stroke in Kahramanmaras, Turkey, and its surroundings. It had a high toll on

This work was supported in part by the Spanish Ministry of Science and Innovation and the European Union FEDER Funds under grants PID2021-122215NB-C32 and PID2021-122215NB-C33.

J. Mikyška et al. (Eds.): ICCS 2023, LNCS 14073, pp. 47–60, 2023.
https://doi.org/10.1007/978-3-031-35995-8_4

deaths in the southeast of Turkey and the neighbour cities of Syria [19]. Other similar episode occurred on 2004 affecting the Indian Ocean and known by the scientific community as the Sumatra-Andaman earthquake [13]. It provoked a massive tsunami with waves up to 30 m high, which destroyed cities and villages in the surrounding coasts of the Indian Ocean. It was one of the deadliest natural disasters in recorded history affecting 14 countries with more than 225 k deaths.

This episode and the 2011 tsunami in Japan showed the necessity to establish advanced earthquake detection systems to inform both population and authorities and specially in subdue zones like the Gulf of Cadiz [16,21]. Results obtained by these works can be used for emergency and decision planners to implement tsunami mitigation measures resulting in tsunami-resilient communities.

These measures include the recommendation of reaching higher ground areas, such as high buildings in cities. Population should concentrate in specific areas and Intelligent Transportation Systems (ITSs) can therefore be implemented to determine routes for emergency and rescue vehicles. In this paper, we propose an ITS for Tsunamis to obtain safe routes considering typical tsunamis conditions, such as flood levels, objects in the pavement and new alerts. The main contributions of this work are the following:

– The design of an ITS in case of tsunamis combining Complex Event Processing (CEP), Fuzzy Inference System (FIS) and Colored Petri Net (CPNs).
– The definition of CEP patterns to determine the tsunami alert level.
– A FIS to deal with the uncertainty.
– A CPN dynamic road model considering the current conditions.
– Obtaining safe routes from the CPN model via simulation techniques.

Figure 1 shows the general diagram, depicting the different steps[1]:

(1a) Several systems provide information on the state of the city roads, and the tsunami alert system information is processed and correlated by a CEP engine, which is responsible for creating the complex events to feed the FIS.
(1b) Cádiz map is modeled as a CPN.
(2) The FIS system provides a set of recommendations stating which city areas are accessible.
(3) The domain expert states whether roads are open, pass-with-precaution or close, on the basis of the FIS recommendations.
(4) The CPN model establishes the mobility restrictions with the information obtained in step (3).
(5) An emergency officer asks for a route, taking the vehicle performance into account.
(6) The CPN model provides this user with a route, which is obtained by simulation.

An ITS is a system which could operate automatically, but some scenarios as the one considered in this work may include situations not considered during the ITS design. Therefore, the domain expert supervision is mandatory.

---

[1] This figure has been designed using images from Flaticon.com.

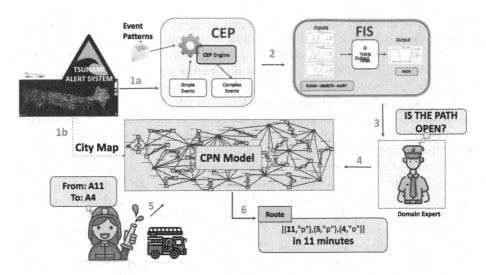

Fig. 1. General diagram of the ITS.

The structure of this work is as follows. Section 2 presents the related work and the scenario of interest is presented in Sect. 3. The CEP patterns are defined in Sect. 4, the proposed FIS inference system is introduced in Sect. 5, and the CPN model and the computation of safe routes are presented in Sect. 6. Finally, the conclusions and lines of future work are presented in Sect. 7.

2   Related Work

Since the post-tsunami situation usually cannot be accurately described, and information is often incomplete, decision makers do not have enough information to perform decision analysis. Therefore, uncertainty is one of the most challenging and important issues in the tsunami management. To deal quantitatively with such an imprecision or uncertainty, the fuzzy theory is adapted to disaster management studies. It is worth highlighting the work presented by Sheu in [24], in which hybrid fuzzy cluster optimization is proposed for the operation of emergency logistics codistribution responding to urgent relief demands in the crucial rescue period. In this line of research, Tsai et al. [26] apply the fuzzy set theory to decision making in a geographic information system for the allocation of disaster shelters. Öztaysi et al. present a very original proposal [22]. They present a fuzzy inference system for the management of spontaneous volunteering developed using MATLAB Fuzzy Logic Toolbox™. It is also worth commenting on the work by Carathedathu et al. [7], in which real-time feedback from sensors is sent to a fuzzy controller that predicts whether a tsunami will strike. To make the prediction it only takes into account two parameters, namely earthquake intensity and wave height. A work closer to our proposal is the study presented by Afandi and Mayasari [3], in which the authors propose an evacuation route using Dijkstra's Algorithm, where the weights of the graph are the fuzzy output obtained based on the length and width of the roads. From our point of view, this work would be useful in a pre-tsunami scenario to know the fastest roads. However, in a post-tsunami scenario any event may have occurred, leaving the roads in difficult conditions due to the presence of obstacles, even being closed or flooded, which are parameters that are taken into account in our model. Moreover, we would like to remark the work by Raj and Sasipraba [10], in which a fuzzy inference rule approach is used to determine which service (namely hospital, fire service or police) is the most appropriate according to the user's needs (in a post-tsunami scenario) and proposes the shortest route to reach that service using Euler's algorithm. Note that we use a different approach, as we compute the route from a fixed location (service) to a certain area, taking the road status into account, so as to avoid flooded areas and considering the times to cross the zones due to the possible obstacles in them.

Some works proposed the use of CEP technology for developing disaster and tsunami management systems [27,28]. However, these systems do not take the advantage of combining CEP with CPN and fuzzy logic. Moreover, event patterns are not defined for automatically detecting tsunami levels, as we do in our work, but for detecting sensor problems such as control error, battery alarm and loss of volts [28].

Regarding the use of Petri nets (PNs) in emergency situations, Tadano et al. [25] proposed the use of stochastic time PNs for disaster victim evacuation, considering road repair sequencing, and Yan et al. [29] used PNs as a methodology for modeling and evaluating the resilience of nuclear power plants.

We can conclude that due to the extreme need for effective disaster management, several works have been developed focusing on different topics related to disaster management, and in particular to the specific case of post-tsunami situations. Mainly, all of them make use of FISs in one way or another. However, we have not found any works that make use of CEP technology combined with fuzzy logic and CPN in an all-in-one solution. Indeed, the major contribution of the proposal presented in this work is the development of a complete system capable of providing an ITS in emergency situations composed of CEP, fuzzy logic and Petri nets which, to the best of our knowledge, has not been yet proposed.

3   Application Scenario

In some previous works [5,9,14], we used CPN models of some Spanish cities for different purposes, such as to obtain routes to reduce pollution, or to avoid close areas in the case of a pandemic, etc. In this paper, we consider a city map for Cádiz city (located in the south of Spain, at latitude 36.52672 and longitude -6.2891). Cádiz is the eighth city more densely populated in Europe due to its location in a small peninsula and it is located close to an active tectonic rift. The city already suffered the devastation of the earthquake occurred in 1755, known as the Lisbon earthquake and future episodes are expected in the area of the Straight between Iberia and north Africa [8]. The hazard planning envisioned for this city includes the recommendation of taking shelter in high grounds. Cádiz

Fig. 2. Map of the Cádiz city.

peninsula is flat, and thus, we have considered its highest building areas shown in Fig. 2. This model consists of 12 areas, taking into account the high buildings in each area, with at least seven stories high[2].

The scenario we consider is the following. After a tsunami, people have sheltered in these buildings. Emergency services, such as paramedical and firefighters, are requested to attend citizens at these areas. A route has to be provided to reach these areas, but objects in the road washed away by the tsunami, flood levels and new incoming alarms can determine whether the mobility is restricted. Emergency services departs from areas 11 and 12, where the firefighting headquarter and the ambulance base are located, respectively. Thus, the routes are obtained from either zone 11 or 12 to any other zone in the city.

We based this scenario on Cádiz, since this is a well studied location for tsunamis and action plans have been provided by authorities [6], but this work can easily be adapted to other locations and the criteria to establish the zones could also be modified in order to consider other parameters of interest.

4   The CEP Event Patterns

CEP is a cutting-edge technology for analyzing and correlating huge amounts of real-time data with the aim of automatically detecting situations of interest (event patterns). These event patterns must be previously implemented using an Event Processing Language (EPL) and then deployed in a CEP engine, the software responsible for data analysis and event pattern detection.

To automatically detect the three tsunami levels according to the Spanish state plan for civil protection against tsunamis [6], we defined three event patterns (*Info, Warning* and *Alert*) by using MEdit4CEP [4]. This is a graphical modeling tool that facilitates domain experts to model event patterns that are transformed automatically into Esper EPL code [2], which is then deployed in the Esper CEP engine.

---

[2] Designed using the mapbox by Raúl Sánchez, elDiario.es (CC BY-NC 4.0).

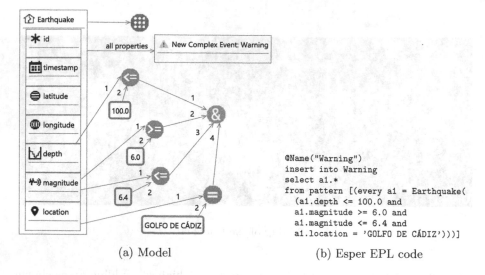

(a) Model

(b) Esper EPL code

Fig. 3. *Warning* event pattern.

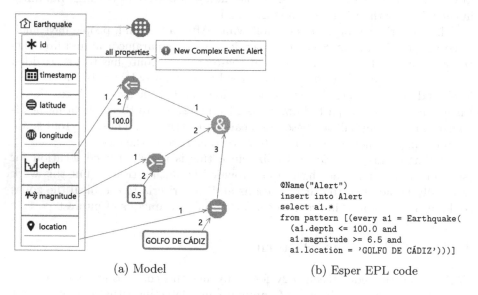

(a) Model

(b) Esper EPL code

Fig. 4. *Alert* event pattern.

As an example, Fig. 3(a) depicts the *Warning* event pattern modeled with MEdit4CEP, while Fig. 3(b) shows the Esper EPL code automatically generated from the designed model. Mainly, this pattern looks for every occurred *Earthquake* event whose depth is less than 100 km., the magnitude is between 6.0 and 6.4, and the location is equal to "GOLFO DE CÁDIZ". When these conditions are satisfied, a *Warning* complex event is created by the CEP engine, which is notified to our proposed FIS. Similarly, Fig. 4(a) illustrates the *Alert* pattern designed with MEdit4CEP, while Fig. 4(b) shows the Esper EPL code automatically generated.

This pattern looks for every *Earthquake* event whose depth is less than 100 km., the magnitude is greater than or equal to 6.5, and the location is equal to "GOLFO DE CÁDIZ". If so, an *Alert* complex event is created and notified to the FIS.

5   The Fuzzy Inference System

FIS is a suitable decision-making system that can be used in the case of natural disasters [22], considering that the inputs and outputs of these systems cannot be sharply defined. We use the Mamdani type FIS [15], considering a set of linguistic rules obtained from domain experts, where the antecedents and the consequents of the rules are expressed as linguistic variables using fuzzy sets [30]. This FIS system provides a highly intuitive knowledge base, being easy to understand and maintain. It is implemented via the Skfuzzy Python toolbox [20], where the `trampmf` and `trimf` membership functions are used to specify the fuzzy sets. We use the default parameters established in the Skfuzzy toolbox, which are: minimum as t-norm (AND operator); t maximum as t-conorm (OR operator); minimum (Implication operator); maximum of consequent fuzzy sets as aggregation method for combining rule consequents; and the defuzzification method used is the centroid (center of gravity).

The fuzzy sets of the input/output variables considered for this FIS are shown in Fig. 5. This FIS only considers three input variables: *flood*, *objects* and *alert*, and a single output variable (*path*). The values of these variables are normalized in a range from 0 to 10, except flood, which is expressed in centimetres, from 0 to 100. In all cases, high values correspond to high values in the registered measurement.

The alert input variable is obtained from the CEP engine taking into account the magnitude property obtained from the detected events. In the case of flood and objects, a low value corresponds to a good condition and a high value means unfavourable conditions for emergency vehicles. We assume that these inputs are

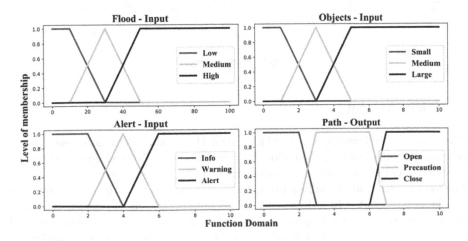

Fig. 5. Input/Output membership functions.

provided by authorities using Machine Learning (ML) technologies, a well known technique used in this domain [18]. Then, they are classified in the linguistic labels: Low, Medium and High for flood; Small, Medium and Large for objects; and Info, Warning and Alert for alerts.

The output variable *path* is also defined in the range 0–10 and the linguistic label classification is the following: *open* (pass without restrictions), *precaution* (can pass with caution, only allowed for vehicles with special performances) and *close* (access is not possible). Table 1 contains the IF-THEN rules defined.

Table 1. IF-THEN Rules for FIS.

1. IF *flood* is Low AND *objects* is Small AND *alert* is No alert THEN *path* is Open
2. IF *flood* is High OR *objects* is Large OR *alert* is Alert THEN *path* is Close
3. IF NOT (*flood* is Low AND *objects* is Small AND *alert* is No alert) AND NOT (*flood* is High OR *objects* is Large OR *alert* is Alert) THEN *path* is Precaution

Figure 6 illustrates the three surfaces obtained considering these rules and the alert levels 2 (Info), 4 (Warning) and 9 (Alert), respectively. For instance, the output is *Precaution* (4.01) taking as input 25 for flood, 2 for objects and 2 for alert, and, for an alert of 9, considering the same input values for flood and objects, the output is *Close* (8.23).

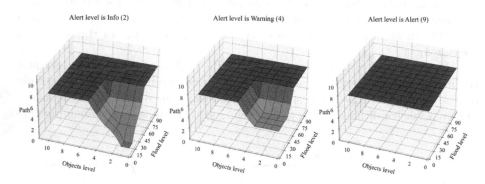

Fig. 6. Fuzzy space.

6   CPN Model

PNs [23] are a well-known graphical formalism that allows us to model concurrent systems and analyze their behavior. PNs are bipartite directed graphs with two types of node: places (drawn as circles) and transitions (drawn as rectangles). The system states are usually represented by the markings of the places (number of tokens on them), while the events producing changes in the system state are modeled as transitions. The arcs can only connect places with transitions and vice versa, and they have a weight associated, which indicates the number of tokens to be removed from the corresponding precondition place or the number of tokens to be added to the corresponding postcondition place of the transition that is executed (*fired* in PN terminology). CPNs [11] extend the plain model with both data and time information in the tokens. CPN Tools [1] is the tool used in this work for editing, simulating, and analyzing CPNs, and therefore, we use the notation of this tool [12]. In CPNs, each place has an associated color set (*colset*), which is similar to a data type, and the tokens on the place must then belong to the colset of the place. Color sets are indicated below the places, in their right bottom part. Transitions can have priorities and guards to restrict their firing. The latter are Boolean expressions constructed using variables, constants and functions, and they are indicated above the transitions, on their left hand-side. Arcs have an expression associated, which must evaluate to a multiset of the color set of the attached place. The *enabling condition* and the *firing rule* are extended to consider the time and data information. Due to the lack of space we omit the technical details, which can be found in [11]. Informally, when a transition is fired under a certain *binding*[3], we remove from its precondition places a number of tokens equal to the resulting value of the expression of the arc connecting the place with the transition, and for the postcondition places we add a number of tokens equal to the resulting value of the corresponding arc expression.

Figure 7 shows a piece of the city map, in which we can see the two color sets that we use: ZoB and I4.

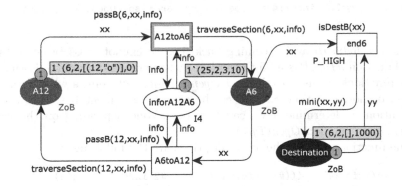

Fig. 7. Detail of the city map.

These color sets are defined as follows:

```
colset INTR=product INT*STRING; | colset ZoB=product INT*INT*INTlistR*INT;
colset INTlistR=list INTR;      | colset I4=product INT*INT*INT*INT;
```

ZoB values consist of 4 fields. The first two elements are integers indicating the destination and the vehicle type, respectively. In this paper, two types of vehicle

---

[3] Bindings are variable assignments, considering the tokens on the precondition places, and the variables labeling the arcs.

are considered: vehicles of type 1, which can only pass under favorable road conditions (Open); and vehicles of type 2, which can also pass under certain unfavorable road conditions (Precaution). The third field represents a route with a list of traversed zones and its mobility conditions ("o" or "p", open and precaution, respectively). The last field indicates the total time to follow the route. As an illustration, let us consider the token in the place $A12$: $1'(6,2,[(12,"o")],0)$. This token indicates a route to zone 6 for a vehicle of type 2, which starts at zone 12, mobility is permitted and the total time is 0.

The color set $I4$ consists of four integer elements, which represent the flood, objects and alert levels, and the last field contains the average time required in normal conditions to cross this zone. As an illustration, the marking of the place *inforA12A6* states that the flood is 25 cm, the object level is 2 and the alert level is 3, with a travel time of 10 min between zones A12 and A6.

The funtions used in the guards are the following:

```
fun noReturnB(i:INT,yy:ZoB)=not( mem (#3yy) (i,"o")) andalso
                       not( mem (#3yy) (i,"p"))
fun isDestB(xx:ZoB)=  hd(rev (#3 xx)) = (#1 xx,"o") orelse
                      hd(rev (#3 xx)) = (#1 xx,"p")
fun isOpenB(i:INT,yy:ZoB)=noReturnB(i,yy) andalso not(isDestB(yy));
fun OpenType(xx:ZoB,info:I4) =
  if (simFIS(info) <3) then true
  else if((simFIS(info)<7) andalso (#2 xx) > 1) then true else false;
passB(i:INT,yy:ZoB,info:I4)=isOpenB(i,yy) andalso  OpenType(yy,info);
```

Function *noReturnB* avoids coming back, i.e. we cannot cross twice the same zone. Function *isDestB* is used to check if we have reached the final destination. These functions are used to define *isOpenB*, which determines whether a route is open or not. Function *passB* uses the expert assessment based on the FIS information to determine if we can traverse a zone depending on the vehicle performance (function *OpentType*).

The functions used in the arc inscriptions are the following:

```
fun tComp(info:I4) = (((#1 info) div 10) + #2 info + #3 info) * 2 div 3;
fun tB(info:I4)= (#4 info) * tComp(info);
fun fis(info:I4)= if  simFIS(info)<=3 then  "o"  else  "p";
fun traverseSection(i:INT,xx:ZoB,info:I4)=
          (#1 xx, #2 xx, (#3 xx)^^[(i,fis(info) )], #4 xx + tB(info));
fun mini(y:ZoB,z:ZoB)= if (#4y < #4z) then 1'y else 1'z
```

Function *tB* is used to compute the times to cross from one zone to another, where the average time to cross between them is proportionally incremented by the level established by the domain expert (*tComp*). The information stored in the *ZoB* places is updated using the traverse function *traverseSection*. It adds to the route the new traversed zone, indicating whether the crossing is open or the

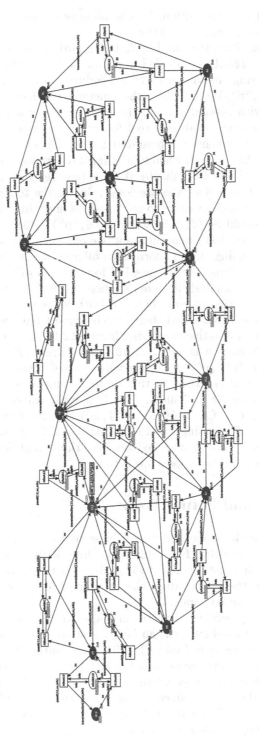

Fig. 8. CPN page of Cádiz Map.

vehicle must proceed with precaution. This is made using the expert assessment (*fis*). It also updates the total travel time.

With the marking indicated in Fig. 7, the guard *passB* is satisfied, and thus the transition *A12toA6* is enabled and can be fired. The token in A12 is removed with its firing, and a new token is produced in place A6, with values (6,2,[(12,"o"),(6,"p")],40). Thus, the emergency vehicle is now in zone 6, it must proceed with precaution, and the total time elapsed in the route is 40 min. Function *tComp* was evaluated to 4 following the expert decision.

Place *Destination* represents the final destination. This place is reached via the transition *end6*, which is fired when it is enabled due to its high priority. It is enabled when the guard *isDestB* indicates that we have reached our destination. Function *mini* is used to replace the previous value of the token on this place with the new information (best route to the previous one). In our case, the new value for this token would be (6,2,[(12,"o"),(6,"p")],40).

Finally, we performed 100 simulations and we obtained the best route from these simulations depending on the information offered by the CEP system, the FIS and the expert decisions. The whole CPN for the map of Cádiz is depicted in Fig. 8. The case under analysis is an ambulance of type 2 traveling from zone A12 to A6. The information to cross between both areas now is (50,6,2,10) with a FIS output closed (8.23) and therefore the expert has decided that this crossing is closed. An alternative is crossing from A12 to A11, and then to A6. In this case, the information is (30,3,2,10) and (3,2,2,2), and the route (6,2,[(12,"o"),(11,"p"),(6,"o")],54) with a total travel time of 54 min. There are other faster alternatives to this route, for instance, the ambulance can proceed via A12 to A8, A7 and A6 with a total travel time of 26 min (6,2,[(12,"o"),(8,"o"),(7,"o"),(6,"o")],26). In this case, the information considered for each crossing is (13,0,2,8) (4,1,2,5) and (7,1,2,7), respectively. The faster time is obtained because the expert has judged that the conditions for crossing these areas are better.

7   Conclusions and Future Work

The ITS proposed in this paper provides a system to obtain safe routes for emergency and rescue vehicles. It complies the three main properties that an ITS must hold, i.e., it must be able to consider the traffic flow, it must be adaptable to capture the system dynamics, and must be concise but able to represent new application scenarios.

The traffic flows represented as routes in maps are modeled via CPNs and tsunami alarms are gathered from the authority systems. These alarms are processed by the CEP technology to feed a FIS with which authorities can assess the road state. These assessments can be used by the CPN to establish the available routes. Simulation allows us to repeat the experiment and thus obtain a short route, as the best of the ones obtained by the simulations. The CPN model has the main advantage that it immediately takes into account the current road and alert conditions, as they are indicated in the markings, and no structural change is required in the model. Thus, this ITS is flexible and the routes are obtained according to the current scenarios.

As future work, we intend to apply Machine Learning and Deep Learning techniques [17] to gather the road situations regarding flood and objects. External communication with the CPN model can be implemented in CPN Tools to achieve this goal. In addition, we plan to perform a usability study through a comparison with existing works in this area.

References

1. CPN Tools home page (2023). http://www.cpntools.org/.
2. Esper - Complex Event Processing (2023). http://www.espertech.com/esper/.
3. Afandi, N., Mayasari, Z.: An evacuation route in Bengkulu city based on fuzzy Dijkstra algorithm. J. Phys: Conf. Ser. **1863**, 012007 (2021). https://doi.org/10.1088/1742-6596/1863/1/012007
4. Boubeta-Puig, J., Ortiz, G., Medina-Bulo, I.: MEdit4CEP: a model-driven solution for real-time decision making in SOA 2.0. Knowl.-Based Syst. **89**, 97–112 (2015). https://doi.org/10.1016/j.knosys.2015.06.021
5. Brazález, E., Macià, H., Díaz, G., Valero, V., Boubeta-Puig, J.: PITS: an intelligent transportation system in pandemic times. Eng. Appl. Artif. Intell. **114**, 105154 (2022). https://doi.org/10.1016/j.engappai.2022.105154
6. Cantavella, J.V., Gaite, B., González, C., Naveiras, F., Ros, E.: Plan Estatal de Protección Civil ante el riesgo de maremotos. Edición comentada. Catálogo de publicaciones de la Administración General del Estado. https://www.ign.es/web/resources/acercaDe/libDigPub/Plan-Estatal-Maremotos.pdf (2021)
7. Carathedathu, M., Jayaraj, N., Vaidyanathan, S.: Artificially intelligent tsunami early warning system, pp. 39–44 (2010). https://doi.org/10.1109/UKSIM.2010.16
8. Chester, D.K.: The 1755 Lisbon earthquake. Prog. Phys. Geogr.: Earth Environ. **25**(3), 363–383 (2001). https://doi.org/10.1177/030913330102500304
9. Díaz, G., Macià, H., Valero, V., Boubeta-Puig, J., Cuartero, F.: An intelligent transportation system to control air pollution and road traffic in cities integrating CEP and colored petri nets. Neural Comput. Appl. **32**(2), 405–426 (2018). https://doi.org/10.1007/s00521-018-3850-1
10. Jeberson Retna Raj, R., Sasipraba, T.: Disaster management system based on GIS web services. In: Recent Advances in Space Technology Services and Climate Change 2010 (RSTS & CC-2010), pp. 252–261 (2010). https://doi.org/10.1109/RSTSCC.2010.5712855
11. Jensen, K.: Coloured Petri Nets: Basic Concepts, Analysis Methods and Practical Use, vol. 2. Springer-Verlag, London (1995)
12. Jensen, K., Kristensen, L.M.: Coloured Petri Nets: Modelling and Validation of Concurrent Systems, 1st edn. Springer, Cham (2009)
13. Lay, T., et al.: The great Sumatra-Andaman earthquake of 26 December 2004. Science **308**(5725), 1127–1133 (2005). https://doi.org/10.1126/science.1112250
14. Macià, H., Díaz, G., Valero, V., Valero, E., Brazález, E., Boubeta-Puig, J.: green-ITS: a proposal to compute low-pollution routes. In: 17th International Conference on Future Networks and Communications / 19th International Conference on Mobile Systems and Pervasive Computing / 12th International Conference on Sustainable Energy Information Technology, vol. 203, pp. 334–341. Elsevier, Niagara Falls (2022). https://doi.org/10.1016/j.procs.2022.07.042

15. Mamdani, E., Assilian, S.: An experiment in linguistic synthesis with a fuzzy logic controller. Int. J. Man-Mach. Stud. **7**(1), 1–13 (1975). https://doi.org/10.1016/S0020-7373(75)80002-2

16. Matias, L.M., Cunha, T., Annunziato, A., Baptista, M.A., Carrilho, F.: Tsunamigenic earthquakes in the Gulf of Cadiz: fault model and recurrence. Nat. Hazards Earth Syst. Sci. **13**(1), 1–13 (2013). https://doi.org/10.5194/nhess-13-1-2013

17. Méndez, M., Ibias, A., Núñez, M.: Using deep learning to detect anomalies in traffic flow. In: Nguyen, N.T., Tran, T.K., Tukayev, U., Hong, T.P., Trawiński, B., Szczerbicki, E. (eds.) Intelligent Information and Database Systems, pp. 299–312. Springer International Publishing, Cham (2022)

18. Mosavi, A., Ozturk, P., Chau, K.W.: Flood prediction using machine learning models: literature review. Water (Switzerland) **10**(11), 1536 (2018). https://doi.org/10.3390/w10111536

19. Naddaf, M.: Turkey-Syria earthquake: what scientists know. Nature **614**(6), 398–399 (2023). https://doi.org/10.1038/d41586-023-00364-y

20. Oliphant, T.P., Jones, P.: SciPy - Skfuzzy (2023). https://pythonhosted.org/scikit-fuzzy/

21. Omira, R., Baptista, M., Miranda, J.: Evaluating Tsunami impact on the Gulf of Cadiz Coast (Northeast Atlantic). Pure Appl. Geophys. **168**(6–7), 1033–1043 (2011). https://doi.org/10.1007/s00024-010-0217-7

22. Öztaysi, B., Behret, H., Kabak, Ö., Sarı, I.U., Kahraman, C.: Fuzzy Inference Systems for Disaster Response. In: Vitoriano, B., Montero, J., Ruan, D. (eds) Decision Aid Models for Disaster Management and Emergencies, pp. 75–94. Atlantis Press, Paris (2013). https://doi.org/10.2991/978-94-91216-74-9_4

23. Peterson, J.L.: Petri Net Theory and the Modeling of Systems. Prentice Hall PTR, Upper Saddle River (1981)

24. Sheu, J.B.: An emergency logistics distribution approach for quick response to urgent relief demand in disasters. Transp. Res. Part E: Logistics Transp. Rev. **43**, 687–709 (2007). https://doi.org/10.1016/j.tre.2006.04.004

25. Tadano, K., Maeno, Y., Carnevali, L.: Road repair sequencing for disaster victim evacuation. Adv. Intell. Syst. Comput. **528**, 401–412 (2017). https://doi.org/10.1007/978-3-319-47253-9_37

26. Tsai, C.H., Chen, C.W., Chiang, W.L., Lin, M.L.: Application of geographic information system to the allocation of disaster shelters via fuzzy models engineering computations. Eng. Comput. **25**, 86–100 (2008). https://doi.org/10.1108/02644400810841431

27. Wächter, J., et al.: Development of tsunami early warning systems and future challenges. Nat. Hazards Earth Syst. Sci. **12**(6), 1923–1935 (2012). https://doi.org/10.5194/nhess-12-1923-2012

28. Whyte, J., Fulton, F., White, A., Putley, A., Paisley, B., Robinson, B.: Realising network intelligence through master data exploitation and dynamic data modelling. CIRED-Open Access Proc. J. **2017**, 2852–2856 (2017). https://doi.org/10.1049/oap-cired.2017.0112

29. Yan, R., Dunnett, S., Tolo, S., Andrews, J.: A petri net methodology for modeling the resilience of nuclear power plants, p. 2426–2432 (2021). https://doi.org/10.3850/978-981-18-2016-8_109-cd

30. Zadeh, L.: Fuzzy sets. Inf. Control **8**(3), 338–353 (1965). https://doi.org/10.1016/S0019-9958(65)90241-X

Downscaling WRF-Chem: Analyzing Urban Air Quality in Barcelona City

Veronica Vidal[1,2]([envelope]) [ID], Ana Cortés[1] [ID], Alba Badia[2] [ID], and Gara Villalba[2] [ID]

[1] Departament d'Arquitectura de Computadors i Sistemes Operatius, Universitat Autònoma de Barcelona, Barcelona, Spain
{veronica.vidal,ana.cortes}@uab.cat
[2] Institut de Ciència i Tecnologia Ambientals, Universitat Autònoma de Barcelona, Barcelona, Spain
{alba.badia,gara.villalba}@uab.cat

Abstract. Improving air quality in highly polluted cities is a challenge for today's society. Most of the proposed strategies include green policies that aim to introduce green infrastructures helping to improve air quality. To design new cities with more green infrastructures, the WRF-Chem model is used to analyze the evolution of the most common pollutants and their dispersion due to the meteorology of the moment. Most of the studies, however, are off an urban scale (hundreds of meters of resolution), and those cases that simulate the meteorology at this resolution need to consider the city's morphology. Using the city of Barcelona as a case study, this paper confirms that the modeling methodology used up to now must be reviewed to design green cities. Certain limitations of the WRF-Chem model have been analyzed, including the BEP-BEM as an urban canopy layer, and the reasons for such limitations are discussed.

Keywords: WRF-Chem · BEP-BEM · Air Quality · Urban scale · Green Infrastructures

1 Introduction

Air quality is a major issue in urban areas where the population's exposure to air pollution directly impacts citizens' health [12]. The main reason for this is that big cities and their metropolitan areas are almost the main contributors to injecting the atmosphere with different kinds of pollutants [2,4]. Therefore, new governmental policies devoted to creating smart and green cities should be designed to transform our current cities into resilient cities that provide citizens with a good and healthy life. One of the trends in this area is to reconfigure the current morphology of the cities by including green infrastructures that help to reduce pollutant concentrations in the underlying urban area. The urban planners, however, must be aware that specific green designs that include urban trees can sometimes produce the opposite effect than the desired because they can obstruct the wind flow and, consequently, lead to higher pollutant concentrations [27]. Therefore, green urban infrastructures should be designed considering those contradictory aspects. Consequently, it is necessary to test different

J. Mikyška et al. (Eds.): ICCS 2023, LNCS 14073, pp. 61–73, 2023.
https://doi.org/10.1007/978-3-031-35995-8_5

vegetation combinations and analyze their implications for the air quality to select alternatives that improve human life quality in the long term. In order to tackle this challenge, one needs to rely on different simulation engines. On the one hand, numerical atmospheric models are required to simulate the atmosphere's behavior in urban areas. Furthermore, the chemical processes associated with pollutant and vegetations interactions must also be considered. Therefore, the models of the atmosphere must be coupled to a chemical model to evaluate the influence of the meteorological variables provided by atmospheric models on these chemical processes involved in the reduction of the air pollution. Moreover, we need to analyze the impact of green infrastructures in urban zones, the city's morphology (also called urban canopy) is the another key factor that will affect the dispersion of the pollutant particles. These three puzzle components (atmosphere, chemistry, and city morphology) should, somehow, be simulated, and their interrelations must also be considered. For that reason, the coupled model WRF-Chem+BEP-BEM has been used in this work. WRF (Weather Research and Forecasting) model is one of the most commonly used atmospheric mesoscale models [21] worldwide. The model offers many physical options and can be coupled with the online numerical atmosphere-chemistry model WRF-Chem [7] for Air Quality applications. Furthermore, the BEP-BEM urban multilayer scheme is also included in the system to analyze the influence of the urban canopy while extending the green infrastructures in the city. However, the main concern when using this multi-model system for air quality evaluation is to deliver high-resolution results useful at urban scales (around 300m). Reaching such a resolution is not easy because there are performance aspects regarding execution time and downscaling numerical processes that should be carefully analyzed. A first step through analyzing the influence of coupling WRF and BEP-BEM urban canopy model into the air quality and simulation execution time was presented in [26]. Other works on the effects of downscaling WRF-Chem at an urban scale are reported in [28]. Wang et al. (2022) claim in those cases that it is possible to achieve a resolution of 100 m to determine air quality aspects. For this purpose, they use WRF-Chem, including the Large-Eddy-Simulator (LES) module, since otherwise, it is not possible to reach out such high-resolution simulations. However, in Wang et al. (2022), there is no consideration of the urban canopy layer, and it should be taken into account because it makes the results useful in urban areas.

On the other hand, one can find works that analyzing the influence of including trees as green infrastructures in urban areas [23]. The main conclusion of these works is that any modification of the urban morphology may have both: the positive and the negative influence on the quality of the urban air. Consequently, the authors conclude that a deep study on both effects must be done appropriately to determine how to configure green areas in urban zones to be sure that, as a result, the air quality is improved instead of worsened. This paper shows several reasons why the current state of the art regarding theair quality simulation at urban resolution must be re-oriented to consider the effect of green infrastructures properly. For this purpose, we have selected the city of Barcelona (Spain) as the case study.

The study case an the air quality model set up is described in Sect. 2. Section 3 includes the experimental results of this paper and, finally, the main conclusions of this work are summarized in Sect. 4.

2 Data, Materials and Methods

2.1 Case Study

The case study used is located in the city of Barcelona (Fig. 1), inside the Catalonia region, in northeastern part of the Iberian Peninsula (Spain). The Metropolitan Area of Barcelona (MAB) with more than 3 million people, is the most populated urban area on the Mediterranean coast. The city of Barcelona reports annually one of the highest air pollution levels in Europe. The most problematic pollutants are NO_2, O_3 and particulate matter (PM_{10} and $PM_{2.5}$) [17]. In Fig. 2, one can observe how in 2021, the levels of NO_2 in the city of Barcelona exceed the recommended WHO (World Health Organization) levels in almost all its territory and, if we take previous years into account, the situation was even worse. In particular, in 2015 the NO_2 annual mean exceeded the 2005 WIIO guideline ($40 \,\mu g/m^3$) in the high traffic urban air pollution ground monitoring stations (*Eixample* and *Gràcia-Sant Gervasi*) [16]. The same year, the mean value for $PM_{2.5}$ and PM_{10} was above the 2005 WHO guideline (20 and $10 \,\mu g/m^3$, respectively) in all urban stations in the city [16]. Exceeding these air quality reference levels is associated with significant risks to public health [13,17].

Fig. 1. Location of the study case, Barcelona, indicated with a yellow mark in the map (Color figure online)

For that reason, in this work, we will focus in the NO_2 and the time period studied belongs to this year and, in particular, we have selected from 16th to 20th of July 2015 for being a part of an episode of high temperatures and high pollution levels in the studied area what could, typically, lead to exceeding the limits recommended by the WHO in 2005. In 2021 the WHO guidelines were updated in order to reduce air pollution levels in European cities and the new air pollution recommendations have been established. For example, recommended value of the NO_2 has been reduced from $40\,\mu g/m^3$ to $10\,\mu g/m^3$. To achieve these levels, the governments of cities like Barcelona, must invest in green policies that have a direct positive the urban air quality. One key point to consider is to redesign cities taking advantage of including green infrastructures (trees, green roofs,...). However, to design it in a smart way, reliable simulations at urban resolution must be done. In the next section the models required for this purpose and, then, the models are used in the experimental section to determine whether they are useful in the way they are currently used or not.

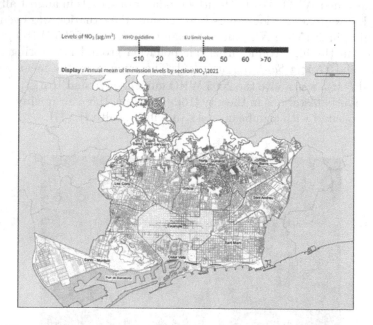

Fig. 2. NO_2 concentration levels in Barcelona in 2021. Data taken from https:// ajuntament.barcelona.cat/mapes-dades-ambientals/qualitataire/ca/

2.2 Model Description, Chemistry and Physics Schemes

The WRF-Chem model is set up with three nested domains covering the Iberian Peninsula, which is the parent domain(D01) with a $9\,km \times 9\,km$ horizontal resolution (WE: 1350 km, NS: 1305 km), followed by the second domain comprehending region of Catalonia at a 3 km × 3 km horizontal resolution and, finally, followed by the third and finer domain of the Metropolitan Area of Barcelona (MAB) at 1 km × 1 km horizontal resolution. Vertically, the three domains are described by 45 vertical layers up to 100 hPa (Fig. 3).

The meteorological and chemical initial and lateral boundary conditions (IC/BCs) were determined using the ERA5 global model [10] and the CAMS-Chem model [5], respectively. The HERMESv3 preprocessor tool [9] was used to create the anthropogenic emissions files from the CAMS-REG-APv3.1 database [6]. Biogenic emissions have been computed online from the Model of Emissions of Gases and Aerosols from Nature v2 (MEGAN; [8]). For the gas-phase chemical scheme, we used the Regional Acid Deposition Model (RADM2, [24]) that accounts for 63 chemical species, 21 photolysis reactions and 136 gas-phase reactions. In WRF-Chem, RADM2 is coupled with MADE/SORGAM aerosol module [3,19]. RADM2 has been broadly used in the air quality studies across Europe [11,25].

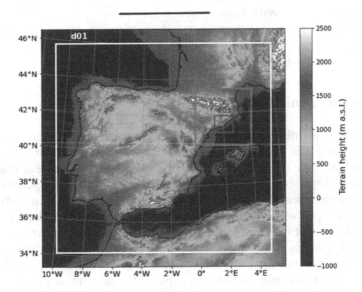

Fig. 3. WRF domains used in our simulations over the Iberian peninsula. Three two-ways nested domains D01: Iberian peninsula (IP), D02: Catalonia region (CAT) and D03: Metropolitan Area of Barcelona (MAB) with 9 km × 9 km, 3 km × 3 km and 1 km × 1 km horizontal resolution, respectively.

In order to represent the urban areas, we used a multi-layer urban canopy scheme, the Building Effect Parameterization (BEP) coupled with the Building

Energy Model (BEP-BEM, [18]). This canopy layer takes into account the energy consumption of buildings and anthropogenic heat, which has previously been validated for the area under study [14,20]. We use the Local Climate Zones (LCZ) classification [22] to associate a specific value of the buildings and ground's thermal, radiative, and geometric parameters in the region of the Metropolitan Area de Barcelona (MAB). The BEP-BEM urban canopy scheme uses 11 urban classes to compute the heat and momentum fluxes in the urban areas (for more details on the use of LCZ and urban morphology refer to [20]). The principal configuration of the model is presented in Table 1.

Table 1. Parameterizations used (same as in [15])

WRF schemes	
Urban scheme	BEP-BEM
Land Surface Model	Noah LSM
PBL scheme	Bougeault-Lacarrère PBL (BouLac), designed to use with urban schemes
Microphysics	WRF Single Moment 6-class scheme
Long- and short-wave radiation	Rapid Radiative Transfer Model for General circulation models (RRTMG) scheme

3   Experimental Results

As it has been mentioned in the previous sections, the experimental study reported in this paper, has been oriented to if the urban air quality simulations must be redefined or not, to be able to play with all the elements that affect the pollutants dispersion in the cities with a high level of the air pollution. The pieces to consider must include the impact of the green urban infrastructures on the chemical processes and also the influence on the meteorological factors. Therefore, the results shown in this section contemplate both: the meteorological and air quality aspects. The results analyzed correspond to the simulation results obtained from $D02$ and $D03$ domains, that is, the results have a resolution of 3 km and 1 km respectively. In order to capture the imprint of two Barcelona's zones that have opposite green morphologies, we have selected *Eixample* and *Ciutadella* areas for the experiments and analysis. The former, is the representative zone of the dense traffic, especially at rush hours and, the second one corresponds to a large green park close to the seaside. Although the meteorological stations and the air quality measurement stations chosen for this work are in different places, the distance between them is irrelevant because we are interested in the station type and its surroundings. However, it is worth noting that the experimentation described below for the selected locations has been carried out for all meteorological and air quality stations in the city of Barcelona. Finally, due to space issues, it was decided to choose two representative areas, one with a no-green and the other with a green urban canopy, since they facilitated the explanation of the objective of this work.

Table 2. Meteorological Stations from the Network of Automatic Weather Stations of Catalonia (XEMA, catalan acronym)

XEMA Station	Measurements	Lat.(°)	Lon.(°)	Alt.msl(m)	Alt.agl(m)
Zoo (Ciutadella)	T, RH	41.38943	2.18847	7	2
Raval (Eixample)	T, RH, WS, WD	41.3839	2.16775	33	40

3.1 Meteorology Results

The observational data used to evaluate the weather variables from the WRF outputs was provided by the Meteorological Service of Catalonia (SMC) [1] through its Network of Automatic Weather Stations (XEMA, catalan acronym). As we have just mentioned, we have collected the data from two meteorologic stations whose location is shown in Table 2. Figure 4 and Figs. 5 show, respectively, the relative humidity and the temperature for the *Eixample* (*Raval*) and *Ciutadella* (**Zoo**) locations respectively. As we may observe, for all cases the results obtained for *D*03 are closer to the real observations than the results provided by *D*02 domain. So, as it is was expected, the higher resolutions the better results. However, there is still an improvement gap between the forecasted 1 km resolution data and the real observations.

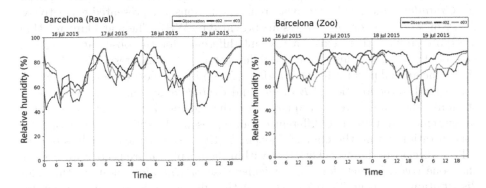

Fig. 4. Evolution of the humidity in the Eixample(Raval) and Ciutadella(Zoo) areas

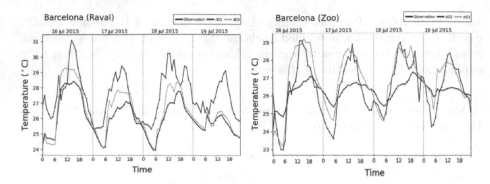

Fig. 5. Evolution of the temperature in the Eixample(Raval) and Ciutadella(Zoo) areas

Table 3. Air Quality Stations from the Atmospheric Pollution Monitoring and Forecasting Network (XVPCA, catalan acronym)

XEMA Station	Measurements	Lat.(°)	Lon.(°)	Alt.msl(m)	Alt.agl(m)
Ciutadella	NO2, O3	41.38640	2.1873	7	2
Eixample	NO2, O3	41.38531	2.15379	26	2

3.2 Air Quality Results

The observational data used to evaluate the WRF-Chem outputs was provided by the *Generalitat de Catalunya* from the *Xarxa de Vigilancia i Previsió de Contaminació Atmosfèrica* (XVPCA). As it has been previously mentioned, for comparison, we have chosen the air quality stations located in *Eixample* and *Ciutadella* (see Table 3). In Figs. 6 and 7, the evolution of the NO_2 pollutant for the *Ciutadella* and *Eixample* measurements stations, are presented respectively. As we may observe, despite the results provided by WRF-Chem at 1 km and 3 km are quite similar, there exists a relevant difference for the observations exists. One of the main reasons for such a difference is the model resolution. Figure 8 shows the annual mean value of emissions (NO_2) in two plots of Barcelona, which correspond to the location of *Eixample* and *Ciutadella* air quality measurements stations. The framed area is a square cell with a side of 1 km around each selected measurement station. The different colors represent the different levels of NO_2 concentration within these areas. Independently of the particular values of each color, it is easy to directly check that within the purple square there appear more than one color, in fact, for the particular case of *Ciutadella* station, the color distribution goes from 20 to $70\,\mu g/m^3$. Although the model set up incorporates the BEP-BEM urban canopy layer with the objective of taking into account the influence of the buildings morphology in the pollutants dispersion, due to the low resolution used, the coupled WRF-Chem model is not able to capture the differences between the parts of the map that are included in the same grid cell. Another aspect to highlight are the differences between the diurnal and night cycles of NO_2 concentration compared to the observations. If one considers as diurnal cycle the time interval that goes from 6:00 to 18:00, we can observe that in the *Ciutadella* station the estimated NO_2 concentration is closer to the observations independently of the day of the week. However, diurnal cycles in the *Eixample* location, are clearly underestimated during business days (2015-07-16 corresponded to a Thursday).

Another issue to analyze is the influence of green areas on air quality. For that purpose, we have cut Figs. 6 and 7 for Friday 2015-07-17, see Fig. 9, but it

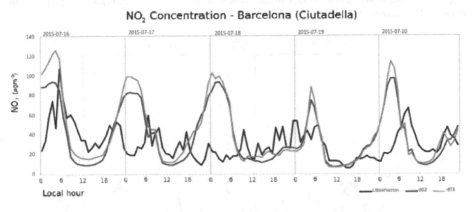

Fig. 6. NO_2 concentration changes in the period of Thursday 2015-07-16 to Monday 2015-07-20 at *Ciutadella* air quality station compared to the results obtained by the WRF-Chem model at $D02$ and $D03$

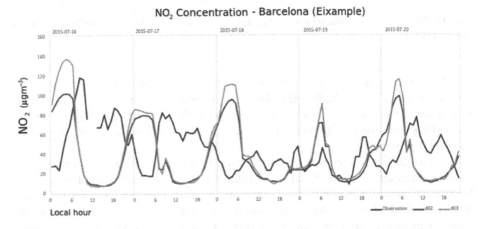

Fig. 7. NO_2 concentration changes in the period of Thursday 2015-07-16 to Monday 2015-07-20 at *Eixample* air quality station compared to the results obtained by the WRF-Chem model at $D02$ and $D03$

Fig. 8. Annual mean of the $NO2$ concentration in 2018 at the *Eixample* and *Ciutadella* areas of Barcelona (Spain)

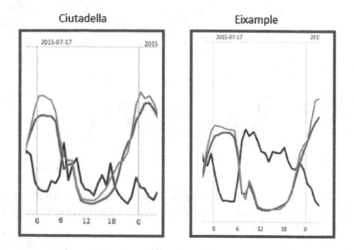

Fig. 9. NO_2 concentration levels for the day 2015-07-17 in the Ciutadella area an in the Eixample

could be done on any other business day. In the cut corresponding to *Ciutadella*, the behavior of NO_2 during the day and at night is quite regular as far as the observations are concerned, while the models at 3 km and 1 km, despite coinciding with the results, cannot capture this behavior as it already has been discussed previously.

In addition, it is very remarkable to keep in mind that this plot corresponds to a green zone with practically no traffic in the entire cell except in certain borders. However, the effect of the park on the air quality of the area is altered if we only look at the models. The main reason for this effect is, not having the coupled models at high resolution, taking into account the green layer existing in *Ciutadella*. From the observed data we can deduce that green infrastructure dampens the harmful effects of pollutants, but this effect cannot be reproduced if it is not worked at high resolution. It is true that in *Ciutadella* area, the emissions produced by traffic are lower, but the model does contemplate this lower injection of pollution into the environment. On the contrary, if we analyze the *Eixample* plot, one may observe that the model cannot capture the behavior of NO_2 concentration in its day or night cycle, as has already been commented on. However, the notable effect is also that the improvement that green spaces' inclusion could produce, nor would it be reproducible if the resolution of all the models involved were not increased. Thus, the main conclusion of this study is that although green areas help to reduce environmental pollution, if complete high-resolution models are not used, it will be difficult to obtain effective green designs.

4   Conclusions

Designing green cities to improve their air quality is one of the planned resilience strategies for many highly polluted cities. To achieve an effective design, it is necessary to test the effect that including certain green areas to the urban morphology would have on the city's air quality. For this, it is necessary to reconsider how to run the proper simulations, since according to the authors' knowledge, there are no proposals that allow to model this influence at a resolution of around 100 m. In this work, the case of the city of Barcelona has been analyzed, emphasizing two areas with opposite characteristics. *Eixample* area has been selected as the area of the high traffic density and many emissions due to the road traffic, and the other area is the *Ciutadella*: Ciutadella: the one that corresponds to a large green park near the sea. The study carried out considering both the meteorological parameters and certain pollutants, has revealed the need of having the complete coupled models (meteorology, chemistry and urban morphology) at urban scale. The main conclusion of this work is that "green resolution matters", that is, downscaling WRF-Chem model without taking green infrastructures into account, may generate designs with the worse air quality for cities than it was before the green modifications.

Acknowledgements. This work has been granted by the Spanish Ministry of Science and Innovation MCIN AEI/10.13039/501100011033 under contract PID2020-113614RB-C21, by the Spanish government under grant PRE2018-085425 and by the Catalan government under contract 2021 SGR 00574.

References

1. Xarxa d'estacions meteorològiques automàtiques. Servei Meteorològic de Catalunya (2017). https://www.meteo.cat/observacions/xema. Accessed 03 Mar 2023
2. La qualitat de l'aire a catalunya. avaluació no2 - 2022. Technical report, Direcció General de Qualitat Ambiental i Canvi Climàtic (2022). https://mediambient. gencat.cat/web/.content/home/ambits_dactuacio/atmosfera/qualitat_de_laire/ avaluacio/balancos_i_informes/documentos/La-qualitat-de-laire-a-Catalunya-NO2-2022.pdf. Accessed 03 Mar 2023

3. Ackermann, I., et al.: Modal aerosol dynamics model for Europe: development and first applications. Atmos. Environ. **32**(17), 2981–2999 (1998). https://doi.org/10. 1016/S1352-2310(98)00006-5. https://www.sciencedirect.com/science/article/pii/ S1352231098000065

4. E., R.M.F.L.A.J.G.A.R.: Informe qualitat de l'aire de barcelona 2021. Technical report, Generalitat de Catalunya (2021). https://mediambient.gencat.cat/ web/.content/home/ambits_dactuacio/atmosfera/qualitat_de_laire/avaluacio/ balancos_i_informes/documentos/estat-qualitat-aire-Catalunya-2021.pdf

5. Gettelman, A., et al.: The whole atmosphere community climate model version 6 (WACCM6). J. Geophys. Res.: Atmos. **124**(23), 12380–12403 (2019). https://doi. org/10.1029/2019JD030943. https://agupubs.onlinelibrary.wiley.com/doi/abs/10. 1029/2019JD030943

6. Granier, C., et al.: The copernicus atmosphere monitoring service global and regional emissions (April 2019 version). Technical report, Copernicus Atmosphere Monitoring Service (2019). https://doi.org/10.24380/D0BN-KX16. https:// atmosphere.copernicus.eu/sites/default/files/2019-06/cams_emissions_general_ document_apr2019_v7.pdf

7. Grell, G., et al.: Fully coupled "online" chemistry within the WRF model. Atmos. Environ. **39**(37), 6957–6975 (2005). https://doi.org/10.1016/j.atmosenv.2005.04. 027

8. Guenther, A.B., et al.: The model of emissions of gases and aerosols from nature version 2.1 (MEGAN2.1): an extended and updated framework for modeling biogenic emissions. Geoscientific Model Dev. **5**(6), 1471–1492 (2012). https://doi.org/ 10.5194/gmd-5-1471-2012. https://gmd.copernicus.org/articles/5/1471/2012/

9. Guevara, M., et al.: HERMESv3, a stand-alone multi-scale atmospheric emission modelling framework - part 1: global and regional module. Geoscientific Model Dev. **12**(5), 1885–1907 (2019). https://doi.org/10.5194/gmd-12-1885-2019. https://gmd.copernicus.org/articles/12/1885/2019/

10. Hersbach, H., et al.: The era5 global reanalysis. Q. J. R. Meteorol. Soc. **146**(730), 1999–2049 (2020). https://doi.org/10.1002/qj.3803. https://rmets.onlinelibrary. wiley.com/doi/abs/10.1002/qj.3803

11. Im, U., et al.: Evaluation of operational online-coupled regional air quality models over Europe and North America in the context of AQMEII phase 2. Part II: particulate matter. Atmos. Environ. **115**, 421–441 (2015). https://doi.org/10. 1016/j.atmosenv.2014.08.072. https://www.sciencedirect.com/science/article/pii/ S1352231014006839

12. Khomenko, S., et al.: Premature mortality due to air pollution in European cities: a health impact assessment. Lancet Planet. Health **5**, 121–134 (2021)

13. World Health Organization: WHO global air quality guidelines: particulate matter (PM2.5 and PM10), ozone, nitrogen dioxide, sulfur dioxide and carbon monoxide. World Health Organization (2021)

14. Ribeiro, I., et al.: Highly resolved WRF-BEP/BEM simulations over Barcelona urban area with LCZ. Atmos. Res. **248**, 105220 (2021). https://doi.org/10. 1016/j.atmosres.2020.105220. https://www.sciencedirect.com/science/article/pii/ S016980952031156X

15. Ribeiro, I., et al.: Highly resolved WRF-BEP/BEM simulations over Barcelona urban area with LCZ. Atmos. Res. **248**(January 2020), 105220 (2021). https:// doi.org/10.1016/j.atmosres.2020.105220

16. Rico, M., Font, L., Arimon, J., Marí, M., Gómez, A., Realp, E.: Informe qualitat de l'aire de Barcelona. Technical report, Agència de Salut Pública de Barcelona (2019)

17. Rivas, I., et al.: Child exposure to indoor and outdoor air pollutants in schools in Barcelona, Spain. Environ. Int. **69**, 200–212 (2014). https://doi.org/10.1016/j.envint.2014.04.009. https://www.sciencedirect.com/science/article/pii/S0160412014001202

18. Salamanca, F., et al.: A study of the urban boundary layer using different urban parameterizations and high-resolution urban canopy parameters with WRF. J. Appl. Meteorol. Climatol. **50**(5), 1107–1128 (2011). https://doi.org/10.1175/2010jamc2538.1

19. Schell, B., et al.: Modeling the formation of secondary organic aerosol within a comprehensive air quality model system. J. Geophys. Res.: Atmos. **106**(D22), 28275–28293 (2001). https://doi.org/10.1029/2001JD000384. https://agupubs.onlinelibrary.wiley.com/doi/abs/10.1029/2001JD000384

20. Segura, R., et al.: Sensitivity study of PBL schemes and soil initialization using the WRF-BEP-BEM model over a Mediterranean coastal city. Urban Climate **39**, 100982 (2021). https://doi.org/10.1016/j.uclim.2021.100982. https://www.sciencedirect.com/science/article/pii/S2212095521002121

21. Skamarock, W., et al.: A Description of the Advanced Research WRF Model Version 4.3. NCAR (2021). https://doi.org/10.5065/1dfh-6p97

22. Stewart, I.D., Oke, T.R.: Local climate zones for urban temperature studies. Bull. Am. Meteorol. Soc. **93**(12), 1879–1900 (2012). https://doi.org/10.1175/bams-d-11-00019.1

23. Fallmann, J., Emeis, S., Suppan, P.: Modeling of the urban heat island and its effect on air quality using WRF/WRF-Chem – assessment of mitigation strategies for a central European city. In: Steyn, D., Mathur, R. (eds.) Air Pollution Modeling and its Application XXIII. SPC, pp. 373–377. Springer, Cham (2014). https://doi.org/10.1007/978-3-319-04379-1_60

24. Stockwell, W., et al.: The second generation regional acid deposition model chemical mechanism for regional air quality modeling. J. Geophys. Res.: Atmos. **95**(D10), 16343–16367 (1990). https://doi.org/10.1029/JD095iD10p16343. https://agupubs.onlinelibrary.wiley.com/doi/abs/10.1029/JD095iD10p16343

25. Tuccella, P., et al.: Modeling of gas and aerosol with WRF/Chem over Europe: evaluation and sensitivity study. J. Geophys. Res.: Atmos. **117**(D3) (2011). https://doi.org/10.1029/2011JD016302. https://agupubs.onlinelibrary.wiley.com/doi/abs/10.1029/2011JD016302

26. Vidal, V., Cortés, A., Badia, A., Villalba, G.: Evaluating WRF-BEP/BEM performance: on the way to analyze urban air quality at high resolution using WRF-Chem+BEP/BEM. In: Paszynski, M., Kranzlmüller, D., Krzhizhanovskaya, V.V., Dongarra, J.J., Sloot, P.M.A. (eds.) ICCS 2021. LNCS, vol. 12746, pp. 516–527. Springer, Cham (2021). https://doi.org/10.1007/978-3-030-77977-1_41

27. Vos, P., et al.: Improving local air quality in cities: to tree or not to tree? Environ. Pollut. **5**(183), 113–122 (2013)

28. Wang, Y., et al.: Coupled mesoscale-LES modeling of air quality in a polluted city using WRF-LES-Chem. EGUsphere, p. 162 (2022). https://doi.org/10.5194/egusphere-2022-1208

Influence of Activation Functions on the Convergence of Physics-Informed Neural Networks for 1D Wave Equation

Paweł Maczuga and Maciej Paszyński[✉]

AGH University of Science and Technology, Kraków, Poland
paszynsk@agh.edu.pl

Abstract. In this paper, we consider a model wave equation. We perform a sequence of numerical experiments with Physics Informed Neural Network, considering different activation functions, and different ways of enforcing the initial and boundary conditions. We show the convergence of the method and the resulting numerical accuracy for different setups. We show that, indeed, the PINN methodology can solve the problem efficiently and accurately the wave-equations without actually solving a system of linear equations as it happens in traditional numerical methods like, e.g., finite element or finite difference method. In particular, we compare the influence of selected activation functions on the convergence of the PINN method. Our PINN code is available on github: https://github.com/pmaczuga/pinn-comparison/tree/iccs.

Keywords: PINN · wave-equations · activation functions · initial conditions · boundary conditions · deep neural network

1 Introduction

Physics Informed Neural Networks (PINN) was introduced by George Karniadakis in 2019 [8]. In PINN, the neural networks represent the solution of PDE

$$u(x,t) = \mathrm{PINN}(x,t) = A_1\sigma(A_2\sigma(A_3\ldots\sigma(A_n\begin{bmatrix}x\\t\end{bmatrix}+y_n)+\cdots+y_3)+y_2)+y_1 \quad (1)$$

where A_k is the weight matrix of layer k and y_k are the biases of this layer. In the training process, PINN learns the residual of the PDE and the residual of boundary and initial conditions by probing the residuals using sample points. The PINN can learn the solution of PDE without actually forming and solving a system of linear equations. The training process is usually performed until we reach a solution with a loss value less than the prescribed accuracy [4], or for the fixed number of epochs [6,8]

J. Mikyška et al. (Eds.): ICCS 2023, LNCS 14073, pp. 74–88, 2023.
https://doi.org/10.1007/978-3-031-35995-8_6

The PINN method can also actually fail. For example, the PINN method, when applied to a problem of advection-diffusion, can only find a solution for small values of convection coefficients, and it fails when the problem is convection dominated. In that case, it is necessary to train the PINN first with the solutions for small values of convection and increase it gradually during the training process [6]. Difficult problems require hard-coding of the boundary conditions into the computational model [2].

Using some references in the PINN literature, we can find some guidelines about the number of layers, the number of neurons per layer, the training rate, the number of epochs during the training, and the number of points used during the training. For example, in [1], the authors use a fully-connected neural network with eight hidden layers and 200 neurons per layer. They also use the training rate of 0.0001 followed by 0.0005 and 0.001, each for 50 epochs. Low number of epochs (for PINNs) comes from the fact, that the authors used huge number of points divided into mini-batches. Usually, increasing the number of points during the training process improves the convergence of the training, but it is also necessary to consider the distribution of points, especially for problems with singularities. From that paper, in different experiment, we can also read an example the number of collocation points for the training of PDE (7000 for the residual of the main PDE and 3000 for the zero divergence equation) as well as for the training of initial and boundary conditions (between 30 to 800 depending on the kind).

Incorporating initial and boundary conditions into PINN also requires some special considerations, often leading to difficulties [11]. The initial conditions and Dirichlet boundary conditions are often enforced in a hard way [10], while the Neumann boundary conditions are usually enforced in a weak way. It is also necessary to weigh the loss functions related to the PDE and to the boundary or initial conditions or use a larger number of samples on the boundary and initial condition residual terms [12].

As observed in [10], the convergence rate of the training process depends strongly on the activation function used. The authors observed that the training rate for the adaptive tanh activation function [3] was much faster than for the Switch activation function [9]. In the end, the accuracy of the obtained solution was the same, but the convergence rates were different.

In this paper, we focus on wave equations, and we employ the manufactured solution technique to check the convergence of the PINN with different activation functions. We refer to the best habits on how to set up the PINN parameters as described in the mentioned literature review.

2 Wave Equation

The main goal of this paper is to demonstrate the influence of different activation functions on the convergence of Physics Informed Neural Networks designed to solve the wave equation, mainly:

$$\frac{\partial u^2(x,t)}{\partial^2 x} = c^2 \frac{\partial u^2(x,t)}{\partial^2 t} \tag{2}$$

for $x \in [0,1]$ and $t \in [0,1]$. With the definition (1) in mind, relating $u(x,t) = \text{PINN}(x,t)$ with a neural network, we define the loss function based on the residual of the PDE.

$$\text{LOSS}_{\text{PDE}}(x,t) = \left(\frac{\partial \text{PINN}^2(x,t)}{\partial^2 x} - c^2 \frac{\partial \text{PINN}^2(x,t)}{\partial^2 t} \right)^2, \tag{3}$$

where we differentiate the neural network, representing the solution, with respect to x and t. Now, we will discuss different possible initial and boundary conditions for the wave equation.

Specifically, we choose one of the following setups:

1) Zero boundary conditions and initial conditions that satisfies the boundary conditions

$$\begin{cases} u(x,0) & = A\sin(\varphi\pi x) \\ \frac{\partial u(x,0)}{\partial t} & = 0 \\ u(0,t) = u(1,t) & = 0 \end{cases} \tag{4}$$

These boundary and initial conditions result in the following family of exact solutions

$$u_{exact}(x,t) = A\sin(\varphi\pi x)\cos(c\ \varphi\pi t), \tag{5}$$

For these boundary and initial conditions, we define the following loss functions

$$\text{LOSS}_{\text{IC}}(x,0) = \left(\frac{\partial \text{PINN}(x,0)}{\partial t} - A\sin(\varphi\pi x) \right)^2,$$

$$\text{LOSS}_{\text{ICdt}}(x,0) = \left(\frac{\partial \text{PINN}(0,t)}{\partial t} - 0.0 \right)^2,$$

$$\text{LOSS}_{\text{BC0}}(0,t) = (u(0,t) - 0.0)^2,$$

$$\text{LOSS}_{\text{BC1}}(1,t) = (u(1,t) - 0.0)^2. \tag{6}$$

2) Reflective boundary condition and corresponding initial solution

$$\begin{cases} u(x,0) & = A\cos(\varphi\pi x) \\ \frac{\partial u(x,0)}{\partial t} & = 0 \\ \frac{\partial u(0,t)}{\partial x} = \frac{\partial u(1,t)}{\partial x} & = 0 \end{cases} \tag{7}$$

which results in the following family of exact solutions:

$$u_{exact} = A\cos(\varphi\pi x)\cos(c\varphi\pi t) \tag{8}$$

With this alternative boundary and initial conditions, we define the following loss functions

$$\text{LOSS}_{IC}(x,0) = \left(\frac{\partial \text{PINN}(x,0)}{\partial t} - A\cdot\cos(\varphi\pi x) \right)^2,$$

$$\text{LOSS}_{ICdt}(x,0) = \left(\frac{\partial \text{PINN}(0,t)}{\partial t} - 0.0 \right)^2,$$

$$\text{LOSS}_{BC0}(0,t) = \left(\frac{\partial u(0,t)}{\partial t} - 0.0 \right)^2,$$

$$\text{LOSS}_{BC1}(1,t) = \left(\frac{\partial u(1,t)}{\partial t} - 0.0 \right)^2. \tag{9}$$

The total loss is

$$\text{LOSS}(x,t) = \text{LOSS}_{PDE}(x,t) + \text{LOSS}_{IC}(x,t) + \text{LOSS}_{ICdt}(x,t) + \text{LOSS}_{BC0}(x,t) + \text{LOSS}_{BC1}(x,t) \tag{10}$$

Now we can choose one of the above setups of initial and boundary conditions and select values of the following parameters:

- c - equation constant.
- A - initial amplitude.
- φ - number of "hills" and "valleys" in the initial condition.

3 Training

Algorithm 1: PINN training

1: Randomly select $x \in (0,1)$
2: Randomly select $t \in (0,1)$
3: Compute all loss functions:

$$\text{LOSS}_{\text{PDE}}(x,t),\ \text{LOSS}_{\text{IC}}(x,t),\ \text{LOSS}_{\text{ICdt}}(x,t),\ \text{LOSS}_{\text{BC0}}(x,t),\ \text{LOSS}_{\text{BC1}}(x,t)$$

4: **for** Number of epochs **do**
5:     **for** Coefficients of matrices from neural network
       layers$\{A^p_{qr}\}_{q=1,\ldots,n_p;r=1,\ldots,m_p;p=1,\ldots,n}$ **do**
6:         Compute the derivatives

$$\frac{\partial \text{LOSS}_{\text{PDE}}(x,t)(x,t)}{\partial A^p_{qr}} \qquad \frac{\partial \text{LOSS}_{\text{IC}}(x,t)}{\partial A^p_{qr}}$$

$$\frac{\partial \text{LOSS}_{\text{BC0}}(x,t)}{\partial A^p_{qr}} \qquad \frac{\partial \text{LOSS}_{\text{BC1}}(x,t)}{\partial A^p_{qr}}$$

7:         Correct the coefficients of matrices from neural network layers

$$A^p_{qr} = A^p_{qr} + \eta_{\text{PDE}}\frac{\partial \text{LOSS}_{\text{PDE}}(x,t)}{\partial A^p_{qr}} \qquad A^p_{qr} = A^p_{qr} + \eta_{\text{IC}}\frac{\partial \text{LOSS}_{\text{IC}}(x,t)}{\partial A^p_{qr}}$$

$$A^p_{qr} = A^p_{qr} + \eta_{\text{BC0}}\frac{\partial \text{LOSS}_{\text{BC0}}(x,t)}{\partial A^p_{qr}} \qquad A^p_{qr} = A^p_{qr} + \eta_{\text{BC1}}\frac{\partial \text{LOSS}_{\text{BC1}}(x,t)}{\partial A^p_{qr}}$$

8:     **end for**
9:     **for** Coefficients of bias vectors from neural network layers$\{b^p_q\}_{q=1,\ldots,n_p;p=1,\ldots,n}$
       do
10:        Compute the derivatives

$$\frac{\partial \text{LOSS}_{\text{PDE}}(x,t)}{\partial b^p_q} \qquad \frac{\partial \text{LOSS}_{\text{IC}}(x,t)}{\partial b^p_q}$$

$$\frac{\partial \text{LOSS}_{\text{BC0}}(x,t)}{\partial b^p_q} \qquad \frac{\partial \text{LOSS}_{\text{BC1}}(x,t)}{\partial b^p_q}$$

11:        Correct the coefficients of matrices from neural network layers

$$y^p_q = y^p_q + \eta_{\text{PDE}}\frac{\partial \text{LOSS}_{\text{PDE}}(x,t)}{\partial y^p_q} \qquad y^p_q = y^p_q + \eta_{\text{IC}}\frac{\partial \text{LOSS}_{\text{IC}}(x,t)}{\partial y^p_q}$$

$$y^p_q = y^p_q + \eta_{\text{BC0}}\frac{\partial \text{LOSS}_{\text{BC0}}(x,t)}{\partial y^p_q} \qquad y^p_q = y^p_q + \eta_{\text{BC1}}\frac{\partial \text{LOSS}_{\text{BC1}}(x,t)}{\partial y^p_q}$$

           {Where $\eta_{\text{PDE}}, \eta_{\text{IC}}, \eta_{\text{BC0}}, \eta_{\text{BC1}}$ are training rates for different loss functions.}
12:        **end for**
13: **end for**

---

4   Numerical Results

In this section, we illustrate the best numerical result obtained from the training of the PINN solver for the 1D wave equation. Figure 3 compares the exact solution with the solution obtained after training of PINN. Figure 2 illustrates how PINN managed to learn the initial condition. Figure 1 shows how the loss function evolved during 60,000 epochs of training. The difference between the exact and PINN solution is computed point-wise. We conclude that the PINN can learn the solution of the wave equation with the accuracy of the order of 0.01. The question if this accuracy is satisfactory will depend on particular applications. The details of the model parameters, the architecture of the neural network, the learning parameters, the loss functions used, and the convergence of the training for different setups are summarized in the next section.

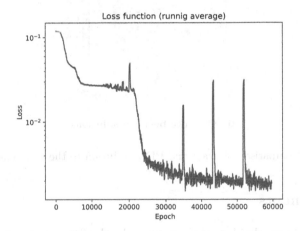

Fig. 1. Convergence of loss function during PINN training.

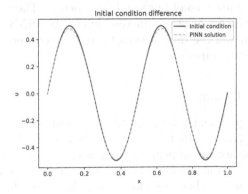

Fig. 2. Learning the initial condition by PINN.

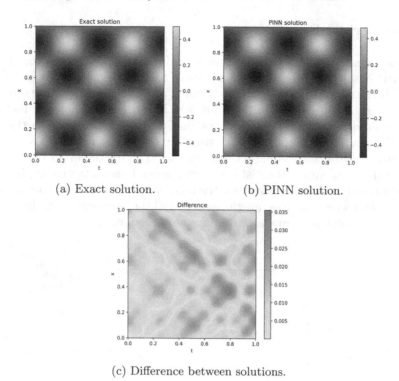

(a) Exact solution.   (b) PINN solution.

(c) Difference between solutions.

Fig. 3. Comparison of exact and PINN solution to the wave equation.

5   Experiments

The experiments are divided into two parts. In the first set of numerical experiments, we created the PINN architecture and trained it to solve the wave equation with various boundary conditions and parameters. The goal of these experiments was to find three sets of parameters where the PINN is converging well to the exact solution and use these setups to test the convergence rate of PINN with different activation functions.

5.1   Parameters Tuning

In our numerical experiments, we referred to setups introduced by authors of six research papers [1,2,6,8,10,11]. Table 1 presents the literature review (the number of layers, the number of neurons per layer, different activation functions, the learning rate (LR), the number of epochs for training (unspecified in some papers) as well as the number of collocation points used for probing the loss functions of the PDE, the initial and boundary conditions).

We chose the PINN architecture, and we solved the wave equation with PINN. The architecture stays the same for all the experiments; only the parameters of

Table 1. Review of PINN architectures. LR - learning rate, N_r, N_i, N_b are number of collocation points for residual, initial condition and boundary condition respectively. Dash represents data that was not specified in the referenced article.

Reference	Layers	Neurons	Activation	LR	epochs	Collocation points
[10]	3	20	Swish ($\beta = 1$)	–	–	–
[2]	5	100	tanh	10^{-3}	–	$N_r = 20,000$ $N_i = 50$ $N_b = 50$
[6]	4	50	tanh	10^{-4}	–	–
[8]	5	100	tanh	10^{-3}	–	$N_r = 20,000$ $N_i = 50$ $N_b = 50$
[1]	8	200	sin	1×10^{-3} 5×10^{-4} 1×10^{-4}	150	$N_r = 3 \times 10_6$
[1]	6	60	tanh	6×10^{-4}	3×10^5	$N_r = 7,000$
[11]	4	50	tanh	10^{-3} adaptive	40,000	–

the equation change. Namely, following Table 1, we selected the following PINN architecture and the training parameters:

- Feed-Forward fully connected network.
- 4 layers.
- 80 neurons per layer.
- learning rate – 0.002.
- 50 000 epochs.
- Equally spaced collocation points.
  - For residual $N_r = 222\,500 = 150 \cdot 150$.
  - For initial conditions $N_i = 150$.
  - For boundary conditions $N_b = 150$.
- tanh activation function.

We train the above PINN with the sets of parameters:

- $c \in \{0.3, 0.5, 1.0, 2.0, 3.0\}$.
- $A \in \{0.5, 1.0, 2.0\}$.
- $\varphi \in \{2, 4, 6\}$.

We employed boundary conditions - zero and reflective.

5.2   Activation Functions

The second set of numerical experiments concerns the investigation of the influence of different loss functions on the convergence of the training. We chose a couple of sets of parameters that yielded interesting results and trained the network again using the different activation functions. The rest of the network architecture stays exactly the same.

For activation functions we choose

- tanh, the most commonly used function in PINNs [2,6,8,11] and indeed it yields very good results.
- sin in [8].
- sigmoid, rarely seen along PINNs.
- swish is adaptive activation function with the following formula: $swish = x \cdot sigmoid(\beta\, x)$ introduced in [9]. β is an adaptive parameter, that changes during training, similar to weights. In the context of PINNs it was used in [10], but with constant β.
- adaptive tanh with the formula atanh $= \tanh(\alpha\, x)$, where α is trainable parameter. Used in [3].

A very popular function for other neural networks other than PINNs is ReLU, which is not present in the points above. That is because, in this (and many other) equations, we need to calculate the second derivative of the neural network. However, ReLU is basically two linear functions merged together, and so its second derivative is always 0, and that, in turn, prevents learning.

6   Results

As seen in Figs. 4, 5 and 6, PINN with that architecture is able to learn simple cases very well. However, it struggles with more difficult ones. The reflective boundary condition is usually harder to learn, but that is not always the case. There are also strange spikes for reflective BC, where increasing A or ϕ actually decreases loss. Figures 7, 8 and 9 shows the influence of different activation functions. The main issue is usually a very large loss connected to the initial condition. This means that PINN is able to learn the equation itself quite well, but for a slightly different initial condition. It is illustrated in Figure 10.

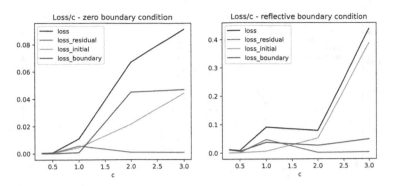

Fig. 4. Influence of equation parameter c on loss function. Other parameters are: $A = 1$, $phi = 4$.

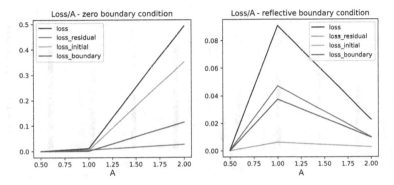

Fig. 5. Influence of initial condition amplitude on loss function. Other parameters are: $c = 1$, $phi = 4$. Strangely loss actually decreases as φ grows. Which means that somehow increasing it makes the equation easier to learn by the neural network. However more experiments are required in order to pin the exact reason, which is outside scope of this work.

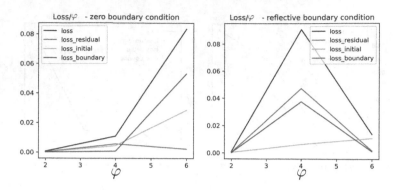

Fig. 6. Influence of *phi* parameter in initial condition on loss function. Other parameters are: $c = 1$, $A = 1$.

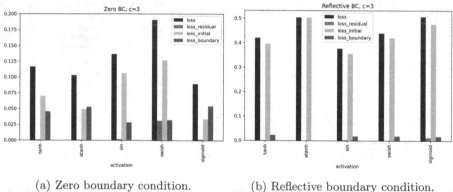

(a) Zero boundary condition.   (b) Reflective boundary condition.

Fig. 7. Loss for different activation functions for zero boundary condition. Parameters are: $c = 3$, $A = 1$, $\phi = 4$.

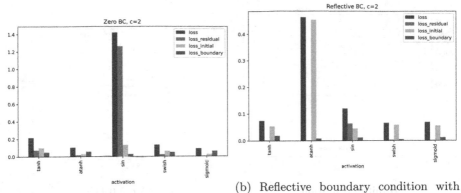

(a) Zero boundary condition with larger c.

(b) Reflective boundary condition with larger c.

Fig. 8. Loss for different activation functions, where one performs significantly worse than the others parameters: $c = 2$, $A = 1$, $\phi = 4$.

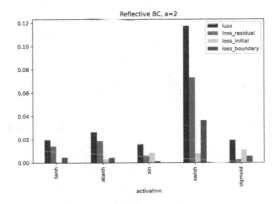

(a) Reflective boundary condition with larger amplitude - A.

Fig. 9. Loss for different activation functions, where one performs significantly worse than the others parameters: $c = 1$, $A = 2$, $\phi = 4$.

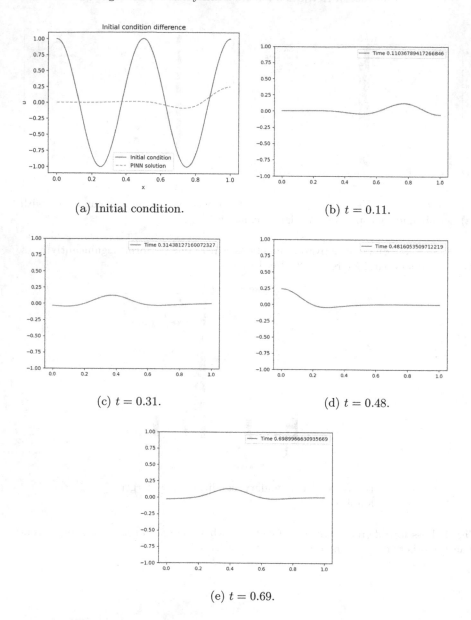

(a) Initial condition.

(b) $t = 0.11$.

(c) $t = 0.31$.

(d) $t = 0.48$.

(e) $t = 0.69$.

Fig. 10. Properly behaving wave, that failed to learned imposed initial condition.

7   Conclusions and Future Work

Surprisingly one of the best result is achieved by the sigmoid function, which is very rarely used for PINNs. As shown in Figs. 8 and 9, there are cases where *atanh*, *sin*, and *swish* achieve much worse results than other activation functions. In the experiment, we performed, it was never the case for either *tanh* or sigmoid. Their effectiveness is as well bound to the specific problem, and there are cases where one or the other achieves better results. Interestingly the same can not be said about adaptive tanh, as it has huge downgrades in performance in some cases. In this case, the problem is connected to a very large loss for the initial condition, and the other two losses are rather low. Both sin and swish failed to learn the equation itself.

Although results in this paper are produced by solving a single PDE they can be in fact generalized to other PDEs as well. The approach is always the same regardless of the specific PDE that the network is trying to solve. That is the minimization of the equation's residual.

The PINNs themselves are likely the future of numerical simulations and worth exploring further. However, they do have some limitations at the moment. Specifically, PINNs are still achieving slightly worse results than state of the art finite element method. Furthermore, there still needs to be a complete theory concerning the Physics Informed Neural Networks. As seen in the results of this paper, there are aspects of PDE that are difficult to learn, most notably the initial condition and more difficult sets of equation parameters. There are ways to mitigate or even eliminate some of them. The most notable are:

- Enforcing the initial condition in a hard way [7].
- Using Variational PINNs (VPINNs), which vastly improves convergence [5].
- Pretraining the PINN for the easier set of parameters and then fine-tuning it using the proper ones [6].

Our future work can be divided in two parts. Firstly we will further explore the three aforementioned ideas and use that to solve 2D problems. Afterwards we plan to use PINNs to solve a real life problem, such as the simulation of tsunami wave caused by underwater earthquake.

Acknowledgment. The Authors are thankful for support from the funds assigned to AGH University of Science and Technology by the Polish Ministry of Science and Higher Education. Research project supported by the program "Excellence initiative - research university" for the AGH University of Science and Technology.

References

1. Cai, S., Mao, Z., Wang, Z., Yin, M., Karniadakis, G.E.: Physics-informed neural networks (PINNs) for fluid mechanics: a review. Acta Mech. Sinica **37**(12), 1727–1738 (2021)
2. Cuomo, S., Di Cola, V.S., Giampaolo, F., Rozza, G., Raissi, M., Piccialli, F.: Scientific machine learning through physics-informed neural networks: where we are and what's next. J. Sci. Comput. **92**(3), 88 (2022)
3. Jagtap, A.D., Kawaguchi, K., Karniadakis, G.E.: Adaptive activation functions accelerate convergence in deep and physics-informed neural networks. J. Comput. Phys. **404**, 109136 (2020)
4. Karniadakis, G.E., Kevrekidis, I.G., Lu, L., Perdikaris, P., Wang, S., Yang, L.: Physics-informed machine learning. Nat. Rev. Phys. **3**(6), 422–440 (2021)

5. Kharazmi, E., Zhang, Z., Karniadakis, G.E.: Variational physics-informed neural networks for solving partial differential equations. CoRR, abs/1912.00873 (2019)
6. Krishnapriyan, A.S., Gholami, A., Zhe, S., Kirby, R.M., Mahoney, M.W.: Characterizing possible failure modes in physics-informed neural networks. arXiv e-prints arXiv:2109.01050 (2021)
7. Lu, L., Pestourie, R., Yao, W., Wang, Z., Verdugo, F., Johnson, S.: Physics-informed neural networks with hard constraints for inverse design. SIAM J. Sci. Comput. **43**, B1105–B1132 (2021)
8. Raissi, M., Perdikaris, P., Karniadakis, G.: Physics-informed neural networks: a deep learning framework for solving forward and inverse problems involving nonlinear partial differential equations. J. Comput. Phys. **378**, 686–707 (2019)
9. Ramachandran, P., Zoph, B., Le, Q.V.: Searching for activation functions. arXiv e-prints arXiv:1710.05941 (2017)
10. Sun, L., Gao, H., Pan, S., Wang, J.-X.: Surrogate modeling for fluid flows based on physics-constrained deep learning without simulation data. Comput. Methods Appl. Mech. Eng. **361**, 112732 (2020)
11. Wang, S., Teng, Y., Perdikaris, P.: Understanding and mitigating gradient flow pathologies in physics-informed neural networks. SIAM J. Sci. Comput. **43**, A3055–A3081 (2021)
12. Yuan, L., Ni, Y.-Q., Deng, X.-Y., Hao, S.: A-PINN: auxiliary physics informed neural networks for forward and inverse problems of nonlinear integro-differential equations. J. Comput. Phys. **462**, 111260 (2022)

Accelerating Multivariate Functional Approximation Computation with Domain Decomposition Techniques

Vijay Mahadevan$^{(\boxtimes)}$ (iD), David Lenz(iD), Iulian Grindeanu(iD),
and Thomas Peterka(iD)

Argonne National Laboratory, Lemont, IL 60439, USA
{mahadevan,dlenz,iulian,peterka}@anl.gov

Abstract. Modeling large datasets through Multivariate Functional Approximations (MFA) plays a critical role in scientific analysis and visualization workflows. However, this requires scalable data partitioning approaches to compute MFA representations in a reasonable amount of time. We propose a fully parallel and efficient method for computing MFA with B-spline bases without sacrificing the reconstructed solution accuracy or continuity. Our approach reduces the total work per task and uses a restricted Additive Schwarz (RAS) method to converge control point data across subdomain boundaries. We also provide a detailed analysis of the parallel approach with domain decomposition solvers to minimize subdomain error residuals and recover high-order continuity with optimal communication cost determined by the overlap regions in the RAS implementation. In contrast to previous methods that generally only recover C^1 continuity for arbitrary B-spline order p or required post-processing to blend discontinuities in the reconstructed data, the accuracy of the MFA remains bounded as the number of subdomains is increased. We demonstrate the effectiveness of our approach using analytical and scientific datasets in 1, 2, and 3 dimensions and show that it is highly scalable (due to bounded outer iteration counts) and that the parallel performance at scale is directly proportional to the nearest-neighbor communication implementations.

Keywords: functional approximation · domain decomposition · B-spline representations · additive Schwarz solvers

1 Introduction

Large-scale discrete data analysis of various scientific computational simulations often require high-order continuous functional representations that have to be evaluated anywhere in the domain. Such expansions described as *Multivariate Functional Approximations* (MFA) [4] in arbitrary dimensions allow the original

---

This work is supported by the U.S. Department of Energy, Office of Science, Advanced Scientific Computing Research under Contract DE-AC02-06CH11357, Program Manager Margaret Lentz.

J. Mikyška et al. (Eds.): ICCS 2023, LNCS 14073, pp. 89–103, 2023.
https://doi.org/10.1007/978-3-031-35995-8_7

discrete data to be modeled, and expressed in a compact form, in addition to supporting higher-order derivative queries (without further approximations such as finite differences) for complex data analysis tasks. MFA utilizes approximations of the raw discrete data using a hypervolume of piecewise continuous functions. One particular option is to use the variations of the B-Spline or NURBS bases [23,24] for the MFA *encoding* of scientific data. The reconstructed data in MFA retains the spatiotemporal contiguity, and statistical distributions, with lesser storage requirements. Due to the potentially large datasets that need to be encoded into MFA, the need for computationally efficient algorithms (in both time and memory) to parallelize the work is critically important. It is also essential to guarantee that the solution smoothness in the reconstructed (or *decoded*) dataset is consistently preserved when transitioning from a single MFA domain to multiple domains during parallelization.

Achieving improved performance without sacrificing discretization accuracy requires an infrastructure that is consistent in the error metrics of the decoded data and an algorithm that remains efficient in the limit of large number of parallel tasks. In this paper, we will utilize domain decomposition (DD) techniques [26] with data partitioning strategies to produce scalable MFA computation algorithms that minimize the reconstruction error when reproducing a given dataset. In such partitioned analysis, it is imperative to ensure that the continuity of the encoded and decoded data across subdomain interfaces is maintained, and remains consistent with the degree of underlying expansion bases used in MFA [23]. This is due to the fact that independently computing MFA approximations in individual subdomains do not guarantee even C^0 regularity in either the MFA space or in the reconstructed data. In order to tackle this issue, we rely on an iterative Schwarz-type DD scheme to ensure that continuity is enforced, and the overall error stays bounded as the number of subdomains are increased (or as the subdomain size decreases).

In addition to remaining efficient, we also require the devised algorithms to extend naturally to arbitrary dimensional settings and to handle large datasets. We next discuss some of the related work in the literature that have been explored for reconstruction of scattered data, and approaches to make these algorithms scalable in order to motivate the ideas presented in the paper.

1.1   Related Work

Domain decomposition (DD) techniques in general rely on the idea of splitting a larger domain of interest into smaller partitions or subdomains, which results in coupled Degrees-of-Freedom (DoF) at their common interfaces. Combining the application of DD schemes and NURBS bases with isogeometric analysis (IGA) [6,7] for high-fidelity modeling of nonlinear Partial Differential Equations (PDEs) [8,10,19] has enjoyed recent success at scale. However, many of these implementations lack full support to handle multiple geometric patches in a distributed memory setting due to non-trivial requirements on continuity constraints at patch boundaries. Directly imposing higher-order geometric continuity in IGA requires specialized parameterizations in order to preserve the approximation properties [16], which can be difficult to parallelize [15] generally.

To overcome some of these issues with discontinuities along NURBS or B-spline patches, Zhang et al. [28] proposed to use a gradient projection scheme to constrain the value (C^0), the gradient (C^1), and the Hessian (C^2) at a small number of test points for optimal shape recovery. Such a constrained projection yields coupled systems of equations for control point data for local patches, and results in a global minimization problem that needs to be solved.

Alternatively, it is possible to create a constrained recovery during the actual post-processing stage i.e., during the decoding stage of the MFA through standard blending techniques [14], in order to recover continuity in the decoded data. However, the underlying MFA representation remains discontinuous, and would become more so with increasing number of subdomains without the ability to recover higher-order derivatives along these boundaries. Moreover, selecting the amount of overlap and resulting width of the blending region relies strongly on a heuristic, which can be problematic for general problem settings.

In contrast, we propose extensions to the constrained solvers used by Zhang et al. [28], and introduce a two-level, DD-based, parallel iterative scheme to enforce the true degree of continuity, independent of the basis function polynomial degree p, unlike the low-order constraints used previously. The outer iteration utilizes the restricted Additive Schwarz (RAS) method [13], with efficient inner subdomain solvers that can handle linear Least-Squares systems to minimize the decoded residual within acceptable error tolerances. Such an iterative solver has low memory requirements that scales with growing number of subdomains, and necessitates only nearest-neighbor communication of the interface data once per outer iteration to converge towards consistent MFA solutions.

2   Approach

With motivations to accelerate the computation of an accurate MFA representation scalably, we utilize a data decomposition approach with overlapping subdomains to create shared layers of piecewise accurate functional reconstructions. This is similar to a multipatch approach typically taken in IGA computations [6,8]. However, in order to ensure that higher-order continuity across domain boundaries are preserved, an outer iteration loop is inevitable to converge the shared unknowns across the interfaces. These global iterations guarantee consistent MFA encodings in parallel, without which the representations will not even ensure C^0 regularity.

In this section, we first provide an illustrative example by formulating the constrained minimization problem to be solved in each subdomain and explain the iterative methodology used in the current work to converge the shared DoFs.

2.1   Numerical Background

A p-th degree NURBS or B-spline curve [24] is defined using the Cox-deBoor functions for each subdomain as

$$C(u) = \sum_{i=0}^{n} R_{i,p}(u)P(i), \quad \forall u \in \Omega \tag{1}$$

$$R_{i,p}(u) = \frac{N_{i,p}(u)W_i}{\sum_{i=0}^{n} N_{i,p}(u)W_i} \tag{2}$$

where $R_{i,p}(u)$ are the piecewise rational functions with P control points of size n, W_i are the control point weights, with the p-th degree B-spline bases $N_{i,p}(u)$ defined on a knot vector u.

Given a set of input points Q that need to be encoded into a MFA, with the weights $W = 1$ (B-spline representations) for simplicity, the unconstrained minimization problem to compute the optimal set of control point locations within a subdomain can be posed as a solution to a linear Least-SQuares (LSQ) system that minimizes the net error of the B-spline approximation.

$$\underset{P \in \mathbb{R}^n}{\arg\min} E = \|Q - RP\|_{L_2},$$

$$\text{where } R \in \mathbb{R}^{m \times n}, \ Q \in \mathbb{R}^m \tag{3}$$

An appropriate LSQ solver such as the one based on Cholesky decomposition or the more efficient ℓ-BFGS scheme [29] can compute the control point solution P that minimizes the residual error E for the given input data Q and MFA representation of degree p. Note that the minimization procedure can be performed independently on each subdomain without dependencies as there are no constraints explicitly specified in Eq. (3). However, in order to recover high-order continuity across subdomain interfaces, computing the unconstrained solution is insufficient. At the minimum, the shared DoFs on subdomain boundaries have to match to recover C^0 continuity in the decoded data (RP).

A straightforward approach to achieve C^0 continuity in the recovered solution is by ensuring that the common control point data P at subdomain interfaces are clamped with repeated knots, in addition to using clamping at the global domain boundaries. In this scheme, the control points exactly interpolate (are clamped to) input data points at the subdomain interface boundaries. Such an approach requires in general a good spatial distribution of Q, and yields only low-order continuous approximations (C^0) when the solution remains smooth across the subdomain interfaces. It should also be noted that as the number of subdomains increases, the global solution being computed becomes further constrained, and more interpolatory due to clamped DoFs. Moreover, the MFA solution computed becomes dependent on the number of subdomains used to decompose the problem; i.e., the global control point data P recovers different reconstructions as a function of number of subdomains ($\mathcal{N}$) used.

While the implementation of the domain decomposed MFA can be much simpler with clamped knots on all subdomain boundaries, ensuring higher-order continuity would require that all $p - 1$ derivatives of the approximation match as well. As a continuous extension, one could relax the interpolatory behavior of clamped knot boundaries by reducing the number of repeated knots, and instead use unclamped knots at internal subdomain boundary interfaces by sharing knot spans between subdomains.

More generally, the constrained minimization problem to recover continuity [24] can be formulated as

$$RP = Q \quad | \quad CP = G, \tag{4}$$

where C is the constraint matrix imposing continuity restrictions on the control points P along with its derivatives, with data exchanged from neighboring domains stored in G, around the neighborhood of the interface $\Omega_{i,j}$ shared by subdomains i and j. With the use of penalized constraints (C) and Lagrange multipliers [12,22], the solution to the constrained LSQ problem can recover optimal control point values. This modification allows us to recover fully consistent (C^0 to C^{p-1}) continuous MFA reconstructions using the solution procedure detailed for the global constrained minimization problem Eq. (4).

2.2 Shared Knot Spans at Subdomain Interfaces

In the current work, in contrast to using clamped knots, we utilize unclamped (floating), shared knot spans at all interior subdomain boundaries such that the high-order continuity and consistency of the decoded solution are recovered, independent of $\mathcal{N}$ (Fig. 1).

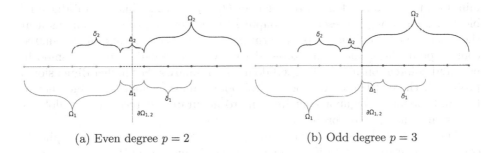

(a) Even degree $p = 2$ (b) Odd degree $p = 3$

Fig. 1. Illustration: 1D parallel partitioned domain with unclamped interior knots and augmented spans ($|\delta| = 2$)

For the purpose of illustration and to explain the proposed solver methodology, let us consider a simple one dimensional domain (Ω) with two subdomains ($\mathcal{N} = 2$) as shown in Fig. (1), where Ω_1 and Ω_2 represent the subdomains that share an interface $\partial\Omega_{1,2}$. In Fig. (1), the layout of the knot spans for both an even degree ($p = 2$) and odd degree ($p = 3$) are shown. For generality, we also introduce here an overlap layer Δ_1 and Δ_2 on each subdomain that represents the set of shared knot spans with its adjacent subdomain (for internal boundaries), and an optional augmented layer δ_1 and δ_2 that has a connotation similar to that of an overlap region in traditional DD schemes [26]. Note that in order to reconstruct the input data in $\Omega_i, \forall i \in [1,2]$, the knot spans must mandatorily include Δ_i regions. This Δ_i overlap region is required by definition to maintain partition of unity of a B-spline curve in order to evaluate Eq. (2). For generality, Δ_i represents the repeated knots along clamped global domain boundaries, and the shared knots between two subdomains in the unclamped interior boundaries. For arbitrary degree p, the number of knot spans in Δ_i is given by $\lfloor \frac{p}{2} \rfloor$, where $\lfloor . \rfloor$ represents the floor operator. In multidimensional tensor product expansions, these shared spans are replaced by shared layers of knot spans along the subdomain interfaces. The δ_i regions are additional, and optional, shared knot spans that can help improve error convergence in a manner similar to overlap regions in DD methods used for PDE solvers [13,26].

Note that the control point DoF vector can be represented by three separate parts based on the local support of the basis expansion. The control point vector is in general given as $P(\Omega \cup \Delta \cup \delta) = [P(\Omega); P(\Delta); P(\delta)]$. However, the owned DoFs represented by $P(\Omega)$ is the only component computed through the LSQ subdomain solver, while the other components, $P(\Delta)$ and $P(\delta)$, are only used to impose the constraints to recover continuity.

Now, the constrained minimization problem for the two subdomain case can be written as

$$\begin{bmatrix} R_1(\Omega_1) & \lambda_{1,2}(\Delta_1 \cup \delta_1) \\ \lambda_{2,1}(\Delta_2 \cup \delta_2) & R_2(\Omega_2) \end{bmatrix} \begin{bmatrix} P_1(\Omega_1 \cup \Delta_1 \cup \delta_1) \\ P_2(\Omega_2 \cup \Delta_2 \cup \delta_2) \end{bmatrix} = \begin{bmatrix} Q_1 \\ Q_2 \end{bmatrix} \qquad (5)$$

where the diagonal operators R_1 and R_2 are the piecewise rational functions that minimize the local subdomain residuals in $\Omega_j, \forall j \in [1,2]$, while the off-diagonal blocks $\lambda_{1,2}$ and $\lambda_{2,1}$ represent the coupling terms between the subdomains near the interface $\partial\Omega_{1,2}$. This coupling term provides the constraints on the shared control point data, and higher-order derivatives as needed to recover smoothness and enforce continuity along subdomain boundaries. For higher dimensional problems, the constraints on the control points must include both face neighbor and diagonal neighbor contributions to accurately determine the globally consistent minimization problem.

The coupling blocks $\lambda_{i,j}$ can be viewed as Lagrange multipliers that explicitly couple the control point DoFs across a subdomain interface ($P_1(\Omega_1 \cup \Delta_1 \cup \delta_1) \cap P_2(\Omega_2 \cup \Delta_2 \cup \delta_2)$) such that continuity is preserved in a weak sense [24]. Using appropriate Schur complements to eliminate the coupled DoF contributions in each subdomain, with $\lambda_{i,j}$ evaluated at *lagged* iterates of adjacent subdomains, the set of coupled constrained equations in Eq. (5) can be completely decoupled for each subdomain. This modified system resembles a block-Jacobi operator of the global system. The scheme illustrated in this section follows ideas similar to the Jacobi-Schwarz method [13] and the overlapping RAS method [27].

In the above description, the coupled control point patches, $P_1(\Delta_1)$ and $P_2(\Delta_2)$ belonging to adjacent subdomains near $\partial\Omega_{1,2}$ are exchanged simultaneously between Ω_1 and Ω_2 before the local domain solves are computed independently. One key advantage with such a DD scheme is that only nearest neighbor exchange of data is required, which keeps communication costs bounded as the number of subdomains $\mathcal{N}$ increase [27], while providing opportunities to interlace recomputation of the constrained control point solution. Note that in such an iterative scheme, nearest neighbor exchanges can be performed compactly per dimension and direction, thereby minimizing communication costs and eliminating expensive global collectives.

Augmenting Knot Spans with Overlap: One of the key metrics of interest is that the parallel solver infrastructure does not amplify any approximation errors unresolved by the tensor product B-spline mesh. Since the local decoupled subdomain solution is encoded accurately to satisfy Eq. (3) in each individual subdomain without any data communication (i.e., embarassingly parallel), imposing the constraints for the shared DoFs in Δ should ensure the error change is bounded. However, as the control point data across subdomains become synchronized, numerical artifacts, especially for high-degree ($p > 2$) basis reconstructions at subdomain interfaces can become dominant sources of error. A key validation metric is to compare the error profiles from the multiple subdomain cases to the single subdomain case, in order to ensure convergence of the solvers to the same unique solution, independent of $\mathcal{N}$.

For many problem domains, overlapping Schwarz solvers [13,18] have been proven to be more stable, efficient and scalable compared to non-overlapping variants [3,27]. We utilize the concept of overlap regions by sharing additional knot spans between subdomains in order to produce better MFA reconstructions of the underlying data. This user-specified, additional overlap is described by $\delta_j, \forall j \in [1,2]$ in Fig. (1). The amount of data overlap utilized for computing the functional approximation can directly affect the accuracy of the subdomain solver, and the scalability of the algorithm. Additionally, the presence of the augmented knot spans in δ keep the residual errors E bounded as the number of subdomain increase with appropriate overlap regions. Note that the optimal range of δ depends on the gradient of the input data near the subdomain interfaces, i.e., $|\delta| = 0$ may suffice for reconstructing smoother solution profiles, while strong gradient reconstructions may require $|\delta| = p$ or $|\delta| = 2p$ in general.

2.3 Solver Workflow

Computing the functional approximation of large datasets require efficient solvers at two levels: first, the local decoupled subdomain problem in Eq. (3), and next, the constrained minimization problem in Eq. (4). Hence, the global problem reduces to a series of minimization problems in each subdomain.

Subdomain Solvers: For the linear LSQ solvers that can be used to compute local subdomain control point solution P, there are a variety of choices available. Direct methods like Singular Value decomposition or Cholesky decomposition operating on the normal equations [2] can compute optimal values. Alternatively, the iterative LSQ solvers such as orthogonal decomposition methods based on

QR and QZ factorizations are more stable, especially when the normal form of the operator, $R^T R$, is ill-conditioned.

Restricted Additive-Schwarz Solvers: The outer RAS iterations work together with nearest neighbor communication procedures to exchange shared DoF data between adjacent subdomains. This is an important step to ensure that P computed through the LSQ procedure is consistent and high-order continuous across subdomain boundaries. The final minimized control point solution is achieved when the interface solutions $P(\Delta)$ match on all $\partial \Omega_{i,j} \in \Omega$.

It is also important to note that unlike the blending approaches that can be directly applied on decoded data [14], the numerical error with the constrained iterative scheme is not bounded by the original partitioned, unconstrained least-squares solution; i,e., imposing subdomain boundary constraints can create artificial numerical peaks (non-monotonic) in reconstructed data as we converge towards continuity recovery. To address this issue, we can increase the subdomain overlap δ to ensure uniform convergence to the true single-subdomain solution error, in the limit of $\mathcal{N} \to \infty$.

2.4   Implementation

The DD techniques presented here for MFA computation are primarily implemented in Python-3, with main dependencies on `SciPy` for B-spline bases evaluations and linear algebra routines. Additionally, the drivers utilize Python bindings (`PyDIY`) for the `DIY` [20] C++ library. `DIY` is a programming model and runtime for block-parallel analytics on distributed-memory machines, built on `MPI-3`. Rather than programming for process parallelism directly in `MPI`, the programming model in `DIY` is based on block parallelism. In `DIY`, data are decomposed into subdomains called blocks. One or more of these blocks are assigned to processing elements (processes or threads) and the computation is described over these blocks, and communication between blocks is defined by reusable patterns. `PyDIY` utilizes `PyBind11`[1] to expose the C++ library in a Pythonic way. In our implementation, `PyDIY` is exclusively used to manage the data decomposition, including specifications to share an interface $\partial \Omega_{i,j}$ and ghost layers that represent the $\Delta \cup \delta$ overlapping domains.

The implementation of non-overlapping and overlapping RAS schemes applied to the computation of MFA exhibits scalable convergence properties in the limit of decreasing subdomain size (i.e., as $\mathcal{N} \to \infty$). This is a favorable property for strong scaling, especially when tackling large datasets, as the net computational cost always remains bounded. This behavior can be explained by the nature of how the RAS iterative procedure resolves the shared DoFs.

---

[1] PyBind11: https://github.com/pybind/pybind11 (last accessed 01/24/2023).

3   Results

To demonstrate the effectiveness of the iterative algorithm for MFA computation, we devised a series of analytical closed form functionals and utilized real-world scientific datasets in both 2 and 3 dimensions obtained from high-fidelity simulations. We also define a *compression ratio* (η), which gives the ratio of total input points in the dataset ($dim\ \boldsymbol{Q}$) to the total control points ($dim\ \boldsymbol{P}$) used in the MFA B-spline representation. Note that as $\eta \to 1$, one can achieve smaller error residuals compared with the reference data for a given degree p, while $\eta \gg 1$ produces smoother approximations with potentially larger pointwise errors.

3.1   Error Convergence Analysis

For a domain $\Omega(x, y, z) = [-4, 4]^3$, we define closed-form synthetic datasets shown in Eq. (6), Eq. (7) and Eq. (8) for 1D, 2D and 3D respectively to verify error convergence and to demonstrate the parallel scalability of MFA computation.

$$F(x) = \mathcal{S}(x + 1) + \mathcal{S}(x - 1), \forall x \in \Omega = [-4, 4], \tag{6}$$

$$F(x, y) = \mathcal{S}(\sqrt{x^2 + y^2}) + \mathcal{S}(2(x - 2)^2 + 2(y + 2)^2), \tag{7}$$

$$F(x, y, z) = \mathcal{S}(\sqrt{x^2 + y^2 + z^2}) + \mathcal{S}(2(x - 2)^2 + (y + 2)^2 + (z - 2)^2), \tag{8}$$

where $\mathcal{S}(x) = \frac{sin(x)}{x}$.

First, to determine the effect of using augmented or overlapped knot span regions (δ) as the number of subdomains $\mathcal{N}$ are increased, we use the 1D function in Eq. (6) on a single subdomain as the reference solution (Fig. (2a)). The error profiles for $\mathcal{N} = 5$ using non-overlapping spans ($|\delta| = 0$) and augmented spans ($|\delta| = 3$) are also shown in Fig. (2c) and Fig. (2d) respectively.

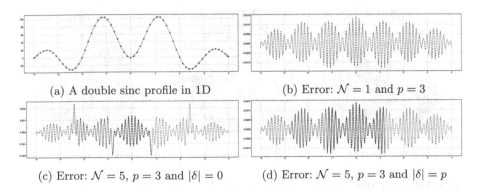

(a) A double sinc profile in 1D   (b) Error: $\mathcal{N} = 1$ and $p = 3$

(c) Error: $\mathcal{N} = 5$, $p = 3$ and $|\delta| = 0$ (d) Error: $\mathcal{N} = 5$, $p = 3$ and $|\delta| = p$

Fig. 2. Demonstration of error convergence, and effect of the overlapping spans to minimize numerical artifacts

It is evident from Fig. (2c) that in the non-overlapping case, the reconstructed data at subdomain boundaries are influenced by contributions from both adjacent domain DoFs, which are enforced to be C^{p-1} continuous by the constrained minimization solver. However, as we increase the number of overlap regions in terms of both the underlying data and the local bases support spans, the error profiles as shown in Fig. (2d) approaches the reference profile (with $\mathcal{N} = 1$) shown in Fig. (2b). Heuristically, for many of the problems tested, using $|\delta| = p$ provides optimal error convergence as number of subdomains increase, even though increasing this parameter to $|\delta| = 2p$ or higher will in general always improve the numerical accuracy at the cost of higher communication costs between neighboring subdomains.

3.2   Real Simulation Datasets

Next, we apply the parallel MFA algorithm to some real world examples cases.

Computational Fluid Dynamics Dataset: In this study, we utilized a 3D dataset from the large-eddy simulation of Navier-Stokes equations for validation of MAX experiments [25] using Nek5000 [11]. The velocity field data is representative of turbulent mixing and thermal striping that occurs in the upper plenum of liquid sodium fast reactors. For the 2D analysis, a slice of the velocity magnitude was resampled onto a 200×200 regular grid [23]. The reference solution and the converged, reconstructed solution with $\mathcal{N} = 5 \times 5 = 25$ subdomains with $p = 6$ and $|\delta| = 2p$ is shown in Fig. (3) for different η. Depending on the use case for MFA reconstruction, the converged error norms with 20 floating control points per subdomain yielding $\eta = 4$ is sufficient to evaluate continuous derivatives everywhere in the domain Ω. The decoded MFA representation (with $\eta = 1$) is also shown in Fig. (3c), which can fully reconstruct the sharp features in the input dataset in contrast to a lossy smoothing shown in Fig. (3b) ($\eta = 4$).

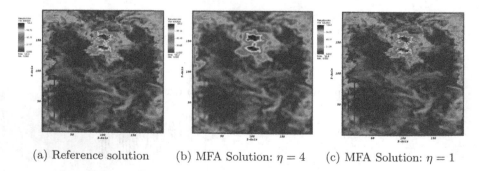

(a) Reference solution   (b) MFA Solution: $\eta = 4$ (c) MFA Solution: $\eta = 1$

Fig. 3. Velocity profile: reference solution (left), B-spline MFA with $p = 6$, $\mathcal{N} = 5 \times 5$, $|\delta| = p$ for $\eta = 4$ (middle), and $\eta = 1$ (right).

Turbulent Combustion Dataset: S3D is a turbulent fuel jet combustion dataset generated from a simulation in the presence of an external cross-flow [5]. The 3D domain has the span $|\Omega| = 704 \times 540 \times 550$, containing three components of the vector field. The magnitude of the velocity field is shown below in Fig. (4a) with 209M points. The converged MFA reconstruction is shown in Fig. (4b) with $8^3 = 512$ subdomains and $n = 62$ per direction in each subdomain yields $\eta = \frac{(704 \times 540 \times 550)}{(8 \times 62)^3} \approx 1.72$. While uniform knot refinement does yield sufficient error reductions in most subdomains, adaptive knot insertion with the DeCasteljau algorithm [24] can reduce the relatively large errors (as shown in Fig. (4c)) to provide better reconstructions everywhere. This was demonstrated on a single subdomain previously [21], but with minor modifications to the communication routines, we can naturally extend the algorithm to handle adaptive knot spans with high-order continuity.

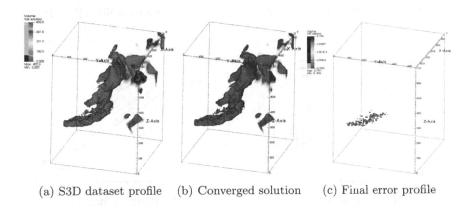

(a) S3D dataset profile   (b) Converged solution   (c) Final error profile

Fig. 4. Volume rendered S3D dataset with $\eta = 1.72$: reference profile (left), converged MFA decoded profile (middle) and the corresponding reconstruction error (right) with $\mathcal{N} = 8 \times 8 \times 8 = 512$, $p = 3$ and $|\delta| = 2p = 6$

3.3 Parallel Scalability

The strong scaling studies using Eq. (7) for 2D and Eq. (8) for 3D were performed on the Theta Cray XC40 supercomputer operated by the Argonne Leadership Computing Facility (ALCF), which provides 4,392 KNL compute nodes with 64 compute cores and 192 GB DDR4 RAM per node. The strong scaling tests in Fig. (5) were performed on 1 to 16,384 tasks in 2D, increasing by a factor of $2^2 = 4$, and the 3D tests were executed on 1 to 32,768 tasks, increasing by a factor of $2^3 = 8$. Note that the 2D studies used 400M input points with $\eta = 4$ and 3D cases used 1.331B points with $\eta = 1.25$ for all runs. In order to better understand the effects of using augmented overlaps (δ) on scalability, two cases with $|\delta| = 0$ and $|\delta| = p$ were also tested. A breakdown of the timings for each task in computing the parallel MFA representation is plotted for all cases. This task-wise breakdown helps us clearly visualize the operations that scale optimally and the ones that do not (Fig. 5).

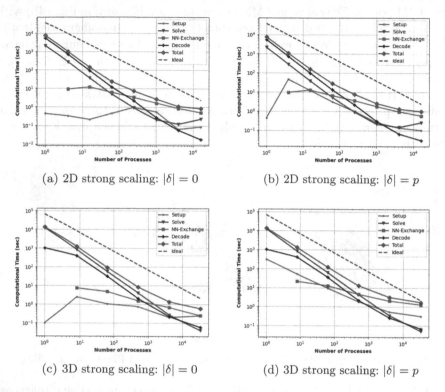

(a) 2D strong scaling: $|\delta| = 0$ (b) 2D strong scaling: $|\delta| = p$

(c) 3D strong scaling: $|\delta| = 0$ (d) 3D strong scaling: $|\delta| = p$

Fig. 5. Strong scaling performance with $p = 3$ and one subdomain per task for non-overlapping (left) and overrlapping (right) RAS schemes

The data partitioned RAS iterative scheme shows good strong scalability for the large dataset experiments, and overall time to compute the MFA in parallel is reduced at a nearly ideal rate up to 8192 MPI tasks in 3D as $\mathcal{N}$ increases, while ensuring C^{p-1} continuity in the subdomain interfaces. It is important to note that the dominant computational time is usually driven by the decoupled LSQ solution computation and the decoding operations, which are embarrassingly parallel as the size of the subdomains decreases in direct proportion to the tasks. Given that the scalability of the linear algebraic LSQ solvers [1,9] and Sparse Matrix Vector (SpMV) products used in the decode tasks are well understood, the bottlenecks potentially occur primarily due to nearest neighbor communication for constraint data exchange between adjacent subdomains. The overall scalability of the algorithm becomes sub-linear when the cost of nearest neighbor exchange cross over the cost of the subdomain solve.

The overall strong scaling efficiency of RAS for MFA remains around 50% for both overlapping and non-overlapping 2D problem cases at 16,384 tasks. However, the added setup cost and sub-optimally scaling nearest neighbor communications reduce the 3D problem efficiency for the overlapping subdomain cases to 25% at 32,768 tasks from 70% in comparison to the non-overlapping cases. The results confirm that the communication cost at scale is driven by the size of the messages being transferred (determined by $|\delta|$) between subdomains.

4 Summary

We have presented a scalable DD approach to tackle the issue of discontinuous B-spline based MFA representations when performing the computations in parallel. The Restricted Additive Schwarz (RAS) method is a natural algorithmic fit for data analysis problems to reconstruct MFA representations in a scalable workflow. Through the use of overlapping Schwarz-based iterative schemes, combined with constrained local subdomain solvers, the two-level iterative technique has been shown to be robust in converging to the functional representation of the given data, without deviating from the single subdomain accuracy metrics.

We have demonstrated that the use of overlap layers δ does preserve the overall MFA accuracy in comparison to the single subdomain case. We determined that for all the problems tested, including real datasets, $|\delta| = p$ to $|\delta| = 2p$ is nearly optimal in terms of error recovery and computational cost even for 3D problems up to 32,768 tasks. A stronger theory for deriving an optimal $|\delta|$ measure will be part of the future work. The MFA-DD scheme applied to both 2D and 3D problems show good parallel scalability and degrades only when the cost of nearest neighbor subdomain data exchanges start to creep up beyond the cost of the locally constrained subdomain solves. Given that scaling characteristics of these sub-processes are well understood in the literature, the parallel speedups behave predictably well at scale on the large computing machines tested.

Another natural extension to improve the accuracy is to utilize a hierarchy of control lattices (multilevel B-spline approximations [17]) to generate a sequence of MFA, the sum of which produces better approximations to datasets with strong gradients. Note that Replacing B-spline bases with NURBS bases ($W \neq 1$) only requires imposing the constraints on the $P_i W_i$ data instead of P_i, and will be explored in the future.

Acknowledgements. This work is supported by Advanced Scientific Computing Research, Office of Science, U.S. Department of Energy, under Contract DE-AC02-06CH11357, program manager Margaret Lentz. This research used the Theta supercomputer operated by the Argonne Leadership Computing Facility (ALCF), a DOE Office of Science User Facility supported under Contract No. DE-AC02-06CH11357, and the Bebop cluster operated by the Argonne Laboratory Computing Resource Center (LCRC) at Argonne National Laboratory.

References

1. Benzi, M.: Preconditioning techniques for large linear systems: a survey. J. Comput. Phys. **182**(2), 418–477 (2002)
2. Björck, Å.: Numerical Methods for Least Squares Problems. SIAM (1996)
3. Bjørstad, P.E., Widlund, O.B.: To overlap or not to overlap: a note on a domain decomposition method for elliptic problems. SIAM J. Sci. Stat. Comput. **10**(5), 1053–1061 (1989)
4. de Boor, C., DeVore, R.: Approximation by smooth multivariate splines. Trans. Am. Math. Soc. **276**(2), 775–788 (1983)
5. Chen, J.H., et al.: Terascale direct numerical simulations of turbulent combustion using s3d. Comput. Sci. Discov. **2**(1), 015001 (2009)
6. Cottrell, J.A., Hughes, T.J., Bazilevs, Y.: Isogeometric Analysis: Toward Integration of CAD and FEA. Wiley, Hoboken (2009)
7. Da Veiga, L.B., Cho, D., Pavarino, L.F., Scacchi, S.: Overlapping Schwarz methods for isogeometric analysis. SIAM J. Numer. Anal. **50**(3), 1394–1416 (2012)
8. Dalcin, L., Collier, N., Vignal, P., Côrtes, A., Calo, V.M.: PetIGA: a framework for high-performance isogeometric analysis. Comput. Methods Appl. Mech. Eng. **308**, 151–181 (2016)
9. Dalcin, L.D., Paz, R.R., Kler, P.A., Cosimo, A.: Parallel distributed computing using python. Adv. Water Resour. **34**(9), 1124–1139 (2011)
10. Dedè, L., Quarteroni, A.: Isogeometric analysis for second order partial differential equations on surfaces. Comput. Methods Appl. Mech. Eng. **284**, 807–834 (2015)
11. Deville, M.O., Fischer, P.F., Mund, E.H.: High-Order Methods for Incompressible Fluid Flow, vol. 9. Cambridge University Press, Cambridge (2002)
12. Dornisch, W., Klinkel, S.: Boundary conditions and multi-patch connections in isogeometric analysis. PAMM **11**(1), 207–208 (2011)
13. Efstathiou, E., Gander, M.J.: Why restricted additive Schwarz converges faster than additive Schwarz. BIT Numer. Math. **43**(5), 945–959 (2003)
14. Grindeanu, I., Peterka, T., Mahadevan, V.S., Nashed, Y.S.: Scalable, high-order continuity across block boundaries of functional approximations computed in parallel. In: 2019 IEEE International Conference on Cluster Computing (CLUSTER), pp. 1–9. IEEE (2019)
15. Hofer, C., Langer, U.: Fast Multipatch Isogeometric Analysis Solvers. Ph.D. thesis, Johannes Kepler University Linz (2018)
16. Kapl, M., Sangalli, G., Takacs, T.: Construction of analysis-suitable G1 planar multi-patch parameterizations. Comput. Aided Des. **97**, 41–55 (2018)
17. Lee, S., Wolberg, G., Shin, S.Y.: Scattered data interpolation with multilevel B-splines. IEEE Trans. Visual Comput. Graph. **3**(3), 228–244 (1997)
18. Lions, P.L.: On the schwarz alternating method. i. In: First International Symposium on Domain Decomposition Methods for Partial Differential Equations, vol. 1, p. 42. Paris, France (1988)
19. Marini, F.: Parallel Additive Schwarz Preconditioning for Isogeometric Analysis. Ph.D. thesis, Università degli Studi di Milano (2015)
20. Morozov, D., Peterka, T.: Block-Parallel Data Analysis with DIY2. In: Proceedings of the 2016 IEEE Large Data Analysis and Visualization Symposium LDAV 2016. Baltimore (2016)
21. Nashed, Y.S.G., Peterka, T., Mahadevan, V., Grindeanu, I.: Rational approximation of scientific data. In: Rodrigues, J.M.F., et al. (eds.) ICCS 2019. LNCS, vol. 11536, pp. 18–31. Springer, Cham (2019). https://doi.org/10.1007/978-3-030-22734-0_2

22. Paul, K., Zimmermann, C., Duong, T.X., Sauer, R.A.: Isogeometric continuity constraints for multi-patch shells governed by fourth-order deformation and phase field models. Comput. Methods Appl. Mech. Eng. **370**, 113219 (2020)
23. Peterka, T., Youssef, S., Grindeanu, I., Mahadevan, V.S., Yeh, R., Tricoche, X., et al.: Foundations of multivariate functional approximation for scientific data. In: 2018 IEEE 8th Symposium on Large Data Analysis and Visualization (LDAV), pp. 61–71 (2018)
24. Piegl, L., Tiller, W.: The NURBS Book. Springer, Heidelberg (2012)
25. Pointer, W., Lomperski, S., Fischer, P.: Validation of CFD methods for advanced SFR design: upper plenum thermal striping and stratification. In: Proc. International Conference on Nuclear Engineering ICONE17, Brussels, Belgium (2009)
26. Smith, B., Bjorstad, P., Gropp, W.: Domain Decomposition: Parallel Multilevel Methods for Elliptic Partial Differential Equations. Cambridge University Press, Cambridge (2004)
27. St-Cyr, A., Gander, M.J., Thomas, S.J.: Optimized multiplicative, additive, and restricted additive Schwarz preconditioning. SIAM J. Sci. Comput. **29**(6), 2402–2425 (2007)
28. Zhang, X., Wang, Y., Gugala, M., Müller, J.D.: Geometric continuity constraints for adjacent NURBS patches in shape optimisation. In: ECCOMAS Congress, vol. 2, p. 9316 (2016)
29. Zheng, W., Bo, P., Liu, Y., Wang, W.: Fast B-spline curve fitting by L-BFGS. Comput. Aided Geom. Des. **29**(7), 448–462 (2012)

User Popularity Preference Aware Sequential Recommendation

Mingda Qian[1,2], Feifei Dai[1(✉)], Xiaoyan Gu[1], Haihui Fan[1], Dong Liu[1], and Bo Li[1]

[1] Institute of Information Engineering, Chinese Academy of Sciences, Beijing, China
{qianmingda,daifeifei,guxiaoyan,fanhaihui,liudong0039,libo}@iie.ac.cn
[2] School of Cyber Security, University of Chinese Academy of Sciences, Beijing, China

Abstract. In recommender systems, users' preferences for item popularity are diverse and dynamic, which reveals the different items that users prefer. Therefore, identifying user popularity preferences are significant for personalized recommendations. Although many methods have analyzed user popularity preferences, most of them only consider particular types of popularity preferences, leading to inappropriate recommendations for users who have other popularity preferences. To comprehensively study user popularity preferences, we propose a User Popularity preference aware Sequential Recommendation (UPSR) method. By sequentially perceiving user behaviors, UPSR captures the type and the evolution of user popularity preferences. Furthermore, UPSR employs contrastive learning to gather similar users and enhance user interest encoding. Then, we can match items and user popularity preferences more accurately and make more proper recommendations. Extensive experiments validate that UPSR not only outperforms the state-of-the-art methods but also reduces popularity bias.

Keywords: Popularity preference · Popularity bias · Sequential recommendation · Neural networks

1 Introduction

In recommender systems, different users have different types of popularity preferences. Many users are sensitive to popularity. For example, some users favor top-selling products, while others prefer less popular items to show their tastes and personalities. Other users are insensitive to popularity and focus on other features, such as price and appearance. Besides, users could have different popularity preferences in different time periods. For example, users' attitudes toward famous brands may change as age grows. As the diverse and personal user popularity preferences reveal the items that users prefer, comprehensively recognizing user popularity preferences are significant for making accurate and personalized recommendations.

User popularity preferences have been widely studied, and existing methods can be separated according to the types of popularity preferences they cover. 1)

J. Mikyška et al. (Eds.): ICCS 2023, LNCS 14073, pp. 104–118, 2023.
https://doi.org/10.1007/978-3-031-35995-8_8

Many methods [3,23] emphasize whether items are popular when making recommendations, which equals to only considering the users prefer popular items. However, these methods may be biased toward hot items and recommend too many popular items to users, and such an issue is called popularity bias. Even worse, popularity bias may bring other critical issues, such as the Matthew effect [29] and echo chamber [13]. 2) To alleviate popularity bias, [1,48] directly reduce the exposure of popular items. However, these methods ignore the users that prefer popular items. 3) Several methods consider more types of popularity preferences to simutanuously alleviate popularity bias and making accurate recommendations. [16,31,45] consider the users either prefer popular items or are insensitive to popularity. However, these methods ignore the users who prefer items with medium or low popularities and can not capture dynamic user popularity preferences. As a result, the users prefer less popular items or largely change their popularity preferences will be wrongly recognized and may receive inappropriate recommendations.

To solve the above issues, we aim to comprehensively capture the diverse and dynamic user popularity preferences. To achieve this, we define user popularity set as $\mathbf{Q}^u = \{\mathbf{P}_i | v_i \in S_u\}$, where item popularity $\mathbf{P}_i$ denotes the number of users who have interacted with item v_i, and S_u is the behavior sequence of user u. Then, we reveal the type and the evolution of user popularity preferences as follows: 1) To identify whether users prefer items with high, medium, or low popularity, we calculate the means μ of $\mathbf{Q}$. Then, we can separate these types of users and make more personalized recommendations. 2) To measure the change amplitude of user popularity preference, we use the standard deviations σ of $\mathbf{Q}$. Hence, we can enhance the learning of user popularity preferences. 3) To obtain the evolution of popularity preferences, we divide user popularity sets according to time periods and capture local user popularity preferences. With the evolution tendencies, we can predict users' future preferences more accurately.

To materialize our ideas, we propose a User Popularity preference aware Sequential Recommendation (UPSR) method. As shown in Fig. 1, UPSR consists of three parts. 1) We apply sequential recommendation methods [7,28,42] as our basic model to extract general features. 2) The sequential popularity perception module enables UPSR to perceive user popularity preferences. To capture the evolution of popularity preferences, this module sequentially processes the subsequences of user behavior sequences. To obtain the types of popularity preferences, we introduce the means and the standard deviations of user popularity sets. Therefore, UPSR can give more personalized recommendations. 3) The popularity contrastive learning module gathers similar users according to the captured popularity preferences and the general features. With the gathered users, this module can further enhance the encoding of user interests, and UPSR can give more proper recommendations.

In summary, we make the following contributions:

- We propose a sequential popularity perception module that comprehensively extracts the type and the evolution of user popularity preference. Therefore, items consistent with user preferences can be recommended.

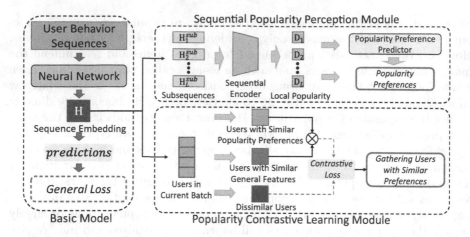

Fig. 1. The overall architecture of UPSR.

- We propose a popularity contrastive learning module that gathers similar users according to popularity preferences and general features. As a result, we can capture user interests more appropriately.
- Extensive experiments are conducted on three popular datasets for evaluation. The results show that UPSR not only outperforms the state-of-the-art methods but also reduces popularity bias.

2   Related Works

This section discusses three lines of works that are closely related to ours: sequential recommendation methods, popularity aware recommendation methods, and contrastive learning methods.

2.1   Sequential Recommendation

General recommendation methods [9,10,24,36,37] model user interests in a static way and gain users' general interests. However, users have dynamic and evolving interests, and recommender systems should rely on users' future interests rather than general interests. To solve this issue, sequential recommendation methods utilize temporal information to capture dynamic user interests and make proper recommendations. With this advantage, sequential recommendation methods receive great attention and are widely used in industry [18,34,40].

The key to sequential recommendation is analyzing user behavior sequences. Early works model the sequences with the Markov Chain [35]. With the development of deep learning [12,41,44,47], Recurrent Neural Networks and its variants are applied [21,22] to analyze the sequences. Recently, the self-attention mechanism [41] has been introduced to model user behavior sequences [25,39] and achieves outstanding performances. Besides, TiSASRec [28] and MEAN-TIME [7] argue that the timestamps of user behavior can not only generate the sequences but also enhance behavior sequence modeling. However, existing methods do not explicitly perceive user popularity preference and may recommend items not align with user popularity preferences. Therefore, we apply UPSR to these methods to capture user popularity preferences and make more accurate recommendations.

2.2 Popularity Aware Recommendation

In recommender systems, popularity has long been recognized as a significant feature. Many existing methods consider utilizing item popularity to improve the performance of recommendations. For example, MostPop is a widely used baseline method that recommends the most popular items. [3, 27] extracts temporal item popularity for recommendations. [23] re-visits and improves MostPop. Graph neural network based methods [5] represent item popularity by the degree of item nodes. However, making recommendations depending on item popularity may expose popular items too frequently, while rarely recommending less popular items. Such an issue is called popularity bias. Popularity bias may cause low-quality recommendations and even echo chamber [13]. Furthermore, popularity bias may bring the Matthew effect [29], which is unfair to small businesses on e-commerce platforms.

To alleviate popularity bias, many existing methods directly reduce the exposure of popular items. For example, [2, 48] take item popularity as prior and reduce the frequency of popular items in recommendation lists. [1] designs regularization according to item popularity. [15] includes random sampling to avoid bias. However, these methods can not perceive user popularity preferences. As a result, the popular items may not be recommended to the users that prefer popular items, and these methods often sacrifice recommendation accuracy to reduce popularity bias.

To solve this issue, several methods consider user popularity preferences when alleviating popularity bias. In these methods, [31] considers the preferences when utilizing item popularity. PDA [45] and CauSeR [16] introduce causal intervention to leverage popularity bias. These methods detect whether a user interacts with an item because of its popularity. However, such a process can hardly accurately model users with niche hobbies or occasionally flow crowds. Meanwhile, these methods ignore that user popularity preferences can be dynamic, leading to inappropriate recommendations. Therefore, we propose UPSR to comprehensively capture user popularity preferences.

2.3 Contrastive Learning

Recently, contrastive learning [6, 14, 33] shows superior ability in self-supervised learning. In recommender systems, existing works construct multiple types of training signals to perform contrastive learning. For example, [46] constructs training signal according to items, attributes, and sequences. CL4SRec [43] makes data augmentation based on user behavior sequences. However, these methods ignore user popularity preferences and may recommend inappropriate items.

3   Proposed Method

As shown in Fig. 1, UPSR consists of three parts. The basic model extracts sequence embeddings from user behavior sequences. The sequential popularity perception module perceives the type and the evolution of user popularity preferences. The popularity contrastive learning module gathers users with similar popularity preferences and general features. By comprehensively capturing popularity preferences, UPSR makes more personalized recommendations.

3.1   Problem Statement

In this work, the behavior sequence of user u is denoted as $S^u = \langle S_1^u, \cdots, S_{|S^u|}^u \rangle$, in which S_i^u represents the i-th item that u has interacted with. Following [7, 25, 39], we regularize user behavior sequences S^u into fixed-length n. If S^u is shorter than n, $n - |S_u|$ special tokens [PAD] are padded before S^u. If S^u is longer than n, the last n behaviors are retained. Given the regularized user behavior sequences s^u, we aim to predict which item will attract user u next.

3.2   Basic Model

To capture the dynamic user interests, we apply regular sequential recommendation methods as our basic model.

Sequential recommendation methods often contain the steps shown in Fig. 1. 1) Regularized user behavior sequence s^u is taken as input. 2) These methods apply neural networks such as RNN, attention mechanism to generate sequence embeddings $\mathbf{H} \in \mathbb{R}^{n \times d}$, where d denotes embedding length and $\mathbf{H}_i$ encodes the user interest when the user interacts with the i-th item in the sequence. 3) The final recommendations are given based on $\mathbf{H}$.

UPSR only requires the basic model to generate sequence embeddings $\mathbf{H}$, and it does not matter what type of neural networks the basic model applied or whether it has other input and output. Therefore, UPSR has promised generality.

3.3   Sequential Popularity Perception Module

Given the sequence embeddings $\mathbf{H}$ from the basic model, the sequential encoder measures local popularity preferences to obtain the evolution of user popularity preferences. Meanwhile, the popularity preference predictor learns the means and the standard deviations of user popularity sets that represent the types of user popularity preferences. Finally, a denoised popularity preference loss is applied to alleviate noise and train the module.

Sequential Encoder. To capture the evolution of user popularity preference, we need to analyze multiple successive local popularity preferences. To obtain local popularity preferences, s^u are divided into $L = \lceil \frac{n}{n_{sub}} \rceil$ subsequences of fixed-length n_{sub}. The subsequence embeddings are generated by averaging the corresponding fragments of the sequence embeddings $\mathbf{H}$:

$$\mathbf{H}_i^{sub} = \frac{\sum_{j \in \phi_i} \mathbf{H}_j}{|\phi_i|} + \mathbf{O}_i, \tag{1}$$

where ϕ_i is the set of behaviors in i-th subsequence, and $\mathbf{O} \in \mathbb{R}^{L \times d}$ is the position embeddings of subsequences. Then, we encode $\mathbf{H}^{sub}$ into local popularity preference embeddings:

$$\mathbf{D}_i = \text{GELU}(\mathbf{H}_i^{sub}\mathbf{W_1} + \mathbf{b}_1)\mathbf{W_2} + \mathbf{b_2}, \tag{2}$$

where $\mathbf{W_1} \in \mathbb{R}^{d \times 2d}, \mathbf{b}_1 \in \mathbb{R}^{1 \times 2d}, \mathbf{W_2} \in \mathbb{R}^{2d \times d}, \mathbf{b}_2 \in \mathbb{R}^{1 \times d}$ are parameters of fully connected layers. Following BERT [11], we utilize GELU [20] as our activation function.

As a result, the local popularity preference embeddings $\mathbf{D}$ perceive the dynamic user popularity preferences.

Popularity Preference Predictor. Given $\mathbf{D}$, we predict the means and the standard deviations of user popularity sets that indicate the type of user popularity preferences. We average the local popularity preference embeddings to generate the overall popularity preference embeddings as:

$$\mathbf{E} = \sum_{i=1}^{L} \mathbf{D}_i / L. \tag{3}$$

Then, the means and the standard deviations are predicted as:

$$\mathbf{G} = \text{GELU}(\mathbf{E}\mathbf{W_3} + \mathbf{b}_3)\mathbf{W_4} + \mathbf{b_4}, \tag{4}$$

where $\mathbf{W_3} \in \mathbb{R}^{d \times d}, \mathbf{b}_3 \in \mathbb{R}^{1 \times d}, \mathbf{W_4} \in \mathbb{R}^{d \times 2}, \mathbf{b_4} \in \mathbb{R}^{1 \times 2}$ are learnable parameters. The two dimensions of $\mathbf{G} \in \mathbb{R}^{1 \times 2}$ are the predicted mean and standard deviation, respectively.

Denoised Popularity Preference Loss. To train the above predictor, we introduce a denoised popularity preference loss. The reason for denoising is that the data of recommender systems are often sparse, and many items have low popularity. As a result, one noisy user behavior could have a non-negligible impact on these items and further influence the reliability of the labels.

For user u, we generate a label $\hat{\mathbf{G}}^u = [\mu^u, \sigma^u]$ that consists of the mean and the standard deviation of the user popularity set $\mathbf{Q}^u$ from the training set. With the labels, the denoised loss function is defined as:

$$\mathcal{L}_{popularity} = \frac{1}{2} \sum_{i=1}^{2} (\text{ReLU}(\mathbf{F}_i - \alpha_p))^2, \tag{5}$$

$$\mathbf{F} = \text{abs}(\mathbf{G} - \log(\hat{\mathbf{G}})), \tag{6}$$

where abs is the absolute value function and $\mathbf{F}$ are the prediction errors. As the values of the labels $\hat{\mathbf{G}}$ are often up to hundreds, we apply the logarithm function to smooth the labels. The ReLU [32] activation discards the gradients when the prediction error is smaller than the threshold α_p. This operation eliminates noise since noisy behavior hardly appears frequently and majorly pollutes small gradients.

With the denoised loss, our predictor can predict the means and standard deviations of user popularity sets, which further informs user embeddings about user popularity preferences.

3.4 Popularity Contrastive Learning Module

This module introduces contrastive learning [6,17] to gather similar users and encode user interests accurately. Two users are identified as similar only if they have close popularity preferences and similar general features. As irrelevant items could have the same popularity, dissimilar users could have similar popularity preferences. Therefore, the identification should not solely depend on popularity preferences.

Following this idea, user similarities are calculated based on the overall popularity preference embeddings $\mathbf{E}$ and the overall sequence embeddings $\widetilde{\mathbf{H}} = \sum_{i=1}^{n} \mathbf{H}_i / n$. Given two users t and u, their similarity is calculated as:

$$\text{sim}(t, u) = \cos(\widetilde{\mathbf{H}}^t + \mathbf{E}^t, \widetilde{\mathbf{H}}^u + \mathbf{E}^u). \tag{7}$$

For each user, its positive samples are the users meet two conditions in the current batch, and the other users are negative samples. The first condition is that positive users should have similar popularity preference types. Given two users t and u, we calculate the distance of their labels:

$$\text{mean}(\text{abs}(\log(\hat{\mathbf{G}}^t) - \log(\hat{\mathbf{G}}^u))) < \alpha_p. \tag{8}$$

The second condition is that positive users should have similar general features. This module records the sequence embeddings in the past epochs and judges positive users based on the records. In epoch l, given the past embeddings $\mathbf{B}^{l-1} \in \mathbb{R}^d$ of user u, the condition is defined as:

$$\cos(\mathbf{B}^{l-1,t}, \mathbf{B}^{l-1,u}) < \alpha_s, \tag{9}$$

where α_s is another threshold. The past embeddings $\mathbf{B}^l$ are generated by a bootstrapping mechanism $\mathbf{B}^l = (1 - \beta)\widetilde{\mathbf{H}} + \beta\mathbf{B}^{l-1}$, where β controls the pace of refreshing.

This module selects samples from current batch and does not introduce additional samples. Hence, UPSR stays light.

Contrastive Learning Loss. The contrastive learning is trained with cross-entropy loss, which is defined as:

$$\mathcal{L}_{contrastive} = - \sum_{i \in \phi_{pos}} \log \frac{\exp(\text{sim}(u, i)/\tau)}{\sum_{j \in \phi_{pos} \cup \phi_{neg}} \exp(\text{sim}(u, j)/\tau)}, \tag{10}$$

where ϕ_{pos} and ϕ_{neg} are the sets of positive and negative samples of user u, and τ is a temperature hyperparameter that scales the values of similarities.

With this loss, UPSR gathers similar users in the embedding space and adjusts user interest appropriately.

3.5 Network Training

UPSR trains the basic model with its original loss function $\mathcal{L}_{basic}$, and the final loss function of UPSR is defined as:

$$\mathcal{L} = \mathcal{L}_{basic} + \gamma_1 \mathcal{L}_{popularity} + \gamma_2 \mathcal{L}_{contrastive}, \tag{11}$$

where γ_1 and γ_2 are trade-off hyperparameters. To avoid overfitting, we apply the widely used dropout technique [38] and layer normalization technique [4] on input embeddings.

4 Experiment

4.1 Datasets

We conduct extensive experiments on three widely used datasets. 1) MovieLens [19] 1M (ML-1M)[1] is a movie rating dataset. 2) Amazon[2] is a series of datasets [30] that contain numerous reviews from Amazon.com, and the datasets are split by product categories. In this work, we adopt the Beauty and the Game category. For dataset preprocessing and splitting, we follow the common procedure in [7,39]. For each user's behavior sequence, we hold out the last behavior for test and the second last behavior for validation. The rest behaviors are for training. In Table 1, we present the statistics of the preprocessed datasets.

4.2 Baselines

We introduce two types of methods as our baselines. 1) We select two sequential recommendation methods. **CL4SRec** [43] introduces contrastive learning, and **MEANTIME** [7] is based on the transformer architecture. Besides, MEAN-TIME takes the exact values of timestamps as additional input. 2) We adopt **PDA** [45] and **CauSeR** [16] that analyze user popularity preferences to alleviate popularity bias and improve recommendation accuracy.

We take CL4SRec and MEANTIME as the basic models of UPSR, PDA, and CauSeR, and we use suffixes _C and _M to indicate which basic model is applied.

---

[1] https://grouplens.org/datasets/movielens/1m/.
[2] http://jmcauley.ucsd.edu/data/amazon/.

Table 1. Dataset statistics. The column Avg. presents the average behaviors per user and the column Pop. presents the average popularity of items.

Dataset	Users	Items	Behaviors	Avg.	Pop.
ML-1M	6,040	3,416	1.00M	165.50	292.63
Beauty	40,226	54,542	0.35M	8.80	6.49
Games	29,341	23,464	0.28M	9.58	11.97

4.3 Implementation Details and Evaluation Metrics

For a fair comparison, the hyperparameters in baselines and UPSR are tuned through a grid search on the validation set. We tune the hyperparameters in UPSR by a grid search, and the final hyperparameters are set as follows: the feature length $d = 128$, the dropout rate is 0.1, the trade-off parameters $\gamma_1 = 1$ and $\gamma_2 = 0.01$, the thresholds $\alpha_p = 0.05$ and $\alpha_s = 0.1$, the number of subsequences $L = 4$, the temperature $\tau = 1$, and the pace $\beta = 0.5$. As the datasets have different densities, the sequence length n is 200 for the ML-1M and 50 for the Amazon datasets. Following the discussion in [8], we obtain the number of training epochs from the validation sets to get more convincing results. Following [26], we treat all the items that a user has not interacted with as negative items. Then, the models are asked to rank the negative items along with the positive items in the test set.

For evaluation, we apply three metrics. To evaluate the recommendation accuracy, we introduce the widely used Hit@K and NDCG@K. K indicates that the method recommends a list of K items for each user, and we select K=10. To evaluate the effect of alleviating popularity bias, we introduce the Average Recommendation Popularity (ARP). ARP $= \sum_{i=1}^{K} \mathbf{P}_i / K$, where $\mathbf{P}_i$ denotes the popularity of the i-th item in the recommdation list.

The ARP metric measures the item popularity of the recommendation list and represents the popularity bias that a method brings. However, a recommender system that achieves a very low ARP is not always good. For example, if a recommender system only recommends unpopular items, it will achieve low ARP but shapely decrease recommendation accuracy. Meanwhile, such a recommender system ignores the users prefer items with medium or high popularities. Therefore, UPSR aims to achieve high performance on Hit and NDCG metrics, while achieving a relatively low ARP.

4.4 Performance Comparison

The overall performance of UPSR and the baselines are presented in Table 2, 3, and 4. The experiments are repeated six times. From the results, we can observe that UPSR not only gives more accurate recommendations but also alleviate popularity bias. The results of Hit@10 and NDCG@10 metrics indicate that UPSR consistently outperforms other methods on three datasets. Meanwhile, the results of ARP show that UPSR can also alleviate popularity bias effectively

Table 2. Overall performance on the ML-1M datasets.

Dataset	Method	Hit@10	NDCG@10	ARP
ML-1M	CL4SRec	0.1943 ± 0.0022	0.1048 ± 0.0017	666.4 ± 12.5
	PDA_C	0.1936 ± 0.0025	0.1041 ± 0.0021	646.5 ± 10.8
	CauSeR_C	0.1957 ± 0.0026	0.1061 ± 0.0019	642.9 ± 13.5
	UPSR_C	$\underline{0.2035} \pm 0.0027$	$\underline{0.1135} \pm 0.0024$	$\underline{642.4} \pm 9.3$
	MEANTIME	0.3304 ± 0.0028	0.1919 ± 0.0022	610.8 ± 11.7
	PDA_M	0.3287 ± 0.0027	0.1909 ± 0.0025	597.9 ± 9.4
	CauSeR_M	0.3337 ± 0.0028	0.1939 ± 0.0022	602.9 ± 8.4
	UPSR_M	$\mathbf{0.3465} \pm 0.0031$	$\mathbf{0.2002} \pm 0.0023$	$\mathbf{597.1} \pm 7.9$

Table 3. Overall performance on the Beauty datasets.

Dataset	Method	Hit@10	NDCG@10	ARP
Beauty	CL4SRec	0.0135 ± 0.0003	0.0070 ± 0.0002	76.6 ± 8.2
	PDA_C	0.0136 ± 0.0003	0.0070 ± 0.0003	63.8 ± 2.5
	CauSeR_C	0.0140 ± 0.0005	0.0073 ± 0.0004	$\underline{63.1} \pm 3.3$
	UPSR_C	$\underline{0.0165} \pm 0.0004$	$\underline{0.0087} \pm 0.0004$	63.2 ± 3.4
	MEANTIME	0.0443 ± 0.0003	0.0245 ± 0.0005	57.7 ± 5.4
	PDA_M	0.0439 ± 0.0003	0.0243 ± 0.0005	49.1 ± 3.8
	CauSeR_M	0.0452 ± 0.0006	0.025 ± 0.0002	$\mathbf{48.2} \pm 3.4$
	UPSR_M	$\mathbf{0.0486} \pm 0.0005$	$\mathbf{0.0266} \pm 0.0003$	48.4 ± 3.7

when making accurate recommendations. These results represent that UPSR learns user popularity preferences more comprehensively, leading to more accurate and proper recommendations.

The results also validate that UPSR has considerable generality. On the three metrics, UPSR achieves considerable improvement compared with the basic models. Therefore, UPSR is compatible with the basic models based on different neural networks or taking additional input.

4.5 Performance on Particular Users

In this section, we validate whether our methods can comprehensively analyze the users prefer items with medium or low popularities or the users largely change their popularity preferences.

Firstly, we sort the users in ascending order of the mean μ^u of the user popularity set $\mathbf{Q}^u$, where μ^u represents the item popularity the user majorly prefers. Then, the users are averagely separated into five groups. We compare the performance on different user groups in Fig. 2. The first two user groups contain users who prefer items with medium or low popularities. On these users, PDA and CauSeR achieve similar Hit and NDCG to the base model MEANTIME. Such

Table 4. Overall performance on the Game datasets.

Dataset	Method	Hit@10	NDCG@10	ARP
Game	CL4SRec	0.0509 ± 0.0012	0.0271 ± 0.0018	9.2 ± 0.18
	PDA_C	0.0511 ± 0.0023	0.0286 ± 0.0018	8.5 ± 0.23
	CauSeR_C	0.0537 ± 0.0016	0.0282 ± 0.0023	8.6 ± 0.26
	UPSR_C	$\underline{0.0570 \pm 0.0027}$	$\underline{0.0292 \pm 0.0023}$	$\underline{8.2 \pm 0.15}$
	MEANTIME	0.1271 ± 0.0017	0.0679 ± 0.0019	9.0 ± 0.16
	PDA_M	0.1299 ± 0.0015	0.685 ± 0.0017	8.6 ± 0.13
	CauSeR_M	0.1275 ± 0.0018	0.686 ± 0.0019	8.7 ± 0.12
	UPSR_M	$\mathbf{0.1331 \pm 0.0018}$	$\mathbf{0.0711 \pm 0.0016}$	$\mathbf{8.1 \pm 0.15}$

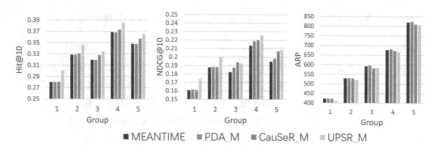

Fig. 2. The performances on users prefer item with different popularities on the ML-1M dataset.

results validate that PDA and CauSeR can hardly handle these users. On the contrary, UPSR outperforms the baselines on all five user groups, representing that UPSR can comprehensively understand the users who prefer items with different popularities.

Secondly, we sort users in ascending order of σ^u / μ^u, where σ^u is the standard deviation of $\mathbf{Q}_u$. The standard deviations represent the change amplitude of user popularity preferences, and we apply the means to normalize the results. The users are also averagely separated into five groups, and the results are presented in Fig. 3. As PDA and CauSeR can hardly capture the users with dynamic popularity preferences, they achieve high Hit and NDCG on users with small amplitudes (i.e., the first two user groups). Meanwhile, the results illustrate that UPSR accurately learns user dynamic popularity preferences regardless of the amplitudes.

The ARP results in the above two experiments have a similar tendency with the other two metrics, although the differences between methods are smaller.

4.6   Ablation Study

To analyze the impact of UPSR's components, we present four variants of UPSR_M in Table 5.

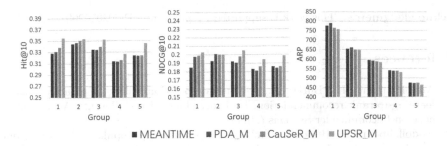

Fig. 3. The performances on users have dynamic popularity preferences on the ML-1M dataset.

Table 5. Hit@10, NDCG@10, and ARP results of the ablation study.

Dataset	ML-1M			Beauty			Game		
Metric	Hit	NDCG	ARP	Hit	NDCG	ARP	Hit	NDCG	ARP
MEANTIME	0.3304	0.1919	610.8	0.0443	0.0245	57.7	0.1271	0.0679	9.0
Mean	0.3407	0.1953	598.0	0.0482	0.0262	49.6	0.1301	0.0698	8.4
Std	0.3384	0.1934	610.4	0.0445	0.0251	56.4	0.1271	0.0707	8.8
Remove_P	0.3358	0.1945	605.5	0.0468	0.0249	51.6	0.1299	0.0696	8.5
Remove_C	0.3414	0.1989	606.4	0.0455	0.0256	51.1	0.1324	0.0705	8.9
UPSR_M	**0.3465**	**0.2002**	**597.1**	**0.0486**	**0.0266**	**48.4**	**0.1331**	**0.0711**	**8.1**

Mean and **Std** are the variants that UPSR only uses the means or the standard deviations of users' popularity sets as labels. These two variants outperform the basic models, indicating that the means and the standard deviations are effective for identifying the types of user popularity preferences.

Remove_P and **Remove_C** are the variants that remove the sequential popularity perception module and the popularity contrastive learning module, respectively. These variants outperform the basic models, denoting that the modules can give more accurate and proper recommendations independently.

5 Conclusion

To comprehensively capture user popularity preferences, we propose to analyze the means and the standard deviations of user popularity sets. Following the idea, we propose a User Popularity preference aware Sequential Recommendation (UPSR) method. On the one hand, UPSR learns the types and the evolutions of user popularity preferences. On the other hand, UPSR employs contrastive learning to gather similar users and encodes user interests more appropriately. Therefore, UPSR gives more accurate and personalized recommendations. Extensive experiment results show that UPSR not only outperforms the state-of-the-art methods but also reduces popularity bias.

116   M. Qian et al.

Acknowledgments. This work is supported by program XDC02050200.

References

1. Abdollahpouri, H., Burke, R., Mobasher, B.: Controlling popularity bias in learning-to-rank recommendation. In: Proceedings of the Eleventh ACM Conference on Recommender Systems (2017)
2. Abdollahpouri, H., Burke, R., Mobasher, B.: Managing popularity bias in recommender systems with personalized re-ranking. In: The Thirty-Second International Flairs Conference (2019)
3. Anelli, V.W., Di Noia, T., Di Sciascio, E., Ragone, A., Trotta, J.: Local popularity and time in top-n recommendation. In: Azzopardi, L., Stein, B., Fuhr, N., Mayr, P., Hauff, C., Hiemstra, D. (eds.) ECIR 2019. LNCS, vol. 11437, pp. 861–868. Springer, Cham (2019). https://doi.org/10.1007/978-3-030-15712-8_63
4. Ba, L.J., Kiros, J.R., Hinton, G.E.: Layer normalization. In: CoRR (2016)
5. van den Berg, R., Kipf, T.N., Welling, M.: Graph convolutional matrix completion. In: KDD (2018)
6. Chen, X., He, K.: Exploring simple Siamese representation learning. In: CVPR (2021)
7. Cho, S.M., Park, E., Yoo, S.: Meantime: mixture of attention mechanisms with multi-temporal embeddings for sequential recommendation. In: RecSys (2020)
8. Dacrema, M.F., Cremonesi, P., Jannach, D.: Are we really making much progress? A worrying analysis of recent neural recommendation approaches. In: RecSys (2019)
9. Dai, F., Gu, X., Li, B., Zhang, J., Qian, M., Wang, W.: Meta-graph based attention-aware recommendation over heterogeneous information networks. In: Rodrigues, J.M.F., et al. (eds.) ICCS 2019. LNCS, vol. 11537, pp. 580–594. Springer, Cham (2019). https://doi.org/10.1007/978-3-030-22741-8_41
10. Dai, F., Gu, X., Wang, Z., Qian, M., Li, B., Wang, W.: Heterogeneous side information-based iterative guidance model for recommendation. In: Proceedings of the 2021 International Conference on Multimedia Retrieval, pp. 55–63 (2021)
11. Devlin, J., Chang, M., Lee, K., Toutanova, K.: BERT: pre-training of deep bidirectional transformers for language understanding. In: Proceedings of NAACL-HLT (2019)
12. Dong, S., Wang, P., Abbas, K.: A survey on deep learning and its applications. Computer Science Review **40**, 100379 (2021)
13. Ge, Y., et al.: Understanding echo chambers in e-commerce recommender systems. In: SIGIR (2020)
14. Gidaris, S., Singh, P., Komodakis, N.: Unsupervised representation learning by predicting image rotations. arXiv preprint arXiv:1803.07728 (2018)
15. Gruson, A., et al.: Offline evaluation to make decisions about playlist recommendation algorithms. In: WSDM (2019)
16. Gupta, P., Sharma, A., Malhotra, P., Vig, L., Shroff, G.: Causer: causal session-based recommendations for handling popularity bias. In: CIKM (2021)
17. Hadsell, R., Chopra, S., LeCun, Y.: Dimensionality reduction by learning an invariant mapping. In: CVPR (2006)
18. Hansen, C., et al.: Contextual and sequential user embeddings for large-scale music recommendation. In: Fourteenth ACM Conference on Recommender Systems (2020)

19. Harper, F.M., Konstan, J.A.: The movielens datasets: history and context. In: ACM TIIS (2015)
20. Hendrycks, D., Gimpel, K.: Bridging nonlinearities and stochastic regularizers with gaussian error linear units. CoRR abs/1606.08415 (2016)
21. Hidasi, B., Karatzoglou, A.: Recurrent neural networks with top-k gains for session-based recommendations. In: CIKM (2018)
22. Hidasi, B., Karatzoglou, A., Baltrunas, L., Tikk, D.: Session-based recommendations with recurrent neural networks. In: ICLR (2016)
23. Ji, Y., Sun, A., Zhang, J., Li, C.: A re-visit of the popularity baseline in recommender systems. In: SIGIR (2020)
24. Kabbur, S., Ning, X., Karypis, G.: FISM: factored item similarity models for top-n recommender systems. In: SIGKDD (2013)
25. Kang, W., McAuley, J.J.: Self-attentive sequential recommendation. In: ICDM (2018)
26. Krichene, W., Rendle, S.: On sampled metrics for item recommendation. In: SIGKDD (2020)
27. Lex, E., Kowald, D., Schedl, M.: Modeling popularity and temporal drift of music genre preferences. Trans. Int. Soc. Music In. Retrieval $3(1)$ (2020)
28. Li, J., Wang, Y., McAuley, J.: Time interval aware self-attention for sequential recommendation. In: WSDM (2020)
29. Liu, Y., Ge, K., Zhang, X., Lin, L.: Real-time attention based look-alike model for recommender system. In: SIGKDD (2019)
30. McAuley, J.J., Targett, C., Shi, Q., van den Hengel, A.: Image-based recommendations on styles and substitutes. In: SIGIR (2015)
31. Nagatani, K., Sato, M.: Accurate and diverse recommendation based on users' tendencies toward temporal item popularity. In: RecSys (2017)
32. Nair, V., Hinton, G.E.: Rectified linear units improve restricted Boltzmann machines. In: ICML (2010)
33. Noroozi, M., Favaro, P.: Unsupervised learning of visual representations by solving Jigsaw puzzles. In: Leibe, B., Matas, J., Sebe, N., Welling, M. (eds.) ECCV 2016. LNCS, vol. 9910, pp. 69–84. Springer, Cham (2016). https://doi.org/10.1007/978-3-319-46466-4_5
34. Qian, M., Gu, X., Chu, L., Dai, F., Fan, H., Li, B.: Flexible order aware sequential recommendation. In: Proceedings of the 2022 International Conference on Multimedia Retrieval, pp. 109–117 (2022)
35. Rendle, S., Freudenthaler, C., Schmidt-Thieme, L.: Factorizing personalized Markov chains for next-basket recommendation. In: WWW (2010)
36. Ricci, F., Rokach, L., Shapira, B., Kantor, P.B. (eds.): Recommender Systems Handbook. Springer, Boston (2015). https://doi.org/10.1007/978-1-4899-7637-6_1
37. Sedhain, S., Menon, A.K., Sanner, S., Xie, L.: AutoRec: autoencoders meet collaborative filtering. In: WWW (2015)
38. Srivastava, N., Hinton, G.E., Krizhevsky, A., Sutskever, I., Salakhutdinov, R.: Dropout: a simple way to prevent neural networks from overfitting. JMLR (2014)
39. Sun, F., et al.: BERT4Rec: sequential recommendation with bidirectional encoder representations from transformer. In: CIKM (2019)
40. Tan, Q., et al.: Sparse-interest network for sequential recommendation. In: Proceedings of the 14th ACM International Conference on Web Search and Data Mining (2021)
41. Vaswani, A., et al.: Attention is all you need. In: NIPS (2017)
42. Wang, S., Hu, L., Wang, Y., Cao, L., Sheng, Q.Z., Orgun, M.A.: Sequential recommender systems: challenges, progress and prospects. In: IJCAI (2019)

43. Xie, X., et al.: Contrastive learning for sequential recommendation. In: ICDE (2022)
44. Zhang, M., Liu, X., Liu, W., Zhou, A., Ma, H., Mei, T.: Multi-granularity reasoning for social relation recognition from images. In: 2019 IEEE International Conference on Multimedia and Expo (ICME), pp. 1618–1623. IEEE (2019)
45. Zhang, Y., et al.: Causal intervention for leveraging popularity bias in recommendation. In: SIGIR (2021)
46. Zhou, K., et al.: S3-rec: self-supervised learning for sequential recommendation with mutual information maximization. In: Proceedings of the 29th ACM International Conference on Information and Knowledge Management, pp. 1893–1902 (2020)
47. Zhu, X., Li, J., Liu, Y., Liao, J., Wang, W.: Operation-level progressive differentiable architecture search. In: 2021 IEEE International Conference on Data Mining (ICDM), pp. 1559–1564. IEEE (2021)
48. Zhu, Z., He, Y., Zhao, X., Zhang, Y., Wang, J., Caverlee, J.: Popularity-opportunity bias in collaborative filtering. In: WSDM (2021)

Data Heterogeneity Differential Privacy: From Theory to Algorithm

Yilin Kang[1,3], Jian Li[1(✉)], Yong Liu[2], and Weiping Wang[1]

[1] Institute of Information Engineering, Chinese Academy of Sciences, Beijing, China
{kangyilin,lijian9026,wangweiping}@iie.ac.cn
[2] Gaoling School of Artificial Intelligence, Renmin University of China,
Beijing, China
liuyonggsai@ruc.edu.cn
[3] School of Cyber Security, University of Chinese Academy of Sciences,
Beijing, China

Abstract. Traditionally, the random noise is equally injected when training with different data instances in the field of differential privacy (DP). In this paper, we first give sharper excess risk bounds of DP stochastic gradient descent (SGD) method. Considering most of the previous methods are under convex conditions, we use Polyak-Łojasiewicz condition to relax it in this paper. Then, after observing that different training data instances affect the machine learning model to different extent, we consider the heterogeneity of training data and attempt to improve the performance of DP-SGD from a new perspective. Specifically, by introducing the influence function (IF), we quantitatively measure the contributions of various training data on the final machine learning model. If the contribution made by a single data instance is so little that attackers cannot infer anything from the model, we do not add noise when training with it. Based on this observation, we design a 'Performance Improving' DP-SGD algorithm: PIDP-SGD. Theoretical and experimental results show that our proposed PIDP-SGD improves the performance significantly.

Keywords: Differential privacy · Machine learning · Data heterogeneity

1 Introduction

Machine learning has been widely applied to many fields in recent decades and tremendous data has been collected. As a result, information disclosure becomes a huge problem. Except for the original data, model parameters can reveal sensitive information in an undirect way as well [15,32].

Differential privacy (DP) [12,13] is a theoretically rigorous tool to prevent sensitive information [10]. It preserves privacy by introducing random noise, to

---

Supplementary: All the proofs are given in https://arxiv.org/pdf/2002.08578.pdf.

J. Mikyška et al. (Eds.): ICCS 2023, LNCS 14073, pp. 119–133, 2023.
https://doi.org/10.1007/978-3-031-35995-8_9

block adversaries from inferring any single individual included in the dataset by observing the machine learning model. As such, DP has been applied to numerous machine learning methods [2,5,6,9,17,21,30,31,33,35,36,39–41,43,44,47] and three main approaches are studied: output perturbation, objective perturbation, and gradient perturbation. However, some problems still exist: First, all data is usually treated equally when training DP model, but in real scenarios, different training data affects the model differently, so treating them all the same lacks 'common sense' and is one of the reasons why low accuracy appears. Meanwhile, previous results always require that the loss function is convex (or even strongly convex), the application scenario is narrow.

To solve the problems, we make the following contributions in this paper: First, we introduce the Polyak-Łojasiewicz (PL) condition [20] to relax the convex (or even strongly convex) assumption. We analyze the excess population risk and give corresponding bounds under PL condition and theoretical results show that our given excess risk bounds are better than previous convex ones. Second, motivated by the definition of DP, we provide a new perspective to improve the performance: treating different data instances differently. In particular, we introduce the Influence Function (IF) [22] to measure the contributions made by different data instances. If the data instance z contributes so little to the machine learning model that the attacker cannot infer anything (represented by the privacy budget ϵ), we do not add noise when training with z, rather than treating all of the data instances as being the same. In this way, we propose a 'Performance Improving' algorithm: PIDP-SGD to improve the model performance, by taking data heterogeneity into account.

The rest of the paper is organized as follows. We introduce some related work in Sect. 2. Preliminaries are presented in Sect. 3. We analyze the excess risk of DP-SGD and give sharper theoretical bounds in Sect. 4. The 'Performance Improving' algorithm is given and analyzed in Sect. 5. In Sect. 6, we compare our proposed method with previous methods in detail. The experimental results are shown in Sect. 7 and we conclude the paper in Sect. 8.

2   Related Work

The first method on DP machine learning is proposed in [9], in which output and objective perturbation methods are introduced. Gradient perturbation is proposed in [33] and DP-SGD is analyzed for the first time. The accuracy of the objective perturbation method is improved by [21]. The excess empirical risk bounds of the methods proposed in [9] and [21] are improved by [5]. An output perturbation method is introduced to DP-SGD by [41], in which a novel ℓ_2 sensitivity is analyzed and better accuracy is achieved. [40] introduces Prox-SVRG [42] to DP and proposes DP-SVRG, in which optimal or near-optimal utility bounds are achieved. Meanwhile, there are also some works concentrated on non-convex analysis. DP is introduced to deep learning by [1], via gradient perturbation method, however, it focuses on the privacy but lacks utility analysis. An output perturbation method is proposed in [44] under non-convex condition.

The Polyak-Łojasiewicz condition is introduced in [40] and the excess empirical risk of gradient perturbation method under non-convex condition is analyzed, however, the excess population risk is not discussed. Aiming to achieve better performance, in [30], more noise is added to those features less 'relevant' to the final model. A Laplace smooth operator is introduced to DP-SGD and a new method: DP-LSSGD is proposed in [36], focusing on non-convex analysis. The excess empirical risk bound and the excess population risk bound of DP model under non-convex condition are analyzed by [38], via Gradient Langevin Dynamics. For non-convex condition, the theoretical results are always unsatisfactory.

All the works mentioned above treat all data instances equally, lack 'common sense' and lead unsatisfactory utility. To solve the problems under both convex and non-convex conditions, we take data heterogeneity into account and propose a 'Performance Improving' algorithm. In this way, our method improves the performance of DP model, superior to previous methods in excess risk bounds.

3   Preliminaries

3.1   Notations and Assumptions

The loss function is $\ell : \mathcal{C} \times \mathcal{D} \to \mathbb{R}$, where $\mathcal{C}$ is the parameter space and $\mathcal{D}$ is the data universe. We assume that the parameter space is bounded, whose radius is r. Supposing there are n data instances in the dataset $D = \{z_1, \cdots, z_n\} \in \mathcal{D}^n$, where z_i are drawn i.i.d from the underlying distribution $\mathcal{P}$. Besides, for each $z = (x, y)$, x is the feature and y is the label. We assume $\|x\|_2 \leq 1$, i.e. $\mathcal{X}$ is the unit ball. Moreover, for a vector $x = [x_1, \cdots, x_d]$, its ℓ_2 norm is defined as: $\|x\|_2 = \left(\sum_{i=1}^d x_i^2\right)^{1/2}$, and the ith element is represented by $[x]_i$.

The population risk over the underlying distribution $\mathcal{P}$ is defined as $L_{\mathcal{P}}(\theta) = \mathbb{E}_{z \sim \mathcal{P}}[\ell(\theta, z)]$. However, we cannot achieve $\mathcal{P}$ in practice, so our goal is to find the optimal model that minimizes the empirical risk $L(\theta; D) = \frac{1}{n}\sum_{i=1}^n \ell(\theta, z_i)$ on dataset D, defined as: $\theta^* = \arg\min[L(\theta; D)]$, and $L(\theta^*; D)$ is represented by L^*. For an algorithm $\mathcal{A} : \mathcal{D}^n \to \mathbb{R}^m$, we denote its output as $\theta_{\mathcal{A}}$. The **excess empirical risk** denotes the gap between $\theta_{\mathcal{A}}$ and θ^*, defined as: $L(\theta_{\mathcal{A}}; D) - L^*$; and the **excess population risk** represents the gap between $\theta_{\mathcal{A}}$ and the optimal model over the underlying $\mathcal{P}$, defined as: $L_{\mathcal{P}}(\theta_{\mathcal{A}}) - \min_{\theta \in \mathcal{C}} L_{\mathcal{P}}(\theta)$. The **generalization error** connects the population risk and the empirical risk, defined as: $L_{\mathcal{P}}(\theta_{\mathcal{A}}) - L(\theta_{\mathcal{A}}; D)$.

Besides, there are some assumptions on the loss function:

Definition 1 (G-Lipschitz). *Loss $\ell : \mathcal{C} \times \mathcal{D} \to \mathbb{R}$ is G-Lipschitz over θ, if for some constant G, any $z \in \mathcal{D}$ and $\theta, \theta' \in \mathcal{C}$, $|\ell(\theta, z) - \ell(\theta', z)| \leq G\|\theta - \theta'\|_2$.*

Definition 2 (L-smooth). *Loss $\ell : \mathcal{C} \times \mathcal{D} \to \mathbb{R}$ is L-smooth over θ, if for some constant L, any $z \in \mathcal{D}$ and $\theta, \theta' \in \mathcal{C}$, $\|\nabla_\theta \ell(\theta, z) - \nabla_\theta \ell(\theta', z)\|_2 \leq L\|\theta - \theta'\|_2$.*

Definitions 1 and 2 upper bound the gradient and the second order gradient, respectively, i.e. $\|\nabla_\theta \ell(\theta, z)\|_2 \leq G$ and $\|\nabla_\theta^2 \ell(\theta, z)\|_2 \leq L$.

3.2 Differential Privacy

Two databases $D, D' \in \mathcal{D}^n$ differing by one single element are denoted as $D \sim D'$, called *adjacent databases*.

Definition 3 (Differential Privacy [13]). *With $S \in range(\mathcal{A})$, the randomized function $\mathcal{A} : \mathcal{D}^n \rightarrow \mathbb{R}^m$ is (ϵ,δ)-differential privacy ((ϵ,δ)-DP) if:*

$$\mathbb{P}\left[\mathcal{A}(D) \in S\right] \leq e^\epsilon \mathbb{P}\left[\mathcal{A}(D') \in S\right] + \delta.$$

Differential privacy requires that adjacent datasets D, D' lead to similar distributions on the output of a randomized algorithm $\mathcal{A}$. This implies that an adversary cannot infer whether an individual participates in the training process because essentially the same conclusions about an individual will be drawn whether or not that individual's data was used. Some kind of attacks, such as membership inference attack, attribute inference attack, memorization attack, can be thwarted by differential privacy [3,7,19].

4 Sharper Utility Bounds for DP-SGD

Considering that SGD naturally fits the condition measuring each data instances differently, before introducing the 'Performance Improving' algorithm in detail, we first analyze the excess risk bounds of DP-SGD.

In DP-SGD, at the tth iteration, we have: $\theta_{t+1} \leftarrow \theta_t - \alpha \left(\nabla_\theta \ell(\theta_t, z_t) + b\right)$, where z_t is the chosen data instance at iteration t, α is the learning rate, and b is the sampled random noise. There is a long list of works to analyze the privacy guarantees of DP-SGD. To the best of our knowledge, the moments accountant method proposed by [1] achieves one of the best results. It claimed that if the Gaussian random noise $b \sim \mathcal{N}(0, \sigma^2 I_m)$ is injected, the loss function is G-Lipschitz, and with

$$\sigma \geq c \frac{G\sqrt{T \log(1/\delta)}}{n\epsilon} \tag{1}$$

for some constant c, then the algorithm satisfies (ϵ, δ)-DP, where T is the total number of training iterations and n is the size of the training dataset.

Previous works always discuss the empirical risk but seldom consider the population risk [1,33,40,41,44]. However, the latter is one of the most concerned terms in machine learning because it demonstrates the gap between the private model and the optimal model over the underlying distribution $\mathcal{P}$.

Excess Empirical Risk. The excess empirical risk measures the gap between θ_{priv} and θ^* over the dataset D, where θ_{priv} denotes the private model. Before the analysis, we first introduce the Polyak-Łojasiewicz (PL) condition [20].

Definition 4 (Polyak-Łojasiewicz condition). *$L(\theta; D)$ satisfies the Polyak-Łojasiewicz (PL) condition if there exists $\mu > 0$ for all θ:*

$$\|\nabla_\theta L(\theta; D)\|_2^2 \geq 2\mu(L(\theta; D) - L^*).$$

PL condition is one of the weakest curvature conditions [26], it does not assume the loss function to be convex and it is commonly used in non-convex optimization. Many non-convex models satisfy the condition, including deep (linear) [8] and shallow neural networks [25].

Remark 1. *If $L(\theta; D)$ satisfies the PL condition, then it satisfies the Quadratic Growth (QG) condition [20], i.e., $L(\theta; D) - L(\theta^*; D) \geq \frac{\mu}{2}\|\theta - \theta^*\|_2^2$, where θ^* denotes the optimal model over dataset D.*

Theorem 1. *Suppose that $\ell(\theta, z)$ is G-Lipschitz, L-smooth, and satisfies PL condition over θ. With learning rate $\alpha = \frac{1}{L}$, σ given in (1) to guarantee (ϵ, δ)-DP, and $T = \mathcal{O}(\log(n))$, then with the dimensions of the model m:*

$$\mathbb{E}\left[L(\theta_{priv}; D) - L^*\right] \leq \mathcal{O}\left(\frac{mG^2 \log(1/\delta)\log(n)}{n^2\epsilon^2}\right),$$

the expectation is taken over the algorithm and dataset D.

Remark 2. *Many researchers discussed the excess empirical risk in previous works. To the best of our knowledge, one of the best results is given by [40], in which T is multiplied 'rudely' to the noise term when summing the loss over T iterations. As a result, the excess empirical risk bound is $\mathcal{O}\left(\frac{mG^2 \log(1/\delta)\log^2(n)}{n^2\epsilon^2}\right)$ in [40]. However, we solve a geometric sequence when summing the loss, and get a tighter bound in this paper. As a result, the excess empirical risk bound is improved by a factor of $\log(n)$ overall.*

Excess Population Risk. To get the excess population risk bound, we first analyze the generalization error, which measures the gap between the performance over the underlying distribution and the dataset D of the private model, connecting the population risk with the empirical risk.

Theorem 2. *If the loss function is G-Lipschitz, L-smooth, and satisfies the PL condition over θ, the generalization error bound of θ_{priv} satisfies:*

$$\mathbb{E}\left[L_{\mathcal{P}}(\theta_{priv}) - L(\theta_{priv}; D)\right]$$
$$\leq \inf_{\tau > 0}\left\{\frac{8(\tau + L)}{\mu}\mathbb{E}[L(\theta_{priv}; D) - L^*] + \frac{16G^2(\tau + L)}{n^2\mu^2} + \frac{L\mathbb{E}[L(\theta_{priv}; D)]}{\tau}\right\},$$
$$\tag{2}$$

where the expectation is taken over the algorithm.

By Theorem 2, one may observe that the generalization error decreases if the optimization error (the excess empirical risk) is smaller, which is in line with the observation in [8,16,25]: 'optimization helps generalization'.

Now, we give the excess population risk bound.

Theorem 3. *If the loss function is G-Lipschitz, L-smooth, and satisfies the PL condition over θ, with learning rate $\alpha = \frac{1}{L}$, the excess population risk of θ_{priv} satisfies:*

$$\mathbb{E}\left[L_{\mathcal{P}}\left(\theta_{priv} \right) - \min_{\theta} L_{\mathcal{P}}\left(\theta \right) \right] \leq (\tau + L)\left(\frac{(8\tau + \mu)}{\mu\tau}\mathbb{E}[L(\theta_{priv}; D) - L^*] + \frac{16G^2}{n^2\mu^2} \right)$$
$$+ \frac{L}{\tau}\mathbb{E}[L^*].$$

Remark 3. *Combining the result given in Theorem 1, if $\mathbb{E}[L^*] = \mathcal{O}(1/n)$, taking $T = \mathcal{O}(\log(n))$, if we ignore constants and $\log(\cdot)$ terms, then for all $\tau > 0$, the excess population risk bound comes to: $\mathcal{O}\left(\frac{m}{\tau n^2 \epsilon^2} + \frac{\tau m}{n^2 \epsilon^2} + \frac{1}{\tau n} \right)$. If $\tau = \mathcal{O}(1)$, it is $\mathcal{O}\left(\frac{m}{n^2 \epsilon^2} + \frac{1}{n} \right)$. If $\tau = \mathcal{O}(\sqrt{n}\epsilon/\sqrt{m})$, the excess population is is $\mathcal{O}\left(\frac{\sqrt{m}}{n^{1.5}\epsilon} + \frac{m^{1.5}}{n^{2.5}\epsilon^3} \right)$. Thus, we get the upper bound of the excess population risk:*

$$\mathbb{E}\left[L_{\mathcal{P}}\left(\theta_{priv} \right) - \min_{\theta} L_{\mathcal{P}}\left(\theta \right) \right] = \mathcal{O}\left(\min\left\{ \frac{m}{n^2\epsilon^2} + \frac{1}{n}, \frac{\sqrt{m}}{n^{1.5}\epsilon} + \frac{m^{1.5}}{n^{2.5}\epsilon^3} \right\} \right).$$

In Remark 3, to get the better result, we assume $\mathbb{E}[L^*] = \mathcal{O}(1/n)$. It is a small value because L^* is the optimal value over the whole dataset. Besides, it is common to assume the minimal population risk $\mathbb{E}[\min L_{\mathcal{P}}(\theta)] \leq \mathcal{O}(1/n)$ [24, 28, 34, 45, 46]. Moreover, under expectation, considering $\theta_{\mathcal{P}}^* = \arg\min L_{\mathcal{P}}(\theta)$ is independent of dataset, so $\mathbb{E}[L^*] \leq \mathbb{E}[\min L_{\mathcal{P}}(\theta)]$ [23]. Thus, the assumption is reasonable.

All the theorems given above only assume that the loss function $\ell(\cdot)$ is G-Lipschitz, L-smooth and satisfies PL inequality, without convex assumption. So the results are general and can be applied to some of the non-convex conditions.

5   Performance Improving DP-SGD

Motivated by the definition of DP, we focus on the contributions made by data instances on the final model. In particular, if the effects caused by a data instance z on the final machine learning model is so little that the attacker cannot realize it (less than e^ϵ), there is no need to add noise to z. Now, only one problem is left: How to measure the impact of the data instances on the model? A classic technique, Influence Function (IF), gives us some inspirations.

5.1   Influence Function and Error Analysis

The contribution of data instance z is naturally defined as $\theta_{-z}^* - \theta^*$, where $\theta_{-z}^* = \arg\min_{\theta} \sum_{z_i \neq z} \ell(\theta, z_i)$. To measure the gap between them, a straight method is to train two models: θ^*, θ_{-z}^*. However, retraining a model for each data instance z is prohibitively slow. To solve the problem, influence [22] measures the contributions on the machine learning model made by data instances:

$$c_z := -\frac{1}{n}\left(-H_{\theta^*}^{-1}\nabla_\theta \ell\left(\theta^*, z \right) \right) \approx \theta_{-z}^* - \theta^*, \tag{3}$$

where $H_{\theta^*} = \frac{1}{n} \sum_{i=1}^{n} \nabla_\theta^2 \ell(\theta^*, z_i)$, assumed positive definite. Via (3), we can measure how the model changes if we 'drop' one data instance, naturally in line with the definition of DP.

The influence function c_z is got by Taylor expansion [27], in which Taylor remainders lead an approximation error. However, [22] only gives an approximation via IF, but not discusses the corresponding error. To fill the gap, we analyze the approximation error in this section, via the definition given below.

Definition 5 (C-Hessian Lipschitz). *A loss function* $\ell : \mathcal{C} \times \mathcal{D} \to \mathbb{R}$ *is C-Hessian Lipschitz over* θ*, if for any* $z \in \mathcal{D}$ *and* $\theta, \theta' \in \mathcal{C}$*, we have:* $\|\nabla_\theta^2 \ell(\theta, z) - \nabla_\theta^2 \ell(\theta', z)\|_2 \leq C\|\theta - \theta'\|_2$.

Remark 4. *C-Hessian Lipschitz means that* $\|\nabla_\theta^3 \ell(\theta, z)\|_2 \leq C$*. For Mean Squared Error,* $\|\nabla_\theta^3 \ell(\theta, z)\|_2 = 0$*. For logistic regression, elements in* $\nabla_\theta^3 \ell(\theta, z)$ *are less than 0.097. The examples above show that the assumption is reasonable.*

Theorem 4. *If* $\ell(\theta, z)$ *is G-Lipschitz, L-smooth, C-Hessian Lipschitz over* θ*, and* $\|H_{\theta^*}\|_2 \geq \zeta$*, then with c_z given in (3) the approximation error satisfies:*

$$E := \|(\theta_{-z}^* - \theta^*) - c_z\|_2 \leq \frac{1}{\zeta^2 n^2}\left(2LG + \frac{CG^2}{\zeta}\right).$$

Remark 5. *Theorem 4 gives a* $\mathcal{O}(1/n^2)$ *approximation error when applying* c_z*, which means that c_z is precise. Besiders, in Theorem 4, we assume that* $\|H_{\theta^*}\|_2 \geq \zeta$*. Because most of the algorithms are regularized, the assumption is easy to hold.*

5.2 Performance Improving DP-SGD

We set a threshold: e^ϵ for adding noise by the following observation: the appearance (or absence) of some data affects the model so little that attackers cannot infer anything from them. Changing those data instances cannot threaten (ϵ, δ)-DP of the model. So, we calculate the 'contribution' of data by IF, only add noise to whom contributes more than e^ϵ.

Details of the Performance Improving algorithm are given in Algorithm 1[1]. Different from traditional DP-SGD algorithm, Algorithm 1 applies a decision process before gradient descent (lines 9 and 10), to decide whether to add random noise or not. If the effect made by the chosen z_t is no more than e^ϵ, SGD runs; otherwise, we sample Gaussian noise b and run DP-SGD. In other words, lines 9 and 10 connects the value of IF with the privacy loss of DP. Meanwhile, we notice that except the training process, the contribution calculating process may also disclose the sensitive information. So we add noise to the $c_t^{(o)}$ to guarantee the claimed DP (line 9). Noting that the contribution given by (3) is based on Taylor expansion, causing an approximation error (discussed in Sect. 5.1), we fix it by adding E to c_t in line 7.

---

[1] In Algorithm 1, $sign(\cdot)$ is the signum function, r is the radius of the parameter space.

Algorithm 1. Performance Improving DP-SGD

Require: dataset D, learning rate α, local iteration rounds T_{local}, global update
   rounds R
1: **function** PIDP-SGD$(D, \alpha, T_{local}, R)$
2:     Initialize $\theta_0^{(g)}, \theta_0^{(l)} \leftarrow \tilde{\theta}$.
3:     **for** $r = 0$ to $R - 1$ **do**
4:         Get $H_{\tilde{\theta}_r^{(g)}}$ and compute its inverse.
5:         **for** $t = 0$ to $T_{local} - 1$ **do**
6:             Choose data instance z_t randomly.
7:             Get contribution of z_t: $c_t^{(o)} = c_{z_t} + E$, via $H_{\tilde{\theta}_r^{(g)}}$.
8:             Sample $b^{(c)} \sim \mathcal{N}(0, \sigma_{(c)}^2 I_m)$.
9:             $c_t = 2c_t^{(o)} + b^{(c)}$, if there exists any i that $|[c_t]_i| \geq \frac{2re^{\epsilon_1}\delta_1}{e^{\epsilon_1}-1}$, jump to line 13;
   otherwise, jump to line 10.
10:             **if** $\ln\left(\frac{2re^{\epsilon_1}\delta_1}{2re^{\epsilon_1}\delta_1-(e^{\epsilon_1}-1)sign([c_t]_i)[c_t]_i}\right) \leq 2\epsilon_1$ for all $i \in [1, m]$, **then**
11:                 $\theta_{t+1}^{(l)} \leftarrow \theta_t^{(l)} - \alpha\nabla_\theta\ell(\theta_t^{(l)}, z_t)$.
12:             **else**
13:                 Sample $b \sim \mathcal{N}(0, \sigma^2 I_m)$,
14:                 $\theta_{t+1}^{(l)} \leftarrow \theta_t^{(l)} - \alpha(\nabla_\theta\ell(\theta_t^{(l)}, z_t) + b)$.
15:             **endif**
16:         **endfor**
17:         $\theta_{r+1}^{(g)} = \theta_{T_{local}}^{(l)}$.
18:     **endfor**
19:     return $\theta_{priv} = \theta_R^{(g)}$.
20: **end function**

It is easy to follow that if the privacy budget ϵ is higher, the constraint of
adding noise is looser, which means that fewer data instances meet the noise. As
a result, the performance of PIDP-SGD will be better if ϵ is higher.

Besides, the method given in this paper will inspire other researchers to apply
it to corresponding fields such as mini-batch gradient descent.

Remark 6. *The time complexity of Algorithm 1 is $\mathcal{O}(Rnm^2 + RT_{local}m)$. Under
the worst case $T_{local} = 1$, it becomes $\mathcal{O}(Rnm^2)$, where R is the total number of
iterations. Fortunately, an efficient approach to calculate the Influence Function
was given in [22], and the time complexity can be reduced to $\mathcal{O}(Rnm)$. For some
other previous performance method, the time complexity also increases, we take
DP-LSSGD as an example here, whose time complexity is $\mathcal{O}(Rm^2)$. Under the
cases $n > m$, our time complexity is larger, but under high dimension cases, when
$n \leq m$, our time complexity is better. However, both theoretical and experimental
results of our method is much better than DP-LSSGD (see Table 1 and Figs. 1
and 2). So under low dimension conditions, the sacrifice on time complexity is
a trade-off against the model performance; under high dimension conditions, our
method is much better on both the model performance and the time complexity.*

5.3   Privacy Guarantees

Theorem 5. *For $\delta_1, \delta_2 > 0$ and $\epsilon_1, \epsilon_2 > 0$, if $\ell(\theta, z)$ is G-Lipschitz over θ, with*
$$\sigma \geq c\frac{G\sqrt{T\log(1/\delta_1)}}{n\epsilon_1}, \, \sigma_{(c)} \geq c'\frac{GR\sqrt{\log(1.25R/\delta_2)}}{n\zeta\epsilon_2}, \text{ where } T = T_{local} * R. \text{ Algorithm 1}$$
is $(\epsilon_1 + \epsilon_2, \delta_1 + \delta_2)$-DP for some constants c, c'.

Theorem 5 shows that the privacy of Algorithm 1 consists of two parts: (1) computing c_t and (2) training model. Specifically, ϵ_1, δ_1 (σ) are for the privacy when training the model and ϵ_2, δ_2 ($\sigma_{(c)}$) are for the privacy when computing c_t.

In Algorithm 1, noise is only added when training with a partition of data instances, which leads better excess risk bounds. In the following, we suppose that there are k data instances affect the model significantly and measure the improvement brought by our proposed 'Performance Improving' algorithm.

5.4   Utility Analysis

We first give the excess empirical risk bound.

Theorem 6. *Suppose that $\ell(\theta, z)$ is G-Lipschitz, L-smooth, and satisfies PL condition over θ. With learning rate $\alpha = \frac{1}{L}$ and k data instances affect the model significantly, the excess empirical risk can be improved to:*

$$\mathbb{E}\left[L(\theta_{priv}; D) - L^*\right] \leq \mathcal{O}\left(\frac{kmG^2 \log(1/\delta_1) \log(n)}{n^3\epsilon_1^2}\right).$$

Noting that $\frac{k}{n} < 1$, the excess empirical risk bound brought by Theorem 6 is better than Theorem 1.

Via Theorem 2, we find that the generalization error is only related to $\|\theta_i^T - \theta^T\|_2$ and $L(\theta_{priv}; D)$, and these terms are only determined by the optimization process, so the generalization error of Algorithm 1 is the same as which given in Theorem 2. Then we come to the excess population risk.

Theorem 7. *If $\ell(\cdot)$ is G-Lipschitz, L-smooth, and satisfies the PL condition over θ. With learning rate $\alpha = \frac{1}{L}$, $T = \mathcal{O}(\log(n))$ and k data instances affect the model significantly, the excess population risk can be improved to:*

$$\mathcal{O}\left(\min\left\{\frac{km}{n^3\epsilon^2} + \frac{1}{n}, \frac{km}{n^{2.5}\epsilon^2} + \frac{1}{n^{1.5}}, \frac{(km)^{1.5}}{n^2\epsilon^3} + \frac{\epsilon}{\sqrt{kmn}}\right\}\right).$$

The proof is similar to Theorem 3 and the discussion given in Remark 3. The first, second, third and last terms are derived from taking $\tau = \mathcal{O}(1), \mathcal{O}(\sqrt{n})$ and $\mathcal{O}(n\epsilon/\sqrt{km})$, respectively. Noting that $\frac{k}{n} < 1$, the result is better than which given by Theorem 3.

In Theorem 7, we theoretically prove that the excess risk of DP models can be better by considering data heterogeneity. It may give new inspirations to the utility analysis in the future work.

Table 1. Comparisons on excess risk bounds between our method and other methods.

	G	L	S.C	C	PL	EPR
[4]	✓	✓	✗	✓	✗	$\mathcal{O}\left(\frac{\sqrt{m}}{n\epsilon} + \frac{1}{\sqrt{n}}\right)$
[14]	✓	✗	✓	✓	✗	$\mathcal{O}\left(\frac{m}{n^2\epsilon^2} + \frac{1}{n}\right)$
[14]	✓	✗	✗	✓	✗	$\mathcal{O}\left(\frac{\sqrt{m}}{n\epsilon} + \frac{1}{\sqrt{n}}\right)$
Ours	✓	✓	✗	✗	✓	$\mathcal{O}\left(\min\left\{\frac{m}{n^2\epsilon^2} + \frac{1}{n}, \frac{\sqrt{m}}{n^{1.5}\epsilon} + \frac{m^{1.5}}{n^{2.5}\epsilon^3}\right\}\right)$
Ours (PIDP-SGD)	✓	✓	✗	✗	✓	$\mathcal{O}\left(\min\left\{\frac{km}{n^3\epsilon^2} + \frac{1}{n}, \frac{km}{n^{2.5}\epsilon^2} + \frac{1}{n^{1.5}}, \frac{(km)^{1.5}}{n^2\epsilon^3} + \frac{\epsilon}{\sqrt{kmn}}\right\}\right)$

6   Comparison with Related Work

Previous works always discuss the excess empirical risk but seldom analyze the excess population risk, so we mainly focus on comparing the excess population risk in this section. Details can be found in Table 1, in which G, L, S.C., C, PL represent G-Lipschitz, L-smooth, strongly convex, convex and PL inequality, respectively and EER, EPR denote the Excess Empirical Risk and the Empirical Population Risk, respectively.

For the excess population risk, the best previous result is $\mathcal{O}\left(\frac{m}{n^2\epsilon^2} + \frac{1}{n}\right)$, given in [14], under strongly convex condition. As shown in Table 1, our result is better by a factor up to $\mathcal{O}\left(\frac{1}{\sqrt{n}}\right)$. When it comes to our proposed PIDP-SGD method, the result is further improved by a factor up to $\mathcal{O}\left(\frac{k}{n}\right)$. Noting that the best result given in [14] requires the loss function to be strongly convex, which means that our result is not only better but also strictly more general than which given in [14]. For the best results under convex condition [4,14]: $\mathcal{O}\left(\frac{\sqrt{m}}{n\epsilon} + \frac{1}{\sqrt{n}}\right)$, our results (both the original one and the one given by PIDP-SGD) are much better. Although it is hard to compare convexity with the PL condition, our results can be applied to some of the non-convex models (shown in Definition 4).

Besides, for the excess empirical risk, our analyzed bound is better than which proposed by [4,5,37,41] and achieves the best result $\mathcal{O}\left(\frac{m}{n^2\epsilon^2}\right)$. For our proposed 'performance improving' method: PIDP-SGD, our analyzed excess empirical risk bound is further tighter by a factor of $\mathcal{O}\left(\frac{k}{n}\right)$. It is worth emphasizing that most of the methods proposed previously assume that the loss function is convex, which is not required in our method. Under this circumstance, we achieve a better result, which is attractive. Additionally, for the non-convex analysis given in [40], the excess empirical risk bound of our analyzed DP-SGD method is better by a factor of $\mathcal{O}\left(\log(n)\right)$ (as discussed in Remark 2) and the PIDP-SGD method is better by a factor of $\mathcal{O}\left(\frac{k\log(n)}{n}\right)$.

7   Experimental Results

Experiments on several real datasets are performed on the classification task. Since our method is based on SGD, we compare our method with previous

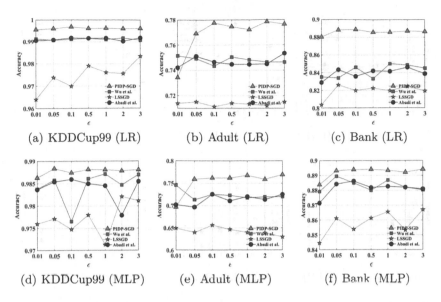

Fig. 1. Accuracy over ϵ, LR denotes logistic regression model and MLP denotes the deep learning model.

DP-SGD methods. Specifically, we compare our method with the gradient perturbation method proposed in [1], the output perturbation method proposed in [41] and the DP-LSSGD method proposed in [36]. The performance is measured in terms of classification accuracy and the optimality gap. The accuracy represents the performance on the testing set, and the optimality gap represents the excess empirical risk on the training set. The optimality gap is denoted by $L(\theta_{priv}; D) - L^*$.

We use both logistic regression model and deep learning model on the datasets KDDCup99 [18], Adult [11] and Bank [29], where the total number of data instances are 70000, 45222, and 41188, respectively. In the experiments, to make the model satisfies the assumptions (such as PL condition) mentioned in the theoretical part, the deep learning model is denoted by Multi-layer Perceptron (MLP) with one hidden layer whose size is the same as the input layer. Training and testing datasets are chosen randomly. In all the experiments, total iteration rounds T is chosen by cross-validation. For PIDP-SGD, we set $RT_{local} = T$. We evaluate the performance of our proposed PIDP-SGD method and some of previous algorithms over the differential privacy budget ϵ. For ϵ, we set it from 0.01 to 3, and in the PIDP-SGD method, we set $\epsilon_1 = 3\epsilon_2 = 3\epsilon/4$ to guarantee $\epsilon_1 + \epsilon_2 = \epsilon$. The results are shown in Fig. 1 and Fig. 2.

Figure 1 shows that as the privacy budget ϵ increases, so does the accuracy, which follows the intuition. When applying the PIDP-SGD algorithm, the accuracy rises on most datasets, which means that our proposed 'performance improving' method is effective. Meanwhile, when ϵ is small, the difference (on accuracy) between traditional methods and 'performance improving' method is

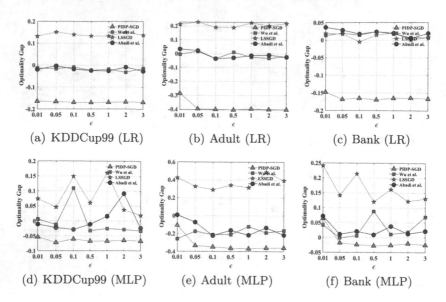

Fig. 2. Optimality gap over ϵ, LR denotes logistic regression model and MLP denotes the deep learning model.

also small. However, as ϵ increases, the 'performance improving' method becomes more and more competitive. The reason is that larger ϵ means that more data instances 'escape' the injected noise, leading to better accuracies.

Figure 2 shows that on some datasets, by applying PIDP-SGD algorithm, the optimality gap of our method is almost 0, which means that it achieves almost the same performance as the model without privacy in some scenarios[2]. Besides, similar to the accuracy in Fig. 1, the optimality gap decreases as ϵ increases, which follows our intuition. Moreover, on some datasets, the performance of some of the methods fluctuates. The reason is that in the setting of differential privacy, random noise is injected into the model, so it is a common phenomenon.

Additionally, on some datasets, the performance of our 'performance improving' method is worse when ϵ is small, the reason is that part of the privacy budget is allocated to c_t, which means 'pure privacy budget' on the model is smaller. Thus, with the increase of ϵ, the 'performance improving' method becomes more competitive, which has been analyzed before in Sect. 5. Experimental results show that our proposed PIDP-SGD algorithm significantly improves the performance under most circumstances.

8   Conclusions

In this paper, we give sharper excess risk empirical and population risk bounds of traditional DP-SGD paradigm. Theoretical results show that our given excess

---

[2] The optimal model (derives L^*) is trained, rather than numerical solutions, and random noise may make the model escape from local minima, so negative optimality gaps appear.

risk bounds are better than previous methods under both convex and non-convex conditions. Meanwhile, based on DP-SGD, we attempt to improve the performance from a new perspective: considering data heterogeneity, rather than treating all data the same. In particular, we introduce the influence function (IF) to analyze the contribution of each data instance to the final model, and the approximation error analysis shows that IF is reasonable to approximate the contribution. In this way, we propose PIDP-SGD: only adding noise to the data demonstrating significant contributions (more than e^ϵ) when training. Detailed theoretical analysis and experimental results show that our proposed PIDP-SGD achieves better performance, without the convexity assumption. Moreover, the new perspective of treating different data instances differently may give new inspirations to future work, including the privacy analysis and the utility analysis. In future work, we will focus on improving the time complexity of PIDP-SGD and applying the algorithm to larger datasets.

Acknowledgements. This work was supported in part by the Excellent Talents Program of IIE, CAS, the Special Research Assistant Project of CAS, the Beijing Outstanding Young Scientist Program (No. BJJWZYJH012019100020098), Beijing Natural Science Foundation (No. 4222029), and National Natural Science Foundation of China (No. 62076234, No. 62106257).

References

1. Abadi, M., et al.: Deep learning with differential privacy. In: ACM SIGSAC Conference on Computer and Communications Security, pp. 308–318 (2016)
2. Arora, R., Upadhyay, J.: On differentially private graph sparsification and applications. In: Advances in Neural Information Processing Systems, pp. 13378–13389 (2019)
3. Backes, M., Berrang, P., Humbert, M., Manoharan, P.: Membership privacy in microrna-based studies. In: ACM SIGSAC Conference on Computer and Communications Security, pp. 319–330 (2016)
4. Bassily, R., Feldman, V., Talwar, K., Guha Thakurta, A.: Private stochastic convex optimization with optimal rates. In: Advances in Neural Information Processing Systems, pp. 11279–11288 (2019)
5. Bassily, R., Smith, A., Thakurta, A.: Private empirical risk minimization: efficient algorithms and tight error bounds. In: IEEE Annual Symposium on Foundations of Computer Science, pp. 464–473 (2014)
6. Bernstein, G., Sheldon, D.R.: Differentially private Bayesian linear regression. In: Advances in Neural Information Processing Systems, pp. 523–533 (2019)
7. Carlini, N., Liu, C., Erlingsson, U., Kos, J., Song, D.: The secret sharer: evaluating and testing unintended memorization in neural networks. In: USENIX Conference on Security Symposium, pp. 267–284 (2019)
8. Charles, Z., Papailiopoulos, D.: Stability and generalization of learning algorithms that converge to global optima. In: International Conference on Machine Learning, pp. 745–754 (2018)
9. Chaudhuri, K., Monteleoni, C., Sarwate, A.D.: Differentially private empirical risk minimization. J. Mach. Learn. Res. **12**, 1069–1109 (2011)

10. Chen, Z., Ni, T., Zhong, H., Zhang, S., Cui, J.: Differentially private double spectrum auction with approximate social welfare maximization. IEEE Trans. Inf. Forensics Secur. **14**, 2805–2818 (2019)
11. Dua, D., Graff, C.: UCI machine learning repository (2017)
12. Dwork, C., McSherry, F., Nissim, K., Smith, A.: Calibrating noise to sensitivity in private data analysis. In: Halevi, S., Rabin, T. (eds.) TCC 2006. LNCS, vol. 3876, pp. 265–284. Springer, Heidelberg (2006). https://doi.org/10.1007/11681878_14
13. Dwork, C., Roth, A., et al.: The algorithmic foundations of differential privacy. Found. Trends® Theor. Comput. Sci. **9**, 211–407 (2014)
14. Feldman, V., Koren, T., Talwar, K.: Private stochastic convex optimization: optimal rates in linear time. In: Annual ACM SIGACT Symposium on Theory of Computing, pp. 439–449 (2020)
15. Fredrikson, M., Lantz, E., Jha, S., Lin, S., Page, D., Ristenpart, T.: Privacy in pharmacogenetics: an end-to-end case study of personalized warfarin dosing. In: USENIX Conference on Security Symposium, pp. 17–32 (2014)
16. Hardt, M., Recht, B., Singer, Y.: Train faster, generalize better: stability of stochastic gradient descent. In: International Conference on Machine Learning, pp. 1225–1234 (2016)
17. Heikkilä, M., Jälkö, J., Dikmen, O., Honkela, A.: Differentially private markov chain monte carlo. In: Advances in Neural Information Processing Systems, pp. 4115–4125 (2019)
18. Hettich, S., Bay, S.D.: The uci kdd archive (1999)
19. Jayaraman, B., Evans, D.: Evaluating differentially private machine learning in practice. In: USENIX Conference on Security Symposium, pp. 1895–1912 (2019)
20. Karimi, H., Nutini, J., Schmidt, M.: Linear convergence of gradient and proximal-gradient methods under the polyak-łojasiewicz condition. In: Joint European Conference on Machine Learning and Knowledge Discovery in Databases, pp. 795–811 (2016)
21. Kifer, D., Smith, A., Thakurta, A.: Private convex empirical risk minimization and high-dimensional regression. In: Conference on Learning Theory, pp. 25–1 (2012)
22. Koh, P.W., Liang, P.: Understanding black-box predictions via influence functions. In: International Conference on Machine Learning, pp. 1885–1894 (2017)
23. Lei, Y., Ledent, A., Kloft, M.: Sharper generalization bounds for pairwise learning. In: Advances in Neural Information Processing Systems (2020)
24. Lei, Y., Ying, Y.: Fine-grained analysis of stability and generalization for stochastic gradient descent. In: International Conference on Machine Learning, pp. 5809–5819 (2020)
25. Lei, Y., Ying, Y.: Sharper generalization bounds for learning with gradient-dominated objective functions. In: International Conference on Learning Representations (2021)
26. Li, S., Liu, Y.: Improved learning rates for stochastic optimization: two theoretical viewpoints (2021)
27. Linnainmaa, S.: Taylor expansion of the accumulated rounding error. BIT Numer. Math. **16**, 146–160 (1976)
28. Liu, M., Zhang, X., Zhang, L., Jin, R., Yang, T.: Fast rates of ERM and stochastic approximation: adaptive to error bound conditions. In: Advances in Neural Information Processing Systems, pp. 4683–4694 (2018)
29. Moro, S., Cortez, P., Rita, P.: A data-driven approach to predict the success of bank telemarketing. Decis. Support Syst. **62**, 22–31 (2014)

30. Phan, N., Wu, X., Hu, H., Dou, D.: Adaptive Laplace mechanism: differential privacy preservation in deep learning. In: IEEE International Conference on Data Mining, pp. 385–394 (2017)
31. Shokri, R., Shmatikov, V.: Privacy-preserving deep learning. In: ACM SIGSAC Conference on Computer and Communications Security, pp. 1310–1321 (2015)
32. Shokri, R., Stronati, M., Song, C., Shmatikov, V.: Membership inference attacks against machine learning models. In: IEEE Symposium on Security and Privacy, pp. 3–18 (2017)
33. Song, S., Chaudhuri, K., Sarwate, A.D.: Stochastic gradient descent with differentially private updates. In: IEEE Global Conference on Signal and Information Processing, pp. 245–248 (2013)
34. Srebro, N., Sridharan, K., Tewari, A.: Optimistic rates for learning with a smooth loss. arXiv preprint arXiv:1009.3896 (2010)
35. Ullman, J., Sealfon, A.: Efficiently estimating erdos-renyi graphs with node differential privacy. In: Advances in Neural Information Processing Systems, pp. 3765–3775 (2019)
36. Wang, B., Gu, Q., Boedihardjo, M., Barekat, F., Osher, S.J.: DP-LSSGD: a stochastic optimization method to lift the utility in privacy-preserving ERM. arXiv preprint arXiv:1906.12056 (2019)
37. Wang, B., Gu, Q., Boedihardjo, M., Barekat, F., Osher, S.J.: DP-LSSGD: a stochastic optimization method to lift the utility in privacy-preserving ERM. CoRR (2019)
38. Wang, D., Chen, C., Xu, J.: Differentially private empirical risk minimization with non-convex loss functions. In: International Conference on Machine Learning, pp. 6526–6535 (2019)
39. Wang, D., Xu, J.: Principal component analysis in the local differential privacy model. Theor. Comput. Sci. **809**, 296–312 (2019)
40. Wang, D., Ye, M., Xu, J.: Differentially private empirical risk minimization revisited: faster and more general. In: Advances in Neural Information Processing Systems, pp. 2722–2731 (2017)
41. Wu, X., Li, F., Kumar, A., Chaudhuri, K., Jha, S., Naughton, J.: Bolt-on differential privacy for scalable stochastic gradient descent-based analytics. In: ACM International Conference on Management of Data, pp. 1307–1322 (2017)
42. Xiao, L., Zhang, T.: A proximal stochastic gradient method with progressive variance reduction. SIAM J. Optim. **24**, 2057–2075 (2014)
43. Xu, C., Ren, J., Zhang, D., Zhang, Y., Qin, Z., Ren, K.: GANobfuscator: mitigating information leakage under GAN via differential privacy. IEEE Trans. Inf. Forensics Secur. **14**, 2358–2371 (2019)
44. Zhang, J., Zheng, K., Mou, W., Wang, L.: Efficient private ERM for smooth objectives. arXiv preprint (2017). arXiv:1703.09947
45. Zhang, L., Yang, T., Jin, R.: Empirical risk minimization for stochastic convex optimization: $o(1/n)$-and $o(1/n^2)$-type of risk bounds. In: Conference on Learning Theory, pp. 1954–1979 (2017)
46. Zhang, L., Zhou, Z.H.: Stochastic approximation of smooth and strongly convex functions: beyond the $o(1/t)$ convergence rate. In: Conference on Learning Theory, pp. 3160–3179 (2019)
47. Zhao, L., et al.: Inprivate digging: enabling tree-based distributed data mining with differential privacy. In: IEEE INFOCOM Conference on Computer Communications, pp. 2087–2095 (2018)

Inference of Over-Constrained NFA of Size $k + 1$ to Efficiently and Systematically Derive NFA of Size k for Grammar Learning

Tomasz Jastrząb[1] , Frédéric Lardeux[2] , and Eric Monfroy[2]

[1] Silesian University of Technology, Gliwice, Poland
Tomasz.Jastrzab@polsl.pl
[2] University of Angers, LERIA, Angers, France
{Frederic.Lardeux,Eric.Monfroy}@univ-angers.fr

Abstract. Grammatical inference involves learning a formal grammar as a finite state machine or set of rewrite rules. This paper focuses on inferring Nondeterministic Finite Automata (NFA) from a given sample of words: the NFA must accept some words, and reject others. Our approach is unique in that it addresses the question of whether or not a finite automaton of size k exists for a given sample by using an over-constrained model of size $k + 1$. Additionally, our method allows for the identification of the automaton of size k when it exists. While the concept may seem straightforward, the effectiveness of this approach is demonstrated through the results of our experiments.

Keywords: grammatical inference · nondeterministic automata · SAT models

1 Introduction

Grammatical inference [4] is the process of learning formal grammars, such as finite automata or production rules, from a given learning sample of words. This method is useful in various fields, such as compiler design, bioinformatics, pattern recognition, and machine learning. The problem we tackle is to learn a finite automaton, specifically a Nondeterministic Finite Automaton (NFA), that can accept a set of positive examples and reject a set of negative examples. The complexity of this problem is determined by the number of states in the automaton. Nondeterministic automata are often smaller in size than deterministic automata for the same language, thus the focus is on learning NFAs. The goal is to minimize the number k of states in the automaton—this is typically done by determining lower (such as 1) and upper bounds (such as the size of a prefix tree acceptor[1]) on the number of states, and using optimization algorithms to find the smallest possible number of states.

---

[1] A prefix tree acceptor (PTA) is a tree-like Deterministic Finite Automaton built from the sample by using each prefix in the sample as a state.

J. Mikyška et al. (Eds.): ICCS 2023, LNCS 14073, pp. 134–147, 2023.
https://doi.org/10.1007/978-3-031-35995-8_10

The problem of learning formal grammars from a given sample of words has been explored from multiple angles. Several algorithms have been proposed, including ad-hoc methods such as DeLeTe2 [2] that focuses on merging states from the prefix tree acceptor (PTA), and newer approaches like the family of algorithms for regular languages inference presented in [18]. Some studies have employed metaheuristics, such as hill-climbing in [16], while others have used complete solvers that can always find a solution if one exists, prove unsatisfiability, and find the global optimum in optimization problems. The problem is often modeled as a Constraint Satisfaction Problem (CSP) and various techniques have been employed, such as Integer Non-Linear Programming (INLP) in [19] and parallel solvers in [6,7]. Additionally, the author of [8] proposed two strategies for solving the CSP formulation of the problem, and [9] presents a parallel approach for solving the optimization variant of the problem.

In this paper, we aim to enhance SAT models for grammatical inference, as opposed to creating a new solver. A SAT (the propositional SATisfiability problem [3]) model consists in defining the problem with Boolean variables and a Boolean formula in Conjunctive Normal Form (CNF). More precisely, we focus here on the definition of over-constrained models of size $k + 1$ with properties that allow deducing information about the classical model of size k. The benefit of these over-constrained models is in terms of spacial complexity. Whereas the complexity of the generation of an NFA of size k of a learning sample S is in $\mathcal{O}(\sigma \cdot k^3)$ variables, and $\mathcal{O}(\sigma \cdot k^3)$ clauses (with $\sigma = \Sigma_{w \in S} |w|$ for most of the models[2]), the generation of our over-constrained NFA of size $k + 1$ is in $\mathcal{O}(\sigma \cdot k^2)$ variables, and $\mathcal{O}(\sigma \cdot k^2)$ clauses.

We made some experiments to test our method. The originality is that we use some of these properties to derive k-state NFAs by building more constrained models of size $k + 1$, where the previous model of size k was unable to provide a solution.

The structure of this paper is as follows. Section 2 gives an overview of the NFA inference problem. Section 3 describes extensions of the classical models. In Sect. 4, we present some properties concerning the extensions. The results of our experiments are discussed in Sect. 5 and we conclude in Sect. 6.

2   The NFA Inference Problem

In this section, we formally introduce the NFA inference problem by utilizing the propositional logic paradigm and providing a generic model. Additionally, we review and highlight some previously established models within the current literature.

---

[2] Only the complexity of the prefix model is in $\mathcal{O}(\sigma \cdot k^2)$ variables, and $\mathcal{O}(\sigma \cdot k^2)$ clauses.

2.1 Notations

Let $\Sigma = \{a_1, \ldots, a_n\}$ be an alphabet of n symbols. A learning sample $S = S^+ \cup S^-$ is given by a set S^+ of so called "positive" words from Σ^* that the inferred NFA must accept, and a set S^- of "negative" words that the NFA must reject.

Let K be the set of the k first non-zero integers, $K = \{1, \ldots, k\}$. We consider the following variables:

- k, an integer, the size of the NFA to be generated,
- a set of k Boolean variables $F = \{f_1, \ldots, f_k\}$ determining whether state i is final or not,
- and, $\Delta = \{\delta_{a,\overrightarrow{i,j}} | a \in \Sigma \text{ and } (i,j) \in K^2\}$, a set of nk^2 Boolean variables representing the existence of transitions from state i to state j with the symbol $a \in \Sigma$, for each i, j, and a.

We define $p_{w,\overrightarrow{i_1,i_{m+1}}}$ as the path $i_1, i_2, \ldots, i_{m+1}$ for a word $w = a_1 \ldots a_m$.

$$p_{w,\overrightarrow{i_1,i_{m+1}}} = \delta_{a_1,\overrightarrow{i_1,i_2}} \wedge \ldots \wedge \delta_{a_m,\overrightarrow{i_m,i_{m+1}}}$$

Although the path is directed from i_1 to i_{m+1} (it is a sequence of derivations), we will build it either starting from i_1, starting from i_{m+1}, or starting from both sides. Thus, to avoid confusion of path and building, we prefer keeping $\overline{i_1, i_{m+1}}$ without any direction.

2.2 A "Meta-model"

A meta-model to define an NFA of size k (noted $k\_$NFA) can be done with the 3 following constraints:

- Special cases for the empty word λ; if it is a word of S^+, the initial state must be final, if it is a negative word, the initial state must not be final:

$$(\lambda \in S^+ \longrightarrow f_1) \wedge (\lambda \in S^- \longrightarrow \neg f_1) \tag{1}$$

- A positive word must terminate on a final state of the $k\_$NFA, i.e., there must be a path from the initial state 1 to a final state i (f_i must be true):

$$\bigvee_{i \in K} p_{w,\overline{1,i}} \wedge f_i \tag{2}$$

- A negative word w must not terminate on a final state of the $k\_$NFA, i.e., either there is no path for w, or each path terminates in a non-final state:

$$\bigwedge_{i \in K} (\neg p_{w,\overline{1,i}} \vee \neg f_i) \tag{3}$$

Of course, the notion of path can be defined and built in many ways. In [10], prefix, suffix, and hybrid approaches are proposed.

2.3   Some Previous Models

As seen in the previous section, several models can be considered for learning an NFA. We skip the direct model (see [9,11]) which has a bad complexity and does not behave well in practice: its space complexity is in $\mathcal{O}(|S^+| \cdot (|\omega_+| + 1) \cdot k^{|\omega_+|})$ clauses, and $\mathcal{O}(|S^+| \cdot k^{|\omega_+|})$ variables with ω_+ the longest word of S^+. We also discard models with 0/1 variables, either from INLP [19] or CSP [14]: we made some tests with various models with [13] and obtained some disastrous results: the NFA inference problem is intrinsically a Boolean problem, and thus, well suited for SAT solvers.

In [12], 7 different models were defined: the prefix model (P), the suffix model (S), and several hybrid models combining prefix and suffix models with different splitting strategies of words, such as the best prefix model $(P^\star)$ to optimize size and use of prefixes, the best suffix model $(S^\star)$ to optimize size and use of suffixes, and 3 hybrid models $(ILS(Init))$ based on a local search optimization [15] of word splittings (starting with an initial configuration $Init$, being either a random splitting of words, the splitting found by the $P^\star$ model, or by the $S^\star$ model).

The main difference is the definition and construction of paths. For example, in the prefix model (P), some extra Boolean variables represent paths for each prefix of the learning sample. Thus, paths are defined with the following two constraints: (each word is represented as $w = va$ with $a \in \Sigma$):

– Case for a word of length 1:

$$\bigvee_{i \in K} \delta_{a,\overrightarrow{1,i}} \leftrightarrow p_{a,\overline{1,i}} \tag{4}$$

– Recursive definition of paths for each prefix w of each word of the sample:

$$\bigwedge_{i \in K} (p_{w,\overline{1,i}} \leftrightarrow (\bigvee_{j \in K} p_{v,\overline{1,j}} \wedge \delta_{a,\overrightarrow{j,i}})) \tag{5}$$

The suffix model (S) is obtained similarly, but with extra variables representing suffixes (i.e., with words represented as $w = av$ with $a \in \Sigma$).

Hybrid models are built with both prefixes and suffixes constructions, each word being considered as the concatenation of a prefix and a suffix. Some constraints are also added to "join" prefix paths and suffix paths.

After transformation in Conjunctive Normal Form (CNF) using Tseitin transformations [17], the spacial complexity of the prefix model is in $\mathcal{O}(\sigma \cdot k^2)$ variables, and $\mathcal{O}(\sigma \cdot k^2)$ clauses with $\sigma = \Sigma_{w \in S}|w|$ (see [11] for details). Although similar, the spacial complexity of the suffix model is in $\mathcal{O}(\sigma \cdot k^3)$ variables and $\mathcal{O}(\sigma \cdot k^3)$ clauses. The reason is that we build prefixes from the initial state 1, whereas suffixes are built from one of the k states (see [11]). Hybrid models are thus also in $\mathcal{O}(\sigma \cdot k^3)$ variables and clauses.

3 $k\_$NFA Extensions

For a given sample, if there is a $k\_$NFA, i.e., an NFA of size k, to recognize words of S^+ and reject words of S^-, there is also an NFA of size $k + 1$. Although obvious and rather useless, this property can be refined to generate k_NFA. To this end, some more constraints are added to the $(k+1)\_NFA$ to build what we call $(k+1)\_NFA$ extensions.

3.1   Building a $(k+1)\_NFA$ from a k_NFA

Let $A = (Q^A, \Sigma, \Delta^A, q_1, F^A)$ be an NFA of size k. Then, there always exists an NFA of size $k+1$, $A' = (Q^{A'}, \Sigma, \Delta^{A'}, q_1, F^{A'})$, such that $Q^{A'} = Q^A \cup \{q_{k+1}\}$, $F^{A'} = \{q_{k+1}\}$ and $\Delta^{A'}$:

$$\forall_{i,j \in (Q^A)^2} \; \delta^A_{a,\overrightarrow{i,j}} \leftrightarrow \delta^{A'}_{a,\overrightarrow{i,j}}$$

$$\forall_{i \in Q^A, j \in F^A} \; \delta^A_{a,\overrightarrow{i,j}} \leftrightarrow \delta^{A'}_{a,\overrightarrow{i,k+1}}$$

In other words, A' has only one final state which is the new state $k+1$, each transition of A also exists in A', and transitions from a state i to a final state of A are duplicated as new transitions from i to state $k+1$. The obvious but important property for the rest of this paper is that the language recognized by A' is the same as the one recognized by A.

In the following, the main idea is to over-constrain a $(k+1)\_NFA$ model (i.e., a model to generate an NFA of size $k+1$) to obtain a model closer or equal to A' as described above. Moreover, the idea is also that the over-constrained $(k+1)\_NFA$ model can be solved and reduced to an NFA of size k more efficiently than solving directly the k_NFA model. We propose two $(k+1)\_NFA$ model extensions for which a new state and new constraints are added. The first model extension, the $(k+1)\_NFA^+$, over-constrains the $(k+1)\_NFA$ to be able to reduce a generated $(k+1)\_NFA$ into a k_NFA with a reduction algorithm (see [10]). However, as shown in [10], we are not always able to reduce a $(k+1)\_NFA^+$ into a k_NFA, and the gain is really poor. In fact, a $(k+1)\_NFA^+$ instance mainly provides information when it is unsatisfiable: in this case, we also know that there is no k_NFA. The second model extension, the $(k+1)\_NFA^\star$ model, does not require any algorithm to reduce a generated $(k+1)\_NFA$ to a $(k+1)\_NFA$: the satisfiability (respectively unsatisfiability) of a $(k+1)\_NFA^\star$ implies the satisfiability (respectively unsatisfiability) of a k_NFA: moreover, in case of satisfiability, the k_NFA can always be directly derived from the $(k+1)\_NFA^\star$ by removing a state and some transitions. This operation has no cost.

3.2 $(k+1)\_NFA^+$ Extension

Let $K = \{1, \ldots, k\}$ be the k first non-zero integers, and $K_+ = \{1, \ldots, k+1\}$ be the $k+1$ first non-zero integers.

A $(k+1)\_NFA^+$ is a $(k+1)\_NFA$ with some extra properties that ensure that it may be reduced to a k_NFA by a reduction algorithm [10]. The extra properties of a $(k+1)\_NFA^+$ are:

- a $(k+1)\_NFA^+$ has one and only one final state, i.e., state $k+1$;

- state $k + 1$ has no outgoing transition;
- each transition from a state i to state $k + 1$ reading the symbol a has an equivalent transition from the state i to a state j ($j \neq k + 1$) with a. State i is called a possibly final state.

A $(k+1)\_NFA^+$ is defined by the same variables as a k_NFA with an addition of those related to state $k + 1$ (a variable for the final state, transitions, and paths). The extra constraints are the following:

- The $(k + 1)\_NFA^+$ has only one final state, state $k + 1$:

$$\bigwedge_{i \in K} (\neg f_i) \wedge f_{k+1} \tag{6}$$

Note that this constraint mainly impacts the suffix model by unit propagation.
- There is no outgoing transition from the $(k + 1)\_NFA^+$ final state:

$$\bigwedge_{a \in \Sigma} \bigwedge_{i \in K_+} \neg \delta_{a,\overrightarrow{k+1,i}} \tag{7}$$

- Each incoming transition of the $(k + 1)\_NFA^+$ final state $k + 1$ must also finish in another state:

$$\bigwedge_{a \in \Sigma} \left(\bigwedge_{i \in K} \left(\delta_{a,\overrightarrow{i,k+1}} \rightarrow \bigvee_{j \in K} \delta_{a,\overrightarrow{i,j}} \right) \right) \tag{8}$$

3.3 $k\_NFA^\star$ Extension

A $(k+1)\_NFA^\star$ is a $(k+1)\_NFA^+$ with some extra properties on words and a new set of Boolean variables representing possibly final states for the corresponding k_NFA ($F^* = \{f_1^*, ..., f_k^*\}$). $(k+1)\_NFA^\star$ may be reduced to a k_NFA by removing state $k + 1$ and its incoming transitions, and fixing the final states among the possible final states, i.e., determining the f_i^* of $\{f_1^*, ..., f_k^*\}$ which are final states of the k_NFA. To determine these final states, we have to ensure:

- A negative word cannot terminate in a possible final state:

$$\bigwedge_{i \in K} \left(f_i^* \rightarrow \bigwedge_{w \in S^-} \neg p_{w,\overline{1,i}} \right) \tag{9}$$

- Each possibly final state validates at least one positive word of S^+:

$$\bigwedge_{i \in K} \left(f_i^* \rightarrow (\bigvee_{va \in S^+} \bigvee_{j \in K} (p_{v,\overline{1,j}} \wedge \delta_{a,\overrightarrow{j,i}} \wedge \delta_{a,\overrightarrow{j,k+1}}) \right) \tag{10}$$

- Each positive word terminates in at least one possible final state:

$$\bigwedge_{w \in S^+} \bigvee_{i \in K} (p_{w,\overline{1,i}} \wedge f_i^*) \tag{11}$$

3.4   Complexity

Extensions $(k+1)\_NFA^+$ and $(k+1)\_NFA^\star$ are of course bigger than k_NFA since the corresponding models contain an extra state, some extra transitions, and some extra constraints. $(k+1)\_NFA^\star$ model needs also k new variables for the possible final states. New constraints to transform $(k+1)\_NFA$ to the extensions also increase the number of clauses and variables. Table 1 provides the cost in terms of variables and clauses. Some details about the arity of the generated clauses are given (unary, binary, and greater). A blank in a cell corresponds to 0.

Table 1. Complexity in terms of variables and clauses for each new constraint allowing the definition of the $(k+1)\_NFA^+$ (Constraints (6–8)) and the $(k+1)\_NFA^\star$ models (Constraints (6–11)).

	Variables	Clauses			
		total	unary	binary	$>$
Constraint 6		$k+1$	$k+1$		
Constraint 7		$n(k+1)$	$n(k+1)$		
Constraint 8		nk			nk
Constraint 9		$k\lvert S^-\rvert$		$k\lvert S^-\rvert$	
Constraint 10	$k^2\lvert S_+\rvert$	$4k^2\lvert S_+\rvert+k$		$3k^2\lvert S_+\rvert$	$k^2\lvert S_+\rvert+k$
Constraint 11	$k\lvert S^+\rvert$	$(3k+1)\lvert S^+\rvert$		$2k\lvert S^+\rvert$	$(k+1)\lvert S^+\rvert$

Both extensions increase the number of variables and clauses of the initial $(k+1)\_NFA$ model. We can observe a lot of unary and binary clauses allowing solvers to perform very well. The global complexity for each extension remains unchanged or decreases, despite the increase in clauses and variables. As we will observe in Sect. 5.2, the implementation of these new constraints significantly lowers the complexity of many of the models ($\mathcal{O}(\sigma \cdot k^3)$ clauses to $\mathcal{O}(\sigma \cdot (k+1)^2)$ clauses) just by simplification.

4   Properties of the Extensions

Extensions $(k+1)\_NFA^+$ and $(k+1)\_NFA^\star$ are over-constrained models for $(k+1)\_NFA$. Indeed, some properties allow them to infer a k_NFA solution. A k_NFA^+ is a "weak" extension because only the existence of a k_NFA can be proved, whereas $k\_NFA^\star$ is a "strong" extension that proves the existence or not of a k_NFA.

4.1 $(k+1)\_NFA^+$

Major Property. Adding Constraints (6), (7), and (8) to k_NFA model leads to the following property:

$$\exists\, k\_NFA \Rightarrow \exists\, (k+1)\_NFA^+ \tag{12}$$

The contraposition allows us to prove the unsatisfiability of k_NFA with $(k+1)$_NFA$^+$:

$$\not\exists\,(k+1)\_\text{NFA}^+ \Rightarrow \not\exists\, k\_\text{NFA} \tag{13}$$

Proof. The main idea is: there always exists a transformation from any k_NFA to a $(k+1)$_NFA$^+$ which recognizes the same language.

1. A new state $k+1$ is added to the k_NFA.
2. Each of incoming transitions of k_NFA final states are duplicated by transitions outgoing to state $k+1$. Thus, a path $p_{w,\overline{1,k+1}}$ exists from the initial state to state $k+1$ if and only if a path exists from the initial state to k_NFA final states.
3. This induces:
   - Each positive word can now finish in state $k+1$,
   - No negative word can finish in state $k+1$.
4. All k_NFA final states are then redundant with state $k+1$ thus state $k+1$ can be considered as the unique final state.
5. The automaton is then a $(k+1)$_NFA$^+$ since it has only one final state, and it validates positive words and rejects negative words.

Minor Property. While the lack of solutions for $(k+1)$_NFA$^+$ allows to conclude the lack of solutions for k_NFA, a solution for $(k+1)$_NFA$^+$ is not sufficient to conclude existence of a k_NFA:

$$\exists\,(k+1)\_\text{NFA}^+ \not\Rightarrow \exists\, k\_\text{NFA} \tag{14}$$

The simple following example (Fig. 1) shows a k_NFA$^+$ solution with $k=3$, whereas there is no k_NFA with $k=2$.

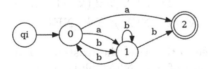

Fig. 1. Example of a $(k+1)$_NFA$^+$ solution for $k=3$ and $S = (\{a, ab, abb, bbb\}, \{aab, b, ba, bab, aaa\})$. It is not possible to find a k_NFA with $k=2$.

It is possible to try to obtain a k_NFA from a $(k+1)$_NFA$^+$ solution with the algorithm presented in [10]. However this algorithm rarely succeeds, and its worst-case complexity is in $O(k \cdot |S|)$.

4.2 $(k+1)$_NFA*

Major Property. Adding Constraints (9), (10), and (11) to $(k+1)$_NFA$^+$ model allows to obtain the equisatisfiability between k_NFA and $(k+1)$_NFA*:

$$\exists\, k\_\text{NFA} \equiv \exists\,(k+1)\_\text{NFA}^* \tag{15}$$

Proof. The main idea is: there are some transformations from $k\_$NFA to $(k + 1)\_$NFA* and from $(k + 1)\_$NFA* to $k\_$NFA preserving acceptance of positive words and rejections of negative words.

$\Rightarrow$ From $k\_$NFA to $(k + 1)\_$NFA*
   1. It is similar to the proof of Sect. 4.1.
$\Leftarrow$ From $(k + 1)\_$NFA* to $k\_$NFA
   1. All transitions $\delta_{a,i,\overline{k+1}}$ are removed, and states j such that $\delta_{a,i,\overline{k+1}} \wedge \delta_{a,i,\vec{j}}$ are now considered as final states.
   2. State $k + 1$ is removed.
   3. Each positive word can terminate in a possible final state because there are some transitions to a possible final state redundant to transitions to state $k + 1$.
   4. No negative word can terminate in a possible final state otherwise, by construction, it would also finish in state $k + 1$.
   5. The automaton is then a $k\_$NFA validating words of S^+ and rejecting words of S^-.

Minor Property. The lack of solutions for $(k + 1)\_$NFA* allows concluding that there are no solutions for the $k\_$NFA model. However, nothing can be deduced for the $(k + 1)\_$NFA model.

$$\nexists \ (k + 1)\_\text{NFA}^* \nRightarrow \nexists \ (k + 1)\_\text{NFA} \tag{16}$$

It is easy to prove Property 16 with its contraposition and its rewritting using Property 15:

$$\exists \ (k + 1)\_\text{NFA} \nRightarrow \exists \ (k + 1)\_\text{NFA}^*$$
$$\exists \ (k + 1)\_\text{NFA} \nRightarrow \exists \ k\_\text{NFA} \tag{17}$$

Property 17 is correct, otherwise, the existence of a $(k + 1)\_$NFA solution would imply the existence of a $k\_$NFA solution. As a counterexample, Fig. 1 proposes a $(k + 1)\_$NFA with $k = 3$ but no $k\_$NFA with $k = 2$ exists.

5   Experimentation

We present an experimental study of two equisatisfiable models, $k\_$NFA and $(k+1)\_$NFA*. Our experiments are conducted on state-of-the-art instances that vary in the sizes of the alphabet and the number of positive and negative words in the learning sample. We present a simplified version of the $(k + 1)\_$NFA* model: indeed, the models briefly presented in Sect. 2.3 can be rewritten and simplified when combined with Constraints (6–11), leading to SAT instances with fewer variables and constraints. We also compare the performance of the two models in terms of the number of solutions and running time.

5.1   Context for Reproductibility

The algorithms were implemented in Python using libraries such as PySAT [5]. The experiments were carried out on a computing cluster with Intel-E5-2695 CPUs, and a fixed limit of 10 GB of memory. Running times were limited to 15 min, including model generation and solving time. We used the Glucose [1] SAT solver with default options.

These experiments were carried out on state-of-the-art instances, described in [12]. These instances can be divided into three categories corresponding to the sizes of the alphabet (2, 5, and 10). The number of positive and negative words is the same in each instance and varies from 10 to 100 in increments of 10 for each category. There are thus 30 instances in total. We test all the possible values of k from 1 to k_{PTA} where k_{PTA} is the upper bound obtained by the size of the computed prefix tree acceptor.

We focus our experiments on k_NFA and $(k+1)\_NFA^\star$ excluding $(k+1)\_NFA^+$ because we want to focus on the comparison of two equisatisfiable models.

Due to the deterministic behavior of Glucose, we run only one execution of each couple (instance, k).

5.2   Simplified Models

The $(k+1)\_NFA^\star$ model is defined by over-constraining a k_NFA model. Mixing and combining all these constraints allows us to obtain a simplified model with fewer variables (f disappear) and fewer constraints (Constraints (1) and (7) are deleted and Constraints (2) and (3) are simplified). The final $(k+1)\_NFA^\star$ models that we use can be defined as follows:

– Variables:
  - a set of k Boolean variables determining whether state i is a possibly final state or not: $F^* = \{f_1^*, \ldots, f_k^*\}$,
  - a set of $nk(k+1)$ Boolean variables representing the transitions from state i to state j with the symbol $a \in \Sigma$: $\Delta = \{\delta_{a,\overrightarrow{i,j}} | a \in \Sigma$ and $i \in K$ and $j \in K_+\}$, and
  - a set of Boolean variables representing the paths (sequence of transitions from a state i to a state j reading a word w):

$$\Pi = \{p_{w,\overrightarrow{i,j}} | (i,j) \in K_+^2, w \in \Sigma^*\}$$

– Constraints:

  - $\bigwedge_{w \in S^+} p_{w,\overline{1,k+1}}$ // *Simplification of Constraint 2*
  - $\bigwedge_{w \in S^-} \neg p_{w,\overline{1,k+1}}$ // *Simplification of Constraint 3*
  - $\bigwedge_{a \in \Sigma} \left(\bigwedge_{i \in K} \left(\delta_{a,\overrightarrow{i,k+1}} \rightarrow \bigvee_{j \in K} \delta_{a,\overrightarrow{i,j}} \right) \right)$ // *Constraint 8*
  - $\bigwedge_{i \in K} \left(f_i^* \rightarrow \bigwedge_{w \in S^-} \neg p_{w,\overline{1,i}} \right)$ // *Constraint 9*

Table 2. Results for $k\_$NFA and $(k+1)\_$NFA* on instances with all values coming from PTA. Each instance corresponds to the average for all samples of words with all k between 1 and the value given by PTA.

Instances				Results			
Model		Vars	Clauses	Sat	Unsat	?	Time
$k\_$NFA	P	56,256	212,229	124	131	158	368
	S	386,508	1,459,936	80	116	217	217
	$ILS(r)$	114,218	446,854	104	133	176	418
	$ILS(P^\star)$	114,781	449,122	110	132	171	423
	$ILS(S^\star)$	51,401	205,593	105	134	174	415
	$P^\star$	412,120	1,574,192	74	126	213	504
	$S^\star$	51,683	206,564	103	134	176	409
$(k+1)\_$NFA*	P	70,362	267,368	128	133	152	356
	S	381,308	1,417,649	86	126	201	169
	$ILS(r)$	122,553	477,081	125	137	151	376
	$ILS(P^\star)$	123,707	479,036	124	138	151	423
	$ILS(S^\star)$	55,192	219,637	125	137	151	362
	$P^\star$	391,773	1,475,492	120	122	171	406
	$S^\star$	55,567	220,898	122	137	154	356

- $\bigwedge_{i \in K} \left(f_i^* \rightarrow (\bigvee_{v.a \in S^+} \bigvee_{j \in K} (p_{v,\overline{1,j}} \wedge \delta_{a,\overrightarrow{j,i}} \wedge \delta_{a,\overrightarrow{j,k+1}})) \right)$ // *Constraint 10*
- $\bigwedge_{w \in S^+} \bigvee_{i \in K} (p_{w,\overline{1,i}} \wedge f_i^*)$ // *Constraint 11*

Additionally, the constraints applied to define the paths either starting from the end, the beginning, or both extremities of words (as outlined in Sect. 2.3) can also be simplified. The most significant simplification is for the suffix model, where the space complexity changes from $\mathcal{O}(\sigma \cdot k^3)$ clauses to $\mathcal{O}(\sigma \cdot (k+1)^2)$ clauses. All other models, which partially use the suffix model, are also affected except for the prefix model.

5.3   Results and Discussions

Table 2 shows results for $k\_$NFA and $(k+1)\_$NFA* with 7 models: the prefix model P, the suffix model S, and 5 hybrid models, the local search model with random initial configuration $ILS(r)$, with $P^\star$ (respectively $S^\star$) initial configuration, the best prefix model $P^\star$, and finally the best suffix model $S^\star$. The combination of learning samples and the number of states (k values) produces 413 instances, and thus, 413 executions for each model. Table 2 is divided into two parts: instances description and results. Instances are described by the model and the average number of variables and clauses. As result, we consider the number of satisfiable instances (Sat), unsatisfiable instances (Unsat), and Unknown solutions (noted ?). Moreover, we give the average running time (in seconds).

Table 3. Difference (in %) from $k\_$NFA to $(k+1)\_$NFA* based on results in Table 2.

Model	Vars	Clauses	Sat	Unsat	?	Times
P	+25.07	+25.98	+3.23	+1.53	−2.35	−3.44
S	−1.35	−2.9	+7.5	+8.62	−8.16	−22.04
$ILS(r)$	+7.3	+6.76	+20.19	+3.01	−10.55	−10.16
$ILS(P^{\star})$	+7.78	+6.66	+12.73	+4.55	−8.26	−0.13
$ILS(S^{\star})$	+7.37	+6.83	+19.05	+2.24	−9.62	−12.9
$P^{\star}$	−4.94	−6.27	+62.16	−3.17	−21.00	−19.5
$S^{\star}$	+7.51	+6.94	+18.45	+2.24	−9.28	−13.01

Table 3 shows the relative difference from $k\_$NFA to $(k+1)\_$NFA* for each model. Column names are the same as in Table 2, but they indicate the percentage difference compared to the results obtained with $k\_$NFA instances.

Tables 2 and 3 show that the number of variables and clauses in $(k+1)\_$NFA* instances increase with respect to $k\_$NFA, but they remain very close; except for the prefix model P with a +25% increase, and for the suffix model S and the best prefix model $P^{\star}$ with a decrease (respectively −1 to −6%), the instances are around +8% larger. Model P increases by 25% because it is the only one that does not use the suffixes (which give a $\mathcal{O}(\sigma k^3)$ complexity). The others, although adding extra Constraints (6–11), keep similar sizes since the $(k+1)\_$NFA* models allow decreasing complexity of suffix constructions from $\mathcal{O}(\sigma k^3)$ to $\mathcal{O}(\sigma k^2)$.

We can observe that the number of Sat solutions increases for all models when we use $(k+1)\_$NFA*. It is similar for Unsat solutions (except for model $P^{\star}$). The number of instances not solved in the given time is therefore logically reduced. The very interesting result is that this improvement in the number of resolved instances is coupled with a diminution of the resolution time.

In terms of global efficiency, best results are obtained by $ILS(r)$, $ILS(P^{\star})$, and $ILS(S^{\star})$ with $(k+1)\_$NFA* (only 151 instances remain unsolved). Adding the running time in the comparison, model $ILS(S^{\star})$ is the best one. Compared to the best model for $k\_$NFA (P), it finds one more Sat and 6 more Unsat solutions with a faster average running time (6 s less).

6   Conclusion

Grammatical inference is the process of learning formal grammars, in our case as a Nondeterministic Finite Automaton. In this paper, we proposed an over-constrained model of size $k+1$ ($(k+1)\_$NFA*) with properties that allow deriving a solution for the classical model of size k ($k\_$NFA). If such an automaton of size $k+1$ exists, it can be freely reduced to an automaton of size k, and if it does not exist, we have the proof there is no automaton of size k. The advantage of using the $(k+1)\_$NFA* model is to obtain shorter resolution times while increasing the rate of solved instances.

Working with a model of size $k + 1$ increases the number of clauses and variables but the additional constraints allow us to simplify the model and limit the combinatorial explosion. Moreover, these constraints allow us to lower the global complexity of most of our models that use a suffix construction (such as the model S).

In the future, we plan to work on adding new constraints to further reduce the complexity of the model $(k + 1)\_NFA^*$. Moreover, a parallel execution of some well-chosen models could lead even more often to solved instances.

References

1. Audemard, G., Simon, L.: Predicting learnt clauses quality in modern SAT solvers. In: Proceedings of IJCAI 2009, pp. 399–404 (2009)
2. Denis, F., Lemay, A., Terlutte, A.: Learning regular languages using RFSAs. Theor. Comput. Sci. **313**(2), 267–294 (2004)
3. Garey, M.R., Johnson, D.S.: Computers and Intractability, A Guide to the Theory of NP-Completeness. W.H. Freeman & Company, San Francisco (1979)
4. de la Higuera, C.: Grammatical Inference: Learning Automata and Grammars. Cambridge University Press, Cambridge (2010)
5. Ignatiev, A., Morgado, A., Marques-Silva, J.: PySAT: a python toolkit for prototyping with SAT oracles. In: Beyersdorff, O., Wintersteiger, C.M. (eds.) SAT 2018. LNCS, vol. 10929, pp. 428–437. Springer, Cham (2018). https://doi.org/10.1007/978-3-319-94144-8_26
6. Jastrzab, T.: On parallel induction of nondeterministic finite automata. In: Proceedings of ICCS 2016. Procedia Computer Science, vol. 80, pp. 257–268. Elsevier (2016)
7. Jastrzab, T.: Two parallelization schemes for the induction of nondeterministic finite automata on PCs. In: Wyrzykowski, R., Dongarra, J., Deelman, E., Karczewski, K. (eds.) PPAM 2017. LNCS, vol. 10777, pp. 279–289. Springer, Cham (2018). https://doi.org/10.1007/978-3-319-78024-5_25
8. Jastrząb, T.: A comparison of selected variable ordering methods for NFA induction. In: Rodrigues, J.M.F., et al. (eds.) ICCS 2019. LNCS, vol. 11540, pp. 741–748. Springer, Cham (2019). https://doi.org/10.1007/978-3-030-22750-0_73
9. Jastrzab, T., Czech, Z.J., Wieczorek, W.: Parallel algorithms for minimal nondeterministic finite automata inference. Fundam. Inform. **178**(3), 203–227 (2021). https://doi.org/10.3233/FI-2021-2004
10. Jastrząb, T., Lardeux, F., Monfroy, E.: Taking advantage of a very simple property to efficiently infer NFAs. In: 34th IEEE International Conference on Tools with Artificial Intelligence, ICTAI 2022, virtual, 1–2 November 2022. IEEE (2022)
11. Lardeux, F., Monfroy, E.: GA and ILS for optimizing the size of NFA models. In: The 8th International Conference on Metaheuristics and Nature Inspired Computing (META), Marrakech, Morocco (2021), https://hal.univ-angers.fr/hal-03284541
12. Lardeux, F., Monfroy, E.: Optimized models and symmetry breaking for the NFA inference problem. In: 33rd IEEE International Conference on Tools with Artificial Intelligence, ICTAI 2021, Washington, DC, USA, 1–3 November 2021, pp. 396–403. IEEE (2021). https://doi.org/10.1109/ICTAI52525.2021.00065
13. Lecoutre, C., Szczepanski, N.: PYCSP3: modeling combinatorial constrained problems in python. CoRR abs/2009.00326 (2020). https://arxiv.org/abs/2009.00326

14. Rossi, F., van Beek, P., Walsh, T. (eds.): Handbook of Constraint Programming, Foundations of Artificial Intelligence, vol. 2. Elsevier, Amsterdam (2006)
15. Stützle, T., Ruiz, R.: Iterated local search. In: Martí, R., Pardalos, P.M., Resende, M.G.C. (eds.) Handbook of Heuristics, pp. 579–605. Springer, Cham (2018). https://doi.org/10.1007/978-3-319-07124-4_8
16. Tomita, M.: Dynamic construction of finite-state automata from examples using hill-climbing. In: Proceedings of the Fourth Annual Conference of the Cognitive Science Society, pp. 105–108 (1982)
17. Tseitin, G.S.: On the complexity of derivation in propositional calculus. In: Siekmann, J.H., Wrightson, G. (eds.) Automation of Reasoning. Symbolic Computation, pp. 466–483. Springer, Heidelberg (1983). https://doi.org/10.1007/978-3-642-81955-1_28
18. de Parga, M.V., García, P., Ruiz, J.: A family of algorithms for non deterministic regular languages inference. In: Ibarra, O.H., Yen, H.-C. (eds.) CIAA 2006. LNCS, vol. 4094, pp. 265–274. Springer, Heidelberg (2006). https://doi.org/10.1007/11812128_25
19. Wieczorek, W.: Grammatical Inference - Algorithms, Routines and Applications. Studies in Computational Intelligence, vol. 673. Springer, Hedelberg (2017). https://doi.org/10.1007/978-3-319-46801-3

Differential Dataset Cartography: Explainable Artificial Intelligence in Comparative Personalized Sentiment Analysis

Jan Kocoń[✉][iD], Joanna Baran[iD], Kamil Kanclerz[iD], Michał Kajstura, and Przemysław Kazienko[iD]

Department of Artificial Intelligence, Wrocław University of Science and Technology, Wrocław, Poland
jan.kocon@pwr.edu.pl

Abstract. *Data Maps* is an interesting method of graphical representation of datasets, which allows observing the model's behaviour for individual instances in the learning process (training dynamics). The method groups elements of a dataset into *easy-to-learn*, *ambiguous*, and *hard-to-learn*. In this article, we present an extension of this method, *Differential Data Maps*, which allows you to visually compare different models trained on the same dataset or analyse the effect of selected features on model behaviour. We show an example application of this visualization method to explain the differences between the three personalized deep neural model architectures from the literature and the HumAnn model we developed. The advantage of the proposed HumAnn is that there is no need for further learning for a new user in the system, in contrast to known personalized methods relying on user embedding. All models were tested on the sentiment analysis task. Three datasets that differ in the type of human context were used: user-annotator, user-author, and user-author-annotator. Our results show that with the new explainable AI method, it is possible to pose new hypotheses explaining differences in the quality of model performance, both at the level of features in the datasets and differences in model architectures.

Keywords: differential data maps · data cartography · personalization · sentiment analysis · explainable artificial intelligence

This work was financed by (1) the National Science Centre, Poland, 2021/41/B/ST6/04471 (PK); (2) the Polish Ministry of Education and Science, CLARIN-PL; (3) the European Regional Development Fund as a part of the 2014-2020 Smart Growth Operational Programme, POIR.01.01.01-00-0288/22 (JK-S140), POIR.01.01.01-00-0615/21 (JK-MHS), POIR.04.02.00-00C002/19 (JB, KK); (4) the statutory funds of the Department of Artificial Intelligence, Wroclaw University of Science and Technology; (5) the European Union under the Horizon Europe, grant no. 101086321 (OMINO).

1   Introduction

Popular approaches in text classification in natural language processing (NLP) assume the development of a general classifier that returns a label based on text-only input. Meanwhile, language and its perception are influenced by many factors: mood, emotions, world view, and sociodemographic conditions. This is a challenge, especially for subjective tasks such as identifying hate speech, aggression, or sentiment. In these tasks, getting a single true label for the text is already difficult at the annotation level and the inter-annotator agreement is low for the same text annotated by several people. Thus, creating a general model that always returns an answer consistent with the user's expectations is difficult.

An example of a subjective task in NLP, among others, is sentiment classification, where the polarity of a text depends on a person's experience and character, both from the perspective of the author of the text and the recipient. It happens that the same text can evoke drastically different reactions. Moreover, it can be argued that a single correct label often simply does not exist, and attempting to enforce it could lead to a model biased against a particular culture or world view, which could lead, e.g., to discrimination against minorities. Obtaining a gold standard from multiple annotations is usually done through aggregation based on instances, such as a mean or majority vote. Still, the more controversial the text, the more difficult it is to find a consensus that satisfies all users.

The aforementioned problems have led to the rapid development of personalized models in NLP, which are trained on datasets annotated by multiple people and using non-aggregated labels. The human context is considered in both the training and inference process, so one gets personalized predictions [9]. In most cases, it improves the prediction quality under the condition of providing human context. Most of the work on this topic is limited to presenting the magnitude of the performance improvement achieved. However, no one focuses on explaining what characterizes the cases for which improvement or deterioration is obtained?

In this work, we propose a new Differential Data Maps (DDM) method from the field of explainable artificial intelligence (XAI), which is an extension of the Data Maps method from the Dataset Cartography [32]. DDM visualizations can help formulate new hypotheses that better explain differences between models, including the impact of individual features and architectures. The capabilities of DDM are demonstrated with an example of personalized sentiment analysis for three datasets in which we have different human contexts.

Our contributions are as follows: (1) we developed a new DDM method that graphically presents the differences in training dynamics on the same datasets for different personalized models; (2) we developed a new HumAnn method that does not require a user ID in the personalized learning/inference process and does not need additional training for new users in the system; (3) we analysed with DDM four personalized methods, including two methods that use the user ID in the model fine-tuning process; (4) we analysed with DDM three datasets with the following human contexts: user-annotator, user-author, and user-author-annotator. (5) through DDM visualizations, we show which situations personalized methods work best and where improvements and deteriorations come from, depending on how the human context is considered.

2   Background

This section briefly describes related works in the area of personalized NLP and provides motivation for the use of XAI methods within this field.

2.1   Personalization in NLP

The existing approaches to human-based NLP can be divided into two groups – based on users' metadata and on using past digital traces such as likes, ratings, or posted texts. Conceptually easier, attempts to adapt personalization in NLP tasks are based on users' metadata and their individual or social-group features. [35] use demographic variables (e.g., gender, age) directly as input into the traditional rule-based model aiming to learn gender differences between users. The demographic adaption is also introduced in the work of [13] and proves that models aware of that features outperform their agnostic counterparts. [36] design a residualized control approach by training a language model over the model's prediction errors using the sociodemographic variables only. Later, the results are combined with factor analysis.

More exploited personalization methods in the literature make advantage of digital traces left by the user. It could be published opinions or ratings. An interesting approach to group-wise sentiment classification is presented by [11]. Taking shared opinions between different people, the authors' solution introduces a non-parametric Dirichlet Process over the individualized models - one for each cluster of users. This lies at the heart of the social comparison theory that humans tend to form groups with others of similar minds and abilities. Inspired by the recommendation systems, the latent factor model can also be used to capture users' specific individuality due to different language habit [30].

The recent works mostly focus on using deep neural networks instead of classical machine learning techniques, especially on the SOTA transformer-based architectures. Those approaches often include global, shared model pre-training, and local, personalized fine-tuning. In the first phase, the model is trained on aggregated, non-personalized data, resulting in a global model unable to incorporate person-level information. After that, the shared model is fine-tuned for each user using their data. There are multiple ways of performing this step. The most basic one is to optimize the whole model, which results in a separate set of weights for each user [29], causing a significant computational and storage overhead. However, there are methods to share a single model between users and learn a unique representation for each person. This representation is combined with the text representation to produce a user-informed prediction [19,37]. Even though these methods mitigate most of the memory-related issues, they continue to require user embedding optimization, which is easier than training the entire model. Still, they can be difficult if the number of users is large or they change frequently. Methods based on an aggregation of user labels [14] do not require training of user embeddings and thus can be easily applied in big-data scenarios. However, in this approach user embedding is fixed and depends only on past texts written or annotated by the user. This results in poor performance if the evaluated text sample introduces a new topic.

2.2   Explainable AI

Modern artificial intelligence (AI) methods are complex. Numerous parameters allow them to learn intricate data patterns. However, there is a risk that the model has memorized specific examples from the training set but does not have general knowledge of the phenomenon it should learn about. To prevent this, explainable artificial intelligence methods should be used to understand the model behaviour [8,34]. Moreover, identifying a missing part can greatly improve the effectiveness of a model [3,18,20,22,25,33]. On the other hand, apart from scientists, there is a growing need for common users to understand AI solutions thoroughly. AI's ethics, trust, and bias are difficult to pinpoint when the algorithm is treated as a black box [1]. Explanations must make the AI algorithm expressive, improving human understanding and confidence that the model makes just and impartial decisions [6]. In addition to curiosity, the need to facilitate better results is growing, especially when the end user is the government [2].

Moreover, it is much more difficult to maintain the transparency, trust, and fairness of the personalized architecture inference process. This requires considering the impact of user context on model behaviour. To the best of our knowledge, no work on methods for analysing the performance of personalized models has been published so far.

3   Datasets

To explore the differences between baselines and personalized models, we used three datasets: (1) **Sentiment140, S140** [10] is a heuristically annotated dataset of 56,557 tweets collected using a predefined set of queries. 1,100 users do binary annotations regarding a positive or negative sentiment based on emoticons contained in the texts. In this dataset, the known user is only the author of the text (user-author). (2) **Internet Movie Database, IMDB** [7] contains 348,415 movie reviews from the IMDB database done by 54,671 users. The labels describe their sentiment in the [1, 10] range. The authors of the corpus randomly selected 50, 000 movies and crawl all their reviews. During the data cleaning procedure, the creators of the dataset filtered out the users whose reviews did not contain numerical ratings and those with less than two reviews. In this dataset, the known user is the author of the text and the author of the evaluation, as it is the same person (user-author-annotator). (3) **Measuring Hate Speech, MHS** [17] consists of 135,556 annotations regarding 39,565 comments retrieved from YouTube, Twitter, and Reddit. They were annotated by 7,912 United

States-based Amazon Mechanical Turk workers in various hate speech-related tasks: sentiment, disrespect, insult, humiliation, inferior status, and others. Here, we focused on sentiment analysis only. This dimension also proved to be difficult for non-personalized methods. In this dataset, the known user is a text annotator (user-annotator).

4   Personalized Architectures

The task of personalized sentiment analysis is approached from many diverse perspectives. We chose four existing methods for comparison: (1) **Baseline** is a conventional fine-tuning of pretrained RoBERTa, without including any user-specific information. (2) **UserIdentifier** [27] takes into account the identity of the text's author. A data augmentation method involves adding a sequence of tokens that identify the user. The string is generated from the username or sampled uniformly from the tokenizer vocabulary, and then appended to the beginning of a text. UserIdentifier uses the same set of parameters to embed both sample content and user identifiers, which is simpler than relying on user-specific embeddings and has been shown to achieve high performance. (3) **UserId** [21, 28] provides the person's identity by appending a user ID to the beginning of the annotated text as a special token before the training procedure. The vector representation of the text with the user ID is obtained via transformer encoding. In this model, all transformer weights are trained to learn the dependencies between the user and the text. (4) **HuBi-Medium** [4, 15, 16, 19] takes inspiration from the collaborative filtering methods. This model learns a personal latent vector for each user during the training procedure. The vector aims to model personal beliefs about the trained task. Similar to the neural collaborative filtering model [12], the personal latent vector is multiplied element-wise with the vector of the text. The resulting vector is fed to a fully connected classification layer.

5   HumAnn

Here, we introduce another personalized architecture, based on **hum**an **ann**otation (HumAnn), which does not require training the whole model for new users. HumAnn is a method combining text representations obtained from a language model, like RoBERTa, and an aggregated user-level score computed by a retrieval module. Texts previously written or annotated by the user are retrieved from the database. Then the text similarity scores are computed and used to calculate a user score representing their preferences. There are various ways of aggregating multiple labels into a single score. In the experiments, we used a KNN-based aggregation that averages the labels of the K most similar samples, where K=3. Textual features are concatenated with a user score. This personalized representation is then passed to a linear classifier for a person-informed prediction. Figure 1 shows components of the entire system.

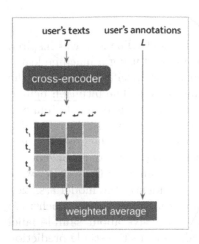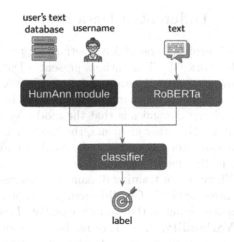

Fig. 1. HumAnn combines text representation from a language model with an aggregated user score. The average user label is weighted by a text similarity computed using a cross-encoder.

In HumAnn, text similarity scores influence the aggregation of previous users' text labels. The labels of samples most similar to the current text impact the final user score. The simplest method of aggregating multiple targets is a weighted arithmetic mean. Similarity score s between a pair of texts plays a role in weighting coefficients. Therefore, if another text is very similar to the sample being evaluated, it has a weight close to 1, and the weight of the dissimilar text is close to 0.

$$s(t_i, T, L) = \frac{1}{N-1} \sum_{n=1}^{N} \mathbb{1}_{n \neq i} \cdot l_n \cdot \text{similarity}(t_i, t_n) \tag{1}$$

where s denotes similarity score, t_i is the text currently predicted, T is a sequence of all user's texts, and L is a sequence of all user's labels. The last two are retrieved from a database of past annotated texts, where N ones belong to the specific user. Similarly, for the KNN-based method, only the K most similar samples are considered during label aggregation.

During the training stage, all the user's texts and labels from the training set, apart from the currently used sample t_i, are utilized to compute an aggregated score. Also, for validation and testing, the method considers only the labels of training examples, preventing data leakage. The model is trained to minimize a standard cross-entropy loss for classification concerning a single, shared parameter set θ.

$$\mathcal{L}_{\text{CE}}(t_i, y_i, T, L; \theta) = -\log Pr(y_i | [t_i; s(t_i, T, L)]) \tag{2}$$

$$\theta = argmin_\theta \mathcal{L}_{\text{CE}}(t_i, y_i, T, L; \theta) \tag{3}$$

where y_i denotes the class of ith text example and θ - model parameters.

6 Differential Data Maps

We propose a novel XAI method for personalized models. The idea was inspired by work [32]. The authors present a Data Maps method using a machine learning model to visualize a dataset. It allows seeing how specific elements of the training set are characterized during the learning process. The intuition behind training dynamics is that the model learns to recognize some elements immediately. For other elements, the model needs more learning epochs, during which it can interchangeably make good or bad decisions relative to the ground truth. Finally, the model cannot learn the ground truth for the last group of elements. Three major training dynamics measures for the ith sample in the dataset were introduced: (1) **Confidence,** $\hat{\mu}_i$ – captures how confidently the model assigned a *true* label to the sample, calculated as a mean probability across epochs; (2) **Variability,** $\hat{\sigma}_i$ – measures how the model was indecisive about sample label during training using standard deviation (low value means the stable prediction of one label, and high value - often change of assigned label); (3) **Correctness,** $\hat{c}_i$ – a fraction of correctly predicted labels for the sample across training epochs E:

$$corr = \frac{\sum_{e=1}^{E}(y_{pred} = y_{true})}{E} \qquad (4)$$

In this work, we extend the idea of Data Maps by proposing visualizing the differences between models in the listed training dynamics measures. Our new method, Differential Data Maps, allows us to interpret differences in the performance of different model architectures and analyse the effect of selected characteristics describing the data on the difference in training dynamics on the same dataset. We define three new metrics based on those presented for Data Maps. Let M1 and M2 be different models trained on the same dataset. Then for ith sample in this dataset, we define new measures: (1) **Confidence change:** $\hat{\mu}_i^C = \hat{\mu}_i^{M2} - \hat{\mu}_i^{M1}$; (2) **Variability change:** $\hat{\sigma}_i^C = \hat{\sigma}_i^{M2} - \hat{\sigma}_i^{M1}$; (3) **Correctness change:** $\hat{c}_i^C = \hat{c}_i^{M2} - \hat{c}_i^{M1}$; where $M1$ is the model whose measures we want to obtain compared to the base model $M2$.

7 Experimental Setup

We evaluated the proposed personalized methods on three sentiment datasets presented in Sect. 3. We used the data split methodology described in the following papers: S140 [23], IMDB [37] and MHS [17]. The datasets contain metadata on the context in which the annotation process occurred. The type of information on users makes it possible to point out some differences between the collections. The MHS dataset provides the most detailed description of the texts' annotators, such as ID, gender, education, income, or severity, but there is no information on the authors of the texts. In contrast, S140 provides information on the texts' authors (namely, the nicknames of the users that tweeted), but there is no information about the annotators. In IMDB, the author and annotator of the text in IMDB are the same person, and we know his or her ID.

We used the RoBERTa-base language model [24] as a baseline and in personalized approaches. For text similarity calculations in HumAnn, we utilized MPNet-based Bi-Encoder, trained on various text-pair datasets [31]. The Cross-Encoder was based on a RoBERTa trained on Semantic Text Similarity Benchmark [5]. Models were fine-tuned using AdamW optimizer with learning rate 1e-5, linear warm-up schedule, batch size 16, and maximum sequence length 512 for 50000 training steps, and the best model was selected according to the validation F-score. For the KNN-based aggregation in HumAnn, K was set to 3. For UserIdentifier, we use 10 tokens drawn from the tokenizer vocabulary as an identifier. This has been shown to enable better differentiation between users than relying on usernames or strings of numbers. Experiments were repeated five times, and the mean F1 score was reported. We conducted statistical tests to measure the significance of differences between each method's performance. Firstly, we checked the assumptions of the t-test for independent samples and conducted it to determine if they were met. Otherwise, the Mann-Whitney U test was used.

8   Results

Table 1. **F1**-macro and **Accuracy** reported for methods. **Bold** values indicate the best performance among all architectures.

	S140		IMDB		MHS	
	F1	Acc	F1	Acc	F1	Acc
Baseline	85.9 ± 0.3	85.1 ± 0.2	43.8 ± 0.8	41.1 ± 1.2	58.1 ± 1.2	48.4 ± 0.7
UserIdentifier	87.4 ± 0.4	86.7 ± 0.4	$\mathbf{47.0 \pm 2.0}$	$\mathbf{44.5 \pm 1.9}$	58.8 ± 0.6	48.5 ± 0.7
UserId	85.2 ± 0.4	86.2 ± 0.2	45.2 ± 1.4	41.8 ± 0.7	59.2 ± 0.6	48.6 ± 0.7
HuBi-Medium	$\mathbf{88.9 \pm 0.2}$	83.8 ± 0.4	43.3 ± 0.6	42.3 ± 2.0	$\mathbf{61.3 \pm 0.5}$	48.3 ± 0.6
HumAnn	87.0 ± 0.2	$\mathbf{88.2 \pm 0.2}$	44.0 ± 0.8	40.5 ± 1.0	58.5 ± 0.8	$\mathbf{51.2 \pm 0.9}$

A comparison of the methods on the three datasets is shown in Table 1. For each dataset, at least one personalized method achieves significantly better results than the baseline. Based on the results, it can be determined that there is no one-size-fits-all architecture. Similarly, depending on the measure, some methods may be better or worse within a particular dataset. In the case of the S140 and MHS datasets, for the F1-macro measure the best method is HuBi-Medium, and for the accuracy measure it is HumAnn. For the IMDB collection, for both quality measures, the best method is UserIdentifier. It is worth noting that the methods belong to two different groups: 1) those that require training the entire model (including finetuning the language model) for new users (UserIdentifier, UserID); 2) those that only require training the representation of the new user (HuBi-Medium, HumAnn). In addition, the datasets represent three different human contexts: 1) user-author (S140), 2) user-annotator (MHS), 3) user-author-annotator (IMDB). From this perspective, it can be assumed that for datasets with a context limited only to the annotator (MHS) or to the author (S140), methods in which the user representation is separated from the language model (HuBi-Medium, HumAnn) are preferable.

It is important to note that the sentiment analysis task is much less subjective than the hate speech detection or emotion recognition tasks [19, 26], for which personalized methods from the literature achieved much better quality gains when annotator context was added relative to the baseline. For some emotions, up to 40 pp of improvement was reported, while for sentiment analysis, the quality gains for F1 and Acc measures are, respectively: 3pp and 3.1pp on S140, 3.2pp and 3.4pp on IMDB, and 3.2pp and 2.8pp on MHS.

The differences in the results of personalized models may not be large for F1 and Acc measures. However, much more interesting conclusions come from analysing differences using Data Cartography. Figure 2 shows the results of the original Data Maps (DM) method for the samples (text, annotation pair), Fig. 3 shows the results of the Differential Data Maps (DDM) method for the samples as well, while Fig. 4 presents the DDM for the users (results for the samples aggregated by user ID) and in both of these Figures the points in each quadrant of the graph are counted. Additionally, in Fig. 4, instead of correctness change, the entropy of user annotation in the set is presented as the colour of a point on the map. Each figure presents results for five architectures in the rows (i.e., baseline and four personalized) and three datasets in the columns. In the case of DM, these are data maps for the pair (model, dataset), and in the case of DDM, these are differential data maps for the pair (baseline-personalized model, dataset). The first interesting observation is that within each set there happen to be data maps for personalized models very similar to the baseline (Fig. 2), for which there are very different ones for differential data maps (Fig. 3 and 4). This means that calculating differences for training dynamic measures gives additional information about differences in the behaviour of models that are not visible at first glance by comparing only the data maps themselves.

Intuitively, it might seem that an increase in the quality of a personalized model relative to the baseline should be associated with a decrease in variability and an increase in confidence for most samples. However, in only three of the six cases among the best models relative to baseline is such a trend observed (S140/HuBi-Medium/F1, MHS/HuBi-Medium/F1, MHS/HumAnn/Acc). In other cases, there is a decrease in confidence and an increase in variability for most samples (S140/HumAnn/Acc, IMDB/UserIdentifier/F1, IMDB/UserIdentifier/Acc). Much more interesting is the observation for DDMs aggregated by user. For example, for the S140 set, for models based on user ID confidence increases

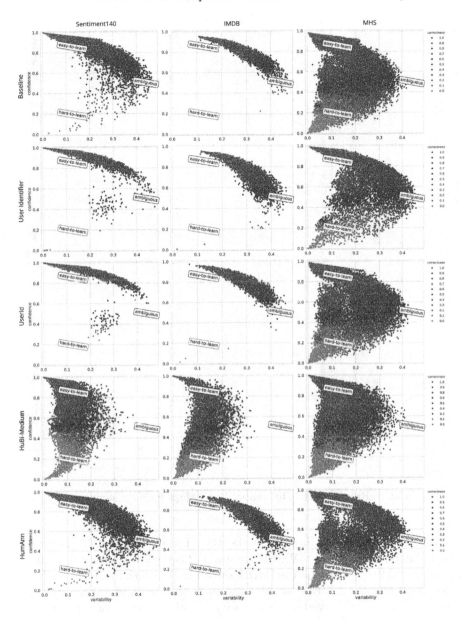

Fig. 2. Results of the original Data Maps method for data samples.

significantly more for authors of similarly rated texts (with low entropy of ratings). For the second group of models, the opposite is true, i.e., these models do better with more diverse people and better model the complex combination of text and human contexts. Models based on user ID better reflect users who behave similarly.

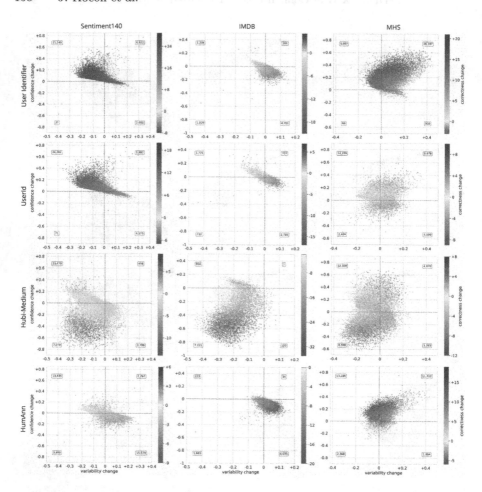

Fig. 3. Results of the Differential Data Maps for data samples.

Even the very similar models on the DM, i.e. UserID and UserIdentifier, behave differently when the DDM is analyzed. They show significant similarity only on the S140 set, while on the IMDB and on the MHS there are no analogous trends regarding the increase or decrease in confidence and variability. Much greater similarity is observed for users with a certain entropy, then, regardless of the dataset, the trends are very similar, i.e. the smallest increase in confidence for users with the lowest entropy on the S140 set, an inverse relationship for the IMDB and no relationship for the MHS.

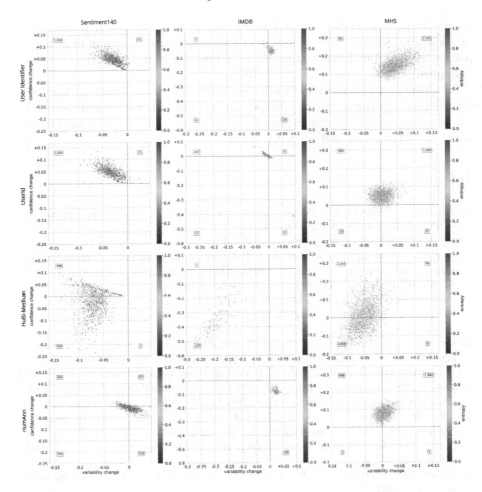

Fig. 4. Results of the Differential Data Maps for data samples aggregated by users.

9   Conclusions and Future Work

In the article, we presented a new Differential Data Maps method that can be used to draw more complex conclusions about either datasets or, more importantly, differences between models. We have presented a few examples of such findings, but further analysis could identify more interesting insights that are not apparent from DM analysis and F1/Acc values.

Our proposed HumAnn model proved to perform on par with other personalized SOTA approach which uses specially trained unique human representation. However, choosing text similarity between past users' written opinions has one major advantage over other methods – frequent retraining is unnecessary. This makes HumAnn easier to deploy in real-world applications. The only limitation of the model is the need to have a certain number of texts from a single person to predict the label in a subjective task correctly. In future, we plan to further train the cross-encoder part of the model to provide similarity scores of even higher quality. Some optimization techniques to reduce compute overhead should also be applied.

The next step in further work will also be to analyse such datasets on which the differences between baseline and personalized models are even greater. This will allow us to further understand for which types of samples the different models perform better and for which worse.

References

1. Adadi, A., Berrada, M.: Peeking inside the black-box: a survey on explainable artificial intelligence (XAI). IEEE Access **6**, 52138–52160 (2018)
2. AI, H.: High-level expert group on artificial intelligence (2019)
3. Baran, J., Kocoń, J.: Linguistic knowledge application to neuro-symbolic transformers in sentiment analysis. In: 2022 IEEE International Conference on Data Mining Workshops (ICDMW), pp. 395–402. IEEE (2022)
4. Bielaniewicz, J., et al.: Deep-sheep: sense of humor extraction from embeddings in the personalized context. In: 2022 IEEE International Conference on Data Mining Workshops (ICDMW), pp. 967–974. IEEE (2022)
5. Cer, D., Diab, M., Agirre, E., Lopez-Gazpio, I., Specia, L.: SemEval-2017 task 1: semantic textual similarity multilingual and crosslingual focused evaluation. In: Proceedings of the 11th Workshop on Semantic Evaluation (SemEval 2017) (2017)
6. Das, A., Rad, P.: Opportunities and challenges in explainable artificial intelligence (XAI): a survey. arXiv preprint arXiv:2006.11371 (2020)
7. Diao, Q., Qiu, M., Wu, C.Y., Smola, A.J., Jiang, J., Wang, C.: Jointly modeling aspects, ratings and sentiments for movie recommendation (JMARS). In: Proceedings of the 20th ACM SIGKDD Conference on Knowledge Discovery and Data Mining (2014)
8. Doshi-Velez, F., Kim, B.: Towards a rigorous science of interpretable machine learning. Stat **1050**, 2 (2017)
9. Flek, L.: Returning the N to NLP: towards contextually personalized classification models. In: Proceedings of the 58th Annual Meeting of ACL (2020)
10. Go, A.: Sentiment classification using distant supervision (2009)
11. Gong, L., Haines, B., Wang, H.: Clustered model adaption for personalized sentiment analysis. In: Proceedings of the 26th Conference on World Wide Web (2017)
12. He, X., Liao, L., Zhang, H., Nie, L., Hu, X., Chua, T.S.: Neural collaborative filtering. In: Proceedings of the 26th Conference on World Wide Web (2017)
13. Hovy, D.: Demographic factors improve classification performance. In: Proceedings of the 53rd Annual Meeting of the ACL&IJCNLP (2015)
14. Kanclerz, K., et al.: Controversy and conformity: from generalized to personalized aggressiveness detection. In: Proceedings of the 59th Annual Meeting of the ACL&IJCNLP (2021)
15. Kanclerz, K., et al.: What if ground truth is subjective? Personalized deep neural hate speech detection. In: Proceedings of the 1st Workshop on Perspectivist Approaches to NLP@ LREC2022, pp. 37–45 (2022)
16. Kazienko, P., et al.: Human-centred neural reasoning for subjective content processing: hate speech, emotions, and humor. Inf. Fusion (2023)

17. Kennedy, C.J., Bacon, G., Sahn, A., von Vacano, C.: Constructing interval variables via faceted Rasch measurement and multitask deep learning: a hate speech application. arXiv preprint arXiv:2009.10277 (2020)
18. Kocoń, J., et al.: Neuro-symbolic models for sentiment analysis. In: Groen, D., de Mulatier, C., Paszynski, M., Krzhizhanovskaya, V.V., Dongarra, J.J., Sloot, P.M.A. (eds.) ICCS 2022. LNCS, vol. 13351, pp. 667–681. Springer, Cham (2022). https://doi.org/10.1007/978-3-031-08754-7_69
19. Kocoń, J., et al.: Learning personal human biases and representations for subjective tasks in natural language processing. In: 2021 IEEE International Conference on Data Mining (ICDM), pp. 1168–1173. IEEE (2021). https://doi.org/10.1109/ICDM51629.2021.00140
20. Kocoń, J., Maziarz, M.: Mapping wordnet onto human brain connectome in emotion processing and semantic similarity recognition. Inf. Process. Manag. **58**(3), 102530 (2021)
21. Kocoń, J., Figas, A., Gruza, M., Puchalska, D., Kajdanowicz, T., Kazienko, P.: Offensive, aggressive, and hate speech analysis: from data-centric to human-centered approach. Inf. Process. Manag. **58**(5) (2021)
22. Korczyński, W., Kocoń, J.: Compression methods for transformers in multidomain sentiment analysis. In: 2022 IEEE International Conference on Data Mining Workshops (ICDMW), pp. 419–426. IEEE (2022)
23. Li, T., Sanjabi, M., Smith, V.: Fair resource allocation in federated learning. CoRR abs/1905.10497 (2019)
24. Liu, Y., et al.: RoBERTa: a robustly optimized BERT pretraining approach. CoRR abs/1907.11692 (2019)
25. Lui, A., Lamb, G.W.: Artificial intelligence and augmented intelligence collaboration: regaining trust and confidence in the financial sector. Inf. Commun. Technol. Law **27**(3), 267–283 (2018)
26. Miłkowski, P., Saganowski, S., Gruza, M., Kazienko, P., Piasecki, M., Kocoń, J.: Multitask personalized recognition of emotions evoked by textual content. In: Pervasive Computing and Communications Workshops (2022)
27. Mireshghallah, F., Shrivastava, V., Shokouhi, M., Berg-Kirkpatrick, T., Sim, R., Dimitriadis, D.: UserIdentifier: implicit user representations for simple and effective personalized sentiment analysis. CoRR abs/2110.00135 (2021)
28. Ngo, A., Candri, A., Ferdinan, T., Kocoń, J., Korczynski, W.: StudEmo: a non-aggregated review dataset for personalized emotion recognition. In: Proceedings of the 1st Workshop on Perspectivist Approaches to NLP@ LREC2022, pp. 46–55 (2022)
29. Schneider, J., Vlachos, M.: Mass personalization of deep learning. CoRR abs/1909.02803 (2019)
30. Song, K., Feng, S., Gao, W., Wang, D., Yu, G., Wong, K.F.: Personalized sentiment classification based on latent individuality of microblog users (2015)
31. Song, K., Tan, X., Qin, T., Lu, J., Liu, T.: MPNet: masked and permuted pretraining for language understanding. CoRR abs/2004.09297 (2020)
32. Swayamdipta, S., et al.: Dataset cartography: mapping and diagnosing datasets with training dynamics. In: Proceedings of the EMNLP2020 (2020)
33. Szołomicka, J., Kocon, J.: MultiAspectEmo: multilingual and language-agnostic aspect-based sentiment analysis. In: 2022 IEEE International Conference on Data Mining Workshops (ICDMW), pp. 443–450. IEEE (2022)
34. Tonekaboni, S., Joshi, S., McCradden, M.D., Goldenberg, A.: What clinicians want: contextualizing explainable machine learning for clinical end use. In: Machine Learning for Healthcare Conference, pp. 359–380. PMLR (2019)

35. Volkova, S., Wilson, T., Yarowsky, D.: Exploring demographic language variations to improve multilingual sentiment analysis in social media. In: Proceedings of the 2013 Conference on Empirical Methods in Natural Language Processing (2013)
36. Zamani, M., Schwartz, H.A., Lynn, V.E., Giorgi, S., Balasubramanian, N.: Residualized factor adaptation for community social media prediction tasks (2018)
37. Zhong, W., Tang, D., Wang, J., Yin, J., Duan, N.: UserAdapter: few-shot user learning in sentiment analysis. In: Findings of the Association for Computational Linguistics: ACL-IJCNLP 2021, pp. 1484–1488 (2021)

Alternative Platforms and Privacy Paradox: A System Dynamics Analysis

Ektor Arzoglou$^{(\boxtimes)}$ and Yki Kortesniemi

Department of Information and Communications Engineering, Aalto University,
Helsinki, Finland
{ektor.arzoglou,yki.kortesniemi}@aalto.fi

Abstract. The term 'privacy paradox' refers to the apparent inconsistency between privacy concerns and actual behaviour that also often leads to the dominance of privacy-careless over privacy-respecting platforms. One of the most important explanations for this phenomenon is based on the concept of social norm, which refers to the influence that an individual's social environment can have on their decisions to accept or reject a specific platform. However, the interdependencies between social norm dynamics and platform adoption have received little attention so far. To overcome this limitation, this article presents a system dynamics simulation model that considers the concept of social norm, shaped by users with diverse privacy concerns, during the adoption process of two alternative social media platforms and identifies the types of situations in which the privacy paradox emerges. The results show a bidirectional minority rule, where (1) the least concerned minority can hinder the more concerned majority from discarding a privacy-careless platform but also (2) the most concerned minority can induce the less concerned majority to adopt a privacy-respecting platform. Both (1) and, to a lesser extent, (2) are types of situations that reflect the privacy paradox.

Keywords: Digital platforms · Privacy · Privacy paradox · Social media · System dynamics

1 Introduction

Digital platforms act as *mediators* of content flows between users. In addition, they typically tailor this content to individual user preferences (i.e. personalisation) based on the processing of accumulated user data. At the same time, the repeated involvement of Big Tech platform owners (e.g. Alphabet and Meta) in cases of user data exploitation has raised privacy concerns, thereby motivating the launch of privacy-respecting platforms, such as the search engine DuckDuckGo and the instant messenger Signal, which were introduced as alternatives to Google Search and WhatsApp, respectively.

Over the last decade, researchers have started to investigate the role of privacy concerns in the adoption of online services, such as digital platforms, by

The original version of this chapter was revised: typesetting mistakes in the Appendix have been corrected. The correction to this chapter is available at
https://doi.org/10.1007/978-3-031-35995-8_49

J. Mikyška et al. (Eds.): ICCS 2023, LNCS 14073, pp. 163–176, 2023.
https://https://doi.org/10.1007/978-3-031-35995-8_12

usually assuming that privacy concerns will likely result in rejection of privacy-careless platforms. However, these studies neglect that, despite expressing high privacy concerns, people may still choose a privacy-careless over a privacy-respecting platform. This inconsistency between privacy concerns and actual behaviour is often referred to as the *privacy paradox* [10].

One of the most important explanations for this phenomenon is based on the concept of *social norm*, which refers to the influence that an individual's social environment can have on their privacy decisions. As a result, individuals may *conform* to the social norm by deciding to accept the use of a platform that they would otherwise reject in order to achieve approval from and harmony with peers and family regardless of privacy preferences and concerns [3]. Social norm is a *dynamic* concept that influences individual behaviour while also being shaped by mass behaviour over time [8]. In addition, it is not necessarily aligned with privacy protection, because it might be shaped by people with less need for privacy and therefore low privacy concerns. Diffusion (i.e. the spread of an innovation through a population) theory [12] and social psychology [1] suggest that more careful attention to the social context should be paid in order to understand the determinants of innovation adoption. However, the concept of social norm has often been modelled as a *static* parameter in innovation diffusion models. This limitation motivates the development of a *system dynamics simulation model* [15] that presents an *endogenous* perspective (i.e. arising from within the system) on the social norm concept in this article.

The research question guiding this article is: *In what types of situations can a social norm outweigh privacy concerns, when choosing from two alternative social media platforms, and how does this help understand the privacy paradox?* The results of the developed system dynamics simulation model show a bidirectional *minority rule*, where (1) the least concerned minority can hinder the more concerned majority from discarding a privacy-careless platform but also (2) the most concerned minority can induce the less concerned majority to adopt a privacy-respecting platform. Both (1) and, to a lesser extent, (2) are types of situations that reflect the privacy paradox. Finally, the contributions of this article also include demonstrating the potential of system dynamics as a tool for analysing privacy behaviour.

The rest of the article is organised as follows. Section 2 reviews literature on social aspects of privacy and privacy paradox. Section 3 describes the applicability of the methodology used, namely system dynamics modelling, to the privacy paradox. Section 4 presents the model of the two alternative social media platforms. The simulation results are discussed in Sect. 5. Finally, Sect. 6 concludes the article.

2   Theoretical Background

The concept of privacy has three main aspects: (1) *territorial privacy*, protecting the close physical area surrounding a person, (2) *privacy of the person*, protecting a person against undue interference, and (3) *informational privacy*, controlling whether and how personal data can be gathered, stored, processed, or selectively disseminated [10,13]. This article focuses exclusively on the third aspect.

2.1 Privacy as a Social Issue

Nissenbaum conceptualises privacy as *contextual integrity*, which is defined as "the appropriate information flows in a given context", and assumes two explicit information norms: (1) *appropriateness*, governing what information is revealed in a given context, and (2) *flow*, governing the recipients of that information in a given context. Contextual integrity is maintained when both norms are upheld and is breached, thus exacerbating privacy concerns, when either of the norms is violated. Unlike previous theories, which often view privacy as a generic and static concept cutting across different contexts, contextual integrity postulates that privacy concerns vary depending on the context [11].

An important implication, which may also explain the privacy paradox, of defining privacy as contextual integrity is that it reveals the *key difference between "giving up" privacy and giving up information*. That is, users do not cede their privacy by sharing their data if they perceive the information flow as appropriate for that specific context. Hence, users may express high privacy concerns before using a service and also retain these concerns while using the service, which they expect to respect the norms governing the recipients of and rights over a certain piece of information.

This implication may apply to single-purpose contexts (e.g. e-commerce platforms, such as Alibaba), in which users enact pre-defined roles and also information sharing is governed by explicit norms. However, it does not apply to multi-purpose contexts (e.g. social media platforms, such as Instagram), in which roles are ever changing and likely unknown a priori (i.e. relationships among users are constantly evolving) and also information norms are implicit (i.e. they encourage behaviour that is consistent with the most common behaviour). In addition, studies show that privacy concerns stem from uncertainty about both data collection by the platform (i.e. violation of appropriateness) and exploitation by third parties (i.e. violation of flow). Finally, feelings of exhaustion and cynicism towards privacy, generated from inability to meet privacy goals, may ultimately lead to a state of resignation about privacy (i.e. "giving up" privacy) and potentially to the inconsistency between privacy concerns and actual behaviour that indicates the privacy paradox [7].

2.2 Social Theory Based Explanations of the Privacy Paradox

Most individuals are not autonomous in their decisions to accept or reject the use of a specific platform, since these decisions are often driven by the need to achieve conformity with the admired peer groups [8]. As such, individuals may conform to the influence of their social environment by neglecting privacy concerns in order to reap the benefits of belonging to a community rather than facing the costs (e.g. social stigmas) of being excluded from the community (i.e. positive net platform value).

In addition, social interactions are categorised into *Gemeinschaften* (communities), determined by internalised emotional ties and implicit rules (i.e. norms), and *Gesellschaften* (societies), determined by rational calculations and explicit rules (i.e. laws) [16]. Certain types of platforms, such as media sharing (e.g. YouTube) and knowledge (e.g. Reddit) platforms, have both a Gemeinschaft side, where people share private information because this is an implicit rule of belonging to a community, but also a Gesellschaft side, where people know explicitly, albeit on an abstract level, and become concerned about the privacy risks based on the platforms' formal rules and policies. Hence, the dominant side is often the Gemeinschaft side, since the concrete and immediate benefits of belonging to a community outweigh the abstract privacy risks of data sharing (i.e. positive net platform value).

3   A System Dynamics Model of the Privacy Paradox

System dynamics is a methodology that uses feedback loops, accumulations, and time delays to understand the *behaviour of complex systems over time* [15]. One of the primary strengths of system dynamics is that it allows for the inclusion of both social and technical elements into the same model and therefore the study of complex sociotechnical systems, such as social media.

Researchers have only recently started to study the privacy paradox by focusing on the interdependencies between social norm dynamics and platform adoption [2,3]. However, these studies focus on the adoption process of a *single* privacy-careless platform, thereby neglecting whether and how a privacy-respecting alternative could at least partially resolve the privacy paradox. To overcome this limitation, this article presents a system dynamics simulation model that considers the concept of social norm, shaped by users with diverse privacy concerns, during the adoption process of *two* alternative social media platforms and identifies the types of situations in which the privacy paradox emerges.

In system dynamics, the model development begins by (1) defining *reference modes*, which are graphs illustrating the problem (e.g. the privacy paradox) as a pattern of behaviour over time, and (2) formulating a *dynamic hypothesis*, which aims to explain the problematic behaviour shown in the reference modes in terms of the underlying *feedback and stock-flow structure* (see Sect. 4) of the system [15].

3.1   Problem Articulation and Dynamic Hypothesis

In order to illustrate the privacy paradox in the context of two alternative social media platforms, this article uses two reference modes relevant to platform adoption: an initial period of growth in adoption of the privacy-careless platform is followed by a decline, during which adoption of the privacy-respecting platform either (1) increases without ultimately dominating (e.g. the privacy-careless platform can maintain a larger fraction of highly concerned users, who are hindered by less concerned users from discarding) or (2) increases and ultimately dominates (e.g. the privacy-respecting platform can obtain a larger fraction of less concerned users, who are induced by highly concerned users to adopt). On one hand, reference mode (1) illustrates a situation in which privacy concerns are inconsistent with adoption of the privacy-careless platform, thus reflecting the privacy paradox. On the other hand, although reference mode (2) illustrates a situation in which privacy concerns are consistent with adoption of the privacy-respecting platform, the privacy paradox is reflected again, but this time to a lesser extent.

The purpose of the model is to explain the types of situations in which a social norm can outweigh privacy concerns using these two modes of dynamic behaviour. As these dynamic behaviours can occur in different settings, the model was built as a generic representation of social media without focusing on any specific platform. Finally, the time horizon of the model is in the order of multiple years, so that the entire platform adoption phase is included in the simulation results.

The dynamic hypothesis guiding the model development is that an extended feedback structure of the Bass model of *innovation diffusion*, which describes the adoption of new products or services (over time) [6], can produce the two modes of dynamic behaviour. As such, platform adoption can be influenced by different factors (e.g. privacy concerns) that have an effect on the feedback loops of the model. The dynamics of the two alternative social media platforms are generated *endogenously* (i.e. from within the system). Conversely, the dynamics of privacy concerns, which can be described as a merely negative concept not bound to any specific context [9,10], are generated *exogenously* (i.e. from without the system).

4   Model Development

In system dynamics, *stock-flow diagrams* consist of variables, shown as named nodes, related by causal links, shown as arrows. *Stocks* are shown as rectangles and represent accumulations of either matter or information. *Flows* are shown as pipes and valves and regulate the rate of change of the stocks. Intermediate variables between stocks and flows indicate *auxiliaries*, which essentially clarify the sequence of events that cause the flows to change the stocks. Finally, a circular sequence of variables related by causal links forms a *feedback loop*, which can be either *reinforcing* (R) (i.e. amplifying change) or *balancing* (B) (i.e. counteracting and opposing change).

4.1   Model Structure

The two alternative social media platforms are modelled by extending the Bass model of innovation diffusion, which considers adoption through exogenous efforts, such as advertising, and adoption through word-of-mouth [6] (Fig. 1). Here, potential users can adopt either one of the two platforms. Both platforms are represented by the same model structure, but each platform is represented by a different subscript (see Appendix). Finally, the model utilises several equations from Ruutu et al. [14].

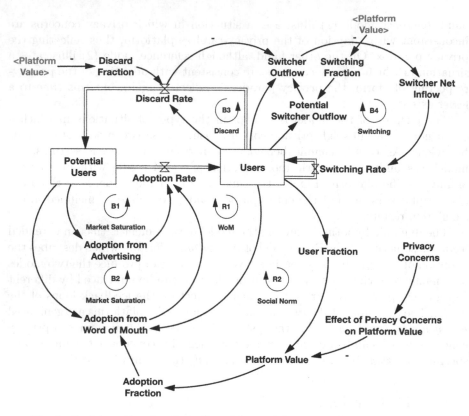

Fig. 1. Social media adoption affected by social norm and privacy concerns

When a platform is launched, the initial number of users is zero, so the only source of adoption are external influences, such as advertising (B1: "Market Saturation"). When the first users enter the platform, the adoption rate increases through word-of-mouth (R1: "WoM"). As the stock of users grows, platform value increases, and the norm related to platform adoption becomes stronger and consequently harder to deviate from. As a result, more potential users conform and adopt the platform (R2: "Social Norm"). The advertising and word-of-mouth effects are largest at the start of the platform diffusion process and steadily diminish as the stock of potential users is depleted (B1, B2: "Market Saturation"). Finally, current users may decide to discard the platform (B3: "Discard") or switch to an alternative (B4: "Switching"), depending on the decrease, caused by privacy concerns, in platform value.

The behaviour of potential and current users is modelled using rules of bounded rationality, which depend on the information available to users at a given point in time. In other words, potential and current users are not assumed to have perfect foresight of how adoption of the two platforms will progress, and they make their decisions regarding platform adoption, discard, and switching based on their perception of platform value to them.

4.2 Model Parameters

The total population (N) considered in the model is 1000 users, divided as per Westin's first privacy segmentation into 550 Pragmatists (mid to high privacy concerns P^*), 250 Fundamentalists (high privacy concerns F^*), and 200 Unconcerned (no or low privacy concerns U^*) [17]. In this regard, F^* is a multiplier of P^*, and U^* can range from zero $(U^* = 1)$ to matching P^* $(U^* = 0)$.

Furthermore, $PC(0)$ determines the initial value of privacy concerns, and $T^0 PC$ determines the time at which privacy concerns start. The effect of privacy concerns on platform value erodes (using exponential smoothing) over time τPC. This erosion essentially indicates the time for users to develop either (1) feelings of exhaustion, resignation, and even cynicism towards privacy (i.e. privacy fatigue) [7] or (2) feelings of privacy safety [5]. As such, privacy concerns are assumed to be boundedly rational.

In addition to the parameters determining privacy concerns, the model includes eight further parameters that have an effect on platform adoption. Initially, an external advertising effort (a), starting at time T^0 and ending at time T, brings the first users in the platform. Thereafter, potential users come into contact (c) with current users, and platform adoption continues only with word-of-mouth. Conversely, it takes some time (τ) for users to process the decrease, caused by privacy concerns, in platform value and react by discarding the platform or switching to an alternative. Moreover, V^* determines the value that users receive from substitutes, and uf^* determines the fraction of users needed in order to obtain the same level of benefits. Therefore, high values of these two parameters make platform adoption harder. Finally, exponent γ determines the strength of social norm (i.e. the dependency of platform value on the number of current users). Hence, in the beginning, when there is a lack of users, high values of γ make platform adoption harder. The model equations and parameter values are listed in the Appendix.

4.3 Model Testing and Validation

The model was built using Vensim DSS for Mac Version 9.0.0 (Double Precision), and the simulation experiments were performed using time step 0.0625 and Euler numerical integration. The validation tests that have been successfully passed to gradually build confidence in the soundness and usefulness of the model, with respect to the purpose presented in Sect. 3, are grouped into *direct structure tests*, which do not involve simulation, and *structure-oriented behaviour tests*, which involve simulation [4]. The results are presented in Table 1.

Table 1. Validation tests applied to the model

Test	Result
Direct structure tests	
Structure confirmation	The feedback structures of the model have been formulated and extended based on the Bass model of innovation diffusion [6]
Parameter confirmation	All parameters in the model (1) have clear and meaningful counterparts in the real world and (2) were set to limited ranges with minimum and maximum values. Since the model was built as a generic representation of social media, the exact parameter values are not significant, and the parameters have not been estimated based on any specific platform
Direct extreme condition	The model includes formulations to ensure that users cannot be added or removed spontaneously (i.e. mass balance) and that stock variables stay non-negative
Dimensional consistency	The units of all variables and parameters have been specified, and the model passes Vensim's dimensional consistency test
Structure-oriented behaviour tests	
Indirect extreme condition	The model behaves as expected when individual variables are subjected to extreme conditions (e.g. no users, no platform value)
Behaviour sensitivity	The model behaves plausibly when individual parameters are set to the limits of their meaningful ranges of variation as well as when several parameters are varied simultaneously in a Monte Carlo experiment

5   Simulation Results

Using the model, it is possible to simulate the two modes of dynamic behaviour presented in Sect. 3.1 and therefore identify the types of situations in which the privacy paradox emerges.

5.1   Simulation Experiment 1

For the first simulation experiment (Fig. 2), platform 1 is launched at $Time = 0$ and platform 2 is launched two years later ($Time = 2$). In platform 1, Fundamentalists are one point two times as concerned as Pragmatists ($P^* = 0.1$, $F^* = 1.2$), on the assumption that privacy preferences of Fundamentalists are somewhat stronger. In addition, Unconcerned are less than one third as concerned as Pragmatists ($U^* = 0.7$), assuming that Unconcerned have significantly less need for privacy than Pragmatists and Fundamentalists. Finally, privacy concerns of Fundamentalists do not erode, on the assumption that this user group is less likely to feel privacy fatigued over time. On the other hand, in platform 2, all three user groups are two fifths as concerned as in platform 1 (P^*, F^*, $U^* = 0.4$), assuming that platform 2 is more than twice as private as platform 1 but still not perfectly private. In addition, erosion of privacy concerns applies to all three user groups, on the assumption that users realise the privacy benefits and feel safer about their data over time.

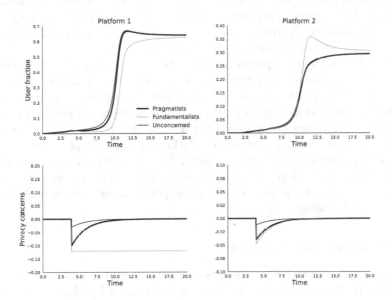

Fig. 2. The social norm created by Pragmatists and Unconcerned results in adoption of platform 1 also for a larger fraction of Fundamentalists, although privacy concerns of the last user group remain constant.

For both platforms, adoption initially takes place through advertising (B1) and word-of-mouth (B2, R1). Advertising efforts (B1) last four years (ending at $Time = 4$ for platform 1 and $Time = 6$ for platform 2), and platform adoption continues only with word-of-mouth (B2, R1) thereafter. Moreover, privacy concerns start at $Time = 4$ in both platforms for all three user groups. At this point, in platform 1, social norm (R2) outweighs privacy concerns of Unconcerned but is outweighed by privacy concerns of Pragmatists and Fundamentalists, thus preserving platform value and adoption only for the first user group. By contrast, the number of Pragmatists and Fundamentalists is starting to decline. On the other hand, in platform 2, social norm (R2) outweighs privacy concerns of all three user groups, and therefore platform adoption increases. As a result, from $Time = 5.5$ until $Time = 10.5$, platform 1 maintains a larger fraction of Pragmatists and Unconcerned, despite the decline of the first user group, while platform 2 obtains a larger fraction of Fundamentalists, who have switched from platform 1. However, as the number of Unconcerned grows and also privacy concerns of Pragmatists erode in platform 1, social norm (R2) outweighs privacy concerns of the second user group, and therefore the number of Pragmatists is once more starting to grow ($Time = 6$). At the same time, adoption of platform 2 increases, but the installed user base remains smaller compared to platform 1.

In other words, the social norm (R2) driving adoption of platform 1 outweighs the social norm (R2) driving adoption of platform 2. Finally, as the number of both Pragmatists and Unconcerned grows in platform 1, social norm (R2) becomes strong enough to eventually outweigh privacy concerns of Fundamentalists too. As a result, the number of Fundamentalists starts to grow again ($Time = 9$), although privacy concerns of this user group do not erode, and platform 1 ultimately dominates platform 2.

The first simulation experiment illustrates a *minority rule that prevents change*, since the smallest user group of Unconcerned initially hinders the largest user group of Pragmatists, before both eventually hinder the user group of Fundamentalists, from switching to platform 2. In addition, although (1) privacy concerns of Pragmatists are not eliminated in platform 1, (2) privacy concerns of Fundamentalists remain constant in platform 1, and (3) platform 2 is more than twice as private as platform 1, a larger fraction of Pragmatists and Fundamentalists eventually adopts platform 1, hence resulting in the privacy paradox. Finally, the results of the first simulation experiment are also consistent with the results of Arzoglou et al. [2].

5.2   Simulation Experiment 2

The setup of the second simulation experiment (Fig. 3) is similar to the first, with the only difference being that privacy concerns of Fundamentalists erode faster in platform 2 ($\tau PC_f = 1$). The assumption is that Fundamentalists are more literate about privacy and therefore able to realise the privacy benefits sooner than Pragmatists and Unconcerned.

As before, privacy concerns start at $Time = 4$ and while the number of Pragmatists and Fundamentalists declines, the number of Unconcerned continues to grow in platform 1. In addition, by $Time = 10.5$, a larger fraction of Fundamentalists has already switched to platform 2, whereas a larger fraction of Pragmatists and Unconcerned remains again in platform 1. However, as the number of Fundamentalists grows in platform 2, the social norm (R2) driving adoption of platform 2 eventually outweighs the social norm (R2) driving adoption of platform 1. As a result, all three user groups switch to platform 2, which ultimately dominates platform 1.

The second simulation experiment illustrates a *minority rule that drives change*, since a smaller user group of Fundamentalists induces a larger user group of Pragmatists and Unconcerned to switch to platform 2. In addition, although privacy concerns of Fundamentalists remain constant, a smaller fraction of this user group eventually adopts platform 1, thus exhibiting again the privacy paradox, but this time to a lesser extent.

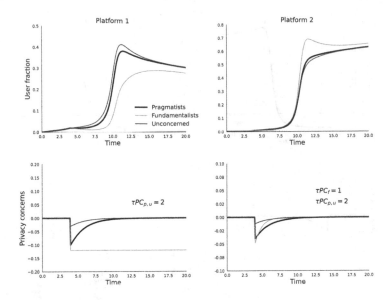

Fig. 3. The social norm created by Fundamentalists results in adoption of platform 2 also for a larger fraction of Pragmatists and Unconcerned.

5.3   Simulation Experiment 3

For the third simulation experiment (Fig. 4), platform 1 is launched at $Time = 0$ and platform 2 is launched at time $Time = 5$, which is one year after privacy concerns start for platform 1 ($Time = 4$). The assumption is that the privacy issues of platform 1 motivate the launch of platform 2. In platform 1, Fundamentalists are two times as concerned as Pragmatists ($P^* = 0.1$, $F^* = 2$), on the assumption that privacy preferences of Fundamentalists are significantly stronger. In addition, Unconcerned are one half as concerned as Pragmatists ($U^* = 0.5$), assuming that Unconcerned have somewhat less need for privacy than Pragmatists and significantly less need for privacy than Fundamentalists. Finally, erosion of privacy concerns applies only to Pragmatists and Unconcerned, once more on the assumption that Fundamentalists are less likely to feel privacy fatigued over time. On the other hand, all three user groups have no privacy concerns in platform 2, which is assumed to be perfectly private.

Again, privacy concerns start at $Time = 4$, and the number of Pragmatists and Fundamentalists declines in platform 1. At the same time, the number of Unconcerned continues to grow, although privacy concerns of this user group are higher compared to the previous two simulation experiments. On the other hand, adoption of platform 2 starts at $Time = 5$ and increases at the highest possible rate. Similar to the first simulation experiment, the number of Unconcerned is sufficient to initially hinder Pragmatists from discarding platform 1. However, contrary to the first simulation experiment, adoption of platform 1 from Pragmatists and Unconcerned becomes easier only when privacy concerns of the two user groups are nearly eliminated. In other words, the social norm (R2) driving adoption of platform 1 outweighs the social norm (R2) driving adoption of plat-

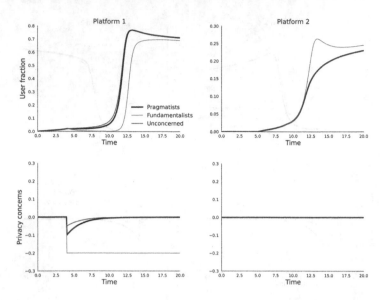

Fig. 4. The social norm created by Pragmatists and Unconcerned results in adoption of platform 1 also for a larger fraction of Fundamentalists, although privacy concerns of the last user group remain constant.

form 2, but this dominance is counterbalanced by the effect of privacy concerns on the value of platform 1. For this reason, adoption of platform 1 increases only when the effect of privacy concerns on the value of platform 1 becomes weaker. Finally, as the number of both Pragmatists and Unconcerned grows in platform 1, social norm (R2) becomes strong enough to eventually outweigh privacy concerns of Fundamentalists too. As a result, the number of Fundamentalists starts to grow again $(Time = 11)$, although privacy concerns of this user group do not erode, and platform 1 ultimately dominates platform 2.

For the third simulation experiment, the minority rule and privacy paradox for all three user groups are similar to the first simulation experiment. In addition, the results of the third simulation experiment are also consistent with the results of Arzoglou et al. [2].

6   Concluding Discussion

This article presents a system dynamics simulation model that considers the concept of social norm, shaped by users with diverse privacy concerns, during the adoption process of two alternative social media platforms and identifies the types of situations in which the privacy paradox emerges. The model illustrates a bidirectional minority rule, where (1) the least concerned minority can hinder the more concerned majority from discarding a privacy-careless platform but also (2) the most concerned minority can induce the less concerned majority to adopt a privacy-respecting platform. Both (1) and, to a lesser extent, (2) are types of situations that reflect the privacy paradox.

Since the model was built as a generic representation of social media, a limitation of the simulation results is that they do not apply exactly to every platform and context. As such, a fruitful topic for future research would be to empirically test and validate the simulation results and thus support the usefulness and applicability of the model to specific platforms across different contexts. Finally, the model could be developed further to present an endogenous perspective on the concept of privacy concerns, determined by e.g. the users' data sharing behaviour and the platform's exploitation of accumulated user data.

Appendix: Model Equations and Parameter Values

The model equations and parameter values are shown in Table 2. In the equations, subscript w refers to the user group (p: Pragmatists, f: Fundamentalists,

Table 2. Model equations and parameter values

Name	Equation/parameter value	Unit	#
Potential users	$\dot{P}_w = \sum_i (DR_{w,i} - AR_{w,i})$ $P_w(0) = 1000$	User	1
Users	$\dot{U}_{w,i} = AR_{w,i} + SR_{w,i} - DR_{w,i}$ $U_{w,i}(0) - 0$	User	2
Adoption rate	$AR_{w,i} = P_w \cdot (a + c \cdot af_{w,i} \cdot U_{w,i}/N_w)$	User/Year	3
Discard rate	$DR_{w,i} = U_{w,i} \cdot df_{w,i}/\tau$	User/Year	4
Switching rate (from j to i)	$SR_{w,i} = \sum_j (U_{w,j} \cdot sf_{w,j,i} - U_{w,i} \cdot sf_{w,i,j})/\tau$	User/Year	5
Adoption fraction	$af_{w,i} = V_{w,i}/(\sum_i V_{w,i} + V^*)$	-	6
Discard fraction	$df_{w,i} = V^*/(V^* + \sum_i V_{w,i})$	-	7
Switching fraction (from i to j)	$sf_{w,i,j} = V_{w,j}/(\sum_i V_{w,i} + V^*)$ (0 if $i = j$)	-	8
Total population	N_w 1000 (divided into 550 Pragmatists, 250 Fundamentalists, and 200 Unconcerned)	User	
Advertising start time	T^0 0, 2 (platform 2 is launched later)	Year	
Advertising end time	T 4, 6	Year	
Advertising effectiveness	a 0.01	1/Year	
Contact rate	c 10	1/Year	
User reaction time	τ 1.5	Year	
User fraction	$uf_{w,i} = U_{w,i}/N_w$	-	9
Reference user fraction	uf^* 0.5	-	
Platform value	$V_{w,i} = (\frac{\sum_w uf_{w,i}}{uf^*})^\gamma + E_{w,i}$	-	10
Reference value	V^* 2.2	-	
Effect of users on platform value	γ 0.7	-	
Privacy concerns (Pragmatists)	$PC_{p,i} = PC(0)_i - \text{Step} (P_i^*, T^0 PC_i)$ Step input function	-	11a
Privacy concerns (Fundamentalists)	$PC_{f,i} = PC(0)_i - \text{Step} (P_i^* \cdot F_i^*, T^0 PC_i)$ Step input function	-	11b
Privacy concerns (Unconcerned)	$PC_{u,i} = PC(0)_i - \text{Step} (P_i^* - (P_i^* \cdot U_i^*), T^0 PC_i)$ Step input function	-	11c
Reference privacy concerns	$PC_{w,i}^* = \text{Smoothi} (PC_{w,i}, \tau PC, PC(0)_i)$ Exponential smoothing function $= PC(0)_i$ (no erosion of privacy concerns)	-	12 12'
Privacy concerns initial value	$PC(0)_i$ 0	-	
Privacy concerns start time	$T^0 PC_i$ 4	Year	
Privacy concerns erosion time	τPC 2	Year	
Reference pragmatism	P^* 0.1, 0.4 (platform 2 is more private)	-	
Reference fundamentalism	F^* 1.2, 0.4	-	
Reference unconcern	U^* 0.7, 0.4	-	
Effect of privacy concerns on platform value	$E_{w,i} = PC_{w,i} - PC_{w,i}^*$	-	13

u: Unconcerned), and subscripts i and j refer to the two alternative social media platforms. For clarity, the equations are shown without the formulations that ensure the validity of stock variables (see Sect. 4.3). For details of the formulations and to ensure the replicability of the simulation results, the simulation model Vensim file is openly available upon request.

References

1. Ajzen, I., Fishbein, M.: The prediction of behavior from attitudinal and normative variables. J. Exp. Soc. Psychol. **6**(4), 466–487 (1970)
2. Arzoglou, E., Kortesniemi, Y., Ruutu, S., Elo, T.: Privacy paradox in social media: a system dynamics analysis. In: Groen, D., de Mulatier, C., Paszynski, M., Dongarra, J.J., Sloot, P.M.A. (eds.) ICCS 2022. LNCS, vol. 13350, pp. 651–666. Springer, Cham (2022). https://doi.org/10.1007/978-3-031-08751-6_47
3. Arzoglou, E., Kortesniemi, Y., Ruutu, S., Elo, T.: The role of privacy obstacles in privacy paradox: a system dynamics analysis. Systems **11**(4), 205 (2023)
4. Barlas, Y.: Formal aspects of model validity and validation in system dynamics. Syst. Dyn. Rev. **12**(3), 183–210 (1996)
5. Bartsch, M., Dienlin, T.: Control your Facebook: an analysis of online privacy literacy. Comput. Hum. Behav. **56**, 147–154 (2016)
6. Bass, F.M.: A new product growth for model consumer durables. Manag. Sci. **15**(5), 215–227 (1969)
7. Choi, H., Park, J., Jung, Y.: The role of privacy fatigue in online privacy behavior. Comput. Hum. Behav. **81**, 42–51 (2018)
8. Cialdini, R.B., Trost, M.R.: Social influence: social norms, conformity and compliance. In: Gilbert, D.T., Fiske, S.T., Lindzey, G. (eds.) The Handbook of Social Psychology, pp. 151–192. McGraw-Hill, New York (1998)
9. Dienlin, T., Trepte, S.: Is the privacy paradox a relic of the past? An in-depth analysis of privacy attitudes and privacy behaviors. Eur. J. Soc. Psychol. **45**(3), 285–297 (2015)
10. Kokolakis, S.: Privacy attitudes and privacy behaviour: a review of current research on the privacy paradox phenomenon. Comput. Secur. **64**, 122–134 (2017)
11. Nissenbaum, H.: Privacy in Context: Technology, Policy, and the Integrity of Social Life. Stanford Law Books, Palo Alto (2009)
12. Rogers, E.M.: Diffusion of Innovations. Free Press, New York (2003)
13. Rosenberg, R.S.: The Social Impact of Computers. Academic Press Inc., Cambridge (1992)
14. Ruutu, S., Casey, T., Kotovirta, V.: Development and competition of digital service platforms: a system dynamics approach. Technol. Forecast. Soc. Chang. **117**, 119–130 (2017)
15. Sterman, J.D.: Business Dynamics: Systems Thinking and Modeling for a Complex World. McGraw-Hill, New York (2000)
16. Tönnies, F.: Community and Society. Dover Publications, New York (2003)
17. Westin, A.F.: Social and political dimensions of privacy. J. Soc. Issues **59**(2), 431–453 (2003)

The First Scientiffic Evidence for the Hail Cannon

Krzysztof Misan[1], Maciej Kozieja[1], Marcin Łoś[1], Dominik Gryboś[1],
Jacek Leszczyński[1], Paweł Maczuga[1], Maciej Woźniak[1],
Albert Oliver Serra[2], and Maciej Paszyński[1(✉)]

[1] AGH University of Science and Technology, Kraków, Poland
maciej.paszynski@agh.edu.pl
[2] Universidad de Las Palmas de Gran Canaria, Las Palmas de Gran Canaria, Spain

Abstract. The hail cannon has been used to prevent hail storms since the 19th century. The idea of the hail cannon is to create a sequence of shock waves to prevent the formation of clouds before the hail storm. Modern hail cannons employ a mixture of acetylene and oxygen to ignite a sequence of explosions in the lower chamber traveling through the neck and into the cone of the cannon, creating shock waves. The shock waves propagate upwards to the cloud, and they are supposed to prevent the formation of the cloud. According to Wikipedia, there is no scientific evidence for the hail cannon, even though it is commonly used in several countries. In this paper, we propose a numerical simulation to verify the idea of the hail cannon. We employ isogeometric analysis and variational splitting methods. We compare our numerical results with the experimental data. We show that our numerical simulation is indeed the scientific evidence for the hail cannon. We also compare our numerical simulations with the experimental measurements performed with a drone before and after a sequence of generated shock waves.

Keywords: Hail cannon · Cloud formation · Advection-Diffusion model · Variational Splitting · Isogeometric Analysis

1 Introduction

A hail cannon is a shock wave generator that is supposed to prevent the formation of hail storms. Modern hail cannons employ a sequence of explosions with an acetylene-oxygen mixture in the combustion chamber. The shock wave created by the explosions travels upward through reversed cone shape pipes. It is supposed to create a shock wave that travels upwards and causes the cloud vapor particles to travel up and to the side. Albert Stiger created the first modern kind of hail cannon in 1895. He was a farmer, and he had a large wine plantation in Austria [6], damaged by local hail storms. The first international congress on hail shooting happened in 1902 in [27]. Despite claims that there is no scientific evidence in favor of hail cannons [30], they are still successfully manufactured [12]. One claim that there is no physical evidence for the hail cannon to work

J. Mikyška et al. (Eds.): ICCS 2023, LNCS 14073, pp. 177–190, 2023.
https://doi.org/10.1007/978-3-031-35995-8_13

is that a thunderstorm is much stronger than the hail cannon, and it does not seem to disturb the hailstorms [8].

In this paper, we perform the first three-dimensional numerical simulations of the hail cannon. We employ the advection-diffusion partial differential equation to model the process. The unknown scalar field is the concentration of the cloud vapor. The shock waves generated by the hail cannon are modeled as the advection vector field. We show that a sequence of generated shock waves can move the cloud vapor up and to the sides (see Fig. 1). After finishing a sequence of shock wave generation, the "hole" in the cloud remains intact for a long time. After that time, if the neighboring clouds are still there, the hole is filled with the cloud vapor particles by the diffusion mechanism. In this sense, we show that generating a sequence of shock waves can produce a hole in the cloud vapor. The argument that the thunderstorm is not removing the hail storm and, thus, the hail cannons are not working is not valid since we need to produce a long sequence of shock waves to obtain the desired effect.

In this paper, we apply the isogeometric finite element method (IGA-FEM) [7] for three-dimensional simulations of the hail cannon. The IGA-FEM employs higher-order and continuity basis functions to approximate different physical phenomena described by Partial Differential Equations. Several researchers applied the IGA-FEM to model different engineering applications. To name a few, IGA-FEM was applied to deformable shell theory [3], phase field modeling [9,10], phase separation simulations with either Cahn-Hilliard [13] or Navier-Stokes-Korteweg higher order models [14], wind turbine aerodynamics [16], incompressible hyper-elasticity [11], turbulent flow simulations [5], transportation of drugs in arterials [15] or the blood flow simulations [4]. In our simulations, we use an explicit dynamics solver, and we employ linear computational cost alternating directions solver [23]. We will use our C++ IGA-ADS code [24] linking LAPACK [2], parallelized into shared-memory multi-core servers using GALOIS library [26]. We developed an interface into the VTK visualization toolkit [28]. Due to the IGA-ADS solver's ability to run fast and accurate three-dimensional simulations on a laptop, it was employed to simulate several phenomena. They include tumor growth [19,21,25], non-linear flow in heterogeneous media with possible applications to CO2 sequestration process [20,24], as well as patogen propagation problem [22].

The structure of the paper is the following. We start with detailed description of the hail cannon and our experimental verification of the cloud vapor reduction by using a drone measurements in Sect. 2. Next, Sect. 3 introduces the Partial Differential Equations modeling the phenomena, together with IGA-FEM discretizations. Section 4 describes details of our simulations of the hail cannon generating a sequence of the shock-waves into the hail cloud. Finally, we conclude the paper in Sect. 5.

Fig. 1. The idea of using shock waves to mix and lift cloud vapor particles.

2   Experimental Verification

In the technological experiments we use the the Inopower anti-hail cannon [17], shown in Fig. 2. The device consists of a container with dimensions of 6.00 m × 2.45 m × 2.60 m with a combustion chamber of 150 dm³, three fuel inlets and a control panel. The shock wave created by the ignition of the acetylene-air mixture is directed upward through the conical outlet pipe. On the other side of the container there are acetylene cylinders and a gas pressure reduction system.

During our experiments, the gauge of the acetylene fed into the explosion chamber was 2.9 bar. During the explosion of the mixture of acetylene and air, a pressure of about 1 MPa was reached. During the experiment, about 300 shock waves were generated in half an hour.

For the experimental verification we measured the temperature, the humidity, and the particular matter concentration in the vertical profile using the equipment placed on the DJI Matrice 200 V2 drone. For each test, a flight was performed immediately before and after the generator was started and 20 min after its completion. During the test, measurements with the drone were made before the experiment and several times after the experiment, in particular, the flight was made 15 and 5 min before starting the generator, 5, 15 and 30 and 40 min after its completion. Figure 3 shows the measurement data of the altitude profile from 0 m to 130 m.

Fig. 2. Hail cannon: Container with acetylene cylinders and control panel. Combustion chamber with the conical outlet pipe.

From the experimental data, we can read the two times reduction of the cloud vapor (possibly with pollution particles) around 15 min after we start the sequence of shock wave generation, and the significant reduction around 30 min of the shock wave generation. Finally, 40 min from the beginning of the

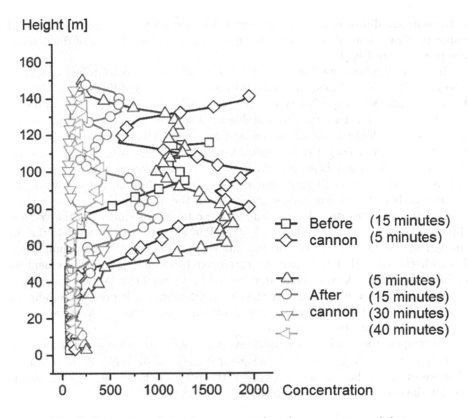

Fig. 3. Reduction of cloud vapor particles after a sequence of shots.

sequence, 10 min after we finish the sequence, we can still see around five times reduction of the cloud vaport. We can conclude that the long sequence of shots can significantly reduce the cloud vapor concentration.

3   Numerical Simulations

We employ advection-diffusion-reaction equations to model the concentration of the water vapor forming a cloud. The equations in the strong form are

$$\frac{\partial u}{\partial t} + (\mathbf{b} \cdot \nabla)u - \nabla \cdot (\mathbf{K}\nabla u) = 0 \text{ in } \Omega \times (0,T],$$
$$\nabla u \cdot n = 0 \text{ in } \Omega \times (0,T], \tag{1}$$
$$u = u_0 \text{ in } \Omega \times 0,$$

where u is the concentration scalar field, $\mathbf{b} = (b_x, b_y, b_z)$ is the assumed "wind" velocity vector field,

$$\mathbf{K} = \begin{pmatrix} K_{11} & 0 & 0 \\ 0 & K_{22} & 0 \\ 0 & 0 & K_{33} \end{pmatrix}, \tag{2}$$

is the isotropic diffusion matrix. The scalar concentration field u represents cloud vapor particles. The initial conditions on the vertical cross-section of the domain are presented in Fig. 4.

In the atmospheric modeling [1, 18, 29] there are two types of diffusion: molecular diffusion and turbulent diffusion (also known as eddy diffusion). In the air, turbulent diffusion is quicker than molecular diffusion. The turbulent diffusion is not isotropic. For example, in a stable atmosphere (all air masses are stratified where the lower ones are denser and less warm than the upper ones that are lighter and warmer), the horizontal turbulence diffusion is greater than the vertical diffusion (there is no transfer between layers). However, when the atmosphere is unstable (some lighter masses are going up through denser ones), the vertical turbulent diffusion can be quicker than the horizontal.

In our model, the diffusion is assumed to be ten times smaller in the vertical direction $K_x = K_y = 1.0$, and $K_z = 0.1$. The advection term models the air movement. We do not investigate the influence of the wind in this simulation. Instead, the advection term will be employed to model the air movement as enforced by the shock-wave generator, according to Eq. (15). We will generate a sequence of shock waves to remove the cloud. Of course, it is possible to add the wind to the model and check how the shock wave generator works if we have a strong wind moving the cloud vapor.

We employ the explicit time integration scheme, where we approximate the time derivative with $\frac{\partial u}{\partial t} = \frac{u^{t+1} - u^t}{dt}$, where dt is the time step size. We also assume that we evaluate the remaining terms in the previous time step. As the result we get the explicit Euler time integration scheme

$$\frac{u^{t+1} - u^t}{dt} = \nabla \cdot (\mathbf{K} \nabla u^t) - (\mathbf{b} \cdot \nabla) u^t = 0. \tag{3}$$

The weak formulation is obtained by testing with B-spline basis functions

$$\left(u^{t+1}, v\right) = \left(u^t, v\right) - dt \left(\mathbf{K} \nabla u^t, \nabla v\right) - dt \left(\mathbf{b} \cdot \nabla u^t, v\right) + \left(c u^t, v\right) \forall v \in V. \tag{4}$$

We discretize with B-spline basis functions defined over the cube shape domain $\Omega = [0, 1]^3$

$$u^{t+1} = \sum_{i=1,\dots,N_x; j=1,\dots,N_y; k=1,\dots,N_z} u_{ij}^{t+1} B_i^x B_j^y B_k^z,$$

$$u^t = \sum_{i=1,\dots,N_x; j=1,\dots,N_y; k=1,\dots,Nz} u_{ij}^t B_i^x B_j^y B_k^z, \tag{5}$$

and we test with B-spline basis functions

$$\sum_{ijk} u_{ijk}^{t+1} \left(B_i^x B_j^y B_k^z, B_l^x B_m^y B_n^z \right) = \sum_{ij} u_{ijk}^t \left(B_i^x B_j^y B_k^z, B_l^x B_m^y B_n^z \right)$$

$$- dt \sum_{ijk} u_{ijk}^t \left(K_x \frac{\partial B_i^x}{\partial x} B_j^y B_k^z, \frac{\partial B_l^x}{\partial x} B_m^y B_n^z \right)$$

$$- dt \sum_{ijk} u_{ijk}^t \left(K_y B_i^x \frac{\partial B_j^y}{\partial y} B_k^z, B_l^x \frac{\partial B_m^y}{\partial y} B_n^z \right)$$

$$- dt \sum_{ijk} u_{ijk}^t \left(K_z B_i^x B_j^y \frac{\partial B_k^z}{\partial z}, B_l^x B_m^y \frac{\partial B_n^z}{\partial z} \right) \tag{6}$$

$$- dt \sum_{ij} u_{ijk}^t \left(b_x(x,y,z) \frac{\partial B_i^x}{\partial x} B_j^y B_k^z, B_l^x B_m^y B_n^z \right)$$

$$+ dt \sum_{ijk} u_{ijk}^t \left(b_y(x,y,z) B_i^x \frac{\partial B_j^y}{\partial y} B_k^z, B_l^x B_m^y B_n^z \right)$$

$$+ dt \sum_{ijk} u_{ijk}^t \left(b_z(x,y,z) B_i^x B_j^y \frac{\partial B_k^z}{\partial z}, B_l^x B_m^y D_n^z \right)$$

$$l = 1, ..., N_x; m = 1, ..., N_y; n = 1, ..., N_z,$$

where $(u, v) = \int_\Omega u(x, y, z)v(x, y, z)dxdydz$.

We separate directions

$$\sum_{ijk} u_{ijl}^{t+1} \left(B_i^x, B_l^x \right)_x \left(B_j^y, B_m^y \right)_y \left(B_k^z, B_n^z \right)_z = \sum_{ij} u_{ijk}^t \left(B_i^x, B_l^x \right)_x \left(B_j^y, B_m^y \right)_y \left(B_k^z, B_n^z \right)_z$$

$$- dt \sum_{ijk} u_{ijk}^t \left(K_x \frac{\partial B_i^x}{\partial x}, \frac{\partial B_l^x}{\partial x} \right)_x \left(K_y B_j^y, B_m^y \right)_y \left(K_z B_j^z, B_n^z \right)_z$$

$$- dt \sum_{ijk} u_{ijk}^t \left(K_x B_i^x, B_l^x \right)_x \left(K_y \frac{\partial B_j^y}{\partial y}, \frac{\partial B_m^y}{\partial y} \right)_y \left(K_z B_k^z, B_n^z \right)_z$$

$$- dt \sum_{ijk} u_{ijk}^t \left(K_x B_i^x, B_l^x \right)_x \left(K_y B_j^y, B_m^y \right)_y \left(K_z \frac{\partial B_k^z}{\partial z}, \frac{\partial B_n^z}{\partial z} \right)_z$$

$$+ dt \sum_{ijk} u_{ijk}^t \left(b_x \frac{\partial B_i^x}{\partial x}, B_l^x \right)_x \left(b_y B_j^y, B_m^y \right)_y \left(b_z B_k^z, B_n^z \right)_z$$

$$+ dt \sum_{ijk} u_{ijk}^t \left(b_x B_i^x, B_l^x \right)_x \left(b_y \frac{\partial B_j^y}{\partial y}, B_m^y \right)_y \left(b_z B_k^z, B_n^z \right)_z$$

$$+ dt \sum_{ijk} u_{ijk}^t \left(b_x B_i^x, B_l^x \right)_x \left(b_y B_j^y, B_m^y \right)_y \left(b_z \frac{\partial B_k^z}{\partial z}, B_n^z \right)_z$$

$$l = 1, ..., N_x; m = 1, ..., N_y; n = 1, ..., N_z.$$

We introduce

$$\mathbf{M}_x = \{(B_i^x, B_l^x)_x\}_{il} = \{\int B_i^x B_l^x \, dx\}_{il}, \tag{7}$$

$$\mathbf{M}_y = \{(B_j^y, B_m^y)_y\}_{jm} = \{\int B_j^y B_m^y \, dy\}_{jm}, \tag{8}$$

$$\mathbf{M}_z = \{(B_k^z, B_n^z)_y\}_{kn} = \{\int B_k^z B_n^z \, dz\}_{kn}. \tag{9}$$

In general, Kronecker product matrix $\mathcal{M} = \mathcal{A}^x \otimes \mathcal{B}^y \otimes \mathcal{C}^z$ over 3D domain $\Omega = \Omega_x \times \Omega_y \times \Omega_z$ is defined as

$$\mathcal{M}_{ijklmn} = \mathcal{A}_{il}^x \mathcal{B}_{jm}^y \mathcal{C}_{kn}^z. \tag{10}$$

Due to the fact, that one-dimensional matrices discretized with B-spline functions are banded and they have $2p + 1$ diagonals (where p stands for the order of B-splines), since

$$(\mathcal{M})^{-1} = (\mathcal{A}^x \otimes \mathcal{B}^y \otimes \mathcal{C}^z)^{-1} = (\mathcal{A}^x)^{-1} \otimes (\mathcal{B}^y)^{-1} \otimes (\mathcal{C}^z)^{-1}. \tag{11}$$

we can solve our system in a linear computational cost. The Kronecker product decomposition on the right-hand-side can also help in developing fast integration algorithms.

4　IGA-ADS Simulation of the Hail Cannon

We perform three-dimensional computer simulations of the hail cannon using our IGA-ADS code [24], employing isogeometric finite element method and linear computational cost solver. In our simulation, the scalar field u represents the water vapor forming a cloud (possibly mixed with the pollution particles). Our initial configuration is the cloud "fixed" at the height of $3/4$ of the domain. We formulate the problem in the domain, with $\Omega = [0, 100\,\text{m}] \times [0, 100\,\text{m}] \times [0, 100\,\text{m}]$. The initial state is presented in Fig. 4. The vertical axis on the left-hand side of the picture represents the vertical dimension of the domain. The vertical axis on the right-hand side represents the cloud vapor concentration field. The display scale is fixed from 0 to 10,000, on all the plots.

We employ the advection-diffusion equations,

$$\frac{\partial u(x, y, z; t)}{\partial t} + b_x(x, y, z; t) \frac{\partial u(x, y, z; t)}{\partial x}$$

$$+ b_y(x, y, z; t) \frac{\partial u(x, y, z; t)}{\partial y}$$

$$+ b_z(x, y, z; t) \frac{\partial u(x, y, z; t)}{\partial z}$$

$$- K_x \frac{\partial^2 u(x, y, z; t)}{\partial x^2} - K_y \frac{\partial^2 u(x, y, z; t)}{\partial y^2} - K_z \frac{\partial^2 u(x, y, z; t)}{\partial z^2} = 0,$$

$$(x, y, z; t) \text{ in } \Omega \times (0, T], \tag{12}$$

$$\nabla u(x, y, z; t) \cdot n(x, y, z) = 0, \quad (x, y, z; t) \text{ in } \Omega \times (0, T], \tag{13}$$

$$u(x, y, z; 0) = u_0 \text{ in } \Omega \times 0, \tag{14}$$

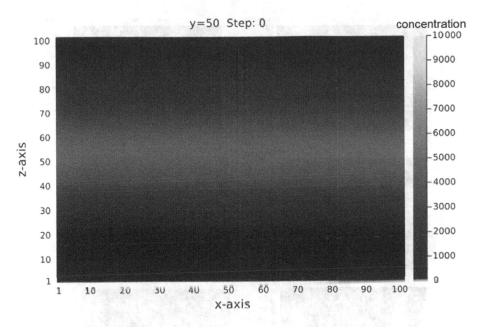

Fig. 4. Initial configuration for the cloud.

where u is the concentration scalar field, where the shock-waves are modeled by the advection field, namely $(b_x(x, y, z; t), b_y(x, y, z; t), b_z(x, y, z; t)) = (0, 0, cannon(x, y, z; t))$ (given by Eq. (15)), $K_x = K_y = 1.0$ are the horizontal diffusion coefficients, and $K_z = 0.1$ is the vertical diffusion.

We employ the implementation of the linear computational cost Kronecker product structure solver as described in [24]. The explicit method formulation is implemented in the `compute_rhs` routine

```
void compute_rhs(int iter)
1     auto& rhs = u;
2     zero(rhs);
3     for(auto e:elements()) {
4       auto U = element_rhs();
5       double J = jacobian(e);
6       for (auto q : quad_points()) {
7         double w = weight(q);
8         for (auto a : dofs_on_element(e)) {
9           auto aa = dof_global_to_local(e, a);
10          value_type v = eval_basis(e, q, a);
11          value_type u = eval_fun(u_prev, e, q);
12          double grad = 1.*u.dx * v.dx +
              1.*u.dy * v.dy +
              0.1*u.dz * v.dz;
13          double val = u.val * v.val - steps.dt * grad+
```

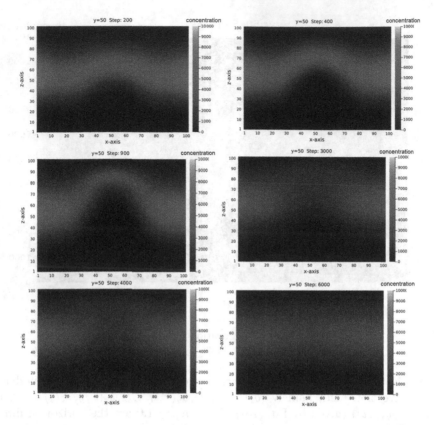

Fig. 5. Reduction of the cloud vapor by generated shock waves. Side view.

```
           steps.dt*cannon(e[1])*u.dz*v.val;
14          U(aa[0], aa[1]) += val * w * J;
15      }
16  }
17  update_global_rhs(rhs, U, e);
18  }
```

In order to simulate the atmospheric cannon, we introduce the shock wave as the advection function in a separable way as

$$
\begin{aligned}
cannon(x, y, z; t) = {}& const * (1 - z) * \\
& sin(10 * \pi * x) * sin(10 * \pi * y) \\
& * max(0, sin(\pi * t/10)),
\end{aligned}
\tag{15}
$$

for $t = s - 100$, where s is the time step size. In other words we run the cannon from time step 100, and we shoot for 10 time steps with a function $(1 - z) * sin(10 * \pi * x) * sin(10 * \pi * y)$ that runs in time like $max(0, sin(\pi * t/10))$.

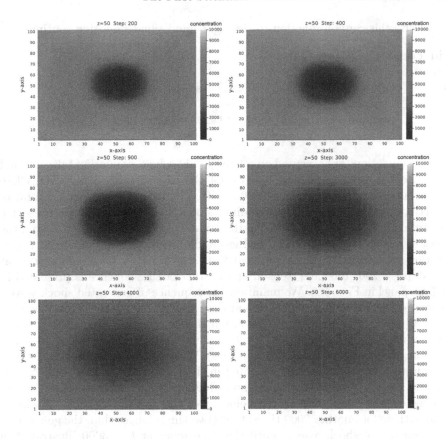

Fig. 6. Reduction of the cloud vapor by generated shock waves. Top view.

```
1    double cannon(double x, double y, double z, int iter) {
2    x=x/40.; y=y/40.; z=z/40.
3    double t=iter;
4    if(x>0.3 && x<0.6 && y>0.3 && y<0.6 &&t>0 && t<1000)
5      return 200.*(1.-z)*
     max(sin(10*PI*x),0) *max(sin(10*PI*y),0)*
     max(0,sin(PI*t/10));
6    else
7      return 0.;
```

We add this cannon function to the right-hand-side computing routine

void compute_rhs(int iter)

...

```
13      double val = u.val * v.val - steps.dt * grad+
        steps.dt*(delta_T(e[1])-
        cannon(e[0],e[1],e[2], iter))*u.dz*v.val;
```

...

The simulations are executed on a laptop. The whole simulation takes around 1 h on Corsair Vengeance LPX, DDR4, 64 GB (2×32 GB), 3200 MHz, CL16 with processor AMD Ryzen 9 - 3900X with 12 physical cores, and a total of 24 virtual cores. We run fully three-dimensional simulations, and we present a horizontal cross-section along OXZ in the middle of the domain in Fig. 5, and the vertical cross-section along OXY in the middle height of the domain in Fig. 6. We start generating the shock waves at time step $t = 100$. The configuration at time step $t = 200$, 100-time steps after the start of the sequence is presented in the first panel in Figs. 5–6. We maintain the sequence of generated shock waves, and we present the next configuration at time step $t = 400$ in the second panel in Figs. 5-6. We can read from this simulation that generating a sequence of shock waves results in a local mixing of the layers and a reduction of the cloud vapor (possibly mixed with the pollution particles). We continue this sequence of shock waves until time step $t = 1000$ presented in the third panel in Figs. 5–6. At this moment, we can observe the maximum reduction of the cloud vapor. Then, we stop the sequence and observe the behavior of the cloud. The "hole" in clouds remains there for another 2000 time steps (two times longer than the generation of a sequence of shock waves) until the time moment $t = 3000$ illustrated in the fourth panel. At this moment, the neighboring cloud vapor particles return slowly to the center by the diffusion mechanism. This is illustrated in the fifth and sixth panels (time moments $t = 4000$ and $t = 6000$) in Figs. 5–6. We conclude that this local water vapor reduction maintains if the cannon creates shocking waves over a repeated period and stays for a significant period. The repetition of the shock waves for a long time is a critical phenomenon for reducing water vapor concentration.

5   Conclusion and Future Work

We claim that the numerical results presented in this paper are the first three-dimensional simulational verification of the hail cannon. We showed that repeating a sequence of shock waves for a long time significantly reduces cloud vapor. Our model is simple and may require several improvements, like computing the real shape of the cannon force modeled by the advection. This will require a solution of the three-dimensional Navier-Stokes equations. This computed field can be introduced into the advection-diffusion model. We can also include different cloud components and the reaction terms between them. The thermal effects that consider the additional movement of the cloud particles can also be incorporated by considering the Navier-Stokes-Boussinesq model. Nevertheless, the presented numerical results are very interesting, and they also confirm our experimental findings.

Acknowledgement. The Authors are thankful for support from the funds assigned to AGH University of Science and Technology by the Polish Ministry of Science and Higher Education. The work of Albert Oliver-Serra is supported by the "Ayudas para la recualificación del sistema universitario español" grant funded by the ULPGC, the Ministry of Universities by Order UNI/501/2021 of 26 May, and the European Union-Next Generation EU Funds.

References

1. Businger, J.A., Arya, S.P.S.: Height of the mixed layer in the stably stratified planetary boundary layer. In: Frenkiel, F., Munn, R. (eds.) Turbulent Diffusion in Environmental Pollution, Advances in Geophysics, vol. 18, pp. 73–92. Elsevier (1975)
2. Anderson, E., et al.: LAPACK Users' Guide, vol. 9. Siam, Delhi (1999)
3. Benson, D.J., et al.: A generalized finite element formulation for arbitrary basis functions: From isogeometric analysis to X-FEM. Int. J. Numer. Methods Eng. **83**, 765–785 (2010)
4. Calo, V., Brasher, N., Bazilevs, Y., Hughes, T.: Multiphysics model for blood flow and drug transport with application to patient-specific coronary artery flow. Comput. Mech. **43**(1), 161–177 (2008)
5. Chang, K., Hughes, T., Calo, V.: Isogeometric variational multiscale large-eddy simulation of fully-developed turbulent flow over a wavy wall. Comput. Fluids **68**, 94–104 (2012)
6. Changnon, S.A., Ivens, J.L.: History repeated: the forgotten hail cannons of Europe. Bull. Am. Meteorol. Soc. **62**(3), 368–375 (1981)
7. Cottrell, J.A., Hughes, T.J.R., Bazilevs, Y.: Isogeometric Analysis: Toward Integration of CAD and FEA. John Wiley & Sons, New York (2009)
8. Curran, J.: Vt. Orchard wakes the neighbors with hail cannon (2018)
9. Dedè, L., Borden, M.J., Hughes, T.: A user's view of solving stiff ordinary differential equations. SIAM Rev. **21**(1), 1–17 (1979)
10. Dede, L., Hughes, T., Lipton, S., Calo, V.: Structural topology optimization with isogeometric analysis in a phase field approach. In: 16th US National Conference on Theoretical and Applied Mechanics, pp. 1–2 (2010)
11. Duddu, R., Lavier, L.L., Hughes, T.J., Calo, V.M.: A finite strain Eulerian formulation for compressible and nearly incompressible hyperelasticity using high-order b-spline finite elements. Int. J. Numer. Methods Eng. **89**(6), 762–785 (2012)
12. Egger, M.: How it works. Hail Cannon Manufacturer (2022)
13. Gomez, H., Calo, V.M., Bazilevs, Y., Hughes, T.J.: Isogeometric analysis of the Cahn-Hilliard phase-field model. Comput. Methods Appl. Mech. Eng. **197**(49–50), 4333–4352 (2008)
14. Gomez, H., Hughes, T.J., Nogueira, X., Calo, V.M.: Isogeometric analysis of the isothermal Navier-Stokes-Korteweg equations. Comput. Methods Appl. Mech. Eng. **199**(25–28), 1828–1840 (2010)
15. Hossain, S.S., Hossainy, S.F., Bazilevs, Y., Calo, V.M., Hughes, T.J.: Mathematical modeling of coupled drug and drug-encapsulated nanoparticle transport in patient-specific coronary artery walls. Comput. Mech. **49**(2), 213–242 (2012)

16. Hsu, M.C., Akkerman, I., Bazilevs, Y.: High-performance computing of wind turbine aerodynamics using isogeometric analysis. Comput. Fluids **49**(1), 93–100 (2011)
17. Inopower: Hail cannon user's manual (2009)
18. Lamb, R., Durran, D.: Eddy diffusivities derived from a numerical model of the convective planetary boundary layer. Il Nuovo Cimento C **1**, 1–17 (1978)
19. Łoś, M., Kłusek, A., Hassaan, M.A., Pingali, K., Dzwinel, W., Paszyński, M.: Parallel fast isogeometric L2 projection solver with GALOIS system for 3D tumor growth simulations. Comput. Methods in Applied Mechanics and Engineering **343**, 1–22 (2019)
20. Łoś, M., Paszyński, M.: Stability of non-linear flow in heterogeneous porous media simulations using higher order and continuity basis functions. J. Comput. Appl. Math. **428**, 115141 (2023)
21. Łoś, M., Paszyński, M., Kłusek, A., Dzwinel, W.: Application of fast isogeometric L2 projection solver for tumor growth simulations. Comput. Methods Appl. Mech. Eng. **316**, 1257–1269 (2017)
22. Łoś, M., Woźniak, M., Muga, I., Paszyński, M.: Three-dimensional simulations of the airborne COVID-19 pathogens using the advection-diffusion model and alternating-directions implicit solver. Bull. Pol. Acad. Sci.: Tech. Sci. **69**(4), e137125 (2021)
23. Łoś, M., Woźniak, M., Paszyński, M., Dalcin, L., Calo, V.M.: Dynamics with matrices possessing Kronecker product structure. Procedia Comput. Sci. **51**, 286–295 (2015)
24. Łoś, M., Woźniak, M., Paszyński, M., Lenharth, A., Hassaan, M.A., Pingali, K.: IGA-ADS: isogeometric analysis finite element method using alternating directions solver. Comput. Phys. Commun. **217**, 99–116 (2017)
25. Paszyński, M., Siwik, L., Dzwinel, W., Pingali, K.: Supermodeling, a convergent data assimilation meta-procedure used in simulation of tumor progression. Comput. Math. Appl. **113**(C), 214–224 (2022)
26. Pingali, K., et al.: The tao of parallelism in algorithms. In: SIGPLAN, vol. 46, no. 6, pp. 12–25 (2011)
27. Plumandon, J.R.: Troisieme congres international de defense contre la grele (in French). Ciel et Terre **22**, 457–471 (1902)
28. Schroeder, W., Martin, K., Lorensen, B.: The Visualization Toolkit, 4th edn, vol. 9. ISBN 978-1-930934-19-1 (2006)
29. Shir, C.C.: A preliminary numerical study of atmospheric turbulent flows in the idealized planetary boundary layer. J. Atmos. Sci. **30**(7), 1327–1339 (1973)
30. Wieringa, J., Holleman, I.: If cannons cannot fight hail, what else? Meteorol. Z. **15**(6), 659–669 (2006)

Constituency Parsing with Spines and Attachments

Katarzyna Krasnowska-Kieraś[ID] and Marcin Woliński[✉][ID]

Institute of Computer Science, Polish Academy of Sciences, Warsaw, Poland
{k.krasnowska,m.wolinski}@ipipan.waw.pl
http://zil.ipipan.waw.pl

Abstract. We propose a hybrid representation of syntactic structures, combining constituency and dependency information. The headed constituency trees that we use offer the advantages of both those approaches to representing syntactic relations within a sentence, with a focus on consistency between them. Based on this representation, we introduce a new constituency parsing technique capable of handling discontinuous structures. The presented approach is centred around head paths in the constituency tree that we refer to as spines and the attachments between them. Our architecture leverages a dependency parser and a large BERT model and achieves 95.96% F1 score on a dataset where ≈10% of trees contain discontinuities.

Keywords: Constituency parsing · Headed constituencies · Syntactic structures

1 Introduction

Syntactic parsing is an important NLP task often used in various text processing pipelines. In this context, syntactic structures are usually represented with constituency or dependency trees. Constituency structures give easy access to phrases that make up sentences. Thus, they are preferred in tasks such as nominal phrase extraction, identification of terminology etc. Dependency trees are closer to predicate-argument structures, so they are the preferred representation for more semantic tasks, involving, e.g., the analysis of events and their actors.

Since constituency and dependency structures are used for different tasks, it seems a practical solution to have both available. This is the task of hybrid parsing. Our goal, however, is to fulfil one more requirement: the structures have to be consistent with each other. Two tokens should be connected with a dependency relation in the dependency tree if and only if there exists a constituent in the constituency tree whose yield contains both tokens.

We plan to create a large parsebank of Polish annotated with both types of structures. The treebank search engine will allow to query both constituency and

Work supported by POIR.04.02.00-00-D006/20-00 national grant (Digital Research Infrastructure for the Arts and Humanities DARIAH-PL).

J. Mikyška et al. (Eds.): ICCS 2023, LNCS 14073, pp. 191–205, 2023.
https://doi.org/10.1007/978-3-031-35995-8_14

dependency structures and we want to be sure that the results of these queries are in line with each other. The goal of the present work is to provide a hybrid parsing method which would provide consistent constituency and dependency structures.

2   Headed Constituencies

To introduce the proposed method, let us take a step back and look at the structures commonly used to represent syntax. In dependency trees, the relations between words are in the focus of interest: the verb's need of its complements, the relation between a noun and its adjectival attribute, and so on. The exact choice of relations is a matter of convention, but, wisely chosen, it leads to binding all words of a sentence in the form of a tree (cf. Fig. 1).

Constituency trees model the hierarchical nature of the natural language by showing how longer fragments are composed of shorter constituents (cf. Fig. 2). The parent-child relation corresponds to the child being a sub-span of the parent.

The grammatical features of words can be generalised to constituents. Thus a constituent *nowego domu*[1] 'new house' can be considered to be in the genitive case and in the singular, since these are the features of the nominal form *domu* 'house' shared by the adjectival form *nowego* 'new'.

Moreover, the relations of dependency syntax can be thought of as occurring between constituents: a verbal phrase *pokazali i opisali* 'showed and described' subcategorises for a subject in the nominative (e.g. *dziennikarze* 'the journalists') and an object in the accusative (e.g. *sytuację* 'the situation').

These relations can be used to create a local dependency structure among the children of a given constituent. For the sentence **S** node, dependencies go from the verbal phrase **VP** (the head) to two nominal phrases (labelled subj and obj) and to a prepositional adjunct. (A technical relation punct is used to take care of punctuation.)

This procedure leads to the creation of a *headed constituency tree* in which each constituent has a head among its children and each non-head child is connected with the head child using a labelled dependency relation (cf. Fig. 3). The constituency and dependency trees can be rather trivially extracted from this structure. However, we prefer to think of headed constituencies as a model of syntax in its own right, which includes both types of syntactic information.

There is one important catch in these considerations: for the joint structure to be possible, the constituency and dependency trees have to be consistent with each other. Typically, constituency trees model surface syntax, that is they reflect purely grammatical interactions within a sentence (e.g. Marcus et al., 1993; Brants et al., 2004). This is often not true in the case of dependency trees, in particular Universal Dependencies (Nivre et al., 2020), which involve more semantic relations. In contrast, an example of a surface syntax oriented dependency scheme is SUD (Gerdes et al., 2018).

---

[1] We use examples in Polish, since its system of 7 grammatical cases makes the grammatical relations more easily visible than in English.

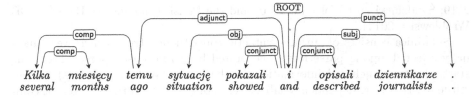

Fig. 1. Dependency tree for the sentence: *Kilka miesięcy temu sytuację pokazali i opisali dziennikarze.* 'The journalists exposed and described the situation several months ago.'

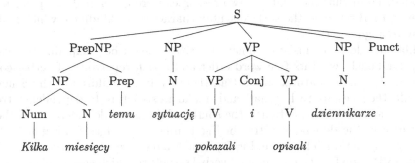

Fig. 2. Constituency tree for the same sentence.

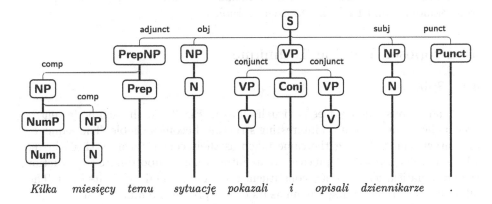

Fig. 3. Headed constituency tree for the same sentence. Bold line joins a constituent with its head child.

3   The Dataset

To perform the experiments we need a constituency treebank consistent with a dependency one. For that reason we chose to work with a Polish dataset built on the *Składnica* constituency treebank[2] (Woliński and Hajnicz, 2021; Woliński,

---

[2] http://zil.ipipan.waw.pl/Sk%C5%82adnica.

2019; Świdziński and Woliński, 2010) and the Polish Dependency Bank[3] (PDB Wróblewska, 2014).

Składnica consists of surface-syntactic constituency trees which were manually selected among parse forests generated by a rule based parser. From the start, the trees included information on the heads (syntactic centres) of constituents. However, dependency labels were not present, so these were not complete headed constituencies as we understand them in the present paper.

The starting point for PDB was converting the trees of Składnica to dependency structures. The result was enriched with dependency labels and manually validated. From that moment the two resources were developed independently. However, PDB retained the surface-syntax character of Składnica, which makes it easy to align the trees.

Składnica does not insist on binary trees. As the authors explain (Woliński, 2019), for Polish, with its free word order and a rich repertoire of verbal complements, it is most natural to treat all these as direct children of the S node. In result, the trees are rather "flat" and similar in structure to dependency trees (cf. Fig. 1 and 2). Coordination is the source of most visible difference in these structures: in the dependency tree in Fig. 1 almost all edges fan out from the node for conjunction. The tree in Fig. 2 provides a more readable structure with a separate **VP** node for the coordinated verbal structure which as a whole becomes a constituent of the sentence **S**.

The size of the dataset, which we were able to create by merging information from Składnica and PDB, is reported in Table 1.[4]

4   Proposed Parsing Technique

4.1   Spines

If a headed constituency tree is visualised as in Fig. 3 – with each node centred over its head constituent an interesting structure becomes visible. The sequences of syntactic units having the same token as their centre form vertical clusters which we call *spines*. The spines are quite intuitive: in a subordinate construction, the grammatical features of a constituent take source (mainly) in its head. Thus all nodes of a spine having a noun as its base represent nominal constructions of various levels of complication. When the nominal construction is, e.g., required by a verb, the nominal spine does not grow higher, but gets attached to a verbal spine. Spines with a conjunction as a base are more context dependent: the higher nodes depend on constituents being coordinated by the conjunction. In both cases the height of a spine is rather limited. It depends on the way modifiers are attached in a given grammar/treebank.

---

[3] http://zil.ipipan.waw.pl/PDB.

[4] See http://git.nlp.ipipan.waw.pl/constituency/spines-attachments for the code and dataset.

Table 1. The sizes of the Polish dataset used in this work

		continuous	discontinuous		total
train	trees	15903	1756	(9.9%)	**17659**
	tokens	239531	40515	(14.5%)	280046
	avg. tokens/tree	15.06	23.07		15.86
validation	trees	1980	231	(10.4%)	**2211**
	tokens	29531	5034	(14.6%)	34565
	avg. tokens/tree	14.91	21.79		15.63
test	trees	1990	215	(9.8%)	**2205**
	tokens	28529	4815	(14.4%)	33344
	avg. tokens/tree	14.34	22.40		15.12

Prediction of a spine for a token can be seen as generalised part of speech tagging. Nodes low in the spine depend mainly on the base token, higher nodes include more contextual information. We think that it is an interesting model in view of the fact that nowadays a large language model is usually used as an encoder for parsing. It seems reasonable to attach the prediction of a constituent to the token in its centre. The mechanism of attention used in current models should be able to provide the necessary data about the token and its context.

More formally, a spine is a maximum path following head edges between the nodes. Each spine ends with some token of the sentence and each token of the sentence is the end of exactly one spine (we assume empty spines for tokens with no preterminal node). Suppose that token t_i is the dependency head of another token t_j, and their corresponding spines are $s_i = n_{i,1} \rightarrow ... \rightarrow n_{i,k}$ and $s_j = n_{j,1} \rightarrow ... \rightarrow n_{j,l}$. This means that $n_{j,1}$ (the topmost nonterminal of s_j) is a non-head child of some nonterminal along s_i.

For example, in the tree shown in Fig. 3, the tokens $t_1 = Kilka$ 'several', $t_2 = miesięcy$ 'months' and $t_3 = temu$ 'ago' have following spines respectively: $s_1 = $ **NP** $\rightarrow$ **Num**, $s_2 = $ **N**, $s_3 = $ **PrepNP** $\rightarrow$ **Prep**. t_2 is a dependent of t_1, and the topmost (and only) node of s_2 (**N**) is a non-head child of the **NP** node of s_1; t_1 is a dependent of t_3, and the topmost node of s_1 (**NP**) is a non-head child of the **PrepNP** node of s_3.

4.2 Spine Based Parsing

Constituency parsing can be decomposed into (1) determining the spines for each token of the input and (2) determining the way their top nodes are attached to other spines. For the latter part, it is necessary to determine which spine is attached to which other spine (2a) and through which node of the head spine (2b). Part (2a) corresponds exactly to determining the dependency structure between tokens. If we use a dependency parser for this part, the resulting constituency trees will be consistent with dependency trees produced by this parser.[5]

A very desirable trait of the proposed method is that if the task (2a) is performed by a dependency parser capable of parsing discontinuous constructions, the resulting constituency parser becomes immune to the problem of discontinuity. Morover, with this technique the other typical problem in constructing constituency parsers, that of unary branches, does not arise at all. Unary branches in the tree are parts of some spines and they get predicted as such.

One issue that needs addressing is that a spine may contain a sequence of consecutive nodes bearing the same syntactic category. For example, the representation of the phrase *jeszcze trzy tygodnie* 'three more weeks' in the tree shown in Fig. 4 reflects its hierarchical structure $(jeszcze(trzy\ tygodnie)_{NP})_{NP}$, with **Part** → *jeszcze* 'more' and **NP** → **N** → *tygodnie* 'weeks' attached to $\mathbf{NP_2}$ → $\mathbf{NP_1}$ → **NumP** → **Num** → *trzy* 'three' as non-head children of $\mathbf{NP_2}$ and $\mathbf{NP_1}$ respectively.[6] Note that such $\mathbf{X_2}$ → $\mathbf{X_1}$ sequence appearing along a spine means that $\mathbf{X_2}$ must have at least one non-head child apart from its head child $\mathbf{X_1}$, since there are no $\mathbf{X}$ → $\mathbf{X}$ unary branchings in our trees. Moreover, the spines resulting from our trees do not contain $\mathbf{X}$ → $\mathbf{Y}$ → ... → $\mathbf{X}$ sequences (where $\mathbf{X}$ ≠ $\mathbf{Y}$).

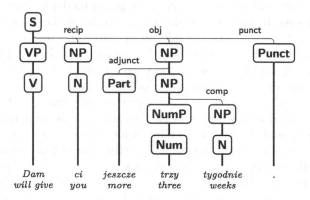

Fig. 4. A tree with an **NP** → **NP** edge along a spine: 'I will give you three more weeks.'

The attachment information consists of the category of the node a spine attaches to and its number in the sequence of identical nodes in its spine, counting from the bottom. Together with the dependencies between tokens, such representation of spines and attachments allows us to encode the complete headed constituency tree.

---

[5] In other words, the tasks (1) and (2b) can be seen as converting the dependency structure to constituencies. Note, however, that the constituency trees are more detailed, so this process adds information.

[6] We use the lower subscript $\mathbf{NP}_i$ to differentiate between two different **NP** nodes, and not to introduce a separate category $\mathbf{NP}_i$.

Table 2. The tree of Fig. 4 encoded with dependency relations, spines and attachments

token	head ID	deprel	spine	attachment
Dam	0	root	S → VP → V	root
ci	1	recip	NP → N	S-1
jeszcze	4	adjunct	Part	NP-2
trzy	1	obj	NP → NP → NumP → Num	S-1
tygodnie	4	comp	NP → N	NP-1
.	1	punct	Punct	S-1

As an example, consider the tree from Fig. 4 and its representation shown in Table 2. The spines for *ci* 'you', *trzy* 'three' and the final punctuation are all attached as children of the same (and only) **S** node along the 'root' spine for *Dam* '(I) will give'. Thererefore, they all have the same **S-1** attachment. The spine for *trzy* contains a sequence of two **NP** nodes. The spine for its one dependent, *tygodnie*, is attached to the first **NP** from the bottom (**NP-1**) and the spine for the other dependent *jeszcze* is attached to the second **NP** from the bottom (**NP-2**). The first two columns of the table contain the dependency relations between tokens following a CoNLL(-U)-like convention.

We also note that, considering the above observations about **X** → **X** head edges, we can alternatively use a 'compressed' representation of spines where any sequence of **X** nodes is collapsed into one **X** node (in practice, there are only sequences of two such nodes in our data). Such modification does not result in any information loss, since any repetitions of same-labeled nodes along spines are retained in the attachment representation. In the example from Table 2, we could represent the spine for *trzy* as **NP** → **NumP** → **Num**, and the information that there are in fact two **NP** nodes is carried by the **NP-2** attachment of *jeszcze*. Such variant of representation reduces the number of different spines. There are 110 distinct spines in our data, producing 70 distinct compressed spines.

Table 3 presents spines ending with **N** and **V** preterminals as an example. The total number of occurrences in our data, as well as the percentage of occurrences among spines with given preterminal, are given for each spine type. In the case of **N** spines, there are 3 types corresponding to types of phrase built around a nominal token: noun phrase with modifiers (hence the repetition of the **NP** node) nested as a non-head in the sentence structure, a nested noun phrase without modifiers, and a noun phrase serving as the syntactic centre of an utterance without a predicate. Figure 5 shows example tree contexts for four **V** spines: main predicate (*wymienia*), relative clause (*zorganizował*), coordinated sentences (*Było, płonęły*) and coordinated verbal phrases (*fascynowały, przerażały*).

5   Parser Architecture

In our approach to constituency parsing, we leverage an already available, well performing dependency parser and combine its analyses with our own constituency component predicting spines and their attachments.

Table 3. Nominal and verbal spines.

spine	occurrences	in data
NP→NP→N	57595	52.84%
NP→N	50864	46.66%
ROOT→NP→NP→N	548	0.50%
ROOT→S→VP→V	16406	36.36%
VP→VP→V	12536	27.78%
S→VP→V	12037	26.68%
CP→S→VP→V	2407	5.33%
VP→V	1725	3.82%
V	7	0.02%
S→S→VP→V	3	0.01%
CP→S→S→VP→V	1	<0.01%
ROOT→S→VP→VP→V	1	<0.01%
S→VP→VP→V	1	<0.01%

As the dependency component, we use the best available dependency parser for Polish which is COMBO[7] (Klimaszewski and Wróblewska, 2021). The published pre-trained models for COMBO operate on UD and are therefore incompatible with the PDB dependency annotation scheme used in our data. Therefore, we trained a dedicated COMBO model on the train portion of our data. The model achieved 96.3% UAS/93.2% LAS on training data and 94.9% UAS/89.8% LAS on the validation portion of our data.

For the constituency component, we used a modified version of the Huggingface Transformers[8] `TFBertForTokenClassification` architecture. The architecture is a simple dense layer classifier added on top of a pre-trained BERT model, predicting a label for each input token. For each token, the classifier is applied to the contextualised vector representation of the token produced by the BERT model, and predicts a vector of logits for all possible classification labels. The whole architecture is then trained, with the classifier being fitted from scratch, and the weights of the BERT model fine-tuned to the specific task. Our modification consisted in adding multiple dense layer classifiers, allowing us to jointly train predictors for several objectives, while fine-tuning the BERT model to all of them. As the BERT model, we use HerBert[9] (Mroczkowski et al., 2021).

The constituency component's model tested in this work involves 3 classifiers, predicting for each token:

---

[7] https://wiki.clarin-pl.eu/pl/nlpws/services/COMBO.
[8] https://huggingface.co/docs/transformers.
[9] https://huggingface.co/allegro/herbert-large-cased.

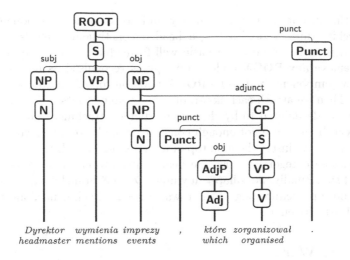

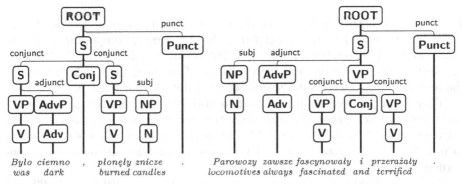

Fig. 5. Examples of verbal spines: 'The headmaster mentions events he organised.', 'It was dark, candles were burning.', 'Steam locomotives have always fascinated and terrified people.'

- its spine,[10]
- the category of its spine's attachment node,
- the number of its spine's attachment node counting from the bottom.

The BERT-style models perform their own tokenisation as a preprocessing step. This tokenisation often splits text words into smaller segments, according to the specific models' tokeniser vocabulary. As a result, for one input sentence token, several BERT vectors can be produced and passed to the final classifiers, leading to multiple (possibly incoherent) predictions associated with the same token. Therefore, we use the common technique of masking the BERT outputs for all but one segment of each token when calculating the loss function. In our experiments, we chose to only leave the first segment unmasked.

---

[10] Since the number of distinct spines is fairly limited, we decided to treat them as atomic labels.

Given the outputs of the dependency and constituency components of our parsing architecture, reconstructing a headed constituency tree is straightforward. First, in order to ensure a basic well-formedness of the produced structure, we remove any **ROOT** nodes from spines recognised as non-root by the dependency component, and add a **ROOT** node on top of the root token's spine if missing. Then we attach each non-root token's spine to its dependency head's spine at the node pointed to by the two predicted attachment labels (category and number). If there are not enough consecutive nodes of given category along the head's spine, we insert them on top of the existing ones. If there are no nodes of the requested category, we use the head spine's topmost node as a fallback attachment site. Finally, we collapse any unary **X** → **X** branches that could result from the above procedure (e.g. if there was an **X**-2 attachment along the spine, but no **X**-1 attachment).

6   Related Work

The proposed method is in a close relation to the attempts at hybrid dependency/constituency parsing. In particular the structure we propose is similar to what Zhou and Zhao (2019); Zhou et al. (2020) call "simplified HPSG". However the authors do not assume full compatibility of the structures, which leads to the necessity of artificial nodes which accommodate for the discrepancies.

There is a substantial line of work devoted to accommodating transition-based constituency parsing to discontinuities, e.g. Coavoux and Cohen (2019). Fernández-González and Martins (2015) present an approach that is very related to ours in that it leverages dependency parsing for (also discontinuous) constituency parsing.

As for continuous constituency parsing, Kitaev et al. (2019) quote 96.36% as F1 score of their neural chart-based parser for Polish. This figure is to some extent suitable for comparison with our experiments since it is calculated on the data of (Seddah et al., 2013) which is in fact an old version of Składnica (including short sentences and simpler grammatical constructions).

Our idea to gather all constituents with the same token being the centre, forming a spine, is, to the best of our knowledge, new. The idea to predict spines at given token positions is similar to the conception of parsing as tagging (e.g. Gómez-Rodríguez and Vilares, 2018). The concept of decorating edges of a constituency tree with relations was present already in the TIGER treebank of German (Brants et al., 2004), but their structures were not strictly headed (the relations did not form a dependency tree).

7   Evaluation

We evaluate our parser in terms of precision (P), recall (R) and F1 metrics over constituents in three variants. The evaluation units are the (possibly discontinuous) yields of all nonterminals of a tree paired with (depending on the metric variant):

Table 4. Results on validation and test data.

	validation data			test data		
	precision	recall	F1	precision	recall	F1
non-compressed						
bracketing	96.40%	96.25%	96.32%	96.56%	96.45%	**96.51%**
constituents	96.46%	96.33%	96.39%	96.64%	96.53%	**96.59%**
headed constit.	95.70%	95.93%	95.81%	95.82%	96.09%	**95.96%**
compressed						
bracketing	96.28%	96.23%	96.26%	96.41%	96.48%	**96.44%**
constituents	96.40%	96.25%	96.32%	96.57%	96.48%	**96.52%**
headed constit.	95.81%	95.67%	95.74%	95.96%	95.88%	**95.92%**

- constituents: the nonterminal's syntactic category,
- headed constituents: the nonterminal's syntactic category and information whether the nonterminal is a head node,
- bracketing only: no additional information (i.e. a constituent needs only to be recognised, regardless of its assigned syntactic category).[11]

All the models for the constituency component were trained using Adam optimiser with learning rate of $2 \cdot 10^{-5}$ and categorical cross-entropy loss summed over all classifiers. The best model was selected using average accuracy on validation data across classifiers. The training was stopped when no better accuracy was achieved for 4 epochs.[12] We trained two variants of the model, using non-compressed and compressed spines. We show the results in Table 4. The difference in performance of both models is negligible, therefore in further analysis we concentrate on the non-compressed version which achieved 95.96% headed constituents F1-score on test portion of our dataset.

In Table 5, we examine how the constituency component of our architecture performs when it comes to predicting spines for particular tokens. The overall accuracy of spine prediction of test data was 97.13%. We calculate aggregate precision, recall and F1[13] on test data for spines ending with particular preterminals in order to check which kinds of spines are more difficult for the parser. We

---

[11] For the bracketings metric, each span is counted only one time, e.g. for the tree in Fig. 4, $(Dam)_{VP}$ and $(Dam)_V$ are treated as the same span (Dam) etc.

[12] We noticed that when validation data loss was used for early stopping and model selection, the accuracies on validation data still exhibited a growing tendency.

[13] Let TP_l, FP_l, FN_l denote the number of true positives, false positives and false negatives respectively for label l in evaluation data. For a set of labels S, we calculate the aggregate precision P_S as $(\sum_{l \in S} TP_l)/(\sum_{l \in S} TP_l + FP_l)$, i.e. the proportion of correctly predicted labels from S to all predicted labels from S. The aggregate recall R_S is $(\sum_{l \in S} TP_l)/(\sum_{l \in S} TP_l + FN_l)$, i.e. the proportion of correctly predicted

Table 5. Aggregate precision, recall and F1 for spine types.

spine	precision	recall	F1	occurrences
... → **Punct**	99.03%	99.36%	99.20%	5158
... → **Prep**	98.31%	98.39%	98.35%	3488
∅	98.78%	97.88%	98.33%	330
... → **Adj**	98.00%	98.41%	98.20%	4079
... → **Part**	98.69%	97.42%	98.05%	1627
... → **N**	98.07%	97.95%	98.01%	10435
... → **Comp**	97.08%	97.51%	97.29%	682
... → **V**	95.47%	95.45%	95.46%	4398
... → **Adv**	95.11%	95.74%	95.42%	1056
... → **Num**	90.87%	90.47%	90.67%	451
... → **Conj**	85.43%	84.68%	85.05%	1364

are not surprised by the **Conj** spines turning out to be the hardest to correctly assign since, in our annotation scheme, they are associated with coordination. One could expect a (near) 100% figure for punctuation which should be very easy for the model to learn. However, some occurences of punctuation are annotated as conjunctions in our data, so their classification is not entirely trivial. We also note that **V** spines are, as a whole, predicted with worse results than the **N** ones.

For a more detailed look into that last osbervation, we also report, in Table 6 precision, recall and F1-score on test data for individual nominal and verbal spines (as shown in Table 3) that appear in the test data portion. For **N** spines, we note that the results for two most common cases (modified vs non-modified noun) the parser performs comparably well. The results for the spine corresponding to noun-centered utterances are visibly lower, but it should be noted that they are two orders of magnitude less frequent in the data, which probably makes learning them substantially harder. As far as the **V** spines are concerned, the results seem to reflect the diversity of syntactic contexts in which particular spines may appear. The **ROOT**-dominated spine corresponds to the main predicate, a **CP** has a specific context, and they are both recognised with a relatively high F1. Meanwhile, the **S**- and **VP**-dominated spines may be associated with a variety of syntactic phenomena: coordination, subordinate infinitival phrase, subordinate clause.

---

labels from S to all gold labels from S. The aggregate $F1_S$ is the harmonic mean of P_S and R_S.

Table 6. Precision, recall and F1 for selected spines.

spine	precision	recall	F1	occurrences
NP → NP → N	98.51%	97.89%	98.20%	5458
NP → N	97.62%	98.17%	97.89%	4926
ROOT → NP → NP → N	95.45%	82.35%	88.42%	51
ROOT → S → VP → V	98.09%	99.10%	98.59%	1661
CP → S → VP → V	96.68%	97.14%	96.91%	210
VP → VP → V	91.89%	95.36%	93.60%	1165
S → VP → V	95.69%	91.18%	93.38%	1168
VP → V	92.47%	88.66%	90.53%	194

Table 7. Comparison with Benepar.

	validation data			test data		
	precision	recall	F1	precision	recall	**F1**
non-compressed, continuous only						
headed	96.05%	96.29%	96.17%	96.13%	96.38%	**96.26%**
Benepar, continuous only						
headed	97.44%	97.43%	97.44%	97.74%	97.75%	**97.75%**

We also performed an experiment comparing our architecture with one of the state-of-the-art constituency parsers, the Berkeley Neural Parser[14] (Benepar, Kitaev et al. 2019). Since Benepar only operates on continuous constituency structures, we trained and evaluated it on the continuous portions of our data, and compared the results with an instance of our architecture trained on the same data (for both the dependency and spine-attachment components). Moreover, as Benepar does not have an explicit mechanism for handling syntactic heads, we represented head nodes by prepending a special character to their labels, thus creating separate labels for head and non-head constituents of each type and roughly doubling the number of possible labels. The results are presented in Table 7. Our architecture performs worse (by 1.5 pp F1-measure) than Benepar on this data. Nevertheless, we find this result encouraging. We believe that, given the inherent ability of our approach to handle discontinuous structures, it is worthwhile to aim at improving its general performance in future work.

---

[14] https://parser.kitaev.io/.

8 Conclusions

In the paper, we have proposed a syntactic structure of headed constituencies and a method for parsing such structures.

We think that this structure, merging constituency and dependency information, is a handy model of syntax. The structure exploits strengths of both representations. In particular constituents provide natural representation for coordinated structures (which are problematic in dependency trees), but the dependencies between tokens (and constituents) are also available in the structure.

The performed experiments show that with present neural models it is possible to express constituency parsing in terms of detecting spines and their attachments. For us, the most important feature of this method is that it generates constituency structures consistent with the dependencies. Another important aspect is the ability to process discontinuous structures, which are hard for many constituency parsers. The method was tested on Polish due to availability of the data, but the approach is not in any way specific to the language in question. Taking into the account that BERT-type models were successfully trained for various languages, we have reasons to believe that the proposed method would work comparably for other languages.

The achieved results of around 96% F1 measure for a dataset with discontinuities show that the proposed method is well in the state-of-the-art zone. Our result for continuous trees is lower than that of Berkeley Neural Parser, which shows that there is room for improvement, which we intend to explore. In this work, we used an external parser as the source of dependency edges. An interesting direction of future work would be to check whether integrating both parts within a joint model could provide some synergy in learning.

References

Brants, S., et al.: TIGER: linguistic interpretation of a German corpus. J. Lang. Comput. **2**, 597–620 (2004). https://doi.org/10.1007/s11168-004-7431-3

Coavoux, M., Cohen, S.B.: Discontinuous constituency parsing with a stack-free transition system and a dynamic oracle. In: Proceedings of the 2019 Conference of the North American Chapter of the Association for Computational Linguistics: Human Language Technologies, Volume 1 (Long and Short Papers), pp. 204–217. Association for Computational Linguistics, Minneapolis, Minnesota (2019). https://doi.org/10.18653/v1/N19-1018

Fernández-González, D., Martins, A.F.: Parsing as reduction. In: Proceedings of the 53rd Annual Meeting of the Association for Computational Linguistics and the 7th International Joint Conference on Natural Language Processing (Volume 1: Long Papers), Beijing, China, pp. 1523–1533. Association for Computational Linguistics (2015). https://doi.org/10.3115/v1/P15-1147

Gerdes, K., Guillaume, B., Kahane, S., Perrier, G.: SUD or surface-syntactic universal dependencies: an annotation scheme near-isomorphic to UD. In: Universal Dependencies Workshop 2018, Brussels, Belgium (2018). https://hal.inria.fr/hal-01930614

Gómez-Rodríguez, C., Vilares, D.: Constituent parsing as sequence labeling. In: Proceedings of the 2018 Conference on Empirical Methods in Natural Language Processing, Brussels, Belgium, pp. 1314–1324. Association for Computational Linguistics (2018). https://doi.org/10.18653/v1/D18-1162

Kitaev, N., Cao, S., Klein, D.: Multilingual constituency parsing with self-attention and pre-training. In: Proceedings of the 57th Annual Meeting of the Association for Computational Linguistics, Florence, Italy, pp. 3499–3505. Association for Computational Linguistics (2019). https://doi.org/10.18653/v1/P19-1340

Klimaszewski, M., Wróblewska, A.: COMBO: State-of-the-art morphosyntactic analysis. In: Proceedings of the 2021 Conference on Empirical Methods in Natural Language Processing: System Demonstrations, Punta Cana, Dominican Republic, pp. 50–62. Association for Computational Linguistics (2021). https://aclanthology.org/2021.emnlp-demo.7

Marcus, M.P., Santorini, B., Marcinkiewicz, M.A.: Building a large annotated corpus of English: the Penn Treebank. Comput. Linguist. **19**(2), 313–330 (1993). https://aclanthology.org/J93-2004

Mroczkowski, R., Rybak, P., Wróblewska, A., Gawlik, I.: HerBERT: Efficiently pretrained transformer-based language model for Polish. In: Proceedings of the 8th Workshop on Balto-Slavic Natural Language Processing, Kiyv, Ukraine, pp. 1–10. Association for Computational Linguistics 2021. https://www.aclweb.org/anthology/2021.bsnlp-1.1

Nivre, J., et al.: Universal dependencies v2: An evergrowing multilingual treebank collection. CoRR, abs/2004.10643 (2020)

Seddah, D., et al.: Overview of the SPMRL 2013 shared task: a cross-framework evaluation of parsing morphologically rich languages. In: Proceedings of the Fourth Workshop on Statistical Parsing of Morphologically-Rich Languages, Seattle, Washington, USA, pp. 146–182. Association for Computational Linguistics (2013). https://aclanthology.org/W13-4917

Świdziński, M., Woliński, M.: Towards a bank of constituent parse trees for Polish. In: Sojka, P., Horák, A., Kopeček, I., Pala, K. (eds.) TSD 2010. LNCS (LNAI), vol. 6231, pp. 197–204. Springer, Heidelberg (2010). https://doi.org/10.1007/978-3-642-15760-8_26

Woliński, M.: Automatyczna analiza składnikowa języka polskiego. Wydawnictwa Uniwersytetu Warszawskiego, Warsaw (2019). https://www.wuw.pl/data/include/cms/Automatyczna_analiza_skladnikowa_Wolinski_Marcin_2019.pdf

Woliński, M., Hajnicz, E.: *Składnica*: a constituency treebank of Polish harmonised with the *Walenty* valency dictionary. Lang. Res. Eval. **55**(1), 209–239 (2021). https://doi.org/10.1007/s10579-020-09511-7

Wróblewska, A.: Polish Dependency Parser Trained on an Automatically Induced Dependency Bank. Ph.D. dissertation, Institute of Computer Science, Polish Academy of Sciences, Warsaw (2014). http://nlp.ipipan.waw.pl/Bib/wro:14.pdf

Zhou, J., Li, Z., Zhao, H.: Parsing all: syntax and semantics, dependencies and spans. In: Findings of the Association for Computational Linguistics: EMNLP 2020, pp. 4438–4449. Association for Computational Linguistics (2020). https://doi.org/10.18653/v1/2020.findings-emnlp.398

Zhou, J., Zhao, H.: Head-driven Phrase Structure Grammar parsing on Penn Treebank. In: Proceedings of the 57th Annual Meeting of the Association for Computational Linguistics, Florence, Italy, pp. 2396–2408. Association for Computational Linguistics (2019). https://doi.org/10.18653/v1/P19-1230

Performing Aerobatic Maneuver with Imitation Learning

Henrique Freitas[1] , Rui Camacho[1,2] , and Daniel Castro Silva[1,3]( )

[1] Department of Informatics Engineering, Faculty of Engineering,
University of Porto, Rua Dr. Roberto Frias, s/n., 4200-465 Porto, Portugal
up201707046@edu.fe.up.pt, {rcamacho,dcs}@fe.up.pt
[2] LIAAD/INESC TEC - Artificial Intelligence and Decision Support
Laboratory/Institute for Systems and Computer Engineering, Technology
and Science, Porto, Portugal
[3] LIACC - Artificial Intelligence and Computer Science Laboratory, Porto, Portugal

Abstract. The work reported in this article addresses the challenge of building models for non-trivial aerobatic aircraft maneuvers in an automated fashion. It is built using a Behavioural Cloning approach where human pilots provide a set of example maneuvers used by a Machine Learning algorithm to induce a control model for each maneuver. The best examples for each maneuver were selected using a set of objective evaluation metrics. Using those example sets, robust models were induced that could replicate (and in some cases outperform) the human pilots that provided the examples (the clean-up effect). Complete complex maneuvers were performed using a meta-controller capable of sequencing the basic ones learned by imitation. This endeavor was rewarded by the results that show several Machine Learning models capable of performing highly complex aircraft maneuvers.

Keywords: Imitation Learning · Behavioural Cloning · Aviation · Autopilot · Aerobatic maneuvers

1 Introduction

With the increase of robot complexity, manually programming the behaviors and actions is becoming costly [1]. It requires excellent knowledge about the robot movement and many resources to develop [15].

In recent years, autonomous driving technologies are improving with Artificial Intelligence methods. One of the used techniques is Imitation Learning (IL) and Behaviour Cloning (BC) [8,24,25]; these have been used for a few decades, are based on Supervised Learning and learn the behaviors based on expert demonstrations through imitation [19,20,22], i.e., learning the mappings between the environment state (input) and actions (output).

The work reported in this paper concerns the application of such Behavioural Cloning methodology to an aviation domain. Currently used autopilots are not helpful in various complex events, being usually only used in specific parts of the flight [6].

J. Mikyška et al. (Eds.): ICCS 2023, LNCS 14073, pp. 206–220, 2023.
https://doi.org/10.1007/978-3-031-35995-8_15

The commercial aviation industry keeps growing, and according to historical data, the passenger count doubles roughly every 15 years [26]. In the 2004–2019 period remarkably, it increased by a factor of 2.4. As stated by the International Air Transport Association in their 20-Year Passenger Forecast, from 2019 to 2040, the global air passenger growth is predicted to be 3.3% annually, which results in a 90% increase in that interval [13].

Although the global airlines market size was over 800B USD in 2019 [14], the airlines anticipate problems in keeping up with the demand, such as pilot shortage [16,27]. The Federal Aviation Administration (FAA) rule that limits airlines to hire pilots with a minimum of 1500 flight hours is one of the possible causes [17]. This rule aggravates the high training cost [17] because it requires pilot trainees to spend even more money to meet this requirement. The United States Air Force is also struggling with pilot training [5].

To solve these problems, an intelligent autopilot system can be built to be responsible for piloting through all phases of a flight or mission and/or help pilot apprentices improve their training with tips and cues on what they are doing wrong and how to improve. This system could also define an evaluation threshold to inform how close to the end one specific trainee is.

Focused on this issue, the first contribution of this work is the implementation of evaluation metrics for three aerobatic maneuvers: Immelmann, Split-S, and Half Cuban Eight. For instance, flight instructors can use these metrics to better understand the eventual difficulties their trainees are having.

These metrics are then used to develop Machine Learning controllers capable of performing the mentioned aerobatic maneuvers. Previous work uses Neural Networks (NN) based on Long Short-Term Memory (LSTM), mostly in autonomous driving [8,24,25] but also in the aerobatics context [18,21,23], where air vehicles are controlled solely by automated systems.

The final contribution is a high-level controller to perform an aerobatic performance show, similar to a Red Bull Air Race performance. This system sequentially invokes the respective controllers to execute the specific maneuvers.

This document is structured as follows: Sect. 2 presents a study and review of the related work using this technology in other use cases. Section 3 briefly describes the data and evaluation metrics. Section 4 explains the controller training process and the results. Section 5 describes the high-level controller for the previously trained controllers. Finally, Sect. 6 concludes this work with possible future work evolution.

2   Related Work

Regarding IL, there are many introductions, studies, and reviews on the subject, such as [2,12,22]. Despite being generic, it has been recently operated thoroughly in robots and similar contexts, such as autonomous driving [1,11].

In particular to aerial vehicles, Müller et al. used a Convolutional Neural Network (CNN) to learn to control a racing drone [21]. The data was collected in a simulator, automatically generating stadium racing tracks and three levels of expert pilots. Joystick controls and images from the Unmanned Aerial Vehicle (UAV) point-of-view were recorded. The control output is divided into throttle, elevator, aileron, and roll. The CNN could fly through the racing tracks at high speed and even outperform state-of-the-art methods. Several skill levels were used to compare how data quality influences the learning pipeline. The authors conclude that better data results in better models and lap times.

Rodriguez-Hernandez at el. also demonstrated how a BC approach could control a Micro Aerial Vehicle (MAV) [23]. The network consisted of a CNN to extract features from the input images and return actions to fly the MAV through gates.

One exciting concept was introduced in [9]: collect drone crashing data to enable better learning through negative examples. One of the real-world fears of autonomous UAV systems is that it hits objects due to the low generalization of the NNs. The authors collected many crashing examples to be used alongside the positive samples, resulting in a robust policy. The negative data were demonstrated to be very important, and the results were comparable to humans in some environments.

As referred above, IL techniques are very used in autonomous driving [8,24,25]. All authors tried end-to-end trainable models, with some variations of CNN architecture connected to LSTM units or Fully Connected Networks (FCN). Curiously, [24] used a UAV to test this approach; nonetheless, the output is context-agnostic as the steering angle is used in any vehicle navigating the environment. Here, the CNN is connected to an LSTM network to ensure a good fit for the temporal-dependent actions. The environment used in [8] and [25] was a simulator, and both architectures output values relative to the vehicle's steering angle, based on the front-road images. Also, in both essays, CNN's last layers were FCN.

On a slightly different effort, [7] focused on a new benchmark to test the scalability of BC. In an autonomous driving setting, they confirmed, besides some well-known limitations, such as dataset bias and overfitting, generalization issues, and training instabilities. The requirement was to further research BC before putting such models in real-world driving.

Regarding aerobatics, [18] compares a standard FCN and an LSTM-based NN. Two controllers were developed, one for the ailerons and another for the elevators. The maneuver performed was an Immelmann turn, and the testing results revealed that the LSTM-based NN is better generalized, with better parameterisable values.

A thorough and complete work is [3], presenting a compilation of four published works and a subsequent article published later [4]. The general goal is to have an intelligent autopilot system, composed of a set of small FCNs, divided into specific flight phases, and a Flight Manager responsible for controlling them and selecting the appropriate ones based on a behavior tree.

With all applications of such techniques, especially to aerial vehicles, it is possible to conclude that applying BC in an airplane-control system is feasible. Also, LSTM-based NNs seem to be a good method to solve this problem.

3   Data Analysis

The data collection phase counted on 25 amateur volunteers from the academic community (mainly BSc., MSc., and Ph.D. students, as well as professors associated with the department), with experience varying from first-time performances to practiced flight simulator enthusiasts. They were tasked with executing the aerobatic maneuvers to collect different examples. The idea was to gather a diverse dataset without the possible bias if collected from only 1 or 2 people. The collected values are relative to aircraft position and attitude, as well as environmental values such as atmospheric pressure and temperature; the airplane used was the aerobatics-capable Extra 300S. Microsoft Flight Simulator X (FSX) was the chosen flight simulator, used both to collect the data and test the controllers.

The complete dataset was published in Zenodo, on https://zenodo.org/record/6803193[1].

3.1   Maneuvers Description

The three maneuvers are composed of specific segments and must obey some 'rules' to be considered good; they are all composed of a main vertical component, upwards for Immelmann and Half Cuban Eight or downwards for Split-S. Figure 1 displays execution diagrams for each one. The Immelmann starts with an upwards semi-loop, followed by a semi-roll to stabilize the airplane. The Split-S comprises a semi-roll for the airplane to become upside down, followed by a semi-loop downwards until straight level flight is achieved. The Half Cuban Eight is similar to the Immelmann, with a 5/8 loop upwards, a semi-roll with a pitch of 45° downwards, and a final 1/8 loop to level the airplane. A good Half Cuban Eight should have the exit point at the same altitude as the entry point; however, as the volunteers are amateurs, this component was not mandatory.

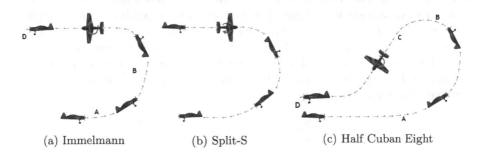

(a) Immelmann   (b) Split-S   (c) Half Cuban Eight

Fig. 1. Aerobatic maneuvers Diagrams

---

[1] Details on the dataset itself are in the *readme.me* file in the repository and specifics on the collection protocol in https://hdl.handle.net/10216/143035.

3.2   Evaluation Metrics

Some metrics were developed to numerically quantify and then sort the collected examples according to their performance; some are maneuver-specific, while others apply to all maneuvers.

Initial and Final Heading (IFH) is the calculated error between the initial and final heading to establish vertical plane consistency; given the initial heading h_0 and the final heading h_f, both in degrees, the difference d is the smaller angle between both values $d = (h_f - h_0 + 180) \% 360 - 180$; normally d should be $180°$ (opposite direction of entry and exit points), so $err = 180 - abs(d)$.

Heading Difference (HD) is the average error of every timeframe heading compared with initial and final values, a metric also related to vertical plane consistency that accounts for mid-execution drifts. Considering the same initial heading h_0 and final heading h_f, h_i is the heading at timeframe $i : 0..f$; d_{i0} and d_{if} are respectively the smaller angle between h_i and h_0 or h_f, calculated as d from the metric above. Then, from both d_{i0} and d_{if}, we consider the smallest (closer to one of the values), $sum = \sum_{i=0}^{f} min(\ abs(d_{i0}),\ abs(d_{if}))$ and finally $err = sum/s$, being s the size of the example.

Semi-Loop (S-L) measures how well the semi-loop trajectory fits a semi-circumference, as intended for the maneuvers; thus, it uses positional data points to find the best circumference curve[2]. Only the timeframes with the airplane pointing upwards (or downwards in Split-S) are considered. After this, the Euclidean distance $d_i, i : 0..f$ from each data point to the closest circumference point is calculated; the error for this metric is the average of the distances, making $sum = \sum_{i=0}^{f} d_i$ and $err = sum/s$.

Semi-Roll Overshoot (S-RO) is one of the metrics that evaluate how well the semi-roll is executed, focusing on whether or not there was some overshooting (more than $180°$ of rotation). Similar to Semi-Loop error calculation, only the timeframes in specific conditions can be considered: pitch near zero and bank value far from straight level flight. This metric works by analyzing the evolution of bank values, looking for any eventual progression over the zero-angle bank value. The implementation uses the multiplication of successive bank values $b_i, i : 0..f - 1$ to find a negative value (when bank goes from positive to negative or vice-versa). If found, future bank differences between consecutive frames d_{b_i} are summed. The formal definition is

$$err = \begin{cases} \sum_{n=i}^{f-1} d_{b_n} & \text{if } \exists i : b_i * b_{i+1} < 0 \\ 0 & \text{otherwise} \end{cases}$$

Semi-Roll Straightness (S-RS) measures the pitch variations during the semi-roll execution within the Half Cuban Eight maneuver, as it is important to

---

[2] Calculation based on the Coope method. Library API documentation available online at https://scikit-guess.readthedocs.io/en/latest/generated/skg.nsphere_fit.html.

keep the pitch constant, near $45°$ downwards. With the same timestamp restriction as Semi-Roll Overshoot, the values considered in this set are relative to the pitch value and calculating standard deviation std, finally $error = std$.

Semi-Roll Altitude Consistency (S-RAC) measures altitude changes during the semi-roll part of a maneuver, as altitude should be kept constant. For this evaluation, all altitudes differences d_{a_i} between two consecutive values $a_i, i : 0..f$ are added in the form $err = \sum_{i=0}^{f-1} abs(d_{d_i})$.

Semi-Roll Completion (S-RC). When collecting Split-S examples, it was frequent for the volunteers to start the semi-loop before the semi-roll was complete, resulting in a less-than-180-degree semi-roll. This caused the airplane to have more significant side drift than supposed; to counteract on and reduce such wrongful actions, this metric was added, calculated as follows: the bank value b_i is correspondent to timeframe $i : 0..f$ from the semi-loop, d_{i0} and d_{i180} is the smaller angle between b_i and 0 (straight level flight) or 180 (upside down), respectively. Then, $sum = \sum_{i=0}^{f} min(\ abs(d_{i0}),\ abs(d_{i180}))$ and $error = sum/s$, s being the size of the loop timeframes set.

Total Evaluation is calculated as the sum of the singular metrics, using a set of weights to manipulate the individual error distribution in order to increase the impact of bigger errors (exponential $- e_i : i = 1..n$) and balance the distributions of the metrics (multiplication $- m_i : i = 1..n$). Thus, the final score is given by $evaluation = \sum_{i=1}^{n} err_i^{e_i} * m_i$.

3.3 Maneuvers Evaluation

Table 1 presents the metrics used for each maneuver, with the respective multiplicative weights; the exponential weights are all set to 1, except S-RS for Half Cuban Eight, which is set to 2.

Besides the two parts that compose the Immelmann (semi-loop and semi-roll), another thing to consider is the vertical plane consistency and the heading deviations from this initial plane, which are considered errors; this is common to the three maneuvers. Figure 2a shows the distribution of the metrics with the weights for all 162 collected examples.

Split-S is quite similar to the Immelmann maneuver, with the exception of some key details: the order of the operations and the direction. Figure 2b displays the metric distribution for the 153 collected examples. We note that Semi-Roll Completion is the component with the highest error, which is congruent with the difficulties felt during data collection, also negatively affecting other metrics.

Again, Half Cuban Eight is similar to Immelmann; the metric distribution for the 157 collected examples is depicted in Fig. 2c.

Table 1. Multiplicative weights used for each maneuver

Metric	Immelmann	Split-S	Half Cuban Eight
IFH	2	0.6	1
HD	2	1	1
S-L	1	1	1
S-RO	1.4	-	1
S-RAC	0.2	0.5	-
S-RC	-	1	-
S-RS	-	-	0.5

4 Controllers Training

Throughout the development of this work, different feature combinations were tried until good results were obtained, with some feature engineering required. Due to space constraints, those experiments are not detailed here.

The used NN configuration is composed of a layer of LSTM cells, with the same amount of units as the number of features. A dropout layer was used with a value of 0.175, a learning rate of 0.005, and a batch size of 64. These values were obtained by a hyperparameter tuning step performed in [18]; also, 15 was the better window size to observe in the LSTM layer.

Since the examples were all executed with full throttle on the airplane and the maneuvers do not require rudder control, only controllers for the elevator and aileron were trained. The specific features for each one are listed in Table 2.

With the evaluation metrics developed, it is possible to select the best examples for training the models. While using only the best examples for training is likely to provide for better results, a balance must be obtained between the quality and the number of samples used for training, as a lower amount of samples is likely to lead to worst results. For this step, several subsets were chosen to train the controllers and compare the results. The chosen subsets are as follows:

- **100% or All**: the control group with all collected samples;
- **Best 90%**: removing the worst 10%, as those are likely to include outliers and bad examples;
- **Best 75%**: mid value between 90 and 60%;
- **Best 60%**: for the most restricted subset, we use 60% of the collected data; considering the variability necessary to train generalizable controllers, any less than that seemed like a considerable reduction in sample size.

4.1 Results

To gather results, the test consisted in positioning the same aircraft in similar conditions in the simulator and having the controllers guide it through a full

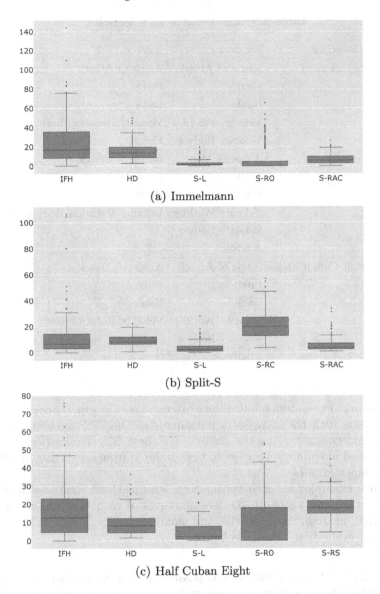

(a) Immelmann

(b) Split-S

(c) Half Cuban Eight

Fig. 2. Metric evaluation distribution for the three maneuvers

execution of the respective maneuver. Hence, five sequential maneuver executions for each previously mentioned version were collected and evaluated using the previously developed metrics; the evaluations for each version were compared with the remaining versions and the human-collected examples (shown as 'Examples' in the graphics).

Table 2. Features used in each controller

Maneuver	Elevator	Aileron
Immelmann	Angle of Attack	Angle of Attack
	Pitch	Pitch
	Bank	Bank
	Velocity World y	Velocity Rotation Body y
	Velocity Body z	Elevator
Split-S	Angle of Attack	Angle of Attack
	Pitch	Pitch
	Bank	Bank
	Velocity World y	Velocity Rotation Body y
	Velocity Body z	
	Aileron	
Half Cuban Eight	Angle of Attack	Angle of Attack
	Pitch	Pitch
	Bank	Bank
	Velocity World y	Velocity Rotation Body y
	Velocity Body z	Elevator
	Aileron	

Immelmann. The results fulfilled the expectations. The graph presents a favorable evolution with the increase in data quality. The 75% model shows great maturity and consistency in the results. The best 60% model (less than 100 examples used in training) appears to have suffered from overfitting or low variability of input examples.

The five examples for each version were grouped and compared in Fig. 3a. The evaluations are visible in the five markings of each box: minimum value, first quartile, median, third quartile, and maximum value. The X axis is relative to the error evaluations – lower is better.

Split-S. Similar to Immelmann, it is possible to notice the increase in performance, although less significant, when training with better data; however, it is most noticeable that the examples have more consistent evaluations, as depicted in Fig. 3b. It is clear the best model is the best 90%, as it gets all five executions with consistent quality relative to the others and even four within the best 25% of the collected data.

Half Cuban Eight. Observing Fig. 3c, the model trained with the best 60% examples presented the best results. However, a strange phenomenon arose: it only started the maneuver with an airspeed close to 200 knots – it was found that the minimum speed to perform a complete semi-loop was 140 knots, as was advised to all volunteers. 200 knots is not an effortless speed to reach in a straight flight with this particular aircraft since its maximum speed is only 240 knots. Therefore, the best 75% model was the best; the base model (100%) was close, but the consistency was decisive.

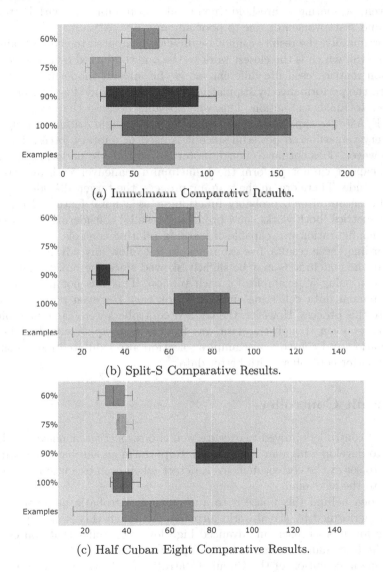

(a) Immelmann Comparative Results.

(b) Split-S Comparative Results.

(c) Half Cuban Eight Comparative Results.

Fig. 3. Maneuvers comparative results. X-axis represents the total evaluation calculated by the metrics - lower is better

4.2 Discussion of Results

The results reflect an expected phenomenon, the difficult balance between using quality data to train the models and the amount and diversity of the data available for the training process. The Immelmann results showcase this phenomenon rather well, presenting a threshold when results stop improving with better data quality and start worsening due to poor amount/diversity.

Unfortunately, the results cannot be directly compared to those obtained by Medeiros [18], which is the closest work to the one conducted by this study. The evaluation metrics used are different, as is the aircraft model used. Medeiros measured the performance by stipulating a required altitude for the Immelmann controller to finish the action. This was possible because the aircraft used, a Boeing F/A-18, is capable of maintaining upwards flight without losing speed – it is very versatile and powerful since it is used in military forces. The Extra 300S, however, does not have enough motor thrust force to perform that ascent, which means it cannot perform the Immelmann maneuver with any specified target altitude. Therefore, in the training process, the target altitude was found not to be accomplished, probably functioning as a noise feature. Despite the different metrics, both works show promising results, considering that a similar network configuration was capable of learning piloting controls.

Regarding these results, five examples do not allow any advanced statistics; therefore, the conclusions may be slightly skewed since the boxes might be different when using 50 examples for each version. This was not possible due to heavily-manual data collection since it is not easy (or even possible) to fully automate this process. However, the trained controllers were capable of piloting the airplane, even though the data was from a non-expert sample. The metric evaluation and consequent examples selection also exhibited partial success, showing better controllers with better data.

5 Circuit Controller

The set of controllers proved to execute well the respective maneuvers, which is enough to develop a high-level controller. To perform an aerobatic performance, it needs to select the correct maneuver, detect when each execution is over, and proceed to the next one.

The idea behind this phase is to automate an aerobatic performance that could be performed by a real-world airplane. Although this is an experimental step, the focus is to reach an advanced high-level controller that can correctly choose the best maneuver suited for the specific flight.

The main execution of the Circuit Controller (CC) was the state machine, where it iterates through the maneuvers' controllers' actions until it finishes all of them. Between maneuvers, the airplane flies in a straight line for 10 s, using FSX built-in autopilot capabilities, so that the maneuvers can be easily isolated and identified, and the aircraft can gain some speed for the next maneuver. For the conducted tests, the order of execution was as follows:

1. **Immelmann**: best 75% trained controller;
2. **Straight Level Flight**: FSX autopilot;
3. **Half Cuban Eight**: best 75% trained controller;
4. **Straight Level Flight**;
5. **Split-S**: best 90% trained controller;
6. **Straight Level Flight**.

Figure 4 shows the trajectory completed when performing the circuit in two different views. The circuit starts at point A, followed by B and C, marking the beginning and end of the Immelmann maneuver. Until point D, the FSX autopilot controlled the airplane in a straight level flight, where the Half Cuban Eight controller started piloting. With E indicating the end, a small deviation is noticeable in heading, both in the semi-loop (curve closer to D) and in the semi-roll (curve closer to E). The Split-S goes from F to G, and it is also showing some signs of deviations when executing the semi-roll, right after point F.

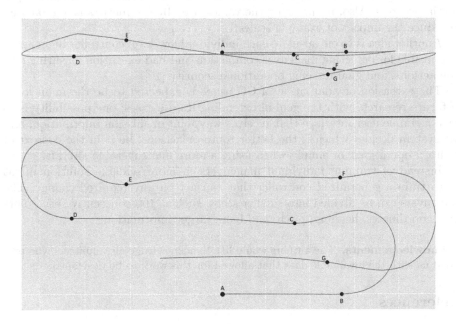

Fig. 4. Circuit views: The top part shows a top (YX) view, while the bottom part shows a side (YZ) view

This system could be evolved and used in automated air vehicle performance shows (something like Automated Red Bull Air Race or Automated Drone Racing League). From what is possible to understand of Baomar's Flight Manager [3], which is the most similar work found, CC is a similar system with an also similar use case, only differing in context: Flight Manager is used in a regular commercial flight, while CC is focused on more complex aerobatic maneuvers.

6 Conclusion

With this work, it is possible to conclude that with better demonstration data the controllers' performance also improves with IL. The technique developed by Medeiros [18] was expanded into other aerobatic maneuvers and a different airplane, showing consistency in the aerial environment.

It was interesting to use non-expert data and still successfully train controllers; besides, after excluding the worst examples by evaluating the amateur data, the controllers' performance also benefited.

We also present the Circuit Controller, an automated system that iterates through the different maneuvers, composing an aerobatic performance show. This system selects between the trained controllers and the built-in autopilot functionalities present in FSX.

This work can be extended or tuned in several ways. Regarding the evaluation metrics, some deeper study about the maneuvers is advised, such as better implementations or relations between the metrics – the multiplicative and exponential weights. Also, normalizing the values used by the metrics is a great way to reduce the impact of example scales.

Another idea is to encapsulate each individual low-level controller in an agent responsible for detecting maneuver completion and bad execution, ensuring failsafe actions, and guaranteeing operational security [10].

The extension or adaptation of CC usage is expected to be the main focus in future research, with the goal of extending its use cases; one possibility is to have a simple user interface that receives waypoint or mission information, and the system decides which is the better route or actions, be it in the context of military operations, or simply when using a more maneuverable aircraft.

Instead of training individual maneuvers, a more scalable solution might be to train a generalized controller that can act on any required change. The maneuvers can be divided into smaller steps, such as 100 consecutive waypoints with coordinates, heading, pitch, and bank target information.

Acknowledgements. The authors would like to acknowledge all volunteers who provided us with the invaluable data that allowed for this work to be developed.

References

1. Argall, B.D., Chernova, S., Veloso, M., Browning, B.: A survey of robot learning from demonstration. Robot. Auton. Syst. **57**(5), 469–483 (2009). https://doi.org/10.1016/j.robot.2008.10.024
2. Attia, A., Dayan, S.: Global overview of Imitation Learning. arXiv preprint, p. 9 (2018). https://doi.org/10.48550/arXiv.1801.06503
3. Baomar, H.: Using Learning from Demonstration to Enable Automated Flight Control Comparable with Experienced Human Pilots. Ph.D. thesis, University College London (2020). https://discovery.ucl.ac.uk/id/eprint/10108999/
4. Baomar, H., Bentley, P.J.: Autonomous flight cycles and extreme landings of airliners beyond the current limits and capabilities using artificial neural networks. Appl. Intell. **51**(9), 6349–6375 (2021). https://doi.org/10.1007/s10489-021-02202-y

5. Caballero, W.N., Gaw, N., Jenkins, P.R., Johnstone, C.: Toward automated instructor pilots in legacy air force systems: physiology-based flight difficulty classification via machine learning. SSRN (2022). https://doi.org/10.2139/ssrn.4170114

6. Claiborne, M.: How does autopilot work on a plane? AeroCorner. https://aerocorner.com/blog/how-does-autopilot-work/ (NA). Accessed Apr 2023

7. Codevilla, F., Santana, E., Lopez, A., Gaidon, A.: Exploring the limitations of behavior cloning for autonomous driving. In: Proceedings of the IEEE/CVF International Conference on Computer Vision (ICCV 2019), 27 October–2 November 2019, Seoul, vol. 2019, pp. 9328–9337 (2019). https://doi.org/10.1109/ICCV.2019.00942

8. Farag, W., Saleh, Z.: Behavior cloning for autonomous driving using convolutional neural networks. In: Proceedings of the 2018 International Conference on Innovation and Intelligence for Informatics, Computing, and Technologies (3ICT), 18–20 November 2018, Sakhier, pp. 189–195 (2018). https://doi.org/10.1109/3ICT.2018.8855753

9. Gandhi, D., Pinto, L., Gupta, A.: Learning to fly by crashing. In: Proceedings of the 2017 IEEE/RSJ International Conference on Intelligent Robots and Systems (IROS 2017), 24–28 September 2017, Vancouver, pp. 3948–3955 (2017). https://doi.org/10.1109/IROS.2017.8206247

10. Garrido, D., Ferreira, L., Jacob, J., Silva, D.C.: Fault injection, detection and treatment in simulated autonomous vehicles. In: Krzhizhanovskaya, V.V., et al. (eds.) ICCS 2020. LNCS, vol. 12137, pp. 471–485. Springer, Cham (2020). https://doi.org/10.1007/978-3-030-50371-0_35

11. Hua, J., Zeng, L., Li, G., Ju, Z.: learning for a robot: deep reinforcement learning, imitation learning, transfer learning. Sensors **21**(4), 21 (2021). https://doi.org/10.3390/s21041278

12. Hussein, A., Elyan, E., Gaber, M.M., Jayne, C.: Deep imitation learning for 3D navigation tasks. Neural Comput. Appl. **29**(7), 389–404 (2018). https://doi.org/10.1007/s00521-017-3241-z

13. IATA: 20 Year Passenger Forecast. IATA: (2021). https://www.iata.org/pax-forecast/. Accessed April 2023

14. IBISWorld: Global Airlines - Market Size 2005–2027. IBISWorld: (2021). https://www.ibisworld.com/global/market-size/global-airlines/. Accessed April 2023

15. Kober, J., Peters, J.: Learning motor primitives for robotics. In: Proceedings of the 2009 IEEE International Conference on Robotics and Automation (ICRA 2009), 12–17 May 2009, Kobe, pp. 2112–2118 (2009). https://doi.org/10.1109/ROBOT.2009.5152577

16. Kotoky, A., Yap, C.: A shortage of pilots looms as the next challenge for airlines. Bloomberg (2021). https://www.bloomberg.com/news/articles/2021-09-21/a-shortage-of-pilots-looms-as-the-next-challenge-for-airlines. Accessed April 2023

17. Mburu, W.: An In-Depth Study of the Pilot Shortage and its Consequences. Oklahoma State University (2017)

18. Medeiros, C.: Learn to Fly: Cloning the Behavior of a Pilot. Master's thesis, Faculty of Engineering, University of Porto (2021)

19. Michie, D., Camacho, R.: Building symbolic representations of intuitive real-time skills from performance data. In: Furukawa, K., Michie, D., Muggleton, S. (eds.) Machine Intelligence 13: Machine Intelligence and Inductive Learning, chap. 15, pp. 385–418. Oxford University Press, Inc. (1994). ISBN: 978-0-19-853850-9

20. Morales, E.F., Sammut, C.: Learning to fly by combining reinforcement learning with behavioural cloning. In: Proceedings of the 21st International Conference on Machine Learning (ICML 2004), 4–8 July 2004, Banff, p. 8 (2004). https://doi.org/10.1145/1015330.1015384

21. Müller, M., Casser, V., Smith, N., Michels, D.L., Ghanem, B.: Teaching UAVs to race: end-to-end regression of agile controls in simulation. In: Proceedings of the European Conference on Computer Vision (ECCV 2018) Workshops, 8–14 September 2018, Munich, pp. 11–29 (2018). https://doi.org/10.1007/978-3-030-11012-3_2

22. Osa, T., Pajarinen, J., Neumann, G., Bagnell, J.A., Abbeel, P., Peters, J.: An algorithmic perspective on imitation learning. Found. Trends Robot. **7**(1–2), 1–179 (2018). https://doi.org/10.1561/2300000053

23. Rodriguez-Hernandez, E., Vasquez-Gomez, J.I., Herrera-Lozada, J.C.: Flying through gates using a behavioral cloning approach. In: Proceedings of the 2019 International Conference on Unmanned Aircraft Systems (ICUAS 2019), 11–14 June 2019, Atlanta, pp. 1353–1358 (2019). https://doi.org/10.1109/ICUAS.2019.8798172

24. Saksena, S.K., Navaneethkrishnan, B., Hegde, S., Raja, P., Vishwanath, R.M.: Towards behavioural cloning for autonomous driving. In: Proceedings of the 2019 Third IEEE International Conference on Robotic Computing (IRC 2019), 25–27 February 2019, Naples, pp. 560–567 (2019). https://doi.org/10.1109/IRC.2019.00115

25. Samak, T.V., Samak, C.V., Kandhasamy, S.: Robust behavioral cloning for autonomous vehicles using end-to-end imitation learning. SAE Int. J. Connected Autom. Veh. **4**(3) (2021). https://doi.org/10.4271/12-04-03-0023

26. The World Bank: Air Transport, Passengers Carried. The World Bank: (2021). https://data.worldbank.org/indicator/IS.AIR.PSGR. Accessed April 2023

27. Wall, R., Tangel, A.: Facing a critical pilot shortage, airlines scramble to hire new pilots. Wall Street J. 1, 1–3 (2018). https://www.wsj.com/articles/pilot-shortage-spurs-hiring-spree-1533720602. Accessed April 2023

An Application of Evolutionary Algorithms and Machine Learning in Four-Part Harmonization

Mikołaj Sikora[✉][ID] and Maciej Smołka[ID]

Institute of Computer Science, AGH University of Krakow, Kraków, Poland
{mikolaj.sikora,smolka}@agh.edu.pl
https://www.informatyka.agh.edu.pl/en/

Abstract. The task of four-voice harmonization of a given melody is one of the most fundamental, but at the same time the most complex problems in functional harmony. This problem can be formulated as a discrete optimization problem with constraints using a set of rules coming from the theory of music. Unfortunately, a straightforward solution of such a problem, i.e., a mere fulfillment of the rules, ensures only the formal correctness of the obtained chord sequences, which does not necessarily imply overall musical quality as perceived by humans. Trying to catch some non-formalized factors of this quality we have decided to utilize artificial intelligence methods with some 'creative' potential that can provide solutions at acceptable level of formal correctness. In this paper we perform the harmonization using a genetic algorithm, an algorithm based on a Bayesian network, as well as a hybrid of these. In a series of experiments we compare the performance of the three algorithms with each other and with a rule-based system that provides chord sequences at a high level of formal correctness. Besides the formal evaluation all obtained solutions were rated by musical experts. The results show that the studied algorithms can generate solutions musically more interesting than those produced by the rule-based system, even if the former are less formally correct than the latter.

Keywords: algorithmic composition · constrained discrete optimization · evolutionary algorithms · machine learning

1 Introduction

Harmony is one of the most important branches of music theory. It governs the co-sounding of several melodies by arranging simultaneous sounds into chords and setting the rules of chord structure and succession. Therefore, it is one of the key factors of musical composition. The foundations of modern harmony date back to the Baroque musical period, i.e., the 17th and the first half of 18th centuries. The main assumptions of the Baroque approach, called the *functional*

This research was supported in part by the funds of Polish Ministry of Education and Science assigned to AGH University of Krakow.

J. Mikyška et al. (Eds.): ICCS 2023, LNCS 14073, pp. 221–236, 2023.
https://doi.org/10.1007/978-3-031-35995-8_16

harmony [11] (a modern description can be found in [13]), were in the mainstream till the beginning of the 20th century and still are foundations of many modern musical genres. In this approach melodies are composed using seven-note *major* or *minor* scales depending on their global key. At every step (i.e., note) of the scale we can build a chord. Basic chords are three-note *triads*, more sophisticated can be built by adding appropriate extra elements, omitting or altering some of existing ones. A key notion in the functional harmony is *harmonic function* of a chord, which determines the role of the chord in a musical composition and in consequence sets the rules for allowed connections with other chords. The basic harmonic functions are: *tonic (T)* based on the first note of the scale, *subdominant (S)* based on the fourth note and *dominant (D)* based on the fifth note. Each of the remaining functions is in a way related to T, S or D, which is then its *base* function, e.g., the chord built on the second note is a kind of subdominant.

The rules of functional harmony can be divided into two groups. The first group determines the validity of connections between subsequent chords and the second one governs the chord structure. In both groups there are *hard rules* which cannot be broken and *soft rules* which can be broken, but breaking these rules degrades the musical quality of the composition.

Four-part harmonization is a problem where the input single-voice melody is transformed into a four-voice choir score using the harmony rules. The score consists of four separate voices (i.e., melodies): the highest soprano with the input melody, alto, tenor and the lowest bass. The harmonization problem is considered fundamental in the musical education and in the professional composition. At the same time it is a complex task, because often it is hard (or even impossible) to satisfy all the rules. Moreover, even compliance with all the rules is not sufficient to obtain a composition that would be satisfactory from the aesthetic point of view.

1.1   State of the Art

The soprano harmonization problem has been deeply studied by computer scientists for the last 30 years. Rule-based expert systems are the most classical method of solving this problem with computers. One of the most famous is *CHORAL* [4] which models 350 functional harmony rules. It was tested on Johann Sebastian Bach's chorales. Rule-based systems perform deterministic solution space search and they are a practical way to satisfy all fundamental rules. In this paper we use a rule-based system described in [2] for three purposes. First, it serves as a reference level for studied stochastic algorithms. Second, it detects broken rules in any solution and provides an evaluator of objectives in our optimization task. Third, it supports the Bayesian network (see below).

The first genetic algorithm solving the harmonization problems was created by McIntyre [7]. It was only able to generate solutions in C-major key. It used the single-point crossover and the binary tournament selection. It turned out that a major challenge is the definition of appropriate genetic (especially mutation) operators. Phon-Amnuaisuk and Wiggins [9] proposed a genetic algorithm with a fitness function based on functional harmony rules. The more rules were broken, the greater (i.e., worse) the fitness value was. The main problem with this solution was the fact that when one connection was fixed, another connection became illegal. De Vega [14] also showed that *divide and conquer* approach does not work in this case. It should be noted that the notion of harmonic function is rarely used in the construction of harmonization algorithms. Paper [8] shows one of notable exceptions.

Neural networks has become very popular machine learning model over the recent years, also in the domain of soprano harmonization. Probably the first neural network to solve the problem was *HARMONET* [5]. Since soprano harmonization can be formulated as a time series prediction task, Long Short-Term Memory networks (LSTMs) and their variations, such as BiLSTMs [6, 16], are currently the most common approaches. For training researchers usually use J. S. Bach's chorales.

Yet another approach to the harmonization is based on Markov models [16, 17]. Some researchers use Markov Decision Processes [17] with awards related to functional harmony rules. Others tried to solve the multi-criteria problem in the case of melodies, chords and tonality [10]. Finally, there exist approaches which use N-gram models, which are solved by Prediction By Partial Match algorithm [15].

1.2   Contribution

As said before, the most of the existing approaches do not use the *harmonic function* term. The main contribution of this paper is the use of this term to model functional harmony rules and create abstraction over the key of the musical pieces, which is useful to model tensions and release them. We propose a model of a chord and of a harmonic function (described in the next section) that covers the whole domain of the Baroque harmony, i.e., allows us to algorithmically handle all chords of this era. This approach can then be used in the construction of algorithms that generate results in more aesthetically pleasing solutions of the soprano harmonization problem. We study three such algorithms: a genetic algorithm, an algorithm based on the Bayesian network and a hybrid of these. We aimed at aesthetically satisfying harmonizations, which is hard to rate automatically. Therefore, the performance of the algorithms was evaluated by human experts, mostly teachers of the Academy of Music in Kraków.

A minor but important contribution concerns the second and the third of the above-mentioned algorithms. Namely, they both make use of machine learning, so they require an appropriate training data. Unfortunately, to the best knowledge of the authors, there are no musical data sets labelled with harmonic functions of chords. Therefore we needed to prepare such a set based on a set of J. S. Bach chorale scores.

2   Soprano Harmonization Problem

In this section we provide a mathematical formulation of the soprano harmonization problem that we shall make use of in the sequel. We put the problem into the framework of discrete optimization with constraints.

We model a chord as a tuple (s, a, t, b, hf, loc), where s, a, t, and b are, respectively, soprano, alto, tenor and bass notes, hf is a harmonic function and loc is the information about the location inside the bar: downbeat, on-beat and off-beat. A harmonic function is a tuple $(base, deg, inv, e, o, down, pos, sys, key)$, where $base$ is the base function (T, S or D), deg is the degree (a note of the scale on which the chord is built), inv is the inversion (chord component in the bass voice), e is a list of extra components, o is a list of omitted components, $down$ is an indication if the degree is lowered by a semitone, pos is the position (a note in the soprano voice), sys is the chord *system* (describing distances between neighboring voices, can be *open*, *close* or *undefined*) and key is chord's key, which is necessary to model some special chords (it is related to the specific step of the scale or its alteration, e.g., a secondary dominant to D has the key equal to the second step of scale). For the attributes of a chord or of a function we adopt the functional notation, e.g., $hf(c)$ shall denote the harmonic function of chord c, $s(c)$ the soprano note of c and $key(f)$ the key of function f.

Let us now make some observations on the set of functional harmony rules. Let C be the set of all chords, let F be a set of all possible harmonic functions. We divide the set R of harmonic rules into disjoint subsets R_c and R_f of rules concerning chords and harmonic functions, respectively. As said before, $R_c = H_c \cup S_c$ and $R_f = H_f \cup S_f$, where H_c, H_f are sets of hard rules and S_c, S_f are sets of soft rules with $H_c \cap S_c = \emptyset$ and $H_f \cap S_f = \emptyset$. Furthermore, we observe that $R_c = R_{c,1} \cup R_{c,2} \cup R_{c,3}$ and $R_f = R_{f,1} \cup R_{f,2}$, where rules $R_{t,i}$ are applied to i consecutive chords or functions for $t \in \{c, f\}$ and $i \in \{1, 2, 3\}$. The final observation is that there are no hard rules applicable to 3 chords, i.e., $H_c \cap R_{c,3} = \emptyset$.

To formulate the harmonization as an optimization problem we introduce the following penalty function.

$$p: \left(\bigcup_{i=1}^{3} R_{c,i} \times C^i \right) \cup \left(\bigcup_{i=1}^{2} R_{f,i} \times F^i \right) \longrightarrow \mathbb{R}_+ \cup \{+\infty\}. \tag{1}$$

We assume that

$$p(r, x_1, \ldots, x_k) = 0 \tag{2}$$

if $(x_1, \ldots, x_k) \in C^k \cup F^k$ comply with $r \in R$,

$$p(r, x_1, \ldots, x_k) = +\infty \tag{3}$$

if $(x_1, \ldots, x_k)$ break hard rule $r \in H_c \cup H_f$ and

$$0 < p(r, x_1, \ldots, x_k) < +\infty \tag{4}$$

if $(x_1, \ldots, x_k)$ break soft rule $r \in S_c \cup S_f$. The penalty function allows us to define two objectives: the one involving rules related to chords

$$V_c(c_1, \ldots, c_n) = \sum_{r \in R_{c,1}} \sum_{j=1}^{n} p(r, c_j) + \sum_{r \in R_{c,2}} \sum_{j=1}^{n-1} p(r, c_j, c_{j+1})$$

$$+ \sum_{r \in R_{c,3}} \sum_{j=1}^{n-2} p(r, c_j, c_{j+1}, c_{j+2}) \tag{5}$$

and the one involving rules related to harmonic functions

$$V_f(f_1, \ldots, f_n) = \sum_{r \in R_{f,1}} \sum_{j=1}^{n} p(r, f_j) + \sum_{r \in R_{f,2}} \sum_{j=1}^{n-1} p(r, f_j, f_{j+1}). \tag{6}$$

Finally, the harmonization problem is as follows: given a sequence of soprano voice notes $S = (s_1, s_2, \ldots s_n)$, a sequence of their locations inside bars $(l_1, \ldots, l_n)$ (that is connected to the *time signature* of a piece) and a key of the whole melody k we seek for a sequence of chords $(c_1^*, c_2^*, \ldots c_n^*)$ and a sequence of harmonic functions $(f_1^*, f_2^*, \ldots f_n^*)$ that are a Pareto solution of two-criteria problem

$$(V_c(c_1^*, \ldots, c_n^*), V_f(f_1^*, \ldots, f_n^*)) = \min\left(V_c(c_1, \ldots, c_n), V_f(f_1, \ldots, f_n)\right) \tag{7}$$

where the minimum is taken over all $(c_1, \ldots, c_n) \in C^n$, $(f_1, \ldots, f_n) \in F^n$ that satisfy constraints

$$hf(c_i) = f_i, \quad s(c_i) = s_i, \quad loc(c_i) = l_i, \quad key(f_i) = k, \tag{8}$$

for $i = 1, \ldots, n$. The input of the soprano harmonization problem consists of a soprano melody, a time signature and a key. The former two are first transformed into sequences $(s_1, \ldots, s_n)$ and $(l_1, \ldots, l_n)$. The main objective is to find the remaining three voices, where each note of every voice corresponds to a single note in the soprano, such that the resulting sequence of chords optimizes total penalties V_c and V_f, i.e., it minimizes the number of broken rules for connections between chords (5) as well as the number of broken rules for connections between harmonic functions (6). As for the constraints, they guarantee that the output soprano melody is the same as the input melody, the output chord c_i^* has function f_i^* and that the output sequence of chords complies with the time signature and the key. Additionally, we would like to achieve solutions that are pleasing and interesting for the human ear. Some of the soft rules are aimed at meeting such expectations, but it is hard (probably even impossible) to cover all aesthetic aspects in a mathematical way.

Below we list soft and hard rules used in this paper. They are collected from the music theory literature [11–13] and discussed with domain experts.

- **Hard rules for chord connections** prohibit: parallel octaves, parallel fifths, overlapping voices, putting alto above soprano, tenor above alto or bass above tenor, exceeding prescribed voice ranges, motion of all voices in one direction, certain jumps in the same voice, false relation, harmonic function repetition without a motion of voices.

- **Soft rules for chords connections** concern: avoiding sum of jumps resulting in a forbidden jump in the same voice, constructing alto, tenor and bass melodies using the smallest possible intervals, preferring some standard movement of voices in particular connections, preferring doubling of the soprano component for simple triad chords, preferring putting the fifth in soprano when there is also the fifth in bass.
- **Hard rules for harmonic function connections** prohibit: (major D)-S connection, placing an unrelated chord after a secondary dominant, using the fifth in bass on an off-beat location, putting together a harmonic function with same function lowered, changing base harmonic function when the previous harmonic function has the same degree, changing the inversion when the degree is increasing by one, doubling the third of a chord unless it is the Neapolitan chord or a VI degree chord following a dominant.
- **Soft rules for harmonic function connections** concern: preferring D-T, S-T and secondary dominants, promoting using non-basic chords, promoting large T-S-D-T connections, discouraging from changing harmonic functions on off-beats, promoting changing functions on every downbeat, preferring the Neapolitan chord over the simple second degree function, preferring T, S and D in meaningful places in harmonization, preferring dominants with the seventh over simple dominants.

As said before, some soft rules are designed to meet the aesthetic expectations of human ears. Their aim is to make final harmonizations more interesting and expressive, but they are not strictly connected to the formal correctness of the resulting sequence of chords. Putting particular non-zero penalty values on such rules can be seen as a mathematical way of preference expression. It is worth noticing that some of the soft rules are conflicting and cannot be fully satisfied simultaneously. In consequence, obtained harmonizations hardly ever can have zero values of the objectives.

3   Algorithmic Approach

In this section we shall describe the algorithms used to solve the harmonization problem.

3.1   Genetic Algorithm

To apply genetic algorithms one needs to define the form of an individual (i.e., a potential solution), a fitness function and kinds of genetic operators transforming sets of individuals. In our case, an individual is simply a sequence of chords, where a component chord fills the role of chromosome with the chord attributes as genes.

Then, we define function g transforming a soprano note into a set of chords harmonizing it. Such a construction is possible and independent on a particular problem, because set F of all harmonic functions is well-known and finite. For every melody note we know the key, its location in the bar, so we can find the subset of F where every function harmonizes the melody note. Finally, we can create all instances of chords for these harmonic functions, where every chord contains the considered melody note in the soprano voice. The construction of g allows us to efficiently compute the fitness (see below) and ensures that all chords generated by genetic operators are valid, which solves a common problem depicted in existing studies [1].

According to Eqs. 5 and 6 we define fitness as the pair (V_c, V_f). We reuse the definition of p from Eq. 1, relaxing it only for broken hard rules, i.e., changing (3) into

$$p(r, x_1, \ldots, x_k) = 1000 \tag{9}$$

for $r \in H_c \cup H_f$. For soft rules we use penalties less than 70, so the above penalty acts as 'soft infinity' and is appropriate for the genetic algorithm used.

We consider two types of single-individual genetic operators, called here *mutators*. The first type is formed by **repair operators**, which are applied with 100% probability and correct errors in specific places where a given rule has been broken. The second type consists of **classic mutation operators** that explore the solution space and are applied with significantly smaller probability.

We use the following repair operators.

i) *DSConnectionSMutator*: in the S-D connection it changes S to T and replaces the chord with a random one with this harmonic function.
ii) *DSConnectionDMutator*: similar as above, but changes D to T.
iii) *SingleThirdMutator*: when a chord contains too many thirds it changes one of them to a different chord component (e.g., fifth).
iv) *DTMutator*: it ensures correctness of the connection D-T.
v) *SeventhToThirdDTMutator*: in the connection of (D with seventh in bass) - T, changes bass chord component in T chord to the third.

Classic mutation operators with probability of their application are as follows.

i) *ChangeBaseFunctionMutator* (0.1): changes base function (T, S, D) to another.
ii) *SwapComponentsMutator* (0.15): swaps alto chord component with tenor chord component.
iii) *ExpandToQuadrupleMutator* (0.1): appends extra chord component to a simple triad chord (e.g., seventh to a pure dominant).
iv) *ChangeInversionMutator* (0.15): changes the inversion of a chord.
v) *ChangeBassOctaveMutator* (0.25): moves bass note an octave up or down.
vi) *AddOmit1ToDominantMutator* (0.05): omits the root in the dominant chord,
vii) *AddOmit5ToDominantMutator* (0.05): omits the fifth in the dominant.
viii) *ChangeSystemMutator* (0.1): changes harmonic function system from open to close or vice versa.
ix) *ChangeDegreeMutator* (0.125): changes the degree of harmonic function (without changing the base harmonic function),
x) *IntroduceModulationMutator* (0.1): introduces a modulated chord.

When a mutator is applied, a new chord is selected randomly from the set generated by g using a probability distribution based on a *chord distance*. The *chord distance* is a sum of many components: difference between MIDI numbers of every voice's note, difference of degrees, difference of base harmonic functions, difference of keys, difference of number of different chord components. Numerous experiments showed that using this metric improves the performance of the genetic algorithm and the mutation process. Moreover, the mutators related to harmonic function changes are disabled after a half of genetic epochs. Therefore, in the final half of computations we focus on the improvement in the correctness of chord connections.

An initial population is chosen randomly. We use a one-point crossover on the bar line with probability 0.2. As the selection operator we use the binary tournament based on scalarized fitness $V_c + V_f$. For the succession we use *NSGA-II* operator [3] choosing non-dominated survivors according to (V_c, V_f) vector value.

3.2   Bayesian Network

Bayesian networks (BNs) are of course machine learning tools and not global optimization methods. Therefore, it cannot be directly applied in the solution of problem (7)–(8). Instead, we use a BN to predict harmonic functions matching given soprano notes. This way we obtain a sequence of harmonic functions that are then passed to the rule-based system, which solves the easier harmonization problem with additional harmonic function labels.

As a training set we used chorales composed by Johann Sebastian Bach. Its usage is very common in learning models for soprano harmonization. Furthermore, these chorales have a homophonic texture with four voices and are excellent examples of functional harmony. They can be found in the *music21* library by MIT[1].

To provide sufficient information to the Bayesian network we need more information than just four melodies. To achieve this, we preprocess chorales with some simple operations to obtain full chordal homophony, where every voice has the same rhythm. Afterwards, the choral contains only chords with no additional notes between them. Then, the harmonic functions of the chords can be identified. Finally, we add remaining information, e.g., about special chord components. The final model of each chorale contains the global key, time signature and the sequence of chords. Each chord contains every voice's note, length of a chord, harmonic function, information about the location inside a bar where the chord is placed, the degree of melody note in the global key. Every note is described by base note (C, D, E, F, G, A, B), MIDI note number and chord component. For example $A\flat 4$ in F minor chord has a MIDI note number equal to 68, base note A and the chord component is the minor third.

After applying the preprocessing algorithm to Bach's chorales we receive the final training set, which contains 168 major and 165 minor chorales[2]. This set

---

[1] https://web.mit.edu/music21.

[2] https://github.com/miksik98/Bach-Chorales-Dataset.

was then used to train the BN. The training was conducted using the maximum likelihood algorithm.

The target of BN is the prediction of a harmonic function based on a specific soprano note and the information about the place of a note inside the bar. The latter is inspired by some rules related to this property. Moreover, we add information about the previous harmonic function, note and place in the bar, as well as information about the next note and place. It is necessary to prevent our model from making invalid predictions such as D-S connection. Figure 1 shows the general concept of a Bayesian network. Notes are processed one by one except the first: we assume that the first harmonic function is T. Then for each note we already know *prevStrongPlace* (flag indicating if note is placed on the on-beat), *prevNote* (degree of the input melody note in the global key), *prevHF* (previous harmonic function), *currentStrongPlace, currentNote, nextStrongPlace, nextNote*. We predict *currentHF* and *nextHF*.

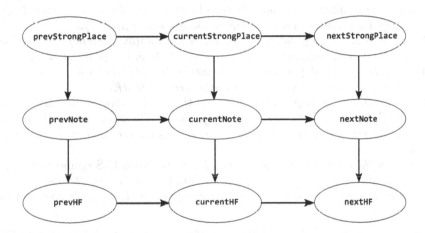

Fig. 1. General concept of the Bayesian network.

Due to the fact that every node of a trained BN has an a posteriori distribution of probability over its values we propose two variants of prediction: *maximum a posteriori (MAP)* (we choose the value with highest probability) and *stochastic (S)* (we draw a random value using a posteriori probability distribution).

Figure 1 does not show the whole picture. Firstly, every harmonic function node is represented by additional nodes: *IsMajor* (flag indicating if chord is major, otherwise it is minor), *Key* (for chords from global key it is null, for others it indicates the key of the modulation), *Base* (base harmonic function - T, S or D), *Extra* (additional components), *Omit* (omitted components), *Degree, Inversion, IsDown* (indication if chord's degree is lowered), *Position* (chord component in the soprano voice). Moreover, we create two separate networks for problems with a major key and for problems with a minor key.

Secondly, there are connections between nodes related to two consecutive functions. As the connections between *prev* (previous) and *current* are identical to those between *current* and *next* here we describe only the former.

i) *currentStrongPlace* - is an observation, no edges incoming.
ii) *currentNote* - is an observation, no edges incoming.
iii) *currentPosition* - is dependent on *currentNote*, *currentKey*, *currentDegree*, because knowing melody note and the root of the chord we can find what kind of chord component is in the soprano.
iv) *currentInversion* - is dependent on *currentStrongPlace* (prevents some rules violation), *currentInversion*, *prevPosition*, *prevInversion* (e.g., prevents parallel fifths), *currentNote*, *currentDegree*.
v) *currentIsDown* - is dependent on *prevDegree*, *currentNote*, *currentBase* (there are few functions which can be lowered and soprano note can imply it).
vi) *currentIsMajor* - is dependent on *currentIsDown* (if chord is lowered, then its mode is minor), *currentNote* (can be the third of chord), *currentDegree*, *currentBase* (there can be different modes, e.g., minor S).
vii) *currentKey* - is dependent on *prevKey* (can imply current), *currentBase*, *currentDegree*, *currentIsMajor*, *currentNote* (can imply modulation).
viii) *currentExtra* - is dependent on *currentPosition*, *currentNote*, *currentInversion* (can be one of extras), *currentDegree*, *currentKey*.
ix) *currentOmit* - is dependent on *currentExtra* (if extra is not empty there should be maybe omit component), *currentPosition*, *currentNote*, *currentInversion* (we cannot skip bass and soprano components), *currentDegree*.
x) *currentBase* - is dependent on *prevBase* (to avoid D-S connections), *prevKey* (if it is nonempty, *currentBase* could be D), *currentDegree*.
xi) *currentDegree* - is dependent on *prevKey*, *prevExtra*, *prevDegree* and *currentNote*.

All edges and dependencies on harmonic function properties are based on functional harmony rules.

We propose a sequential algorithm for predicting the properties of a sought harmonic function. After every step we update the probability distribution for every unobserved node. Firstly we start with *currentDegree (MAP)*, then *currentKey (S)*, *currentisDown (MAP)*, *currentIsMajor (S)*, *currentBase (MAP)*, *currentPosition (MAP)*, *currentInversion (S)*, *currentExtra (MAP)*, *currentOmit (MAP)*. With the above schema we are sure that every predicted harmonic function is valid because the training set contains only valid harmonic functions.

If the training set covers the domain of harmony, thus defined Bayesian network can generate interesting harmonic function sequences. To solve the whole soprano harmonization problem based on these sequences we use a rule-based system that produces chords complying with the functions generated by the BN.

3.3 Hybrid Algorithm

The hybrid algorithm is a combination of genetic and Bayesian network algorithms. A part of the initial population of the genetic algorithm is created using the network, and a second part is created randomly. Then we proceed as in the genetic algorithm. This approach mixes already correct solutions from the network with the random, which produces new higher-quality results faster. In our tests, we used an initial population with a probability of 0.5% of taking a solution from a Bayesian network algorithm.

4   Test Results

For the implementation of algorithms we used the Scala language. For the genetic algorithm we used additional *jenetics* library[3] and for Bayesian network we used *SMILE* library[4]. All details can be found in the open-source code repository[5].

The main objective of the tests is to compare the performance of all three algorithms, i.e., genetic algorithm (GA), algorithm using Bayesian network (BN) and hybrid algorithm (HA), with the rule-based system (RS). The tests were carried out on a machine with Intel Xeon CPU E5-2680 v4 (2.40GHz) processor with 28 cores and 64 GB RAM. We selected 10 input melodies from the Polish harmony book by Targosz [12]. Four of them have major keys, another four have minor keys and the remaining two contain alterations. The average length of tasks is 9 bars. Moreover, these melodies are varied in terms of keys (keys with flats, keys with sharps) and time signature $(\frac{2}{4}, \frac{3}{4}, \frac{4}{4}, \frac{6}{8})$. Every solution obtained with a specific algorithm was judged in two categories:

1. evaluation of broken rules using scalarized fitness from GA, i.e., $V_c + V_f$,
2. evaluation of musical quality by domain experts using scale $\{1, 2, 3, 4, 5, 6\}$, where 6 is the highest mark.

Every algorithm was run on each melody, so the domain experts (professional musicians, harmony teachers) received 10 input melodies and 40 harmonizations as MP3 files[6]. For nondeterministic algorithms we chose one of the results with the lowest value of rule metric, i.e., total penalty (scalarized fitness).

We started tests with the parametrization of the genetic algorithm. After a set of tests we found final parameters: 2000 epochs, 2000 individuals in population, probability of crossover - 0.2, mutation probability - proposed in the previous chapter, survivors percentage - 30%. In the hybrid algorithm we altered the above parameters with a population size of 1000 and 1500 epochs. Every task was solved 30 times. For the genetic algorithm we found confirmation of the mutators disabling strategy, because the final solution was always found in the second half of epochs. As shown in the Table 1 the best performance by the rule metric received the rule-based system. It confirmed the RS ability to generate formally valid harmonizations as expected. The second algorithm in most tasks was the hybrid. In the Fig. 2 we can see that HA was very stable for the minor key group

---

[3] https://jenetics.io.

[4] https://www.bayesfusion.com/smile.

[5] https://github.com/HarmonySolverTeam/HarmonySolverBackend/tree/mgr.

[6] Available at https://youtube.com/playlist?list=PLysSbLi9j6PjOqNIzTdekEowPFix 2bOlU.

Table 1. Rule metric results.

Group	Melody	GA		HA		RS	BN
		Mean	IQR	Mean	IQR		
Major	1	2703.5	1001.5	1994.5	2064.5	390	3970
	2	3650.5	1399.5	3095.5	1869.5	479	3369
	3	1676	1018	700.5	952.5	479	2834
	4	858.5	1044	930.5	963	370	1956
Minor	5	851.5	768	804	31.25	379	811
	6	522	68.75	619	201.5	379	4846
	7	5465.6	1227.75	3353.5	947.5	644	5505
	8	760	262.75	756	1041.75	472	1820
With alterations	9	7605	2022.5	4416	48	914	5624
	10	7350	1195.25	6925	1962.75	1013	12544

(tasks 5–8). The poorest results were received for the group with alterations: sometimes more than 8 broken hard rules. What is surprising, BN's performance received the inter-quartile range (IQR) close to 0. RS is deterministic and we always get the same harmonization for a given melody. For these reasons, in Table 1 IQR for RS and BN is not reported and the only value shown is the mean result for the rule metric.

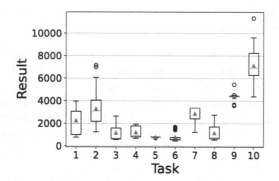

Fig. 2. Rule metric results for HA.

The average expert grades are aggregated in the Table 2. The extreme ratings number per algorithm is shown in the Fig. 3. The highest number of 6 grades received a hybrid algorithm and it was twice better than the rule-based system, which seemed to be the worst algorithm in the group with alterations. The genetic algorithm and rule-based system received the highest number of 1 and 2 grades. RS generated simple harmonic structures in harmonizations, which was not appreciated by the experts. In contrast, GA generated too bizarre harmonic structures. BN and HA were more consistent in this aspect. What is more, BN received the second lowest number of 1 and 2 grades, because experts appreciated Bach's harmonization style, which was used to train the network and could be observed in resulting harmonizations.

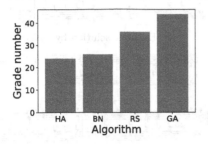 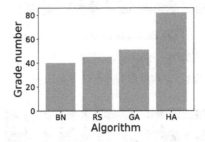

Fig. 3. Number of lowest (left) and highest (right) experts' grade.

Example solutions for the tenth task are shown in Fig. 4. In the first bar in the GA and HA solutions there are consecutive secondary dominants, which build tension at the beginning of the final harmonization. RS is very typical and not advanced. BN uses some chord inversions as well as interesting static connections, like in the second bar. The solutions generated by genetic algorithms are more colorful than those of other algorithms. It was sometimes viewed negatively by experts.

Table 2. Comparison of average expert metric results of algorithms.

Group	Melody	GA	BN	RS	HA
Major	1	4.40	4.05	4.13	4.25
	2	3.73	3.40	4.60	4.83
	3	4.28	4.75	4.43	4.85
	4	4.53	4.18	4.60	4.33
Minor	5	4.68	4.43	4.10	4.90
	6	3.88	4.68	4.13	4.50
	7	3.454	3.80	4.95	4.35
	8	4.70	4.05	4.43	4.20
With alterations	9	4.35	4.35	3.43	4.83
	10	3.83	4.25	3.80	4.10

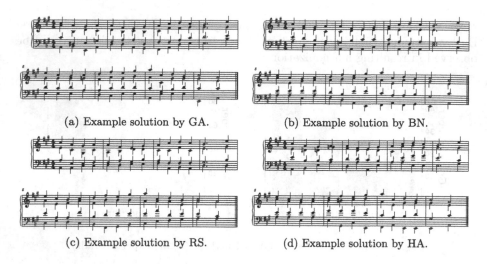

(a) Example solution by GA.   (b) Example solution by BN.

(c) Example solution by RS.   (d) Example solution by HA.

Fig. 4. Example solutions for 2nd test melody.

5   Conclusions

The results of experiments shown in the previous section and comments of domain experts confirm that a rule-based system produces harmonizations which are the best in terms of rules of functional harmony, but it is achieved by using uncomplicated harmonic structures. These results were rated as safe and boring by experts. The Bayesian network algorithm reached low ranks in both metrics, but this is probably because the test melodies were not similar to chorales that the network was trained on. Maybe if the melodies had been taken from Bach's chorales the algorithm would perform better. The domain experts said there were many interesting harmonic connections, but they did not have any sensible consequences. They also noticed many typical Baroque harmonic functions (e.g., minor S in major key), so we can assume training was successful. The genetic algorithm generated the most interesting harmonic connections related with melody phrases, especially creating tensions and releasing them. On the other hand there were many random parts. The bass line was illogical and there were many jumps inside its melody. The most common mistakes were false relation and repetition of the same function through the bar line. The hybrid algorithm generated the best solutions. It received twice as many *6 grades* as the rule-based system, and the highest average grade. It contains elements of both algorithms it combines, so we can see features of Bach's style which were enriched by the genetic algorithm.

To sum up, the results confirm that the use of evolutionary algorithms and machine learning can produce more interesting harmonizations than rule-based systems. They can also satisfy most of the functional harmony rules. A straightforward future research should probably focus on the hybrid algorithm, which uses domain knowledge for learning and mutation and which was highly rated by the experts. Another promising option is the use of Bayesian networks trained on a more general musical data and with a more complex structure, e.g., considering sequences of chords with length greater than 3. Finally, there are some other promising methods for soprano harmonization to explore like LSTM and BiLSTM neural networks. A more comprehensive idea is the composition of algorithmic harmonizers with techniques based on the counterpoint theory that would result in more complex polyphonic musical compositions.

References

1. Allan, M., Williams, C.: Harmonising chorales by probabilistic inference. In: Advances in Neural Information Processing Systems, vol. 17. MIT Press (2004)
2. Dajda, J., et al.: Current trends in software engineering bachelor theses. Comput. Inform. **40**(4), 930–956 (2021)
3. Deb, K., et al.: A fast and elitist multiobjective genetic algorithm: NSGA-II. IEEE Trans. Evol. Comput. **6**(2), 182–197 (2002)
4. Ebcioğlu, K.: An expert system for harmonizing four-part chorales. Comput. Music. J. **12**(3), 43–51 (1988)
5. Hild, H., Feulner, J., Menzel, W.: HARMONET: a neural net for harmonizing chorales in the style of J. S. Bach. In: Advances in Neural Information Processing Systems, vol. 4. Morgan-Kaufmann (1991)
6. Lim, H., Rhyu, S., Lee, K.: Chord generation from symbolic melody using BLSTM networks. In: Proceedings of the 18th International Society for Music Information Retrieval Conference, ISMIR 2017, Suzhou, China, 23–27 October 2017, pp. 621–627. ISMIR (2017)
7. McIntyre, R.A.: Bach in a box: the evolution of four part Baroque harmony using the genetic algorithm. In: Proceedings of the First IEEE Conference on Evolutionary Computation. IEEE World Congress on Computational Intelligence, vol. 2, pp. 852–857. IEEE (1994)
8. Mycka, J., Żychowski, A., Mańdziuk, J.: Human-level melodic line harmonization. In: Groen, D., de Mulatier, C., Paszynski, M., Krzhizhanovskaya, V.V., Dongarra, J.J., Sloot, P.M.A. (eds.) ICCS 2022. LNCS, vol. 13350, pp. 17–30. Springer, Cham (2022). https://doi.org/10.1007/978-3-031-08751-6_2
9. Phon-Amnuaisuk, S., Wiggins, G.A.: The four-part harmonisation problem: a comparison between genetic algorithms and a rule-based system. In: Proceedings of the AISB 1999 Symposium on Musical Creativity, pp. 28–34. AISB London (1999)
10. Raczynski, S., Fukayama, S., Vincent, E.: Melody harmonisation with interpolated probabilistic models. J. New Music Res. **42**, 223–235 (2012)
11. Rameau, J.P.: Treatise on Harmony (1722). Dover Publications, Inc. (1971)
12. Targosz, J.: Podstawy harmonii funkcyjnej. PWM (2004). In Polish
13. Toutant, W.: Functional Harmony. Wadsworth Publishing Company (1985)
14. de Vega, F.F.: Revisiting the 4-part harmonization problem with GAs: a critical review and proposals for improving. In: 2017 IEEE Congress on Evolutionary Computation (CEC), pp. 1271–1278. IEEE (2017)
15. Whorley, R.P., et al.: Multiple viewpoint systems: time complexity and the construction of domains for complex musical viewpoints in the harmonisation problem. J. New Music Res. **42**, 237–266 (2013)

16. Yeh, Y.C., et al.: Automatic melody harmonization with triad chords: a comparative study. J. New Music Res. **50**(1), 37–51 (2021)
17. Yi, L., Goldsmith, J.: Automatic generation of four-part harmony. In: Proceedings of the Fifth UAI Bayesian Modeling Applications Workshop, CEUR Workshop Proceedings, vol. 268, pp. 81–86. CEUR-WS.org (2007)

Predicting ABM Results with Covering Arrays and Random Forests

Megan Olsen[1]([✉]), M. S. Raunak[2], and D. Richard Kuhn[2]

[1] Loyola University MD, Baltimore, MD 21210, USA
mmolsen@loyola.edu
[2] NIST, Gaithersburg, MD 20899, USA
{raunak,kuhn}@nist.gov

Abstract. Simulation is a useful and effective way to analyze and study complex, real-world systems. It allows researchers, practitioners, and decision makers to make sense of the inner working of a system that involves many factors often resulting in some sort of emergent behavior. The number of parameter value combinations grows exponentially and it quickly becomes infeasible to test them all or even to explore a suitable subset. How does one then efficiently identify the parameter value combinations that matter for a particular simulation study? In addition, is it possible to train a machine learning model to predict the outcome of an agent-based model (ABM) with a systematically chosen small subset of parameter value combinations? We explore these questions in this paper. We propose utilizing covering arrays to create t-way ($t = 2$, 3, etc.) combinations of parameter values to significantly reduce an ABM's parameter value exploration space. In our prior work we showed that covering arrays are useful for systematically decreasing an ABM's parameter space. We now build on that work by applying it to Wilensky's Heat Bugs model and training a random forest machine learning model to predict simulation results by using the covering arrays to select our training and test data. Our results show that a 2-way covering array provides sufficient training data to train our random forest to predict three different simulation outcomes. Our process of using covering arrays to decrease parameter space to then predict ABM results using machine learning is successful.

Keywords: agent-based modeling · machine learning · calibration

1 Introduction

Modeling and simulation is a useful and effective way to study complex, real world systems. Through modeling we can examine the inner working of an intricate system, and ask questions about how a change to one aspect of the system affects other aspects or the system as a whole. Scenarios such as the spread of a

---

Supported by NSF MRI grant 1626262.

pandemic, the operations of a self-driving car, or the flow of patients in an emergency department can be studied with simulation models. Simulations allow us to study these complex scenarios without building the physical system, which can be costly, dangerous, or simply infeasible.

Agent-based modeling (ABM) is a popular modeling technique for studying these types of systems and phenomena through simulation. Agents are individual autonomous entities that make decisions about their actions and interactions within the environment. ABM can be a bottom-up approach, where the global behavior of a system emerges out of the individual decisions and actions of agents. A change to parameter values can affect model outcomes, sometimes in unexpected ways, as parameters affect the algorithms for how the agents make decisions, and how the system is updated over time or in response to those agent decisions. Thus it is important to understand what parameter values to use and explore while simulating and studying an agent based model. However, a typical ABM will include many parameters, each with a potentially very large set of possible values. The number of parameter value combinations grows exponentially with the addition of each parameter, quickly becoming infeasible to test all parameter combinations, or even to explore a suitable subset of them. This problem is not unique to ABM, and is present in any reasonably large system. How does one then efficiently identify the parameter value combinations that matter for a particular simulation study, or are likely to influence the output the most? These questions are crucial for developing a well-calibrated, valid simulation model.

There are many approaches for calibrating, testing, and validating ABMs, most of which rely on choosing a suitable subset of parameter values to examine. In this paper, we explore using covering arrays from the software testing literature to systematically cover an effective subset of the parameter space to choose the parameter value to test. We have previously shown the usefulness of covering arrays for reducing the parameter value space of simulation models [16,18]. Building on that work, after using covering arrays to thoughtfully reduce the parameter space, we ask the question: can we use machine learning (ML) to train a model on data from running an agent-based model on a small subset of its parameter space, and then predict the results for a larger set of new parameter combinations? This approach could significantly decrease the number of simulation experiments required to fully understand the impact of parameter changes on a model. We also analyze the relationships between parameter value combinations and simulation outputs. As far as we can tell, we are the first to explore this approach on simulation models. We test the approach by training random forest models on data from Wilensky's Heatbugs Netlogo agent based model [20], and our results show that it is feasible to make these predictions.

2   Related Work

One of the objectives of modeling and simulation is to understand the behavior of large, complex systems under different environmental conditions. Through the selection of parameter values, it can also allow us to perform predictive analysis of a system. For the predictions to be accurate and useful, the model has to reliably represent the real system and robust against parameter value changes such that it isn't too sensitive, resulting in uncertain output swings. Thus the rate and magnitude of changes in model output when the input parameter values change is an important aspect of developing useful simulation models. Researchers and practitioners perform uncertainty or risk analysis, i.e., estimation of output variance, parameter calibration, and sensitivity analysis of simulation models to gain better understanding of model accuracy and robustness [3,5]. These analyses require running the model under many parameter value combinations.

Model calibration adjusts the initial model to a reference system by tuning parameter values. Hoffmann showed that for large, complex models as well as for model federations with interdependent parameters, model calibration is an NP-complete problem [7], and consequently is too costly in many cases. Researchers are thus left to find more pragmatic approaches of exploring a small subset of parameter values. Sensitivity analysis towards validation and optimization of the model is another well studied area [9]. In his 2004 article Kleijnen commented, "Few statisticians have studied random simulations. And only some simulation analysts have focused on strategic issues, namely which scenarios to simulate, and how to analyze the resulting I/O data" [8]. Kleijnen used his argument to motivate the need for using a Design of Experiment (DOE) study of the meta-models, using mathematical equations as a metamodel of computer simulation models. Multiple research works have studied DOE for deterministic metamodels such as the polynominal regression metamodel, Krigin models, and more [1,8]. None of these research studies, however, looked into systematically reducing the parameter value space to tackle the challenge of dealing with an exponentially large search space, especially on ABMs.

Covering arrays also have roots in DOE, and represent an approach to effectively cover the overall parameter value combination space [11]. These arrays include all t-way combinations of parameter values (typically for small t, i.e., 2 to 6), and are effective in designing software testing solutions [11,13]. In our earlier work, we showed that covering arrays can also be useful in focusing the parameter value space and in choosing a useful subset of values to test an ABM [16]. In this study we build on this approach by applying it on a different ABM, and investigating the effectiveness of using a Random Forest ML model to predict simulation outputs on unseen combinations of parameter values. In a similar vein, Lanus et. al. used covering array based approaches to describe how two different data sets used in training and testing of an ML model differ [15]. Their study used combinatorial set difference metrics to identify which features are most influential as a determining factor of how successful an ML model is going to be on a new data set. The objective and application of their study, however, was different than our work presented here.

We are unaware of prior work that attempts to predict the outcome of an ABM after training a machine learning model on a subset of that model's possible parameter combinations. Machine learning is instead used for processes such as to de-feature (feature simplification, removal, etc.) CAD models for simulation studies [6], or as part of building the model itself [21]. There are also numerous approaches for trying to find optimal parameter combinations for a model, such as genetic algorithms [4,17,19] and Robust Parameter Estimation [10]. However, this work is not focused on determining optimal combinations, but instead in making it easier to understand the type of behavior caused by any particular set of parameter values, and how a specific type of change to parameters would change model output, without needing to run the ABM on those values.

3   Approach

Our goal is to explore the feasibility and usefulness of using a combination of covering arrays and machine learning models for predicting results of an agent-based simulation model (ABM) within the vast parameter value combination space. The challenge is to select parameter values that are representative of the model's overall behavior, so that we can train the machine learning (ML) model to be able to correctly predict behavior on previously untested areas of the parameter space. We have chosen Wilensky's Heatbugs ABM in NetLogo [20] for our study. It is a simple model, amenable to quick data generation, with a limited number of outputs to predict, and with emergent behavior. This model therefore allows exploration of this new approach.

We utilize covering arrays to reduce the parameter value space systematically, run the model for each parameter set in the 2-way and 3-way covering arrays, train a random forest model on the 2-way data (33,351 parameter combinations), and test its ability to predict the outcome of the simulation on the significantly larger 3-way data that was not seen during the training of the model (3, 971, 955 parameter combinations). This section provides details on each step.

3.1   Heatbugs Model

In Wilensky's Heatbugs ABM [20], agents move in a 2D grid in an attempt to find a location with their ideal temperature. Only one agent can exist in each location, and they have a random chance of moving at any given time step. Agents give off heat to the local environment, which is then diffused to neighboring squares. The heat also dissipates from each square at a given rate. The result is a level of unhappiness for each agent ($|ideal\_temp - current\_temp|$), with the goal of low unhappiness for all agents. The behavior of the agents and their environment are defined by the eight parameters in Table 1.

3.2   Choosing Parameters via Covering Arrays

As seen in Table 1, there are 5.6386e15 valid combinations for this model. If we hold the number of agents steady, we reduce this number to approximately 11 trillion parameter value combinations (1.1277e13). If we needed to test every combination this is the minimum number of times the model would need to run; however, it is infeasible to test more than a tiny fraction of these combinations.

Table 1. Heatbugs model parameters and their valid values. There are 2.28939e16 potential combinations. Given that each pair of min/max parameters must have the relationship min < max, there are 5.6386e15 valid combinations.

Parameter	Min Value	Max Value	Increment
Agent Number	1	500	1
Min Ideal Temp	0	200	1
Max Ideal Temp	0	200	1
Min Output Heat	0	100	1
Max Output Heat	0	100	1
Evaporation Rate	0.01	1	0.01
Diffusion Rate	0	1	0.1
Random Move Chance	0	100	1

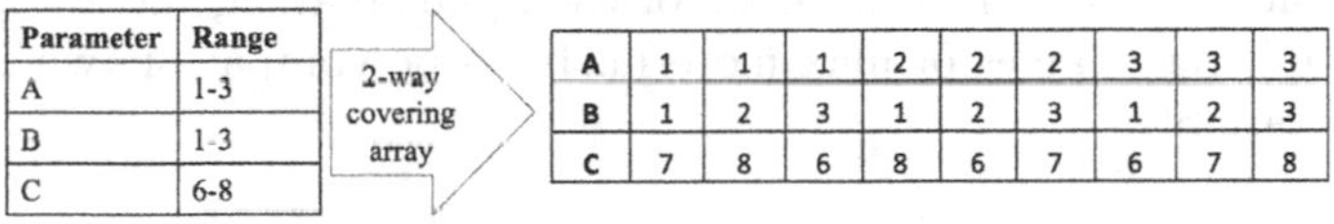

Parameter	Range
A	1-3
B	1-3
C	6-8

2-way covering array ⟹

A	1	1	1	2	2	2	3	3	3
B	1	2	3	1	2	3	1	2	3
C	7	8	6	8	6	7	6	7	8

Fig. 1. Example of converting a simple 27 combination parameter space to a 2-way covering array of 9 combinations (each column is a combination).

An efficient approach for reducing the parameter space is to use covering arrays. A *t-way* covering array is a matrix of values that includes all *t-way* combinations of parameter values. Suppose we have three parameters A, B, and C, and each of these parameters can take three values. Figure 1 shows the parameter values and a 2-way covering array constructed from those values. A 2-way covering array [11] as shown in the figure, includes every pair of parameter values at least once. As a larger example, if we instead have 10 binary parameters, there are $2^{10} = 1024$ possible value combinations one can use to run the model. However, all 3-way interactions of parameter values are included in a covering array of only 13 rows. Those 13 rows include every possible 3-way combination of parameter values, substantially reducing the search space. From the perspective of a simulation model study, that covering array allows systematic exploration of the parameter value space with only 13 simulation runs instead of 1024 runs.

We use the ACTS tool [2] to generate covering arrays of 2-way and 3-way combinations of parameter values to significantly reduce the parameter value exploration space while ensuring broad coverage of possible parameter interactions. When agent count is held steady for the Heatbugs model, there are 1.1277e13 valid parameter value combinations across the seven remaining parameter values. This number is reduced to 33,551 parameter value combinations where every possible 2-way interactions of the parameter values are present. Similarly, all 3-way interactions are captured by 3,972,000 value combinations.

Both 2-way and 3-way covering arrays reduce the parameter value space while maintaining good coverage of parameter interactions. We ensure that none of the 2-way rows are present in the 3-way data. Figure 2 shows the frequency of each

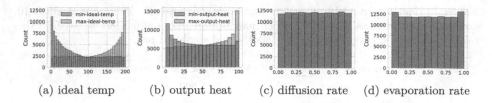

(a) ideal temp   (b) output heat   (c) diffusion rate   (d) evaporation rate

Fig. 2. Histograms of the feature values as they exist in the 2-way covering array.

parameter's values in the 2-way covering array. The diffusion and evaporation rates are evenly distributed across all potential values, while the pairs of ideal temperature and output heat parameters have higher rates on the lower and higher ends of their scales. This uneven application of values is expected due to the relationship between min and max for each pair, making the lower values of min and the higher values of max more likely to be compatible with values of the related variable.

3.3 Machine Learning

We learn random forest models to attempt to predict the outcome of the ABM. Random forests are a decision tree ensemble method. A decision tree is a supervised learning method that at each node branches based on the value of a particular feature, such as whether or not diffusion rate is less than 0.2. Each branch takes you closer to a prediction, and how many levels the tree needs will depend on the data. For a random forest, essentially multiple decision tree models are learned, and then the prediction that is most common among all learned trees is the prediction of the overall ensemble. This ensemble method is more likely to correctly predict outcomes than a single decision tree. Details of our implementation are in Sect. 4.

4   Experimental Setup

4.1   Data Gathering and Preparation

We run the simulation for each parameter combination in the 2-way and 3-way covering arrays four times, for 25,000 simulation steps. We run with four different random number seeds due to the stochasticity in the model, to make it more likely that the overall results produced are due to parameters instead of randomness. After each simulation ends we calculate the following metrics[1]:

- avg: the average unhappiness of the heat bugs
- $avgF$: the average unhappiness across the final 500 time steps

---

[1] The processed data used in this paper can be found at https://data.nist.gov/.

- $stdF$: standard deviation of unhappiness across the final 500 time steps
- $minF$ and $maxF$: minimum and maximum unhappiness of all agents across the final 500 time steps.

After all data are gathered, we prepare it for machine learning. Each set of four rows, one for each model run of the same parameters but different random number seeds, is combined into a single row of data and labeled based on whether that set of parameter values appears to achieve the outcome to be predicted. Each of our experiments prepare the data for a different type of prediction:

A) the model reaches low unhappiness with low variation of unhappiness across agents and time;
B) the model reaches a steady state; and
C) the average final unhappiness level of the agents.

In all experiments, each parameter combination appears once in the prepared data. We determine the thresholds for creating the class labels based on the 2 way (training) data, and then apply those thresholds to the 3-way (test) data. Each threshold creates a set of data with a different level of imbalance between the classes, so that we can test how imbalance affects predictive ability. In each experiment a threshold has been included that leads to balanced class labels on the 2-way data. In experiments A and B we predict a binary class, e.g. Class 1 represents meeting the criteria, and Class 0 represents not meeting that criteria. Experiment C is instead a multi-class classification problem.

Experiment A: Low Unhappiness and Low Variation: Experiment A attempts to predict if the model reaches low overall unhappiness with a low variation of unhappiness across agents during the final 500 time steps. In these experiments, the class of the parameter set is determined by $stdF < threshold$ for a particular percentage of the four simulations run for a particular parameter set. Class 1 label is applied to data below the threshold, as it represents low variation in standard deviation, which also correlates to relatively low overall unhappiness level for this ABM. We test our predictive ability on thresholds of $0.1, 0.15, 0.2, 0.5, 0.75, 1$. For each threshold we test two approaches for combining the four simulations of the same parameter combination: either 50% or 100% of the four simulation runs must meet the threshold criteria to be Class 1; otherwise, the parameter set is labeled as Class 0.

Experiment B: Steady State: Experiment B attempts to predict if the unhappiness level in the Heatbugs model reaches a steady state by the end of the simulation. We define steady state as $stdF/avgF < threshold$. The standard deviation on its own is insufficient as the overall averages vary widely, between 0 to 45,000, from run to run. The high average values are due to certain parameter combinations leading to overwhelming heat in the system well above the bugs' ideal temperature; low values occur when the situation keeps heat from building up too high. When the average is high, a relatively small standard deviation could still be in the hundreds; whereas when average is low, it would be single digits at most. Thus we found that the value of standard deviation can be misleading for this model. By dividing the standard deviation by the average, we can observe how much the unhappiness level is varying in terms of the overall average at that point in the simulation. We test our predictive ability on thresholds of $0.001, 0.002, 0.005, 0.0075, 0.01$. For each threshold we test three approaches for combining the four simulations of the same parameter combination: either 50%, 75%, or 100% of the four simulation runs must meet the threshold criteria for the parameter set to be labeled Class 1; otherwise, it is labeled Class 0.

Experiment C: Average Unhappiness: Experiment C attempts to predict the overall average unhappiness during the final 500 time steps, $avgF$. We create classes defined by quantiles of the $avgF$ column of the 2-way data:

1. 4 quantiles: <30.91, <74.08, <125.19, ≥125.195
2. 6 quantiles: <12.27, <38.45, <74.08, <113.84, <164.66, ≥164.66
3. 10 quantiles: <12.27, <24.24, <38.45, <55.60, <74.08, <92.76, <113.84, <137.98, <164.66, ≥164.66

In this experiment we determine if it is possible to predict the general region of the average unhappiness for agents in the model. Although quantiles do not need to be used, it is a natural way to define the categories. To ensure a proper machine learning process is followed, we do not analyze quantiles of the 3-way data but apply the thresholds developed on the 2-way data to the 3-way data.

4.2   Machine Learning in All Experiments

We use the simulation results from the 2-way covering array data to generate training and validation data for each random forest model, then use the results from the 3-way covering array to test each model. This setup allows us to see if we can train a random forest model on a very small subset of the overall parameter space ($3.3551e4$ combinations), and then predict results for a significantly larger amount of new parameter combinations ($3.972e6$). In each case, the test data is processed in the same way as the equivalent training data. Each combination of thresholds and approaches to combining results of the same parameters results in a new set of data and a random forest model.

We use scikit-learn's implementation of random forests, and learn using a randomized grid search. A randomized grid search takes all of the various hyperparameter (e.g. parameters to the random forest) options and randomly chooses a subset of all possible hyperparameter combinations to search for the best model. We use 10 fold cross validation, with balanced accuracy as the scoring function on 100 hyperparameter combinations. Cross validation is the standard best practice to help reduce the chance of overfitting on the training data, with 10 folds being the generally agreed upon best choice. Balanced accuracy is used as our datasets are mostly imbalanced, which can affect how well the model learns.

We test the following hyperparameters: number of decision trees in ensemble (200, 288, 377, 466, 555, 644, 733, 822, 911, 1000); maximum number of features

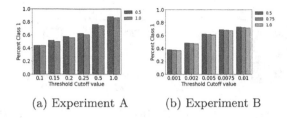

(a) Experiment A   (b) Experiment B

Fig. 3. The classes are generally unbalanced, with the most balanced data occurring when the cutoff threshold is 0.15 for Exp. A and 0.002 for Exp. B.

to consider at a branching (2, 7); maximum depth of learned trees (10, 20, 30, 40, 50, 60, 70, 80, 90, 100, 110); minimum samples required to branch (2, 5, 10); minimum samples required at a leaf node, e.g. at the point at which the decision tree is making a prediction, how many rows from the data should be represented by the prior branch in this direction (1, 2, 4); Gini impurity to determine quality of split; and bootstrapping either on or off. The grid search tests 100 randomly chosen combinations of these hyperparameters. After completing the search, it determines the best hyperparameters and generates a final random forest model using those hyperparameters. The 3-way data is tested on that model.

5   Results

We analyze *balanced accuracy*, *precision*, and *recall* as they are standard ML metrics that are particularly useful for unbalanced data. Unlike regular accuracy that only determines what percentage of predictions are correct, balanced accuracy is the sum of the number correct in each class divided by the number of classes. Balanced accuracy is thus more accurate when a large number of examples are from one class. As can be seen in Fig. 3, our classes are imbalanced in most of our training data, which is to be expected. The recall and precision helps us to understand how that imbalance affects the predictions for each class. The recall tells us how likely we are to predict a given class if it actually is that class; so recall for Class 1 in Experiment B is how likely we are to predict that a simulation run with a specific set of parameter values will be steady when it actually is steady. Precision tells us how often a prediction is correct; so a high precision on Class 1 means that if we predict Class 1 then it is likely to actually be Class 1. As a reminder, in all scenarios, 2-way data trained the model; the results are from using that learned model to predict the 3-way results. High accuracy, precision, and recall implies that our overall process works for training a random forest to predict the results of an agent-based model by only running a small percentage of potential parameter combinations.

5.1   Experiment A: Low Unhappiness and Low Variation

In Experiment A we predict whether the standard deviation over the last 500 time steps ($stdF$) crosses a given threshold. Class 1 represents situations where

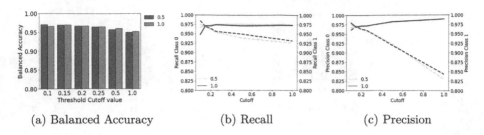

(a) Balanced Accuracy   (b) Recall   (c) Precision

Fig. 4. Recall and Precision from Experiment A. The solid line is for Class 1 (e.g. below threshold), and the dotted line is Class 0 (e.g. at or above threshold). The colors denote what percentage of the four runs needed to be below the threshold for the parameter combination to be labeled as Class 1.

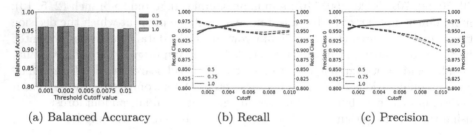

(a) Balanced Accuracy   (b) Recall   (c) Precision

Fig. 5. Recall and Precision from Experiment B. The solid line is for Class 1 (e.g. below threshold), and the dotted line is Class 0 (e.g. at or above threshold). The colors denote what percentage of the four runs needed to be below the threshold for the parameter combination to be labeled as Class 1.(Color figure online)

it is below the threshold, e.g. is steady and the overall unhappiness values are likely small. As can be seen in Fig. 4, balanced accuracy varies between 95% and 97% on the 3-way data, with accuracy decreasing as we become more imbalanced with a higher percentage of the data in Class 1. Our recall for both classes is actually quite good in all versions of the problem, meaning that we are getting most examples correct. The precision for Class 0 is closer to 80% for the more imbalanced situations, meaning that when we predict Class 0 we are less likely to be correct than when we predict Class 1. Overall, many values for standard deviation cutoff are predictable, even when the training data has significantly fewer examples in one class. It is feasible to predict if standard deviation will be low in the simulation by training on a small subset of the parameter space, although best results occur in a relatively balanced training set. How the four results from each parameter set are combined does not affect the result.

5.2   Experiment B: Steady Unhappiness

In this experiment we predict how steady the unhappiness level is across agents during the last 500 time steps. We defined *steadiness* as standard deviation divided by average, to take into account that some average unhappiness levels are

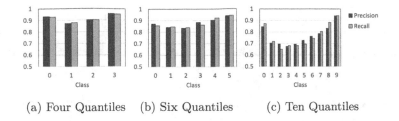

(a) Four Quantiles   (b) Six Quantiles   (c) Ten Quantiles

Fig. 6. Recall and Precision from Experiment C for each class in each experiment.

orders of magnitude higher than others. As seen in Fig. 5, the balanced accuracy score is high across all situations tested, always at least 95%. The precision and recall is always at least 90% as well, which is better than in Experiment A. We are very likely to get each class correct, although in the more imbalanced data we again see a dip in precision for Class 0 meaning that predictions for Class 0 are less trustworthy than predictions for Class 1. However, even for imbalanced data our models' predictions are generally accurate. Overall we seem able to predict steadiness of unhappiness level, with minimal impact by how steadiness is defined or how the four runs from each parameter set are combined.

5.3 Experiment C: Average Unhappiness

In experiment C we predict the overall average unhappiness level in categories defined by quantiles. Our balanced accuracy for four quantiles is 91.6%, for six quantiles is 87.8%, and for ten quantiles is 76.7%. We can see from balanced accuracy that the more fine grained our categories, the harder it is to predict. This result is not surprising, but is a confirmation of what may be most difficult to predict via machine learning. Figure 6 confirms that our prediction ability is high in the four quantile scenario, and steadily decreases as we increase the number of quantiles. As the number of quantiles increases we are increasing the difficulty of the prediction, as fewer categories means more precision on average. For ten quantiles, the random forest is not able to adequately predict any classes other than the smallest and largest categories. In all quantile experiments the highest average values are the most likely to be accurately predicted, meaning that it's easiest to determine if a high average unhappiness level will occur.

5.4 Feature Importance

One of the benefits of learning a random forest model is that the model can readily tell us how much influence each parameter had on the final result. Feature importance is a useful piece of information, as a goal in this process is to be able to better understand the parameter space and how those parameters affect the final outcomes in the simulation. We use the feature_importance feature of scikit-learn's implementation of Random Forests to compute this result.

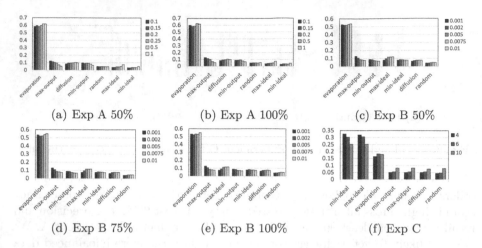

(a) Exp A 50%   (b) Exp A 100%   (c) Exp B 50%

(d) Exp B 75%   (e) Exp B 100%   (f) Exp C

Fig. 7. Feature Importance. In Experiments A and B, evaporation rate has the strongest influence on classification. For all quartiles in Experiment C, the min and max ideal temp play the strongest role in determining classification. Neither percent nor threshold heavily influence this ranking.

In Fig. 7 we see importance of each feature for determining which class a set of parameter combinations predict. A feature (e.g. parameter) importance of 1 would imply that the single feature alone can predict the results. As expected, no feature has that high an impact, nor does any pair of features. However, in each experiment a single feature has a majority importance. For Experiments A and B, evaporation rate is the most important feature and is significantly more important than all of the other features with an importance over 0.5 (Fig. 7). For Experiment C where we are predicting the averages, the minimum and maximum ideal temperature parameters have the most influence on the results, and are almost equally important with each other with an importance of at least 0.25 in each scenario (Fig. 7f). The thresholds for determining the class label do not affect which feature is most important for making correct predictions; the feature importance seems tied instead to what we are trying to predict.

For a closer look at the importance of features, we can consider 2-way and 3-way combinations of feature values in the covering arrays, without machine learning. The notion of *combination frequency differences* (CFD) [14] identifies combinations of feature values that are more strongly associated with one class than another, an approach useful in explainable AI [13] and vulnerability analysis for physically unclonable functions [12]. The CFD value for a single attribute value combination is the difference between the rate of occurrence of a particular value combination in one class versus another class. For example, a particular combination of fur and eye color may occur in 75% of one dog breed as compared with 10% of another breed, for a difference of 0.65. Differences are computed for all attribute value combinations, for $v^t \binom{n}{t}$ t-way combinations of n attributes with v values each. Graphing these differences for every value combination in

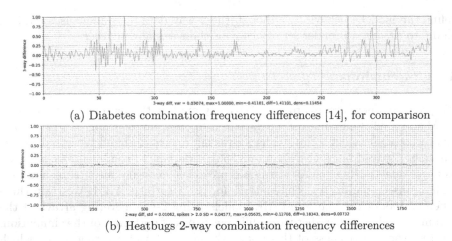

(a) Diabetes combination frequency differences [14], for comparison

(b) Heatbugs 2-way combination frequency differences

Fig. 8. Combination frequency differences to identify combinations of features values that are more strongly associated with one class versus another. In (a) we see a comparison graph where some feature combinations strongly affect the outcome. In (b) we see that no parameter combinations are strongly associated with the outcome for our 2-way data.

a data set produces a graph such as shown in Fig. 8a, which shows a machine learning data set for clinical values related to diabetes. The graph shows the difference for all 3-way combinations in this data set. As can be seen in Fig. 8a, three combinations occur in 100% of the positive cases (above center line), indicating a strong association with these combinations of clinical values [14].

Applying this method to the Heatbugs 2-way covering array for Experiment A produces a graph as shown in Fig. 8b. A separate report (not shown due to space limitations) shows that, consistent with the random forest model, evaporation rate is the single most important attribute. However, the 2-way combination frequency analysis also shows that low values of *minimum ideal temp* in combination with evaporation rate are also strongly associated with class predictions. Similarly, a few other combinations are also found to be significant. This additional useful information is not apparent from the random forest model. Also note that the range of difference values is much narrower in Fig. 8b than in Fig. 8a, indicating that the Heatbugs problem is in some sense "harder" for machine learning than the diabetes prediction problem (which has >99% accuracy), as there are no combinations that clearly indicate a particular class on their own. We plan to investigate this CFD approach further in future work, to determine how well it can aid in determining the impact of parameters on model results, and/or how likely a model is to be predictable via machine learning.

Both the ML feature importance results and the combination frequency difference results confirm that the ABM outcomes cannot be fully explained by 2-way combinations alone, reconfirming that training our ML model on 2-way data to predict 3-way data does mean we are able to predict situations previously unseen in the training (2-way) data. Pairs of variables are not enough to fully define the model's behaviors, although the 2-way data does allow us to train a model to predict a broader set of behavior than was already seen.

6   Conclusions

We propose a new process to systematically explore the parameter space of an ABM and predict outcomes of simulations using machine learning (ML) and combinatorial analysis. This process uses covering arrays to significantly reduce the parameter space from $\sim$11 trillion to 3.3551e4 parameter combinations, allowing us to systematically explore, learn, and predict simulation outputs for almost 4 million unseen parameter combinations for the Heatbugs ABM. We train three different sets of random forest ML models using 2-way covering arrays of the parameter values, to predict three different types of outcomes of the simulation. We test the effectiveness of this approach on the 3-way covering array, which represents significantly more parameter combinations than the training data, to see if it is possible to train the model on a small part of the parameter space and then predict the results on a much larger part of the parameter space. We perform combinatorial analysis and feature importance analysis for insights into simulation output predictability, and the role of parameters in the predictions.

Overall our results show that this process works. All three experiments on Heatbugs had high success in predicting a) if the standard deviation of the average unhappiness level will be low, b) if the standard deviation of unhappiness will be low in relation to the average unhappiness, and c) the average unhappiness. We tested many different thresholds for our categories, all with high success. Our best success was when our classes were most balanced, although learning was still successful even with unbalanced data. These results indicate that the process could be effective on other ABMs as well, and that further work should be done on the predictability of ABM results.

In future studies we plan to explore what types of predictions can be made about ABMs, if the process will work on all types of models or only on models with specific properties, and if a different ML algorithm will garner better results. Random forests were a good fit for this model's data and our goals, but another ML algorithm may better fit other models. We also plan to further explore the use of CFD in analyzing results from the covering arrays in determine parameter importance in results prediction. We plan to continue this research to further define how and when our approach can be successfully applied, so that eventually one could use our process to explore a parameter space more effectively on many ABMs.

Disclaimer. Commercial products may be identified in this document, but such identification does not imply recommendation or endorsement by NIST, nor that the products identified are necessarily the best available for the purpose.

References

1. Kleijnen, J.P., Sargent, R.G.: A methodology for fitting and validating metamodels in simulation. Eur. J. Oper. Res. **120**(1), 14–29 (2000)
2. Borazjany, M., Lei, Y., Kacker, R., Kuhn, R.: Combinatorial testing of acts: a case study. In: First International Workshop on Combinatorial Testing, IEEE Fifth International Conference on Software Testing, Verification and Validation (ICST 2012), pp. 591–600 (2012)
3. Cacuci, D., Ionescu-Bujor, M., Navon, I.M.: Sensitivity and Uncertainty Analysis: Applications to Large-Scale Systems. Taylor & Francis Group, CRC Press (2005)
4. Calvez, B., Hutzler, G.: Automatic tuning of agent-based models using genetic algorithms. In: Sichman, J.S., Antunes, L. (eds.) MABS 2005. LNCS (LNAI), vol. 3891, pp. 41–57. Springer, Heidelberg (2006). https://doi.org/10.1007/11734680_4
5. Calvez, B., Hutzler, G., et al.: Adaptive dichotomic optimization: a new method for the calibration of agent-based models. In: Proceedings of the 2007 European Simulation and Modelling Conference (ESM 2007), pp. 415–419 (2007)
6. Danglade, F., Pernot, J.P., Véron, P.: On the use of machine learning to defeature CAD models for simulation. Comput.-Aided Design Appl. **11**, 358–368 (2014)
7. Hofmann, M.: On the complexity of parameter calibration in simulation models. J. Defense Model. Simul. **2**(4), 217–226 (2005)
8. Kleijnen, J.: An overview of the design and analysis of simulation experiments for sensitivity analysis. Eur. J. Oper. Res. **164**(2), 287–300 (2005)
9. Kleijnen, J.: Sensitivity analysis of simulation models: an overview. In: 6th International Conference on Sensitivity Analysis of Model Output (2010)
10. Krauße, T., Cullmann, J.: Towards a more representative parametrisation of hydrologic models via synthesizing the strengths of particle swarm optimisation and robust parameter estimation. Hydrol. Earth Syst. Sci. **16**, 603–629 (2012)
11. Kuhn, D.R., Kacker, R., Lei, Y.: Introduction to Combinatorial Testing. Chapman & Hall/CRC (2013)
12. Kuhn, D.R., Raunak, M., Prado, C., Patil, V.C., Kacker, R.N.: Combination frequency differencing for identifying design weaknesses in physical unclonable functions. In: 2022 IEEE IEEE International Conference on Software Testing, Verification and Validation Workshops (ICSTW), pp. 110–117. IEEE (2022)
13. Kuhn, D., Kacker, R., Lei, Y.: Advanced combinatorial test methods for system reliability. Technical report, 2010, IEEE Reliability Society (2011)
14. Kuhn, D., Raunak, M.S., Kacker, R.: Combination frequency differencing. Technical report, National Institute of Standards and Technology (2021). https://nvlpubs.nist.gov/nistpubs/CSWP/NIST.CSWP.12062021-draft.pdf
15. Lanus, E., Freeman, L.J., Richard Kuhn, D., Kacker, R.N.: Combinatorial testing metrics for machine learning. In: 2021 IEEE International Conference on Software Testing, Verification and Validation Workshops (ICSTW), pp. 81–84 (2021)
16. Maalouf, C., Olsen, M., Raunak, M.S.: Combinatorial testing for parameter ealuation. In: Proceedings of the 2019 Winter Simulation Conference (WSC19). Society for Computer Simulation International (2019)
17. Olsen, M., Laspesa, J., Taylor-D'Ambrosio, T.: On genetic algorithm effectiveness for finding behaviors in agent-based predator prey models. In: Proceedings of the Summer Simulation Multi-Conference (SummerSim 2018). Society for Computer Simulation International, Bordeaux (2018)
18. Olsen, M., Raunak, M.S.: Efficient parameter exploration of simulation studies. In: 2022 IEEE 29th Software Technology Conference, pp. 190–191 (2022)

19. Stonedahl, F.: Genetic algorithms for the exploration of parameter spaces in agent based models. Ph.D. thesis, Northwestern University (2011)
20. Wilensky: Netlogo heatbugs model (2004). http://ccl.northwestern.edu/netlogo/models/Heatbugs
21. Zhang, W., Valencia, A., Chang, N.B.: Synergistic integration between machine learning and agent-based modeling: a multidisciplinary review. IEEE Trans. Neural Netw. Learn. Syst. 34(5), 2170–2190 (2023)

Vecpar – A Framework for Portability and Parallelization

Georgiana Mania[1,2]([✉])[ID], Nicholas Styles[1][ID], Michael Kuhn[3][ID],
Andreas Salzburger[4][ID], Beomki Yeo[5,6], and Thomas Ludwig[2,7]

[1] Deutsches Elektronen-Synchrotron DESY, Notkestr. 85, 22607 Hamburg, Germany
{georgiana.mania,nicholas.styles}@desy.de
[2] University of Hamburg, Hamburg, Germany
[3] Otto von Guericke University Magdeburg, Magdeburg, Germany
michael.kuhn@ovgu.de
[4] CERN, 1211 Geneva, Switzerland
andreas.salzburger@cern.ch
[5] Lawrence Berkeley National Laboratory, Berkeley, CA 94720, USA
[6] Department of Physics, University of California, Berkeley, CA 94720, USA
beomki.yeo@berkeley.edu
[7] Deutsches Klimarechenzentrum, Bundesstraße 45a, 20146 Hamburg, Germany
ludwig@dkrz.de

Abstract. Complex particle reconstruction software used by High
Energy Physics experiments already pushes the edges of computing
resources with demanding requirements for speed and memory throughput, but the future experiments pose an even greater challenge. Although
many supercomputers have already reached petascale capacities using
many-core architectures and accelerators, numerous scientific applications still need to be adapted to make use of these new resources. To
ensure a smooth transition to a platform-agnostic code base, we developed a prototype of a portability and parallelization framework named
vecpar. In this paper, we introduce the technical concepts, the main features and we demonstrate the framework's potential by comparing the
runtimes of the single-source vecpar implementation (compiled for different architectures) with native serial and parallel implementations, which
reveal significant speedup over the former and competitive speedup versus the latter. Further optimizations and extended portability options
are currently investigated and are therefore the focus of future work.

Keywords: Performance portability · Parallel computing ·
Heterogeneous computing

1 Introduction

High Energy Physics experiments, such as those at the Large Hadron Collider
(LHC) at CERN [2], use complex algorithms to resolve the huge amounts of

This work was supported by DASHH under grant number HIDSS-0002.

data from their detectors into accurate descriptions of the particles of interest to be studied. The number of particles traversing the detector within a given time interval is frequently increasing, resulting in a significant rise in the multiplicity of discrete measurements. This will be further exacerbated in future, when the High-Luminosity LHC will provide events containing up to 10 000 charged particles to be reconstructed.

Multi-threading and accelerator support could be one possibility to alleviate this challenge. GPUs, many-core CPUs or architectures like ARM are efficient solutions in terms of FLOPS per watt, which makes them the perfect candidates for the world's most powerful supercomputers. As listed in the November 2022 edition of the TOP500, the ten highest-ranked clusters use a variety of processing units provided by different vendors including AMD, NVIDIA, IBM, Intel and ARM [23]. This great diversity comes with a major challenge for programmers: how to ensure code portability and maintainability, when each targeted platform requires different low-level assembly code that can be generated by a multitude of programming languages, language extensions, libraries or tools, some of which are vendor-proprietary.

Furthermore, there is also the question of how much the currently commonly used particle reconstruction software could benefit from the heterogeneous architectures, keeping in mind that the parallelization potential is limited in a number of places by its sequential nature. Thus, we designed *vecpar* to be an easy-to-use framework for parallelization targeting CPU and GPU with single-source C++ code, compiled for different platforms. Domain scientists can easily implement an algorithm without any previous knowledge of dedicated language extensions and test its performance on heterogeneous architecture, knowing that vecpar adds minimal overhead to native implementations. In the end, they can choose to either gradually port more complex algorithms to vecpar abstractions or to implement a specific native solution based on the performance results obtained by the prototype vecpar implementation.

This paper is organized as follows. In Sect. 2 we present an overview of the state of the art and related work. Then we introduce the vecpar framework in Sect. 3. Preliminary performance results are discussed in Sect. 4, while the conclusions and future work are summarized in Sect. 5.

2   State of the Art and Related Work

In the last 15 years, the HPC community has included support for co-processors, GPUs and FPGAs into existing state-of-the-art parallel programming standards like OpenMP [18] and developed new ones like OpenACC [17], OpenCL [7], and SYCL [20]. Even the C++ standard added concepts like parallel execution policies, polymorphic allocators, increased support for compile-time polymorphism and many others in a first step to unify the execution environment on heterogeneous devices. Additionally, each vendor provides native interfaces and dedicated compiler tools, which are optimized to ensure the best performance out of their hardware. Examples include CUDA [16], the parallel computing platform for NVIDIA GPUs and Data Parallel C++ [19], the multi-architecture programming model proposed by Intel to target all their platforms: CPU, GPU and FPGA.

Unsurprisingly, these developments offer a trade-off between performance and portability [9], a difficult decision that needs to be made having several factors in mind: the available hardware for production environment, the programming skills needed to write the code, the effort to maintain potentially several native implementations targeting different platforms and last, but certainly not least, the parallelization potential of the scientific application. Consequently, this led to the development of several heterogeneous libraries and frameworks which offer the flexibility of having single-source C++ code compiled for different platforms while still ensuring good performance. With complex abstractions for memory and compute layouts, Kokkos [25] and Alpaka [26] deliver top performance but with some drawbacks in term of productivity considering that application developers are usually scientific domain experts rather than computer scientists. Other libraries like GrPPI [11] and SkePU [8] define similar abstractions to vecpar, but the former does not ensure GPU offloading support, while the latter is missing some features (like the filter skeleton) to accommodate some steps of the reconstruction flow (e.g. filtering the measurements above a specific threshold). Similar to these two libraries, vecpar aims to decouple scientific algorithms from parallelization strategies while automatically handling memory transfers, which can increase development productivity significantly.

Particle physicists from major experiments at CERN already started to adapt algorithms and event data models for parallel execution and GPU offloading in software frameworks like AthenaMT [13], CMSSW [6], ALICE O^2 [21] and Allen [3]. A Common Tracking Software Project (ACTS) is an experiment-independent toolkit for track[1] reconstruction [4], which offers a realistic and thread-safe test-bed for R&D projects that explore heterogeneous architectures; these include the following three libraries. Firstly, *algebra-plugin* [1] is a linear algebra library with two available heterogeneous backends: *cmath*, which provides custom implementations in C++, CUDA and SYCL, and *eigen*, uses Eigen library [12] for data types and mathematical functions. Secondly, *vecmem* [24] is a heterogeneous memory management library with C++, CUDA, HIP and SYCL support, which defines several memory resources, iterable containers based on polymorphic allocators (e.g. `vector` and `jagged_vector`) and utility functions to support vector-related operations similar to the ones provided by the C++ Standard Library but enhanced with GPU-friendly features. Lastly, *detray* [22] is a header-only detector surface intersections based on algebra-plugin and vecmem.

While these libraries increase portability to different platforms and vendors by implementing the core algorithms inline to the GPU's restrictive requirements (e.g. no dynamic memory allocations and polymorphism), they move away from the concept of single-source repository because they provide dedicated backends for different architectures; this translates into a substantial development effort for domain scientists to maintain all of them. Moreover, the applications which

---

[1] A track is a charged particle trajectory through a detector.

use these tools, must assemble a reconstruction chain for CPU and a different one for GPU. Vecpar is intended to close this gap by delivering comparable levels of performance portability when using a single-source implementation, without requiring any knowledge of parallelization strategies.

3   Proposed Approach

In this section we first introduce vecpar's underlying programming concepts and then we describe the framework's design, the Application Programming Interface (API) and associated compiler support.

Vecpar[2] is implemented as an open-source header-only library. To address the most common scenarios in particle reconstruction (but not necessarily limited to them), a series of abstractions were defined which allow the inversion of control of the execution flow from the application itself to the framework. This enables the separation of concerns between the scientific code and the parallelization strategies which can be extended, modified or replaced with minimum or even no impact on the invoking code.

Conceptually, vecpar is based on the *map-filter-reduce* notations and calculus from functional programming for specifying and manipulating computable functions over lists [5] and ensures a thread-safe environment by handling immutable data structures protected by *const correctness*[3]. It defines the main operators: `parallel_map`[4], `parallel_filter` and `parallel_reduce`, and the composed versions `parallel_map_filter` and `parallel_map_reduce`, which are implemented by each of the vecpar backends using different languages. Additionally, a generic `parallel_algorithm` operator is provided, which dispatches the execution to the appropriate implementation based on C++20 concepts and partial specializations, which were chosen because they employ compile-time polymorphism. The operators apply user-defined functions wrapped in SPSVERBc9s on C++ vectors allocated either in host, device or CUDA unified memory using the vecmem library support for heterogeneous resources. For example, by extending the `vecpar::parallelizable_map` algorithm class which is templated on (a) the number of iterable collections and (b) the data types for input and output, the user has to provide an implementation for `mapping_function`, which is called by the framework at run-time by invoking a wrapper lambda function that will eventually be executed by each parallel thread. The algorithm classes are stateless and have no virtual functions. This ensures a straight-forward mapping to the GPU's memory, if needed. The infrastructure supports up to five collections of the same length that can be iterated over in the same time.

Currently, vecpar fully supports two parallel execution backends: CPU using OpenMP threads and GPU using CUDA threads. An experimental backend

---

[2] https://github.com/wr-hamburg/vecpar.
[3] As an exception motivated by performance optimization purposes, the vecpar API provides a limited number of mutable versions as well.
[4] A `parallel_mmap` operator which allows mutable data structures is also defined.

based on OpenMP target was recently implemented, in order to extend portability to AMD GPUs. Similarly, new backends could be easily added to the generic dispatch system. The execution flow is summarized in Fig. 1. To run a vecpar algorithm in a parallel (and potentially GPU offloaded) manner, it has to be passed to a specific API function like `vecpar::parallel_map(algorithm,...)` or the generic `vecpar::parallel_algorithm(algorithm,...)` which would delegate the execution to the parallel implementation using OpenMP or CUDA, respectively. In this case, the decision is made implicitly based on evaluating internal flags like `__CUDA__` or `_OPENMP` which are set by the compiler based on information provided either by the user (e.g. through compilation flags) or inferred at compile/link time (e.g. available support for OpenMP). There is also the option to invoke a specific backend directly; for example calling `vecpar::cuda::parallel_algorithm(algorithm,...)` bypasses the dispatch system. This could be particularly useful when two imbricated levels of parallelism are required, for instance when reading multiple event data files from disk in a CPU multi-threaded environment while offloading the computations associated with each event to the GPU.

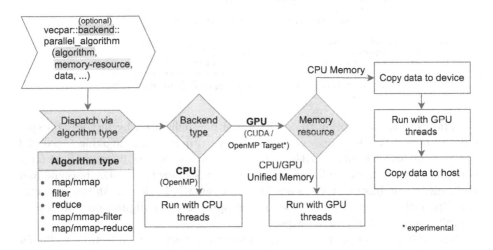

Fig. 1. Decision flow in vecpar

If an algorithm doesn't necessarily fit into one of the map-filter-reduce patterns or the input data structures are different from the ones expected by the predefined base classes, vecpar offers a generic `vecpar::parallel_map` function which can be invoked with a lambda that defines the behavior, provided by the user. While this offers more flexibility and ensures that the anonymous function can be executed as a GPU kernel, it passes the responsibility of data management and parallelization to the user implementation.

In the case of a GPU parallelization, the framework handles not only the distribution of independent work items to different threads, but also the required memory transfers between host and device, using vecmem library support. Vec-par's CUDA backend provides separate implementations for executing parallel algorithms handling collections stored in managed memory and host memory. This is an important distinction since the former requires only a retrieval of a pointer to an unified memory address, while the latter requires explicit memory copy between host and device for both input and output.

The algorithm chaining functionality simplifies the implementation of a use case which requires several algorithms as intermediate steps to reach a final result. As a precondition, these algorithms need to be mathematically compos-able or otherwise a compile-time error will be shown. Vecpar's implementation calls `parallel_algorithm(algorithm,...)` on each algorithm, in order, using the result from the previous one as input for the current one. The execution flow of the chain is then reduced to the one described above in Fig. 1. To use this feature, an instance of a `vecpar::chain` templated on the input and output data types is needed as a first step. Then the algorithm instances and the input data associated with the first algorithm are passed to the chain through calls to its member functions as showed in Listing 1.1.

```
1  vecpar::chain<vecmem::host_memory_resource,
2              double,
3              vecmem::vector<int>> chain(mr);
4
5  chain.with_config(config) // optional
6      .with_algorithms(alg_1, alg_2, alg_3)
7      .execute(vecmem_vector, context_object);
```

Listing 1.1. Algorithm chaining definition example

The user has the option to configure the parallelization by providing a spe-cific number of workers in a `vecpar::config` object when invoking a parallel function or parallel chain. Without passing it or by leaving it empty, a default configuration is generated based on the problem size for the CUDA backend and the runtime environment variables for the OpenMP backend. At the moment, the OpenMP Target backend does not support user configuration and uses a default one for optimization purposes.

The GNU `gcc` compiler can be used to compile vecpar code to target `x86_64` and `aarch64` architectures and when built with offloading capabilities, it can also target NVIDIA and AMD GPUs through vecpar's OpenMP target backend. In addition to what is offered by `gcc`, LLVM/`clang` compiler can additionally gener-ate NVIDIA's Parallel Thread Execution assembly code (NVPTX) and therefore target NVIDIA GPUs with native assembly, which makes it more versatile for building heterogeneous code. `ROCm/aomp` can also be used to compile for CPU and AMD GPUs using the OpenMP target backend. An NVIDIA compiler which supports C++20 is required to build vecpar sources.

4   Evaluation

In this section, we demonstrate the benefits of using vecpar by evaluating it in comparison to native state-of-the-art parallel solutions like OpenMP (OMP),

OpenMP Target (OMPT) and CUDA, and to similar approaches like Kokkos, in several use cases, starting from trivial kernels to more complex scenarios from particle reconstruction software.

Four environments were used for running the experiments: **Env1** – Intel Core i7-10870H CPU and NVIDIA GeForce RTX 3060 GPU, **Env2** (HPC cluster node) – 20-cores Intel Xeon Gold 5115 CPU and NVIDIA Tesla V100 GPU, **Env3** (Raspberry Pi Cluster node) – 4-cores ARM Cortex A72 CPU, and **Env4** – Intel i5-8600K CPU and AMD Radeon RX6750 XT. Clang 14 (with offload support) was used as default C++ compiler for the Env1 and Env2, gcc 12 was used for the third one, while aomp 16 was used for Env4. CUDA 11.6 (driver v.510.47.03), ROCm 5.3.3 and OpenMP 4.5 ensured the support for parallelism.

4.1   BabelStream Benchmark

The BabelStream benchmark [10] defines a framework to evaluate (a) the wall clock execution times and (b) the memory throughput of simple mathematical operations when using different parallelization APIs and/or compilers. To achieve this, different abstractions are required for memory allocations (and GPU transfers if needed) and kernel execution. This clear separation is not transparent in a vecpar algorithm implementation since memory handling is a built-in feature of the library, as previously shown in Fig. 1; nevertheless, vecpar offers a lambda-based feature, which allows more flexibility to the user by trading off strictly single-source nature of the implementation. This lambda implementation is used for the GPU-CUDA benchmark while single-source vecpar algorithms were used for all the other scenarios. The vecpar branch for BabelStream is available open-source on github[5]. For the present results, we used the *triad* benchmark which performs $a_i = b_i + scalar \times c_i, \forall 0 \leq i < 2^{25}$, with double-precision operands. This kernel is run 100 times and the mean execution time (which does not include the necessary time for memory transfers in the case of the GPU) is computed. The measurements shown in Fig. 2 prove that vecpar achieves comparable performance with native parallel implementations; for the CPU, there is a deviation of up to $\approx 10\%$ for both vecpar and kokkos OMP backends, while the GPU targeting implementations (using OMPT and CUDA) are within $\pm 1\%$ of the other benchmarks.

4.2   Vecpar Internal Benchmark

To evaluate the performance and portability of a single-source vecpar algorithm compiled for different platforms, we designed a custom benchmark, available in the vecpar repository. In this case, we include CPU-GPU memory transfer times in the total execution time of a given scenario. Each test was repeated 20 times.

The Single and Double Precision $a \times X + Y$ (SAXPY/DAXPY) benchmarks provides three implementations: OpenMP, CUDA and vecpar (using different backends), which perform the same operation: $y_i = a \times x_i + y_i, x_i \in X, y_i \in$

---

[5] https://github.com/wr-hamburg/BabelStream/tree/vecpar.

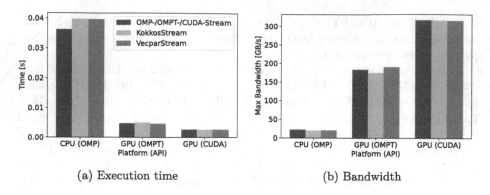

(a) Execution time   (b) Bandwidth

Fig. 2. BabelStream `triad` kernel, using vectors of 2^{25} FP64, on Env1

$Y, \forall\, 0 \leq i < N$ and the code is compiled with `clang` for the NVIDIA GPU and with the `aomp` compiler for the AMD GPU. Figure 3 shows that the overhead of vecpar in comparison to the native CUDA implementations on NVIDIA hardware is up to 0.5%. The vecpar OpenMP target code compiled for AMD GPU seems to be slower for smaller problem size while showing a 4× speedup for vectors of one million elements; this is most likely credited to the link-time optimization of the LLVM compiler.

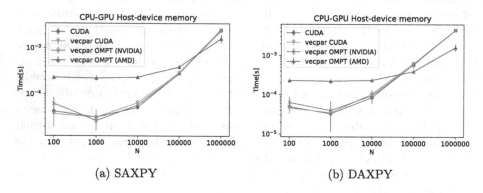

(a) SAXPY   (b) DAXPY

Fig. 3. Mean and standard deviation for SAXPY and DAXPY kernels, using vectors of different sizes N, on Env1 and Env4

4.3   Track Reconstruction Use Cases

In order to establish the position and momentum of charged particles in a magnetic field, the second order differential equation of motion is integrated to provide a numerical solution which is later corroborated with detector measurements to obtain a realistic estimate. The 4th order Runge-Kutta-Nyström (RKN) algorithm is widely used for its precision but it has the downside of being inherently sequential, with each stage depending on the calculations of the previous one. The *detray* project features an implementation of an algorithm based on adaptive

RKN methods [14,15] (*RKN stepper*) which provides estimates for the transport of track parameters[6] and their covariance matrices, either in global coordinates (i.e. free parameters) or in local coordinates based on an intersection surface (i.e. bound parameters) through the detector layers. For each experiment a different executable is built for every linear algebra backend from *algebra-plugin* project and then ran on different platforms. For all the tests, the wall-clock times include host-device transfers when the data is in host memory and a GPU is used to compute the results. In case the data is already in managed memory, the time spent for initialization is not factored in.

Track Parameters Estimation. The first case that we explored is a simplified STEP algorithm which provides estimates for track parameters only (without computing the Jacobians) and limits the numerical integration to 100 steps in both forward and backward direction. Moreover, the magnetic field is assumed constant in z direction. For each test, we simulate 10 000 (free) tracks starting from the center of the detector, with a negative unit charge and a magnetic field of (0, 0, 2T).

```
__global__ void rk_stepper_test_kernel (vecmem::data::
    vector_view<free_track_parameters> tracks_data,
    const vector3 B) {
    int gid = threadIdx.x + blockIdx.x * blockDim.x;
    vecmem::device_vector<free_track_parameters>
    tracks(tracks_data);
    // Prevent overflow
    if (gid >= tracks.size()) {
        return;
    }
    // Get a track
    auto& traj = tracks.at(gid);
    // Define RK stepper
    rk_stepper_type rk(B);
    // Index for stepping
    unsigned int i_s=0;
    // Forward direction
    rk_stepper_type::state forward_state(traj);
    for (i_s=0; i_s<rk_steps; i_s++) {
        rk.step(forward_state);
    }
    // Backward direction
    traj.flip();
    rk_stepper_type::state backward_state(traj);
    for (i_s=0; i_s<rk_steps; i_s++) {
        rk.step(backward_state);
    }
}
```

```
struct rk_stepper_algorithm :
    public vecpar::algorithm::parallelizable_mmap<
    free_track_parameters, vector3>{

    TARGET free_track_parameters& mapping_function(
        free_track_parameters& traj, vector3 B) override
    {
        // Define RK stepper
        rk_stepper_type rk(B);
        // Index for stepping
        unsigned int i_s=0;
        // Forward direction
        rk_stepper_type::state forward_state(traj);
        for (i_s=0; i_s<rk_steps; i_s++) {
            rk.step(forward_state);
        }
        // Backward direction
        traj.flip();
        rk_stepper_type::state backward_state(traj);
        for (i_s=0; i_s<rk_steps; i_s++) {
            rk.step(backward_state);
        }
        return traj;
    }
};
```

Listing 1.2. CUDA   **Listing 1.3.** C++/vecpar

To evaluate the development productivity, we first look at the implementation details. The CUDA and vecpar RKN kernels in Listing 1.2 and Listing 1.3[7] show that while the former requires knowledge about accessing the data from unified memory and thread parallelization based on global index, the latter fits the code into vecpar's *map* abstraction templated on the input and output data types and can just focus on the actual hardware-agnostic algorithm which is **identical** in

---

[6] A set of parameters describing the helical trajectory followed by a charged particle moving within a magnetic field.

[7] TARGET is a vecpar macro which adds extra qualifiers at compile time.

both cases, as highlighted in the rectangle. Moreover, the vecpar implementation can be compiled with `clang` for x86_64 and NVIDIA GPU platforms without further code changes, which ensures portability and productivity by avoiding to maintain distinct code sources for CPU and GPU.

In terms of performance, we compared the vecpar single-source implementation compiled for different platforms (`vecpar_cpu` and `vecpar_gpu`) with the initial sequential version (`seq_cpu`), a hard-coded OpenMP version (`omp_cpu`) and a CUDA version (`cuda_nvcc`) and we evaluated the impact of different factors like the problem size, the precision of the operands and the number of parallel workers on Env1 and Env2. Figure 4 shows the results for the tests focusing on single precision; both OpenMP and vecpar versions provide close to ideal strong scaling performance, when using up to 40 threads, on Env2.

A closer analysis shows that the parallel versions are faster than the sequential one for double precision operands on both linear algebra backends (cmath and eigen), for both CPU and GPU, as depicted in the speedup diagrams from Fig. 5. For the GPU, the advantage of the vecpar solution over the native CUDA

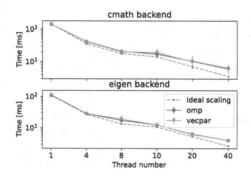

Fig. 4. Strong-scaling evaluation: Multi-threading implementations of the RKN stepper, in single precision, cmath/eigen backends, on Env2

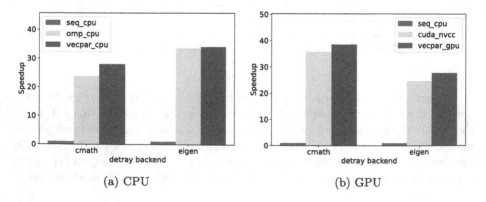

Fig. 5. Speedup factors of the vecpar single-source implementation compiled for CPU and GPU over other implementations, on Env2

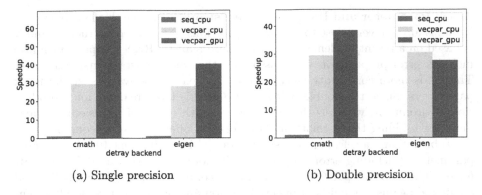

(a) Single precision   (b) Double precision

Fig. 6. Speedup for vecpar implementation over the sequential CPU implementation, using cmath/eigen math backends, in single/double precision, on Env2

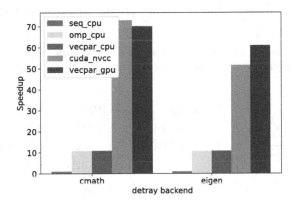

Fig. 7. Speedup diagram for simplified RKN stepper using cmath/eigen backends, in single precision, one million tracks, without fastmath support, on Env1

is assumed to be due to NVPTX optimizations done by `clang` since compiling the same RKN kernel with `clang` already shows a slight speedup over the executable produced by `nvcc`.

Figure 6 summarizes these experiments with the conclusion that for this given scenario, the vecpar single-source implementation ensures a significant speedup of 28–65× over the initial sequential implementation, comparable to native OpenMP and CUDA implementations.

Extreme Load Use Case. We evaluated the performance of the vecpar implementation in an extreme-load test of one million tracks using 16 OpenMP threads and 3907 × 256 CUDA threads. The results in Fig. 7 show that the vecpar implementation is comparable with native ones, showing up to 12× and 70× speedups for CPU and GPU respectively over the initial sequential one.

Track Parameter and Its Covariance Estimation. The third set of experiments is already very close to common operations in track reconstruction and it is based on a recent implementation of a more realistic RKN stepper which also includes error-propagation while assuming the same constant magnetic field. This adds more complexity since it computes partial derivatives of the track parameters. Since the double precision use case is more realistic for a production environment, we focus the experiments on this level of precision; this also includes disabling the fastmath support features offered by some compilers.

Table 1 shows a selection of results obtained by estimating 10 000 free track parameters (including error propagation) in double precision on different platforms. The same vecpar C++ source file is compiled for CPU and GPU, and compared against the initial sequential version and against native parallel solutions (using OpenMP threads for CPU and CUDA threads for GPU). The measurements involve using the cmath backend for detray, but similar results were obtained with the eigen backend, but they are omitted due to space restrictions.

Table 1. Mean time (in seconds) and standard deviation for estimating 10 000 tracks using sequential, parallel and vecpar implementations on different CPU/GPU platforms

Platform	Sequential	OpenMP/CUDA	Vecpar
ARM aarch64 (Env3)	4.54 ± 0.04%	1.16 ± 4.11%	1.27 ± 9.78%
Intel x86_64 (Env2)	1.69 ± 5.51%	0.1058 ± 4.75%	0.1052 ± 5.41%
NVIDIA V100 (Env2)	–	0.004 ± 6.32%	0.02 ± 1.86%

Firstly, we explored the extended portability to aarch64; the vecpar implementation is within ±9% of the native implementation and 3.55× faster than the sequential one. This is close to the maximum theoretical speedup of 4× which is limited by the number of OpenMP threads (which is 4 in this case). For the x86_64 platform, the vecpar implementation is within ±3% of the OpenMP implementation while being 15× faster than the sequential one. Secondly, for the GPU case, regardless of the storage location, the vecpar implementation takes longer than native CUDA compiled with nvcc and data in managed memory. Despite much better memory and computation throughput, there are several uncoalesced global memory accesses due to unnecessary duplication of the offloaded algorithmic code. We expect this overhead to be alleviated by using a CUDA 11.7 feature: __grid_constant__; this will store the algorithm object into GPU constant cache which provides much better latency than global memory. Experimental validation should be possible as soon as LLVM/clang enables support for it.

Nevertheless, there is still a benefit of having a single parallel implementation which can be executed on different platforms as it is shown in Fig. 8. Speedups up to 19× and 108× can be observed for CPU and GPU respectively.

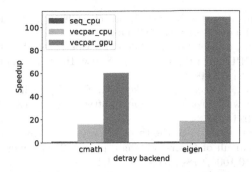

Fig. 8. Speedup of the vecpar implementation for RKN stepper (with error propagation) for 10 000 tracks over the sequential implementation for cmath and eigen backends, in double precision, on Env2

5   Conclusions and Future Work

In this paper, we presented a new framework for efficient parallelized execution of charged particle tracking[8]. It relies on new C++ features, OpenMP and CUDA to ensure improved performance and portability while decoupling the scientific algorithms from the parallelization strategies. Although vecpar is still in an early development phase, it was successfully used to port a simplified step of the track reconstruction flow and demonstrated its potential by obtaining speedups up to 108× over sequential implementations using single-source C++ code. Moreover, it provides speedups competitive with those obtained by both native and related APIs, while being easier to implement and port to different architectures. We showed its portability by testing it on different platforms like `x86_64`, `ARM`, `NVIDIA Volta` and `Ampere`, and `AMD Radeon`.

In future versions, we plan to improve the memory access patterns for the CUDA backend, to finalize the development of the OpenMP target backend and to extend the abstractions to allow more complex patterns like hierarchical parallelism and uneven distributed workloads.

Acknowledgment. We acknowledge the support by DASHH (Data Science in Hamburg - HELMHOLTZ Graduate School for the Structure of Matter) with the Grant-No. HIDSS-0002. The National Analysis Facility (NAF) at Deutsches Elektronen-Synchrotron (DESY), the University of Hamburg (UHH) and Deutsche Klimarechenzentrum (DKRZ) provided the hardware resources for the experiments.

References

1. Algebra-plugin. https://github.com/acts-project/algebra-plugins/. Accessed 8 Dec 2022

---

[8] While the initial goal was to contribute to the increase of the performance portability of open-source track reconstruction software, other scientific use cases are welcome in the future.

2. European Organization for Nuclear Research (CERN). Nature **184**(4702), 1844 (1959). https://doi.org/10.1038/1841844b0
3. Aaij, R., Albrecht, J., Belous, M., Billoir, P., Boettcher, T., et al.: Allen: a high-level trigger on GPUs for LHCb. Comput. Softw. Big Sci. **4**(1) (2020). https://doi.org/10.1007/s41781-020-00039-7
4. Ai, X., Allaire, C., Calace, N., Czirkos, A., Ene, I., Elsing, M., et al.: A common tracking software project. Comput. Softw. Big Sci. (2022). https://doi.org/10.1007/s41781-021-00078-8
5. Bird, R.S.: An introduction to the theory of lists. In: Broy, M. (ed.) Logic of Programming and Calculi of Discrete Design, pp. 5–42. Springer, Heidelberg (1987). https://doi.org/10.1007/978-3-642-87374-4_1
6. Bocci, A., Kortelainen, M., Innocente, V., Pantaleo, F., Rovere, M.: Heterogeneous reconstruction of tracks and primary vertices with the CMS pixel tracker (2020). https://doi.org/10.48550/ARXIV.2008.13461, http://arxiv.org/2008.13461
7. Breitbart, J., Fohry, C.: OpenCL - an effective programming model for data parallel computations at the Cell Broadband Engine. In: 24th IEEE International Symposium on Parallel and Distributed Processing, IPDPS 2010, Atlanta, Georgia, USA, 19–23 April 2010 - Workshop Proceedings, pp. 1–8. IEEE (2010). https://doi.org/10.1109/IPDPSW.2010.5470823
8. Dastgeer, U.: Skeleton programming for heterogeneous GPU-based systems (2011)
9. Deakin, T., Poenaru, A., Lin, T., McIntosh-Smith, S.: Tracking performance portability on the yellow brick road to exascale. In: 2020 IEEE/ACM International Workshop on Performance, Portability and Productivity in HPC (P3HPC), pp. 1–13 (2020). https://doi.org/10.1109/P3HPC51967.2020.00006
10. Deakin, T., Price, J., Martineau, M., McIntosh-Smith, S.: Evaluating attainable memory bandwidth of parallel programming models via BabelStream. Int. J. Comput. Sci. Eng. **17**(3), 247–262 (2018). https://doi.org/10.1504/IJCSE.2018.095847. Special Issue on Novel Strategies for Programming Accelerators
11. del Rio Astorga, D., Dolz, M.F., Fernández, J., García, J.D.: A generic parallel pattern interface for stream and data processing. Concurr. Comput. Pract. Exp. **29**(24), e4175 (2017). https://doi.org/10.1002/cpe.4175, https://onlinelibrary.wiley.com/doi/abs/10.1002/cpe.4175
12. Guennebaud, G., Jacob, B., et al.: Eigen v3 (2010). https://eigen.tuxfamily.org
13. Leggett, C., et al.: AthenaMT: upgrading the ATLAS software framework for the many-core world with multi-threading. J. Phys.: Conf. Ser. **898**, 042009 (2017). https://doi.org/10.1088/1742-6596/898/4/042009
14. Lund, E., Bugge, L., Gavrilenko, I., Strandlie, A.: Track parameter propagation through the application of a new adaptive Runge-Kutta-Nystrom method in the ATLAS experiment. Technical report (2009). https://cds.cern.ch/record/1113528/files/ATL-SOFT-PUB-2009-001.pdf
15. Lund, E., Bugge, L., Gavrilenko, I., Strandlie, A.: Transport of covariance matrices in the inhomogeneous magnetic field of the ATLAS experiment by the application of a semi-analytical method. Technical report (2009). https://cds.cern.ch/record/1114177/files/ATL-SOFT-PUB-2009-002.pdf
16. Nickolls, J., Buck, I., Garland, M., Skadron, K.: Scalable parallel programming with CUDA. In: International Conference on Computer Graphics and Interactive Techniques, SIGGRAPH 2008, Los Angeles, California, USA, 11–15 August 2008, Classes, pp. 16:1–16:14. ACM (2008). https://doi.org/10.1145/1401132.1401152
17. Organization, O.: The OpenACC application programming interface, version 3.2. https://www.openacc.org/sites/default/files/inline-images/Specification/OpenACC-3.2-final.pdf

18. van der Pas, R., Stotzer, E., Terboven, C.: Using OpenMP - The Next Step: Affinity, Accelerators, Tasking, and SIMD. MIT Press (2017). https://mitpress.mit.edu/ books/using-openmp-next-step

19. Reinders, J., Ashbaugh, B., Brodman, J., Kinsner, M., Pennycook, J., Tian, X.: Data Parallel C++. Apress, Berkeley (2021). https://doi.org/10.1007/978-1-4842-5574-2

20. Reyes, R., Lomüller, V.: SYCL: single-source C++ accelerator programming. In: Joubert, G.R., Leather, H., Parsons, M., Peters, F.J., Sawyer, M. (eds.) Parallel Computing: On the Road to Exascale, Proceedings of the International Conference on Parallel Computing, ParCo 2015, 1–4 September 2015, Edinburgh, Scotland, UK. Advances in Parallel Computing, vol. 27, pp. 673–682. IOS Press (2015). https://doi.org/10.3233/978-1-61499-621-7-673

21. Rohr, D., Gorbunov, S., Schmidt, M.O., Shahoyan, R.: Track reconstruction in the ALICE TPC using GPUs for LHC Run 3 (2018). https://doi.org/10.48550/ ARXIV.1811.11481, https://arxiv.org/abs/1811.11481

22. Salzburger, A., Niermann, J., Yeo, B., Krasznahorkay, A.: Detray: a compile time polymorphic tracking geometry description. J. Phys.: Conf. Ser. **2438**(1), 012026 (2023). https://doi.org/10.1088/1742-6596/2438/1/012026

23. Strohmaier, E., Dongarra, J., Simon, H., Meuer, M.: Top500 List. https://www. top500.org. Accessed 01 Dec 2022

24. Swatman, S.N., Krasznahorkay, A., Gessinger, P.: Managing heterogeneous device memory using C++17 memory resources. J. Phys.: Conf. Ser. **2438**(1), 012050 (2023). https://doi.org/10.1088/1742-6596/2438/1/012050

25. Trott, C.R., Lebrun-Grandié, D., Arndt, D., Ciesko, J., Dang, V., et al.: Kokkos 3: programming model extensions for the exascale era. IEEE Trans. Parallel Distrib. Syst. **33**(4), 805–817 (2022). https://doi.org/10.1109/TPDS.2021.3097283

26. Zenker, E., et al.: Alpaka - an abstraction library for parallel kernel acceleration. IEEE Computer Society (2016). https://arxiv.org/abs/1602.08477

Self-supervised Deep Heterogeneous Graph Neural Networks with Contrastive Learning

Zhiping Li[1,2], Fangfang Yuan[1(✉)], Cong Cao[1(✉)], Dakui Wang[1], Jiali Feng[1,2], Baoke Li[1,2], and Yanbing Liu[1,2]

[1] Institute of Information Engineering, Chinese Academy of Sciences, Beijing, China
{lizhiping,yuanfangfang,caocong,wangdakui,fengjiali,
libaoke,liuyanbing}@iie.ac.cn
[2] School of Cyber Security, University of Chinese Academy of Sciences,
Beijing, China

Abstract. Heterogeneous graph neural networks have shown superior capabilities on graphs that contain multiple types of entities with rich semantic information. However, they are usually (semi-)supervised learning methods which rely on costly task-specific labeled data. Due to the problem of label sparsity on heterogeneous graphs, the performance of these methods is limited, prompting the emergence of some self-supervised learning methods. However, most of self-supervised methods aggregate meta-path based neighbors without considering implicit neighbors that also contain rich information, and the mining of implicit neighbors is accompanied by the problem of introducing irrelevant nodes. Therefore, in this paper we propose a self-supervised deep heterogeneous graph neural networks with contrastive learning (DHG-CL) which not only preserves the information of implicitly valuable neighbors but also further enhances the distinguishability of node representations. Specifically, (1) we design a cross-layer semantic encoder to incorporate information from different high-order neighbors through message passing across layers; and then (2) we design a graph-based contrastive learning task to distinguish semantically dissimilar nodes, further obtaining discriminative node representations. Extensive experiments conducted on a variety of real-world heterogeneous graphs show that our proposed DHG-CL outperforms the state-of-the-arts.

Keywords: Heterogeneous graph neural networks · Self-supervised learning · Contrastive learning

1 Introduction

Heterogeneous graphs are widely present in real-world networks, which can model various types of entities and their relations, such as academic networks, social networks, etc. Figure 1(a) shows an example of a heterogeneous graph. Recently, heterogeneous graph neural networks (HGNNs) have become an emerging tool for mining heterogeneous graph-structured data by aggregating the properties of neighbors. Precisely because of preserving both attribute and structural

© The Author(s), under exclusive license to Springer Nature Switzerland AG 2023
J. Mikyška et al. (Eds.): ICCS 2023, LNCS 14073, pp. 268–282, 2023.
https://doi.org/10.1007/978-3-031-35995-8_19

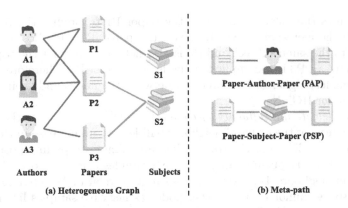

Fig. 1. A toy example of a heterogeneous graph (ACM) and its meta-path. (a) A heterogenous graph ACM consists three types of nodes (author (A), paper (P) and subject (S)) and two types of relations. (b) Two meta-paths involved in ACM (i.e., Paper-Author-Paper and Paper-Subject-Paper).

information, HGNNs show superior performance in various graph-data mining tasks such as node classification, link prediction, and graph classification. In fact, HGNNs are usually trained under the (semi-)supervised learning paradigm, which means that the process requires the task-specific labeled data. In most real-world scenarios, it is very difficult and expensive to obtain labeled data. For example, in the case of microbiome networks, labeling these data requires domain-specific knowledge which is often scarce or non-existent [33].

Motivated by methods [19,20], self-supervised learning capable of training deep models on unlabeled data promises to be a solution. Contrastive learning as a typical method of self-supervised learning has drawn massive attention for its outstanding performance [1,4,23,29], exploiting instance discrimination as a pretext task to learn more discriminative representations. Recently, some efforts have been devoted to investigating the potential of contrastive self-supervised learning on heterogeneous graphs. For example, DMGI [17] using mutual information and HeCo [28] using co-contrastive learning achieve desirable results on heterogeneous graphs. However, they only focus on aggregating meta-path based neighbors without considering implicit neighbors that also contain rich information.

To fully utilize the information of implicit neighbors and reduce the influence of irrelevant neighbors on heterogeneous graphs, in this paper, we take an attempt to perform a contrastive self-supervised learning on heterogeneous graphs. However, mining the valuable neighbors on heterogeneous graphs is a non-trivial problem, presenting us with two key challenges:

1. *How to fully mine the relevant neighbors?* It is well known that meta-paths [22] can describe different semantic contents between the nodes and their neighbors. However, due to the complexity of heterogeneous graphs, there are many deep implicit semantics on heterogeneous graphs and it may be informative for the learning of node representations. For example, in Fig. 1,

it only uses the local neighbors of the paper P1 to predict the topic of P1 through the meta-path PAP. However, the paper P3 and P1's neighbor P2 belong to the same subject and P3 is also helpful to predict the topic of P1. That is to say, P3 is the implicit valuable neighbor of P1. Therefore, modeling these implicit higher-order semantics between the nodes and their neighbors is critical for HGNNs.

2. *How to alleviate the influence of irrelevant neighbors?* We note that for each node, increasing the order of neighbors will lead to an increase in the number of irrelevant neighbors. Contrastive learning aiming to pulling positive samples together and pushing apart negative samples can be used to distinguish irrelevant neighbors. However, there is a high correlation between nodes in the graph, so we cannot treat all other nodes as negative samples like in images [1] and sentences [4]. Therefore, it is imperative to construct a graph-based contrastive learning task on heterogeneous graphs.

To address the above challenges, we propose a self-supervised deep heterogeneous graph neural networks with contrastive learning, named DHG-CL, which not only preserves the information of implicitly valuable neighbors but also further enhances the distinguishability of node representations. Specifically, we first design a projection function on heterogeneous graphs that maps different types of nodes into the same low-dimensional space. Next, to address the first challenge, we design a cross-layer semantic encoder, including a semantic-aware attention mechanism and a cross-layer message passing mechanism, which focuses on implicitly aggregating high-order neighbors in different semantic spaces and explores the deep semantics. Then, to address the second challenge, we introduce a graph-based contrastive learning task to distinguish irrelevant neighbors, further learning discriminative representations. More specifically, we construct a graph-based sampling strategy: (1) Inspired by the data augmentation [4], we pass the same node to the encoder twice with standard dropout to obtain a positive pair, aiming to generate a different contrastive object without changing the graph topology, so that we can get a strong supervision signal. (2) Because of the high correlation between nodes on the graph, we only treat nodes that are semantically irrelevant as negative samples. To delineate semantic irrelevance, we take the nodes with the number of meta-paths less than the threshold to the target node as negative samples. To summarize, the main contributions of this paper are as follows:

- We propose a self-supervised deep heterogeneous graph neural networks with contrastive learning. Unlike previous meta-path based methods, DHG-CL can fully mine the higher-order valuable neighbors on heterogeneous graphs to achieve a more informative node representation.
- We propose a graph-based contrastive learning task on heterogeneous graphs. Specifically, we introduce a dropout mask and semantic based negative sampling strategy to further learn discriminative representations.
- We conduct rich experiments on four real-world datasets and the results demonstrate that the proposed DHG-CL significantly outperforms the state-of-the-arts.

2 Related Work

In this section, we review recent developments in heterogeneous graph neural networks and contrastive learning.

2.1 Heterogeneous Graph Neural Networks

Graph neural networks have attracted extensive attention due to their superior performance in modeling graph-structured data. Many researchers [26,30] have made a detailed summary of them. Some works [11,24] propose to use convolutional neural networks and attention mechanisms for homogeneous graph. In recent years, some HGNNs [2,6–8,27] are proposed to learn node representations on heterogeneous graphs. For example, Wang et al. [27] propose a heterogeneous graph attention network named HAN consisting of node-level and semantic-level attention. Ji et al. [8] propose a deeper architecture and improve the node-level aggregating process to alleviate the semantic confusion. However, since the above HGNNs usually rely on a large amount of labeled data, they fail to fully perform well in sparse labeled scenarios. In this paper, we perform a self-supervised learning on heterogeneous graphs to learn the deep structural properties of graphs.

2.2 Contrastive Learning

The methods of contrastive learning have shown their success in self-supervised learning by distinguishing positive and negative samples [13,14]. In CV [1,5] and NLP [4,16], many excellent contrastive learning methods have emerged. For example, Chen et al. [1] propose SimCLR, which utilizes data augmentation techniques to generate two related views as a positive pair by randomly transforming a given image. Gao et al. [4] propose SimCSE to perform data augmentation through dropout operations. In graph learning, studies on contrastive learning [9,25,28] are proposed to learn node representations. Veličković et al. [25] propose DGI to maximize the mutual information between global representation and local patches. On heterogeneous graphs, Park et al. [17] propose DMGI to integrate the node embeddings by introducing the consensus regularization framework and the universal discriminator. Wang et al. [28] propose HeCo, which constructs a cross-view contrastive learning task after learning node embeddings from two views (network schema and meta-path views). However, both of them ignore the mining of deep implicit semantics, failing to explore neighbors other than the meta-path based neighbors. To go one step further, we implement a cross-layer semantic encoder to explore implicitly valuable neighbors, and then design graph-based contrastive learning to reduce the influence of irrelevant neighbors.

3 Preliminary

In this section, we will present several basic concepts of heterogeneous graphs.

Definition 1. Heterogeneous Graph. *A heterogeneous graph is defined as a graph* $\mathcal{G} = (\mathcal{V}, \mathcal{E}, \mathcal{A}, \mathcal{R})$, *where* $\mathcal{V}$ *and* $\mathcal{E}$ *denote the sets of nodes and edges. It is also associated with type mapping functions, including a node type mapping function* $\phi : \mathcal{V} \rightarrow \mathcal{A}$ *and an edge type mapping function* $\varphi : \mathcal{E} \rightarrow \mathcal{R}$. $\mathcal{A}$ *and* $\mathcal{R}$ *denote a set of node types and a set of edge types, where* $|\mathcal{A}| + |\mathcal{R}| > 2$.

Definition 2. Meta-path. *A meta-path* $\mathcal{P}$ *is defined as a path which consists of a set of nodes and edges. It is also in the form of* $A_1 \xrightarrow{R_1} A_2 \xrightarrow{R_2} \cdots \xrightarrow{R_l} A_{l+1}$ *(abbreviated as* $A_1 A_2 \cdots A_{l+1}$*), which can also be described as* $R_1 \circ R_2 \circ \cdots \circ R_l$, *where* $\circ$ *denote a combination operator on relations.*

Definition 3. Semantic context. *Given a meta-path* $\mathcal{P}$, *the semantic context* $\mathcal{N}_i^{\mathcal{P}}$ *of node* i *is defined as a set of neighbors connected to node* i *via meta-path* $\mathcal{P}$.

4   The Proposed DHG-CL Model

In this section, we will introduce the proposed DHG-CL in detail and Fig. 2 presents the overview of DHG-CL. We encode deep implicit semantics on heterogeneous graphs through a semantic-aware attention mechanism and a cross-layer message passing mechanism. To further enhance the distinguishability of embeddings, we introduce dropout mask and semantic based negative sampling strategy to construct a graph-based contrastive learning task. Finally, DHG-CL iteratively updates node embeddings via optimizing the contrastive loss.

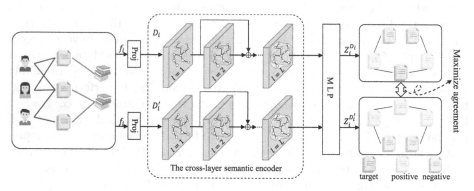

Fig. 2. The overview of the proposed DHG-CL.

4.1   Node Transformation

On heterogeneous graphs, different types of nodes lie in different feature spaces due to the heterogeneity of nodes. Therefore, we first apply a type-specific transformation to project different types of nodes into the same feature space. Specifically, we design a transformation matrix W_{ϕ_i} for node i with type ϕ_i:

$$h_i = \sigma \left(W_{\phi_i} \cdot f_i \right), \tag{1}$$

where h_i is the projected embedding of node i, σ is the activation function and f_i is the feature of node i. h_i can also be viewed as the node embedding h_i^0.

4.2 Cross-Layer Semantic Encoder

To capture the deep implicit semantics, we design a cross-layer semantic encoder, illustrated as Fig. 3. Now, we give a brief introduction to it, including neighbor aggregation and message combination:

$$h_i^l \leftarrow \text{Combine}^l \left(\text{Aggregate}^l \left(h_j^{l-1} \right), h_i^{l-1} \right), \tag{2}$$

where $\text{Aggregate}^l (\cdot)$ is the neighbor aggregation function in layer l, $\text{Combine}^l (\cdot)$ is the message combination function in layer l, h_j^{l-1} is the output embedding of node j in layer $l-1$, node $j \in \mathcal{N}_i$, denoted as $\left\{ \mathcal{N}_i^{\mathcal{P}_1}, \ldots, \mathcal{N}_i^{\mathcal{P}_m} \right\}$, $\mathcal{N}_i^{\mathcal{P}_n}$ is the set of neighbors of node i in meta-path $\mathcal{P}_n$.

Neighbor Aggregation. Since there are multiple semantic contexts, in order to obtain sufficient expressive ability, we design a semantic-aware attention mechanism. To be specific, we do not simply fuse them in the same space [27], but instead maintain its own semantic space for the different semantics. Therefore, we design a semantic projection to map semantic context into their own space:

$$h_{j,\mathcal{P}}^l = W_\mathcal{P}^l \cdot h_j^{l-1}, \tag{3}$$

where $W_\mathcal{P}^l$ is the semantic-specific transformation matrix, node $j \in \mathcal{N}_i^\mathcal{P}$.

After that, different neighbors will make different contributions to the target node, so we design an intra-semantic attention to preserve the important information of neighbors as much as possible:

$$e_{ij,\mathcal{P}}^l = \text{LeakyReLU} \left({a_\mathcal{P}^l}^\mathrm{T} \cdot [h_{i,\mathcal{P}}^l \| h_{j,\mathcal{P}}^l] \right), \tag{4}$$

where $e_{ij,\mathcal{P}}^l$ indicates the importance of node j to node i in meta-path $\mathcal{P}$, $a_\mathcal{P}^l$ is the semantic-specific attention vector for meta-path $\mathcal{P}$, $\|$ is the concatenate operation, $\text{LeakyReLU} (\cdot)$ is the activation function (with negative slope $a = 0.01$).

Then we perform mask attention which means we only compute $e_{ij,\mathcal{P}}^l$ for nodes $j \in \mathcal{N}_i^\mathcal{P}$ to inject the graph structure into the model. After obtaining the different importance of nodes, we normalize them to get the weight $\alpha_{ij,\mathcal{P}}^l$ by softmax function:

$$\alpha_{ij,\mathcal{P}}^l = \frac{\exp \left(\text{LeakyReLU} \left({a_\mathcal{P}^l}^\mathrm{T} \cdot [h_{i,\mathcal{P}}^l \| h_{j,\mathcal{P}}^l] \right) \right)}{\sum\limits_{k \in \mathcal{N}_i^\mathcal{P}} \exp \left(\text{LeakyReLU} \left({a_\mathcal{P}^l}^\mathrm{T} \cdot [h_{i,\mathcal{P}}^l \| h_{k,\mathcal{P}}^l] \right) \right)}. \tag{5}$$

Next, we aggregate the context information with the corresponding weights:

$$\tilde{h}_{i,\mathcal{P}}^l = \sigma \left(\sum_{j \in \mathcal{N}_i^\mathcal{P}} \alpha_{ij,\mathcal{P}}^l \cdot h_{j,\mathcal{P}}^l \right), \tag{6}$$

where $\tilde{h}^l_{i,\mathcal{P}}$ is the learned embedding of node i in semantic $\mathcal{P}$. For example, as shown in Fig. 3, we learn the embedding of node i with semantics "co-author" and "co-subject" in ACM. Then, we design a concatenation operation to fuse the semantics in different spaces, and then map the fused embedding of node i back to its original space:

$$\tilde{h}^l_i = W^l_o \cdot \text{Concatenate}\left(\tilde{h}^l_{i,\mathcal{P}_1}, \ldots, \tilde{h}^l_{i,\mathcal{P}_m}\right), \tag{7}$$

where W^l_o is the transformation matrix, $\tilde{h}^l_i$ is the aggregated embedding of node i in layer l.

Message Combination. For meta-path based HGNNs, they only need one layer to aggregate the information of neighbor j into node i. However, in this way the model can only learn a single manually designing semantic, ignoring the implicit higher-order neighbors. To capture implicit higher-order proximity information, we need a deeper model. Therefore, we stack our model to L layers, which enables nodes to reach a large proportion of nodes on the graph and mines the deeper semantics. In addition, to preserve the semantic information of each layer as much as possible, we design a cross-layer message passing mechanism:

$$h^l_i = h^{l-1}_i + \tilde{h}^l_i, \tag{8}$$

where h^l_i is the output embedding of node i in layer l.

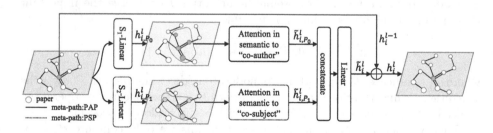

Fig. 3. The cross-layer semantic encoder.

4.3   Graph-Based Contrastive Learning

On heterogeneous graphs, there are various implicit relationships between nodes, which usually contain rich semantics. With the proposed encoder, we successfully fuse the rich information among nodes. However, as the order of neighbors increases, the target node will fuse the information of many irrelevant nodes. To enhance the distinguishability of representations, we need to push away nodes that have little semantic relationship with the target node.

The customized meta-path is usually based on domain knowledge, which reflects the known high-order semantics of nodes under the specific task. For example, the meta-path PAP contains a semantic: two papers published by the same author are more similar in topic, which is more helpful to distinguish the category of the paper. Therefore, the number of meta-paths connected between two nodes reveals their similarity. That is, the higher the number is, the more similar they are. Given nodes i and j, if node j has C_n^{ij} instances to reach node i through meta-path $\mathcal{P}_n$, we define a function to measure the number of meta-path connections between nodes i and j:

$$C_{ij} = \sum_{n=1}^{m} C_n^{ij}. \tag{9}$$

Motivated by SimCSE [4], the key factor for us to obtain positive pairs is to independently sample x_i and x_i^+ by using the dropout mask. Specifically, we pass the same node to the pre-trained encoder twice. By exploiting the standard dropout twice, we can obtain two different representations as a positive pair without changing the graph topology. Then, we take nodes with C_{ij} less than T_{neg} as negative samples. The model learns the contrastive information by predicting the positive sample among the negative samples.

Before calculating the contrastive loss, we feed embeddings we get from the encoder into a multilayer perceptron to project them into the contrastive learning space:

$$Z_i = W_2 \cdot \sigma \left(W_1 \cdot h_i^L + b_1 \right) + b_2, \tag{10}$$

where W_1 and W_2 are the transformation matrices, h_i^L is the output embedding of node i from encoder. For the contrastive learning task, we simply feed the same input to the encoder twice. In this way, we can obtain two embeddings with different dropout masks and we define them as $Z_i^{D_i}$ and $Z_i^{D_i'}$, where D_i and D_i' are the different dropout masks. Finally, we calculate the contrastive loss as follows:

$$\mathcal{L}_i = -\log \frac{e^{\text{sim}(Z_i^{D_i}, Z_i^{D_i'})/\tau}}{\sum_{j \in S_i \setminus \{i\}} e^{\text{sim}(Z_i^{D_i}, Z_j^{D_j})/\tau} + \sum_{j \in S_i} e^{\text{sim}(Z_i^{D_i}, Z_j^{D_j'})/\tau}}, \tag{11}$$

where S_i is the set of positive and negative samples of node i, $S_i = \{i, j | j \in V \text{ and } C_{ij} < T_{neg}\}$, T_{neg} is the threshold for sampling negative samples, τ is a temperature parameter, $\text{sim}(Z_1, Z_2)$ is the cosine similarity $\frac{Z_1^T \cdot Z_2}{||Z_1|| \cdot ||Z_2||}$.

5 Experiments

5.1 Experimental Setup

Datasets. We conduct experiments on four real-world networks. The detailed descriptions are summarized in Table 1.

- **ACM** [32]. ACM is extracted from KDD, SIGMOD, SIGCOMM, Mobi-COMM, and VLDB. The target nodes are papers which are divided into three classes: Database, Wireless Communication, and Data Mining.

Table 1. The statistics of four datasets.

Dataset	Nodes	Edges	Meta-path
ACM	paper (P): 4,019	P-A: 13,407	PAP
	author (A): 7,167	P-S: 4,019	PSP
	subject (S): 60		
DBLP	author (A): 4,057	P-A: 19,645	APA
	paper (P): 14,328	P-C: 14,328	APCPA
	conference (C): 20	P-T: 85,810	APTPA
	term (T): 7,723		
Freebase	movie (M): 3,492	M-A: 65,341	MAM
	actor (A): 33,401	M-D: 3,762	MDM
	direct (D): 2,502	M-W: 6,414	MWM
	writer (W): 4,459		
IMDB	movie (M): 3,676	M-A: 11,028	MAM
	actor (A): 4,353	M-D: 3,676	MDM
	direct (D): 1,678		

- **DBLP** [3]. DBLP is extracted from the computer science bibliography website. The target nodes are authors which are divided into four classes: Database, Data Mining, Artificial Intelligence, and Information Retrieval.
- **Freebase** [12]. Freebase is a dataset about movies extracted from Freebase. The target nodes are movies which are divided into three classes: Action, Comedy and Drama.
- **IMDB** [32]. IMDB is a subset of dataset IMDB. The target nodes are movies which are divide into three classes: Action, Comedy, and Drama.

Baselines. We compare the proposed DHG-CL with three categories of baselines, including: two unsupervised homogeneous methods (i.e., DeepWalk, DGI), four unsupervised heterogeneous methods (i.e., Mp2vec, HERec, DMGI, HeCo), and a semi-supervised heterogeneous method (i.e., HAN), to verify the effectiveness of DHG-CL.

- **DeepWalk** [18]. It performs random walk on homogeneous graphs and learns the representations of nodes through the skip-gram model.
- **DGI** [25]. It maximizes the mutual information between the graph-level summary and the local patches.
- **Mp2vec** [2]. It performs meta-path based random walk on heterogeneous graphs and learns the representations of nodes through the skip-gram model.
- **HERec** [21]. It utilizes a meta-path based random walk strategy to filter node sequences and applies DeepWalk to embed the heterogeneous graphs.
- **DMGI** [17]. It minimizes the disagreements among node embeddings and employs a universal discriminator to discriminate the graph-level summary and local patches.
- **HeCo** [28]. It adopts network schema and meta-path structure as two views to perform contrastive learning on heterogeneous graphs.
- **HAN** [27]. It adopts hierarchical attention mechanism to model both node-level and semantic-level importance.

Implementation Details. For all baselines, we use the settings in their original paper and modify a few parameters. Specifically, for random walk based methods like DeepWalk, Mp2vec and HERec, we set the walk length to 20, the window size to 5, the number of walks to 10 and the number of negative samples to 5. For Deepwalk, we ignore the heterogeneity of nodes and test its performance on the whole heterogeneous graph. For DGI, Mp2vec and HERec, we test all the meta-paths for them and report the best performance.

For the proposed DHG-CL, we adopt Adaptive Moment Estimation (Adam) [10] optimizer, set the dimension of node embeddings to 64, tune the learning rate from 0.0005 to 0.001, and tune the patience for early stopping from 20 to 50. For the cross-layer semantic encoder, we set the dimension of the semantic space to 64, and the number of layers to 2. For the contrast task, we tune dropout including feature dropout from 0.1 to 0.9, attention dropout from 0.1 to 0.9, temperature coefficient τ from 0.5 to 0.9, and negative threshold from 3 to 5. To reduce randomness, we perform all methods 10 times and report the average results.

Table 2. Comparison results ($\%\pm\sigma$) for node classification.

Datasets	Metrics	Split	HAN	DeepWalk	DGI	Mp2vec	HERec	DMGI	HeCo	DHG-CL
ACM	Ma-F1	20	87.61±0.5	66.04±2.1	82.24±3.4	68.05±1.0	68.67±0.7	87.99±0.5	88.33±0.2	**89.64±0.3**
		40	87.68±0.6	69.70±1.2	84.01±2.6	66.30±0.4	66.64±0.6	88.23±0.4	87.28±0.4	**90.71±0.2**
		60	86.28±0.8	63.64±2.0	84.79±2.5	67.77±0.8	68.25±1.5	89.48±0.3	88.97±0.4	**90.88±0.1**
	Mi-F1	20	87.53±0.5	62.46±3.9	81.79±3.9	67.98±1.9	68.82±1.2	87.30±0.6	88.38±0.3	**89.28±0.4**
		40	87.89±0.6	71.87±2.8	83.66±3.1	67.81±0.3	68.87±2.1	87.78±0.4	87.02±0.4	**90.51±0.3**
		60	86.67±0.5	67.69±4.7	84.36±2.9	70.53±1.0	71.15±1.7	89.00±0.4	88.44±0.5	**90.55±0.3**
DBLP	Ma-F1	20	89.28±2.4	82.14±0.7	88.95±0.3	89.56±0.8	87.74±1.3	89.46±0.1	91.06±0.3	**92.24±0.7**
		40	89.86±1.9	85.01±0.4	87.98±0.4	90.71±0.6	89.24±0.9	89.44±0.1	89.90±0.5	**91.07±0.8**
		60	89.55±1.5	85.69±0.3	90.87±0.4	91.52±0.8	89.90±0.7	88.68±1.0	90.90±0.4	**91.96±0.5**
	Mi-F1	20	89.84±2.4	83.00±0.5	90.04±0.3	90.12±0.8	88.51±1.2	90.28±0.1	91.76±0.3	**92.74±0.7**
		40	90.24±1.9	85.26±0.4	88.74±0.3	91.12±0.6	89.81±0.9	90.11±0.1	90.22±0.6	**91.46±0.9**
		60	90.72±1.1	86.55±0.4	91.85±0.3	92.28±0.7	90.86±0.6	89.20±0.8	91.77±0.3	**92.71±0.6**
Freebase	Ma-F1	20	57.75±0.9	50.15±2.5	55.76±0.9	52.26±1.9	55.48±1.5	54.56±0.5	59.33±0.9	**61.74±0.8**
		40	55.51±1.2	50.87±2.1	54.33±1.4	53.59±1.2	57.31±1.8	51.93±0.8	61.15±0.5	**61.28±1.1**
		60	56.13±0.9	51.50±1.3	53.48±0.7	51.42±1.8	53.69±1.5	50.21±1.0	**60.93±0.5**	60.62±1.4
	Mi-F1	20	61.22±1.7	59.67±1.7	63.85±2.2	54.36±2.1	57.89±1.7	61.60±2.1	61.95±1.1	**66.56±0.5**
		40	57.60±1.5	61.94±1.7	61.76±3.6	55.86±1.5	59.84±1.9	62.86±3.2	63.88±0.8	**65.91±0.9**
		60	58.94±1.2	61.39±1.4	64.47±2.1	53.78±1.9	56.34±1.6	58.85±2.9	63.94±0.8	**66.40±1.2**
IMDB	Ma-F1	20	39.38±1.8	35.94±0.5	36.20±0.6	36.47±1.5	34.38±1.4	36.86±0.9	39.98±0.9	**45.13±1.0**
		40	40.30±1.9	36.01±1.0	39.56±0.4	36.88±1.3	36.89±1.5	37.93±0.7	40.76±0.8	**41.49±0.7**
		60	**44.79±1.3**	33.63±1.0	40.82±0.5	37.05±1.9	35.25±1.1	37.22±1.2	41.79±0.8	41.90±0.9
	Mi-F1	20	39.98±1.4	35.38±1.1	36.39±0.6	36.73±1.4	35.93±1.7	38.07±0.7	40.14±1.0	**44.74±1.0**
		40	40.50±1.8	36.71±1.2	39.49±0.5	36.97±1.3	37.49±1.3	37.92±0.7	40.96±0.8	**41.75±1.1**
		60	**45.27±1.3**	35.32±0.5	40.96±0.5	37.08±1.9	37.15±1.1	37.87±1.2	41.87±0.8	43.80±1.3

5.2   Node Classification

After obtaining the learned embeddings of nodes by pre-training model, we feed them into a linear classifier to predict their labels. Following [28], we choose 20, 40, 60 labeled nodes for each class as the training set, 1000 nodes as the validation set, and 1000 nodes as the test set. Then, we use common evaluation metrics: Macro-F1, Micro-F1 and report the test performance when it performs best in the validation set. The comparison results are shown in Table 2.

From Table 2, we can see that DHG-CL achieves the best performance in most cases. Comparing DeepWalk and Mp2vec, DGI and DMGI, we can see that unsupervised methods for heterogeneous graphs generally outperform unsupervised methods for homogeneous graphs, which demonstrates the benefits of graph heterogeneity. We can also see that DHG-CL outperforms HeCo and DMGI, indicating that encoding deep semantic information and introducing data augmentation technology have brought certain performance improvements. Moreover, DHG-CL outperform the semi-supervised method HAN, demonstrating the potential capabilities of self-supervised HGNNs.

5.3 Node Clustering

To more comprehensively evaluate the embeddings of nodes learned by DHG-CL, we also conduct a node clustering task. In this task, we utilize the K-means algorithm to perform clustering of nodes and set the number of clusters as the number of classes. Then, we adopt normalized mutual information (NMI) and adjusted rand index (ARI) to assess the quality of the clustering results. Since initializing the centroids will affect the performance of K-means, we repeat the process for 10 times and report the average, as shown in Table 3. Similarly, the proposed DHG-CL significantly outperforms other methods in most cases, which further proves the effectiveness of DHG-CL. Since HAN is trained with label guidance, we do not compare with it.

Table 3. Comparison results for node clustering.

Methods	ACM		DBLP		Freebase		IMDB	
	NMI	ARI	NMI	ARI	NMI	ARI	NMI	ARI
DeepWalk	36.59	28.77	66.17	71.25	13.47	14.92	0.35	0.08
DGI	50.13	45.27	69.88	75.63	16.26	17.25	0.09	0.08
Mp2vec	37.75	30.21	73.74	78.81	16.50	17.24	0.47	0.44
HERec	35.97	28.90	72.19	77.75	16.30	17.23	0.44	1.04
DMGI	50.05	40.38	67.05	71.72	16.10	16.29	0.73	**1.17**
HeCo	57.92	56.45	71.11	76.64	15.23	16.88	0.81	0.84
DHG-CL	**60.57**	**59.07**	**77.37**	**81.57**	**16.74**	**17.93**	**0.88**	0.96

5.4 Visualization

For a more intuitive evaluation, we conduct the task of visualization on ACM dataset. We use t-SNE [15] algorithm to visualize embeddings and choose different colors for different labels.

From Fig. 4, we can observe that Deepwalk and DGI designed for homogeneous graphs do not perform well and have blurry boundaries, which may

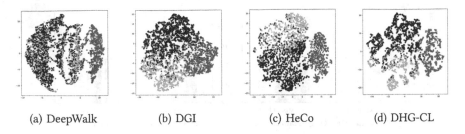

| (a) DeepWalk | (b) DGI | (c) HeCo | (d) DHG-CL |

Fig. 4. Visualization paper embedding on ACM. Each point indicates one paper and its color indicates the topic.

lead to confusion among nodes. In contrast, the heterogeneous graph methods perform much better than the above homogeneous graph methods. It demonstrates that graph heterogeneity contains rich information for node classification, node clustering tasks, etc. For heterogeneous graph method, DHG-CL has higher intra-class similarity and a clearer boundary than HcCo to distinguish nodes.

5.5 Variant Analysis

In order to verify the effectiveness of each part of DHG-CL, we design two variants of DHG-CL as follows:

- DHG-CL$_{shallow}$. To verify the effectiveness of cross-layer semantic encoder, we remove the cross-layer message passing mechanism in the encoder and set the number of layers to 1. Therefore, the DHG-CL$_{shallow}$ can only preserve shallow graph information and cannot capture higher-order semantics.
- DHG-CL$_{noise}$. To verify the effectiveness of contrasting learning strategy in DHG-CL, we introduce other common data augmentation techniques. In specific, we maintain our negative sampling strategy, and then we sample positive sample from the new graph constructed by removing a portion of the edges [31].

We use the same parameters for them and conduct comparison between them and DHG-CL on ACM and DBLP. The comparison results of 20 labeled nodes per class are shown in Fig. 5. Obviously, DHG-CL significantly outperforms their variants. DHG-CL performs better than DHG-CL$_{shallow}$, indicating that it is necessary to preserve both shallow and implicit higher-order semantics and some helpful information is also in higher-order semantics. DHG-CL performs better than DHG-CL$_{noise}$, which proves that dropout mask as a novel method is more robust than data augmentation technique which modifies (e.g. add, remove nodes/edges) the inherent structure of the graph.

5.6 Parameter Analysis

In this section, we investigate the sensitivity of parameters and report the Macro-F1 values of node classification on ACM dataset with different parameters.

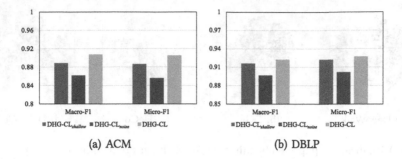

Fig. 5. The comparison of DHG-CL and its variants.

Number of Layers L. L describes the depth of the heterogeneous graph, which determines the high-order proximity of the nodes. We vary the value of it and report the results in Fig. 6(a). We can see that with the increase of the number of layers, the performance of DHG-CL firstly goes up and then decreases drastically when L is higher than 2. This is because as the number of layers increases, the unhelpful information will bring more negative gains.

Threshold T_{neg}. T_{neg} describes the minimum number of meta-paths between nodes, which affects the number of negative samples. We explore the performance of DHG-CL with various thresholds. From Fig. 6(b), we can see that with the growth of the threshold, the performance increases first and then starts to decrease when we set $T_{neg} = 5$. The reason is that the larger the threshold is, the more negative samples are selected, which will include some semantically similar samples as negative samples.

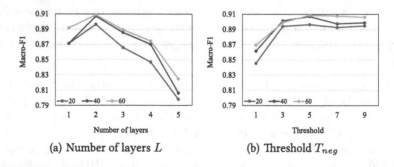

Fig. 6. Parameter sensitivity of DHG-CL w.r.t. number of layers L and threshold T_{neg}.

6   Conclusion and Future Work

In this work, we propose a novel self-supervised deep heterogeneous graph neural networks with contrastive learning, named DHG-CL. DHG-CL employs a cross-layer semantic encoder and the graph-based contrastive learning to preserve the information of implicitly valuable neighbors and further enhance the distinguishability of representations. In cross-layer semantic encoder, we design a semantic-aware attention mechanism and a cross-layer message passing mechanism to capture deep implicit semantics. In graph-based contrastive learning, we innovatively introduce dropout mask and semantic based negative sampling strategy to further obtain discriminative representation. Experimental results on four real-world heterogeneous graphs demonstrate the effectiveness of DHG-CL.

Acknowledgements. This work is supported by Key Research and Development Program Projects of Xinjiang (No. 2022B03010-2), Strategic Priority Research Program of the Chinese Academy of Sciences (No. XDC02030400).

References

1. Chen, T., Kornblith, S., Norouzi, M., Hinton, G.: A simple framework for contrastive learning of visual representations. In: International Conference on Machine Learning, pp. 1597–1607. PMLR (2020)
2. Dong, Y., Chawla, N.V., Swami, A.: Metapath2vec: scalable representation learning for heterogeneous networks. In: Proceedings of the 23rd ACM SIGKDD International Conference on Knowledge Discovery and Data Mining, pp. 135–144 (2017)
3. Fu, X., Zhang, J., Meng, Z., King, I.: MAGNN: metapath aggregated graph neural network for heterogeneous graph embedding. In: Proceedings of The Web Conference 2020, pp. 2331–2341 (2020)
4. Gao, T., Yao, X., Chen, D.: SimCSE: simple contrastive learning of sentence embeddings. arXiv preprint arXiv:2104.08821 (2021)
5. He, K., Fan, H., Wu, Y., Xie, S., Girshick, R.: Momentum contrast for unsupervised visual representation learning. In: Proceedings of the IEEE/CVF Conference on Computer Vision and Pattern Recognition, pp. 9729–9738 (2020)
6. Hong, H., Guo, H., Lin, Y., Yang, X., Li, Z., Ye, J.: An attention-based graph neural network for heterogeneous structural learning. In: Proceedings of the AAAI Conference on Artificial Intelligence, vol. 34, pp. 4132–4139 (2020)
7. Hu, Z., Dong, Y., Wang, K., Sun, Y.: Heterogeneous graph transformer. In: Proceedings of The Web Conference 2020, pp. 2704–2710 (2020)
8. Ji, H., Wang, X., Shi, C., Wang, B., Yu, P.: Heterogeneous graph propagation network. IEEE Trans. Knowl. Data Eng. **35**, 521–532 (2021)
9. Jiang, X., Lu, Y., Fang, Y., Shi, C.: Contrastive pre-training of GNNs on heterogeneous graphs. In: Proceedings of the 30th ACM International Conference on Information & Knowledge Management, pp. 803–812 (2021)
10. Kingma, D.P., Ba, J.: Adam: a method for stochastic optimization. arXiv preprint arXiv:1412.6980 (2014)
11. Kipf, T.N., Welling, M.: Semi-supervised classification with graph convolutional networks. arXiv preprint arXiv:1609.02907 (2016)
12. Li, X., Ding, D., Kao, B., Sun, Y., Mamoulis, N.: Leveraging meta-path contexts for classification in heterogeneous information networks. In: 2021 IEEE 37th International Conference on Data Engineering (ICDE), pp. 912–923. IEEE (2021)
13. Liu, X., Zhang, F., Hou, Z., Mian, L., Wang, Z., Zhang, J., Tang, J.: Self-supervised learning: generative or contrastive. IEEE Trans. Knowl. Data Eng. **35**, 857–876 (2021)
14. Liu, Y., Pan, S., Jin, M., Zhou, C., Xia, F., Yu, P.S.: Graph self-supervised learning: a survey. arXiv preprint arXiv:2103.00111 (2021)

15. Van der Maaten, L., Hinton, G.: Visualizing data using t-SNE. J. Mach. Learn. Res. **9**(11) (2008)
16. Meng, Y., et al.: COCO-LM: correcting and contrasting text sequences for language model pretraining. Adv. Neural Inf. Process. Syst. **34** (2021)
17. Park, C., Kim, D., Han, J., Yu, H.: Unsupervised attributed multiplex network embedding. In: Proceedings of the AAAI Conference on Artificial Intelligence, vol. 34, pp. 5371–5378 (2020)
18. Perozzi, B., Al-Rfou, R., Skiena, S.: DeepWalk: online learning of social representations. In: Proceedings of the 20th ACM SIGKDD International Conference on Knowledge Discovery and Data Mining, pp. 701–710 (2014)
19. Radford, A., Wu, J., Child, R., Luan, D., Amodei, D., Sutskever, I., et al.: Language models are unsupervised multitask learners. OpenAI Blog **1**(8), 9 (2019)
20. Razavi, A., van den Oord, A., Vinyals, O.: Generating diverse high-fidelity images with VQ-VAE-2. In: Wallach, H., Larochelle, H., Beygelzimer, A., d'Alché-Buc, F., Fox, E., Garnett, R. (eds.) Advances in Neural Information Processing Systems, vol. 32. Curran Associates, Inc. (2019). https://proceedings.neurips.cc/paper/2019/file/5f8e2fa1718d1bbcadf1cd9c7a54fb8c-Paper.pdf
21. Shi, C., Hu, B., Zhao, W.X., Philip, S.Y.: Heterogeneous information network embedding for recommendation. IEEE Trans. Knowl. Data Eng. **31**(2), 357–370 (2018)
22. Sun, Y., Han, J., Yan, X., Yu, P.S., Wu, T.: PathSim: meta path-based top-k similarity search in heterogeneous information networks. Proc. VLDB Endow. **4**(11), 992–1003 (2011)
23. Tian, Y., Krishnan, D., Isola, P.: Contrastive multiview coding. In: Vedaldi, A., Bischof, H., Brox, T., Frahm, J.-M. (eds.) ECCV 2020. LNCS, vol. 12356, pp. 776–794. Springer, Cham (2020). https://doi.org/10.1007/978-3-030-58621-8_45
24. Veličković, P., Cucurull, G., Casanova, A., Romero, A., Lio, P., Bengio, Y.: Graph attention networks. arXiv preprint arXiv:1710.10903 (2017)
25. Velickovic, P., Fedus, W., Hamilton, W.L., Liò, P., Bengio, Y., Hjelm, R.D.: Deep graph infomax. In: ICLR (Poster), vol. 2, no. 3, p. 4 (2019)
26. Wang, X., Bo, D., Shi, C., Fan, S., Ye, Y., Yu, P.S.: A survey on heterogeneous graph embedding: methods, techniques, applications and sources. arXiv preprint arXiv:2011.14867 (2020)
27. Wang, X., Ji, H., Shi, C., Wang, B., Ye, Y., Cui, P., Yu, P.S.: Heterogeneous graph attention network. In: The World Wide Web Conference, pp. 2022–2032 (2019)
28. Wang, X., Liu, N., Han, H., Shi, C.: Self-supervised heterogeneous graph neural network with co-contrastive learning. In: Proceedings of the 27th ACM SIGKDD Conference on Knowledge Discovery & Data Mining, pp. 1726–1736 (2021)
29. Wu, Z., Wang, S., Gu, J., Khabsa, M., Sun, F., Ma, H.: CLEAR: contrastive learning for sentence representation. arXiv preprint arXiv:2012.15466 (2020)
30. Wu, Z., Pan, S., Chen, F., Long, G., Zhang, C., Philip, S.Y.: A comprehensive survey on graph neural networks. IEEE Trans. Neural Netw. Learn. Syst. **32**(1), 4–24 (2020)
31. You, Y., Chen, T., Sui, Y., Chen, T., Wang, Z., Shen, Y.: Graph contrastive learning with augmentations. Adv. Neural. Inf. Process. Syst. **33**, 5812–5823 (2020)
32. Zhao, J., Wang, X., Shi, C., Liu, Z., Ye, Y.: Network schema preserving heterogeneous information network embedding. In: International Joint Conference on Artificial Intelligence (IJCAI) (2020)
33. Zitnik, M., Sosič, R., Leskovec, J.: Prioritizing network communities. Nat. Commun. **9**(1), 1–9 (2018)

First-Principles Calculation to N-type Beryllium Related Co-doping and Beryllium Doping in Diamond

Delun Zhou[1], Jinyu Zhang[1], Ruifeng Yue[1,2(✉)], and Yan Wang[1,2,3(✉)]

[1] School of Integrated Circuit, Tsinghua University, Beijing, China
yuerf@mail.tsinghua.edu.cn
[2] Beijing National Research Center for Information Science and Technology, Beijing, China
[3] Beijing Innovation Center for Future Chips (ICFC), Beijing, China

Abstract. The Beryllium-doped (Be-doped) diamond and Beryllium related (Be-X) co-doped diamond have been carefully investigated by the density functional theory (DFT) to explore the possibility to achieve effective and shallow n-type doping in diamond. Although the ionization energy and formation energy of interstitial/substitutional Be-doped diamond is not ideal, the introduction of Be-related co-doping techniques (Be-N/O/S) greatly improves the electrical properties in diamonds. We found, for the first time, n-type diamond doping can be realized in Be-N, Be-O and Be-S co-doped systems, among which Be-N_3 has the best performance. Be-N3 has the advantages of low ionization energy (0.25 eV), low formation energy (-1.59 eV), and direct bandgap. The N-2p states play a crucial role in the conduction band edge of Be-N3 co-doped diamond. Hence, the Be-N_3 could be expected to become a promising alternative for N-type shallow doping in diamond.

Keywords: Beryllium-related Co-doping · First-principles Calculation · Diamond

1 Introduction

Diamond has been extensively studied due to its outstanding properties since the end of the last century. Diamond is not only a superhard material for mechanical cutting, but also a semiconductor material with full potential, and it is even expected to become the ultimate semiconductor material. Diamond has large bandgap (5.5 eV) [1], unparalleled thermal conductivity (22 W/cm·K) [2], and its electric breakdown field (10 MV/cm) [3] and carrier mobility (4500 cm^2/Vs for electrons [4]; 3800 cm^2/Vs for holes [3]) show obvious advantages compared with other semiconductor materials. These unique properties make it a promising option to be used in novel electronic device, especially in the field of high temperature, high pressure and high power.

Although diamond material has tremendous advantages, a series of fundamental problems remain to be resolved. In addition to the immense difficulty of growing large-scale single crystal diamonds, achieving effective diamond doping is one of the most

© The Author(s), under exclusive license to Springer Nature Switzerland AG 2023
J. Mikyška et al. (Eds.): ICCS 2023, LNCS 14073, pp. 283–294, 2023.
https://doi.org/10.1007/978-3-031-35995-8_20

significant obstacles. In order to construct basic diamond electronic devices, it is necessary to solve the n-type and p-type doping of diamond. At present, the p-type doping of diamond has been basically solved through boron doping [5], however, the problem of n-type doping has not been properly resolved. Diamond n-type doping requires the introduction of appropriate dopants to achieve effective shallow doping. Currently, research on n-type diamond doping can be mainly divided into single element doping and co-doping. The research of single doping includes interstitial doping of the first group elements (Li, Na) [6, 7] and the substitutional doping of the fifth and sixth group elements (N, P, O, S) [8–12]; co-doping includes Li-N [13], B-S [14], Li-P [15], B-N [16, 17] in diamond. Although the formation energy of B-S co-doping is lower than that of sulfur doping, it is still relatively high. Li-N co-doping has outstanding performance but due to the existence of barrier energy, the progress in experimental preparation is slow. Though researchers all over the world have made continuous attempts and efforts, there is no recognized technical route to achieve ideal diamond N-type doping yet. Therefore, exploring the method of realizing n-type doping in diamond is the focus of current research.

Other second-period elements in the periodic table (Li, B, N, O) [5, 6, 8, 11] have been extensively studied in diamond doping. Beryllium element, which is also in the second period, also has research potential due to its close atomic radius to C atom. Beryllium-doped diamond has been successfully prepared by MPCVD method in experiments [18], proving the feasibility of incorporating beryllium into diamond. However, there is no theoretical/experimental research on beryllium and other elements co-doped diamond research, investigations on Be-related co-doping on the realization of N-type diamond doping is still rare, worthy of in-depth research.

In this article, we calculated the doping of beryllium in diamond and managed to achieve excellent performance of diamond n-type doping by co-doping Beryllium and other elements (Nitrogen, Oxygen, and Sulfur). We systematically calculated the formation energy, ionization energy, band structure, density of states (TDOS and PDOS) of these structures to carefully analyze the electrical properties of these structures, and finally found that some beryllium co-doped structures may have outstanding performance.

2   Calculation Methods

This publication follows the same computational methodology as our earlier work [13, 14, 16]. All calculations are done using density functional theory (DFT) to optimize the geometry, calculate the density of states (DOS), and determine the band structure. VASP through Perdew-Burke-Ernzerhof (PBE) are implemented to perform exchange-correlation function within the generalized gradient approximation (GGA). [19] Projected augmented plane wave (PAW) potentials are employed to describe core-valence interactions. In this study, the cut-off energy of the plane wave is set to 500 eV. The convergence criterion of the electronic structure relaxation calculation is set to 1×10^{-5} eV, and its criterion of the interatomic force is adjusted to 1×10^{-4} eV after the convergence verification [20]. For calculations, the 216-atom supercell of diamond ($3 \times 3 \times 3$) and the Monkhorst-Pack grid of KPOINTS ($9 \times 9 \times 9$) are employed. Larger cell sizes are deemed essential to guarantee the reliability of output results, particularly for shallow doping studies in diamond. The atomic positions where the lattice constant of bulk diamond converges to 3.573 Å. We found that result was consistent with previous studies [22].

3 Results and Discussion

3.1 Impurity Formation Energy (Ef)

The impurity formation energy is calculated to judge the possibility and stability of impurity doping into the material. The lower the value, the easier and more stable the impurity doping is, which is conducive to effective doping.

The formation energy of impurity X in charge state q in the doped diamond $E_f [X^q]$ is defined as [23]:

$$E_f[X^q] = E_{tot}[X^q] - E_{tot}[C, bulk] - \sum nX \mu X - q(E_F + \Delta V) \tag{1}$$

where $E_{tot}[X^q]$ is the total energy of the whole structure including the charged dopant X and $E_{tot}[C, bulk]$ is the total energy of the perfect diamond bulk. n_X is denoted as the number of atom of element X (host atoms or impurity atoms) that has been added to ($n_X > 0$) or removed ($n_X < 0$) from the supercell to form the defect. μ_X is the chemical potential of impurity X, and the summation is conducted when various impurities coexist in the supercell at the same time. In this paper, we investigate the chemical potential of Beryllium from the hexagonal close-packed structure of pure Be bulk, while the chemical potential of Oxygen is based on O_2, Sulfur from H_2S, and Nitrogen from NH_3. In addition, E_F is the Fermi level of the doped diamond, and we assume the ΔV as a correction term to unify the electrostatic potentials between the diamond bulk and the supercell with impurities. In the following analysis, we would focus on the neutral charge state of the impurity and therefore, the last term is zero.

The impurity solubility is proportional to $N \exp(-E_f/kT)$, where E_f represents the impurity formation energy, k denotes the Boltzmann's constant, T denotes the temperature, and N denotes the density of states in the supercell that the dopant atom occupies. Therefore, the solubility is strongly correlated to the formation energy. Smaller formation energy indicates better doping effectivity.

We first calculated the formation energies of Be interstitial doping (Be$_i$) and Be substitutional doping (Be$_s$). The formation energies of Be$_i$ and Be$_s$ are 10.07 eV and 4.81 eV. Similar to the doping of the first group elements (Li, Na), the formation energy characteristics of Be doping alone are not ideal. Due to its high formation energy, it can be concluded that Be's solid solubility is low, and it is difficult to achieve effective doping by directly incorporating Be into the diamond.

Co-doping in diamond may cause smaller lattice distortion, thereby reducing the impurity formation energy. Therefore, we tried to co-dope Be with other element atoms into the diamond. We calculated the formation energy of BeX$_n$ (X denotes N, S, O, n = 1–4) in a supercell, where Be occupies the position of the original carbon atom in a diamond structure in the form of substitution, and the X element atoms replace those carbon atoms that are nearest neighbors of the central Be atom.

As shown in Table 2, some Be-X_n (X denotes N, O, S, n = 1–4) structures have a smaller formation energy than Be single doping, that is, the co-doping method is effective for reducing the formation energy of Be impurities in diamonds. The formation energy of Be-S co-doped diamond is generally higher. One possible explanation is that the atomic radius of sulfur atoms is larger (much larger than that of nitrogen and oxygen atoms), and the lattice distortion caused by the co-doped structure is correspondingly larger. The formation energy of Be-N_3 and Be-N_4 is comparatively low, which are −1.59 eV and − 1.42 eV accordingly. In general, the formation energy of Be-O co-doping is lower than Be-S co-doping but higher than Be-N co-doping, and it is better than most single-doping structures in terms of defect formation energy.

The length of the C-C bond in the pure diamond structure is 1.54 Å, which is smaller than those of the C-Be bond in substitutional Be doped (1.68 Å). The reason for the longer C-Be bond length may be due to the larger radius of Be atoms, which squeezes the surrounding C atoms to the outside after incorporation. Selecting the structure with the lowest formation energy under each co-doping system, we found that the distance of the Be-S bond is 1.79 Å, the Be-O bond is 1.61 Å, and the Be-N bond is 1.64 Å. The bond length and the internal lattice distortion are related to the defect formation energy. Generally speaking, the larger the lattice distortion is, the longer the bond length is, and the larger the impurity formation energy is. The combination of bond length and impurity formation energy of Be-Xn (X denotes N, S, O, n = 1–4) is consistent with the above theory, and also verifies that Be-N co-doping is the result of better performance in Be-X_n co-doping structures in diamond (Figs. 1, 2 and Table 1).

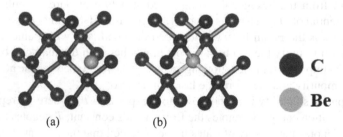

(a)   (b)

Fig. 1. The doping structures of Be_i (a), Be_s (b).

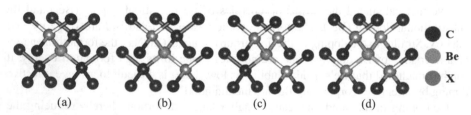

(a)   (b)   (c)   (d)

Fig. 2. The doping structures of Be-X_1 (a), Be-X_2 (b), Be-X_3 (c), and Be-X_4 (d).

Table 1. Impurity formation energy (E_f) of interstitial Be (Bei), substitutional Be (Bes), and Be-related dopants in diamond

Compound	Position	E_f /eV
$C_{216}Be$	Be_i	10.08
$C_{215}Be$	Be_s	4.81
$C_{214}Be_1N_1$	$Be-N_1$	1.57
$C_{213}Be_1N_2$	$Be-N_2$	−1.38
$C_{212}Be_1N_3$	$Be-N_3$	−1.59
$C_{211}Be_1N_4$	$Be-N_4$	−1.42
$C_{214}Be_1O_1$	$Be-O_1$	3.30
$C_{213}Be_1O_2$	$Be-O_2$	4.72
$C_{212}Be_1O_3$	$Be-O_3$	5.79
$C_{211}Be_1O_4$	$Be-O_4$	8.45
$C_{214}Be_1S_1$	$Be-S_1$	7.23
$C_{213}Be_1S_2$	$Be-S_2$	11.41
$C_{212}Be_1S_3$	$Be-S_3$	14.74
$C_{211}Be_1S_4$	$Be-S_4$	18.67

3.2 Ionization Energies

Ionization energy is a key parameter for judging the performance of doping. The lower the ionization energy, the easier it is to be activated. In this paper, the thermodynamic transition level $\varepsilon(q1,q2)$ is adopted to estimate the ionization energy with the PBE methods. The equation of thermodynamic transition level is provided as follows [24]

$$\varepsilon(q_1, q_2) = \frac{E_{tot}[X^{q_1}] - E_{tot}[X^{q_2}]}{q_2 - q_1} - E_V - \Delta V \tag{2}$$

where the connotation of $E_{tot}[X^{q_1}]$ and $E_{tot}[X^{q_2}]$ are consistent with the previous part of formation energy. E_V is the valence band maximum of the pure diamond. The transition level $\varepsilon(0/-)$ is equal to the ionization energy (acceptor level) E_A, and the ionization energy of the donor is $E_D = E_g - \varepsilon(0/+)$ [25], where E_g is the DFT calculation bandgap of pure diamond of 4.1 eV.

We have verified the correctness of our calculation results by calculating the structure of some existing experimental/theoretical results and comparing them with our own results (Table 2).

The ionization energies of Be-doped (substitutional and interstitial) and Be-co-doped diamond calculated according to the formula are shown in Table 3.

Substitutional Be-doped diamond is a p-type semiconductor, while interstitial Be-doped diamond is n-type. The ionization energy of interstitial Be-doped semiconductors is xx, which is better than N single-doping and S single-doping, and not as good as Na and Li single-doping. Part of the Be-related co-doped structure can achieve n-type doping. BeN_3 and BeN_4 in the Be-N co-doped system, BeO_2, BeO_3, BeO_4 in the Be-O co-doped system, and BeS_2, BeS_3, BeS_4 in the Be-S co-doped system may have donor energy levels. The ionization energies of BeN_3, BeO_2, and BeS_2 structures are 0.25 eV, 0.86 eV, and 0.56 eV, respectively, the best in each co-doped system. Combined with the previous calculation results of impurity formation energy, we believed that BeN_3, BeS_2, and BeO_2 are the structures with outstanding performance among Be-related co-doped structures, worthy of more in-depth exploration.

Table 2. Ionization energies of some n-type dopants in diamond.

Dopant	E_D/eV (Our work)	E_D/eV (Ref.)
P	0.59	$0.43^b - 0.56^a$ [9, 26]
S	1.44	1.4^b [10]
O	0.45	0.32^b[11]
N	1.43	$1.4^{a,b}$[6, 27]
Li (interstitial)	0.04	0.1^a[7]
BS	0.55	$0.39^a - 0.52^b$[28, 29]
Li-N_4	0.232	0.271^a[30]

[a] Theoretical values [b] Experimental values

3.3 Electronic Structure

We performed further calculation and analysis on Be doping and Be-related co-doping in diamond. As shown in the Fig. 3, the substitutional Be-doped diamond demonstrated p-type characteristics with an acceptor level. The Fermi level of interstitial Be-doped diamond is close to the bottom of the conduction band, indicating n-type diamond characteristics. These calculation results are consistent with the ionization energy calculation results above.

Table 3. Ionization energies (eV) of Be-related co-doped defects in diamond

Dopant	E_D/eV
Be_i	1.76 eV(E_D)
Be_s	1.16 eV(E_A)
Be-N_1	0.68 eV(E_A)
Be-N_2	0.43 eV(E_A)
Be-N_3	0.25 eV(E_D)
Be-N_4	0.33 eV(E_D)
Be-O_1	0.77 eV(E_A)
Be-O_2	0.86 eV(E_D)
Be-O_3	0.91 eV(E_D)
Be-O_4	0.88 eV(E_D)
Be-S_1	0.62 eV(E_A)
Be-S_2	0.56 eV(E_D)
Be-S_3	0.71 eV(E_D)
Be-S_4	1.03 eV(E_D)

ED refers to the donor level of n-type semiconductor
EA refers to the acceptor level of p-type semiconductor

Combined with the previous results of formation energy and ionization energy, we further calculated the structures with better performance in various co-doped systems. We investigated the total density of states (TDOS) of Be-N_3, Be-S_2, and Be-O_2 co-doped diamond, and the partial density of states (PDOS) of Be-N_3 so as to identify their donor properties. The calculation results (TDOS and PDOS) are displayed in Fig. 3 and Fig. 4. According to our calculation results, the Fermi level of Be-N_3, Be-S_2, and Be-O_2 doped diamond is close to the conduction band minimum (CBM). These results are in agreement with those of previous band calculations. The mutually corroborated calculation results prove that these structures are capable of N-type diamond doping.

Among Be co-doped diamonds, the performance of the Be-N co-doped system is more advantageous, and after further comparative analysis, BeN$_3$ has the most outstanding performance, since its properties of low formation energy and low ionization energy at the same time. The atomic radii of Be and N atoms are the closest to C atoms, and the lattice distortion caused by doping is the smallest, which may be a possible explanation of the calculation results.

Calculation results shown in Fig. 4, demonstrate that N-2p states play a crucial role in the conduction band edge of Be-N$_3$ doped diamond, which explains the reason why Be-N$_3$ could achieve n-type diamond doping. Co-doped atoms not only lower the formation energy of impurity (Be) but are also a decisive factor for the realization of n-type doping. According to the calculation results of the TDOS/PDOS, it is evident that these Be-related co-doping can achieve n-type diamond doping, which are consistent with the previous discussion of ionization energy and band structure.

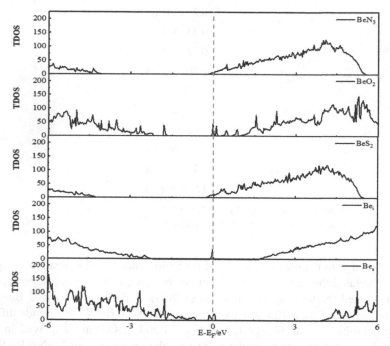

Fig. 3. TDOS of the Be-N$_3$, Be-O$_2$, Be-S$_2$, Be$_i$, and Be$_s$ doped diamond.

3.4 Band Structure

We have calculated the band structure of a series of Be co-doped diamond, mainly for those with better performance in each co-doped system. While credible n-type doping is achieved through these doping structures, conduction band minimums (CBM) vary dramatically. As is shown in Fig. 5, the CBM structure of Be-N$_3$ and Be-S$_2$ doped diamond is clear, which is consistent with the CBM of typical n-type semiconductors. However, the CBM structure of Be-O$_2$ doped diamond is relatively complex, in which the intermediate energy level and CBM almost overlap, which is also consistent with the results of DOS calculations.

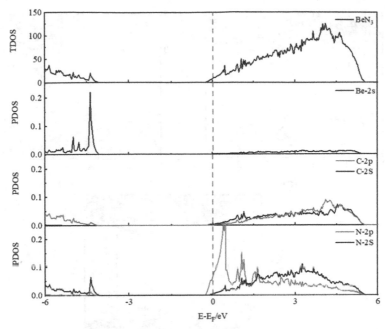

Fig. 4. TDOS and PDOS of the Be-N$_3$-doped diamond. (a) TDOS of the Be-N$_3$-doped diamond; PDOS of the Be-N$_3$-doped diamond for (b) Be atoms, (c) C atoms, (d) N atom.

It is worth noting that Be-X doping has a direct band gap, which is very different from pure diamond and also different from traditional single-doped diamond. The conduction band minimums (CBM), and valence band maximums (VBM) in this structure are all on the same KPOINTS, which is the characteristic of direct bandgap semiconductor.

There is doubt that the band structure can be more precisely calculated by using non-local hybrid functionals of HSE06 [5]. However, given the limited calculation resources, standard PBE functional was adopted in our research. Although it underestimates the bandgap of diamond, its negative effects on the energy difference of charge state are not evident, and our analysis of formation energy is not affected as well.

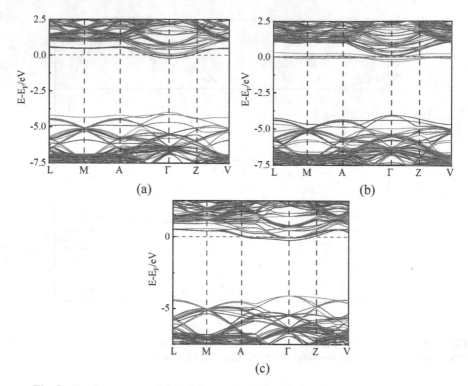

Fig. 5. Band structures of doped diamond with Be-N$_3$ (a), Be-O$_2$ (b), and Be-S$_2$ (c)

4 Conclusions

In conclusion, Be-related co-doping is expected to achieve N-type doping in diamond. Co-doped N, O, and S atoms not only reduce the formation energy, but are also a key factor in achieving n-type doping. Be-N, Be-O, and Be-S co-doped systems are expected to achieve diamond n-type doping, and the performance of Be-N co-doping is the best overall. We systematically calculated the formation energy, ionization energy, band structure, and DOS of various doped structures. These calculation results all prove that BeN3 has outstanding performance among Be-related co-doping structures and is expected to achieve high concentration and shallow doping of n-type diamond. Our calculation results need to be further verified by experiments.

References

1. Shen, S., Shen, W., Liu, S., et al.: First-principles calculations of co-doping impurities in diamond. Mater. Today Commun. **23**, 100847 (2019)
2. Li, Y., Liao, X., Guo, X., et al.: Improving thermal conductivity of epoxy-based composites by diamond-graphene binary fillers. Diam. Relat. Mater. **126**, 126 (2022)
3. Zhang, Z., Lin, C., Yang, X., et al.: Solar-blind imaging based on 2-inch polycrystalline diamond photodetector linear array. Carbon **173**(42), 427–432 (2021)

4. Liu, X., et al.: Boron–oxygen complex yields n-type surface layer in semiconducting diamond. Proc. Natl. Acad. Sci. **116**(16), 7703–7711 (2019)
5. Czelej, K., Śpiewak, P., Kurzydłowski, K.J.: Electronic structure and n-type doping in diamond from first principles. MRS Adv. **1**(16), 1093–1098 (2016). https://doi.org/10.1557/adv.201 6.87
6. Kajihara, S.A., et al.: Nitrogen and potential n-type dopants in diamond. Phys. Rev. Lett. **66**(15), 2010–2013 (1991)
7. Goss, J.P., Briddon, P.R.: Theoretical study of Li and Na as n-type dopants for diamond. Phys. Rev. B **75**(7), 2978–2984 (2007)
8. Shah, Z.M., Mainwood, A.: A theoretical study of the effect of nitrogen, boron and phosphorus impurities on the growth and morphology of diamond surfaces. Diam. Relat. Mater. **17**(7–10), 1307–1310 (2008)
9. Kato, H.: Diamond bipolar junction transistor device with phosphorus-doped diamond base layer. Diam. Relat. Mater. **27**, 19–22 (2012)
10. Sque, S.J., Jones, R., Goss, J.P., et al.: Shallow donors in diamond: chalcogens, pnictogens, and their hydrogen complexes. Phys. Rev. Lett. **92**(1), 017402 (2004)
11. Prins, J.F.: n-type semiconducting diamond by means of oxygen-ion implantation. Phys. Rev. B **61**(11), 7191–7194 (2000)
12. Kato, H., Makino, T., Yamasaki, S., et al.: n-type diamond growth by phosphorus doping on (001)-oriented surface. MRS Proc. **1039**(40), 6189 (2007)
13. Delun, Z., Tang, L., Geng, Y., et al.: First-principles calculation to N-type LiN Co-doping and Li doping in diamond. Diam. Relat. Mater. **110**, 108070 (2020)
14. Tang, L., Yue, R., Wang, Y.: N-type BS co-doping and S doping in diamond from first principles. Carbon **130**, 458–465 (2018)
15. Shao, Q.Y., Wang, G.W., Zhang, J., et al.: First principles calculation of lithium-phosphorus co-doped diamond. Condens. Matter Phys. **16**(1), 1 (2013)
16. Zhou, D., Tang, L., Zhang, J., Yue, R., Wang, Y.: n-type B-N co-doping and N doping in diamond from first principles. In: Groen, D., de Mulatier, C., Paszynski, M., Krzhizhanovskaya, V.V., Dongarra, J.J., Sloot, P.M.A. (eds.) Computational Science (ICCS 2022). LNCS, pp. 530–540. Springer, Cham (2022). https://doi.org/10.1007/978-3-031-08751-6_38
17. Sun, S., Jia, X., Zhang, Z., et al.: HPHT synthesis of boron and nitrogen co-doped strip-shaped diamond using powder catalyst with additive h-BN. J. Cryst. Growth **377**(15), 22–27 (2013)
18. Ueda, K., Kasu, M.: Beryllium-doped single-crystal diamond grown by microwave plasma CVD. Diam. Relat. Mater. **18**(2–3), 121–123 (2009)
19. Sancho-García, J.C., Brédas, J.L., Cornil, J.: Assessment of the reliability of the Perdew–Burke–Ernzerhof functionals in the determination of torsional potentials in π-conjugated molecules. Chem. Phys. Lett. **377**(1–2), 63–68 (2003)
20. Jones, R., Goss, J.P., Briddon, P.R.: Acceptor level of nitrogen in diamond and the 270-nm absorption band. Phys. Rev. B: Condens. Matter **80**(3), 1132–1136 (2009)
21. Zongbao, L., Yong, L., Ying, W., et al.: Synergistic effect in B and N co-doped Ib-type diamond single crystal: a density function theory calculation. Can. J. Phys. **94**(9), 929–932 (2016)
22. Rilby, D.P.: Lattice constant of diamond and the C-C single bond. Nature **153**(3889), 587–588 (1944)
23. Freysoldt, C., Grabowski, B., Hickel, T., et al.: First-principles calculations for point defects in solids. Rev. Mod. Phys. **86**(1), 253 (2014)
24. Kajihara, S.A., Antonelli, A., Bernholc, J., Car, R.: Nitrogen and potential n-type dopants in diamond. Phys. Rev. Lett. **66**(15), 2010–2013 (1991)
25. Goss, J.P., Briddon, P.R., Eyre, R.J.: Donor levels for selected n-type dopants in diamond: a computational study of the effect of supercell size. Phys. Rev. B: Condens. Matter **74**(24), 4070–4079 (2006)

26. Miyazaki, T., Okushi, H.: A theoretical study of a sulfur impurity in diamond. Diam. Relat. Mater. **10**(3–7), 449–452 (2001)
27. Schwingenschlögl, U., Chroneos, A., Schuster, C., et al.: Doping and cluster formation in diamond. J. Appl. Phy. **110**(V110N5), 162 (2011)
28. Jing, Z., Li, R., Wang, X., et al.: Study on the microstructure and electrical properties of boron and sulfur codoped diamond films deposited using chemical vapor deposition. J. Nanomater. **2014**(21), 4338–4346 (2014)
29. Eaton, S.C., Anderson, A.B., Angus, J.C., et al.: Diamond growth in the presence of boron and sulfur. Diam. Relat. Mater. **12**(10–11), 1627–1632 (2003)
30. Moussa, J.E., Marom, N., Sai, N., Chelikowsky, J.R.: Theoretical design of a shallow donor in diamond by lithium-nitrogen codoping. Phys. Rev. Lett. **108**(22), 226404 (2012)

Machine Learning Detects Anomalies in OPS-SAT Telemetry

Bogdan Ruszczak[1,3]([✉]) [iD], Krzysztof Kotowski[3] [iD], Jacek Andrzejewski[3],
Alicja Musiał[3], David Evans[4], Vladimir Zelenevskiy[4,5], Sam Bammens[4],
Rodrigo Laurinovics[4,6], and Jakub Nalepa[2,3]([✉]) [iD]

[1] Faculty of Electrical Engineering, Automatic Control and Informatics, Department
of Informatics, Opole University of Technology, Prószkowska 76, 45-758 Opole, Poland
b.ruszczak@po.edu.pl
[2] Faculty of Automatic Control, Electronics and Computer Science,
Department of Algorithmics and Software, Silesian University of Technology,
Akademicka 16, 44-100 Gliwice, Poland
jnalepa@ieee.org
[3] KP Labs, Konarskiego 18C, 44-100 Gliwice, Poland
{bruszczak,kkotowski,jandrzejewski,amusial,jnalepa}@kplabs.pl
[4] European Space Agency/ESOC, Robert-Bosch-Str. 5, 64293 Darmstadt, Germany
{David.Evans,Sam.Bammens}@esa.int,
{Vladimir.Zelenevskiy,Rodrigo.Laurinovics}@ext.esa.int
[5] Telespazio Germany GmbH, Europaplatz 5, 64293 Darmstadt, Germany
[6] IrbGS ltd, "Viraki", Ance Parish, Irbene LV-3601, Latvia

Abstract. Detecting anomalies in satellite telemetry data is pivotal in
ensuring its safe operations. Although there exist various data-driven
techniques for the task of determining abnormal parts of the signal, they
are virtually never validated over real telemetries. Analyzing such data
is challenging due to its intrinsic characteristics, as telemetry may be
noisy and affected by incorrect acquisition, resulting in missing parts of
the signal. In this paper, we tackle this issue and propose a machine
learning approach for detecting anomalies in single-channel satellite
telemetry. To validate its capabilities in a practical scenario, we build
a dataset capturing the nominal and anomalous telemetry data captured on board OPS-SAT—a nanosatellite launched and operated by the
European Space Agency. Our extensive experimental study showed that
the proposed algorithm offers high-quality anomaly detection in real-life
satellite telemetry, reaching 98.4% accuracy over the unseen test set.

Keywords: machine learning · anomaly detection · feature
engineering · satellite telemetry

This work was partially supported by the: "On-board Anomaly detection from the OPS-
SAT telemetry using deep learning" project funded by the European Space Agency
under contract No. 4000137339/22/NL/GLC/ov. JN was supported by the Silesian
University of Technology Rector's grant (02/080/RGJ22/0026).

J. Mikyška et al. (Eds.): ICCS 2023, LNCS 14073, pp. 295–306, 2023.
https://doi.org/10.1007/978-3-031-35995-8_21

1   Introduction

Many critical systems are being monitored using various sensors for detecting their malfunction [11, 20]. This is extremely important for satellites to ensure their safe operations and to appropriately respond to any abnormal events that are reflected in the telemetry data. In general, there are three types of anomalous events which are commonly observed in satellite telemetry. In point anomalies, telemetry signal values fall outside the nominal operational range, and hence can be easily detected using out-of-limit checks. On the other hand, collective anomalies correspond to the overall sequences of consecutive telemetry values that are abnormal, whereas in contextual anomalies, the single telemetry values are anomalous within their local neighborhood [17]. There exist data-driven algorithms for detecting such types of abnormal events [18], spanning across classic techniques exploiting expert systems [25], unsupervised approaches [8, 14] and deep learning models [10] (often benefiting from recurrent neural networks which are especially well-fitted to process time-series data [1, 9]). Such algorithms, however, are virtually never validated over real-life telemetry data. Also, they often require long time-series data to build a model reflecting the nominal operation of the satellite—capturing it on board is tedious and time-consuming, thus data-level digital twins have been blooming to simulate the correct telemetry [2]. Benchmark datasets commonly exploited to validate detection algorithms contain time-series data, where each time series is split into its training and test parts, presenting similar characteristics. Such data is not affected by the practical challenges commonly observed in on-board telemetry, such as data noisiness or missing data, due to e.g., inappropriate signal acquisition. Therefore, the estimated anomaly detection capabilities of data-driven techniques may easily become over-optimistic, and the experimental scenarios are often flawed by methodological issues in the field [23].

In this paper, we tackle the issue of thorough validation of machine learning techniques for anomaly detection in satellite telemetry data. We not only propose an end-to-end pipeline for detecting such abnormal events in signal data. We also investigate its capabilities over a dataset capturing real-life telemetry data captured on board OPS-SAT—a nanosatellite launched and operated by the European Space Agency (ESA). OPS-SAT is a *flying laboratory*, providing a platform for on-board experiments (e.g., those which may be too risky to be executed on other operational satellites [7]), with one of them being anomaly detection from telemetry data using the pipeline discussed in this work. Other experiments performed on-board OPS-SAT include, among others:

- on-board unsupervised machine learning for spacecraft autonomy [13, 22],
- the compression of housekeeping telemetry [5],
- assessing the stability of the attitude during inertial pointing mode using on-board images of the sky [21],
- deploying and maintaining the MO/MAL ground infrastructure with this aircraft [15],
- testing the on-board thermal vacuum [12],
- and addressing fail-safety and redundancy design challenges [24].

Here, we build upon the available OPS-SAT telemetry data to build a representative dataset that is used to train and validate our anomaly detection techniques in a practical setting, reflecting real on-board telemetry challenges, e.g., missing data and aperiodicity of the signal [6]. Our experimental validation performed over a carefully bundled OPS-SAT telemetry dataset revealed that our approaches exploiting hand-crafted feature extractors (which are independent of the length of the signal and can effectively operate with missing data) and classic supervised learners offer high-quality anomaly detection, reaching 98.4% classification accuracy over the unseen test set.

This paper is structured as follows. In Sect. 2, we discuss the main objective tackled in this paper and present the OPS-SAT telemetry channels of interest, together with our end-to-end machine learning pipeline for detecting abnormal events from time-series data. Our experimental study is reported and discussed in Sect. 3. Section 4 concludes the paper and highlights the most exciting future research pathways which may emerge from our work.

2   Materials and Methods

In this section, we discuss the investigated OPS-SAT telemetry (Sect. 2.1), together with our approach for detecting anomalies in such data (Sect. 2.2).

2.1   Dataset

In this work, we investigate the telemetry signals indicated by the ESA OPS-SAT team as the most interesting from the operational point of view, and we downloaded such data from the WebMUST client telemetry directory available for OPS-SAT [4]. Those telemetry channels include the magnetometer readouts, alongside the photo diode (PD) values—the following signals are collected:

- **Magnetometer telemetry channels:** I_B_FB_MM_0 (CADC0872) (see its fragment rendered in Fig. 1), I_B_FB_MM_1 (CADC0873), I_B_FB_MM_2 (CADC0874),
- **PD channels:** I_PD1_THETA (CADC0884), I_PD2_THETA (CADC0886), I_PD3_THETA (CADC0888), I_PD4_THETA(CADC0890), I_PD5_THETA (CADC0892), I_PD6_THETA (CADC0894).

It is of note that there are several practical challenges can be identified while exploring telemetry data captured by experimental satellites. They encompass—but are not limited to—the following issues:

- **High fragmentation of telemetry signals.** The data was registered mainly during the important stages of the OPS-SAT mission.
- **Missing data**, reflected as a high number of "gaps" in the telemetry readouts.
- Presence of the **recurring parts with noisy fragments** or with an unusual number of peaks (not necessarily abnormal) in the signal.

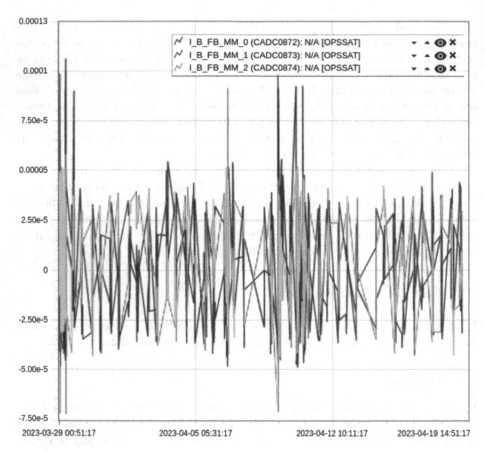

Fig. 1. A preview of three different OPS-SAT telemetries for a long (several weeks) timeframe. We can observe that the signal can easily become aperiodic and noisy, hence exploiting the approaches which focus on predicting the nominal signal and confronting it with the actual readouts [9] can easily become infeasible.

- **Non-uniform acquisition frequency rates** (the signal may be acquired hourly, daily, or even monthly), and there might be some radical changes in the signal characteristics along the analyzed months (due to e.g., various changes in satellite operations or hardware characteristics).
- **Lack of periodic and consistent nominal data** which would be captured over a long period of time.

The above-discussed telemetry channels were split into sub-parts, roughly corresponding to their periods, and they were labeled as nominal or anomalous by a human rater, and then the assigned labels were verified by the OPS-SAT

Operations Team using the OXI labeling system[1]. Overall, we collected **1273 nominal and 328 anomalous telemetry fragments for training**, and **416 nominal and 117 anomalous fragments for testing**, with the latter fragments, never seen during training (this training-test dataset split is stratified and it was generated randomly over all available telemetry fragments, to include approximately 25% of all telemetry fragments in the test set).

2.2 Detecting OPS-SAT Anomalies Using Machine Learning

In our machine learning anomaly detection pipeline, we **(1) extract an array of discriminative features from the telemetry fragments**[2], and such feature vectors are later **(2) fed to a supervised learner which performs prediction** (here, we tackle a binary classification task of determining abnormal vs. nominal telemetry fragments). For each telemetry segment, we extract the following 18 features (referred to as $(a$–$p)$), with some of them already proven to be discriminative for handling the classification tasks in telemetry data [16]:

- The features reflecting the **basic statistics** within the telemetry fragment (those include its *(a)* duration, *(b)* length, *(c)* mean value, *(d)* variance, and *(e)* standard deviation of a segment).
- The **number of peaks** within the fragment. More specifically, we sum the number of peaks with prominence of at least 10% of the segment value, and check this statistic for several segment representations, i.e., for *(f)* raw segment, alongside for its two variants after applying a smoothing transformation, using the "lighter" *(g)*, and "stronger" smoothing*(h)*, and for its *(i)* first and *(j)* second derivative.
- The **segment's variance** using its *(k)* first and *(l)* second derivative.
- The squared sum of *(m)* the **number of missing readouts**.
- The segment's *(n)* **weighted length** (relative to its sampling that also differs across different telemetry segments in the dataset), and *(o)* the **variance relative to segment's length** and *(p)* **to the segment's duration**.

Our extractors are independent of the length of the segment, and they can be effectively exploited to extract features from the segments with missing readouts—this may not be possible while operating over the raw telemetry data.

To tackle the issue of the relatively small amount and limited representativeness of OPS-SAT training samples, we introduce several data augmentation techniques specifically targeting the telemetry segments. Although they can be effectively used to synthesize both nominal and anomalous segments based on the actual samples, we augment the nominal examples only, in order to ensure that the abnormal signal segments were validated by the OPS-SAT Operations

---

[1] The OXI labeling system has been developed by KP Labs and it is available at https://oxi.kplabs.pl. OXI allows for not only investigating time-series data, together with the ground-truth information but also for generating ground truth.

[2] It is of note, however, that our approach can be applied to any time-series data, not only satellite telemetry.

Team (and they are *not* the result of synthetic data augmentation). We introduce the following data transformations to augment such telemetry samples:

- **Mirroring a segment over the OX axis** (ω_1)—a segment is flipped along the horizontal axis.
- **Mirroring a segment over the OY axis** (ω_2)—in a single telemetry channel, two consecutive nominal segments are flipped along their median value, starting from the half length of the first segment to the half of the second one.
- **Shifting a segment** (ω_3)—in a single telemetry channel, two consecutive nominal segments are shifted for a certain number of steps (steps of 15% and 25% of the segments' length were applied). The resulting segment starts at the shifted (by a given number of steps) starting position, and it finishes at the shifted ending position of the first segment. Hence, the last point of the first (shifted) segment will overlap the second segment—to avoid information leaks across the training and test sets, both neighboring segments that undergo this augmentation step must be originally included in the training set.

Finally, once the features are extracted and the training set has been potentially augmented, such nominal and anomalous examples are fed to the classifier to perform training. We may easily exploit any supervised learner in the proposed approach—this flexibility of our processing chain is shown in Sect. 3.

3   Experimental Validation

The main objective of our experimental study is to investigate the classification performance of the proposed classification engine for detecting anomalies in real-life OPS-SAT telemetry data. On top of that, we are aimed at verifying the flexibility of the processing chain, alongside the impact of improving the classification part of the pipeline working on a stable set of extracted features. We focus on classic classification models, including widely-adopted random forests, multi-layer perceptrons, adaptive boosting algorithms, k-nearest neighbors and support vector machines (SVMs) with a linear kernel function. To quantify the performance of the models, we calculate their accuracy, precision, recall and F_1 score, as well as the Matthews correlation coefficient (r_ϕ) over the unseen test set, with r_ϕ being the measure commonly used in imbalanced classification tasks [3]. All metrics should be maximized, with one indicating their perfect score. We split the experimental study into two experiments: in Experiment 1 (Sect. 3.1), we train the classification models over the original data only, whereas in Experiment 2 (Sect. 3.2) we benefit from data augmentation of nominal training samples.

3.1   Experiment 1: Exploiting Original Training Dataset

In the preliminary experiment, we optimized the most important hyperparameters of the classification models using a five-fold cross-validation procedure over the training set [19]. Thus, we ultimately had the maximum depth of random forests of 20 and a maximum allowed number of features of 50% and 50 estimators; the size of the hidden layer in the multi-layer descriptor was 400; for adaptive boosting, we used 500 estimators; for SVMs, the regularization (C) parameter was set to 8; and $k = 5$ for k-nearest neighbors. The results obtained for such optimized models are gathered in Table 1 (they are sorted according to r_ϕ). Although precision exceeds 0.92 for the k-nearest neighbor classifier, we can observe rather low r_ϕ values across all investigated algorithms, indicating that the number of false positives (i.e., abnormal telemetry segments incorrectly classified as nominal) is large. This kind of error is, however, unacceptable in practice, as it could lead to missing anomalous events, e.g., in the on-board fault detection, isolation, and recovery system. On the other hand, a too large number of false negatives can deteriorate the usability of the system, if such incorrectly classified nominal segments are to be reviewed by the operations team before taking action in response to potential on-board anomalies. In practice, the latter issue is commonly a more severe problem, especially if the number of false alarms gets (very) large. In this work, we hypothesize that deploying data augmentation routines to synthesize nominal training examples may help deal with those issues by generating a larger and more representative training set that will allow us to elaborate well-generalizing models.

Table 1. The classification results obtained over the test set using machine learning models with default parameterization (sorted by r_ϕ). The best results are **boldfaced**.

Model	Accuracy	Precision	Recall	F_1 score	r_ϕ
Random forest	**0.9294**	**0.8826**	0.8050	**0.8233**	**0.7830**
Multilayer perceptron	0.8958	0.7814	**0.8077**	0.7688	0.7215
Adaptive boosting	0.9101	0.8138	0.7774	0.7646	0.7151
k-nearest neighbors	0.9051	0.9204	0.7495	0.7643	0.7133
SVM with linear kernel	0.8957	0.8249	0.7132	0.7369	0.6765

3.2 Experiment 2: Augmenting Training Datasets

Table 2. The summary of the segment data set for OPS-SAT telemetry anomalies.

Class	Training set	ω_1	ω_2	ω_3	Validation set
Nominal	1273	2677	703	1406	416
Anomalous	328	—	—	—	117

The augmentation methods can substantially increase the size of the training set (Table 2). Several examples of augmented training samples are visualized in Fig. 2, where each column represents one telemetry segment (marked with a solid line) and its augmented version (rendered as a dashed line). In this experiment, we initially trained a random forest classifier (as it achieved the largest F_1 score in the previous experiment across all supervised models) using all versions of the augmented training sets. To better understand the impact of specific data augmentation routines on the overall performance of the classifier, we employed them separately (and then collectively), hence obtaining four augmented training sets. In Table 3, we gather the performance gain for each augmented dataset separately, and for all of them collectively (see the last row in this table). We can appreciate that the biggest improvement in the random forest's accuracy is obtained while exploiting ω_1 (mirroring training segments over the OX axis), as well as while utilizing all augmentation techniques ($\omega_1 \wedge \omega_2 \wedge \omega_3$). Here, the results are reported for the very same test set to ensure fair comparison.

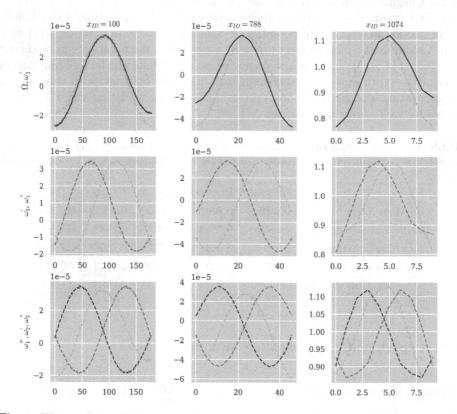

Fig. 2. The examples of the nominal telemetry segments (Ω, green solid lines), with their augmented versions with different parameterizations ($\omega_1, \omega_2, \omega_3$, blue and violet dashed lines). Each column provides one selected telemetry segment and its augmented versions. Some augmentations have been applied more than once, thus we have synthesized segments denoted as: $\omega_1', \omega_1'', \omega_1'''$. The presented segments are extracted from the following OPS-SAT telemetry channels: $x_{ID} = 100$ from $CADC0872$, $x_{ID} = 788$ from $CADC0874$, and $x_{ID} = 1074$ from $CADC0888$. (Color figure online)

Table 3. The benefit of exploiting data augmentation, quantified as the gain in a metric when compared to the classifier trained over the original training set. The positive and negative values indicate an increase or decrease in the model's performance.

Augmentation	Accuracy	Precision	Recall	F_1 score	r_ϕ
ω_1	0.0491	0.1055	−0.0764	0.0109	0.0535
ω_2	0.0064	−0.0363	−0.0615	−0.0592	−0.0535
ω_3	0.0347	0.0870	−0.0521	−0.0054	0.0243
$\omega_1 \wedge \omega_2 \wedge \omega_3$	0.0548	0.1174	−0.0978	−0.0079	0.0393

Knowing that the latter approach results in the largest accuracy and precision gains (hence, we minimize the number of false positive alarms), we selected this augmented version of the training set for further experiments with all other supervised models. Here, we optimized their hyperparameters following the five-fold cross-validation procedure over the augmented training set, resulting in 400 neurons in the hidden layer of the multi-layer perceptron, 500 estimators in adaptive boosting, $C = 8$ for an SVM, and $k = 5$ for k-nearest neighbors. In Table 4, we summarize the classification measures obtained using all optimized classifiers, with the random forest model (max. depth of 25) exceeding the accuracy of 0.98, and the corresponding r_ϕ amounting to 0.826, significantly outperforming the random forest model trained over the original training set (Table 1).

To provide a comparison for those metrics, a deep learning algorithm [9] obtained precision: 0.855 and recall: 0.855 (it was evaluated for the SMAP Spacecraft data), and precision: 0.926 and recall: 0.694 (for the Curiosity telemetry dataset). Although we are aware that comparing the algorithms over different sets may easily become biased, we can indeed observe that our classification models were able to exceed the reported metrics for very challenging OPS-SAT telemetry data which indicates their significant generalization capabilities and robustness against noisy and difficult time-series data.

Table 4. The results (obtained over the test set) elaborated using all investigated classification models with optimized hyperparameters and trained over the augmented training set ($\omega_1 \wedge \omega_2 \wedge \omega_3$), sorted by r_ϕ. The best results are **boldfaced**.

Model	Accuracy	Precision	Recall	F_1	r_ϕ
Random forest	**0.9843**	**1.0000**	0.7317	**0.8239**	**0.8261**
Multi-layer perceptron	0.9792	0.8716	**0.7437**	0.7795	0.7815
Adaptive boosting	0.9812	0.9422	0.6858	0.7817	0.7893
SVM with linear kernel	0.9729	0.9097	0.6154	0.6683	0.6802
k-nearest neighbors	0.9491	0.8662	0.6637	0.7186	0.7008

Finally, for the best model (random forest), we present the confusion matrix in Fig. 3, which—besides presenting all measures (the number of true positives, true negatives, false positives, and false negatives: N_{TP}, N_{TN}, N_{FP}, and N_{FN}, respectively) brings the examples of correctly and incorrectly classified test segments. The classifier correctly identified 82% abnormal segments (96/117), and 99% (413/416) nominal segments, showing its potential practical utility for onboard anomaly detection. Additionally, we can observe a significant heterogeneity of the test segments, highlighting the difficulty of the classification task.

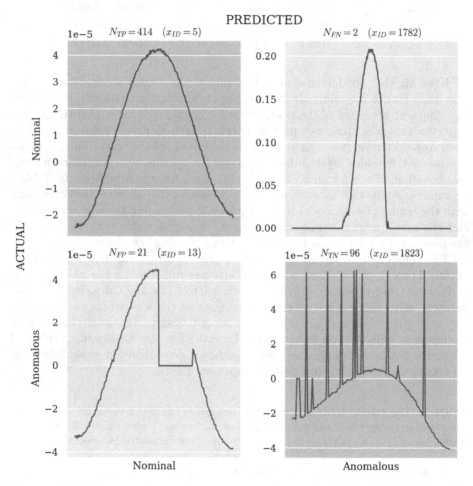

Fig. 3. The confusion matrix for the best classifier (random forest trained over the augmented training set with optimized hyperparameters), and the examples of correctly and incorrectly classified test segments (with the green and orange background, respectively). These example segments were extracted from the following OPS-SAT telemetries: $x_{ID} = 1782$ from $CADC0894$, $x_{ID} \in \{5, 13, 1823\}$ from $CADC0872$. (Color figure online)

4 Conclusions and Future Work

Detecting anomalies in spacecraft telemetry data is of paramount practical importance to appropriately respond to various unexpected events that may happen on-board an operational satellite. Albeit there exist data-driven approaches toward this task, they are virtually never validated over telemetry channels captured by a real spacecraft. In this paper, we tackled this problem and proposed an end-to-end machine learning pipeline for detecting such abnormalities. Our approach benefits from hand-crafted feature extractors which are independent of the length of the telemetry segments which are later classified by a supervised learner as nominal or anomalous. To understand its generalization capabilities, we validated it over a curated (and validated by the ESA Operations Team) set of nominal and abnormal telemetry channels acquired on-board OPS-SAT—a nanosatellite operated by ESA. The experiments indicated that our technique offers high-quality detection of anomalies from OPS-SAT telemetry. Also, we showed that exploiting the suggested data augmentation routines allows for significantly improving the generalization capabilities of the classification models.

Our current research efforts are focused on deploying the proposed anomaly detection system on-board OPS-SAT. It is of note that the model has been already trained on the ground, and we are in the process of uplinking it to the satellite. Also, we are working on utilizing our method for other missions, to further prove its generalizability. Finally, it would be interesting to exploit unsupervised clustering over large-scale telemetry datasets while utilizing our feature extractors. This may help accelerate the process of generating ground truth for supervised models (as pre-classified telemetry segments would have to be verified by humans), and possibly uncover intrinsic characteristics of such time-series data that might have not been observed by the Operations Teams.

References

1. Benecki, P., Piechaczek, S., Kostrzewa, D., Nalepa, J.: Detecting anomalies in spacecraft telemetry using evolutionary thresholding and LSTMs. In: Proceedings of the GECCO, GECCO 2021, ACM, New York, NY, USA, pp. 143–144 (2021)
2. Castellani, A., Schmitt, S., Squartini, S.: Real-world anomaly detection by using digital twin systems and weakly-supervised learning. CoRR abs/2011.06296 (2020). https://arxiv.org/abs/2011.06296
3. Chicco, D., Jurman, G.: The advantages of the Matthews correlation coefficient (MCC) over F1 score and accuracy in binary classification evaluation. BMC Genomics **21**(1), 6 (2020)
4. ESA: WebMUST - web client for OPS-SAT directory (2021). https://opssat1.esoc.esa.int/webclient-must
5. Evans, D., et al.: Implementing the New CCSDS Housekeeping Data Compression Standard 124.0-B-1 (Based on POCKET+) on OPS-SAT-1. Small Satellite Conference (2022). https://digitalcommons.usu.edu/smallsat/2022/all2022/133
6. Evans, D.J., Ortega, M., Zeif, R., Sergert, T.: OPS-SAT: FDIR design on a mission that expects bugs - and lots of them. In: SpaceOps 2016 Conference. American Institute of Aeronautics and Astronautics (2016). https://doi.org/10.2514/6.2016-2481

7. Fratini, S., Policella, N., Silva, R., Guerreiro, J.: On-board autonomy operations for OPS-SAT experiment. Appl. Intell. **52**(6), 6970–6987 (2022)
8. Gao, Y., et al.: An unsupervised anomaly detection approach for spacecraft based on normal behavior clustering. In: Proceedings of the ICICTA, pp. 478–481 (2012)
9. Hundman, K., Constantinou, V., Laporte, C., Colwell, I., Söderström, T.: Detecting spacecraft anomalies using LSTMs and nonparametric dynamic thresholding. In: Proceedings of the SIGKDD, pp. 387–395. ACM (2018)
10. Kacker, S., Meredith, A., Cahoy, K., Labrèche, G.: Machine learning image processing algorithms onboard OPS-SAT. In: Small Satellite Conference (2022). https://digitalcommons.usu.edu/smallsat/2022/all2022/65
11. Kopp, M., Pevný, T., Holeňa, M.: Anomaly explanation with random forests. Expert Syst. Appl. **149**, 113187 (2020)
12. Kubicka, M., Zeif, R., Henkel, M., Hörmer, A.J.: Thermal vacuum tests for the ESA's OPS-SAT mission. Elektrotech. Informationstechnik **139**(1), 16–24 (2022)
13. Labreche, G., et al.: OPS-SAT spacecraft autonomy with TensorFlow lite, unsupervised learning, and online machine learning. In: Proceedings of the IEEE Aerospace Conference. IEEE Computer Society (2022)
14. Li, K., et al.: A novel method for spacecraft electrical fault detection based on FCM clustering and WPSVM classification with PCA feature extraction. Proc. Inst. Mech. Eng. **231**(1), 98–108 (2017)
15. Marszk, D., Evans, D., Mladenov, T., Labrèche, G., Zelenevskiy, V., Shiradhonkar, V.: MO services and CFDP in action on OPS-SAT. In: Small Satellite Conference (2022). https://digitalcommons.usu.edu/smallsat/2022/all2022/67
16. Nalepa, J., et al.: Toward On-Board Detection Of Anomalous Events From Ops-Sat Telemetry Using Deep Learning (2022). https://www.researchgate.net/publication/367091759
17. Nalepa, J., Myller, M., Andrzejewski, J., Benecki, P., Piechaczek, S., Kostrzewa, D.: Evaluating algorithms for anomaly detection in satellite telemetry data. Acta Astronaut. **198**, 689–701 (2022)
18. Nassif, A.B., Talib, M.A., Nasir, Q., Dakalbab, F.M.: Machine learning for anomaly detection: a systematic review. IEEE Access **9**, 78658–78700 (2021)
19. Pedregosa, F., et al.: Scikit-learn: machine learning in Python. J. Mach. Learn. Res. **12**, 2825–2830 (2011)
20. Primartha, R., Tama, B.A.: Anomaly detection using random forest: a performance revisited. In: Proceedings ICoDSE, pp. 1–6 (2017)
21. Segret, B., et al.: On-board images to characterize a CubeSat's ADCS. In: Small Satellite Conference (2022). https://digitalcommons.usu.edu/smallsat/2022/all2022/174
22. Toussaint, F., Thomassin, J., Laurens, S.: ASTERIA in-orbit testing on OPSSAT: an on-board autonomous orbit control solution including collision risks avoidance. In: Small Satellite Conference (2022). https://digitalcommons.usu.edu/smallsat/2022/all2022/103
23. Wu, R., Keogh, E.J.: Current time series anomaly detection benchmarks are flawed and are creating the illusion of progress (extended abstract). In: 2022 IEEE 38th International Conference on Data Engineering (ICDE), pp. 1479–1480 (2022)
24. Zeif, R., Henkel, M., Hörmer, A., Kubicka, M., Wenger, M., Koudelka, O.: The redundancy and fail-safe concept of the OPS-SAT payload processing platform. In: 69th International Astronautical Congress, IAC 2018 (2018)
25. Zeng, Z., Jin, G., Xu, C., Chen, S., Zhang, L.: Spacecraft telemetry anomaly detection based on parametric causality and double-criteria drift streaming peaks over threshold. Appl. Sci. **12**(4), 1–24 (2022)

Wildfire Perimeter Detection via Iterative Trimming Method

Li Tan[1], Yangsheng Hu[1], Shuo Tan[1], Raymond A. de Callafon[1]([⊠]),
and Ilkay Altıntaş[2]

[1] Department of Mechanical and Aerospace Engineering,
University of California San Diego, La Jolla, CA, USA
{ltan,yah071,sht001,callafon}@eng.ucsd.edu
[2] San Diego Supercomputer Center, University of California San Diego,
La Jolla, CA, USA
ialtintas@ucsd.edu

Abstract. The perimeter of a wildfire is essential for prediction of the spread of a wildfire. Real-time information on an active wildfire can be obtained with Thermal InfraRed (TIR) data collected via aerial surveys or satellite imaging, but often lack the actual numerical parametrization of the wildfire perimeter. As such, additional image processing is needed to formulate closed polygons that provide the numerical parametrization of wildfire perimeters. Although a traditional image segmentation method (ISM) that relies on image gradient or image continuity can be used to process a TIR image, these methods may fail to accurately represent a perimeter or boundary of an object when pixels representing high infrared values are sparse and not connected. An ISM processed TIR image with sparse high infrared pixels often results in multiple disconnected sub-objects rather than a complete object. This paper solves the problem of detecting wildfire perimeters from TIR images in three distinct image processing steps. First, Delaunay triangulation is used to connect the sparse and disconnected high-value infrared pixels. Subsequently, a closed (convex) polygon is created by joining adjacent triangles. The final step consists of an iterative trimming method that removes redundant triangles to find the closed (non-convex) polygon that parametrizes the wildfire perimeter. The method is illustrated on a typical satellite TIR image of a wildfire, and the result is compared to those obtained by traditional ISMs. The illustration shows that the three image processing steps summarized in this paper yield an accurate result for representation of the wildfire perimeter.

Keywords: Wildfire Perimeter · Thermal Infrared Image · Delaunay Triangulation · Iterative Algorithm

---

Work is supported by WIFIRE Commons and funded by NSF 2040676 and NSF 2134904 under the Convergence Accelerator program.

J. Mikyška et al. (Eds.): ICCS 2023, LNCS 14073, pp. 307–320, 2023.
https://doi.org/10.1007/978-3-031-35995-8_22

1   Introduction

As unprecedented wildfire activity has occurred in recent years, the reliable prediction of the spread of an active wildfire has proven to be challenging task. Ensemble Kalman filtering [9] is often used as the data assimilation technique for wildfire spread prediction [15]. Many applications have combined a data-driven fire model such as FARSITE [11] with ensemble Kalman filtering to improve the prediction accuracy [10,20,21]. Such data assimilation techniques do rely on the availability of (past) fire perimeter measurements to predict (future) wildfire perimeters for characterizing wildfire progression.

For the moment, manual delineation is still frequently used to obtain the wildfire perimeter. Due to the significant advances in sensor technology for wildfire monitoring, new real-time data sources such as Thermal InfraRed (TIR) imaging can be used for a data-driven approach to predict wildfire progression. For example, wildfires can be monitored via MODIS data [6], satellite heat images [19] or thermal infrared imaging (TIR) on aerial flight systems [23] that characterize ground temperature. In case of TIR or heat images, image pixels represent temperature information or strength of the infrared band. In particular for an RGB image, the R value is chosen as the data for the image pixels.

Many contributions can be found that provided automated extraction of wildfire perimeters by applying classic edge detection algorithm, such as global intensity thresholding algorithm, Sobel gradient operator, and Canny edge detector [17,24]. Among them, Canny method has the ability to outperform the others [24]. Such edge detection methods might be good tools to automatically delineate the wildfire perimeter, and the application of other edge detection methods, such as graph-cut method and level set method, are beneficial for the detection of wildfire perimeter. However, due to limited resolution of the image, discontinuity of the two-dimensional image data, and possibly partial activity of a wildfire along its boundary, identification of the most recent fire perimeter remains a challenge. Furthermore, different image pixels may be independent and subjected to noise or temporary fire inactivity. As a result, the burn area in the wildfire image can be disconnected or even sparse.

Although combining Machine Learning (ML) methodologies with real-time image data has been recognized to advance in wildfire science and management [13], there is an important restriction that a huge amount of training sets are required by ML techniques to learn the characteristics of the heat map of the burn area. Although computer vision is applied more and more on wildfire detection and measurement [22], few contributions can been found that provide a closed polygon for the parametrization of the wildfire perimeter on the basis of TIR or heat images. In [24], unsupervised edge detectors were applied to obtain the wildfire perimeter automatically, but performance in case of sparse TIR data had not been demonstrated.

The main contribution of the paper is to provided a novel TIR image processing technique to characterize a closed polygon for the wildfire perimeter. To solve the problem of discontinuity of a heat image, the basic concept of Delaunay triangulation is used to obtain a convex polygon of a burn area. Similar ideas are explored in [6] where the so-called α-shape algorithm is used to determine the wildfire perimeter using information on hot spots. Unfortunately, the α-algorithm can only adjust the detected wildfire perimeter globally by changing the value of α, and is barely able to distinguish spot fires from the main burn area in an TIR image. To solve this problem, the novelty of the TIR image processing lies in the iterative trimming of triangular objects created by the Delaunay triangulation to obtain a closer match of the wildfire perimeter.

Both rough trimming and fine trimming are included in the iterative trimming method. For the rough trimming, after obtaining the convex polygon covering all pixels of the burn area by applying the Delaunay triangulation, two threshold values are created for this step. One is related to the longest side of the triangle created by Delaunay triangulation, and the other one is related to the relative burn area surrounding a vertex in a chosen domain. Based on these two threshold values, the iterative trimming method will first delete the redundant abnormally large triangle created by the vertex of the polygon, and then delete the isolated pixels of burn area caused by spot fire. As a result, a new convex polygon can be obtained relying on the remaining pixels of the burn area, and this process will be repeated iteratively until all the pixels of spot fires are removed. When the rough trimming is finished, another two threshold values related to the longest side of triangles connected to the vertex of the polygon and the relative burn area surrounding a vertex are created for the fine trimming. The wildfire perimeter can be finally obtained by tuning the threshold values. The performance of the iterative trimming method is illustrated by comparing the wildfire perimeter created by the iterative trimming method to those created by some classical edge detection methods, such as Canny edge detection, graph-cut method, and level set method.

2   Thermal Infrared Image of a Wildfire

One typical discontinuous RGB TIR image of wildfire is presented in Fig. 1, and the corresponding R-value TIR image is presented in Fig. 2. It can be observed from Fig. 1 and Fig. 2 that the bright area alternates with the dark area. Although the different bright areas are close, they are not connected. Delaunay triangulation is applied in this paper to link up those bright areas.

To prepare for the Delaunay triangulation, the active pixels and inactive pixels are defined as

$$y_{i,j} = f([i,j], b) = \begin{cases} 0, & \text{if } b < b_t, \\ 1, & \text{if } b \geq b_t, \end{cases} \tag{1}$$

where i, j describe the location of the pixel, b is the R-value of the TIR image, and b_t is the threshold value to distinguish between active pixel and inactive pixel. $y_{i,j} = 1$ means the pixel at i, j is identified as an active pixel. It is part of burn area and can be used to detect the wildfire perimeter; $y_{i,j} = 0$ means the pixel at i, j is considered to be an inactive pixel. It represents either inactive

Fig. 1. Example of an RGB TIR image, courtesy of DigitalGlobe WorldView-3 satellite data of the Happy Camp Complex fire. The size of the image is 850 × 550 pixels.

wildfire or unburned area and should not be used for the detection of wildfire perimeter. In this paper, b_t is chosen to be 50. The reason is that the pixels with the R-value smaller than 50 belong to nearly completely dark area as shown in Fig. 2.

3 Delaunay Triangulation and Iterative Trimming

3.1 Delaunay Triangulation

In 1934, Boris Delaunay introduces the Delaunay triangulation [7] that is well known for maximizing the minimum of all the angles of the resulting triangles. The Delaunay triangulation has been used in many applications due to the property of providing connectivity information for a given set of points [12]. This property is also adopted in this paper to solve the problem of missing connection between the burn area.

By applying the Delaunay triangulation on the active pixels determined from Fig. 2 in Sect. 2, the connecting triangles (red) can be established as shown in Fig. 3(a). The union set of all the resulting triangles created by Delaunay triangulation is a convex hull depicted in Fig. 3(b). It can be observed that the convex hull contains many redundant (sliver) triangles that hide the shape of the wildfire perimeter. These redundant large triangles are mostly caused by the vertices of the polygon characterizing the wildfire perimeter. Delaunay Triangulation connects these vertices as it did for the internal disconnected burn area. In addition, some active pixels are caused by spot fires and should also be removed to discover the main perimeter of the main wildfire. Therefore, it is important to distinguish the triangles outside the main wildfire perimeter from those inside the wildfire perimeter. To this end, an iterative trimming method is set up to trim the convex hull and reveal the wildfire perimeter.

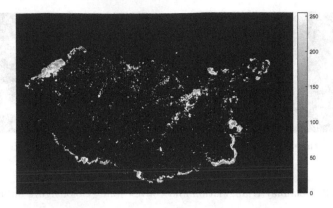

Fig. 2. R-value of TIR image depicted in Fig. 1.

3.2 Iterative Trimming Method

To cut out the redundant triangles established by Delaunay triangulation and the active pixels caused by spot fires, two iterative trimming steps are used: rough trimming and fine trimming. In the process of rough trimming, the abnormally large triangles and the pixels of spot fire will be removed iteratively. After that, the remaining polygon is further trimmed in the fine trimming process. Details on both trimming processes are as follows.

Rough Trimming. As mentioned, the goal of rough trimming is to remove the abnormally large triangles created by Delaunay triangulation and some small groups of active pixels caused by spot fire to obtain a coarse shape of the main fire. The abnormally large triangles are selected by using the histogram of the longest side of all triangles. If the longest side of a triangle appears just once in a bin of the histogram and larger than the upper boundary of the last bin with count larger than two, then the triangle is categorized into the abnormally large triangle. The detailed procedure to recognize the abnormally large triangle is summarized in Algorithm 1.

Figure 4 shows the histogram calculated on the basis of the triangles depicted earlier in Fig. 3(a). From the histogram in Fig. 4, it can be observed that most of the triangles have a longest side smaller than 50, and abnormally large triangles can be recognized by choosing the last few triangles with one count in the histogram.

(a) Delaunay triangulation on active pixels (b) Polygon (red) by the union set of the
of Figure 2.   triangles

Fig. 3. Delaunay triangulation and convex hull. (Color figure online)

Algorithm 1. Identifying abnormally large triangle

Input: Longest side of each triangle, s
Output: Abnormally large triangle

1: Construct a histogram about the longest side of each triangles, h=histogram(s).
2: Find the index of the last triangle with count larger than one,
   k=find(h.count > 1, 'last').
3: Abnormally large triangles are identified as the triangles with longest side larger
   than $l_r = k \times h$.binwidth.

Due to the fact that spot fire is outside the main body of the wildfire, it is
only necessary to check whether the vertices of the currently established polygon
belong to the main fire or a spot fire. Spot fires are defined as a tiny connected
burn area that is isolated from the main fire. Therefore, the active pixels of spot
fires can be distinguished by the relative surface area within a chosen domain.
The chosen domain can be a square patch of the TIR image centered by the
vertex of the current polygon, and the relative surface area can be calculated by
the ratio of the number of the connected active pixels containing the vertex of
the current polygon and the number of pixels inside the whole square patch. The
approach to select the active pixels of spot fires is summarized in Algorithm 2.

Algorithm 2. Identifying active pixels of spot fire

Input: Vertices of currently established polygon, n; size of square image patch, $m \times m$;
threshold value, d_u.
Output: Active pixels of spot fires.

1: For each vertex, calculate the summation, p_q, $q = 1, 2, \ldots, n$, of the number of all
   connected active pixels starting from the vertex inside the $m \times m$ square patch.
2: Compute the ratio $\frac{p_q}{m \times m}$, for $q = 1, 2, \ldots, n$
3: If $\frac{p_q}{m \times m} \leq d_u$, all the connected active pixels with respect to the q_{th} vertex of the
   polygon are regarded as part of the active pixels of spot fires.

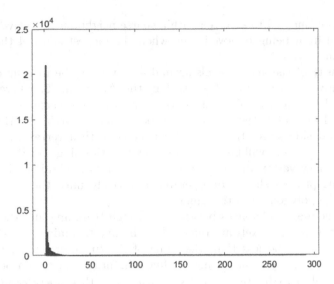

Fig. 4. Histogram of the longest side of the triangles in Fig. 3(a). The unit of the side length is (one) pixel length.

As a consequence of removing the active pixels of spot fires, the set of points for Delaunay triangulation are affected. Hence, an iterative process of Delaunay triangulation, removing the abnormally large triangles, and removing the active pixels of spot fires is operated until all the active pixels of spot fires are eliminated. During the iteration, the number n of the vertices of the polygon will also change accordingly. After the rough trimming, a coarse shape of the main wildfire can be acquired, and the fine trimming will further remove the redundant triangles produced by Delaunay triangulation.

Fine Trimming. Instead of just removing the abnormally large triangles coarsely as done in rough trimming, fine trimming is more rigorous. With the goal of obtaining the perimeter of the main wildfire, the fine trimming is focused on the triangles connected with the vertices of the current polygon to avoid making the polygon disconnected. The redundant triangles can also be determined by the histogram of the longest side, or specific requirement on the wildfire perimeter.

Furthermore, considering the fact that the TIR image is sparse, the density of the active pixels around a vertex should also be taken into account. Heavier trimming can be done on the triangles connected to a vertex having a denser neighbouring active pixels, and those triangles connected to a vertex that has a relatively sparse neighbouring active pixels should be discarded more carefully. As a result, two threshold values are established for the fine trimming. The first one is the threshold value l_f, set for the longest side of the triangles that should be removed, and the second one is the threshold value d_l, set for the density of the neighbouring active pixels around a vertex. The value of d_l is chosen so that the triangles connected to a vertex with a sparse neighbouring active pixels will be protected from being removed even when the longest sides of the triangles are longer than l_f.

The density of the active pixels around a vertex can be measured similarly by the relative surface area $\frac{p_q}{m \times m}$ used in the Algorithm 2, where p_q is the number of all the connected pixels starting from the q_{th} vertex of the current polygon, and $m \times m$ is the size of the chosen square patch. All the triangles with a longest side larger than l_f and connected with a vertex that meet the requirement $\frac{p_q}{m \times m} > d_l$ will be removed in the fine trimming. During the process of trimming, new vertex will appear as the triangles are eliminated. Therefore, the trimming process should be operated iteratively until the number of the vertices of the polygon stays the same.

One of the main differences between the rough trimming and the fine trimming is that no active pixels are removed in the fine trimming. For this reason, the value of d_l can be less than the value of d_u. In other words, new vertex with $\frac{p_q}{m \times m} \leq d_u$ may appear during the fine trimming, and d_l can be chosen as $d_l < \frac{p_q}{m \times m} \leq d_u$ to trim the triangles connected to this newly created vertex. The complete procedure of the iterative trimming method including the step of Delaunay triangulation is summarized in Algorithm 3.

Algorithm 3. Iterative trimming method

Input: TIR image of wildfire.

Output: Wildfire perimeter.

1: Determine the set of active pixels.

2: Find the locations of the active pixels, and apply the Delaunay triangulation on the active pixels.

3: Remove the abnormally large triangles and the active pixels caused by spot fires summarized in Algorithm 1 and Algorithm 2 respectively.

4: If the number of active pixels changes, go back to step 2.

5: Remove the triangle connected to the the q_{th} vertex with $\frac{p_q}{m \times m} > d_l$, if the longest side of this triangle is larger than l_f.

6: If the number of vertices of the polygon changes in step 5, repeat step 5.

The performance of the iterative trimming method based on the TIR data given earlier in Fig. 1 is illustrated in the next section. In addition, the established polygon of the main wildfire perimeter obtained by the iterative trimming method on the basis of the Delaunay triangulation is compared to those obtained by the Canny edge detector, the graph-cut method, and the level set method.

4   Results and Discussion

4.1   Iterative Trimming Method

Considering the distribution of active pixels in Fig. 2, the computed longest side of the triangles in Fig. 3(a) and the resulting histogram depicted in Fig. 4, the values of l_f, m, d_u, d_l are chosen as $50, 11, 0.16, 0.08$ respectively. The polygons obtained after the rough trimming and the fine trimming are shown in Fig. 5(a) and Fig. 5(b). It is worthwhile to note that, although the high-value infrared pixels are sparse and disconnected, the iterative trimming method can obtain a closed polygon of the wildfire automatically. It can also be noticed that some isolated active pixels outside the red polygon are regarded as the spot fires, and are not used to establish the polygon of the main wildfire. The computation time is around one second. As reference for the computation time, all calculations were performed on an Intel Core i7-7500U CPU with 16 GB RAM.

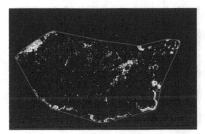 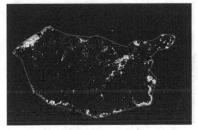

(a) Polygon (red) obtained by rough trimming. (b) Polygon (red) obtained by fine trimming.

Fig. 5. Results of iterative trimming method. (Color figure online)

4.2   Canny Edge Detector

The Canny edge detector is developed by John F. Canny in 1986 [5]. Although Canny edge detection is a traditional edge detection method, it has been widely applied and improved in more recent researches [1,8,18]. It was also utilized in the study of wildfire monitoring [24] and was shown to be one of the most effective unsupervised detection algorithms. Three performance criteria: good detection, good localization, and unique response to a single edge, form the basis of Canny edge detector.

A Canny edge detection has five steps including smoothing the image by Gaussian filter, calculating the gradient of the image, deleting spurious response to true edges, using double threshold to find out prospective edge, and tracking edge by preserving strong edges and weak edges that are connected to strong edges. The image processing toolbox of MATLAB provides the standard edge detection algorithm, and Canny edge detector can be applied by calling the function `edge(I,'canny',threshold,sigma)`, where I is the image, `threshold` is used to ignore the unnecessary edges, and `sigma` decides the standard deviation of the Gaussian filter.

It is clear from the procedure of Canny edge detector that the image gradient is the basis for this edge detection method. For the TIR image given earlier in

Fig. 2, the Canny edge detector is more likely to work on detecting the boundaries of all isolated clusters of pixels instead of the main wildfire. To blur the image to a greater extent so that the effect of the isolation is reduced, a larger standard deviation of the Gaussian filter can be applied.

The results of Canny edge detection with default value for the `threshold` and two different standard deviations of the Gaussian filter are shown in Fig. 6. Although increasing the standard deviation of Gaussian filter leads to a slightly better result, the Canny edge detector is still not able to detect the main boundary of the main wildfire.

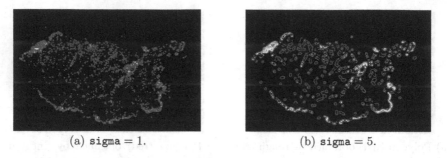

(a) `sigma` = 1.   (b) `sigma` = 5.

Fig. 6. Boundaries (red) of the wildfire generated by Canny edge detector. (Color figure online)

4.3   Graph-Cut Method

The graph-cut method is a widely used method in the field of computer vision and details can, for example, be found in [2,4,25]. Here a short summary of the graph-cut method is given. An image is first transformed to a graph consisting of nodes and edges, where edges are used to connect every two neighbour nodes, and nodes are composed of two terminal nodes, source node and sink node, and all pixels. Each edge is assigned with a weight or cost, and the goal of the graph-cut method is to find a minimum cut of the graph by using the max-flow min-cut theorem [3]. With the minimum cut, the image is divided into a foreground and a background.

The graph-cut method has a good ability to produce an optimal solution to the image segmentation of a binary problem, which is similar to distinguishing between the burn area and unburned area in a TIR image of a wildfire. Therefore, the graph-cut method is applied on Fig. 2 to obtain the boundary of the wildfire via the image segmenter application in MATLAB. The seeds required by graph-cut method are highlighted as Fig. 7(a), and the result of the graph-cut method is shown in Fig. 7(b).

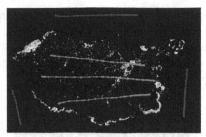 

(a) Seeds for foreground (green) and back-   (b) Boundaries (red) generated by graph-
ground (red).   cut methods..

Fig. 7. Results of graph-cut method. (Color figure online)

It can be observed that graph-cut method only works well on identifying part of the wildfire boundary near the provided seeds. Although a good wildfire perimeter can be created by graph-cut method when more detailed and compli cated seeds are provided, too much human interaction is required and the work involved is almost the same as a manual delineation, limiting the application in automatic wildfire perimeter detection.

4.4   Level Set Method

The level set method is an impressive tool for image segmentation by exploiting the information of regions and boundaries of the object [16]. It applies level sets for numerical analysis of surfaces, and the application of level sets makes it beneficial to track the change of the topology, such as the development of a hole. In addition, the level set method provides an implicit description of the object without the need of parameterizing the object. Due to the fact that a level set method has a good ability to separate two regions, it might be a solution to the binary problem of detecting the burn area from the unburned area.

Unfortunately, for a disconnected TIR image of a wildfire as depicted in Fig. 2, it is infeasible to decide whether a hole exists, or what the size of the hole is. Moreover, continuous image gradient and intensity of the pixels are important components of level set method to determine the speed of the evolving and the shape described by each level set. Another potential problem of the level set method is that the final result is dependent on the choice of the initial contour, but the shape of the wildfire can be arbitrary. Hence, an algorithm of automatically generating an initial zero-level contour is required, or a fire expert needs to provide an initial contour for each image of a wildfire to apply the level set method.

To test the performance of the level set method on Fig. 2, the level set method introduced by [14] is adopted. By choosing the parameters $\alpha = 2$, $\lambda = 5$, $\mu = 0.01$, the result can be achieved after 200 iterations as Fig. 8(b), with the initial contour established as Fig. 8(a). It can be observed that the boundary captured by level set method passes through the disconnected part of the burn area of the TIR image. The reason for that is level set method also relies on the image gradient or pixel intensity to calculate the level sets, and there is a huge change in the image gradient between the disconnected areas.

(a) Initial zero level contour.   (b) Final zero level contour

Fig. 8. Results of level set method.

5   Conclusions

This paper introduces an iterative trimming method (ITM) based on Delaunay triangulation with a goal to establish a closed polygon of the main wildfire perimeter automatically for a TIR image of wildfire. Although the burn area caused by spot fire is deleted in the ITM, they can be captured respectively by treating each burn area of a spot fire as a main wildfire perimeter. The performance of the iterative trimming method is validated by providing a study that compares the result of the ITM with those of the various edge detection methods based on a satellite generated TIR image. The comparison study shows that the various edge detection methods fail to provide a single closed polygon that parametrizes the main wildfire perimeter. Often, disconnected burn areas will be detected separately. The proposed ITM shows good performance with a single closed polygon for the wildfire perimeter. For further studies, more information of the wildfire can be used in the iterative trimming method. For example, the wind direction and wind speed can be used to predict the location of the spot fire, and a priori knowledge of the spot fire can be included in the iterative trimming method to better capture and remove the active pixels caused by spot fires.

References

1. Bao, P., Zhang, L., Wu, X.: Canny edge detection enhancement by scale multiplication. IEEE Trans. Pattern Anal. Mach. Intell. **27**(9), 1485–1490 (2005)
2. Boykov, Y., Funka-Lea, G.: Graph cuts and efficient nd image segmentation. Int. J. Comput. Vis. **70**(2), 109–131 (2006)

3. Boykov, Y., Kolmogorov, V.: An experimental comparison of min-cut/max-flow algorithms for energy minimization in vision. IEEE Trans. Pattern Anal. Mach. Intell. **26**(9), 1124–1137 (2004)

4. Boykov, Y.Y., Jolly, M.P.: Interactive graph cuts for optimal boundary & region segmentation of objects in nd images. In: Proceedings of the Eighth IEEE International Conference on Computer Vision. ICCV 2001, vol. 1, pp. 105–112. IEEE (2001)

5. Canny, J.: A computational approach to edge detection. IEEE Trans. Pattern Anal. Mach. Intell. **6**, 679–698 (1986)

6. Chiaraviglio, N., et al.: Automatic fire perimeter determination using MODIS hotspots information. In: 2016 IEEE 12th International Conference on e-Science (e-Science), pp. 414–423 (2016)

7. Delaunay, B., et al.: Sur la sphere vide. Izv. Akad. Nauk SSSR, Otdelenie Matematicheskii i Estestvennyka Nauk **7**(793–800), 1–2 (1934)

8. Ding, L., Goshtasby, A.: On the canny edge detector. Pattern Recogn. **34**(3), 721–725 (2001)

9. Evensen, G.: The ensemble Kalman filter: theoretical formulation and practical implementation. Ocean Dyn. **53**(4), 343–367 (2003)

10. Fang, H., Srivas, T., de Callafon, R.A., Haile, M.A.: Ensemble-based simultaneous input and state estimation for nonlinear dynamic systems with application to wildfire data assimilation. Control Eng. Pract. **63**, 104–115 (2017)

11. Finney, M.A.: FARSITE, Fire Area Simulator-model development and evaluation. No. 4, US Department of Agriculture, Forest Service, Rocky Mountain Research Station (1998)

12. Engquist, B. (ed.): Encyclopedia of Applied and Computational Mathematics. Springer, Heidelberg (2015). https://doi.org/10.1007/978-3-540-70529-1

13. Jain, P., Coogan, S.C., Subramanian, S.G., Crowley, M., Taylor, S., Flannigan, M.D.: A review of machine learning applications in wildfire science and management. arXiv preprint arXiv:2003.00646 (2020)

14. Li, C., Xu, C., Gui, C., Fox, M.D.: Distance regularized level set evolution and its application to image segmentation. IEEE Trans. Image Process. **19**(12), 3243–3254 (2010)

15. Mandel, J., et al.: Towards a dynamic data driven application system for wildfire simulation. In: Sunderam, V.S., van Albada, G.D., Sloot, P.M.A., Dongarra, J.J. (eds.) ICCS 2005. LNCS, vol. 3515, pp. 632–639. Springer, Heidelberg (2005). https://doi.org/10.1007/11428848_82

16. Mumford, D., Shah, J.: Boundary detection by minimizing functionals. In: IEEE Conference on Computer Vision and Pattern Recognition, vol. 17, pp. 137–154. San Francisco (1985)

17. Ononye, A.E., Vodacek, A., Saber, E.: Automated extraction of fire line parameters from multispectral infrared images. Remote Sens. Environ. **108**(2), 179–188 (2007)

18. Rong, W., Li, Z., Zhang, W., Sun, L.: An improved canny edge detection algorithm. In: 2014 IEEE International Conference on Mechatronics and Automation, pp. 577–582. IEEE (2014)

19. Schroeder, W., Oliva, P., Giglio, L., Quayle, B., Lorenz, E., Morelli, F.: Active fire detection using landsat-8/oli data. Remote Sens. Environ. **185**, 210–220 (2016). landsat 8 Science Results

20. Srivas, T., Artés, T., de Callafon, R.A., Altintas, I.: Wildfire spread prediction and assimilation for FARSITE using ensemble Kalman filtering. Procedia Comput. Sci. **80**, 897–908 (2016)

21. Subramanian, A., Tan, L., de Callafon, R.A., Crawl, D., Altintas, I.: Recursive updates of wildfire perimeters using barrier points and ensemble Kalman filtering. In: Krzhizhanovskaya, V.V., et al. (eds.) ICCS 2020. LNCS, vol. 12142, pp. 225–236. Springer, Cham (2020). https://doi.org/10.1007/978-3-030-50433-5_18

22. Toulouse, T., Rossi, L., Campana, A., Celik, T., Akhloufi, M.A.: Computer vision for wildfire research: an evolving image dataset for processing and analysis. Fire Saf. J. **92**, 188–194 (2017)

23. Valero, M.M., Rios, O., Mata, C., Pastor, E., Planas, E.: An integrated approach for tactical monitoring and data-driven spread forecasting of wildfires. Fire Saf. J. **91**, 835–844 (2017). fire Safety Science: Proceedings of the 12th International Symposium

24. Valero, M., Rios, O., Pastor, E., Planas, E.: Automated location of active fire perimeters in aerial infrared imaging using unsupervised edge detectors. Int. J. Wildland Fire **27**(4), 241–256 (2018)

25. Yi, F., Moon, I.: Image segmentation: a survey of graph-cut methods. In: 2012 International Conference on Systems and Informatics (ICSAI2012), pp. 1936–1941. IEEE (2012)

Variable Discovery with Large Language Models for Metamorphic Testing of Scientific Software

Christos Tsigkanos[1,2]($\boxtimes$) (iD), Pooja Rani[3] (iD), Sebastian Müller[4] (iD), and Timo Kehrer[1] (iD)

[1] University of Bern, Bern, Switzerland
`christos.tsigkanos@inf.unibe.ch`
[2] University of Athens, Athens, Greece
[3] University of Zurich, Zürich, Switzerland
[4] Humboldt-Universität zu Berlin, Berlin, Germany

Abstract. When testing scientific software, it is often challenging or even impossible to craft a test oracle for checking whether the program under test produces the expected output when being executed on a given input – also known as the oracle problem in software engineering. Metamorphic testing mitigates the oracle problem by reasoning on necessary properties that a program under test should exhibit regarding multiple input and output variables. A general approach consists of extracting metamorphic relations from auxiliary artifacts such as user manuals or documentation, a strategy particularly fitting to testing scientific software. However, such software typically has large input-output spaces, and the fundamental prerequisite – extracting variables of interest – is an arduous and non-scalable process when performed manually. To this end, we devise a workflow around an autoregressive transformer-based Large Language Model (LLM) towards the extraction of variables from user manuals of scientific software. Our end-to-end approach, besides a prompt specification consisting of few examples by a human user, is fully automated, in contrast to current practice requiring human intervention. We showcase our LLM workflow over three case studies of scientific software documentation, and compare variables extracted to ground truth manually labelled by experts.

Keywords: Scientific Software · Metamorphic Testing · Large Language Models · Natural Language Processing

1 Introduction

With software being the most important driver for research in many scientific disciplines, the functional correctness of scientific software in particular is of utmost importance [8,27,28,30]. Scientific software that has bugs may produce wrong outcomes leading to erroneous evidence, which in turn may have severe

© The Author(s), under exclusive license to Springer Nature Switzerland AG 2023
J. Mikyška et al. (Eds.): ICCS 2023, LNCS 14073, pp. 321–335, 2023.
https://doi.org/10.1007/978-3-031-35995-8_23

consequences in terms of research costs, scientific reputation, or even human well-being (e.g., when relying on invalid theories) [33,42]. Software testing is the predominant way for ensuring functional correctness of a program under test. Traditional testing techniques rely on a test oracle for checking whether the program under test produces the expected output when being executed on a given input. In some cases, however, obtaining reliable oracles is challenging or even impossible – this is generally known as the test oracle problem in software engineering [4].

Scientific software is particularly affected by the oracle problem [23]. A major reason for this is that the scientific process is often exploratory in nature, with software written to find answers that are previously unknown [9,47,50], while lacking functional requirements being specified up-front [35]. Moreover, the underlying scientific theory may involve complex computations or inherent uncertainties, making it hard to determine the expected output for a given input [12,26,27,50].

Metamorphic Testing [43] is a property-based software testing approach that mitigates the oracle problem by relying on so-called metamorphic relations [44]. Roughly speaking, a metamorphic relation specifies how the output changes according to a change made to the input. This way, a huge amount of test cases may be generated based on a *single* input-output-pair, invalidating the traditional prerequisite of being able to accurately determine the expected output for *any* given input of a program under test. By mitigating the oracle problem, metamorphic testing qualifies as a promising approach for testing scientific software [11,12,14,21,31,32,34]. However, scientific software is often characterized by having large input-output (I/O) spaces which are hard to cover by metamorphic relations when being specified ad-hoc, calling for methods to support scientific software engineers in systematically deriving metamorphic relations.

In recent work on metamorphic testing of scientific software, Peng et al. [36,37] proposed to use auxiliary artifacts such as user manuals as a potential source for extracting metamorphic relations. The fundamental prerequisite consists of extracting input-output variables of interest; subsequently, relations can be devised with the overall goal of enabling metamorphic testing. The problem is illustrated in Fig. 1 over an excerpt of a scientific software manual: certain variables (e.g., "status", "startup") appear in the text, along with some hints on metamorphic relations (shaded in Fig. 1). Variable extraction has been initially performed manually [36], an arduous and non-scalable process. Following that, Peng et al. [37] proposed to semi-automate the variable extraction using supervised machine learning algorithms and manually crafted natural language processing-based patterns. However, the supervised learning methods still demand manual work in creating a ground truth, crafting the NLP features.

To further increase the level of automation, we devise a workflow around an autoregressive transformer-based Large Language Model (LLM) towards the first critical step of extracting I/O variables for metamorphic testing from documentation of scientific software. In contrast to manual extraction, our end-to-end approach is fully automated, besides a prompt specification consisting of few examples by a human user. We showcase variable extraction over documentation of scientific software and compare the variables extracted by our workflow to a ground truth that has been manually labelled by experts [37].

We extend recent previous work [46] which appeared as a short paper at the Early Research Achievements track of the 30th IEEE International Conference on Software Analysis, Evolution and Reengineering, where we outlined the underlying research problem and illustrated emerging results. In this paper, we present our approach for variable discovery with LLMs for metamorphic testing in detail. We devote extensive discussion of evaluation aspects of the approach – investigating partial and exact discovered variables, performance of the implied binary classification test, as well as threats to validity. Specifically, our contributions are:

- We present a concrete instantiation of our approach for variable discovery from scientific software manuals for metamorphic testing with LLMs;
- We evaluate the advocated LLM workflow experimentally over three different case studies, demonstrating its feasibility and investigating its performance in tandem with operationalization options, and finally;
- We provide a replication package allowing our results to be reproduced by the research community.

The remainder of this paper is structured as follows. In Sect. 2 we contextualize our approach within the state of the art. In Sect. 3, we describe our solution in the form of a workflow revolving around an LLM, and discuss implementation particulars. Section 4 presents our evaluation and discusses results along with threats to validity. Finally, Sect. 5 concludes the paper and provides an outlook on future work.

2   State of the Art

A number of previous works has demonstrated the feasibility of metamorphic testing for testing scientific software, yet relying on metamorphic relations that have been manually specified in a largely ad-hoc manner [11,12,14,21,31,32].

A major research stream on supporting the discovery of metamorphic relations is devoted to observing behavior of a running program. Su et al. [45] present KABU, a tool to automatically find metamorphic relations by generating new inputs for the program under test and then inferring relations in a rule-based manner. Kanewala et al. [17,22,24] investigate the applicability of different machine learning approaches on the task of metamorphic relation discovery. Hiremath et al. [18] apply machine learning to identify all possible metamorphic relations on oceanographic software that are then minimized according to a cost function. With these approaches, metamorphic relations are learned directly from the behavior of the program under test, which means that all found relations can only be used for regression testing of future program versions. Our approach is to learn metamorphic relations from auxiliary documents such as user manuals, and is thus more general.

The on/off status$_\ominus$ of pumps can be controlled dynamically by specifying startup$_\ominus$ and shutoff water depths at the inlet node or through user-defined Control Rules. (...) For a Type 5 pump, its operating curve shifts position such that flow$_\ominus$ changes in direct proportion to the controlled speed setting while head$_\ominus$ changes in proportion to the setting squared. The principal input parameters for a pump include: =" names of its inlet and outlet$_\ominus$ nodes = name of its pump curve (or * for an Ideal pump) = initial on/off status$_\ominus$ = startup$_\ominus$ and shutoff depths (optional). 3.2.9 Flow Regulators are structures or devices used to control and divert flows$_\ominus$ within a conveyance system. They are typically used to: =" control releases from storage$_\ominus$ facilities =" prevent unacceptable surcharging =" divert flow to treatment facilities and interceptors. SWMM can model the following types of Flow Regulators: Orifices$_\ominus$, Weirs$_\ominus$, and Outlets$_\ominus$. Orifices$_\ominus$ are used to model outlet and diversion structures in drainage systems (...) Orifices$_\ominus$ can be used as storage$_\ominus$ unit outlets$_\ominus$ under all types of flow routing$_\ominus$. If not attached to a storage unit node, they can only be used in drainage networks that are analyzed with Dynamic Wave flow$_\ominus$ routing.

Fig. 1. Excerpt of a page from the Storm Water Management Model [41] scientific software manual by the U.S. Environmental Protection Agency, showing words that the LLM workflow correctly$_\ominus$ classified as variables, and ones that it misclassified. Words not marked were correctly classified as non-variables. An instance of a metamorphic relation is shown shaded.

In recent work on metamorphic testing of scientific software, Peng et al. [36, 37] proposed to use auxiliary artifacts such as user manuals, discussion forums, or documentation as a potential source for extracting metamorphic relations. As a first step towards metamorphic relation discovery from these artifacts, their goal is to identify input-output variables from natural language descriptions. Variable extraction has been initially performed manually [36], an arduous and non-scalable process; variables occurring throughout a scientific software manual were identified in a laborious task involving two researchers and amounting to 40 human-hours. Following that, Peng et al. [37] proposed to semi-automate the variable extraction using supervised machine learning algorithms and manually crafted natural language processing-based patterns. Although these methods are promising, they are limited as (i) they require considerable human interventions in preparing the ground truth and crafting features, and (ii) they rely on how similar variables are written in scientific software. Our approach aims at end-to-end automation and a more general applicability of extracting input-output variables.

Recent research has shown interest in exploring LLMs to overcome the general limitations of classical ML and NLP techniques. LLMs are trained on billions of parameters retrieved from large-scale natural language sources, e.g., web pages. Despite learning a specific task on a particular dataset, they have been shown to learn tasks without external supervision [38]. They can leverage learned features to work on a variety of other problems, such as in machine translation or spelling correction [7]. Specific to software engineering, LLMs have been explored e.g., for classifying issue reports [13], generating code from docstrings [6], generating docstrings from code [10], or synthesizing programs [3]. Given these results, we adopt LLMs towards our overall goal of automating MR discovery which has not yet been investigated for applications of LLMs in software engineering.

3   Discovering I/O Variables with an LLM

To extract I/O variables, we devise a data processing workflow revolving around a generative Large Language Model. Given a scientific software manual in PDF format, the target task consists of inferring the metamorphic variables inherent to it. Due to the underlying deep learning model being an LLM, special attention is given to prompt construction, discussed subsequently.

3.1   LLM-Based Workflow

The data processing workflow we adopt for variable extraction is illustrated in Fig. 2 and revolves around an LLM and two stages: (i) pre-processing, where the scientific software manual is prepared to be submitted to the LLM, and (ii) post-processing, where I/O variables are extracted from its output. In the following, we detail key steps as components, noting that given a source document, all steps described are automated, except for a few human-specified examples used for the prompt.

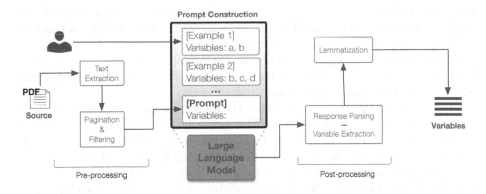

Fig. 2. Data analysis workflow adopted for inferring variables for metamorphic testing from a source document, with an LLM having the central role.

Text Extraction. This first step concerns extracting text from a source document, which is in PDF format, typical for scientific software documentation. Faithful to the automation objective pursued, no filtering by a human user is assumed. The input includes the *entirety* of the document including auxiliary content (e.g., tables of contents, title pages, etc.). For this initial step, to ensure that text is correctly extracted and PDF artifacts are kept minimal, we employ optical character recognition (OCR) on screenshots of pages – several options exist, e.g., tesseract and EasyOCR[1]. This is in contrast to extracting text directly from the PDF, which is prone to produce irregular results and depends strongly on the document encoding. We note that the realization of the OCR pipeline may affect the quality of text extracted.

Pagination and Filtering. The source text is subsequently filtered for irregular characters (e.g., special characters, OCR noise) and maintained in pagewise chunks according to the source document. At this step, pages with less than some specified number of characters can be omitted – such pages assumed to correspond to e.g., full-page figures, separator pages, etc., may not contain I/O variables.

LLM Invocation. A prompt is constructed consisting of few-shot examples provided by the expert user (discussed subsequently in Sec. 3.2). Afterward, the prompt is tokenized according to the target model architecture and the LLM is subsequently invoked. Several autoregressive transformer-based LLMs can be utilized, open ones such as GPT-J-6B [49], GPT-NeoX-20B [5], or closed ones[2] such as OpenAI's GPT-3 or Google's PaLM.

Response Parsing and Variable Extraction. After invoking an inference operation over the prompt to the LLM, its textual response is parsed. The model in essence has completed the prompt given to it, by appending a list of potential variables. Those are extracted from the corresponding response and deduplicated.

Lemmatization. Observe that variables (as words), may occur in several forms in the source text, and as such they should be as uniquely identifiable as possible. This step refers to the process of turning a word into its lemma. A lemma is the "canonical form" of a word, commonly corresponding to its dictionary version – for instance, "flow rates" would be transformed to "flow rate". We perform such lemmatization for each variable extracted.

The result of the workflow is succinctly illustrated in the example of Fig. 1 over an excerpt of a page of the SWMM scientific manual [41]: certain words have been classified as variables, while others have been classified as not being variables. Furthermore, note that certain words may be misclassified – either as false positives (instances which are not relevant but which the model incorrectly identified as relevant), or false negatives (instances which are relevant but which the model incorrectly identified as not relevant).

3.2   Prompt Construction and LLM Particulars

A prompt for the LLM is compiled consisting of the few-shot labelled page examples provided by the user, along with the target page being processed appended

---

[1] tesseract-ocr.github.io, github.com/JaidedAI/EasyOCR – uses ResNet, CTC, and beam-search-based decoder.

[2] openai.com/api, deepmind.com/publications, goo.gle/palm-paper.

to them – Fig. 3 illustrates a fragment of such a prompt, which starts with an instruction to the model, defining the task that it is required to do. The form of the prompt consists of labelled instances of "Text: $\mathcal{T}_i$ and Variables: $\mathcal{V}_i$"; where $\mathcal{T}_i$ corresponds to the text of page i of the source document and $\mathcal{V}_i$ to a comma-separated list of variables occurring in page i, in order to steer the model to what output (and structure of the output) is expected. Subsequently, the text of the current page being processed is appended, and the prompt ends with an (empty) "Variables:" directive.

Your task is to extract Variables from the Text.
Text: (...) If no value for P_UPDIS is entered, the model will set P_UPDIS = 20. (...) NPERCO controls the amount of nitrate removed from the surface layer in runoff relative to the amount removed via percolation. The value of NPERCO can range from 0 to 1.0. (...) If no value for PHOSKD is entered, the model will set PHOSKD = 175.0. (...)
Variables: P_UPDIS, NPERCO, PPERCO, PHOSKD
Text: [**current page being processed**]
Variables:

Fig. 3. Fragment of the prompt given to the LLM, reflecting (part of) one example page labelled by the expert user, along with the current page being processed by the workflow. Note the initial instruction (first line) and that the prompt string ends with 'Variables:', that the LLM should complete by appending.

The response of few-shot LLMs can be unstable and be strongly dependent on the prompt format, the given specific examples, as well as their order. As such, prompt construction in LLMs plays a central role [40]. In our case, user input consists of few examples of text and certain words occurring in the text, which the expert user labels as variables. This user input provides a threat to *majority label bias*, where the model may suggest responses which are more frequent in the examples given by the user, throughout invocations. The intuition is that variable examples initially specified should be as descriptive, unique and prominent in the text as possible. However, as we discuss in future work, investigation into more effective prompt construction is a direction that warrants further consideration – a highly active topic in current LLM research. Since LLMs are sensitive to examples' ordering, we randomize the ones given by the user in each invocation (*recency bias*). *Common token bias*, where the model tends to yield tokens more common in its pre-training data, is an issue as well – this means that variables having more common names may be suggested with higher frequency.

Parameters common in LLM models are temperature, top-p, maximum length (in tokens, of the input and output), and frequency/presence penalties. Informally, lower temperature values render the model increasingly confident in its top choices, while higher decrease confidence while encouraging creative outputs. Top-p has a similar effect, while frequency/presence penalties can be used to suppress repetition.

4   Evaluation

To concretely support evaluation and investigate feasibility, we realized a proof-of-concept implementation [1] reflecting the proposed workflow of Fig. 2. Thereupon, we assess our approach for variable inference. Specifically, we target its accuracy as a binary classifier (labelling words as variables or non-variables), as well as its performance as true positive rates. Subsequently, we discuss our findings, operationalization options and threats to validity.

4.1   Ground Truth and Experiment Setup

Experimental Subjects. Our evaluation target consists of three scientific software user manuals, in line with the state of the art: in foundational works in metamorphic testing [36,37], I/O variables occurring throughout scientific software manuals were manually identified, in a laborious task. We acknowledge this significant effort and treat this result as the ground truth, against which we compare our LLM-based workflow.

SWMM. The Storm Water Management Model (SWMM [41]) by the U.S. Environmental Protection Agency (EPA), is a dynamic rainfall-runoff simulation software that computes runoff quantity and quality within mostly urban areas. The scientific users of SWMM include physicists, hydrologists and engineers involved in planning, analysis, and design related to storm water runoff, combined and sanitary sewers, and other drainage systems. Its user manual [41] is a 353-page PDF document; the ground truth [36] consists of 1005 I/O variables.

SWAT. The Soil and Water Assessment Tool (SWAT [2]) is a watershed to river basin-scale model used to simulate the quality and quantity of surface and groundwater as well as predicting the environmental impact of land use, land management practices, and climate change. SWAT is widely adopted to evaluate soil erosion prevention and control, non-point source pollution and overall regional management of watersheds. Its user manual [2] is a 649-page PDF document; the ground truth [37] consists of 1461 I/O variables.

MODFLOW. The Modular Hydrologic Model (MODFLOW [29]) targets groundwater-flow simulation, including groundwater/surface-water coupling, solute transport, land subsidence, and others. It is widely used for over 30 years by scientists, consultants and governmental organizations. Its user manual [29] is a 188-page PDF document; the ground truth [37] consists of 772 I/O variables.

Instantiation of the Workflow. We realized the workflow of Fig. 2 by employing the (medium-sized) GPT-J-6B [49] – an open-source transformer model trained using JAX [48], known to be trained on a mix of code and natural language text from several programming languages. Specifically, GPT-J-6B was trained on the Pile [15], a large-scale dataset curated by EleutherAI[3]. For our experiments, we deployed the workflow on an NVIDIA RTX A6000 (CUDA 12.0,

---

[3] www.eleuther.ai

PyTorch 1.13), over model GPT-J-6B (float32), with each invocation taking less than 10 s. Regarding configuration, a few-shot prompt consisting of 3 examples of text – variable pairs is used at each invocation, followed by the source document page currently processed. For extracting text from the document, we employ EasyOCR. We ignore pages with less than 400 characters, to filter out pages with, e.g., full-sized figures, tables of contents, etc. Other than that, the respective document is provided to the processing pipeline page by page. We set presence penalty to 0.9, top-p to 0.7, and limit the number of new tokens to 35. Within each invocation, the order of the human-specified examples is randomized at the respective prompt. Thereupon, we invoke the workflow for different values of the critical temperature parameter. Variable discovery (excluding OCR) for the entirety of the source software manuals is in the range of 19 min for SWMM, 28 min for SWAT and 8 min for MODFLOW.

Evaluation Metrics. To assess our approach, we compare the workflow-extracted variables for each case study to the respective ground truth [36]. We first investigate *accuracy* of the approach, against the implied binary classification test. Subsequently, and for a more refined metric, we assess *performance* achieved as true positives over different values of the critical temperature parameter.

Accuracy. We evaluate the advocated LLM workflow as a binary classifier, where the task is to classify every word appearing in the source document as a variable – or not a variable. Following previous works in this direction [16,20,39], we compute accuracy to evaluate our results. For a binary classifier, recall that accuracy is defined as the proportion of correct predictions (both true positives and true negatives) among the total number of cases examined. To this end, we compare the output of the workflow given the source document against ground truth; words of the source PDF are defined as unique instances of characters occurring between spaces, which are subsequently lemmatized.

True Positive Rate. Recall the example of Fig. 1, which illustrates the workflow's outputs for an example page fragment. We consider two types of *true positives*: (i) exact true positives, where the LLM workflow derived variable exactly matches one in ground truth, and (ii) partial true positives, where the workflow derived a variable which is part of one in ground truth. The latter is because variables are often phrasal, and the LLM workflow may label part of the phrase – for example, derived variable "water depth" partially occurs within "maximum water depth" specified in ground truth. We treat as partial, (standalone) variables identified by the workflow which are comprised of at least 2 characters, occurring within a ground truth variable. We specifically investigate partial variables due to the high potential they have to be integrated within an interactive human-machine labelling process.

4.2 Results

Experiment results are summarized in Table 1. We obtain a best accuracy of 0.88 (SWMM case), 0.91 (SWAT case) and 0.90 (MODFLOW case). Accuracy, up to the second digit is the same for both partial and exact matches. Table 1 further shows in detail the number of true positives for exact and partial matches and accuracy achieved over the unique lemmatized words of the source document, for best and worst temperature values.

Table 1. Accuracy of the LLM workflow as a binary classifier. Illustrated is accuracy achieved for best ($\star$) and worst temperature for exact variable matches.

	Temp	True P. (exact)	True P. (partial)	Unique Lemmas	Accuracy
SWMM $\star$	0.4	250	415	11154	0.88
SWMM	0.9	228	403	11154	0.87
SWAT $\star$	0.2	692	1019	18013	0.91
SWAT	0.9	672	992	18013	0.91
MODFLOW $\star$	0.7	223	271	4874	0.90
MODFLOW	0.9	220	263	4874	0.90

Observe that the critical temperature parameter does not significantly affect accuracy – as such, we subsequently investigate performance as manifested in the true positive rate. Specifically, we assess performance as true positives over the ground truth – that is, variables correctly identified against ground truth – for different temperature parameters, and for both exact and partial matches. Our results are illustrated in Fig. 4. Notably, the workflow successfully derives 58% (SWMM), 83% (SWAT) and 59% of the ground truth as partial matches, and 33% (SWMM), 58% (SWAT) and 49% (MODFLOW) as exact matches. Overall results imply a certain trend: moderate to medium temperature values yield the best true positive rate (Fig. 4), while deviation in accuracy between best and worst (Table 1) remains relatively low.

Conversely, Fig. 5 illustrates false positives, as the number of variables discovered for each case which are not relevant and which the model incorrectly identified as relevant. False positive rate generally increases with temperature – behavior in line with the definition of the temperature parameter, where higher values render the LLM more creative with its choices. Notably, for each case, there is a (low) temperature value with which false positives are (relatively) low.

4.3   Discussion and Threats to Validity

We especially note that besides a few examples given for the prompt, this is the output of an automatic procedure utilizing a model not fine-tuned; we believe it illustrates significant future potential and warrants further investigation. However, we acknowledge that due to the generic generative model used, there is a high amount of false positives (Fig. 5), especially against tailored approaches [36]. The juxtaposition of Figs. 5 and 4 shows that a balance between true positives and false positives needs to be achieved regarding the choice of temperature. Fine-tuning the LLM is the next major conceptual direction to investigate to mitigate this.

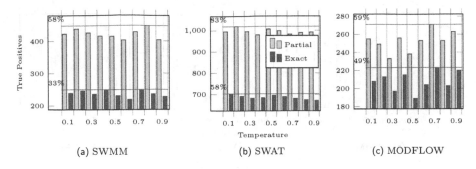

Fig. 4. Overall performance for the three case studies considered as true positives correctly derived by the advocated LLM workflow against ground truth. Illustrated over different temperature parameters, for variables partially identified (gray bars) and exactly identified (black bars). The maximum percentage of ground truth correctly identified in both cases is also marked. (Color figure online)

Regarding operationalization, an invocation of the workflow against a page of the source document takes seconds, rendering the approach fitting to an application where a user interacts with the workflow in an online fashion. Such a human-in-the-loop approach can target a trade-off between manual labelling and full automation. While leveraging partial variable matches for this represents an obvious way forward, one can envision the human expert user correcting model output, especially regarding false positives, and steering the model towards better overall task performance.

Threats to validity of our investigation revolve around the use of the generative LLM, which is the pillar of our approach. LLMs, by design, are trained against huge corpuses of general text [15] presenting a threat to construct validity. Regarding external validity, we note the difference of performance between, e.g., the SWAT case and the MODFLOW case. The conjecture is that natural language, language structure, and variable name choice inherent in a specific document against the LLM (and its training) leads to varying performance. As such, our results may not generalize to other scientific software documentation, although we performed 3 case studies. Additionally, we treated the scientific software document page by page; this represents a first approach, and making use of context of particular text (e.g., same paragraph, same section) is likely to lead to better results. As for threats to conclusion validity and repeatability, we release our analysis data and implementation in the form of a reproduction kit [1].

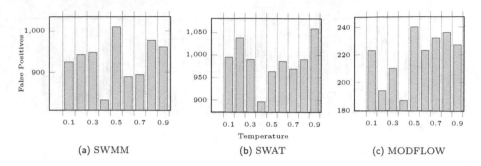

(a) SWMM   (b) SWAT   (c) MODFLOW

Fig. 5. False positive rate illustrated over different temperature parameters.

5   Conclusion and Future Work

Metamorphic testing involves reasoning on necessary properties that a program under test should exhibit regarding multiple input and output variables. A general approach consists of extracting metamorphic relations from auxiliary artifacts such as user manuals or documentation, a strategy particularly fitting to testing scientific software. However, large input-output spaces are common and relations can be complex, hindering the application of metamorphic testing. The fundamental prerequisite consists of extracting input-output variables of interest; subsequently, relations can be devised with the overall goal of enabling testing. By virtue of our workflow design and the preliminary results presented, we believe to have demonstrated the potential that LLMs have for this critical step of variable discovery. Thereupon, we identify key research directions towards model refinement, prompt construction, and human-in-the-loop configurations.

Firstly, larger and more advanced models are very likely to perform better [25]. Naturally, we identify model fine-tuning as the key driver for achieving higher performance [19]. A systematic investigation into more effective prompt construction is warranted, especially towards issues raised by recency bias, common token bias, and majority label bias is a priority (as occurring in the particular problem tackled). Optimizations can include mixing previous model responses in the few-shot prompt, along with the one that is human-specified. This can steer the model into the current context, by providing more information about the topics inherent in neighboring text. Additionally, instead of autoregressive language models, Masked Language Models (MLMs [51]) can be employed. MLMs can be used to predict masked text parts based on its neighboring context [51]. Finally, leveraging partial variable matches represents an obvious way forward for operationalization; one can envision the human expert user correcting model output, especially regarding false positives, and steering the model towards better overall task performance.

Acknowledgements. Funded in part by the Deutsche Forschungsgemeinschaft (DFG, German Research Foundation) - SFB 1404 FONDA.

References

1. Implementation of the LLM-based workflow and reproduction kit (2023). https://seg.inf.unibe.ch/papers/mt-varextract-gpt-0.7.tar.gz
2. Arnold, J.G., Kiniry, J.R., Srinivasan, R., Williams, J.R., Haney, E.B., Neitsch, S.L.: United States department of agriculture. Soil and Water Assessment Tool (SWAT). Texas Water Resources Institute (2012)
3. Balog, M., Gaunt, A.L., Brockschmidt, M., Nowozin, S., Tarlow, D.: DeepCoder: learning to write programs. arXiv preprint arXiv:1611.01989 (2016)
4. Barr, E.T., Harman, M., McMinn, P., Shahbaz, M., Yoo, S.: The oracle problem in software testing: a survey. IEEE Trans. Softw. Eng. **41**(5), 507–525 (2014)
5. Black, S., et al.: GPT-NeoX-20B: an open-source autoregressive language model. arXiv preprint arXiv:2204.06745 (2022)
6. Brown, T., et al.: Language models are few-shot learners. Adv. Neural. Inf. Process. Syst. **33**, 1877–1901 (2020)
7. Carlson, A., Fette, I.: Memory-based context-sensitive spelling correction at web scale. In: International Conference on Machine Learning and Applications, pp. 166–171. IEEE (2007)
8. Carver, J.C., Hong, N.P.C., Thiruvathukal, G.K.: Software Engineering for Science. CRC Press (2016)
9. Carver, J.C., Kendall, R.P., Squires, S.E., Post, D.E.: Software development environments for scientific and engineering software: a series of case studies. In: International Conference on Software Engineering, pp. 550–559. IEEE (2007)
10. Chen, M., et al.: Evaluating large language models trained on code. arXiv:2107.03374 (2021)
11. Chen, T.Y., Feng, J., Tse, T.H.: Metamorphic testing of programs on partial differential equations: a case study. In: International Computer Software and Applications, pp. 327–333. IEEE (2002)
12. Chen, T.Y., Ho, J.W.K., Liu, H., Xie, X.: An innovative approach for testing bioinformatics programs using metamorphic testing. BMC Bioinform. **10**(1), 1–12 (2009)
13. Colavito, G., Lanubile, F., Novielli, N.: Issue report classification using pre-trained language models. In: International Workshop on Natural Language-Based Software Engineering, pp. 29–32 (2022)
14. Ding, J., Zhang, D., Hu, X.-H.: An application of metamorphic testing for testing scientific software. In: International Workshop on Metamorphic Testing, pp. 37–43 (2016)
15. Gao, L., et al.: The pile: An 800gb dataset of diverse text for language modeling. arXiv preprint arXiv:2101.00027 (2020)
16. Han, C., Fan, Z., Zhang, D., Qiu, M., Gao, M., Zhou, A.: Meta-learning adversarial domain adaptation network for few-shot text classification. arXiv preprint arXiv:2107.12262 (2021)
17. Hardin, B., Kanewala, U.: Using semi-supervised learning for predicting metamorphic relations. In: International Workshop on Metamorphic Testing, pp. 14–17. IEEE (2018)
18. Hiremath, D.J., Claus, M., Hasselbring, W., Rath, W.: Towards automated metamorphic test identification for ocean system models. In: International Workshop on Metamorphic Testing, pp. 42–46. IEEE (2021)
19. Howard, J., Ruder, S.: Universal language model fine-tuning for text classification (2018)

20. Jiang, X., et al.: On the importance of attention in meta-learning for few-shot text classification. arXiv preprint arXiv:1806.00852 (2018)
21. Kanewala, U., Bieman, J.M.: Techniques for testing scientific programs without an oracle. In: International Workshop on Software Engineering for Computational Science and Engineering, pp. 48–57. IEEE (2013)
22. Kanewala, U., Bieman, J.M.: Using machine learning techniques to detect metamorphic relations for programs without test oracles. In: International Symposium on Software Reliability Engineering, pp. 1–10. IEEE (2013)
23. Kanewala, U., Bieman, J.M.: Testing scientific software: a systematic literature review. Inf. Softw. Technol. **56**(10), 1219–1232 (2014)
24. Kanewala, U., Bieman, J.M., Ben-Hur, A.: Predicting metamorphic relations for testing scientific software: a machine learning approach using graph kernels. Softw. Test. Verif. Reliab. **26**(3), 245–269 (2016)
25. Kaplan, J., et al.: Scaling laws for neural language models. arXiv preprint arXiv:2001.08361 (2020)
26. Kelly, D., Sanders, R.: The challenge of testing scientific software. In: Annual Conference of the Association for Software Testing, pp. 30–36 (2008)
27. Kelly, D., Sanders, R., et al.: Assessing the quality of scientific software. In: International Workshop on Software Engineering for Computational Science and Engineering (2008)
28. Kelly, D., Smith, S., Meng, N.: Software engineering for scientists. Comput. Sci. Eng. **13**(05), 7–11 (2011)
29. Langevin, C.D., Hughes, J.D., Banta, E.R., Provost, A.M., Niswonger, R.G., Panday, S.: MODFLOW 6 modular hydrologic model version 6.2. 1: US geological survey software release (2021). https://doi.org/10.5066/F76Q1VQV
30. Leser, U., et al.: The Collaborative Research Center FONDA. Datenbank-Spektrum, (1610–1995) (2021)
31. Lin, X., Simon, M., Niu, N.: Exploratory metamorphic testing for scientific software. Comput. Sci. Eng. **22**(2), 78–87 (2018)
32. Lin, X., Simon, M., Niu, N.: Hierarchical metamorphic relations for testing scientific software. In: International Workshop on Software Engineering for Science, pp. 1–8 (2018)
33. Miller, G.: A scientist's nightmare: software problem leads to five retractions. Science **314**(5807), 1856–1857 (2006)
34. Müller, S., Gogoll, V., Vu, A.D., Kehrer, T., Grunske, L.: Automatically finding metamorphic relations in computational material science parsers. In: International Workshop on Software Engineering for eScience (2022)
35. Nguyen-Hoan, L., Flint, S., Sankaranarayana, R.: A survey of scientific software development. In: International Symposium on Empirical Software Engineering and Measurement, pp. 1–10 (2010)
36. Peng, Z., Lin, X., Niu, N., Abdul-Aziz, O.I.: I/O associations in scientific software: a study of SWMM. In: Paszynski, M., Kranzlmüller, D., Krzhizhanovskaya, V.V., Dongarra, J.J., Sloot, P.M.A. (eds.) ICCS 2021. LNCS, vol. 12747, pp. 375–389. Springer, Cham (2021). https://doi.org/10.1007/978-3-030-77980-1_29
37. Peng, Z., Lin, X., Santhoshkumar, S.N., Niu, N., Kanewala, U.: Learning I/O variables from scientific software's user manuals. In: Groen, D., de Mulatier, C., Paszynski, M., Krzhizhanovskaya, V.V., Dongarra, J.J., Sloot, P.M.A. (eds.) ICCS 2022. LNCS, vol. 13353, pp. 503–516. Springer, Cham (2022). https://doi.org/10.1007/978-3-031-08760-8_42
38. Radford, A., et al.: Language models are unsupervised multitask learners. OpenAI Blog **1**(8), 9 (2019)

39. Ren, X., He, W., Qu, M., Huang, L., Ji, H., Han, J.: AFET: automatic fine-grained entity typing by hierarchical partial-label embedding. In: Conference on empirical methods in natural language processing, pp. 1369–1378 (2016)
40. Reynolds, L., McDonell, K.: Prompt programming for large language models: beyond the few-shot paradigm. In: Extended Abstracts, CHI, pp. 1–7 (2021)
41. Rossman, L.A.: Storm water management model user's manual, version 5.0. Cincinnati: National Risk Management Research Laboratory, Office of Research and Development, US Environmental Protection Agency (2010)
42. Sanders, R., Kelly, D.: Dealing with risk in scientific software development. IEEE Softw. **25**(4), 21–28 (2008)
43. Segura, S., Fraser, G., Sanchez, A.B., Ruiz-Cortés, A.: A survey on metamorphic testing. IEEE Trans. Softw. Eng. **42**(9), 805–824 (2016)
44. Segura, S., Towey, D., Zhou, Z.Q., Chen, T.Y.: Metamorphic testing: testing the untestable. IEEE Softw. **37**(3), 46–53 (2018)
45. Su, F-H, Bell, J, Murphy, C., Kaiser, G.: Dynamic inference of likely metamorphic properties to support differential testing. In: International Workshop on Automation of Software Test, pp. 55–59. IEEE (2015)
46. Tsigkanos, C., Rani, P., Müller, S., Kehrer, T.: Large language models: the next frontier for variable discovery within metamorphic testing? In: International Conference on Software Analysis, Evolution and Reengineering, Early Research Achievements (ERA) track. IEEE Computer Society (2023)
47. Vu, A.D., Kehrer, T., Tsigkanos, C.: Outcome-preserving input reduction for scientific data analysis workflows. In: 37th IEEE/ACM International Conference on Automated Software Engineering, ASE 2022, Rochester, MI, USA, 10–14 October 2022, pp. 182:1–182:5. ACM (2022)
48. Wang, B.: Mesh-transformer-JAX: model-parallel implementation of transformer language model with JAX. https://github.com/kingoflolz/mesh-transformer-jax
49. Wang, B., Komatsuzaki, A.: GPT-J-6B: a 6 billion parameter autoregressive language model (2022). https://github.com/kingoflolz/mesh-transformer-jax
50. Weyuker, E.J.: On testing non-testable programs. Comput. J. **25**(4), 465–470 (1982)
51. Xu, F.F., Alon, U., Neubig, G., Hellendoorn, V.J.: A systematic evaluation of large language models of code. In: ACM SIGPLAN International Symposium on Machine Programming, pp. 1–10 (2022)

On Irregularity Localization for Scientific Data Analysis Workflows

Anh Duc Vu[1]([✉])[ID], Christos Tsigkanos[2,3][ID], Jorge-Arnulfo Quiané-Ruiz[4][ID],
Volker Markl[5,6][ID], and Timo Kehrer[2][ID]

[1] Humboldt-Universität zu Berlin, Berlin, Germany
vuducanh@informaXk.hu-berlin.de
[2] University of Bern, Bern, Switzerland
[3] University of Athens, Athens, Greece
[4] IT University of Copenhagen, Copenhagen, Denmark
[5] Technical University of Berlin, Berlin, Germany
[6] DFKI Berlin, Berlin, Germany

Abstract. The paradigm shift towards data-driven science is massively transforming the scientific process. Scientists use exploratory data analysis to arrive at new insights. This requires them to specify complex data analysis workflows, which consist of compositions of data analysis functions. Said functions encapsulate information extraction, integration, and model building through operations specified in linear algebra, relational algebra, and iterative control flow among these. A key challenge in these complex workflows is to understand and act upon irregularities in these workflows, such as outliers in aggregations. Regardless whether irregularities stem from errors or point to new insights, they must be localized and rationalized, in order to ensure the correctness and overall trustworthiness of the workflow. We propose to automatically reduce a workflow's input data while still observing some outcome of interest, thereby computing a minimal reproducible example to support workflow debugging. In essence, we reduce the problem to the determination of the input relevant to reproducing the irregularity. To that end, we present a portfolio of different strategies being tailored to data analysis workflows that operate on tabular data. We investigate their feasibility in terms of input reduction, and compare their effectiveness and efficiency within three characteristic cases.

1 Introduction

Computationally-intensive research methods exploiting large amounts of data are becoming ubiquitous in numerous scientific disciplines [12]. During the scientific process, researchers leverage exploratory data analysis to analyze, manipulate, and investigate data sets in order to apply statistical techniques, spot anomalies, test hypotheses, or check assumptions. Typically, this involves the composition of heterogeneous collections of data processing functions (e.g., data integration, normalization, and filtering) into complex data analysis workflows [16]. Exploratory analysis in scientific computing often yields results for

© The Author(s), under exclusive license to Springer Nature Switzerland AG 2023
J. Mikyška et al. (Eds.): ICCS 2023, LNCS 14073, pp. 336–351, 2023.
https://doi.org/10.1007/978-3-031-35995-8_24

which it is hard to judge whether they are correct or not [23]. Irregularities in workflow results could point to new interesting insights, or may be caused by errors in the workflow or corrupt input data [5]. Either way, one has to perform data debugging [3] with the goal of finding the cause of irregularities in order to increase the trustworthiness in the workflow [21].

Many researchers pursued debugging and validation in data processing systems from different angles, notably by allowing the inspection of intermediate [8] or sub-results [6], or tracking data provenance [2,13,14] for dataflow systems like Apache Spark or Flink. What is common among these data provenance and debugging approaches is that they assume that the workflow is a white-box: they must be able to follow and label tuples through the entire process. However, this is far from reality in scientific data analytics, where workflows are mostly composed of black-box processes. Moreover, classical fault localization techniques, e.g., [1,9,26], assume that suspicious behavior manifests in an error that can be identified by a pre-defined test. Nevertheless, scientific data analytics is explorative by nature, making it often impossible to define tests that specify correct behavior in terms of expected outcomes [15,19].

Conversely, we aim at finding the cause of an irregularity in a workflow's input data by computing a *minimal reproducible example* that produces the irregularity. The general idea is to iteratively reduce the input data to ease the task of finding the cause of the output's irregularity. As opposed to classical fault localization, we refer to this process as *irregularity localization*, emphasizing the fact that the outcome investigated may or may not be faulty.

Recently, we proposed *outcome-preserving input reduction* as a generic approach to support irregularity localization in data analysis workflows [24]. The idea is to iteratively reduce the workflow's input data while still observing some outcome of interest in its results. While we do not ask developers for pre-defined tests, we assume that, by looking at the workflow's result, they may specify a debugging question that characterizes the outcome of interest. For instance, the expressions "($result.Time$ = "07h00" $\land result.Count \geq 166$)" may be questioning a counted number in an aggregation. Then, the workflow may be executed on a reduced input, and the debugging question is evaluated on its output data in order to check whether a reduction is outcome-preserving and thus permitted. For example, whether "($result.Time$ = "07h00" $\land result.Count \geq 166$)" still holds in the new result. We intend to perform this in iterations until we reach a certain fixpoint (e.g., a degree of reduction or the consumption of a given resources budget), thus obtaining a minimal reproducible example as a result.

While the framework advocated in [24] is generally applicable, as it abstracts from workflow implementations and execution engines, it needs to be instantiated by providing implementations of its components. In particular, implementing a reducer facility is challenging. The large amount of data and the limited information about the workflow's underlying mechanism make it difficult to distinguish the relevant parts of the input data and require careful selection strategies for the reduction attempts. In this paper, we present a portfolio of reduction strategies tailored to scientific data analysis workflows, making the following contributions:

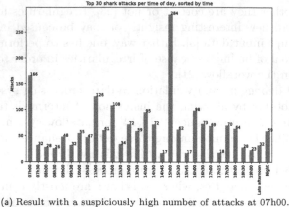

(a) Result with a suspiciously high number of attacks at 07h00.

(b) Data profile of a minimal reproducible example.

Fig. 1. An example workflow that aggregates shark attacks per time of day.

- We present a concrete instantiation of a general framework for irregularity localization in data analysis workflows (Sect. 3);
- We introduce a portfolio of strategies targeting the reduction of tabular data. In particular, we propose Similarity-based Isolation, a data reduction strategy that isolates those tuples, together with all their similar tuples, having an effect on the outcome of interest (Sect. 4).
- We evaluate the proposed strategies experimentally over three different cases with respect to feasibility and their individual performance (Sect. 5);
- We provide a replication package allowing our results to be reproduced.

2   Motivation and Background

Consider a data analysis workflow implemented as a computational notebook taken from Kaggle[1] over a dataset comprising shark attacks worldwide in the last 100 years, where the workflow seeks to figure out the number of attacks per time of the day. Figure 1a shows a plot of the workflow's result. Notably, for 07h00 the number of attacks appears to be quite high compared to other times of frequent attacks, which are around noon and afternoon – although the peak at 15h00 is also high, the attacks at 07h00 are at a suspicious time. It may certainly be the case that the data indeed show a high amount of attacks for that time; however, there may also be an error or wrong assumption in the analysis performed to produce this result, or the data may be corrupt. Either way, there is an aspect of the result which is suspicious, i.e., showing an *irregularity* which we should investigate further. In [24] we argued that such investigation of suspicious results within data analysis workflows must accommodate the specific needs of the scientists developing these workflows. Like the scientific discovery

---

[1] kaggle.com/mysarahmadbhat/shark-attacks.

process itself, the investigation should be done in an explorative manner. Our example represents a characteristic case where exploratory analysis is required to deduce if the data indeed support shark attacks occurring early in the morning. Intuitively, one would seek to spot the cause of the suspicious outcome with the help of a minimal example of the input, referred to as *irregularity localization*.

The dataset of our motivating example contains about 25k rows and 24 columns, rather small compared to what is typically processed within scientific data analytics, but still overwhelming for a human being. Thus, it is desirable to reduce the input dataset to the minimum set of rows and columns of the input table that reproduce the outcome of interest. We refer to this task as *outcome-preserving input reduction*. For our example case, it turns out that we require only a minor fraction of the input dataset to enable the workflow to reproduce the suspicious result. Figure 1b depicts a profile of such a reduced input dataset, comprising only 166 tuples and one column that are needed to reproduce the peak of irregularly many shark attacks at 07h00. Observe that only column 'Time' has been deemed relevant, and the relevant tuples take the values of 'Morning', 'Evening' and '07h00', which is surprising and would have been difficult to find manually. This *minimal reproducible example* provides a strong hint to the user; some data transformation applied within the workflow did yield these irregular values, which needs to be investigated. For this particular example workflow, the seemingly high number of attacks is caused by a data cleaning function that attempts to map textual reports to clock time. Apparently, the workflow interprets the string values "Morning" and "Evening" to 7 o'clock (which is a questionable assumption made by the workflow developer), and it misses to distinguish a.m. from p.m. (which is clearly an error in the workflow's implementation).

Without this automated reduction, users may try to query the input data by selecting the tuples where 'Time = 07h00' to investigate their suspicion on the workflow's output. The result, however, would be that only 21 tuples of the input that actually have the value '07h00' would show up – as seen in Fig. 1b – giving the user no indication where the remaining 145 attacks come from. More generally, the key challenge addressed by our approach is to support irregularity localization in cases where the suspicious parts of the output are a result of the workflow's execution, but they cannot be easily spotted in the input data, rendering any ad-hoc analysis of the input data infeasible – especially when the workflow contains tasks that change the data before processing it.

2.1 General Framework for Outcome-Preserving Input Reduction

A high-level vision of how to systematically support irregularity localization through outcome-preserving input reduction for data analysis workflows is illustrated in Fig. 2. A scientist initiates a workflow by submitting some input data, yielding an output (marked as (1)). On further inspection by the scientist, certain parts or aspects of the output data may be of particular interest, perhaps because they are perceived as suspicious or exhibit irregularities that the scientist did not expect and lacks a reasonable explanation for (2).

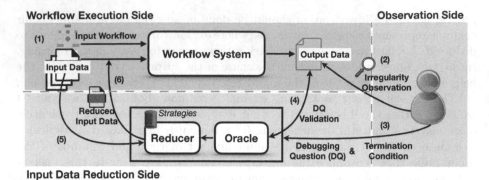

Fig. 2. Outcome-preserving input data reduction for irregularity localization.

A debugging question – in essence a query specified by the user that formally describes the outcome of interest – together with a termination condition – e.g. minimality of the input – is then supplied to the input reduction facility (3). The debugging question is used by the Oracle component to generate an assertion to answer whether the outcome specified by the question holds on the output data (4). The difference to classical test oracles in software testing is that, in general, there is no assessment on whether the specified outcome is correct or not. Moreover, whereas tests are specified beforehand to define expected outcomes, debugging questions are meant to be used ad-hoc on any observed outcome, making them tools for explorative investigation.

The Reducer component operates upon the input data facilitating its reduction strategies in order to yield a subset of it (5). The analysis workflow may be triggered again with the reduced input, producing a new output (6). Thereupon, again the oracle is employed to decide if the outcome of interest is still observable in the new output data. This process is repeated iteratively until the termination condition holds (4)–(6). The result (at each iteration) is a reduced input intended to aid understanding the circumstances involving the irregularity that is sought to be investigated.

3   Instantiation of the General Framework

To serve as a practical solution, the general framework previously introduced needs to be instantiated by providing concrete implementations of (i) the formalism for specifying debugging questions to be interpreted by an oracle, and (ii) the strategy of the reducer. Both heavily depend on the structure of the workflow's input and output. In this paper, we focus on tabular data, widely used in data science and scientific computing. We assume that a workflow takes a single dataset as input and produces a single output dataset. Both input and output data are in the shape of a table, comprising arbitrary rows over a fixed number of columns whose data types are defined by a schema.

Debugging Question and Oracle. When a user wishes to investigate their suspicion regarding a result, the manner on how they express and formalize the observed irregularity poses a challenge. Assuming the output is in the form of a table, we propose to use an "SQL-like" query to characterize the suspicious parts of the table. With ease of use and familiarity in mind, we select the query syntax of the python software library pandas[2]. This allows for a smoother adoption of the technique advocated, as it eliminates the need for most users to learn yet another new syntax. In our motivating example, for the unusual high number of shark attacks at 7 o'clock in the workflow's result, we may formulate the pandas query '(Time == "07h00" & Count == 166)'. An oracle evaluates the debugging question on the output table, obtaining a certain number k of output tuples. We store this amount of output tuples when the debugging question is evaluated on the output table obtained from running the workflow on the original input data set. Later on, after every iteration, the same query is then evaluated on the output table obtained from running the workflow on a reduced input data set. The outcome of interest is preserved if the query yields the same amount of k output tuples.

Outcome-Preserving Input Data Reduction. Reducing an input dataset to a minimal one where any smaller subset does not yield the outcome of interest anymore presents a conceptual challenge. Without any assumptions on the workflow behavior, finding such a minimal input dataset requires an exhaustive search, which is infeasible for real-world datasets comprising a large number of tuples. We can ease the problem by only requiring a local minimum which, however, still requires testing its exponentially many subsets. A similar problem that suffers from such a combinatorial explosion has been described by Zeller et al. [26] in the context of test minimization. To keep the search for a local minimum tractable, even when the search space is large, an approximation based on the notion of n-minimality is defined which we adapt for our setting as follows:

Definition 1 (n-minimality). *Let I be a set of tuples, A be a subset of I, and is_preserving : $2^I \to \{True, False\}$ be an evaluation function with is_preserving(A) = True. A is n-minimal if for all $B \subset A$ it holds that if $|B| \leq n$ then is_preserving(A - B) = False.*

In the use case of test minimization, [26] argues for *1-minimality* being sufficient to aid the developer to investigate what exactly causes the test case to fail. We adopt a similar viewpoint for irregularity localization. If the user is presented with a reduced input dataset where each tuple is necessary to reproduce results obeying the outcome of interest, they can either: (i) surmise that the parts of the workflow processing the data in question are not behaving as intended, or (ii) improve their understanding of the workflow for further validation.

4   Investigated Reduction Strategies

We now describe the algorithms employed to find *1-minimal* outcome preserving sets. We first review some baseline strategies: leave-one-out, dd-min, prob-dd;

---

[2] https://pandas.pydata.org/.

and afterwards describe a novel strategy based on similarity search and fault isolation (`similarity-iso`).

4.1 Baseline (Leave-One-Out)

To find a 1-minimal set, a naive strategy consists of iteratively removing a single data element from the input data set, running the workflow, and consulting the oracle of whether the outcome of interest is preserved. This step is applied to all the elements from the input data set until an element removal changes the outcome, and the next iteration starts until no more elements can be removed. In our approach, it serves as the baseline upon which the strategies outlined in the following are to be compared.

4.2 Delta Debugging (dd-min)

Our second strategy is based on the principle of *Delta Debugging*. Specifically, we adapt `ddmin` [26], a classical fault localization technique originally proposed for test case minimization. `ddmin` is based on a binary search: we start by splitting the input space into two halves and testing each of them whether they can be removed from the input without changing the outcome. If one of the halves can be removed, we found a new smaller dataset that still preserves the outcome. We continue halving the smaller but still outcome preserving input partition. If neither of the halves preserve the outcome, the granularity is increased to remove quarters of the dataset. This is done until the granularity is so small that we try to remove single elements from the dataset, at which point it becomes the leave-one-out strategy.

4.3 Probabilistic Delta Debugging (prob-dd)

Wang et. al. [25] pointed to the weakness of the `ddmin` algorithm in not taking advantage of past executions while iterating, but only following a pre-determined schedule of testing subsets. To ameliorate this, they devised *Probabilistic Delta Debugging*; the improvement consists of using a probabilistic model to guide the selection which is updated after each test. In such a model, each element of the input is assigned a random variable that decides whether that element will be in the reduced dataset or not. We refer to the original paper [25] for technical details.

4.4 Similarity-Based Isolation (similarity-iso)

The next strategy is constructed with the objective of better localizing elements of interest than `ddmin`. It is based on the intuition that similar elements – by means of a distance function – will cause similar effects on the outcomes of a computation. This strategy is a derivative of the *dd* fault isolation algorithm [26], which computes an *n-minimal-difference* between two sets.

Definition 2 (n-minimal-difference). *Let I be a set of data tuples; A, B are subsets of I, and is_preserving : $2^I \rightarrow \{True, False\}$ is an evaluation function. Given that $B \subset A$, is_preserving$(A) = True$, and is_preserving$(B) = False$, then $A - B$ is an n-minimal-difference if for all $C \subseteq (A - B)$ it holds that if $|C| \leq n$ then is_preserving$(A - C) = False$ and is_preserving$(B \cup C) = True$.*

Algorithm 1: Similarity-based Isolation

    Data: input
    Result: reduced_input

1 min_preserving ← input
2 isolated ← ∅
3 **while** *True* **do**
      /* Isolate a 1-minimal difference */
4   max_changing ← ∅
5   Δ ← (min_preserving − max_changing) − isolated
6   **while** $|\Delta| > 1$ **do**
7   test ← (*get_subset*$(\Delta, \frac{1}{2}|\Delta|)$ + max_changing) + isolated
8   **if** *is_preserving(test)* $= True$ **then**
9 | min_preserving ← test
10   **else**
11 | max_changing ← test
12   Δ ← (min_preserving − max_changing) − isolated
13   **if** $\Delta = \emptyset$ **then**
        /* No new tuples can be isolated */
14   *break*
15   **else**
        /* Isolate Δ and all similar tuples in min_preserving */
16   isolated ← isolated + Δ + *get_similar_elements*(min_preserving, Δ)

17 *return* ddmin(isolated)

Our strategy exploiting the notions of a 1-minimal difference and similarity of data tuples is illustrated in Algorithm 1. The outer while loop (lines 3 to 16) maintains two sets. min_preserving refers to the smallest input dataset that preserves the outcome of interest; it is initialized with the original input dataset (line 1) and constantly reduced within each iteration. isolated is initialized with the empty set (line 2) and accumulates those input tuples which, presumably, have an effect on the outcome. In each iteration, min_preserving and isolated are updated in two steps: First, we try to find a single tuple that is not yet included in isolated (lines 4–12) but causes a different outcome if removed. Second, if such a tuple can be found, this tuple as well as all similar – by some distance function, e.g. levenshtein – tuples in min_preserving that presumably also change the outcome are added to isolated (line 16). This process of alternating isolation and similarity search is repeated until no new elements can be isolated (line 14) and a fixpoint is reached. The procedure finishes by running any of the other minimization strategies – currently ddmin – to find a *1-minimal* solution (line 17).

Isolation: The isolation step (lines 4–12) is realized by isolating a 1-minimal difference, referred to as Δ, between min_preserving and max_changing, a set that accumulates the currently largest set of input tuples changing the outcome of interest. Since we can safely assume that the empty set produces a radically different result than the original input data set and thus changes the outcome of interest, max_changing is initialized with the empty set (line 4). The set Δ is initialized as the set difference between min_preserving and max_changing, from which we also remove those tuples that have been isolated in previous iterations (line 5). Then, in each iteration of the inner while loop (lines 6–12), half of the elements are taken from Δ, and unified with the currently largest outcome-changing set max_changing and the already isolated elements in isolated (line 7). For the obtained set of input tuples, referred to as test, we run the workflow and let our oracle decide whether the outcome of interest is preserved (line 8). If so, a smaller outcome-preserving set has been located (line 9). Otherwise, a new largest outcome changing set has been located (line 11). We proceed with the next iteration of the inner while loop, as long as Δ contains more than one tuple (line 6). The procedure takes logarithmic many steps to isolate an *1-difference*.

Similarity Search: The similarity search (line 16) follows the intuition that similar data is likely to have similar effects on the outcome of a computation. Given our assumption of tabular data, we define the similarity of two data records as the average similarity over each column value For each column value, we employ classical distance metrics, depending on the column's data type. In the base case, we currently use normalized Euclid distance for numerical values, Levenshtein distance for strings, and equality for categorical data.

5   Evaluation

We conducted experiments to assess how suitable it is to perform outcome-preserving input reduction with our framework and the aforementioned strategies. Our evaluation[3] addresses the following research questions: (**RQ1**) How feasible is a search for 1-minimal datasets for outcome-preserving input reduction?, and; (**RQ2**) How do the reduction strategies compare to each other with respect to their resource consumption? To answer RQ1, we consider measures from the user's perspective – namely the degree of reduction and the overall time it takes to do so. For RQ2, we investigate the runtime properties of the reductions, measuring the number of times the workflow is executed by a strategy, and the time a strategy requires between each of these iterations.

---

[3] Our replication package can be found at https://osf.io/fk2x4/?view_only=442434edaec94c2b8172a759699d0886.

5.1   Experimental Setup

We adopt three workflows derived from publicly available computational note-books, referred to as Shark (described in Sect. 2), NBA (a workflow analyzing NBA players), and FEC (a workflow analyzing the 2012 US federal election commissions data).

The NBA case is an erroneous notebook from a study about failure identification strategies in Jupyter notebooks [20], which entails explorative analysis on a dataset about NBA players – study participants were tasked with fixing various errors. At some point the NBA players are grouped into point and shooting guards, and their average height must be calculated. The study mentions that one of the participants remarked that a "mean height of 12 doesn't seem to make a lot of sense", likely referring to an expected length unit of centimeters or meters. As such, we adopt the parts of the NBA notebook calculating the average height, and apply our search strategies with the debugging query (Position == "PG" ∧ Height < 100) to check why point guards appear with such low height. The dimensions of this dataset are 1184 rows, and 13 columns.

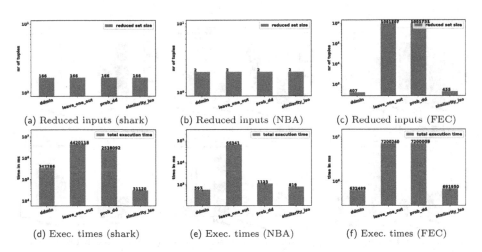

(a) Reduced inputs (shark) (b) Reduced inputs (NBA) (c) Reduced inputs (FEC)

(d) Exec. times (shark) (e) Exec. times (NBA) (f) Exec. times (FEC)

Fig. 3. Size of reduced input in no. of tuples and total execution time in ms (log scale).

The third case concerns the US federal election commissions data from 2012 and is taken from the book "Python for Data Analysis, 3rd Edition"[4]. The input dataset consists of about 1M rows and 16 columns. One of the results of this analysis yields a table listing the donations by state to candidates Barack Obama and Mitt Romney. Of interest is that for the state Arkansas, Obama received about 77% of the donations from individuals, more than three times as much as Romney. However, in the election of 2012, Arkansas was won by

---

[4] wesmckinney.com/book/data-analysis-examples.html.

the Republican party with more than 60% of the votes – a possible irregularity. Therefore, we use the following debugging question to investigate: '*contbr_st*' == "*AR*" ∧ '*Obama, Barack*' > 0.77 ∧ '*Obama, Barack*' < 0.78.

We proceed to apply the reduction strategies of Sect. 4. Experiments were performed on an Intel E7-4880 2.5GHz CPU. Each experiment was repeated 10 times; measurements reported in the sequel are average values.

5.2   Results

The issue that causes the Shark workflow to calculate a surprisingly high number of attacks has been already discussed in Sec. 2. A minimal example to reproduce this result is a dataset that comprises the 166 tuples taking the values "7h00", "Morning", and "Evening" in the column 'Time'.

The low average height calculated by the NBA workflow is caused by a data cleaning step which neglects that players' heights are given not only in centimeters but also in inches. So any input set of size 2 where at least one of the elements has a height value in inches is sufficient to reproduce the outcome specified.

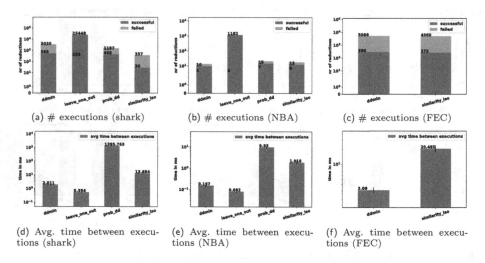

(a) # executions (shark)   (b) # executions (NBA)   (c) # executions (FEC)

(d) Avg. time between executions (shark)   (e) Avg. time between executions (NBA)   (f) Avg. time between executions (FEC)

Fig. 4. Number of workflow executions and average time between executions (log scale).

The FEC workflow was obtained with no fault injection – a reduced input set should confirm that the analysis is as intended, illustrating that the presented approach is not only useful for irregularity localization but also for validation by providing a smaller dataset.

RQ1. Sizes of the reduced datasets and the time needed to conclude the reductions are reported in Fig. 3. Note that the y-axes are scaled logarithmically. For the Shark and NBA workflow, each strategy found 1-minimal input sets of the same size, namely 166 and 2 tuples, which corresponds to reductions of 99,5% and 99.8% compared to the original input sizes of 25614 and 1184, respectively. Regarding the overall time needed to find these 1-minimal sets, `leave-one-out` took the longest, with a factor of almost 100 compared to the other startegies in case of the NBA workflow. Also `Prob-dd` was clearly outperformed by `ddmin` and `similarity-iso` on both cases. For the shark workflow, `prob-dd` took around 40 min, `ddmin` required about 6 min, and `similarity-iso` was the clear winner requiring less than a minute. For the NBA case, `ddmin` requires around 600ms, which is about 25% less than the second best strategy `similarity-iso`. In the FEC case, we observe that `ddmin` with a runtime of 8 min is about 1 min faster than `similarity-iso` and produces a dataset about 10% smaller. `leave_one_out` and `prob-dd` are not competitive at all within a time-out of 2 h. Because the number of tuples in the FEC dataset is so high, `leave_one_out` is getting timed out because it is only removing a single tuple per iteration, and `prob-dd` takes so long to update probabilities for each element that it also times out before it can finish. In the sequel, we will thus not consider these strategies for the FEC workflow evaluation, and they are omitted from Fig 4f and Fig. 4c, respectively.

RQ2. As observed, there are differences in how long the various strategies require to find a solution. Figure 4 illustrates RQ2-related measurements, again noting that the y-axes are log-scaled. Figure 4a shows the number of reductions that were attempted by each strategy for the shark workflow, divided by successful and failed attempts – the sum represents the total number of workflow executions. Unsurprisingly, the longest running strategy `leave-one-out` also executed the workflow the most, while the fastest strategy `similarity_iso` required the least amount of attempts. Interestingly, even though `prob-dd` only took about half as many reduction attempts than `ddmin`, it took nearly 7 times as long. For an explanation, observe Fig. 4d, depicting the average execution time between workflow executions. This metric records the average time each strategy requires to decide on the next reduction candidate based on the result of the previous execution. For this metric, we observe that `prob_dd` requires considerably more time to choose the next candidate, but with the payoff of needing only half as many executions than `ddmin`. The same can be observed with `similarity-iso`, which takes more than 6 times as much to choose the next potential solution than `ddmin`, but requiring 10 times less workflow executions. For the NBA case, Fig. 4e illustrates that the longest running strategy also executed the workflow the most, with `leave_one_out` doing so almost more than 100 times as much as the others. Again, `prob-dd` takes considerably more time between workflow executions. `ddmin` requires the least amount of executions. Paired with a low overhead between the executions, this allows `ddmin` to be faster. The FEC workflow shows that even though `similarity-iso` did less iterations than `ddmin`, the higher time between iterations cause `similarity-iso` to perform a bit slower.

5.3   Summary and Threats to Validity

Our results, when viewed *ex post facto*, illustrate some key findings. Firstly the baseline strategy `leave-one-out` performs the worst and is impractical, especially for a dataset like FEC that has millions of tuples – as expected for a strategy that removes only a single element each iteration. Secondly, when relevant data (and resp., irrelevant data) is spatially grouped together, a simple partitioning strategy like ddmin can easily remove unnecessary elements. However, in the shark example, where relevant tuples are scattered, ddmin underperforms. Instead, our novel strategy similarity-iso is able to isolate the scattered data quickly, vastly outperforming ddmin in those cases, while still being competitive with ddmin's best cases by only adding minimal overhead. prob-dd is performant at choosing promising candidates while simultaneously being generically applicable. However, the amount of time and space needed to update element probabilities can be costly when the number of elements is large, which we observed in the FEC workflow, where the high number of tuples caused prob-dd to timeout. Overall, our results show that similarity search is rather balanced and performant, concluding up to 10 times faster in cases that pose difficulties for the others, while where ddmin outperforms, similarity-iso is only marginally slower while still requiring less iterations. The above strengthen the argument that similarity-iso offers the best tradeoff and has the strongest potential for further big data refinements.

Regarding threats to validity, risk of internal validity mainly lies in the correctness of implementation (which we release along with a reproduction kit). The core implementations of ddmin and prob-dd are based on their respective original sources with instrumentation for measurements and time-outs. Risk to construct validity lies in the fact that the debugging questions, which are central to the approach, have been specified by us. But, since this technique is mainly supposed to be a tool for the exploration of the workflow behaviour, any debugging question that specifies a property of the result could have been taken to showcase the reduction strategies. As such, finding the "correct" debugging question, is out of the scope of this work. Another possible point is the number of experiment repetitions; three of the four strategies however are deterministic, and prob-dd's main hurdle is the amount of processing between iterations. The threat to external validity lies in the subject selection – vastly different cases, e.g., where tuple similarity is difficult to be defined, may yield different results. However, it may be argued that the cases presented are representative of typical workflows utilizing tabular data.

6   Related Work

Much of the research on scientific workflows has focused on optimizing for speed and resource utilization, led by high-performance computing [17,22]. However, with workflows becoming ubiquitous, there is a changing mindset that human productivity arguably still is the most expensive resource [4].

Artemis [11] and Nautilus [10] are systems that provide explanations of why or why not certain tuples are in the result, by modifying sub-query operators or by inserting tuples into the data. In our black box setting we cannot modify the individual operators of a workflow and we do not address why a result is missing but rather why a particular result is there using only the input data.

So-called why-provenance [2] is popular to answer why a specific tuple appears in the result, representing the set of input tuples that are responsible for its computation. Ikeda et al. [13] demonstrate how to utilize provenance to debug workflows by enabling forward tracing of input tuples and backward tracing of result tuples. Titian [14] enables collection of provenance for Apache Spark by tracking records through the various data operators. TagSniff [3] is a data debugging model for big data that allows users to efficiently capture data provenance. However, given the black-box nature of scientific workflow tasks, the above works are not applicable because they require white-box access to either track or reverse calculate the relevant tuples. Gulzar et. al. [7] also remark that provenance often returns excessively much data, and in the worst case the whole input. To tackle this, their BigSift approach combines provenance with delta debugging to find a minimal set that leads to a test failure, which relies on a test suite that cannot be generally assumed to be available.

BugDoc [18] can find the responsible component in a pipeline when given an oracle, by iteratively re-executing it with different instantiations. We note that our technique can be used after BugDoc identifies the responsible input data, to reduce it to a minimal reproducible example for further cause localization.

7   Conclusion and Future Work

Motivated by the need to support exploratory analysis within scientific computing, we proposed to iteratively reduce a workflow's input data while still observing some outcome of interest to produce a minimal reproducible example – in essence, we determine the input relevant to reproducing the irregularity. To that end, we presented a portfolio of reduction strategies applicable to tabular data, the shape most often encountered by scientist users. We further investigated the strategies' input reduction and resource consumption over three case studies. To realize an end-to-end framework, we identify open challenges at different levels of abstraction, which provide avenues for future work. Firstly, we considered tabular data; in general, data shape as well as processing and size naturally affect choice and development of reduction strategies, yielding tradeoffs that should be assessed. Thus, future work should investigate the applicability on larger sizes of data, and also data across different domains like imaging and genomic data for example. Secondly, regarding the debugging questions, it is true that the burden of formulating them still lies on the user. We argue that our use case is slightly easier than the well known oracle problem from software testing, where a user has to specify correct behaviour upfront. In the presented use case, the user observes already computed results from the workflow and may formulate properties that this data has right now, as opposed to properties that all results should have. Thus, instead of thinking about all possible results, the user only has to consider the currently observed data. Still, the important step of the debugging question specification should be investigated; users should be supported effectively in their formulation in subsequent works.

Acknowledgements. Funded in part by the Deutsche Forschungsgemeinschaft (DFG, German Research Foundation) - SFB 1404 FONDA.

References

1. Abreu, R., Zoeteweij, P., Van Gemund, A.J.: On the accuracy of spectrum-based fault localization. In: Testing: Academic and Industrial Conference Practice and Research Techniques. IEEE (2007)
2. Buneman, P., Khanna, S., Wang-Chiew, T.: Why and where: a characterization of data provenance. In: Van den Bussche, J., Vianu, V. (eds.) ICDT 2001. LNCS, vol. 1973, pp. 316–330. Springer, Heidelberg (2001). https://doi.org/10.1007/3-540-44503-X_20
3. Contreras-Rojas, B., Quiané-Ruiz, J., Kaoudi, Z., Thirumuruganathan, S.: TagSniff: simplified big data debugging for dataflow jobs. In: ACM Symposium on Cloud Computing, pp. 453–464. ACM (2019)
4. Deelman, E., et al.: The future of scientific workflows. J. High Perform. Comput. Appl. **32**(1) (2018)
5. Galhotra, S., Fariha, A., Lourenço, R., Freire, J., Meliou, A., Srivastava, D.: DataExposer: exposing disconnect between data and systems. arXiv preprint arXiv:2105.06058 (2021)
6. Grust, T., Kliebhan, F., Rittinger, J., Schreiber, T.: True language-level SQL debugging. In: International Conference on Extending Database Technology (2011)
7. Gulzar, M.A., Interlandi, M., Han, X., Li, M., Condie, T., Kim, M.: Automated debugging in data-intensive scalable computing. In: Symposium on Cloud Computing (2017)
8. Gulzar, M.A., et al.: BigDebug: debugging primitives for interactive big data processing in spark. In: ICSE. IEEE (2016)
9. Heiden, S., et al.: An evaluation of pure spectrum-based fault localization techniques for large-scale software systems. Softw. Pract. Exp. **49**(8), 1197–1224 (2019)
10. Herschel, M., Eichelberger, H.: The nautilus analyzer: understanding and debugging data transformations. In: International Conference on Information and Knowledge Management, pp. 2731–2733 (2012)
11. Herschel, M., Hernández, M.A.: Explaining missing answers to SPJUA queries. Proc. VLDB Endow. **3**(1–2), 185–196 (2010)
12. Hey, A.J., Tansley, S., et al.: The Fourth Paradigm: Data-intensive Scientific Discovery, vol. 1. Microsoft Research (2009)
13. Ikeda, R., Cho, J., Fang, C., Salihoglu, S., Torikai, S., Widom, J.: Provenance-based debugging and drill-down in data-oriented workflows. In: International Conference on Data Engineering. IEEE (2012)
14. Interlandi, M., et al.: Titian: Data provenance support in spark. In: Proceedings of VLDB, vol. 9 (2015)
15. Kanewala, U., Bieman, J.M.: Testing scientific software: a systematic literature review. Inf. Softw. Technol. **56**(10), 1219–1232 (2014)
16. Leser, U., et al.: The Collaborative Research Center FONDA. Datenbank-Spektrum (1610–1995) (2021)
17. Lin, B., et al.: A time-driven data placement strategy for a scientific workflow combining edge computing and cloud computing. IEEE Trans. Industr. Inform. **15**(7), 4254–4265 (2019)
18. Lourenço, R., Freire, J., Shasha, D.: BugDoc: a system for debugging computational pipelines. In: Proceedings of the 2020 ACM SIGMOD (2020)

19. Pimentel, J.F., Murta, L., Braganholo, V., Freire, J.: A large-scale study about quality and reproducibility of Jupyter notebooks. In: Internatonal Conference on Mining Software Repositories. IEEE (2019)
20. Robinson, D., Ernst, N.A., Vargas, E.L., Storey, M.A.D.: Error identification strategies for python Jupyter notebooks. arXiv preprint arXiv:2203.16653 (2022)
21. Sanders, R., Kelly, D.: Dealing with risk in scientific software development. IEEE Softw. **25**(4), 21–28 (2008)
22. Shirvani, M.: A hybrid meta-heuristic algorithm for scientific workflow scheduling in heterogeneous distributed computing systems. Eng. Appl. Artif. Intell. **90**, 103501 (2020)
23. Vogel, T., Druskat, S., Scheidgen, M., Draxl, C., Grunske, L.: Challenges for verifying and validating scientific software in computational materials science. In: International Workshop on SE for Science. IEEE (2019)
24. Vu, A.D., Kehrer, T., Tsigkanos, C.: Outcome-preserving input reduction for scientific data analysis workflows. In: International Conference on Automated Software Engineering, New Ideas and Emerging Results (2022)
25. Wang, G., Shen, R., Chen, J., Xiong, Y., Zhang, L.: Probabilistic delta debugging. In: ESEC/FSE (2021)
26. Zeller, A., Hildebrandt, R.: Simplifying and isolating failure-inducing input. IEEE Trans. Softw. Eng. **28**(2), 183–200 (2002)

RAFEN – Regularized Alignment Framework for Embeddings of Nodes

Kamil Tagowski[(✉)] , Piotr Bielak , Jakub Binkowski ,
and Tomasz Kajdanowicz

Department of Artificial Intelligence, Wrocław University of Science and Technology,
Wrocław, Poland
kamil.tagowski@pwr.edu.pl

Abstract. Learning representations of nodes has been a crucial area of the graph machine learning research area. A well-defined node embedding model should reflect both node features and the graph structure in the final embedding. In the case of dynamic graphs, this problem becomes even more complex as both features and structure may change over time. The embeddings of particular nodes should remain comparable during the evolution of the graph, what can be achieved by applying an alignment procedure. This step was often applied in existing works after the node embedding was already computed. In this paper, we introduce a framework – RAFEN – that allows to enrich any existing node embedding method using the aforementioned alignment term and learning aligned node embedding during training time. We propose several variants of our framework and demonstrate its performance on six real-world datasets. RAFEN achieves on-par or better performance than existing approaches without requiring additional processing steps.

Keywords: Dynamic Graphs · Graph Embedding · Embedding Alignment · Graph Neural Networks · Link prediction · Machine Learning

1 Introduction

In recent years, representation learning on structured objects, like graphs, gained much attention in the community. The wide application range, which includes recommender systems, molecular biology, or social networks, motivated a rapid development of new methods for learning adequate representations of such data structures. Especially real-world data often involves accounting for temporal phenomena, which requires handling structural changes over time (i.e., dynamic graphs).

Researchers widely use the embedding of dynamic graphs in a discrete-time manner, i.e., the graph data is aggregated into snapshots and embeddings are computed for each one of them. In opposite to the continuous-time approach, one does not require to provide an online update mechanism, thus allowing one

© The Author(s), under exclusive license to Springer Nature Switzerland AG 2023
J. Mikyška et al. (Eds.): ICCS 2023, LNCS 14073, pp. 352–364, 2023.
https://doi.org/10.1007/978-3-031-35995-8_25

to utilize any classical static embedding approaches or temporal embedding. On the other hand, the discrete-time approach requires an embedding aggregation method that transforms the sequence of node snapshot embeddings into a single node embedding matrix which captures the whole graph evolution. Unfortunately, due to the stochastic properties of node embedding methods, such embeddings are algebraically incomparable. To solve this problem, most existing approaches apply a post-hoc correction [1,14–16].

In this work, we consider a different scenario where we build aligned node embeddings by introducing an additional regularization term to a model's loss function. This term explicitly aligns embeddings of corresponding nodes from consecutive timesteps, which eventually helps to improve results on a downstream task. In particular, we summarize the contributions of the paper as follows:

C1. We propose a novel framework RAFEN for learning aligned node embeddings in dynamic networks. It can be used with any existing node embedding method (both for static and dynamic networks) and, contrary to existing post-hoc methods, it does not rely on a time-consuming matrix factorization approach.

C2. We propose three different versions of the alignment function, including one that treats all nodes equally and one that utilizes temporal network measures for weighting the alignment process.

C3. We evaluate the proposed framework variants in a link prediction task over six real-world datasets and show that our approach allows to improve results on the downstream task compared to other existing methods.

2   Related Work

Static Node Embedding. To date, plethora of node embedding methods have been introduced. The seminal work of Grover et al. [4] introduced Node2Vec method, which enables learning structural embeddings of nodes in an unsupervised manner. It pushes node embeddings apart when nodes are away from each other but makes embeddings closer for nearby nodes, leveraging nodes co-occurrences on common random walks as a measure of node "distance". Recently, the main body of literature was focused on Graph Neural Networks, among which GCN [8], GIN [18], and GAT [17] are the most notable ones. In particular, the Graph AutoEncoder (GAE) [7] architecture leverages GCNs to learn unsupervised node structural embeddings, using reconstruction of the adjacency matrix as the training task. As static node embeddings can be adopted for time-discrete dynamic graphs, we utilize Node2Vec and GAE as base for computing dynamic node representations in this work.

Dynamic Node Embedding. The problem of dynamic node embedding has recently gained more focus from researchers. We can distinguish meta approaches that could utilize any static embedding method to produce dynamic snapshot embeddings [1,14–16]. However, most of them utilize only the Node2Vec [4] embeddings method. Another group of models are the ones dedicated for dynamic node embedding, such as Dyngraph2vec [3], where the authors utilize the Autoencoder (AE) architecture to capture the temporal network's evolving structure and provide three variants of the model which differ in the representation method of a neighbor vector.

Learning of Compatible Representation. This problem has already been studied in literature [5,10,13]. In [5], the authors present the BC-Aligner method that allows obtaining previous embedding models based on a learnable transformation matrix. [10] achieves model compatibility with an alignment loss that aligns class centers between models with a boundary loss to constraint new features to be more centralized to class centers. [13] proposes backward compatibility training for image classification by adding an influence loss that can capture dataset changes along with new classes.

Post-hoc Alignment for Node Embeddings. There are already approaches for alignment of node embeddings based on the Orthogonal Procrustes method [1,14–16]. In [14,16], the authors use all common nodes between snapshots to perform the alignment. In [1], it has been found that some nodes change their behavior too significantly. Therefore, they use only a subset of common nodes, which is selected based on activity and selection method. This approach in further explored in [15] by introducing a wide range of activity functions that can be used for the alignment process.

3   The Proposed RAFEN Framework

Definition 1. *A dynamic network (graph) is a tuple $G_{0,T} = (V_{0,T}, E_{0,T})$, where $V_{0,T}$ denotes a set of all nodes (vertices) observed between timestamp 0 and T, and $V_{0,T}$ is a set of edges in the same timestamp range. We model such a network as a sequence of graph snapshots $G_{0,1}, G_{1,2}, G_{2,3}, \ldots, G_{T-1,T}$.*

Definition 2. *Node embedding is a function $f : V \rightarrow \mathbb{R}^{|V| \times d}$ that maps a set of nodes V into low-dimensional (i.e., $d \ll |V|$) embedding matrix F, where each row represents the embedding of a single node.*

Definition 3. *Node activity scoring function $s : V_{com} \rightarrow \mathbb{R}^{|V_{com}|}$ (with $V_{com} = V_{i,j} \cap V_{k,l}$ and $i < j < k < l$) assigns a scalar score to each node from the set of common nodes V_{com}, which reflects the change in a node's activity between two snapshots $G_{i,j}$ and $G_{k,l}$.*

Definition 4. *Reference nodes V_{ref} is a subset of common nodes between two snapshots retrieved by a selection function. We use the percent selection function as defined in [1], with chooses the top p% of nodes – i.e., $select(S, V) = V_{ref} \subseteq sort_S(V)$, s.t. $|V_{ref}| = \boldsymbol{p}|V|$.*

Problem Statement. Given a dynamic graph $G_{0,T}$ in the form of discrete graph snapshots (see Definition 1), the objective is to find at any given timestamp $t \in \{1, 2, \ldots, T\}$, the node embedding $F_{t-1,t}$ of the current snapshot $G_{t-1,t}$, such that $F_{t-1,t}$ will be algebraically compatible (aligned) with the node embedding $F_{t-2,t-1}$ of the previous snapshot $G_{t-2,t-1}$. Such compatibility is required to properly aggregate snapshot embeddings $F_{0,1}, F_{1,2}, \ldots, F_{t-1,t}$ into a single node embedding matrix $F_{0,t}$ that summarizes the whole graph evolution (see [1] for more details about the aggregation mechanism).

RAFEN. As a solution for the alignment problem, we propose a novel framework RAFEN that enhances the loss function of any existing node embedding method L_{model} by means of an alignment regularization term $L_{alignment}$. It allows to learn node embeddings that are aligned with a given anchor embedding (in our case: the previous graph snapshot). Our framework can be used along with expert knowledge by setting a model hyperparameter α or using temporal network measures to automatically determine the alignment coefficient of each node. We will discuss all framework details in the sections below. Moreover, we present a general overview of our pipeline in Fig. 1.

How to Incorporate the Alignment Term? We enhance the loss function of the node embedding model L_{model} by adding an alignment term $L_{alignment}$. In order to balance between the information from the previous graph snapshot (through the optimization of the $L_{alignment}$ loss) and the information from the current snapshot (through the optimization of the L_{model} loss), we propose to combine both terms in the following way:

$$L = (1 - \alpha) \cdot L_{model} + \alpha \cdot L_{alignment} \tag{1}$$

The α parameter controls the trade-off between the model's loss and the alignment term, which are both critical factors in the model's learning behavior. We can distinguish two boundary conditions of the α parameter:

- $\alpha = 0$ – the model uses only model loss,
- $\alpha = 1$ – the model uses only alignment term.

Choosing the right α value is essential for the model's performance. It can be done by a simple grid search or through expert knowledge. However, performing an α grid search might be time-consuming and computationally expensive. To address this, we propose a simplified version of our framework, where we set equal importance to both loss terms, which is equivalent to setting $\alpha = 0.5$

$$L = L_{model} + L_{alignment} \tag{2}$$

Alignment Loss Term ($L_{alignment}$). The foundation of our proposed RAFEN framework is the alignment loss term $L_{alignment}$. We build upon the mean squared error (MSE) between node v's embedding from the current snapshot $F_{t-1,t}^{(v)}$ and its representation from the previous snapshot $F_{t-2,t-1}^{(v)}$. We apply such regularization to a subset of the common nodes $V_{com} \in V_{t-2,t-1} \cap V_{t-1,t}$.

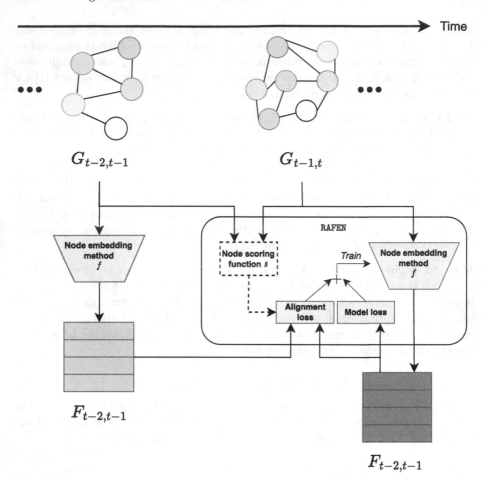

Fig. 1. Overview of the RAFEN pipeline. The pipeline can be decomposed into two steps: (1) For the snapshot $G_{t-2,t-1}$ that we want align to, we compute node embeddings $F_{t-2,t-1}$ using the vanilla node embedding method f. (2) For the next snapshot $G_{t-1,t}$, we additionally compute the alignment loss term, which we add to the node embedding method model loss – Eq. (2). The alignment loss term is calculated as an MSE (mean squared error) – Eq. (3) between the representation of common nodes of currently trained node embeddings $F_{t}-1, t$ and embeddings from the previous snapshot $F_{t-2,t}$. There is no need to take all common nodes to calculate the alignment loss. Instead, we apply node scoring function s that computes the difference in node's activity between snapshots and selects a subset of common nodes – Eq. (5) or use the output as weights in alignment loss term – Eq. (4)

Let us now define three RAFEN variants:

– RAFEN_ALL – we use all common nodes between consecutive snapshots:

$$L_{\text{alignment}} = \frac{1}{|V_{com}|} \sum_{v \in V_{com}} \left(F_{t-1,t}^{(v)} - F_{t-2,t-1}^{(v)} \right)^2 \tag{3}$$

– `RAFEN_Weighted` – similarly to the previous case, we use all common nodes, but instead of treating each one of them equally, we assign weights that reflect the node activity change (node activity scoring function $s(\cdot)$ - see Definition 3) of a particular node between the graph snapshots;

$$L_{\text{alignment}} = \frac{1}{|V_{com}|} \sum_{v \in V_{com}} \left(\left(F_{t-1,t}^{(v)} - F_{t-2,t-1}^{(v)} \right)^2 \cdot s(v) \right) \tag{4}$$

– `RAFEN_REF` – in [1,15], the authors show that utilizing all common nodes in the alignment process may lead to degraded performance in the downstream tasks due to the fact that some nodes change their behavior too much. They use $V_{ref} \subset V_{com}$ subset of common nodes, which is built according to the node activity scores (the top $p\%$ of the common nodes are selected).

$$L_{\text{alignment}} = \frac{1}{|V_{\text{ref}}|} \sum_{v \in V_{\text{ref}}} \left(F_{t-1,t}^{(v)} - F_{t-2,t-1}^{(v)} \right)^2 \tag{5}$$

4 Experimental Setup

We evaluate all methods for the link prediction task. Following, we discuss the details and hyperparameters of the experiments.

4.1 Datasets

We conducted experiments on six real-world datasets. We followed the graph snapshots split procedure defined in [15], such that we split graphs based on the timestamp frequency (monthly or yearly). Moreover, the first four snapshots in *ppi*, and *ogbl-collab* datasets were ignored, as merging them would result in a too broad timespan of such merged snapshots. Also, the two last snapshots of *bitcoin-alpha* and *bitcoin-otc* were merged to provide more data for validation. We present an overview of datasets in Table 1. Despite the similarities with the experimental setup in [15], we reproduced all experiments due to a refined evaluation protocol on link prediction. Also, our method differs in alignment procedure, i.e., it aligns embedding to the previous snapshot instead of the first.

Table 1. Statistics of graph datasets. $|V|$ - number of nodes, $|\mathcal{E}|$ - number of edges, Dir - whether the graph is directed or not, T - number of snapshots

Dataset	$|V|$	$|\mathcal{E}|$	Dir	Timespan	T	Snapshot timespan	Network domain
fb-forum [12]	899	33 720	✗	5.5 months	5	1 month	social
fb-messages [12]	1 899	61 734	✗	7.2 months	7	1 month	social
bitcoin-alpha [12]	3 783	24 186	✓	5.2 years	5	1 year	crypto
bitcoin-otc [12]	5 881	35 592	✓	5.2 years	5	1 year	crypto
ppi [12]	16 386	141 836	✗	24 years	5	5 years	protein
ogbl-collab [6]	233 513	1 171 947	✗	34 years	7	5 years	citation

4.2   Node Embeddings

We selected two different methods as our base models for computing node embeddings. First, we used Node2Vec, a widely renowned node structural embedding method that relies on random walks. To show our framework's universality, we included a second method – GAE, an encoder-decoder model that utilizes the graph convolution operation. Both methods are capable of capturing meaningful structural features of the graph and are trained in an unsupervised manner. Nonetheless, due to the flexibility of our framework, other node embedding models could also be utilized. For both models, each snapshot representation was recomputed 25 times to account for randomness in these methods (e.g., random walks in Node2Vec, weight initialization in the encoder of the GAE model).

Node2Vec. We trained the Node2Vec model, reusing the hyperparameters configuration from [15], as these were obtained from a hyperparameter search procedure. In particular, we use 128-dimensional embeddings, except for the *bitcoin-alpha* dataset where we use 32 as the embedding size. Further, we use the same configurations in our *RAFEN* models, as we desire that our models perform well without additional hyperparameter search. We utilize the Node2Vec implementation from the PyTorch Geometric library [2].

GAE. We use a Graph Autoencoder with two Graph Convolutional layers (GCN) as an encoder and the Inner-Product decoder. Since datasets do not come with node features required by the GCN layers, we use an additional trainable lookup embedding layer, which serves as a node feature matrix for the model (we evaluated different strategies in preliminary experiment – this one turned out to work the best). We train these models for 100 epochs using the Adam optimizer with a learning rate of 0.01, and a hidden layer size of 128.

4.3   Aligned Models

We tested variants of our RAFEN framework with loss functions introduced in Sect. 3, namely RAFEN_ALL, RAFEN_WEIGHTED, and RAFEN_REF. In addition, RAFEN_WEIGHTED involves scoring function, denoted as $s(\cdot)$ in Eq. (4). For scoring function, we selected *Edge Jaccard* (EJ) [15], and *Temporal Betweenness* (TB) [19] due to their good performance in post-hoc alignment methods in previous work. Also, these two functions provide desired variability, such that they are from a different family of methods: *Edge Jaccard* is based only on edge changes in the static scheme, while *Temporal Betweenness* takes into account the time aspect. We denote the two variants as RAFEN_WEIGHTED_EJ and RAFEN_WEIGHTED_TB, for *Edge Jaccard* and *Temporal Betweenness* respectively.

It is worth noting that L_{model} and $L_{alignment}$ are in different scales. Therefore, we additionally scale them with the loss value of the model from the first batch.

4.4   Link Prediction

We define the link prediction task as edge existence prediction on the last snapshot of the graph $G_{T-1,T}$, based on the previously learned representations from previous snapshots $F_{0,1}, \ldots, F_{T-2,T-1}$. We take existing links in the graph $G_{T-1,T}$ as positive examples. To avoid out-of-distribution nodes in evaluation, we filter edges wherein one of the nodes was not observed previously. The number of negative edges (non-existing edges) is equal in size to the positive ones. In the negative sampling process, we also applied a simple edge reject criterion that prevents adding already existing edges as negative ones. Having edges sampled, we split the datasets into train, val and test splits in the proportion: 60%, 20%, 20%, respectively. This split is both leveraged during grid-search hyperparameter optimization and the method evaluation. Finally, using optimal hyperparameters, we performed the final evaluation with training and validation subsets merged. The results of the evaluation are reported in Sect. 5.

To fuse the node embeddings from all snapshots into a single one that reflects the graph evolution, we evaluate 4 different aggregation schemes:

- mean – node embeddings are averaged across snapshots,
- last – only the node embedding from the last snapshot is taken,
- FILDNE – weighted incremental combination of previous snapshot embeddings as in [1],
- k-FILDNE – weighted incremental combination of previous snapshot embeddings with automated weight estimation as in [1].

Further, we use a logistic regression classifier, which takes the Hadamard product of the two aggregated node embeddings. We utilize the implementation from the scikit-learn [11] library with *liblinear* solver, keeping other hyperparameters values default.

Summarizing, the experiments were performed on link prediction with embeddings trained with 3 variants of loss function and 4 embeddings aggregation schemes. We also report results for the two embedding methods (Node2Vec, GAE) trained without alignment. In addition, we used 3 post-hoc alignment methods from [15] as our baselines. To sum up, we evaluate the following models:

- **(whole graph)** – embeddings are trained on whole graph $G_{0,T-1}$,
- **(last snapshot only)** – embeddings are trained on last graph snapshot $G_{T-2,T-1}$,
- **Posthoc-PA** – Procrustes Aligner (PA),
- **Posthoc-EJ** – Edge Jaccard Aligner (EJA),
- **Posthoc-TB** – Temporal Betweenness (TB),
- RAFEN_ALL – RAFEN variant with all common nodes between consecutive snapshots and simplified loss importance term,
- RAFEN_Weighted_EJ – RAFEN_Weighted variant with Edge Jaccard node activity scoring method and simplified loss importance term,
- RAFEN_Weighted_TB – RAFEN_Weighted variant with Temporal Betweenness node activity scoring method and simplified loss importance term.

5   Results

This section presents the results of our experiments, divided into three subsections presenting different framework characteristics. First, we compare the embedding aggregation methods. Next, we show the results for the two formulations of RAFEN loss function, i.e., with and without hyperparameter α. Finally, we present the results of several variants of RAFEN models compared to the baselines approaches.

5.1 Embeddings Aggregation Method Comparison

Foremost, we evaluated the embedding aggregation methods, namely last, mean, FILDNE, k-FILDNE. We summarize these results in Table 2. We chose the best aggregation method that performs well on both GAE and Node2Vec-based models based on mean ranks for each dataset separately, i.e., bitcoin-alpha: k-FILDNE; bitcoin-otc: last; fb-forum: FILDNE; fb-messages: k-FILDNE; ogbl-collab: last, ppi: last. Please note that for the bitcoin-alpha dataset, the best aggregation method for Node2Vec and GAE differs. However, we chose k-FILDNE due to a better combination of mean ranks (i.e., 1.22 and 1.56 vs 3.44 and 1.44).

Counterintuitively to previous dynamic graph representation learning papers [1] for four out of six datasets, the best aggregation method was last, which does not aggregate all historical representations. That behavior led us to an additional study that we conducted to further explore both Posthoc and RAFEN methods. We contrasted AUC on the link-prediction of RAFEN_ALL and Posthoc-PA with plain Node2Vec embeddings and computed their relative performance. In particular, we evaluated two scenarios:

- **Prev** – we take each node's v embedding $F_{t-1,t}^{(v)}$ at time step t and evaluate it on link-prediction using the previous snapshot's graph structure $G_{t-2,t-1}$ as a target,

Table 2. Mean ranks of aggregation methods on different datasets. Ranks were calculated on the link prediction task (evaluated on last snapshot), using the AUC metric over 25 runs. Then we took mean ranks of all the RAFEN and Posthoc models for Node2Vec and GAE models, seperately.

Dataset	Model	Aggregation method			
		FILDNE	k-FILDNE	last	mean
bitcoin-alpha	Node2Vec	2.11	**1.22**	3.22	3.44
	GAE	3.00	1.56	4.00	**1.44**
bitcoin-otc	Node2Vec	1.89	3.00	**1.11**	4.00
	GAE	1.67	2.89	**1.44**	4.00
fb-forum	Node2Vec	**1.22**	2.11	3.11	3.56
	GAE	**1.33**	2.22	2.78	3.67
fb-messages	Node2Vec	2.00	3.00	**1.00**	4.00
	GAE	1.83	2.50	**1.67**	4.00
ogbl-collab	Node2Vec	2.00	3.00	**1.00**	4.00
	GAE	2.00	3.00	**1.00**	4.00
ppi	Node2Vec	2.33	2.67	**1.00**	4.00
	GAE	2.89	2.06	**1.11**	3.94

– **Next** – we take each node's v embedding $F_{t-1,t}^{(v)}$ at time step t and evaluate it on link-prediction using next snapshot's graph structure $G_{t,t+1}$ as a target.

We report the link prediction AUC ratio between alignment methods and the non-aligned Node2Vec version. Results are shown in Fig. 2.

Our proposed RAFEN method increased the evaluation performance on previous snapshots for all of the benchmarked datasets. The gain ranges from 3.5% on the first snapshot of the ppi dataset up to a 63% difference on the third snapshot of bitcoin-otc. In the case of the **Next** scenario, we observe that the last snapshot evaluation performance increased on four our of six datasets – ranging from 1.2% in the case of the ogbl-collab dataset up to 7.4% in fb-forum.

5.2   Comparison of RAFEN's α-Based Variants

In Sect. 3, we proposed two formulations of the loss function for our RAFEN framework. Since the first variant defined in Eq. (1) involves an α hyperparameter, which requires additional domain knowledge or hyperparameter search procedure (e.g., grid search), we investigate whether its simplified variant with L_{model} and $L_{alignment}$ treated equally important provides competitive performance. Likewise, RAFEN_REF models require defining a reference node selection function with hyperparameters, like top $p\%$ in our case. Therefore, we compare it to RAFEN_Weighted models that rely only on scoring function and do not involve additional hyperparameters. Results of the comparisons are shown in Table 3.

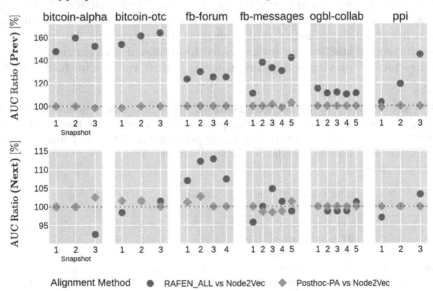

Fig. 2. Single snapshot representation evaluation study. In the first figure row, the previous snapshot evaluation is shown (denoted as **Prev**), and in the second row next snapshot evaluation (denoted as **Next**). Each column represents a different dataset. Y-Axis contains a link prediction AUC metric Ratio calculated between Alignment Method and vanilla Node2Vec. X-Axis presents snapshot IDs on which node representation was trained.

Across almost all the shown comparisons, we can observe that percentage AUC difference in *alpha*-based models to simplified/weighted versions is in the range (-1.61% to 1.71%). We can observe outlier values in both **bitcoin-alpha** and **ppi** favoring α hyperparameter. Conversely, there is an outlier in **fb-forum** favoring the weighted method. As in most cases, the differences are negligible, we decided to use simplified/weighted versions for the comparison with baselines. In the aftermath, RAFEN requires only one additional hyperparameter compared to plain GAE or Node2Vec models, which is the node activity scoring function.

5.3   RAFEN Comparison to Baselines

The results of the comparison between our proposed RAFEN framework and baseline models are presented in Table 4. Compared to snapshot-based Node2Vec and GAE baselines, RAFEN was better on four datasets: bitcoin-otc, fb-forum, ppi, and ogbl-collab – only losing to the variant trained on the full graph data. For the remaining two datasets, namely fb-messages, and bitcoin-alpha, we observe the superior performance of post-hoc methods. However, as the previous study (see Sect. 5.2) shows, these datasets provided the most significant differences between α based and weighted. When the models were tuned, they would have also improved performance compared to RAFEN_Weighted on these datasets.

In the case of RAFEN_Weighted, the temporal betweenness scoring function was barely better than Edge Jaccard, showing notable differences only in the case of GAE on ogbl-collab and Node2Vec on fb-forum. When comparing RAFEN_ALL and RAFEN_Weighted, RAFEN_ALL approach served as a strong baseline, showing better performance and higher rank in most of the experiments. We hypothesize that the scoring functions we leveraged led to decreased performance. Hence, it requires further investigation and refinements, which we leave for future work.

Table 3. Comparison of α-based RAFEN with simplified/weighted versions. We present the percentage difference of averaged link prediction AUC metric on the last snapshot over 25 runs. Values below zero denote better performance of the simplified/weighted variant, and values above zero denote better α variant performance. We mark outlier values with **bold**.

Model	Selector	RAFEN variant	bitcoin alpha	bitcoin otc	fb forum	fb messages	ogbl collab	ppi
Node2Vec	All	α vs Simplified	-0.55%	0.42%	-0.58%	1.71%	0.59%	0.16%
	TB	REF_α vs Weighted	**4.84%**	0.00%	**-2.59%**	1.34%	0.35%	-0.49%
	EJ	REF_α vs Weighted	**5.01%**	-0.42%	1.04%	**2.79%**	0.47%	-0.49%
GAE	All	α vs Simplified	0.49%	1.16%	-0.12%	-0.54%	0.76%	0.50%
	TB	REF_α vs Weighted	**2.07%**	0.91%	-1.53%	**2.12%**	0.25%	0.17%
	EJ	REF_α vs Weighted	**3.96%**	0.52%	0.81%	-1.64%	1.27%	0.00%

Table 4. Last snapshot link prediction evaluation results. We report AUC metric values with mean and standard derivation over 25 model retrains. We mark the top three results with **bold**. Methods mean ranks represent the ranking established upon mean AUC values over retrains.

Model	Alignment method	bitcoin alpha	bitcoin otc	fb forum	fb messages	ogbl collab	ppi	Mean rank
Node2Vec	(whole graph)	70.60 ± 5.80	71.70 ± 2.60	$\mathbf{91.40 \pm 0.80}$	71.70 ± 4.70	$\mathbf{89.00 \pm 0.10}$	$\mathbf{63.80 \pm 0.20}$	**7.00**
	(last snapshot only)	72.70 ± 4.40	69.30 ± 2.70	80.00 ± 2.40	74.30 ± 3.90	84.50 ± 0.10	60.30 ± 0.40	10.42
	Posthoc-PA	74.60 ± 7.00	69.70 ± 2.20	84.40 ± 2.10	73.60 ± 6.50	84.50 ± 0.20	60.30 ± 0.30	10.00
	Posthoc-EJ	76.50 ± 7.50	69.60 ± 1.80	84.20 ± 2.10	73.30 ± 6.00	84.50 ± 0.20	60.30 ± 0.30	10.33
	Posthoc-TB	72.70 ± 7.60	70.20 ± 2.80	84.10 ± 1.90	74.00 ± 5.40	84.50 ± 0.20	60.30 ± 0.30	10.00
	RAFEN_ALL	73.10 ± 5.70	70.80 ± 2.50	$\mathbf{87.20 \pm 1.70}$	74.60 ± 5.40	$\mathbf{84.80 \pm 0.10}$	61.90 ± 0.40	**5.75**
	RAFEN_Weighted_EJ	70.20 ± 7.20	71.10 ± 2.60	85.80 ± 1.80	73.20 ± 2.90	84.70 ± 0.20	$\mathbf{61.90 \pm 0.40}$	9.17
	RAFEN_Weighted_TB	70.80 ± 8.80	71.40 ± 3.00	$\mathbf{87.00 \pm 1.70}$	73.70 ± 3.80	$\mathbf{84.80 \pm 0.10}$	61.90 ± 0.30	7.25
GAE	(whole graph)	72.10 ± 4.30	$\mathbf{77.80 \pm 1.70}$	68.20 ± 6.30	65.60 ± 4.50	80.70 ± 0.20	61.30 ± 0.60	10.00
	(last snapshot only)	77.80 ± 2.00	74.50 ± 1.90	79.60 ± 1.70	$\mathbf{75.80 \pm 3.80}$	78.20 ± 0.30	58.00 ± 1.00	9.92
	Posthoc-PA	$\mathbf{82.70 \pm 2.80}$	74.40 ± 2.30	84.80 ± 1.70	$\mathbf{75.70 \pm 3.80}$	78.20 ± 0.30	58.00 ± 0.80	8.42
	Posthoc-EJ	$\mathbf{82.60 \pm 3.00}$	74.50 ± 2.20	84.90 ± 1.80	75.40 ± 3.30	78.20 ± 0.30	58.10 ± 0.70	7.92
	Posthoc-TB	$\mathbf{82.30 \pm 3.50}$	74.80 ± 2.00	84.80 ± 1.60	$\mathbf{75.90 \pm 3.50}$	78.20 ± 0.30	58.10 ± 0.60	7.58
	RAFEN_ALL	81.50 ± 2.60	$\mathbf{76.80 \pm 1.40}$	86.60 ± 1.10	74.00 ± 3.50	78.50 ± 0.40	59.30 ± 0.50	**6.75**
	RAFEN_Weighted_EJ	80.10 ± 2.60	76.30 ± 1.50	86.10 ± 1.20	74.50 ± 2.60	78.00 ± 0.30	59.40 ± 0.60	8.00
	RAFEN_Weighted_TB	80.50 ± 3.30	$\mathbf{76.40 \pm 1.90}$	86.30 ± 1.20	73.90 ± 2.80	78.50 ± 0.30	59.30 ± 0.50	7.50

6 Conclusion and Future Work

In this paper, we tackled the problem of learning aligned node embedding for dynamic graphs modeled as a series of discrete graph snapshots. Contrary to existing approaches, such as post-hoc methods, we perform the alignment step during the embedding training. We proposed a novel framework – RAFEN – which allows to enrich any existing node embedding method with the aforementioned alignment capabilities. We conducted experiments on six real-world datasets and showed that our approach, even in its simplified version (with minimum set of hyperparameters), achieves better or on-par performance compared to existing approaches. For future work, we want to explore more advanced mechanisms for the actual alignment step, including better investigation of the scoring function in RAFEN_Weighted setting and the applicability of RAFEN for continuous-time dynamic graphs.

Our code with experiments and data, all enclosed in DVC pipelines [9], is available publicly at https://github.com/graphml-lab-pwr/rafen.

Acknowledgments. This work was financed by (1) the Polish Ministry of Education and Science, CLARIN-PL; (2) the European Regional Development Fund as a part of the 2014–2020 Smart Growth Operational Programme, CLARIN - Common Language Resources and Technology Infrastructure, project no. POIR.04.02.00-00C002/19; (3) the statutory funds of the Department of Artificial Intelligence, Wroclaw University of Science and Technology, Poland; (4) Horizon Europe Framework Programme MSCA Staff Exchanges grant no. 101086321 (OMINO).

References

1. Bielak, P., Tagowski, K., Falkiewicz, M., Kajdanowicz, T., Chawla, N.V.: FILDNE: a framework for incremental learning of dynamic networks embeddings. Knowl.-Based Syst. **236**, 107453 (2022)
2. Fey, M., Lenssen, J.E.: Fast graph representation learning with PyTorch Geometric. In: ICLR Workshop on Representation Learning on Graphs and Manifolds (2019)
3. Goyal, P., Chhetri, S.R., Canedo, A.: dyngraph2vec: capturing network dynamics using dynamic graph representation learning. Knowl.-Based Syst. **187**, 104816 (2020)
4. Grover, A., Leskovec, J.: node2vec: scalable feature learning for networks. In: Proceedings of the 22nd ACM SIGKDD International Conference on Knowledge Discovery and Data Mining, pp. 855–864 (2016)
5. Hu, W., Bansal, R., Cao, K., Rao, N., Subbian, K., Leskovec, J.: Learning backward compatible embeddings. In: Proceedings of the 28th ACM SIGKDD Conference on Knowledge Discovery & Data Mining (2022)
6. Hu, W., et al.: Open graph benchmark: datasets for machine learning on graphs. arXiv preprint arXiv:2005.00687 (2020)
7. Kipf, T.N., Welling, M.: Variational graph auto-encoders. arXiv preprint arXiv:1611.07308 (2016)
8. Kipf, T.N., Welling, M.: Semi-supervised classification with graph convolutional networks. arXiv (2017). https://doi.org/10.48550/arXiv.1609.02907, http://arxiv.org/abs/1609.02907. [cs, stat]
9. Kuprieiev, R., et al.: DVC: data version control - git for data & models (2022). https://doi.org/10.5281/zenodo.7093084
10. Meng, Q., Zhang, C., Xu, X., Zhou, F.: Learning compatible embeddings. In: Proceedings of the IEEE/CVF International Conference on Computer Vision, pp. 9939–9948 (2021)
11. Pedregosa, F., et al.: Scikit-learn: machine learning in Python. J. Mach. Learn. Res. **12**, 2825–2830 (2011)
12. Rossi, R., Ahmed, N.: The network data repository with interactive graph analytics and visualization. In: Twenty-Ninth AAAI Conference on Artificial Intelligence (2015)
13. Shen, Y., Xiong, Y., Xia, W., Soatto, S.: Towards backward-compatible representation learning. In: Proceedings of the IEEE/CVF Conference on Computer Vision and Pattern Recognition, pp. 6368–6377 (2020)
14. Singer, U., Guy, I., Radinsky, K.: Node embedding over temporal graphs. In: Proceedings of the 28th International Joint Conference on Artificial Intelligence, pp. 4605–4612. AAAI Press (2019)
15. Tagowski, K., Bielak, P., Kajdanowicz, T.: Embedding alignment methods in dynamic networks. In: Paszynski, M., Kranzlmüller, D., Krzhizhanovskaya, V.V., Dongarra, J.J., Sloot, P.M.A. (eds.) ICCS 2021. LNCS, vol. 12742, pp. 599–613. Springer, Cham (2021). https://doi.org/10.1007/978-3-030-77961-0_48
16. Trivedi, P., Büyükçakır, A., Lin, Y., Qian, Y., Jin, D., Koutra, D.: On structural vs. proximity-based temporal node embeddings (2020)
17. Veličković, P., Cucurull, G., Casanova, A., Romero, A., Lio, P., Bengio, Y.: Graph attention networks. arXiv preprint arXiv:1710.10903 (2017)
18. Xu, K., Hu, W., Leskovec, J., Jegelka, S.: How powerful are graph neural networks? (2019). http://arxiv.org/abs/1810.00826. [cs, stat]
19. Yu, E.Y., Fu, Y., Chen, X., Xie, M., Chen, D.B.: Identifying critical nodes in temporal networks by network embedding. Sci. Rep. **10**(1), 1–8 (2020)

CLARIN-Emo: Training Emotion Recognition Models Using Human Annotation and ChatGPT

Bartłomiej Koptyra[(✉)], Anh Ngo, Łukasz Radliński, and Jan Kocoń

Department of Artificial Intelligence, Wrocław University of Science and Technology,
Wrocław, Poland
bartlomiej.koptyra@pwr.edu.pl

Abstract. In this paper, we investigate whether it is possible to automatically annotate texts with ChatGPT or generate both artificial texts and annotations for them. We prepared three collections of texts annotated with emotions at the level of sentences and/or whole documents. CLARIN-Emo contains the opinions of real people, manually annotated by six linguists. Stockbrief-GPT consists of real human articles annotated by ChatGPT. ChatGPT-Emo is an artificial corpus created and annotated entirely by ChatGPT. We present an analysis of these corpora and the results of Transformer-based methods fine-tuned on these data. The results show that manual annotation can provide better-quality data, especially in building personalized models.

Keywords: ChatGPT · Emotion recognition · Automatic annotation

1 Introduction

Emotions play an important role in human lives. Expressing and detecting them is a vital skill in social interactions, including professional careers. We even try to measure our abilities in this aspect, calling them emotional intelligence. Therefore it is no wonder that many studies concern themselves with trying to detect and use publicly expressed emotions to improve on a wide range of tasks such as stock market prediction [14], understanding user-preferences [28], public health monitoring and surveillance [27].

Understanding human emotions is one of the more challenging tasks in natural language processing. Not only are they a very subjective topic, but humans

This work was financed by the European Regional Development Fund as a part of the 2014–2020 Smart Growth Operational Programme: (1) POIR.04.02.00-00C002/19 (AN,ŁR); (2) POIR.01.01.01-00-0615/21 (BK); (3) POIR.01.01.01-00-0288/22 (JK); (4) the statutory funds of the Department of Artificial Intelligence, Wroclaw University of Science and Technology; (5) the European Union under the Horizon Europe, grant no. 101086321 (OMINO).

J. Mikyška et al. (Eds.): ICCS 2023, LNCS 14073, pp. 365–379, 2023.
https://doi.org/10.1007/978-3-031-35995-8_26

can also lack the capability to express themselves in written language. This inaccuracy of expressed emotions is because more natural, multimodal forms of emotional expression, such as speech, body language, and facial expressions, can, unlike text, portray peculiar cues used to recognize emotions [1]. The difference between formal and informal writing styles, accounting for grammatical errors (especially in short texts) [1], use of figurative language like sarcasm [22], and many more are aspects that contribute to the difficulty. This emphasizes a need for good training data. This is not a trivial task, as creating a dataset with emotion labels requires a set of guidelines for annotators to follow for annotations to be consistent. Examining the publicly available datasets for emotion recognition in the Polish language reveals the lack thereof. CLARIN-Emo dataset was created as a subset of a Polish sentiment annotated consumer reviews corpus PolEmo [3, 15, 18–20, 23, 37] with additional emotion annotations to fill this gap.

With the recent rise in popularity of large language models, we took this opportunity to explore the possibility of creating consistent emotion annotations using such a model, namely chatGPT, on a different data set and compare it against a model trained and tested on the newly created human-annotated emotions dataset. This allows checking the consistency of such annotations by training a model on them. We also explore the possibility of not only annotating but also generating already annotated data using this multi-purpose dialog system created by OpenAI. Such a system requires dialogue breadth (coverage) and dialog depth (complexity). Emotion recognition and expression are important features that can increase dialog depth [5]. Checking the consistency of emotion recognition of such a system helps to recognize at what level of dialog depth are modern, popular large language models. Our findings suggest that such models are still not up to human standards.

We use three created datasets to train and evaluate multilabel emotion classification models, on the level of both documents and sentences, by fine-tuning pre-trained transformers. We treat sentences as a part of the document to train sequential sentence classification models. We also perform a personalized setup for sequence sentence classification in the new CLARIN-Emo dataset, similar to [13, 17, 26], and find that personalization has a positive impact on model performance in emotion recognition.

2   Related Work

There are many non-Polish datasets for emotion recognition that follow different emotion taxonomies. The most basic approach is to use a distinction proposed by Ekman [11], which states six basic human emotions: anger, surprise, disgust, enjoyment, fear, and sadness. Such an approach can be found, for example, in the EmotionLines dataset [5], which assigns emotions to utterances in a dialog setting to achieve the contextual flow of emotions, and in ArmanEmo [25], which is a sentence-level human-labeled annotated dataset based on Twitter, Instagram, and comments from an Iranian e-commerce company. Plutchnik's "wheel of emotions" [30] extended this notion into four contrasting emotion pairs: joy vs sadness, anger vs fear, trust vs disgust, and surprise vs anticipation. It is a popular approach used in many datasets because of its relatively small number of classes and the use of the wheel as representation. This representation allows using angles in the wheel as distances between emotions. Plutchnik's set of emotions can be found used in creating a word-emotion association lexicon [4], even in crowdsourcing setting, assessing the psychology of story characters [31], emotion detection on social media [35], and many more. Plutchink's set of emotions can also be expanded into 24 emotions with inter-annotator agreement based on distances on the wheel [10]. There are also other datasets with more emotions, like 28, used in the GoEmotions dataset [9], which trade F1-score for finer granularity.

Text classification has been a prominent field of research for many years, with recent years bringing a shift towards more advanced deep learning algorithms [24]. The latter dominated the field over the last few years with particular transformer models such as RoBERTa [21], XML-RoBERTa [8] and DistilBERT [34] achieving impressive results for many text classification tasks.

When it comes to research on ChatGPT for text processing tasks, most of the studies conducted so far mainly focus on text summarisation [12, 29] and question answering [2, 36]. Very few explore such areas as sentiment analysis [38] or humor recognition and generation [6]. However, there is a large-scale evaluation of Chat-GPT that has been recently published [16], which mainly focused on exploring ChatGPT performance when compared to SOTA solutions for many different text processing tasks, including text classification and sentiment analysis. Chat-GPT didn't outperform any SOTA solutions. However, it performed significantly worse on most emotion and sentiment processing tasks. There haven't been any major studies conducted that would explore the potential usage of the chatGPT for data annotation or generation of synthetic datasets for text processing tasks.

3 Datasets

As part of our work, we have prepared several datasets that will be used to compare different approaches to emotion recognition. The first dataset (CLARIN-Emo) contains Polish reviews written by humans and manually annotated with emotions, also by humans. The second dataset (Stockbrief-GPT) contains Polish business articles written by humans and annotated with emotions using Chat-GPT. The third dataset contains English reviews, where ChatGPT generated both the text and the emotion annotations.

3.1 CLARIN-Emo: Human Texts and Annotations

The CLARIN-Emo dataset contains a subset of consumer reviews from the PolEmo 2.0 corpus. The reviews belong to one of four domains: *hotels, medicine, products,* and *school.* The collection also includes non-opinion texts, which are informative, belong to the same domains, and are mostly neutral. The CLARIN-Emo corpus contains a total of 1110 opinions composed of 8891 sentences. Annotators labeled the sentiment and emotions of the opinions at the level of each sentence and the whole opinion. The emotion labels are taken from Plutchik's model. There are eight emotion dimensions: *joy, trust, anticipation, surprise, fear, sadness, disgust,* and *anger.* In addition, three sentiment dimensions were used: positive, negative, and neutral. The corpus was annotated by six people who did not see each other's decisions. The annotation was multi-labeled – several labels could describe each sentence and each opinion. Table 1 shows the agreement of annotations for each emotion dimension and the total agreement for all dimensions. Those labels that were annotated at least twice by different annotators were included in the final corpus. The same Table 1 also summarizes the number of annotations in the final corpus for each label, assigned at the sentence and text levels.

Table 1. Annotations statistics in CLARIN-Emo dataset.

Emotion	Number of instances		Krippendorff's alpha coefficient	
	Document-level	Sentence-level	Document-level	Sentence-level
Joy	626	4222	0.6497	0.7190
Trust	274	1972	0.3514	0.3990
Anticipation	92	1078	0.1258	0.3307
Surprise	72	582	0.1118	0.3497
Fear	50	370	0.1214	0.3436
Sadness	604	3892	0.5145	0.6481
Disgust	313	1664	0.2359	0.4276
Anger	274	1406	0.4238	0.4877
Positive	665	4687	0.7116	0.7317
Negative	618	4122	0.7510	0.8195
Neutral	176	2364	0.5908	0.5990
Total	3764	26359	0.4171	0.5323

3.2   Stockbrief-GPT: Human Texts Annotated by ChatGPT

Stockbrief-GPT is a corpus of Polish articles on economics containing 227 articles and 3795 sentences. We used ChatGPT to annotate these articles at the sentence and whole-text levels. Below is a translation of the sample prompt and the system's response.

Task: Stockbrief-GPT (example translated from Polish to English).

Prompt

Company: cd project
===========
Rate the following sentences against the Company given earlier using the following label identifiers from the specified categories: "S:sentiment" (-3:strongly negative, -2:negative, -1:weakly negative, 0:neutral, 1:weakly positive, 2:positive, 3:strongly positive), "E:emotion" (fear, stress, sadness, anger, joy, trust, surprise, anticipation).

Additionally, evaluate the entire text using the same labels. Present the result as a JSON object containing the following fields: { "sentences": { "sentence_number": {"S": identifier_label_sentence, "E": list_label_emotion}}, "evaluation_text": { "S": identifier_label_sentiment, "E": list_label_emotion}}.

1. The CD Project's poker play. 2. On Monday, after the trading session, the world of listed game companies was flooded with information that all investors interested in domestic gaming had been waiting for - CD Projekt officially confirmed that it was working on the fourth installment of the "Witcher" game series. 3. However, the developers have indicated that it will not be a game called "The Witcher 4", as the market had expected. 4. As originally expected, the new installment of the Witcher saga was to be titled 'The Witcher 4' and tell the story of Ciri - Geralt's ward, wearing a medallion from the Witcher School of the Cat. (...)

ChatGPT answer

```
{ "sentences": { "1": {"S": 1, "E": ["anticipation"]},
"2": {"S": 2, "E": ["anticipation"]},
"3": {"S": -1, "E": ["surprise"]},
"4": {"S": 0, "E": ["anticipation"]}, ...}, "evaluation_text": {"S": 0, "E": [ "anticipation" ] }}
```

Table 2 how many occurrences of each label are in the Stockbrief-GPT dataset. As strong positive and negative labels were very sparse in the Stockbrief-GPT dataset for our experiments, we cast those labels as positive and negative, respectively.

Table 2. Number of annotations of different emotions in Stockbrief-GPT dataset.

Emotion	Document-level	Sentence-level
Fear	6	97
Stress	21	171
Sadness	47	415
Anger	35	347
Joy	75	895
Trust	44	403
Suprise	12	219
Anticipation	123	1247
Strong negative	0	32
Negative	18	481
Weak negative	45	415
Neutral	40	904
Weak positive	94	797
Positive	30	1135
Strong positive	0	31
Total	590	7589

3.3   ChatGPT-Emo: ChatGPT Texts and Annotations

ChatGPT-Emo dataset contains opinions and reviews with corresponding domains and emotional labels of the opinion. The initially generated dataset comprised 6684 texts with 3486 unique emotional labels and 613 domains. Chat-GPT fully generated the dataset. Below is an example of the prompt and the system's answer:

Task: ChatGPT-Emo.

Prompt

Create example dataset containing 3 long opinions (at least 5-10 sentences each) form different domains and labels described emotions evoked by these opinions. The output should be an array of JSON objects containing the following keys: "text" is a text of the opinion, "domain" is the domain of the opinion, and "labels" is an array of emotions evoked by the opinion. Do not explain yourself at all, ONLY array of JSONs should be you answer. Do not make textual introduction before the array. Start the texts differently, make them as real opinions of users from the web. Each opinion should start with different word. If you know real opinions from the web, use them. Use from 5 to 10 emotional labels for each example.

ChatGPT answer

[{"text": "I recently attended a concert by this band and it was an unforgettable experience. The energy and passion they brought to the stage was electrifying, and the crowd was completely enthralled. The music was amazing, and I was dancing and singing along the entire time. I would highly recommend this band to anyone looking for a high-energy and entertaining concert experience. The stage production was also top-notch, and I was blown away by the light show and special effects. I can't wait to see them again.",
"domain": "Entertainment", "labels": ["Excitement", "Thrills", "Engagement", "Entertainment", "Admiration", "Satisfaction", "Anticipation", "Awe", "Enthrallment", "Unforgettable"]}, ...]

Many of the initially generated emotional labels were synonymical to each other e.g. *joy, joyful, happy, happiness* etc., thus, it was important to map the similar labels to each other. In order to achieve that, a three-step approach was used. Firstly, all labels with the same lemma recognized by the spaCy English language model were mapped to the lemma of the labels. Then lemmas with similar meaning were grouped manually into 8 major categories: *disappointed, angry, disgusted, excited, happy, satisfied, sad* and *rare* where rare included all labels that didn't have synonymous labels in the dataset and appeared 50 or less times in the entire dataset. Then all duplicate texts and all texts with only rare labels assigned were removed from the dataset. Domains were manually maped into 13 categories: emphScience, Games & Technology, Art, Culture & Entertainment, Food & Dining, Health & Fitness, Travel, Cities & Tourism, Cars, Workplace & Economics, Home, Animals & Nature, Shopping & Products, Personal & Lifestyle, Politics & Social issues, Education. Final dataset consists of 5959 texts of 13 domains with 7 unique emotional labels assigned.

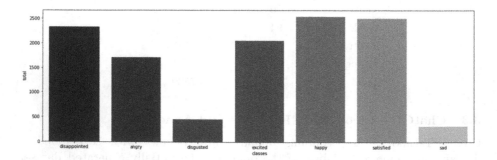

Fig. 1. ChatGPT-Emo distribution of texts into 7 classes: disappointed, angry, disgusted, excited, happy, satisfied, sad

Figure 1 shows the distribution of different classes. It can be observed that most 'positive labels' occur with similar frequency and the 'negative' labels have a greater variety of occurrence. An argument could be made since categories *satisfied* and *happy* have similar meanings and almost identical frequency of occurrence. However, the category *satisfied* was used to group labels that were positive but not explicitly emotional such as *fulfill* and *content*. *Happy* was used to more explicitly emotional labels such as *joy* and *pleased*.

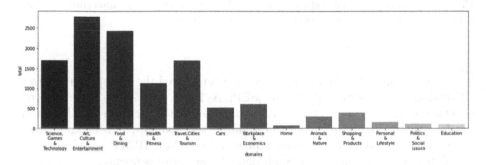

Fig. 2. ChatGPT-Emo distribution of texts into 13 domains: Science, Games & Technology, Art, Culture & Entertainment, Food & Dining, Health & Fitness, Travel, Cities & Tourism, Cars, Workplace & Economics, Home, Animals & Nature, Shopping & Products, Personal & Lifestyle, Politics & Social issues, Education

Figure 2 shows the distribution of texts into 13 distinguished domains. As seen in the figure most common categories are broad categories such as *Art, Culture & Entertainment* and *Food & Dining* and more narrow ones such as *Home* or *Education* are among the most rarely occurring ones. Surprisingly many reviews regarded *cars* even though it is arguably the most narrow category of all listed.

4   Models

4.1   Sequential Sentence Classification

For the Sequential Sentence Classification (SSC) task, given a sequence of sentences $X = x_1, x_2, ..., x_N$ where N is the number of sentences, the goal is to predict the set of labels for each sentence based on the context in which the sentences appear. Our model is closely derived from the approach proposed by [7], in which the authors proposed a novel input representation and utilized the BERT model by fine-tuning the target SSC tasks to classify sentences in scientific abstracts into their rhetorical roles. This section describes the model along with our modifications.

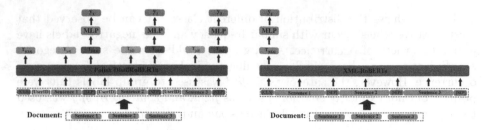

Fig. 3. Overview architecture of Sequential Sentence Classification

Fig. 4. Overview architecture of Text Classification

Figure 3 illustrates the model overview. It employed the two special tokens of BERT-based architecture, [CLS] and [SEP]. As input to BERT, a given sequence of sentences $X = x_1, x_2, ..., x_N$ is concatenated by separator tokens [SEP], resulting in a whole long sequence, before adding the standard classification token [CLS] at the beginning. However, unlike the standard BERT, which uses the [CLS] token as the input representation of the entire sequence, the proposed method uses the [SEP] token encodings as the input representation fed into a multi-layer feedforward network for classification. The author's idea behind using [SEP] to represent a sequence of sentences is that the [SEP] tokens learn sentence structure and the relations between consecutive sentences through BERT's pre-training, allowing the BERT's self-attention layer to leverage contextual information from all the words in the entire document, while still benefitting from BERT's pre-trained weights.

Loss Function. The original model is trained on multi-class classification tasks and aims to minimize cross-entropy loss between the predicted probability distribution and the target distribution of the classes. In addition, when we conducted some initial experiments, we noticed that the model struggled severely with unbalanced label distributions. Since each sentence in our dataset could have multiple labels, we introduced the positive weights to the BCEWithLogitsLoss, which are calculated by weighing the samples as the inverse of the class frequency for its classes in the training set. The formula followed the previous study from [33]:

$$weight_{class_n} = \frac{\sum_{i=1}^{n} x_n}{x_n}, \qquad (1)$$

where x_n is the count of samples of class n in the training set

Personalized Approach. Since emotion recognition is a subjective task [17], each annotator could have a different perspective on the same sentence. We decided to combine the SSC model with a personalized approach called UserID proposed by [26]. In this method, each annotator is represented by a special token in text embedding, by which the model is provided with information about each annotator.

4.2   Text Classification

The main goal of the text classification task is to assign a set of labels to the given record of text. In the case of our paper, the main goal is to assign the set of emotional labels from the given set to each text in the given dataset. The model we use is the base version of XML-RoBERTa [8]. This version of the BERT model is trained in multiple languages, which is important in this case as two of the tested datasets are in Polish, and one is in English. As seen in Fig. 4 the architecture of the model is way simpler than for sequence classification, as there is no need to generate separate outputs for each sentence. Opposite to the first architecture in this case the [CLS] token is used as the input to MLP which then assigns the set of labels to the given text.

5   Experiments

5.1   Sequential Sentence Classification

We formulated experiments as multi-label classification on two datasets Clarin-Emo and Stockbrief-GPT, which aim to predict emotional labels for each sentence in a given document. The macro F1-score and micro F1-score were used for model evaluation. Both datasets were split into Train (70%), Dev (15%), and Test (15%). Except for the experiment with the personalized method UserID where the labels from six annotators are considered individually, in other experiments concerning CLARIN-Emo, these labels are aggregated using majority voting.

Language Models. All experiments utilized the Polish DistilRoBERTa multilingual pretrained model proposed by [32]. We also run a longer experiment on a CLARIN-Emo dataset, finetuning a larger model, namely the large version of XLM-RoBERTa model from [8] using FocalLoss with weighted BCEWithLogitsLoss function as the cross-entropy part of FocalLoss, and included stochastic weight averaging.

Hyperparameter Settings. In the experiments using the Polish DistilRoBERTa language model on both datasets, the optimal learning rate was 1e−5. We used AdamW optimizer with a weight decay of 0.01 and a learning rate scheduler with the fraction of the steps to increase the learning rate of 0.1. The training epoch numbers were obtained for each experiment scenario separately using early stopping with a patience of 20 epochs. The SSC multilabel task and the SSC UserID task on Clarin-Emo trained for 50 epochs, and for the SSC multilabel task on Stockbrief-GPT, it was 70 epochs. To prevent overfitting, a dropout layer with a rate of 0.1 was added. To train the larger model we use the same optimizer and its parameters alongside a linear schedule with a warmup scheduler with 10% warmup steps. We use early stopping of 10 epochs and do 20 annealing epochs for the stochastic weight averaging, with a learning rate of 1e−6.

5.2   Text Classification

All of our datasets contain labels concerning text as a whole. We evaluate separately the model's performance on all three of them. All three datasets have been split into the train (70%), dev (15%), and test (15%). In the case of CLARIN-Emo and Stockbrief, the documents in each split are the same as in sequential sentence classification.

Language Model. To achieve at least somewhat comparable results all three models are fine-tuning the same model, the base version of XLM-RoBERTa. We chose a multilingual model for the text classification comparison as two of our datasets are in Polish and one is in English.

Hyperparameter Settings. Similarly to sequential sentence classification, we use AdamW optimizer with a learning rate of $1e-5$ and weight decay of 0.01 and linear schedule with warmup scheduler with 10% warm up steps. For CLARIN-Emo and ChatGPT-Emo we find patience of 10 enough but we set it as 30 for StockBrief.

6   Results

Table 3 summarizes the results of the sequential sentence classification task on the CLARIN-Emo dataset. Figure 5 illustrates the results of the Polish DistilRoBERTa-based model on the Stockbrief-GPT dataset.

For the CLARIN-Emo dataset, regarding the macro F1-score, the personalized approach using UserID with Polish DistilRoBERTa surpassed the others, about 6 pp better than the XLM-RoBERTa-Large. However, concerning the micro F1-score, both models performed comparably, with 85.68% and 84.08% respectively. Overall, the UserID model achieved good performance on almost emotional labels, followed by the XLM-RoBERTa-Large, and in the final position is the SSC multilabel based on Polish DistilRoBERTa. *disgust* is the only emotion that Polish DistilRoBERTa got slightly higher than the other two models. Still, the difference is not significant, just about 2 pp. Besides, *sadness* is another emotion that UserID performed worse than the other two models, with a difference of about 6–7 pp. In addition, the lowest F1-scores were from two emotions, *surprise* and *fear* for all models. However, there is a notable difference between the highest (UserID) and the lowest (Polish DistilRoBERTa), around 24 pp and 11 pp for *surprise* and *fear*, respectively. Since these are two emotions with the lowest inter-annotator agreement and the lowest frequency in label distribution, it could explain this phenomenon and strengthen the benefit of personalized approaches.

Regarding the results from the Stockbrief-GPT dataset, while both micro F1-score (53.64%) and macro F1-score (47.72%) were not considered high and the lowest score being for *fear* (29.9%), it achieved considerably better scores on some emotions, such as *anticipation* and *joy*.

In the case of text classification, we achieved the highest results on ChatGPT-Emo (Fig. 8). It achieves a macro-F1 score of 71.65%, with an abnormally high

Table 3. Sequential Sentence Multilabel Classification F1-score (%) on CLARIN-Emo.

F1-scores	PL-DistilRoBERTa	PL-DistilRoBERTa UserID	XLM-RoBERTa-Large
Micro	76.73	**85.68**	84.08
Macro	68.74	**76.56**	70.31
Joy	84.36	**91.18**	88.18
Trust	70.44	**78.41**	59.29
Anticipation	55.94	**66.37**	60.73
Surprise	34.14	**58.85**	45.19
Fear	41.97	**52.82**	43.04
Sadness	85.82	79.85	**87.36**
Disgust	**67.17**	64.57	64.96
Anger	65.22	**69.63**	68.06
Positive	86.20	**92.06**	88.80
Negative	87.15	**93.00**	90.04
Neutral	77.72	**95.46**	77.75

F1-score for *disappointed* label (98.05%). We achieve satisfactory results on the CLARIN-Emo dataset (Fig. 6). As the text classification setup did not account for class balancing we can see the classifier struggle with rare labels but otherwise achieve a good performance. As presented in Fig. 7, our classifier struggled on many of the emotion labels in Stockbrief-GPT, achieving only 22.86% macro F1-score. One of the possible explanations for this is too few training examples combined with a big class imbalance.

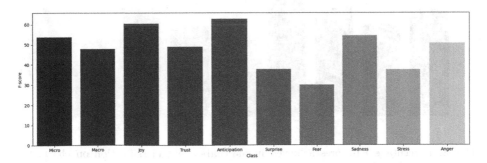

Fig. 5. Percentage F1-scores achieved in Sequential sentence multilabel classification on Stockbrief-GPT.

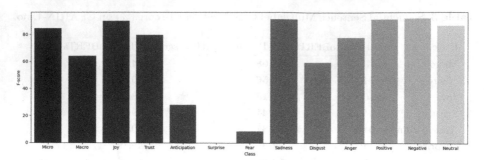

Fig. 6. Percentage F1-scores achieved in multilabel text classification on CLARIN-Emo.

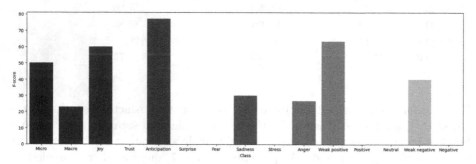

Fig. 7. Percentage F1-scores achieved in multilabel text classification on Stockbrief-GPT

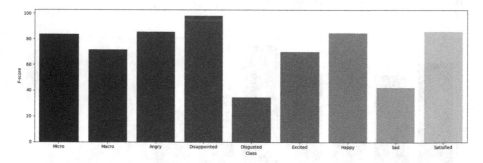

Fig. 8. Percentage F1-scores achieved in multilabel text classification on chatGPT-Emo.

7   Conclusions and Future Work

In this paper, we analyzed several datasets annotated with emotions. The CLARIN-Emo dataset was the only dataset where both the text and the annotations came from a human. The ChatGPT-Emo dataset was generated entirely by the ChatGPT model, while the Stockbrief-GPT dataset contained text written by a human and annotated by ChatGPT. We acknowledge that the results are not directly comparable. At the same time, we note that emotion recognition is a difficult task. Manual annotation at the sentence level for the CLARIN-Emo set yielded a high-quality model trained on human labels (F1-macro at 68.74%). In paper [16], we showed that ChatGPT achieves only 53.23% on the same task (a decrease of about 15pp). In addition, due to the subjective nature of the task, we showed that adding human information to this set further increases F1-macro to 76.56% (an increase of about 8pp). If we train the model on the labels generated by the system at the sentence level, as in the case of the stock letter GPT set, we can obtain a model with an F1-macro quality of 47.72%, which requires further research as it remains to be seen what is the source of the uncertainty: the underdeterminism of the ChatGPT model or the lack of world knowledge to solve such a pragmatic task by other transformer-based models. Intuitively, however, one can assume that ChatGPT was not specifically trained to solve this type of task, hence the lower quality of both the annotation and the system trained on such an annotation.

The third dataset tested was built entirely with ChatGPT, i.e. both text and labels were generated by the model. Such a setup ensured a high consistency of emotion labeling for the text, yielding an F1-macro of 71.64% and an F1-micro of 83.71%. This is a much better result than the annotation evaluation performed by ChatGPT on human-written text. However, this may be due to the fact that the text generation prompt does not allow to enforce the diversity of the generated content, and many texts are semantically similar to each other. In further work, we plan to build a text generator based on random words from different domains, so that we can partially force a greater diversity of generated opinions.

References

1. Acheampong, F.A., Wenyu, C., Nunoo-Mensah, H.: Text-based emotion detection: advances, challenges, and opportunities. Eng. Rep. **2**(7), e12189 (2020)
2. Antaki, F., Touma, S., Milad, D., El-Khoury, J., Duval, R.: Evaluating the performance of ChatGPT in ophthalmology: an analysis of its successes and shortcomings. medRxiv (2023)
3. Baran, J., Kocoń, J.: Linguistic knowledge application to neuro-symbolic transformers in sentiment analysis. In: 2022 IEEE International Conference on Data Mining Workshops (ICDMW), pp. 395–402. IEEE (2022)
4. Bravo-Marquez, F., Frank, E., Mohammad, S.M., Pfahringer, B.: Determining word-emotion associations from tweets by multi-label classification. In: 2016 IEEE/WIC/ACM International Conference on Web Intelligence (WI), pp. 536–539. IEEE (2016)
5. Chen, S.Y., Hsu, C.C., Kuo, C.C., Ku, L.W., et al.: EmotionLines: an emotion corpus of multi-party conversations. arXiv preprint arXiv:1802.08379 (2018)
6. Chen, Y., Eger, S.: Transformers go for the LOLs: generating (humourous) titles from scientific abstracts end-to-end (2022)
7. Cohan, A., Beltagy, I., King, D., Dalvi, B., Weld, D.: Pretrained language models for sequential sentence classification. In: Proceedings of the 2019 Conference on Empirical Methods in Natural Language Processing and the 9th International

Joint Conference on Natural Language Processing (EMNLP-IJCNLP), Hong Kong, China, pp. 3693–3699. Association for Computational Linguistics (2019)

8. Conneau, A., et al.: Unsupervised cross-lingual representation learning at scale. CoRR abs/1911.02116 (2019)

9. Demszky, D., Movshovitz-Attias, D., Ko, J., Cowen, A., Nemade, G., Ravi, S.: GoEmotions: a dataset of fine-grained emotions. arXiv preprint arXiv:2005.00547 (2020)

10. Desai, S., Caragea, C., Li, J.J.: Detecting perceived emotions in hurricane disasters. arXiv preprint arXiv:2004.14299 (2020)

11. Ekman, P.: Are there basic emotions? Psychol. Rev. **99**(3), 550–553 (1992)

12. Gao, C.A., et al.: Comparing scientific abstracts generated by ChatGPT to original abstracts using an artificial intelligence output detector, plagiarism detector, and blinded human reviewers. bioRxiv (2022)

13. Kazienko, P., et al.: Human-centred neural reasoning for subjective content processing: hate speech, emotions, and humor. Inf. Fusion (2023)

14. Kim, Y., Jeong, S.R., Ghani, I.: Text opinion mining to analyze news for stock market prediction. Int. J. Adv. Soft Comput. Appl. **6**(1), 2074–8523 (2014)

15. Kocoń, J., et al.: Neuro-symbolic models for sentiment analysis. In: Groen, D., de Mulatier, C., Paszynski, M., Krzhizhanovskaya, V.V., Dongarra, J.J., Sloot, P.M.A. (eds.) ICCS 2022. LNCS, pp. 667–681. Springer, Cham (2022). https://doi.org/10.1007/978-3-031-08754-7_69

16. Kocoń, J., et al.: ChatGPT: jack of all trades, master of none. arXiv preprint arXiv:2302.10724 (2023)

17. Kocoń, J., et al.: Learning personal human biases and representations for subjective tasks in natural language processing. In: 2021 IEEE International Conference on Data Mining (ICDM), pp. 1168–1173. IEEE (2021)

18. Kocoń, J., Miłkowski, P., Kanclerz, K.: MultiEmo: multilingual, multilevel, multidomain sentiment analysis corpus of consumer reviews. In: Paszynski, M., Kranzlmüller, D., Krzhizhanovskaya, V.V., Dongarra, J.J., Sloot, P.M.A. (eds.) ICCS 2021. LNCS, vol. 12743, pp. 297–312. Springer, Cham (2021). https://doi.org/10.1007/978-3-030-77964-1_24

19. Kocoń, J., Miłkowski, P., Zaśko-Zielińska, M.: Multi-level sentiment analysis of PolEmo 2.0: extended corpus of multi-domain consumer reviews. In: Proceedings of the 23rd Conference on Computational Natural Language Learning (CoNLL), pp. 980–991 (2019)

20. Korczyński, W., Kocoń, J.: Compression methods for transformers in multidomain sentiment analysis. In: 2022 IEEE International Conference on Data Mining Workshops (ICDMW), pp. 419–426. IEEE (2022)

21. Liu, Y., et al.: RoBERTa: a robustly optimized BERT pretraining approach (2019)

22. Maynard, D.G., Greenwood, M.A.: Who cares about sarcastic tweets? Investigating the impact of sarcasm on sentiment analysis. In: LREC 2014 Proceedings. ELRA (2014)

23. Miłkowski, P., Gruza, M., Kazienko, P., Szołomicka, J., Woźniak, S., Kocoń, J.: Multi-model analysis of language-agnostic sentiment classification on multiemo data. In: Nguyen, N.T., Manolopoulos, Y., Chbeir, R., Kozierkiewicz, A., Trawiński, B. (eds.) ICCCI 2022. LNCS, pp. 163–175. Springer, Cham (2022). https://doi.org/10.1007/978-3-031-16014-1_14

24. Minaee, S., Kalchbrenner, N., Cambria, E., Nikzad, N., Chenaghlu, M., Gao, J.: Deep learning based text classification: a comprehensive review (2020)

25. Mirzaee, H., Peymanfard, J., Moshtaghin, H.H., Zeinali, H.: ArmanEmo: a Persian dataset for text-based emotion detection. arXiv preprint arXiv:2207.11808 (2022)

26. Ngo, A., Candri, A., Ferdinan, T., Kocoń, J., Korczynski, W.: StudEmo: a non-aggregated review dataset for personalized emotion recognition. In: Proceedings of the 1st Workshop on Perspectivist Approaches to NLP@ LREC2022, pp. 46–55 (2022)

27. Paul, M.J., et al.: Social media mining for public health monitoring and surveillance. In: Biocomputing 2016: Proceedings of the Pacific Symposium, pp. 468–479. World Scientific (2016)

28. Păvăloaia, V.D., Teodor, E.M., Fotache, D., Danileţ, M.: Opinion mining on social media data: sentiment analysis of user preferences. Sustainability 11(16), 4459 (2019)

29. Phillips, T., Saleh, A., Glazewski, K.D., Hmelo-Silver, C.E., Mott, B., Lester, J.C.: Exploring the use of GPT-3 as a tool for evaluating text-based collaborative discourse. In: Companion Proceedings of the 12th, p. 54 (2022)

30. Plutchik, R.: A general psychoevolutionary theory of emotion. In: Theories of Emotion, pp. 3–33. Elsevier (1980)

31. Rashkin, H., Bosselut, A., Sap, M., Knight, K., Choi, Y.: Modeling Naive psychology of characters in simple commonsense stories. arXiv preprint arXiv:1805.06533 (2018)

32. Reimers, N., Gurevych, I.: Making monolingual sentence embeddings multilingual using knowledge distillation. In: Proceedings of the 2020 Conference on Empirical Methods in Natural Language Processing. Association for Computational Linguistics (2020)

33. Rotsztejn, J., Hollenstein, N., Zhang, C.: ETH-DS3Lab at SemEval-2018 task 7: effectively combining recurrent and convolutional neural networks for relation classification and extraction. In: Proceedings of the 12th International Workshop on Semantic Evaluation, New Orleans, Louisiana, pp. 689–696. Association for Computational Linguistics (2018)

34. Sanh, V., Debut, L., Chaumond, J., Wolf, T.: DistilBERT, a distilled version of BERT: smaller, faster, cheaper and lighter (2019)

35. Sprugnoli, R., et al.: Multiemotions-it: a new dataset for opinion polarity and emotion analysis for Italian. In: Proceedings of the Seventh Italian Conference on Computational Linguistics (CLiC-it 2020), pp. 402–408. Accademia University Press (2020)

36. Susnjak, T.: ChatGPT: the end of online exam integrity? arXiv preprint arXiv:2212.09292 (2022)

37. Szołomicka, J., Kocon, J.: MultiAspectEmo: multilingual and language-agnostic aspect-based sentiment analysis. In: 2022 IEEE International Conference on Data Mining Workshops (ICDMW), pp. 443–450. IEEE (2022)

38. Tabone, W., de Winter, J.: Using ChatGPT for human-computer interaction research: a primer. Manuscript Submitted for Publication (2023)

B^2-FedGAN: Balanced Bi-directional Federated GAN

Ali Anaissi[1]([✉]) [iD] and Basem Suleiman[1,2] [iD]

[1] School of Computer Science, University of Sydney, Sydney, Australia
{ali.anaissi,basem.suleiman}@sydney.edu.au
[2] School of Computer Science and Engineering, The University of New South Wales, Sydney, Australia
b.suleiman@unsw.edu.au

Abstract. In Federated Learning (FL), a shared model is learned across dispersive clients each of which often has small and heterogeneous data. As such, datasets in FL setting may suffer from the non-IID (Independent and identically distributed) problem. In this paper, we propose a BAGAN as machine learning model which has the ability to create data for minority classes, and a Bi-FedAvg model as a new approach to mitigate non-IID problems in FL settings. The performance comparison between FedAvg and Bi-FedAvg in both IID and Non-IID environments will be shown in terms of accuracy, converge stability and category cross-entropy loss. On the other hand, the training and testing performance among FedAvg, FedAvg with a conditional GAN model, and FedAvg with BAGAN-GP model, on IID and Non-IID environments with three imbalanced datasets will be compared and discussed. The results indicate that Bi-FedAvg fails to outperform Fed-Avg, for Bi-FedAvg suffers from model quality loss or even divergence when running on non-IID data partitions. In addition to that, our experiments demonstrate that higher quality images for complex image datasets can be generated by BAGAN and combining federated learning and Balancing GAN model together is conducive to obtaining a high-level privacy-preserving capability and achieving more competitive model performance. The project will give an inspired further exploration of the implementation of a combination between Federated learning and BAGAN on image classification in real-world scenarios.

Keywords: GAN · Federated Learning · Imbalanced dataset

1 Introduction

Deep learning methods have been employed extensively in the area of image classification. Due to its complexities, deep learning models require a large number of samples for learning process. In such domain such as medical applications it is often not possible to collect large datasets in one single hospitals (a.k.a clients). At the same time, it's also hard to combine data from different clients at one

J. Mikyška et al. (Eds.): ICCS 2023, LNCS 14073, pp. 380–392, 2023.
https://doi.org/10.1007/978-3-031-35995-8_27

single central location because of their privacy regulations [7]. In this sense, Federated Learning (FL) became a dominant solution to this kind of problem [9]. In Federated Learning, only local client data will be utilized to train a local model rather than be gathered to a central server, and the central server will subsequently aggregate only the local model coefficients into the global model. The privacy of users is well safeguarded in Federated Learning since no other parties will ever have access to client data.

Nevertheless, in real situations, data from different clients are very heterogeneous. In other words, data of different classes are not evenly distributed in multiple clients [3]. The clients may have too few samples of certain classes, or missing some classes. This is the non independent and identically distributed, or Non-IID data problem which is currently an open research question in federated learning. For instance, the predicting accuracy of standard federated learning models is lower than 55% with highly skewed Non-IID data, where each client only contains one class of data, compared with IID type of data [20].

This paper proposes a novel federated learning method based on a balance GAN (BGAN) model to handle the problem of imbalanced classes in datasets, and a bi-directional FedAvg method to mitigate the effect of Non-IID problem. The adversarial module in BGAN will learn the pattern of both majority and minority classes in the dataset, and its generative model will generate images for the minority classes.

The rest of this paper is organized as follows. In Sect. 2, various related studies are discussed. In Sect. 3, we present our approach and data pre-processing methodology. Section 4 presents our experiment and results. Finally, the conclusions are drawn in Sect. 5.

2 Related Work

Many studies underscore the potential of deep learning will help to identify complex patterns in medical industry. For that, sufficiently large and diverse datasets are required for training. However, as multi-institutional collaborations centrally share the patient data and as a result, there will be privacy and ownership challenges [17]. Federated Learning have been recently considered as a potential solution for building a predictive model in medical industry without sharing patient data to the Central model [14].

FedAvg algorithm is the original version of federated aggregation algorithm which was initially proposed by McMahan et al. [19]. It uses the local SGD updates to build a global model by taking average model coefficient from a subset of clients with non-IID data. In FedAvg, a central server is used to communicate between clients without accessing the data in local server. For Federated Learning, three main challenges have attracted several researches to improve the FedAvg algorithm. (1) how to handle non-identically distributed (Non-IID) data across the network (statistical heterogeneity); (2) how to aggregate the coefficient at the central model; and (3) who and when to communicate client's model coefficients. Accordingly, several derived models of FedAvg have been developed with theoretical guarantee. For instance, a method called FedProx is proposed by LiTian [18] to tackle the Non-IID problem. FedProx can be viewed as a generalization and re-parametrization of FedAvg. Despite very little changes to the method itself, but there will be significant implications both in theory and in practice. Compared with FedAvg, one key improvement for FedProx is that it introduces an additional proximal term to the local training, which essentially restricts the local updates to be closer to the latest global model, and that will help the federated training to converge faster [18]. Subsequently, pFedBayes [19] is also proposed to mitigate the non-Non-IID problem. The idea for this model is that each client uses the aggregated global distribution as prior distribution and updates its personal distribution by balancing the construction error over its personal data and the KL divergence with aggregated global distribution. Another algorithm called PC-FedAvg [1,2], it personalizes the resulting support vectors to addresses the problem of Non-IID distribution of data in FL. Simlarly, pFedMe [6] formulates a new bi-level optimization problem by using the Moreau envelope as a regularized loss function. The idea behind that is allowing clients to pursue their own models with different directions, but not stay far away from the reference point. In this sense, the statistical diversity among clients will be smaller. Therefore, the above algorithms could help us to think about how to improve the current model or make some changes into the model for the Non-IID problem and ultimately get a better performance.

A very recent method propsed by LiZheng [12] called FedFocus. The idea behind this method is that the training loss of each model is taken as the basis for parameter aggregation weights, and as training layer deepens, a constantly updated dynamic factor is designed to stabilize the aggregation process, which could improve the training efficiency and accuracy. On the other hand, the paper applies it into COVID-19 detection on CXR images. The author claimed that the training efficiency, accuracy and stability was significantly improved by using FedFocus. In this paper the authors utilise a GAN model on client side for learning process. However, datasets are often class-imbalanced, which will negatively affect the accuracy of deep learning classifiers [13]. It is pointed that the existing GANs and their training regimes only work well on balanced datasets but fail to be effective in case of imbalanced datasets.

This paper proposes a bi-directional version of the traditional FedAvg model to alleviate the problem of Non-IID data with a better aggregation method on the client side. Furthermore, our method used an improved version of GAN which has the ability to deal with class-imbalanced datasets.

3   Bi-directional FedAvg Based Balanced GAN for Learning Process

Federated learning uses a distributed framework which allows multiple clients to collaboratively train a machine learning model using their own unique dataset. This is to say that the client trains local data to obtain a local model and aggregates the local model by updating parameters to the central server to obtain a global model. After multiple iterations until meeting a terminating condition, the final model converges to the centralised machine learning result. Based on the research work we reviewed, the accuracy of model decreases when the diversity of data acroos the clients increases [5]. Hence, our new method i.e. Bi-FedAvg (Bi-directional FedAvg) adds an extra computation layer on the client side after the central server has published the aggregated global model. Figure 1 presents our Bi-FedAvg framework architecture.

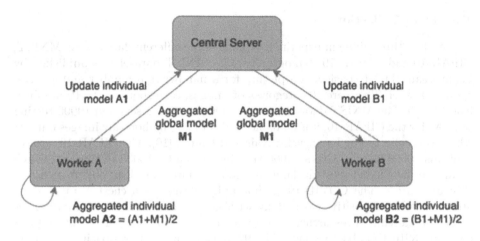

Fig. 1. Core Architecture of Bi-FedAvg.

As can be seen from Fig. 1, Bi-FedAvg averages the current aggregated weights from the global model with the local client model's weights to alleviate the problem that training data differs from user to user as well as not population-specific characteristics.

We implement a sample MLP model which includes two fully-connected layers and one softmax layer. After the training iterations, it is expected that local data contributions will become less divergent. However, it still cannot mitigate the Non-IID problem, which means the model may be negatively affected by Non-IID data among clients. In reality, the majority of clients' data categories are unevenly distributed. Thus, we will use the BAGAN to solve the imbalanced dataset. The BAGAN is a methodology to restore the balance of an imbalanced dataset by using generative adversarial networks. Additionally, it has a higher accuracy of deep-learning classifiers trained over the augmented dataset where the balance has been restored [15]. However, it might also suffer from some problems. For instance, it is unstable when images in different classes look similar. For example, the imbalanced Flowers dataset has many similar classes so BAGAN performs not well. On the other hand, it is hard to train and sensitive to its architecture and hyper-parameters [8]. Therefore, we have adopted the improved version of BAGAN-GP, which can improve the loss function of BAGAN with gradient penalty. Moreover, Huang & Jafari (2021) propose an architecture of autoencoder with an intermediate embedding model, which helps the autoencoder learn the label information directly. In our model architecture, the BAGAN model will run on each client to augment the local data and restore class balance, then federated learning will be applied to the combination of augmented data and original data of the client.

4   Experimental Setup

4.1   Data Collection

We conduct three different experiments using three different datasets i.e. MNIST, CIFAR-10 and COVID-19 Radiography. The MNIST dataset was published by Lecun team [11], it consists of a picture for a handwritten number and a corresponding label. There are 10 categories of images, corresponding to 10 numbers from 0 to 9. The MNIST dataset contains 60000 training sets and 10000 testing sets. As for the CIFAR-10, it is a labelled subset of 80 million tiny images dataset which was collected by Krizhevsky, Nair and Hinton [10]. The CIFAR-10 dataset contains 60000 images with a 50000 training set and a 10000 testing set, which include 10 labelled classes: airplane, automobile, bird, cat, deer, dog, frog, horse, ship and truck. The COVID-19 Radiography dataset is a chest X-ray images about COVID-19 positive cases along with normal and viral pneumonia images created by a team of researchers along with medical doctors [4,16]. This dataset contains 3616 COVID-19 images, 10192 normal, 6012 lung opacity and 1345 viral pneumonia images. Figure 2 shows the distribution of classes in COVID-19 dataset. As can be seen, the COVID-19 dataset is highly imbalanced. The Normal class has 10192 images, which is 3 times as many as COVID-19 class (3616). The details of three datasets, including image resolution, number of classes, and minimum/maximum/total number of images in classes, were shown in Table 1.

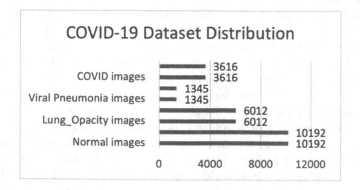

Fig. 2. The distribution of classes in COVID-19 dataset.

Table 1. Datasets information including resolution, number of classes, and min/max/total number of images in classes.

Dataset	Resolution	Classes	Min	Max	Total
MNIST	28 * 28	10	5421	6742	60000
CIFAR-10	32 * 32	10	5000	5000	50000
COVID-19	299 * 299	4	1345	10192	21165

4.2   Results

Comparison Experiments on FedAvg and Bi-FedAvg: The first experiment we conducted was to evaluate the performance of Bi-FedAvg algorithm in both IID and Non-IID settings, and to compare the resultant accuracies with FedAvg using the three different image datasets MNIST [11], CIFAR-10 [10] and COVID-19 Radiography dataset. Initially, we apply the following data pre-processing procedures on the three datasets. RGB images in CIFAR-10 and COVID-19 datasets were converted to grayscale. Images in the COVID-19 dataset were resized to 64 * 64 pixels to avoid large time consumption and insufficient memory problems in the testing environment. After scaling feature vectors to 0 to 1 scale, training and test datasets with a test size of 0.30 were created. Using RELU as the activation function, a multilayer perceptron with two hidden layers and 200 hidden units on each layer was constructed. Each client contained the same number of data points. However, clients in the IID and Non-IID settings had different arrangements of the data. The data was further processed and batched with a batch size of 32 for each client. For FedAvg, an initial global model's weight was set and would serve as the initial weights for all local models. In each communication round for a total of 30 to 300 communication rounds, 10 randomly selected clients were fitted into local models. Local weights for each client were averaged, updated to a global model, and this process was iterated to find the performance on the test dataset. For Bi-FedAvg, an additional average process was implemented between the global model weight and local models' weights before assigning the global weight to each local model.

The experiment was implemented in both IID and Non-IID settings. IID-imbalanced setting indicates that each client has an imbalanced dataset containing all classes. Non-IID setting means that each client has an imbalanced dataset with a random number of classes obtained. Figure 3 showed the training accuracy and categorical cross entropy loss plot on three datasets in the IID environment. It was evident that Bi-FedAvg had upward trends for training accuracy performance on all datasets, however FedAvg only had an upward trend for training accuracy performance on the MNIST dataset, whereas CIFAR-10 and COVID-19 datasets exhibited a brief upward trend for 10 communication rounds followed by a decline trend. The early stopping callback approach was used to terminate training when the validation loss reached a stopping decrease;

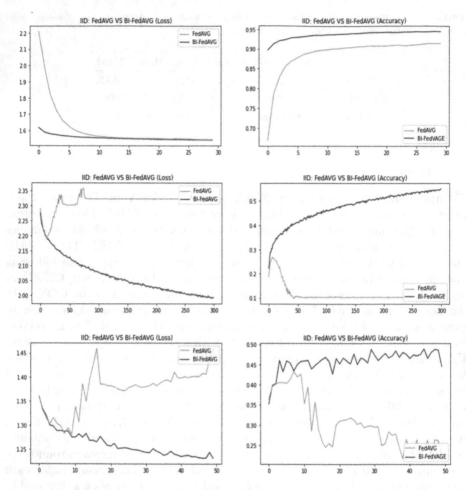

Fig. 3. Training accuracy and categorical cross entropy loss plot of three datasets in IID environment. Top: MNIST. Middle: CIFAR-10. Bottom: COVID-19 dataset.

nevertheless, the performance of Bi-FedAvg outperformed FedAvg on all selected datasets in the IID setting.

Experiments on Bi-FedAvg with CGAN, BAGAN-GP, and Without GAN Model: The second experiment was designed to assess the training and testing performance of several experiment settings, including FedAvg, FedAvg with a conditional GAN model, and FedAvg with BAGAN-GP model, on IID and Non-IID environments with three imbalanced datasets. This experiment utilized the same datasets as the previous one, including the MNIST, CIFAR-10, and

COVID-19 Radiography datasets. We applied the same pre-processing and the structure of multilayer perceptrons as in the first experiment. Before distributing data samples into each client, BAGAN-GP or CGAN were used to restore class balance by generating images for classes with insufficient numbers of images. The data were then combined with real and generated images to create additional data for each client, which was subsequently fitted into local models. The hyperparameters for BAGAN-GP are defined as follows: The optimizer was set to the Adam algorithm with a learning rate of 0.0002 and momentum (0.5, 0.9); the default batch size and dimension of the latent vector were set to 128, and the training ratio of the discriminator to the generator was set to 5 [8]. Each client's images were fitted to BAGAN-GP with 100 learning steps and two epochs per learning step. For CGAN, the discriminator and generator were set to multilayer perceptron with three hidden layers and LeakReLU as activation function. The optimizer was the same as BAGAN-GP and the batch size was set to 512. To restore balance, a random half-batch of images was selected for each epoch, and 2000 epochs were run to generate images. For the IID environment, an additional parameter was set to the randomly selected step to allocate each client with data points from all classes with a proportion of minimum data points in each class. This ensured that there were no situations where certain classes only contain one or two data points.

Figure 4 shows the training accuracy and category cross-entropy loss of several experiments using three datasets in an IID environment. For the performance on the MNIST and COVID-19 datasets, all three experiments had similar training and testing performances. The accuracy plots on the right-hand side increased rapidly on the first 50 communication rounds and turned to increase mildly after the rest of the communication rounds. FedAvg and FedAvg with BAGAN-GP performed slightly better on test datasets, achieving 96.967% accuracy and 1.496 category cross-entropy loss on the MNIST dataset, while FedAvg with CGAN performed slightly better with 81.17% accuracy and 0.931 category cross-entropy loss on COVID-19 dataset. On the Cifar-10 dataset, the growth trend of accuracy plots was relatively more gradual than that on the MNIST dataset. FedAvg with BAGAN-GP had the best performance on the test dataset with 42.41% accuracy and 2.052 categorical cross-entropy loss. Table 2 demonstrated the test accuracy and category cross-entropy loss of three experiments using three datasets in the IID environment.

Table 2. Test accuracy and category cross-entropy loss of three experiments using three datasets in the IID environment.

	MNIST-Accuracy	MNIST-Loss	CIFAR-10-Accuracy	CIFAR-10-Loss	COVID-Accuracy	COVID-Loss
FedAvg	96.97%	1.496	41.15%	2.062	80.42%	0.936
FedAvg+CGAN	96.83%	1.498	40.15%	2.070	81.17%	0.931
FedAvg+BAGAN-GP	96.97%	1.496	42.41%	2.052	80.79%	0.937

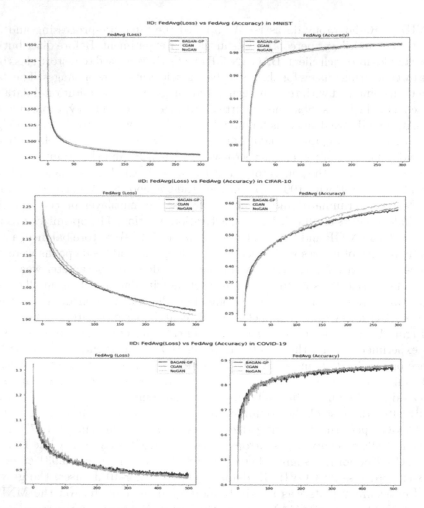

Fig. 4. Training accuracy and category cross-entropy loss of three experiments using three datasets in the IID environment.

For Non-IID environments, a random selection function was assigned to each client in order to distribute data points from a random number of classes with a minimum percentage of data points in each class.

Figure 5 depicted the training accuracy and category cross-entropy loss of several experiments using three datasets in the non-IID environment. For the performance on the MNIST dataset, the training and testing performances of all three experiments are similar. Before 50 communication rounds, the curve displayed a few mild undulations. FedAvg with the BAGAN-GP model gets the greatest performance on the test dataset, with 96.23 percent accuracy and 1.508 percent category cross-entropy loss. On the CIFAR-10 dataset, the performance plots of all training experiments demonstrated large fluctuations, but an overall upward

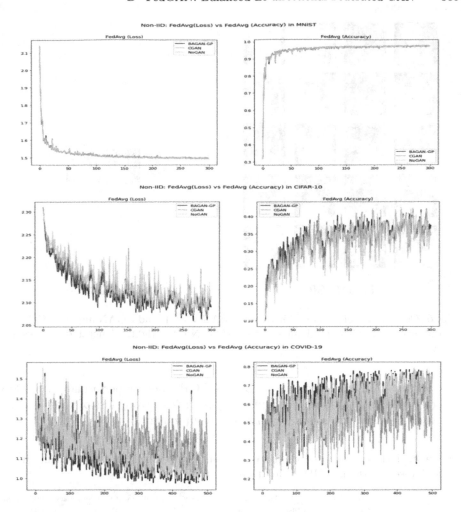

Fig. 5. Training accuracy and category cross-entropy loss of three experiments using three datasets in the non-IID environment

trend. FedAvg with the BAGAN-GP model achieved the best performance on the test dataset, with 32.29% accuracy and 2.131 category cross-entropy loss. On the COVID-19 dataset, the performance plots of all training experiments demonstrated an extreme fluctuation, which indicated that the model did not converge within 500 communication rounds. On the test dataset, FedAvg performed the best, followed by FedAvg with BAGAN-GP and FedAvg with CGAN. FedAvg and FedAvg with BAGAN-GP had a similar upward trend; however, FedAvg with BAGAN-GP had a greater peak for the same variation during the training process as FedAvg without GAN model, indicating that BAGAN-GP should have a superior performance. Possible mistakes or volatility in the model's learning processes may have impacted the final accuracy or category cross-entropy

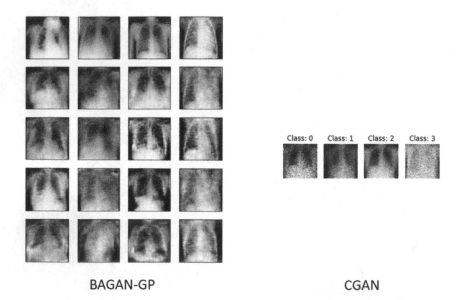

BAGAN-GP   CGAN

Fig. 6. Comparison of generated images for BAGAN-GP and CGAN model in COVID-19 dataset. Left: Generated images for BAGAN-GP. Right: Generated images for CGAN.

loss on test datasets, which led to BAGAN-GP's poorer performance. Table 3 demonstrated the test accuracy and category cross-entropy loss of three experiments using three datasets in the non-IID environment.

Table 3. Test accuracy and category cross-entropy loss of three experiments using three datasets in the non-IID environment.

	MNIST-Accuracy	MNIST-Loss	CIFAR-10-Accuracy	CIFAR-10-Loss	COVID-Accuracy	COVID-Loss
FedAvg	96.17%	1.508	29.04%	2.156	73.30%	1.028
FedAvg+CGAN	95.64%	1.513	29.05%	2.158	65.53%	1.090
FedAvg+BAGAN-GP	96.23%	1.508	32.29%	2.131	67.47%	1.063

Furthermore, Fig. 6 demonstrated the comparison of generated images for the BAGAN-GP and CGAN model in the COVID-19 dataset. It was evident that BAGAN-GP generated higher-quality images than CGAAN, especially when images had minor differences between classes. The conditional GAN model did not help to increase the performance due to its relatively simple architecture that could not generate high quality images for complex image datasets.

5 Conclusion

We have also introduced BAGAN as a data augmentation tool to resolve the heterogeneous and imbalanced data distribution problem in federated learning. The purpose of using BAGAN is to restore class balance in client data and improve local data quality. Experiment results showed that BAGAN outperforms our baseline model and CGAN in most of the scenarios. Different from other data augmentation methods in federated learning where client data are shared with the server to train a good generator, our GAN model is trained specifically on each client using only client data. While the majority of data augmentation methods used data sharing and may raise the risk of data privacy leakage, our approach can protect client data privacy because no local data or labelling information is shared with the central server. Our research offers a new data augmentation framework for the generative adversarial network (GAN) based method that can enhance model performance while maintaining client data privacy. In the field of bio-medics, gathering and distributing large amounts of medical photographs appears to be impossible, in part because of the lack of sufficient public access to sensitive data and patient privacy concerns. However, research has found that most patients are willing to volunteer their data for research purposes if sufficient precautions have been taken to protect their privacy. Therefore, our framework provides a potential solution to break the boundary of data sharing limitations without leaking patient-sensitive data.

Acknowledgement. We would like to thank Zhibin Ye, Zekai Zhang, Yinxuan Ding, and Yiting Li who helped in carrying out data analysis and model implementation in this project.

References

1. Anaissi, A., Suleiman, B., Alyassine, W.: Personalised federated learning framework for damage detection in structural health monitoring. J. Civ. Struct. Health Monit. **13**, 1–14 (2022)
2. Anaissi, A., Suleiman, B., Alyassine, W.: A personalized federated learning algorithm for one-class support vector machine: an application in anomaly detection. In: Groen, D., de Mulatier, C., Paszynski, M., Krzhizhanovskaya, V.V., Dongarra, J.J., Sloot, P.M.A. (eds.) ICCS 2022, Part IV. LNCS, vol. 13353, pp. 373–379. Springer, Cham (2022). https://doi.org/10.1007/978-3-031-08760-8_31
3. Cao, L.: Beyond I.I.D.: non-IID thinking, informatics, and learning. IEEE Intell. Syst. **37**(4), 5–17 (2022). https://doi.org/10.1109/MIS.2022.3194618
4. Chowdhury, M., et al.: Can AI help in screening viral and COVID-19 pneumonia? IEEE Access **8**, 132665–132676 (2020). https://doi.org/10.1109/ACCESS.2020.3010287
5. Deng, Y., Kamani, M.M., Mahdavi, M.: Adaptive personalized federated learning (2020). https://doi.org/10.48550/ARXIV.2003.13461, https://arxiv.org/abs/2003.13461
6. Dinh, C.T., Tran, N.H., Nguyen, T.D.: Personalized federated learning with moreau envelopes (2020). https://doi.org/10.48550/ARXIV.2006.08848, https://arxiv.org/abs/2006.08848
7. Ha, T., Dang, T.K., Le, H., Truong, T.A.: Security and privacy issues in deep learning: a brief review. SN Comput. Sci. **1**(5), 1–15 (2020). https://doi.org/10.1007/s42979-020-00254-4

8. Huang, G., Jafari, A.: Enhanced balancing GAN: minority-class image generation. Neural Comput. Appl. **35**, 5145–5154 (2021). https://doi.org/10.1007/s00521-021-06163-8
9. Kaissis, G., Makowski, M.R., Rückert, D., Braren, R.F.: Secure, privacy-preserving and federated machine learning in medical imaging. Nat. Mach. Intell. **2**, 305–311 (2020). https://doi.org/10.1038/s42256-020-0186-1
10. Krizhevsky, A., Hinton, G.: Learning multiple layers of features from tiny images. Technical report 0, University of Toronto, Toronto, Ontario (2009)
11. Lecun, Y., Bottou, L., Bengio, Y., Haffner, P.: Gradient-based learning applied to document recognition. Proc. IEEE **86**(11), 2278–2324 (1998). https://doi.org/10.1109/5.726791
12. Li, Z., et al.: Integrated CNN and federated learning for COVID-19 detection on chest X-ray images. IEEE/ACM Trans. Comput. Biol. Bioinform. 1–11 (2022). https://doi.org/10.1109/TCBB.2022.3184319
13. Li, Z., Shao, J., Mao, Y., Wang, J.H., Zhang, J.: Federated learning with GAN-based data synthesis for non-IID clients (2022). https://doi.org/10.48550/ARXIV.2206.05507, https://arxiv.org/abs/2206.05507
14. Liu, Y., Zhang, L., Ge, N., Li, G.: A systematic literature review on federated learning: from a model quality perspective (2020). https://doi.org/10.48550/ARXIV.2012.01973, https://arxiv.org/abs/2012.01973
15. Mariani, G., Scheidegger, F., Istrate, R., Bekas, C., Malossi, C.: BAGAN: data augmentation with balancing GAN (2018). https://doi.org/10.48550/ARXIV.1803.09655, http://arxiv.org/abs/1803.09655
16. Rahman, T., et al.: Exploring the effect of image enhancement techniques on COVID-19 detection using chest X-rays images (2020). https://doi.org/10.48550/ARXIV.2012.02238, https://arxiv.org/abs/2012.02238
17. Sheller, M.J., et al.: Federated learning in medicine: facilitating multi-institutional collaborations without sharing patient data. Sci. Rep. **10** (2020). https://doi.org/10.1038/s41598-020-69250-1
18. Su, L., Xu, J., Yang, P.: A non-parametric view of FedAvg and FedProx: beyond stationary points (2021). https://doi.org/10.48550/ARXIV.2106.15216, https://arxiv.org/abs/2106.15216
19. Zhang, X., Li, Y., Li, W., Guo, K., Shao, Y.: Personalized federated learning via variational Bayesian inference. In: International Conference on Machine Learning, pp. 26293–26310. PMLR (2022)
20. Zhao, Y., Li, M., Lai, L., Suda, N., Civin, D., Chandra, V.: Federated learning with non-IID data (2018). https://doi.org/10.48550/ARXIV.1806.00582, https://arxiv.org/abs/1806.00582

Graph-Level Representations Using Ensemble-Based Readout Functions

Jakub Binkowski[✉][iD], Albert Sawczyn[iD], Denis Janiak[iD], Piotr Bielak[iD],
and Tomasz Kajdanowicz[iD]

Department of Artificial Intelligence, Wroclaw University of Science and Technology,
Wrocław, Poland
jakub.binkowski@pwr.edu.pl

Abstract. Graph machine learning models have been successfully deployed in various application areas. One of the most prominent types of models – Graph Neural Networks (GNNs) – provides an elegant way of extracting expressive node-level representation vectors, which can be used to solve node-related problems, such as classifying users in a social network. However, many tasks require representations at the level of the whole graph, e.g., molecular applications. In order to convert node-level representations into a graph-level vector, a so-called *readout* function must be applied. In this work, we study existing readout methods, including simple non-trainable ones, as well as complex, parametrized models. We introduce a concept of ensemble-based readout functions that combine either representations or predictions. Our experiments show that such ensembles allow for better performance than simple single readouts or similar performance as the complex, parametrized ones, but at a fraction of the model complexity.

Keywords: Graph Neural Networks · Readout · Pooling · Graph Classification · Graph Regression · Machine Learning

1 Introduction

In recent years, machine learning has seen dramatic progress in areas where data is more complex, irregular, or structured. The abundance of methods treating objects of various structured domains eventually led to the movement of Geometric Deep Learning [3], which tries to characterize neural network architectures through the lens of data geometry and symmetries. In particular, research advancements in graph neural networks resulted in multiple novel architectures being able to address symmetries present in graph objects. One of the most important milestones was the development of Graph Neural Networks (GNNs), which rely on the convolution operation adapted to the irregular graph structure. GNNs can learn node representations by transforming initial input features throughout several message-passing layers. The final node representations might be further utilized in a task of interest and are often learned end-to-end using a task-related objective.

J. Mikyška et al. (Eds.): ICCS 2023, LNCS 14073, pp. 393–405, 2023.
https://doi.org/10.1007/978-3-031-35995-8_28

In the realm of Graph Machine Learning, one can distinguish several types of tasks. Among them, the most prevalent are link prediction, node-level prediction (classification or regression) and graph-level prediction (classification or regression). It turns out that for node-level tasks, using node representations obtained from message-passing GNNs is straightforward as those representations could be immediately passed to the layer producing final predictions per each node. However, in the case of graph-level tasks, before the classification stage, one has to obtain a global graph-level representation vector, which summarizes the entire object. The common practice is to aggregate node representations from the last or all GNN layers through a so-called **readout function** [12]. Since graph-level representations must be invariant to node permutations, a readout function has to aggregate node embeddings regardless of their ordering, i.e., for each possible node permutation, it must return the same value. This enforces a significant constraint and poses a real challenge. Since graph-level tasks are of great importance, e.g., in fields like chemistry or biology, and their performance strongly depends on the readout function, this work focuses on the evaluation of various approaches to perform readout. Additionally, we include proposed ensemble methods and discuss obtained results to give an intuition for choosing the proper readout for a problem of interest.

In the simplest scenario, one might leverage permutation-invariant functions, such as: sum, mean, or max. However, as shown by Xu et al. [28], these approaches might be suboptimal and lead to significant data loss. Along with the advancements in various aggregation schemes for GNN message-passing, many works attempted to refine the readout function for graph-level tasks. Recently, Buterez et al. [5] performed a large-scale evaluation of various readouts, giving several unexpected insights. In this work, we aim to perform a follow-up study of selected readouts. In particular, we introduce an ensemble approach to readouts, in which we first perform several readout functions in parallel and then aggregate results in two scenarios: (1) representations of all readouts are aggregated, and the result is passed for prediction, (2) prediction is performed over embeddings from each readout individually, and then outcomes are aggregated.

Contributions. This paper is aimed at bringing the following contributions:

1. We introduce ensemble-based readout models at representation and prediction levels, which help obtain rich graph representation from nodes' representations and outperform non-invariant SOTA results obtained with MLP and GRU models.
2. We perform a comparative study of readouts on datasets with varying sizes, characteristics, and target tasks.
3. We analyse the obtained results concerning the computational burden of considered readout functions, giving guidelines for readout function selection.

2   Related Work

2.1   Graph Neural Networks

The early approaches, which enabled to perform tasks, like graph classification, involved graph kernels [16]. However, due to their limitations, e.g., the necessity of manually designing combinatorial features of a graph, as well as big advancements in deep learning, Graph Neural Networks were proposed. GNNs arose as a generalization of convolution to irregular graph structures. The groundbreaking works in graph machine learning proposed to perform graph convolution in a spectral domain, leveraging eigendecomposition of the graph Laplacian [4], or by employing polynomial spectral filters [8]. Gilmer et al. [12] proposed to unify GNNs through the message-passing framework, showing that previous convolution operators are its special cases. The rapid growth in the field led to many variants of Message Passing Neural Networks (MPNNs), among which the most renowned ones are GCN [15], GAT [25], GIN [28]. Such MPNNs were widely used in tasks like graph property prediction (classification or regression), node property prediction (node classification or regression), or link prediction.

2.2   Graph-Level Prediction

While MPNNs can transform input features into nodes' representations, predicting at a graph level requires a readout function that summarises a graph's node embeddings into a single vector, further passed to a prediction model (classifier or regressor). Such function should be invariant to node order permutations [3], and it is desired to be injective for the sake of expressivity [28].

The primary approach for readouts leveraged simple permutation-invariant functions, including sum, mean, or max. Xu et al. [28] proved the properties of these functions, showing their advantages and limitations. Although the work considered these functions in the context of neighbourhood aggregation in graph convolution, the same applies to them being readouts. Also, the authors raised the issue of function injectivity and stated that in certain situations, the sum operator might satisfy this property.

Recently, researchers developed more advanced approaches considering permutation invariance with respect to inputs. Most of these works aimed to perform various tasks on sets, which resemble the scenario for readouts. Zaheer et al. [31] presented the DeepSets architecture as the main framework to obtain permutation invariant representations of sets. We adopted this approach in our experimental scenario, and we will discuss it in more detail in the next section. Moreover, the SetTransformer [17] model enabled the efficient application of Transformer [24] architecture to sets, which scales to large-sized inputs.

It is also worth noting that several works proposed to leverage local pooling, which attempts to group similar nodes together iteratively. For instance, Diff-Pool [30] clusters together similar nodes in several iterations, such that in each iteration number of clusters decreases up to only 1 at the last layer, which serves as a graph representation. Besides, there are other local pooling methods, like

Graph Memory Networks [14], or based on GRACLUS [22]. However, as shown by Mesquita et al. [19], there are no significant performance benefits from using local pooling, hence in this work, we do not consider them in the experiments.

Recently, Buterez et al. [5] showed results of various readouts based on large-scale evaluation, stating that in some situations losing permutation-invariance and employing models, like MLP or GRU recurrent neural network [6], might bring superior performance. Moreover, they showed that general adaptive (i.e. parametrized and learnable) readouts often lead to better efficacy with sufficient data. Our work also employs MLP and GRU, however, as opposed to Buterez et al., we also propose to use node-shared parameters (which, in hand with a prediction layer, forms the DeepSets architecture [31]). Moreover, we propose leveraging an ensemble approach that utilizes several readouts in a single model, similar to the aggregation scheme in PNA graph convolution [7].

3   Graph Neural Network Readouts

Graph Neural Networks operate on graph input data, transforming initial node features into rich node representations. Formally, we denote a graph as $\mathcal{G} = (\mathbf{X}, \mathcal{A})$, where $\mathbf{X} \in \mathbb{R}^{|\mathcal{V}| \times d}$ are node features, and A is an adjacency matrix describing the connection between nodes, such that $a_{uv} = \mathbb{1}((u, v) \in \mathcal{E})$ for $\mathcal{E}$ is a set of edges and $u, v \in \mathcal{V}$ are nodes. In this work, we consider unweighted edges. Further, a GNN layer is a function taking a graph as input and transforming it into a latent space – see: Eq. (1). Such formulation describes a generic framework of Message Passing Neural Network. First, it aggregates the neighborhood of a node through a permutation-invariant $\oplus$ function to compute a message, and then it combines the message with the node's representation through the Θ function. Both functions could be parameterized by neural networks. Particular instances of GNNs, like GCN or GIN, vary only in the Θ and $\oplus$ functions. The whole network often comprises several such layers, and their number should be adjusted to a particular dataset.

$$\mathbf{h}_u^{(l+1)} = \Theta^{(l)} \left(\mathbf{h}_u^{(l)}, \oplus^{(l)} \left(\{ \mathbf{h}_v^{(l)}, \forall v \in \mathcal{N}(u) \} \right) \right) \tag{1}$$

3.1   Problem Statement

Formally, a readout function, denoted as $\mathcal{R}$, is an aggregation function that takes a matrix of stacked node embeddings (usually taken from the last GNN layer) at the input and returns a single vector at its output, i.e., $\mathcal{R}(\mathbf{H}^{(L)}) \to \mathbf{z}$, where $\mathbf{H}^{(L)} \in \mathbb{R}^{|\mathcal{V}| \times d_{\mathcal{V}}}$, and $\mathbf{z} \in \mathbb{R}^{d_{\mathcal{G}}}$, where L is the number of layers, $d_{\mathcal{V}}$ dimension of a hidden node representation, and $d_{\mathcal{G}}$ dimension of graph representation. In many scenarios, the dimension of the graph representation vector is equal to the dimension of the nodes' representation, i.e., $d_{\mathcal{G}} = d_{\mathcal{V}}$. Figure 1 visually depicts the concept of the readout function.

Readout function has to be invariant to the permutation of the node order in a graph, meaning that $\mathcal{R}(\mathbf{H}^{(L)}) = \mathcal{R}(\mathbf{P}\mathbf{H}^{(L)})$, where $\mathbf{P}$ is a row-permutation

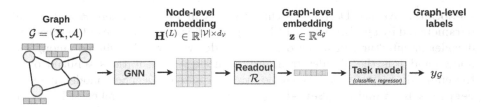

Fig. 1. A readout function in the whole graph-level prediction pipeline.

matrix. As showed by Zaheer et al. [31], permutation-invariant functions over the elements could be expressed as in Eq. (2), where ρ and ϕ could be functions parametrized by neural networks. Zaheer et al. considered such parametrization with MLP networks as the DeepSets architecture. However, it might be hard to train representations in a DeepSets setting, which was theoretically discussed by Wagstaff et al. [27]. Nonetheless, we include DeepSets approach in our experiments as well. Besides, there are several different approaches to performing prediction on sets, e.g., SetTransformer [17], or Set2Set [26].

$$f(X) = \rho \left(\sum_{x \in X} \phi(x) \right) \tag{2}$$

3.2 Non-parametrized Readout Functions

This work considers three basic non-parametrized functions: sum, mean, and max. These are the simplest realizations of readouts which satisfy the necessary invariance conditions. However, there are certain drawbacks to such aggregations. The concerns were raised by Xu et al. [28], who characterized each of the three aggregators with respect to their expressivity, stating that sum could capture the most information from the node embeddings, yet the effectiveness of such readout function might depend on the underlying data. However, as proved by Corso et al. [7], a single aggregation might not satisfy injectivity over real numbers, hence they propose to concatenate results of several aggregations at each convolutional layer. Motivated by these results, we propose an ensemble approach for building readout function. In particular, we evaluate the effectiveness of a readout function which composes sum, mean, and max aggregator into one vector.

3.3 Parametrized Readout Functions

In contrast to sum, mean, and max, parametrized readout functions could be optimized during training. In this work, we consider three such functions: DeepSets, Virtual Node, and an ensemble approach. Also, motivated by the surprising performance of permutation-sensitive methods [5], we include MLP- and GRU-based readouts. Now, let us discuss each parametrized readout in more detail.

DeepSets. We use DeepSets architecture as in Eq. (2), where ϕ and ρ are parametrized by an MLP network. This approach is a special case of the encoder-decoder architecture, in which ϕ is the encoder with weights shared among all nodes, and ρ is the decoder transforming node representations summed over the embedding dimension. We employ two variants of DeepSets model, namely `DeepSets-Base` and `DeepSets-Large`, which differ by the number of layers in the encoder. Both versions have the same 2-layer MLP at the decoder, which predicts the final output.

Virtual Node. In this approach, no explicit readout function is present in the architecture. Instead, the graph adjacency matrix is altered such that it includes one extra node connected with all other nodes in the graph [12]. Then, the graph-level representation is the embedding of this single node trained along with the remaining node embeddings (virtual node features are initialized with zeros). The aggregation function of the graph convolution might be seen as an implicit readout, and due to its permutation-invariance, this approach is also insensitive to the modification in the node ordering.

Dense and GRU. In their recent work, Buterez et al. [5] empirically showed that on many datasets loosening permutation-invariance and leveraging MLP (Dense) or GRU architectures might lead to outperformance over other readouts, including the learnable ones. However, they argue that this approach is particularly suitable for data with consistent node ordering like canonical-SMILES [20] in molecular datasets.

3.4 Ensemble Readouts

In this work, we propose to evaluate ensemble approaches to readout functions that combine results from many readouts. As described by Xu et al. [28], each aggregation from `sum`, `mean`, `max` has specific desirable properties, hence using three of them might bring a performance improvement. We propose to use these three readout functions and consider three approaches to aggregate representation computed with each of them.

1. First, we propose to simply concatenate the results of each aggregation, which are then forwarded to a target predictor. We denote that approach as `ConcatR`, where $\|$ denotes the concatenation operator:

$$\mathbf{z} = \mathcal{R}_1(\mathbf{H}^{(L)}) \;\|\; \mathcal{R}_2(\mathbf{H}^{(L)}) \;\|\; \dots \;\|\; \mathcal{R}_N(\mathbf{H}^{(L)}), \tag{3}$$

2. In the second scenario, we build the final graph-level representation as a weighted sum of each readout's output. This approach is formally expressed by Eq. (4), where $\{\mathcal{R}_1, \dots, \mathcal{R}_N\}$ is a set of N readout functions, $(\mathbf{W}_r^{(\mathbf{p})}, \mathbf{b}_r^{(\mathbf{p})})$ are weights and biases of a linear projection layer specific to the readout r,

and $(w_r^{(c)}, b_r^{(c)})$ are weights and biases of a combination layer. Note that representations from each readout might result in different scales, so we experiment with turning the projection layer on and off.

$$\mathbf{z} = \sum_{r \in \{\mathcal{R}_1,...,\mathcal{R}_N\}} w_r^{(c)} \left(\mathbf{W}_{\mathbf{r}}^{(\mathbf{P})} \mathbf{z}_r + \mathbf{b}_{\mathbf{r}}^{(\mathbf{P})} \right) + b_r^{(c)} \tag{4}$$

We propose two variants of such representation-level ensembles:

- WMeanR (*Weighted Mean of Readouts*) – the representation from each readout is multiplied by a learnable weight and summed together (the projection layer is replaced with an identity map: $\mathbf{W}_{\mathbf{r}}^{(\mathbf{P})} = \mathbb{I}$, $\mathbf{b}_r^{(p)} = \mathbf{0}$),
- WMeanR+Proj (*Weighted Mean of Readouts with Projection layer*) – the representation from each readout is first transformed by the projection layer, and then weighted and summed together (as in the previous model).

3. The last scenario involves the aggregation of results from the three readouts at the prediction level, as shown in Eq. (5). The symbols are the same as in Eq. (4), and $\psi(\cdot)$ denotes a predictor model. Again, we include models with and without a projection layer in our experiments.

$$\hat{\mathbf{y}} = \sum_{r \in \{\mathcal{R}_1,...,\mathcal{R}_N\}} \left(w_r^{(c)} \psi(\mathbf{W}_{\mathbf{r}}^{(\mathbf{P})} \mathbf{z}_r + \mathbf{b}_{\mathbf{r}}^{(\mathbf{P})}) + b_r^{(c)} \right) \tag{5}$$

- MeanPred (*Mean of readouts' Predictions*) – predictions are computed for each readout separately and then averaged over all readouts (the projection layer is replaced with an identity map: $\mathbf{W}_{\mathbf{r}}^{(\mathbf{P})} = \mathbb{I}$, $\mathbf{b}_r^{(p)} = \mathbf{0}$; we also used fixed values for the combination parameters: $w_r^{(c)} = 1, b_r^{(c)} = 0$),
- WMeanPred (*Weighted Mean of readouts' Predictions*) – predictions are computed for each readout separately, then aggregated by learnable weights (the projection layer is replaced with an identity map: $\mathbf{W}_{\mathbf{r}}^{(\mathbf{P})} = \mathbb{I}$, $\mathbf{b}_r^{(p)} = \mathbf{0}$),
- WMeanPred+Proj (*Weighted Mean of readouts' Predictions with Projection layer*) – projection layer is applied to representations from each readout, then predictions are computed and aggregated by learnable weights

Altogether, we introduce **6 different ensemble-based variants of readout functions**.

4 Experiments

Here, we describe experiments and obtained results in detail. First, we elaborate on the datasets and splits, further, we specify the experimental setting, and finally, we discuss the results.

4.1 Datasets

In our experiments, we utilized four datasets, which are MUTAG [16], ENZYMES [1], ZINC [2] (12K subset of the dataset, as proposed in [9]), and REDDIT-MULTI-12K [29]. Datasets were selected to cover various tasks, domains, and graph sizes. Table 1 summarizes datasets statistics. For each dataset from the TUD repository (MUTAG, ENZYMES, REDDIT-MULTI-12K), we performed random splits to *train/val/test* subsets in proportions 80%/10%/10%. In the case of the ZINC dataset, we leverage pre-defined random splits without any further modifications. In the case of binary classification on the MUTAG dataset, we measure performance with the $F1$ score, for multi-class classification on ENZYMES and REDDIT-MULTI-12K datasets, we leveraged the macro-averaged $F1$ score, and for ZINC the R^2 metric was used.

Table 1. Tasks and statistics for each of four datasets used in the experiments.

Dataset	Task	# graphs	Avg. nodes	Avg. edges
ENZYMES	multi-class	600	32.9	125.4
MUTAG	binary	188	18.0	39.9
ZINC	regression	12 000	23.2	49.8
REDDIT-MULTI-12K	multi-class	11 929	391.0	913.3

4.2 Evaluation Protocol

We evaluated three different and most prevalent graph convolutions, i.e., GCN, GAT and GIN, for each readout described in Sect. 3. We set the hyperparameters of each model based on the ones found in [9] with slight modifications when necessary. All hyperparameters specific to each dataset are presented in Table 2. For Dense and GRU readout, we adopted the architectures from Buterez et al. [5]. For DeepSets-Base, we used MLP with two layers of size 128 followed by batch normalization [13] and ReLU activation, and the last layer additionally contains dropout [23] with probability set to 0.4. For DeepSets-Large, we leveraged 6 such layers, followed by dropout with probability set to 0.4. The two considered DeepSets architectures were arranged to resemble Dense approach. It is worth noting that the number of trainable parameters of Dense method is relatively large, even when compared to DeepSets-Large. For the final predictor, i.e., classifier or regressor, the resultant representation of the readout function is always passed to an MLP with one hidden layer of size 128 (three layers including input and output).

Based on the well-established protocol schemes, we apply five times repeated bootstrap evaluation of each combination of graph convolution and readout. The models for each dataset were trained for a minimum of 10 epochs with early stopping on validation set loss and the patience set to 25 epochs. We leverage dynamic learning rate starting from a value of 1×10^{-3}, which was multiplied by 0.5 on each plateau with patience set to 10 epochs (minimum learning rate was set to 1×10^{-6}). Models were optimized with AdamW optimizer [18], following default PyTorch parameters (except for learning rate). In the case of classification, we used cross-entropy loss, and for regression - mean-squared error. Experiments were implemented with, `PyTorch` [21], `torch-geometric` [11] and `pytorch-lightning` [10] libraries. To ensure reproducibility, we provide the full implementation of our experiments, containing entire hyperparameters configurations, in our public git repository https://github.com/graphml-lab-pwr/ensemble-readouts.

Table 2. Hyperparameters specification for each dataset, where L is number of layers (last layer index), $d_{\mathcal{V}}$ is the hidden graph representation dimension, and $d_{\mathcal{G}}$ is the graph representation dimension.

Dataset	L	$d_{\mathcal{V}}$	$d_{\mathcal{G}}$
ENZYMES	3	128	128
MUTAG	3	64	128
ZINC	3	128	128
REDDIT-MULTI-12K	3	128	128

4.3 Results

Following the experimental setup described in the previous section, we obtain results presented in Table 3, which shows the achieved efficacy for each dataset, split by graph convolution and readout (for all metrics *greater = better*). We observe that in most cases the best results are obtained either by *parametrized* or *ensemble* methods. Among all parametrized readouts, non-invariant methods turned out to perform best only on MUTAG, which is the smallest dataset used in our experiments (containing only 188 graphs overall, and only *19 graphs* in the test set). We hypothesize that such results might be caused by overfitting due to a relatively large number of trainable parameters in these models and a fixed node ordering. Therefore, we consider such methods as the worst choice for the readout function. On the other hand, the `DeepSets` architecture outperformed other models for ENZYMES using all types of GNNs, and for ZINC using GCN and GAT. The usage of a `Virtual Node` did not lead to any improvement.

Furthermore, the results showed that **ensemble** approaches outperformed other readouts for REDDIT-MULTI-12K dataset using GCN and GIN layers, as well as using the GIN layer on the ZINC dataset. It is worth noting that when using GIN convolution, the best ensemble approach is consistently the best overall or nearly the best. We did not observe such consistent results for GCN and GAT, yet the **ensemble** approaches provide better or comparative metric values.

Whether to use ensembles over representations or predictions depends on the dataset and graph convolution. Each ensemble enables an additional interpretability property, since the trained weights might be further analyzed to assess the contribution of each readout method.

Regarding the projection layer considered in these models, we observe that projection layers in ensembles do not guarantee better performance and could be omitted for the sake of decreasing model size. Nonetheless, when using a combination of various readouts from non-aligned embedding spaces, one should always verify the necessity of additional projection layers. We also tested a combination of DeepSets with *non-parameterized* methods in an *ensemble* setting, but we did not observe an increase in performance, hence we omitted results for brevity.

Table 3. Results of the conducted experiments. Each cell contains mean and standard deviation in percent of the metric over 5 runs. Metric $F1$ score was used for ENZYMES, MUTAG, REDDIT-MULTI-12K, and R^2 for ZINC (for all metrics *greater = better*). Best results are presented in **bold**, best within a class (*non-parametrized, parametrized, ensemble*) are presented with underline. NON-PAR, PAR and ENS denote non-parametrized, parametrized and ensemble-based readout functions, respectively.

		ENZYMES (F1 ↑)			MUTAG (F1 ↑)		
		GAT	GCN	GIN	GAT	GCN	GIN
NON-PAR	max	56.68 ± 7.43	54.63 ± 4.86	48.29 ± 4.33	77.55 ± 7.40	78.57 ± 6.94	82.61 ± 9.39
	mean	57.60 ± 7.31	45.62 ± 6.74	50.46 ± 2.77	75.99 ± 9.52	76.60 ± 8.59	77.70 ± 8.66
	sum	51.61 ± 12.71	49.23 ± 8.08	51.19 ± 9.65	84.65 ± 6.87	82.62 ± 6.28	81.47 ± 8.83
PAR	DeepSets-Base	62.32 ± 10.09	64.39 ± 6.26	55.66 ± 9.77	66.63 ± 28.74	75.42 ± 13.24	84.68 ± 6.86
	DeepSets-Large	63.65 ± 7.01	58.08 ± 11.49	42.44 ± 11.31	78.00 ± 3.93	72.59 ± 25.46	79.66 ± 5.97
	Dense	59.70 ± 7.58	54.24 ± 5.32	40.43 ± 4.15	83.91 ± 3.89	88.61 ± 8.58	85.77 ± 3.26
	GRU	43.72 ± 7.39	47.56 ± 3.75	27.96 ± 6.42	86.41 ± 5.74	85.90 ± 5.30	86.76 ± 4.02
	Virtual Node	55.49 ± 10.04	45.54 ± 3.99	51.37 ± 8.42	75.77 ± 8.74	78.41 ± 6.73	84.46 ± 7.14
ENS	ConcatR	56.38 ± 8.12	44.76 ± 10.68	46.06 ± 9.24	85.65 ± 7.03	85.63 ± 8.36	85.63 ± 8.36
	WMeanR	50.53 ± 8.03	44.82 ± 7.43	49.04 ± 3.65	85.65 ± 7.03	83.32 ± 8.73	85.26 ± 8.27
	WMeanR+Proj	57.90 ± 5.75	47.09 ± 11.09	49.77 ± 9.11	84.09 ± 6.93	85.65 ± 7.03	86.31 ± 6.08
	MeanPred	50.95 ± 9.59	47.56 ± 11.81	54.49 ± 5.51	83.21 ± 5.61	80.18 ± 9.37	86.20 ± 8.39
	WMeanPred	51.89 ± 10.17	50.06 ± 9.43	48.52 ± 8.90	82.10 ± 7.16	79.62 ± 10.43	84.58 ± 8.13
	WMeanPred+Proj	58.35 ± 9.43	52.59 ± 7.55	50.67 ± 10.82	84.18 ± 8.35	84.95 ± 8.26	86.25 ± 6.95
		REDDIT-MULTI-12K (F1 ↑)			ZINC (R^2 ↑)		
		GAT	GCN	GIN	GAT	GCN	GIN
NON-PAR	max	37.66 ± 1.56	38.15 ± 2.95	39.09 ± 2.12	65.74 ± 0.76	65.88 ± 1.43	71.62 ± 3.41
	mean	34.89 ± 2.73	37.77 ± 1.77	36.53 ± 0.55	68.10 ± 0.57	67.84 ± 0.79	78.29 ± 2.85
	sum	39.02 ± 1.16	40.38 ± 0.97	39.86 ± 0.92	67.88 ± 0.89	68.21 ± 0.26	81.68 ± 1.65
PAR	DeepSets-Base	36.59 ± 2.35	35.42 ± 2.75	39.08 ± 1.82	64.42 ± 9.36	73.82 ± 2.39	75.40 ± 2.17
	DeepSets-Large	37.61 ± 1.71	37.29 ± 2.06	16.90 ± 10.11	70.09 ± 2.77	83.69 ± 3.26	81.96 ± 1.24
	Dense	32.12 ± 1.67	32.65 ± 1.10	34.91 ± 1.36	61.93 ± 1.73	62.57 ± 1.31	69.65 ± 1.39
	GRU	27.49 ± 1.38	26.62 ± 1.49	29.21 ± 2.40	58.23 ± 8.00	65.26 ± 1.20	63.81 ± 2.91
	Virtual Node	37.26 ± 1.60	40.39 ± 0.44	40.37 ± 1.21	67.43 ± 1.44	69.16 ± 0.31	76.67 ± 3.01
ENS	ConcatR	38.62 ± 1.90	41.54 ± 0.75	41.06 ± 1.05	68.08 ± 0.56	67.24 ± 1.33	83.23 ± 2.56
	WMeanR	38.83 ± 0.94	41.02 ± 0.75	39.22 ± 2.58	67.66 ± 0.88	67.08 ± 0.82	81.04 ± 2.00
	WMeanR+Proj	37.70 ± 1.04	40.65 ± 1.17	40.37 ± 1.15	67.97 ± 0.63	67.80 ± 0.26	82.18 ± 2.31
	MeanPred	38.46 ± 1.52	40.59 ± 1.79	40.20 ± 1.25	68.22 ± 0.58	67.90 ± 0.47	78.14 ± 2.24
	WMeanPred	37.56 ± 1.30	39.91 ± 0.98	38.62 ± 2.02	67.28 ± 1.90	68.22 ± 0.37	80.06 ± 3.41
	WMeanPred+Proj	38.18 ± 1.11	40.65 ± 1.95	38.92 ± 1.34	67.77 ± 1.19	66.98 ± 2.22	85.66 ± 2.70

In order to discuss the results from the perspective of model size, we plotted the efficacy achieved by models on each dataset against the number of trainable parameters, presented in Fig. 2. Without accounting for a specific GNN type, the plots reveal that we can often benefit from **ensemble** or **parametrized** approaches which are comparable in size and efficacy. The exception is the dense model represented by the rightmost outlier points, which in connection with a lack of permutation-invariance position it as a poor choice for a readout function.

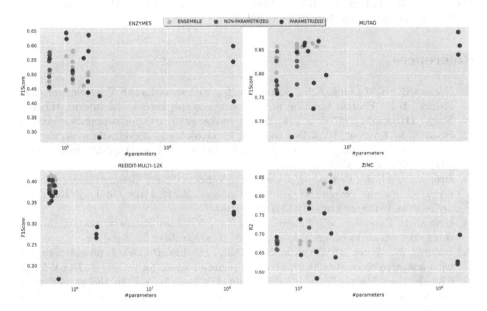

Fig. 2. Comparison of the number of parameters with achieved efficacy for the four considered datasets.

5   Conclusion

In this work, we evaluated various neural network readouts on several datasets covering different domains, cardinalities, graph properties, and tasks. We introduced **ensemble**-based approach that combines basic *non-parametrized* readout functions in several ways and evaluated them against other *parametrized* methods known in the literature, including DeepSets, Virtual Node, and non-invariant MLP and GRU. We showed that DeepSets and **ensemble** approaches outperform other methods in the majority of experiments and hence should be a first choice when designing models for graph-level predictions. We also do not recommend using non-invariant readouts, as they introduce larger models, might lead to overfitting, and not necessarily lead to better generalization. For future work, we suggest investigating probabilistic modelling of node representations, including Bayesian approach, to prevent information loss, which might happen while using deterministic readouts presented in this work.

Acknowledgments. This work was financed by (1) the Polish Ministry of Education and Science, CLARIN-PL; (2) the European Regional Development Fund as a part of the 2014–2020 Smart Growth Operational Programme, CLARIN - Common Language Resources and Technology Infrastructure, project no. POIR.04.02.00-00C002/19; (3) the statutory funds of the Department of Artificial Intelligence, Wroclaw University of Science and Technology, Poland; (4) Horizon Europe Framework Programme MSCA Staff Exchanges grant no. 101086321 (OMINO).

References

1. Borgwardt, K.M., Ong, C.S., Schönauer, S., Vishwanathan, S.V.N., Smola, A.J., Kriegel, H.P.: Protein function prediction via graph kernels. Bioinform. (Oxford Engl.) **21**(Suppl 1), i47–56 (2005). https://doi.org/10.1093/bioinformatics/bti1007
2. Bresson, X., Laurent, T.: A Two-step graph convolutional decoder for molecule generation. arXiv, Vancouver, Canada (2019). https://doi.org/10.48550/arXiv.1906.03412, arXiv:1906.03412 [cs, stat]
3. Bronstein, M.M., Bruna, J., Cohen, T., Veličković, P.: Geometric deep learning: grids, groups, graphs, geodesics, and gauges (2021). https://doi.org/10.48550/arXiv.2104.13478, arXiv:2104.13478 [cs, stat]
4. Bruna, J., Zaremba, W., Szlam, A., LeCun, Y.: Spectral networks and locally connected networks on graphs. arXiv (2014). arXiv:1312.6203 [cs]
5. Buterez, D., Janet, J.P., Kiddle, S.J., Oglic, D., Liò, P.: Graph neural networks with adaptive readouts (2022). https://openreview.net/forum?id=yts7fLpWY9G
6. Cho, K., van Merrienboer, B., Bahdanau, D., Bengio, Y.: On the properties of neural machine translation: encoder-decoder approaches (2014). arXiv:1409.1259 [cs, stat]
7. Corso, G., Cavalleri, L., Beaini, D., Liò, P., Velickovic, P.: Principal neighbourhood aggregation for graph nets (2020)
8. Defferrard, M., Bresson, X., Vandergheynst, P.: Convolutional neural networks on graphs with fast localized spectral filtering. arXiv (2017). arXiv:1606.09375 [cs, stat]
9. Dwivedi, V.P., Joshi, C.K., Luu, A.T., Laurent, T., Bengio, Y., Bresson, X.: Benchmarking graph neural networks. J. Mach. Learn. Res. **23** (2022). arXiv:2003.00982 [cs, stat]
10. Falcon, W., team, T.P.L.: PyTorch lightning (2019). https://doi.org/10.5281/zenodo.7545285, https://zenodo.org/record/7545285
11. Fey, M., Lenssen, J.E.: Fast graph representation learning with PyTorch geometric. arXiv (2019). arXiv:1903.02428 [cs, stat]
12. Gilmer, J., Schoenholz, S.S., Riley, P.F., Vinyals, O., Dahl, G.E.: Neural message passing for quantum chemistry. In: Proceedings of the 34th International Conference on Machine Learning, vol. 70. arXiv, Sydney (2017). https://doi.org/10.48550/arXiv.1704.01212, arXiv:1704.01212 [cs]
13. Ioffe, S., Szegedy, C.: Batch normalization: accelerating deep network training by reducing internal covariate shift (2015). https://doi.org/10.48550/arXiv.1502.03167, arXiv:1502.03167 [cs]

14. Khasahmadi, A.H., Hassani, K., Moradi, P., Lee, L., Morris, Q.: Memory-based graph networks. arXiv (2020). https://doi.org/10.48550/arXiv.2002.09518, arXiv:2002.09518 [cs, stat]
15. Kipf, T.N., Welling, M.: Semi-supervised classification with graph convolutional networks. arXiv (2017). https://doi.org/10.48550/arXiv.1609.02907, arXiv:1609.02907 [cs, stat]
16. Kriege, N., Mutzel, P.: Subgraph matching kernels for attributed graphs. In: Proceedings of the 29th International Conference on International Conference on Machine Learning. Omnipress, Edinburgh (2012)
17. Lee, J., Lee, Y., Kim, J., Kosiorek, A.R., Choi, S., Teh, Y.W.: Set transformer: a framework for attention-based permutation-invariant neural networks. In: Proceedings of the 36th International Conference on Machine Learning, pp. 3744–3753. arXiv (2019). https://doi.org/10.48550/arXiv.1810.00825, arXiv:1810.00825 [cs, stat]
18. Loshchilov, I., Hutter, F.: Decoupled weight decay regularization (2019). https://doi.org/10.48550/arXiv.1711.05101, arXiv:1711.05101 [cs, math]
19. Mesquita, D., Souza, A.H., Kaski, S.: Rethinking pooling in graph neural networks (2020). arXiv:2010.11418 [cs]
20. O'Boyle, N.M.: Towards a Universal SMILES representation - a standard method to generate canonical SMILES based on the InChI. J. Cheminform. 4(1), 22 (2012). https://doi.org/10.1186/1758-2946-4-22
21. Paszke, A., et al.: PyTorch: an imperative style, high-performance deep learning library (2019). https://doi.org/10.48550/arXiv.1912.01703, arXiv:1912.01703 [cs, stat]
22. Rhee, S., Seo, S., Kim, S.: Hybrid approach of relation network and localized graph convolutional filtering for breast cancer subtype classification (2018). https://doi.org/10.48550/arXiv.1711.05859, arXiv:1711.05859 [cs]
23. Srivastava, N., Hinton, G., Krizhevsky, A., Sutskever, I., Salakhutdinov, R.: Dropout: a simple way to prevent neural networks from overfitting. J. Mach. Learn. Res. 15(56), 1929–1958 (2014). http://jmlr.org/papers/v15/srivastava14a.html
24. Vaswani, A., et al.: Attention is all you need (2017). arXiv:1706.03762 [cs]
25. Veličković, P., Cucurull, G., Casanova, A., Romero, A., Liò, P., Bengio, Y.: Graph attention networks (2018). arXiv:1710.10903 [cs, stat]
26. Vinyals, O., Bengio, S., Kudlur, M.: Order matters: sequence to sequence for sets (2016). https://doi.org/10.48550/arXiv.1511.06391, arXiv:1511.06391 [cs, stat]
27. Wagstaff, E., Fuchs, F.B., Engelcke, M., Posner, I., Osborne, M.: On the limitations of representing functions on sets (2019). arXiv:1901.09006 [cs, stat]
28. Xu, K., Hu, W., Leskovec, J., Jegelka, S.: How powerful are graph neural networks? (2019). arXiv:1810.00826 [cs, stat]
29. Yanardag, P., Vishwanathan, S.: Deep graph kernels. In: Proceedings of the 21th ACM SIGKDD International Conference on Knowledge Discovery and Data Mining, KDD 2015, pp. 1365–1374. Association for Computing Machinery, New York (2015). https://doi.org/10.1145/2783258.2783417
30. Ying, R., You, J., Morris, C., Ren, X., Hamilton, W.L., Leskovec, J.: Hierarchical graph representation learning with differentiable pooling (2019). https://doi.org/10.48550/arXiv.1806.08804, arXiv:1806.08804 [cs, stat]
31. Zaheer, M., Kottur, S., Ravanbakhsh, S., Poczos, B., Salakhutdinov, R., Smola, A.: Deep sets (2018). https://doi.org/10.48550/arXiv.1703.06114, arXiv:1703.06114 [cs, stat]

Parallel Adjoint Taping Using MPI

Markus Towara[✉][iD], Jannick Kremer[iD], and Uwe Naumann[iD]

Software and Tools for Computational Engineering, RWTH Aachen University,
Aachen, Germany
towara@stce.rwth-aachen.de
https://www.stce.rwth-aachen.de

Abstract. The adjoint reversal of long evolutionary calculations (e.g. loops), where each iteration depends on the output of the previous iteration, is a common occurrence in computational engineering (e.g. computational fluid dynamics (CFD) simulation), physics (e.g. molecular dynamics) and computational finance (e.g. long Monte Carlo paths). For the edge case of a scalar state, the execution, as well as adjoint control flow reversal, are inherently serial operations, as there is no spatial dimension to parallelize. Our proposed method exploits the run time difference between passive function evaluation and augmented forward evaluation, which is inherent to most adjoint AD techniques. For high dimensional states, additional parallelization of the primal computation can and should be exploited at the spatial level. Still, for problem sizes where the parallelization of the primal has reached the barrier of scalability, the proposed method can be used to better utilize available computing resources and improve the efficiency of adjoint reversal.

We expect this method to be especially useful for operator-overloading AD tools. However, the concepts are also applicable to source-to-source transformation and handwritten adjoints, or a hybrid of all approaches. For illustration, C++ reference implementations of a low dimensional evolution (lorenz attractor) and a high dimensional evolution (computational fluid dynamics problem in OpenFOAM) are discussed. Both theoretical bounds on the speedup and run time measurements are presented.

Keywords: Algorithmic Differentiation · Adjoints · MPI

1 Introduction

Motivation

Adjoint sensitivities are desired in a variety of academic and industrial applications due to their high accuracy and low computational cost compared to forward methods (e.g. finite differences), for functions which map a high number n of input parameters to a low (if not scalar) number of outputs m ($f : \mathbb{R}^n \to \mathbb{R}^m$ with $m \ll n$). Such sensitivities can be efficiently computed with Algorithmic Differentiation (AD, commonly also referred to as Automatic Differentiation). With adjoint AD, the full Jacobian $\nabla f(x) : \mathbb{R}^n \to \mathbb{R}^{m \times n}$ can be obtained at a computational complexity $\mathcal{O}(m) \cdot cost(f)$.

© The Author(s), under exclusive license to Springer Nature Switzerland AG 2023
J. Mikyška et al. (Eds.): ICCS 2023, LNCS 14073, pp. 406–420, 2023.
https://doi.org/10.1007/978-3-031-35995-8_29

The goal of this paper is to propose a novel way to better utilize parallel computing resources in the context of adjoint control flow reversal. Our proposed method exploits the run time difference between passive function evaluation and augmented forward evaluation (see definitions in Sect. 2).

The computation of adjoints typically consists of two distinct phases (outlined in more detail in the main matter):

Augmented forward evaluation. Execution of function f, but all intermediate values required for the reverse propagation of adjoints are cached. Efficient AD tools will pre-accumulate [10,32] partial Jacobians required for the reversal process already during this step, reducing the amount of data to be stored and the number of operations to be performed during reverse propagation.

Reverse propagation of adjoints. Evaluation of the adjoint model [11], restoring cached values as necessary and utilizing pre-accumulated Jacobians. Sensitivities will be propagated from the outputs back to the inputs and parameters.

The augmented forward evaluation is typically significantly slower than the passive (i.e. regular) evaluation of the same code, due to added memory bandwidth usage for the caching of values, added computational operations for the pre-accumulation of Jacobians, and (for some AD tools) added overhead due to the overloading of operators. The proposed method exploits this run time factor between *augmented forward* and *passive* evaluation by distributing the work to multiple processors and only performing a specific part of the calculation in *augmented forward* mode and remaining in *passive* mode for the remaining part. Depending on the implementation, this introduces additional *passive* computations, however, these are hidden in the parallelization and do not influence the wall clock time.

We mainly focus on C++ codes, where the overhead for operator-overloading is low, due to inlining and compile-time optimization. For interpreted languages (e.g. Python, MATLAB) the overhead can be much higher. General purpose operator-overloading tools include ADOL-C [33] (C,C++), CoDi-Pack [27] (C++), dco/c++ [16], Sacado [25], and AdiMat [2] (Matlab).

General purpose AD tools which use the source code transformation approach [11] include TAPENADE [12] (C and Fortran), dcc [8] (C) and Tangent [9] (python). Recently, a lot of research has focused on providing source code transformation for languages based on the LLVM stack [15] (e.g. C++, Rust, but also interpreted languages like Julia). This research has produced frameworks such as Enzyme [20], Zygote [13], and Diffractor [6]. Furthermore, domain-specific tools exist, e.g. DolfinAdjoint [5] for the Dolfin [17]/FEniCS [1] packages for solving differential equations using FEM. Especially in the context of machine learning, a variety of new tools, often geared toward specific linear algebra operations, have been developed (e.g. JAX [3]).

For high dimensional states (e.g. discretization points in numerical simulations), additional parallelization of the primal computation can and should be exploited on the spatial level. Still, for problem sizes where the parallelization of the primal has reached the barrier of scalability, this method can be used to utilize remaining computing resources and improve the efficiency of adjoint reversal.

We expect the proposed method to be especially useful for operator-overloading AD tools, however, the concepts are also applicable to source-to-source transformation and handwritten adjoints.

Limitation of State-of-the-Art Approaches

Many parallelization schemes utilized in computational engineering face challenges once the number of involved processes gets very large due to data access patterns, communication overheads, and other factors [7]. Existing strategies for parallelizing adjoint computations often focus on assembling individual stencils (e.g. [19]), or involve reversing the communication patterns in MPI (e.g. [27,28]). Both of these approaches require significant code changes in order to incorporate adjoint computations. Our approach, though limited to evolution-style algorithms, does not require many changes below the outer iteration level, as mostly a black box differentiation [11] approach is retained.

Limitations of the Proposed Approach

The current implementation of the proposed method is limited to MPI parallelization. Conceptually, it can also be extended to OpenMP or GPU parallelism (e.g. CUDA), however, memory bandwidth limitations might reduce the effectiveness of the approach. As shown in Sect. 4.2, the scaling of the proposed method is bound by a constant factor. To achieve scaling onto a high number of processors it needs to be combined with a parallelization strategy for the underlying problem.

Data Availability

The code for the case study in Sect. 5 and further illustrative code is available at github.com/STCE-at-RWTH/parallel_taping.

2 Notation and Foundations

We denote scalar variables in italics (e.g. $y \in \mathbb{R}$) and vectors in bold (e.g. $\mathbf{x} \in \mathbb{R}^n$). The corresponding adjoints are identified by $\bar{\mathbf{x}} \in \mathbb{R}^n$ and $\bar{y} \in \mathbb{R}$ (Notation of [11]). For a multivariate function $y = f(\mathbf{x})$ the adjoint model reads

$$\bar{\mathbf{x}} = \bar{y} \cdot \nabla f(\mathbf{x}), \tag{1}$$

where $\nabla f(\mathbf{x})$ denotes the gradient of f evaluated at location $\mathbf{x}$. For reference, we also introduce the tangent model with tangents $\dot{\mathbf{x}} \in \mathbb{R}^n$ and $\dot{y} \in \mathbb{R}$

$$\dot{y} = \nabla f(\mathbf{x}) \cdot \dot{\mathbf{x}}. \tag{2}$$

Equations (1) and (2) can be evaluated for a given code implementation of f explicitly (by writing down the tangent/adjoint statements line by line, see code in GitHub) or implicitly by means of Algorithmic Differentiation (AD).

The gradient $\nabla f(\mathbf{x})$ can be calculated at cost $\mathcal{O}(1) \cdot cost(f)$, where $cost(f)$ is a measure proportional to the run time needed to evaluate f, by choosing $\bar{y} = 1$ and evaluating the adjoint model. To acquire the same gradient we need n calls to the tangent model (by letting $\dot{\mathbf{x}}$ range over the unit vectors e_i), yielding cost $\mathcal{O}(n) \cdot cost(f)$. Thus, the adjoint method is the natural choice for calculating gradients whenever $1 \ll n$.

For the adjoint mode, the data flow of the program has to be reversed to propagate adjoints from the program outputs back to the inputs. With the adjoint model, the program execution decomposes into a distinct forward and reverse section.

Conceptually, all computer codes can be transformed into the execution of elemental functions φ_j operating on a single vector of data $\mathbf{v} \in \mathbb{R}^{n+p+m}$, using n inputs (v_0 to v_{n-1}), p intermediate values (v_n to v_{n+p-1}) and m outputs (v_{n+p} to $v_{n+p+m-1}$) [10, 21]. The corresponding adjoints can then be accumulated in $\bar{\mathbf{v}}$. In the following procedure $i \prec j$ denotes a (direct) dependence of variable v_j on v_i (that is $\frac{\partial \varphi_j}{\partial v_i} \neq 0$). Then the forward section computes the intermediates and output variables; Only after this is finished the reverse section starts to propagate the adjoints of outputs and intermediates back to the inputs.

$$
\left.
\begin{aligned}
&\text{for } j = n, \ldots, n + p + m - 1 \\
&\quad v_j = \varphi_j(\mathbf{v}_i)_{i \prec j}
\end{aligned}
\right\} \text{ forward section,}
$$

$$
\left.
\begin{aligned}
&\text{for } i = n + p - 1, \ldots, 0 \\
&\quad \bar{v}_i = \sum_{j : i \prec j} \frac{\partial \varphi_j(\mathbf{v}_i)}{\partial v_i} \cdot \bar{v}_j
\end{aligned}
\right\} \text{ reverse section.}
\tag{3}
$$

Note, that the derivatives $\frac{\partial \varphi_j}{\partial v_i}$ required in the reverse section still potentially depend on $\mathbf{v}_i$ (if φ_j is non-linear in v_i). Thus, either the values $\mathbf{v}_i$, or the Jacobian $\frac{\partial \varphi_j}{\partial v_i}$ (pre-accumulation), need to be cached until they are consumed in the reverse section. This caching is where the majority of the adjoint methods' memory penalty stems from. Figure 1 illustrate the concept of Jacobian pre-accumulation on a per-statement level. This technique reduces the amount of data that needs to be stored for the reverse section and also allows to perform most of the calculations already during the forward phase. Statement level pre-accumulation can efficiently be implemented by C++ expression template techniques [16].

3   Serial Adjoint Reversal of Evolution

First, we consider a serial evolution with state $\mathbf{x} \in \mathbb{R}^{n_x}$ (states are repeatedly overwritten by an iteration formula) and parameter $\mathbf{p} \in \mathbb{R}^{n_p}$ (parameters remain constant throughout the computation). Each iteration step only depends on the state $\mathbf{x}$ from the previous iteration and global parameters $\mathbf{p}$ (within each iteration additional intermediate variables might be used). To uniquely identify the states $\mathbf{x}$ during the different phases of the computation we introduce an upper index for each iteration. In practice, the state may use only a single memory location (thus overwriting the existing memory). However, for the adjoint reversal, it may be required to cache the states $\mathbf{x}$.

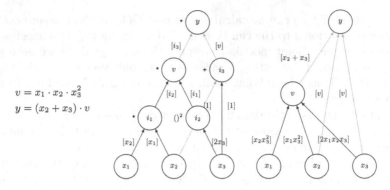

$$v = x_1 \cdot x_2 \cdot x_3^2$$
$$y = (x_2 + x_3) \cdot v$$

Fig. 1. Example for pre-accumulation of two statements. Elimination of intermediate nodes and edges from the middle graph saves memory for three nodes and four edges. Edges are labeled with partial derivatives.

```
1    double f(double x, const double p){ return p*sin(x); }
2
3    double evolution(int n, double x, const double p){
4        for(int i=0; i<n; i++)
5            x = f(x,p);
6        return x;
7    }
```

Listing 1.1. Example of evolution with function f, scalar input x, and scalar parameter $p$xrightmargin

First, we consider scalar evolutions $(n_x = 1, n_p = 1)$ of the form:

$$x^n = f^n(x^{n-1}, p) \circ f^{n-1}(x^{n-2}, p) \circ \ldots \circ f^2(x^1, p) \circ f^1(x^0, p)$$

with arbitrary (differentiable) functions $f^i : \mathbb{R} \times \mathbb{R} \to \mathbb{R} \in C^1$. An example of such an evolution, where f is always the same iteration procedure $x = p \cdot \sin x$, is given in Listing 1.1. This can easily be generalized to vector-valued evolutions of the form:

$$y = \mathcal{J} \circ f^n(\mathbf{x}^{n-1}, \mathbf{p}) \circ f^{n-1}(\mathbf{x}^{n-2}, \mathbf{p}) \circ \ldots \circ f^2(\mathbf{x}^1, \mathbf{p}) \circ f^1(\mathbf{x}^0, \mathbf{p})$$

with arbitrary (differentiable) functions $f^i : \mathbb{R}^{n_x} \times \mathbb{R}^{n_p} \to \mathbb{R}^{n_x} \in C^1$ and a final reduction $\mathcal{J} : \mathbb{R}^{n_x} \to \mathbb{R} \in C^1$ that reduces the state $\mathbf{x}$ to a scalar, such that the adjoint model only needs to be run once to get a full gradient of $\mathcal{J}$ with respect to $\mathbf{x}$ and $\mathbf{p}$. A graphical representation of the vector evolution, as well as the corresponding derivatives, are given in Fig. 2.

3.1   Reversal of Serial Evolution

We define the augmented forward run $\hat{f}$ of a function f as a function evaluation, which produces the same outputs, but caches all information required for the adjoint data flow reversal.

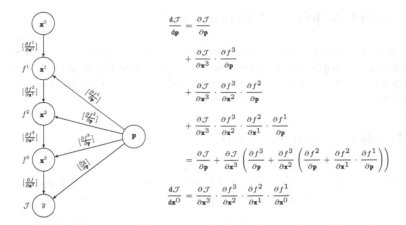

Fig. 2. Directed acyclic graph (DAG) representation of a three-step evolution (left), and the total derivative of J w.r.t. $\mathbf{p}$ (right). Note how the derivative can be found by multiplying the partial derivatives along paths from $\mathbf{p}$ to J, adding up parallel paths.

Conceptually, the adjoint reversal of a (scalar) evolution loop always consists of the following steps: The evolutionary loop, where all variables which are required for the adjoint reversal are cached

$$x^n = \hat{f}^n(x^{n-1}, p) \circ \hat{f}^{n-1}(x^{n-2}, p) \circ \ldots \circ \hat{f}^2(x^1, p) \circ \hat{f}^1(x^0, p)$$

is followed by the calculation of adjoints of the inputs $\bar{x}^0$, $\bar{p}$, as well as all intermediate adjoints $\bar{x}^i$ (either implicitly using an AD tool, or by explicitly calculating the required gradients):

$$\bar{x}^0 = \frac{\mathrm{d}}{\mathrm{d}x} \left[f^n(x^{n-1}, p) \circ f^{n-1}(x^{n-2}, p) \circ \ldots \circ f^1(x^0, p) \right] \cdot \bar{x}^n$$

as well as the adjoint of parameters p:

$$\bar{p} = \frac{\mathrm{d}}{\mathrm{d}p} \left[f^n(x^{n-1}, p) \circ f^{n-1}(x^{n-2}, p) \circ \ldots \circ f^1(x^0, p) \right] \cdot \bar{x}^n.$$

An example code on how to implement such a reversal using handwritten adjoints is given in the code on GitHub. For a general-purpose AD tool the workflow is always similar, where the adjoint statements are built automatically during the second step:

1. Mark initial values x^0 and parameters p as active inputs.
2. Execute augmented forward section for full n-step evolution, storing required intermediate values.
3. Seed the outputs, e.g. final state x^n.
4. Propagate adjoints back from $\bar{x}^n$ to $\bar{x}^0$ and $\bar{p}$.
5. Harvest adjoints $\bar{x}^0$ and $\bar{p}$.

4   Parallel Adjoints of Evolution

Building on the serial reversal procedure, we now expand it to a parallel setting. For notational simplicity, we focus on scalar states and parameters, however, the procedure can be applied to vector-valued states and parameters virtually unchanged.

4.1   Parallel Reversal Procedure

To transition from a serial evolution to a parallel adjoint evolution, we define an evolutionary loop where the first $n-1$ iterations are calculated in passive mode and only the last $\hat{f}^n$ is calculated in augmented forward mode:

$$x^n = \hat{f}^n(x^{n-1}, p) \circ f^{n-1}(x^{n-2}, p) \circ \ldots \circ f^2(x^1, p) \circ f^1(x^0, p)$$

As only one iteration step is cached, *only* the following adjoints can be calculated:

$$\bar{x}^{n-1} = \frac{\mathrm{d}f(x^{n-1}, p)}{\mathrm{d}x^{n-1}} \cdot \bar{x}^n$$

as well as a partial adjoint update for p:

$$\bar{p}^n = \frac{\mathrm{d}}{\mathrm{d}p} f(x^{n-1}, p) \cdot \bar{x}^n$$

where, due to the incremental nature [11] of the adjoint:

$$\bar{p} = \sum_{i=0}^{n} \bar{p}^i.$$

For the following discussion of parallel adjoint propagation, we assume that the n iteration steps can be distributed onto n processors. If less processors are available $(P < n)$, one can, without loss of generality (for evenly divisible n), redistribute the loop iterations, such that each processor reverses n/n_p steps of the iterations.

As before, each processor calculates only one iteration step $\hat{f}_i$ in augmented primal mode, such that it can later calculate the adjoint reversal $\bar{x}^{i-1} = \bar{x}^i \cdot \frac{\partial f^i}{\partial x^{i-1}}$. If we have n processors, then all steps can be reversed. However, all processors, except the last, have to wait for adjoint $\bar{x}^i$ to be calculated and sent to it by processor $(i+1)$. Thus, the reverse propagation phase remains a serial operation, as the propagation of the next iteration can not start until the current one has finished. All gains come from the fact that each processor has to perform only one active forward evaluation and can rely on faster passive evaluations for the remaining iterations up to the active one. Each processor except the first has to perform some amount of redundant passive computation (f^0 to f^{i-1}) before it can execute $\hat{f}^i$. A graphical representation of the procedure is given in Fig. 3.

In the following we outline the procedure for each processor i:

1. Execute passive evolution up to step $(i-1)$:
   $x^{i-1} = f^{i-1}(x^{i-2}, p) \circ \ldots \circ f^2(x^1, p) \circ f^1(x^0, p)$;
2. Execute augmented forward section $x^i = \hat{f}^i(x^{i-1}, p)$;
3. Seed outputs $\bar{x}^i$ (received from processor $(i+1)$ if $i < n$, else $\bar{x}^n = 1$);
4. Propagate adjoints back from $\bar{x}^i$ to $\bar{x}^{i-1}$ and $\bar{p}^i$;
5. Harvest adjoints $\bar{x}^{i-1}$ and $\bar{p}$;
6. Send calculated adjoints $\bar{x}^{i-1}$ to processor $(i-1)$ (if $i > 1$);
7. Reduce partial parameter adjoints $\bar{p} = \sum_{i=0}^{n} \bar{p}^i$ (e.g. MPI Allreduce).

A possible implementation for the parallel adjoint reversal of the previous evolution is available on GitHub.

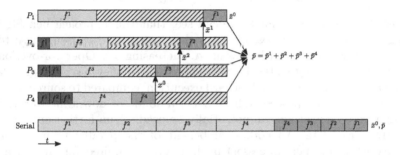

Fig. 3. Illustration of serial and parallel adjoint reversal for $n = 4, \lambda = 5, \alpha = 2$. Adjoints for the state x are passed along from processor to processor, as necessary requirements for the further reversal of the evolution. Adjoints for parameter p can be calculated thread local and are summed up with an `MPI_Allreduce` operation at the end. For n = 4 Eq. (4) predicts a speedup of $28/18 \approx 1.55$ and an upper limit for the speedup of $7/3 \approx 2.33$.

4.2   Run Time Analysis

For the following analysis of the run time of the algorithm, we assume the following:

- MPI communication cost (Send, Receive, Allreduce) is negligible compared to computation cost;
- Cost of each *active* function evaluations $\hat{f}^i$ is constant:
  $cost(\hat{f}^i) = cost(\hat{f}) =: c_{\hat{f}}$;
- Cost of each *passive* function evaluations f^i is constant:
  $cost(f^i) = cost(f) =: c_f$;
- Active (augmented forward) evaluation is *slower* than passive by a factor of $\lambda > 1$: $c_{\hat{f}} = \lambda c_f$;
- Adjoint propagation has cost $c_{\bar{f}}$, with a run time factor $\alpha > 0$: $c_{\bar{f}} = \alpha c_f$.

With the above assumptions we obtain the following costs:

- Total serial execution cost (n augmented forward evaluations and consecutive reverse propagations):

$$T_1 = n \cdot c_{\hat{f}} + n \cdot c_{\bar{f}} = n \cdot (c_{\hat{f}} + c_{\bar{f}}) = n \cdot (\lambda + \alpha) \cdot c_f$$

- Total parallel execution cost ($n - 1$ passive evaluations, one augmented forward evaluations, and n reverse propagations)

$$T_n = (n - 1) \cdot c_f + 1 \cdot c_{\hat{f}} + n \cdot c_{\bar{f}} = (n + \lambda + \alpha n - 1) \cdot c_f$$

With T_1 and T_n we can calculate the possible speedup (for $n \to \infty$) as:

$$S_\infty = \frac{\lim_{n\to\infty} T_1}{\lim_{n\to\infty} T_n} = \lim_{n\to\infty} \frac{\lambda + \alpha}{1 + \alpha + \frac{\lambda - 1}{n}} = \frac{\lambda + \alpha}{1 + \alpha} \tag{4}$$

Looking at the derived formula, we can see, that this approach benefits from shifting as much work as possible from the reverse propagation phase to the augmented forward phase (decreasing α, increasing λ). Operator-overloading AD tools are especially suited for this approach, as they have a higher factor between passive and augmented forward execution, compared to source-to-source tools. All except one processor have some amount of downtime during their blocking receive. This time can potentially be used to run further optimizations on the recorded data. An additional benefit of the parallel taping approach is the reduction of (per processor) peak memory required for the caching of the augmented forward section (only one instead of n steps have to be held in memory). While the overall memory consumption (for the cache) stays the same, using MPI it can be straightforwardly spread over multiple compute nodes.

5   Case Study: Lorenz Attractor

To demonstrate the feasibility of the proposed method we model a simple Lorenz attractor [18] using explicit forward Euler time integration of the form:

$$\mathbf{x}^{i+1} = \mathbf{x}^i + \Delta t \begin{pmatrix} \sigma * (x_1^i - x_0^i) \\ x_0^i * (\rho - x_2^i) - x_1^i \\ x_0^i * x_1^i - \beta * x_2^i \end{pmatrix}.$$

Thus, we have state $\mathbf{x} = (x_0, x_1, x_2) \in \mathbb{R}^3$ and parameters $\mathbf{p} = (\beta, \rho, \sigma) \in \mathbb{R}^3$. We run the iteration with a small time-step Δt for $2 \cdot 10^7$ iterations. In terms of floating point operations, this computation is not very demanding, as only multiplications and subtractions are executed.

The iteration is implemented in C++ and differentiation is handled by the operator-overloading AD tool dco/c++. We obtain a scalar output by evaluating the distance of the final iterate from the origin: $\mathcal{J} = \sqrt{x_0^2 + x_1^2 + x_2^2}$. The gradient $\frac{d\mathcal{J}}{d\mathbf{p}}$ is then calculated by one evaluation of the adjoint model.

Timing the execution on 8 of 36 cores of an *Intel Xeon E5-2695 v4* [4] system we get the following run-time factors (as defined in Sect. 4):

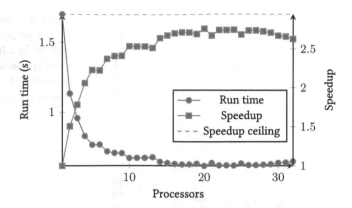

Fig. 4. Run time of $2 \cdot 10^7$ iterations of the (adjoint) Lorenz attractor (left axis) and speedup compared to serial execution (right axis). The limit predicted by Eq. 4 is marked as a dashed line. Run time average of 20 executions.

Augmented forward to passive ratio: $\lambda = 6.83$
Adjoint propagation to passive ratio: $\alpha = 2.05$

Thus, Eq. (4) limits the theoretical speedup to approximately three. For $P = 20$ we see an actual speedup of 2.75. The run time and speedup for up to 32 processors are presented in Fig. 4. After a certain number of processors no more improvement is achieved and even a slight slowdown is observed. As the problem dimension is not scaled with the number of processors, the amount of (traced) work per processor shrinks, making communication and memory allocation overheads more visible.

The time the MPI parallel program spends in the different phases of the computation is outlined in Fig. 5. All processors, except the last, spend time waiting for incoming adjoints from the next processor (gray parts of the bars). This time can conceivably be used for productive waiting, e.g. optimizing the recorded tape, such that once the reverse propagation starts (green part) it is executed faster. In the current implementation, passive calculations of the same iteration are performed on multiple processors (f_0 on all but the first, f_1 on all but the first and second, and so on). The passive computations could be bundled on a single processor, at the expense of additional communication and some further implementation work (processors need to be able to advance their state forward without actually performing the iteration). The resulting code would not necessarily be faster, but certainly more energy efficient (even more if idle processors are used in some other way or are allowed to enter a low power state). The approach of utilizing a dedicated process for all passive calculations is used in the following example.

6 Case Study: Parallel Taping in OpenFOAM

Here we present the reversal of evolutions as encountered in OpenFOAM [22]. OpenFOAM is an open-source computational fluid dynamics solver suite, based

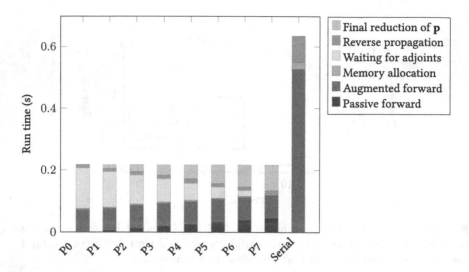

Fig. 5. Time spent in different phases of the computation for eight processors. For reference also the serial execution without any MPI communication is given on the right. Reverse interpretation starts on P_7 and finishes on P_0.

on the finite volume method. A version of OpenFOAM augmented with features of algorithmic differentiation has been developed by the author's institution [29].

For this case study, we use a variation of the `simpleFoam` solver, which implements the SIMPLE algorithm [24] with added penalty parameters to facilitate topology optimization [23]. The solver is run for a fixed number of time steps, all of which are traced and reversed by AD (except for linear solver calls, which are differentiated symbolically [30]).

Two different versions of the OpenFOAM solver, implementing the (MPI-) parallel taping approach, were developed [14]:

1. A version that operates with AD data types throughout (even if a time step does not need to be reversed, the AD data types are retained, but the tape recording is switched off).
2. A version where the time steps required to kick-start the adjoint recording are calculated on a different thread, running a different set of binaries.

Due to architectural limitations within OpenFOAM, we can currently not instantiate the full codebase with different floating-point types at the same time (e.g. with a passive floating point type `double` and an adjoint AD type `ad::adjoint_t⟨double⟩`). Using the first approach, the execution time for evaluating a single SIMPLE iteration step is not as fast as with fully passive (e.g. `double`) data types. That is, the observed run time factor lies somewhere in between c_f and $c_{\dot{f}}$.

However, it is possible to compile different sets of binaries for different types. In the second implementation, we use this to calculate all time steps on a single processor running a binary with true passive `double` data types. This data is

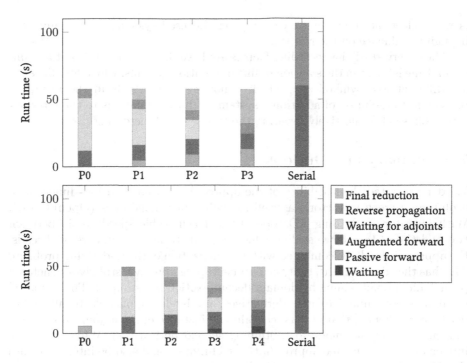

Fig. 6. Run time (in s) for a run of code variant 1.) on 4 nodes (top) and a run of code variant 2.) on 4+1 nodes (bottom). Run times are averaged over 10 runs.

used to kick-start the calculation of augmented forward and reverse propagation phases on the remaining processes, which are running the version of OpenFOAM instantiated with adjoint data types. As most processors have idle time after the recording of the tape, it is possible to move the extra process calculating with true double data types onto a processor with such idle time, eliminating the need for an extra processor.

On the investigated hardware (Intel Skylake Platinum 8160), the passive run time is about $\lambda_1 = 2.5$ times lower than the active run time (4.4 s vs 11.3 s per 64 time steps). With an observed interpretation factor of $\alpha_1 = 1.54$ we get the scaling limit according to Eq. (4) as $S_1 \approx 1.6$. As expected, for version 2 the factor $\lambda_2 = 6.0$ is higher (1.86 s vs 11.3 s). With an observed interpretation factor of $\alpha_2 = 3.6$ we get the scaling limit according to Eq. (4) as $S_2 \approx 2.1$.

Figure 6 shows an example run with AD tool dco/c++ on the common pitz-Daily [26] OpenFOAM case (flow over a backward-facing step). The case is run and adjoined over 256 time steps, obtaining the gradient w.r.t. roughly 50 000 design parameters. Here we distribute the calculation over four processors, thus each processor is responsible for recording and adjoining 64 time steps. We observe, that each recorded time step requires about 400 MB of memory. The state that needs to be communicated by MPI is just about 600 kB big, which is way below the interconnect capacity on the used system to have any non-negligible influence on the run time.

The observed adjoint recording factors are lower because a significant amount of run time is spent in the solution of linear equation systems, which for efficiency are differentiated symbolically, which removes complexity from the recording phase and the solution of additional systems of linear equations to the interpretation phase [31], effectively lowering the factor λ and increasing α.

7   Summary and Outlook

We demonstrated the feasibility of the approach to scalar and low-dimensional evolutions (lorenz attractor), as well as high dimensional ones (OpenFOAM). With operator-overloading AD tools, typical achievable speedups lie between two and three. To achieve scaling which extends to larger numbers of threads, the approach should be combined with parallelization of the underlying problem. This has the added benefit, that the parallel taping approach still gives benefits if primal and adjoint scaling by domain decomposition has stopped. The proposed method is not limited to first derivatives but is also applicable to all higher derivatives schemes, that at least contain one instance of the adjoint model (e.g. tangent over adjoint mode for obtaining second derivatives). While currently implemented as a manual approach, this technique could also be integrated into the AD tool itself, also removing the limitation to evolutions.

Acknowledgements. The authors would like to thank the reviewers for their detailed comments, which we hope we addressed adequately. OpenFOAM simulations were performed with computing resources granted by RWTH Aachen University under project thes1094.

References

1. Alnæs, M., et al.: The FEniCS project version 1.5. Arch. Numer. Softw. **3**(100), 9–23 (2015)
2. Bischof, C.H., Bücker, H.M., Vehreschild, A.: A macro language for derivative definition in ADiMat. In: Bücker, M., Corliss, G., Naumann, U., Hovland, P., Norris, B. (eds.) Automatic Differentiation: Applications, Theory, and Implementations. LNCSE, vol. 50, pp. 181–188. Springer, Heidelberg (2006). https://doi.org/10.1007/3-540-28438-9_16
3. Bradbury, J., et al.: JAX: composable transformations of Python+NumPy programs (2022)
4. Intel Corporation. Intel xeon prozessor e5–2695 v. 4 (2022). https://ark.intel.com/content/www/de/de/ark/products/91316/intel-xeon-processor-e52695-v4-45m-cache-2-10-ghz.html
5. Farrell, P.E., Ham, D.A., Funke, S.W., Rognes, M.E.: Automated derivation of the adjoint of high-level transient finite element programs. SIAM J. Sci. Comput. **35**(4), C369–C393 (2013)
6. Fischer, K., et al.: Diffractor - next generation ad (2022). https://github.com/JuliaDiff/Diffractor.jl

7. Fischer, P.F.: Scaling limits for PDE-based simulation. In: 22nd AIAA Computational Fluid Dynamics Conference, p. 3049 (2015)
8. Förster, M.: Algorithmic Differentiation of Pragma-Defined Parallel Regions: Differentiating Computer Programs Containing OpenMP. Springer, Wiesbaden (2014). https://doi.org/10.1007/978-3-658-07597-2
9. Google LLC. Tangent: Source-to-source debuggable derivatives (2017). https://research.googleblog.com/2017/11/tangent-source-to-source-debuggable.html
10. Griewank, A., Juedes, D., Utke, J.: Algorithm 755: ADOL-C: a package for the automatic differentiation of algorithms written in C/C++. ACM Trans. Math. Softw. **22**(2), 131–167 (1996)
11. Griewank, A., Walther, A.: Evaluating Derivatives: Principles and Techniques of Algorithmic Differentiation. SIAM (2008)
12. Hascoët, L., Pascual, V.: The Tapenade automatic differentiation tool: principles, model, and specification. ACM Trans. Math. Softw. **39**(3):20:1–20:43 (2013)
13. Innes, M.: Don't unroll adjoint: differentiating SSA-form programs. CoRR, abs/1810.07951 (2018)
14. Kremer, J.: Parallel adjoint taping approaches with OpenFOAM. Bachelors thesis, RWTH Aachen University (2022)
15. Lattner, C., Adve, V.: LLVM: a compilation framework for lifelong program analysis and transformation, San Jose, CA, USA, pp. 75–88 (2004)
16. Leppkes, K., Lotz, J., Naumann, U.: Derivative Code by Overloading in C++ (DCO/C++): Introduction and Summary of Features. Technical Report AIB-2016-08, RWTH Aachen University (2016)
17. Logg, A., Wells, G.N.: DOLFIN: automated finite element computing. ACM Trans. Math. Softw. (TOMS) **37**(2), 20 (2010)
18. Lorenz, E.N.: Deterministic nonperiodic flow. J. Atmos. Sci. **20**(2), 130–141 (1963)
19. Luporini, F., et al.: Architecture and performance of Devito, a system for automated stencil computation. ACM Trans. Math. Softw. (TOMS) **46**(1), 1–28 (2020)
20. Moses, W.S., et al.: Reverse-mode automatic differentiation and optimization of GPU kernels via enzyme. In: Proceedings of the International Conference for High Performance Computing, Networking, Storage and Analysis, pp. 1–16 (2021)
21. Naumann, U.: The Art of Differentiating Computer Programs. An Introduction to Algorithmic Differentiation. SIAM (2012)
22. OpenFOAM Ltd., OpenFOAM - The Open Source Computational Fluid Dynamics (CFD) Toolbox. http://openfoam.com/
23. Othmer, C., de Villiers, E., Weller, H.G.: Implementation of a continuous adjoint for topology optimization of ducted flows. In: 18th AIAA Computational Fluid Dynamics Conference (2007)
24. Patankar, S.V., Spalding, D.: A calculation procedure for heat, mass and momentum transfer in three-dimensional parabolic flows. Int. J. Heat Mass Transfer **15**(10), 1787–1806 (1972)
25. Phipps, E.T., Bartlett, R.A., Gay, D.M., Hoekstra, R.J.: Large-scale transient sensitivity analysis of a radiation-damaged bipolar junction transistor via automatic differentiation. In: Bischof, C.H., Bücker, H.M., Hovland, P., Naumann, U., Utke, J. (eds.) Advances in Automatic Differentiation. LNCSE, vol. 64, pp. 351–362. Springer, Heidelberg (2008). https://doi.org/10.1007/978-3-540-68942-3_31
26. Pitz, R.W., Daily, J.W.: Combustion in a turbulent mixing layer formed at a rearward-facing step. AIAA J. **21**(11), 1565–1570 (1983)
27. Sagebaum, M., Albring, T., Gauger, N.R.: High-performance derivative computations using codipack. ACM Trans. Math. Softw. (TOMS) **45**(4), 1–26 (2019)

28. Schanen, M.: Semantics driven adjoints of the message passing interface. Dissertation, RWTH Aachen University (2014)
29. Towara, M.: Discrete adjoint optimization with OpenFOAM. Dissertation, RWTH Aachen University, Aachen (2018)
30. Towara, M., Naumann, U.: Simple adjoint message passing. Optim. Methods Softw. **33**, 1232–1249 (2018)
31. Towara, M., Schanen, M., Naumann, U.: MPI-parallel discrete adjoint OpenFOAM. Procedia Comput. Sci. **51**, 19–28 (2015). 2015 International Conference on Computational Science
32. Utke, J., Lyons, A., Naumann, U.: Efficient reversal of the intraprocedural flow of control in adjoint computations. J. Syst. Softw. **79**(9), 1280–1294 (2006). Selected Papers from the Fourth Source Code Analysis and Manipulation (SCAM 2004) Workshop
33. Walther, A., Griewank, A.: Getting started with Adol-C. In: Naumann, U., Schenk, O. (eds.) Combinatorial Scientific Computing, pp. 181–202. Chapman-Hall CRC Computational Science (2012)

A Moral Foundations Dictionary for the European Portuguese Language: The Case of Portuguese Parliamentary Debates

Mafalda Zúquete[1]([✉])(iD), Diana Orghian[2], and Flávio L. Pinheiro[1](iD)

[1] NOVA IMS - Universidade Nova de Lisboa, Lisbon, Portugal
mzuquete@novaims.unl.pt
[2] FeedZai, Lisbon, Portugal

Abstract. Moral Foundations Theory (MFT) has shown that American liberals and conservatives rely on fundamentally different moral principles, offering a different perspective on the deepening political divide in US politics. However, results outside the US have been less clear, particularly in countries with a more diverse political landscape that does not fall into the traditional Liberal/Conservative dichotomy. Here, we expand the Moral Foundations Dictionary to European Portuguese, which we then use to analyze 10 years of transcripts of parliamentary sessions using standard Data Science and Text Mining techniques. Despite a larger number of represented parties, we show that no traditional parties fall into the Conservative or Liberal characterization and that the political landscape in Portugal is relatively homogeneous with the major difference observed concerning the dichotomy between Government and the parliament.

Keywords: Moral Foundations Theory · Moral Foundations Dictionary · Political Discourse · Text Mining

1 Introduction

How can opposite beliefs co-exist, and which side is "morally right"? When trying to understand the co-existence of disparate moralities, Haidt and Joseph [18] proposed a structured way to study morality based on our intuitions and how they adapt to the culture we inhabit. This would give origin to what is now called *Moral Foundations Theory* (MFT) that assumes the existence of innate pillars which capture society's moral variety [12].

According to MFT, human morality stands atop five different foundational dimensions: **Harm** (Care/Harm, it makes us sensitive to signs of suffering and need; it makes us despise cruelty and want to care for those who are suffering); **Fairness** (Fairness/Cheating, it makes us sensitive to indications that another person is likely to be a good (or bad) partner for collaboration and reciprocal altruism. It makes us want to shun or punish cheaters); **Ingroup** (Loyalty/Betrayal, it makes us sensitive to signs that another person is (or is not) a

J. Mikyška et al. (Eds.): ICCS 2023, LNCS 14073, pp. 421–434, 2023.
https://doi.org/10.1007/978-3-031-35995-8_30

team player. It makes us trust and reward such people, and it makes us want to hurt, ostracize, or even kill those who betray our group or us); **Authority** (Authority/Subversion, it makes us sensitive to signs of rank or status, and to signs that other people are (or are not) behaving properly, given their position); and **Purity** (Sanctity/Degradation, it includes the behavioral immune system, which can make us wary of a diverse array of symbolic objects and threats. It makes it possible for people to invest in objects with irrational and extreme positive and negative values, which are essential for binding groups together). Moreover, Graham et al. [12] considered the possibility of more than five moral foundations, namely Liberty/Oppression, but claimed that there was not enough evidence to support their foundationhood at this stage.

Graham et al. [13] developed a Moral Foundations Dictionary (MFD) to measure the moral load in political texts. By using the MFD in tandem with *Linguistic Inquiry and Word Count* (LIWC), Graham et al. [13] showed that liberals and conservatives relied on different sets of moral foundations: liberals tended to value the Harm and Fairness foundations primarily; where conservatives' moral discourse was more evenly distributed across foundations. When it comes to the creation of an MFD in languages other than (American) English, it is relevant to highlight the works of Matsuo et al. [26] (Japanese), Carvalho at al. [6] (Brazilian Portuguese), Alper et al. [1] (Turkish), Carvalho and Guedes [5] (Spanish), and Wan et al. [33] (Chinese). Here, we propose the creation of an MFD in European Portuguese, its validation, and an example analysis of the moral loading in Portuguese politics by studying the transcripts of the Diary of the Assembly of the Republic from 2011 to 2021. From this analysis, we noticed how the morality profile of the parties was quite similar among them, while being relatively different from the morality embedded in the Government discourse. In that sense, we also glimpsed a dimension in parliamentary dynamics that we found relevant: the parties morality profile reflected their proximity to governmental power.

2   Related Work

Across the literature, the relationship between moral foundations and American political ideology shows a consistent pattern: liberals primarily value the Harm and Fairness foundations, and conservatives' moral beliefs are based on all moral foundations [13,16,27].

We highlight the work of Graham et al. [13], which used methods from text mining to the study of morality under the paradigm of MFT, by creating the MFD to detect moral words in church sermons. Despite the replicability of these results when the study is precisely reproduced, extensions to include sermons from other churches, political transcripts, and texts from media outlets, meant that the conclusions did not hold [9]. This outcome highlights the shortcomings of this dictionary-based methodology [2]:

1. A limited amount of lemmas and stems of words;
2. "Radical" lemmas rarely used in everyday language;

3. An association with a moral binary scale, but lacking of a measure of "strength".

New, more complex, algorithms have been developed to detect the underlying morality in the text, such as Garten et al.'s [10] *Distributed Dictionary Representations* (DDR) of the MFD, or Araque et al.'s [2]'s *MoralStrength*, which was obtained as an extension of the MFD, based on WordNet synsets. The creation of the *Moral Foundation Twitter Corpus* (MFTC), a corpus of tweets classified with regards to morality [21], is likely to promote the development of more MFD-related algorithms, following previous studies of morality in tweets [8,24].

Indeed, the framework of MFT has been used to explore political chasms. The work of Koleva et al. [25] noted that the strongest unique predictor of a "culture war" opinion was often a sub-scale of the Moral Foundations Questionnaire (MFQ) (often Purity) instead of political ideology, interest in politics, religious attendance, or any demographic variable.

Despite the evidence for a binary view of American politics, a cluster analysis of answers to the MFQ resulted in four main clusters – secular liberals, libertarians, religious leftists, and social conservatives – which bound themselves to the two main political parties [17]. Furthermore, a deeper look at the differences between two groups of liberals – Clinton and Obama supporters during the 2008 Democratic primaries – showed how, despite both groups identifying themselves as equally liberal, the moral foundations that they valued were not the same: Clinton supporters showing more substantial support of the binding moral foundations, closer to the typical conservative voter; Obama supporters defaulting to the individualizing foundations [22]. Libertarians proved to be an interesting outlier to the bipartisan dynamic of American politics, as they appeared to reject all of the five moral foundations [17]. This led to the proposal of a sixth moral foundation, Liberty/Oppression, which would help characterize the moral profile of libertarians [23]. Still, Graham et al. [12] discarded this foundation, claiming that the evidence for foundationhood was still lacking.

While contemporary American politics shows a high appeal when studying how politics intersects with morality, there are works outside the American context. Parker et al. [30] focused on Australian politics, looking at how time and moral foundations were related, showing a more complex landscape to Australian politics than a myopic view of liberal/conservative would suggest. Moreover, outside the political sphere, relevant insights can be achieved by studying societal dynamics from the lenses of MFT. Examples include, but are not limited to, topics such as sports fandom [35], religion [11], education [14], health habits [32], social cognition [19], or media [31].

3 European Portuguese Moral Foundations Dictionary

3.1 Semi-supervised Translation of the English MFD

The English MFD created by Graham et al. [13] consisted of a set of full words and word stems related to each of the five moral foundations. The dictionary included both foundation-supporting words, or *Virtues*, and foundation-violating words, *Vices*. Here, we adapted the methodology proposed by Matsuo et al. [26] to develop a European Portuguese Moral Foundations Dictionary. The proposed approach involves six steps, some automated and some manual (Fig. 1).

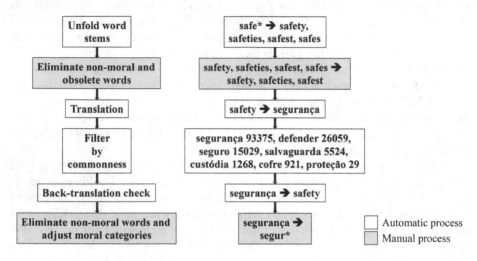

Fig. 1. Illustrative workflow of the translations process (left) using the stem "safe*" as an example (right).

Starting from the MFD developed by Graham [13], all the word stems were unfolded with the aid of *OneLook* [7] to find all words that started with each stem and then, using the *Common Words* filter, we selected the most common words associated with each stem.

The *Merriam-Webster* [29] and the *wikionary* [34] dictionaries were used to find and filter out words flagged as *obsolete* or *archaic*. We then used the *Cambridge University's English (US) - Portuguese* [3] to translate each word to Portuguese. Translations that were expressions rather than words were discarded. In the case of words that translated to reflexive verbs, the verb in the infinitive was manually added to the list of translations. Since the MFD was meant to represent current speech, we then filtered out uncommon words using the *Corpus de Referência do Português Contemporâneo* (CRPC) [28]. For words that were radicals on the original dictionary, the 10 most common translations were kept, and the 5 most common translations were kept in the case of plain words. To ensure that translations kept the moral load of the original words, a backtranslation check was performed using the *Cambridge University's Portuguese - English (US)* [4] dictionary. All the words for which the backtranslation differed from the original were removed. Finally, the words were checked to ensure that they were all moral, and the moral categories were adjusted to reflect the Portuguese language's current use better.

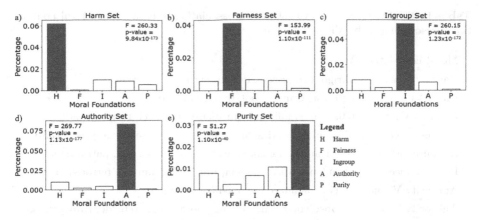

Fig. 2. Mean percentage of dictionary words in examples for each Moral Foundation after corrections to the MFD.

The outcome of the above steps is a list of words, unlike the original MFD, which was a list of words and word stems. As such, words that started the same way and were associated with the same moral foundations were manually gathered into lemmas, checking the CRPC to ensure that using the lemma on a word would not yield unwanted results. Also, because of the Portuguese Language Orthographic Agreement of 1990, double orthography was considered for some words and word stems. The European Portuguese MFD consists of a list of words and word stems, each associated with one or more moral foundations and the corresponding Vice/Virtue loading (except for the words associated with Morality General).

3.2 Experimental Validation

To test the validity of the resulting dictionary, we ran a survey-based study on Prolific that counted with the participation of 324 individuals who provided valid responses. The pool of participants were all native Portuguese speakers, including 36% females and 64% males, who were, on average, 24 years old.

During the survey, each participant was initially briefed on the definition of each moral foundation (which consisted of the translation of the definitions presented by Haidt [15]) and was asked to describe scenarios where the foundations were validated or violated. Participants who submitted short answers (under 20 words) or failed to answer for any moral foundations were considered to have invalid answers and, thus, were ignored in the subsequent analysis.

From the survey answers, we then determined the validity of the dictionary by conducting a one-way analysis of variance (ANOVA) on the main effect of the moral foundations. That is, we checked if there was a significantly higher frequency of words from the moral foundation in question than the words from the remaining moral foundations.

Table 1. The five words with the strongest association to each of the Moral Foundations, descriminated by Virtue and Vice.

Moral Foundation	Words
Harm Virtue	abrigo, compaixão, defesa, empatia, segurança
Harm Vice	agressor, bruto, cruel, dano, indefeso
Fairness Virtue	balanço, equilíbrio, equivalente, razoável, reciprocidade
Fairness Vice	discriminação, desigual, desonesto, intolerância, preconceito
Ingroup Virtue	coletivamente, colónia, cooperativa, nacional, patriotismo
Ingroup Vice	abandonar, discordância, imigração, inimigo, terrorismo
Authority Virtue	cargo, comando, lei, obediência, ordem
Authority Vice	desobediência, infrator, insubordinação, rebelde, transgressão
Purity Virtue	igreja, puro, sagrado, santo, virgem
Purity Vice	mancha, obsceno, pecado, profano, sujo

Noteworthy to mention that after a pre-test of the survey materials we found an unexpected overabundance of Authority-related words in the Ingroup foundation. Under closer inspection, we found some ambiguity in the moral categorization of words starting with "trai*" and "lea*". As such, we revised the MFD so that these words would only be Ingroup, rather than being both Authority and Ingroup. We also changed the category of "deslea*" in the same way, as this word was the direct antonym of "lea*" and it would not make sense to have them be classified differently. Furthermore, we noticed how "cumpr*" and "viol*" were consistently misclassified across all Moral Foundations, which was related to the presence of such words in the survey questions that primed participants to use them by default. As such, we decided to exclude those words from the MFD altogether.

Results show that, for most moral foundations, the most commonly identified MFD words were associated with the correct foundation, see Fig. 2. Moreover, the MFD also passed the ANOVA, which supports the main effect of moral foundations (see Fig. 2), and that these words effectively discriminate between different dimensions.

We can also measure which words show a stronger association with each moral dimension, based on which had the lowest entropy (seed words). Table 1 shows for each moral dimension the five words with the strongest association.

The European Portuguese Moral Foundations Dictionary is available for download [36].

4   Morality in the Portuguese Assembly of the Republic, a Case Study

4.1   Parliamentary Transcripts

As an example a possible application of the European Portuguese MFD, we studied the moral load embedded in the political discourse of the Portuguese

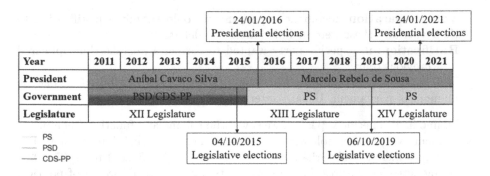

Fig. 3. A brief timeline of the parties in power in Portuguese politics.

Assembly of the Republic. In particular, how do the parties currently in parliament relate to each other with regards to morality?

To that end, we used the publicly available transcripts [20] of parliamentary sessions and debates between 20/06/2011 and 22/07/2021 from the *Diário da Assembleia da República* (DAR). This period covers the harsh austerity mandated by *Troika*, the economic growth that followed, the first *superavit* in 40 years, the slow but steady growth and normalization of the far right, and the COVID-19 pandemic. Figure 3 illustrates which parties were in power during the time period under study. This includes both presidential and governmental power.

Since the transcripts were in raw text format, some pre-processing steps were necessary to obtain a suitable structured working dataset. These included eliminating blank lines, replacing em-dashes with hyphens, removing page headers, adding name and party information to government members, identifying unidentified speakers, and checking for spelling errors.

Moreover, members of the government were often only identified by their function. Given that different people could occupy the same position over time, name and party information were manually added in those cases. If a line did not have an identified speaker, it was determined it was the same speaker as in the previous line, unless the line was referring to: applause, protests, laugh, pauses, votes, or changes in the presidency of the assembly. This way, lines that were not interventions by members of parliament were not wrongly attributed to a speaker. Finally, transcripts were manually checked for orthographic errors with the help of *Microsoft Word*'s spell check, as the text files contained some errors.

The edited transcripts were then processed line by line. Each line was classified as one of 6 categories if it did not have a specified speaker:

- **Summary:** the debate's summary;
- **Note:** notes at the end of the transcripts;
- **Proposals:** the transcripts featured legislative proposals that could be attributed to political parties;

- **Vote declaration:** members of parliament could submit a justification to their vote and these were included after the debate;
- **Ratification:** if a specific party wanted to correct a mistake that appeared in a previous diary;
- **Voices:** interventions during the debate that were not attributed to a single member of parliament. Usually interruptions and interjections;

If the line had a specified speaker, we saved their name and function (member of parliament (MP) or their role within the government) and their intervention. The final dataset included all parliamentary sessions from 20/06/2011 to 22/07/2021, spanning three legislatures, four governments, and ten years' worth of political interventions on the parliament floor.

4.2   Measuring Moral Load

To study the underlying morality in parliamentary speech we looked at the deviations in the use of moral words by MPs, compared to the expected number of moral words given the number of words spoken. This approach deviated from past works [13,30]. Frimer [9] argued that American liberals and conservatives were not so different, as his replication and expansion of previous experiments [13] found no evidence of differences between the moral languages of liberals and conservatives. This informed our choice of how to encode the moral speech of Portuguese political parties, leading us to focus on how political speech would differ from expected.

In order to encode the moral load in the transcripts, we took the following steps:

1. Group the speech of all MPs for each legislative session and sum the total number of words and the number of words in each moral foundation;
2. Create log-log regressions for all moral foundations, where the x-axis was the number of words in a moral foundation and the y-axis was the total number of words used;
3. Calculate the residuals for each moral foundation, for each MP, for each legislative session, using the previously defined regressions;
4. Group the residuals by party, by year, and calculate the average residuals for each moral foundation, to get the morality profile of each party during each legislative session.

Only MPs who spoke more than 100 words in the entire legislative session were considered in step 2, and that threshold was raised to 1000 words when the party data was grouped in step 4. This was due to the fact that there were many MPs who had minimal interventions throughout the legislative session, and thus their contributions were not considered significant with regards to the morality of the party.

These regressions (see Fig. 4) had R^2 scores that could be considered quite high, lending confidence to the results from these regressions.

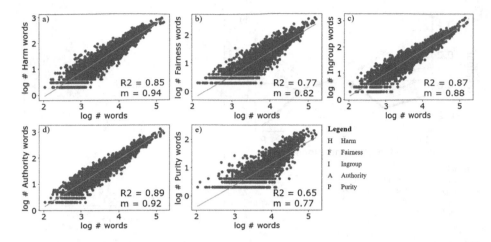

Fig. 4. Log-log regressions of the total number of words by the total number of foundation words.

4.3 Political Analysis

Figure 5a shows the moral load of each foundation per party in the 2020/21 legislative session. While we were comparing parties, the sample sizes for each are quite varied. Furthermore, some of the parties in this analysis only had one MP and others had plenty. As such, the samples of some parties, like Chega! (CH) or Iniciativa Liberal (IL), do not fully represent that party but only one, or very few MPs.

We could see that most political forces had relatively similar moral profiles, with the government and Partido Ecologista "Os Verdes" (PEV) demonstrating some of the most extreme values. It appears that the parties do not differ much with regards to Harm, Ingroup, or Authority, and show the most disparate values in the Fairness and Purity foundations. More importantly we see that the government shows a substantially different profile from the remaining parties in the Parliament.

To better understand the dynamic between parties we ran a Factor Analysis (FA) on the party profiles for this year. Since the Bartlett sphericity test had a p-value of 0.00017 (under 0.05) and the Kaiser-Meyer-Olkin score was 0.76 (over 0.60) we could consider we could perform an FA on our dataset.

We chose to run our analysis with two factors as determined by the Kaiser criterion. The eigenvalues of the first two factors were 3.72 and 0.91. Though a strict application of the Kaiser criterion would dictate that only one factor was chosen, as the eigenvector of the second factor is lower than 1, since 0.91 is very close to 1 we decided to keep two factors. To confirm our choice of two factors, we analyzed the cumulative variance explained by one factor versus two. The cumulative variance went from 0.72 to 0.86, which we considered relevant enough to keep the two factors in our analysis. These values, as well as the factor loadings, are in Table 2.

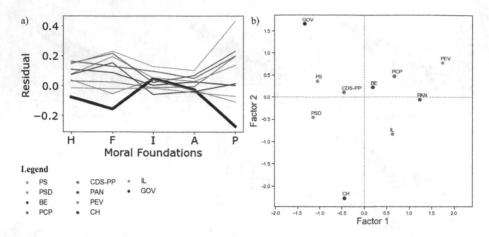

Fig. 5. Morality profile of every party currently in parliament (a) and a visualization of its factor analysis (b).

Though these factors are not easy to interpret, they allow us to better understand the dynamics of the parties. Parties in the first quadrant of Fig. 5b exhibit high values for all moral foundations (PEV and Partido Comunista Português (PCP)), whereas parties in the second quadrant have lower values for most values but comparatively high Ingroup values (government and Partido Socialista (PS)).

Though these results do not reflect clearly the right/left-wing dynamic as its commonly understood, the visualization of the factors shows that the parties closer to the government are the ones which have been in power up to this point (PS, Partido Social-Democrata (PSD), and CDS-Partido Popular (CDS-PP)), and the ones farthest away are more recent parties, which elected MPs for the first time in 2019 (CH and IL).

The above conclusion is further reinforced when we look at the evolution of parliamentary dynamics over time. We excluded the most recent parties, IL and

Table 2. Factor loadings and corresponding variance.

	Factor 1	Factor 2
Harm	0.88	−0.18
Fairness	0.90	−0.43
Ingroup	0.55	0.61
Authority	0.96	0.28
Purity	0.89	−0.08
Variance	0.72	0.14
Cumulative variance	0.72	0.86

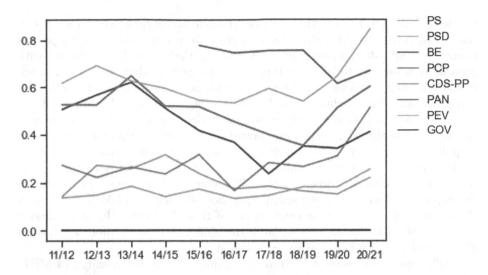

Fig. 6. Evolution of the distance between parties and the government.

CH, from our analysis, since they have only been in Parliament for two legislative sessions. As such, to study the evolution of parliamentary speech we plotted the euclidean distance between each party, in each year, and the government, over time (Fig. 6).

In Fig. 6 the dynamic we saw before is quite clear, with PS and PSD being quite close to the government throughout the entire period we considered. However, after the 2015 elections, when the PSD/CDS-PP coalition lost its power and PS became the governing party (see Fig. 3), CDS-PP started drifting away from the government.

In opposition to the behavior of CDS-PP, we remark on how the moral profile of Bloco de Esquerda (BE) and PCP have evolved over time. These parties started getting closer to the government following the 2015 election, when their parliamentary support was instrumental for the PS minority government to be able to stand, but after the 2019 election, as PS won reelection but no longer had a formal parliamentary agreement with these parties, they started distancing themselves again.

Farthest away from the government we find the two environmental parties, PEV and Pessoas-Animais-Natureza (PAN).

5   Conclusions

We presented the steps employed in the development of a Moral Foundations Dictionary for European Portuguese. These dictionaries are key to support text-based analysis based on LIWC approaches.

As an example, we then studied the case of Political Discourse in the Portuguese Parliament transcripts between 2011 and 2021. Here, not only did the parties not fit into the Liberal and Conservative archetypes determined previously, but there was not a clear distinction between right and left wing parties either. Instead, after running an FA we found that there is a dynamic with regards to proximity to the government, as the parties which have been traditionally in power are closer to the government, and less mainstream parties are farther away. This insight is consolidated through a temporal analysis of the distance between parties and the government over the years. Indeed, the parties which have shared governmental power since the establishment of the Third Republic have a moral profile which is very similar to that of the government.

Our work is not without limitations. For instance, we did not consider the different sample sizes between parties. This created limitations with regard to the representativeness of some samples, as not all parties and MPs speak with the same frequency or for the same length of time. Moreover, the methodology used here is relatively simple and lacks the ability to extract information from the context in which moral words are present.

Still, some of the most advanced algorithms used in the research of MFT, which are currently not available in European Portuguese, were built on top of the original MFD. This means that better and more accurate algorithms to detect moral foundations in the text could be created by expanding on the work we presented. Furthermore, the proposed MFD was a translation of an MFD created specifically for the American context. As such, this MFD likely failed to capture some nuances and peculiarities of the Portuguese language and Portugal's moral context, which could be better grasped by an MFD created from scratch, by linguists or specialists in MFT. Nonetheless, our work provides a baseline upon which future work could be developed and improvements could be made.

While this project looked at 10 years' worth of parliamentary data, our work could still be expanded by taking an even broader look to the Portuguese parliament, or even expanding the type of text we are taking into consideration. Close analysis of morality in speech during elections or referendums would undoubtedly create interesting insights. Another exciting avenue of research would be to consider texts extracted from the Internet, namely tweets. Studying this type of text would mean the expansion of the MFD via supervised learning, as well as a corpus of annotated tweets in Portuguese to study the accuracy of these methodologies.

Acknowledgements. MZ and FLP acknowledge the financial support provided by FCT Portugal under the project UIDB/04152/2020 – Centro de Investigação em Gestão de Informação (MagIC). FLP acknowledges the financial support of the Portuguese Foundation for Science and Technology ("Fundação para a Ciência e a Tecnologia") through grant DSAIPA/DS/0116/2019. The authors are thankful to César Hidalgo and Cristian Candia for the useful discussions and insightful comments.

References

1. Alper, S., Bayrak, F., Us, E.Ö., Yilmaz, O.: Do changes in threat salience predict the moral content of sermons? The case of Friday Khutbas in Turkey. Eur. J. Soc. Psychol. **50**(3), 662–672 (2020)
2. Araque, O., Gatti, L., Kalimeri, K.: MoralStrength: exploiting a moral lexicon and embedding similarity for moral foundations prediction. Knowl.-Based Syst. **191**, 105184 (2020)
3. Cambridge University Press: Cambridge Dictionary: Translations from English to Portuguese. https://dictionary.cambridge.org/us/dictionary/english-portuguese/
4. Cambridge University Press: Cambridge Portuguese-English Dictionary: Translations from Portuguese to English. https://dictionary.cambridge.org/us/dictionary/portuguese-english/
5. Carvalho, F., Guedes, G.: Dicionário de fundamentos morais em espanhol (2022)
6. Carvalho, F., Okuno, H.Y., Baroni, L., Guedes, G.: A Brazilian Portuguese moral foundations dictionary for fake news classification. In: 2020 39th International Conference of the Chilean Computer Science Society (SCCC), pp. 1–5. IEEE (2020)
7. Datamuse: OneLook Dictionary Search. https://www.onelook.com/
8. Dehghani, M., et al.: Purity homophily in social networks. J. Exp. Psychol. Gen. **145**(3), 366 (2016)
9. Frimer, J.A.: Do liberals and conservatives use different moral languages? Two replications and six extensions of Graham, Haidt, and Nosek's (2009) moral text analysis. J. Res. Pers. **84**, 103906 (2020)
10. Garten, J., Hoover, J., Johnson, K.M., Boghrati, R., Iskiwitch, C., Dehghani, M.: Dictionaries and distributions: combining expert knowledge and large scale textual data content analysis. Behav. Res. Methods **50**(1), 344–361 (2018)
11. Graham, J., Haidt, J.: Beyond beliefs: religions bind individuals into moral communities. Pers. Soc. Psychol. Rev. **14**(1), 140–150 (2010)
12. Graham, J., et al.: Moral foundations theory: the pragmatic validity of moral pluralism. In: Advances in Experimental Social Psychology, vol. 47, pp. 55–130. Elsevier (2013)
13. Graham, J., Haidt, J., Nosek, B.A.: Liberals and conservatives rely on different sets of moral foundations. J. Pers. Soc. Psychol. **96**(5), 1029 (2009)
14. Graham, J., Haidt, J., Rimm-Kaufman, S.E.: Ideology and intuition in moral education. Int. J. Dev. Sustain. **2**(3), 269–286 (2008)
15. Haidt, J.: The Righteous Mind: Why Good People are Divided by Politics and Religion. Vintage (2012)
16. Haidt, J., Graham, J.: When morality opposes justice: conservatives have moral intuitions that liberals may not recognize. Soc. Just. Res. **20**(1), 98–116 (2007)
17. Haidt, J., Graham, J., Joseph, C.: Above and below left-right: ideological narratives and moral foundations. Psychol. Inq. **20**(2–3), 110–119 (2009)
18. Haidt, J., Joseph, C.: Intuitive ethics: how innately prepared intuitions generate culturally variable virtues. Daedalus **133**(4), 55–66 (2004)
19. Hirsh, J.B., DeYoung, C.G., Xu, X., Peterson, J.B.: Compassionate liberals and polite conservatives: associations of agreeableness with political ideology and moral values. Pers. Soc. Psychol. Bull. **36**(5), 655–664 (2010)
20. Histórico-Parlamentar, A.: Debates Parlamentares. https://debates.parlamento.pt
21. Hoover, J., et al.: Moral foundations twitter corpus: a collection of 35k tweets annotated for moral sentiment. Soc. Psychol. Pers. Sci. **11**(8), 1057–1071 (2020)

22. Iyer, R., Graham, J., Koleva, S., Ditto, P., Haidt, J.: Beyond identity politics: moral psychology and the 2008 democratic primary. Anal. Soc. Issues Public Policy **10**(1), 293–306 (2010)
23. Iyer, R., Koleva, S., Graham, J., Ditto, P., Haidt, J.: Understanding libertarian morality: the psychological dispositions of self-identified libertarians (2012)
24. Kaur, R., Sasahara, K.: Quantifying moral foundations from various topics on twitter conversations. In: 2016 IEEE International Conference on Big Data (Big Data), pp. 2505–2512. IEEE (2016)
25. Koleva, S.P., Graham, J., Iyer, R., Ditto, P.H., Haidt, J.: Tracing the threads: how five moral concerns (especially purity) help explain culture war attitudes. J. Res. Pers. **46**(2), 184–194 (2012)
26. Matsuo, A., Sasahara, K., Taguchi, Y., Karasawa, M.: Development and validation of the Japanese moral foundations dictionary. PLoS One **14**(3), e0213343 (2019)
27. McAdams, D.P., Albaugh, M., Farber, E., Daniels, J., Logan, R.L., Olson, B.: Family metaphors and moral intuitions: how conservatives and liberals narrate their lives. J. Pers. Soc. Psychol. **95**(4), 978 (2008)
28. Mendes, A., Bacelar do Nascimento, F.: Corpus de Referência do Português Contemporâneo. http://alfclul.clul.ul.pt/CQPweb/
29. Merriam-Webster: Dictionary by Merriam-Webster: America's most-trusted online dictionary. https://www.merriam-webster.com/
30. Parker, P., Sahdra, B.K., Ondaatje, M.: Individual differences are a better predictor of moral language use in Australian prime ministers' speeches than is their political party (2019)
31. Tamborini, R.: Moral intuition and media entertainment. J. Media Psychol. **23**(1), 39–45 (2011)
32. Van Leeuwen, F., Park, J.H., Koenig, B.L., Graham, J.: Regional variation in pathogen prevalence predicts endorsement of group-focused moral concerns. Evol. Hum. Behav. **33**(5), 429–437 (2012)
33. Wan, Y., Peng, Z., Wu, F., Li, M., Gao, J.: Understanding public ethical acceptance and its antecedents and moderators of AI surveillance technology: analysis from social media data. Available at SSRN 4246846
34. Wikimedia Foundation: Wiktionary. https://www.wiktionary.org/
35. Winegard, B., Deaner, R.O.: The evolutionary significance of red sox nation: sport fandom as a by-product of coalitional psychology. Evol. Psychol.: Int. J. Evol. Approaches Psychol. Behav. **8**(3), 432–446 (2010)
36. Zúquete, M., Pinheiro, F.L., Orghian, D.: Moral foundations dictionary in European portuguese. https://github.com/mafalda-zuquete/mfd-pt

Towards Automatic Generation of Digital Twins: Graph-Based Integration of Smart City Datasets

Sebastian Ernst[(✉)] [iD], Leszek Kotulski [iD], and Igor Wojnicki [iD]

Department of Applied Computer Science, AGH University of Science and Technology, Al. Mickiewicza 30, Kraków, Poland
{ernst,kotulski,wojnicki}@agh.edu.pl

Abstract. This paper presents a graph-based approach to modelling and analysis of spatial (GIS) datasets supporting the deployment of *smart city* solutions. The presented approach is based on the *spatially-triggered graph transformations* (STGT) methodology, which allows for *materialisation* of spatial relationships detected using suitable tools, as well as performing measurements and modifications of geometries. The theory is illustrated using a real-world example which concerns street lighting. It shows how an existing traffic sensor network can be used to enable dynamic dimming of lamps, which can result in significant energy usage savings. Also, network analysis is applied to broaden the coverage of such systems, even in case of sensor sparsity. The presented results have been obtained in a real-world project and are due for larger-scale validation in the near future.

Keywords: graph transformations · GIS · smart cities · energy saving

1 Introduction

Reduction of energy usage is an important goal for governments and has implications in numerous areas, including economy, sustainability, protection of the environment and assuring the safety of citizens. It has also been one of the underlying goals of the *smart city* concept.

While there are many definitions of a *smart city* [1,8], practically all of them include the use of ICT solutions based on data collection and analysis. Hence, many projects supporting the development of smart cities involved deployment of sensor networks (or integration of data registered by existing devices). Data availability is an enabler for the data to be utilised in day-to-day operations. For instance, real-time vehicle traffic intensity data can be used to dim the streetlights [7,12], as the EN 13201 standard used throughout Europe and in other parts of the World allows for dynamic changes of the road lighting class [4].

However, a big problem lies in the *integration* of individual datasets. Defining the relationships between lamps, the areas each of them illuminates, how these areas constitute a road segment and which road segments each traffic sensor

© The Author(s), under exclusive license to Springer Nature Switzerland AG 2023
J. Mikyška et al. (Eds.): ICCS 2023, LNCS 14073, pp. 435–449, 2023.
https://doi.org/10.1007/978-3-031-35995-8_31

applies to is a time-consuming task. Another problem arises when not all streets are covered by sensors, which may reduce the applicability of dimming and, as a result, diminish the positive impact of dimming. These issues are further elaborated in Sect. 2. Such challenges are addressed during establishing so-called Digital Twins [5] which are actual data-based representations of physical infrastructural objects, policies, as well as sociological or technology bound behaviors.

This paper proposes a twofold solution to such problems. First, a graph structure, the Digital Twin Graph, able to store spatial data originating from various sources is introduced. Then, graph transformations, including those triggered by spatial relations (see STGT in Sect. 3.1), are used to integrate the objects from separate datasets as well as to derive added value by interpreting the data in each dataset.

Building upon the example of lamp dimming based on traffic intensity, as a result of this phase, the sensors are associated with the road segments they monitor. Later, the graph model is further analysed to estimate the coverage of the existing sensors in order and expand the applicability of dynamic lighting, thus aiding the city officials in the decision-making process. This shows that the proposed approach can be used to both enable day-to-day operations and support long-term evolution of smart cities.

2   Motivation

As mentioned in Sect. 1, collection of smart city data is the first step, but actual benefits come from proper utilisation of such data. Following up on the example of dynamically-dimmed street lighting, savings can be generated by just providing appropriate control logic for existing devices. These savings are also significant, as street lighting is one of the main contributors of cities' energy bills [9]; as shown in [11], implementing such dynamic control can reduce power consumption by 34%.

A common problem with dataset integration is that often they are logically linked only by the spatial (GIS, Geographic Information System) component. For instance, having a dataset modelling streetlight locations (as points) and one modelling streets (as lines or polygons), there usually won't be an attribute relationship between them. Therefore, to determine that a lamp illuminates a given street, one must resort to analysing spatial relationships, such as the distance. This process is not straightforward, as we've shown e.g. in [3]: while it was possible to perform large-scale (Washington D.C.) lamp-to-street assignment programatically, the process involved a lot of fine-tuning and was error-prone.

During a pilot project for intelligent street lighting, jointly executed by the city of Kraków, Poland, and AGH University of Science and Technology, there were two challenges related to traffic intensity sensors:

1. the sensor locations were provided as points, without any relationship with streets,
2. the sensor network was rather sparse, which limited the applicability of street lighting.

The first problem was solved by simply performing a manual assignment of sensors to the street segments they monitor. The second issue required more work and resulted in development of the concept of *virtual sensors*, based on the Dual Graph Grammar formalism [13]. However, the formulae used by the virtual sensors to compute the estimated traffic intensity based on the neighbouring real sensors still had to be developed by hand.

This paper presents a graph model, along with the necessary formalisms and operations, to resolve both of the aforementioned challenges. It uses publicly-available OpenStreetMap data as the basis for integration of other datasets and proposes an extensible graph model to store all of its data. It allows for automatic assignment of camera-based sensors to street segments, and by inferring the possible traffic flows, it can also estimate the coverage of neighbouring streets automatically.

3   Formal Background

We assume that the digital twin of city will have a multi-layer graph structure. Its flat representation will be generated with the help of methods described in [6,10]. Every layer will be generated by another graph grammar, with transformations fired by a dedicated expert system considering both spatial and semantic relations.

3.1   Spatially-Triggered Graph Transformations

The Spatially-Triggered Graph Transformations (STGT) mechanism was first introduced in [2] and involves the usage of specialised tools to detect spatial relationships among objects in different datasets. This allows for *materialisation* of such relationships in the graph, which, in turn, enables their usage for further analysis.

In essence, the methodology uses an external tool, called the *expert system* (Ξ), to:

- detect spatial relationships (such as vicinity, intersection, parallelism) between geographic objects modelled by graph nodes,
- perform spatial measurements and/or transformations of existing geometries, e.g. to supplement the detected relationships with attributes (such as distance, angle, etc.) or to generate new geometries based on existing ones.

In practice, to implement the concept of STGT, we use a programmatic procedure based on spatial (GIS) analysis tools, such as a spatial database (PostGIS[1], Spatialite[2]) or a spatial analysis library, such as GeoPandas[3]. The latter choice seems preferable in this case, as it is well-suited to ad-hoc analysis.

---

[1] https://postgis.net.

[2] https://www.gaia-gis.it/fossil/libspatialite/index.

[3] https://geopandas.org/en/stable/.

A practical example of such application can be found in Sect. 4.3, and the formal notations of the STGT concept and its prerequisites are presented below.

3.2 Notations

The Digital Twin Graph will represent the entire structure of the city data and the semantic relations among their elements, and formally it is a eight-tuple.

Definition 1 (Digital Twin Graph). *The DTG is an attributed graph over the set of node labels Σ and the set of edge labels Γ, such that:*

$$DTG = (V, E, \Delta, lab_{X\,(X=V,E)}, att_{X\,(X=V,E)}, \Pi)$$

where:

- *V is a finite, nonempty set of graph nodes identified unambiguously by some injective indexing function $\text{Index} : V \to \mathbb{N}$,*
- *$E \subseteq V \times V$ is a set of edges,*
- *$\Delta \subseteq V$ is a set of nonterminal nodes.*
- *$lab_V : V \to \Sigma$ is a node labelling function,*
- *$lab_E : E \to \Gamma$ is a edge labelling function,*
- *$att_X : X \to 2^{Att\_name_X \times Att\_val_X}$ $(X = V, E)$, is a node/edge attributing function, such that for $a \in X$ return a set of pairs (an, av) where $an \in Att\_name_X$ is an attribute name, and $av \in Att\_val_X$ is an attribute value.*

Comments: the set of available attributes in the attributing depends on the label of the node or edge; however, formally, the notation which is a subset of all possible name and values combinations is sufficient.

The transformations are formally defined as a part of the graph grammars notation presented below.

Definition 2 (Graph grammar). *A graph grammar Ω is a tuple:*

$$\Omega = (\Sigma, \Gamma, \Delta, \Phi, S, \Pi)$$

where:

- *Σ is the set of node labels,*
- *$\Delta \subset \Sigma$, is the set of terminal node labels,*
- *Γ is the set of edge labels,*
- *Φ is the set of transformation rules (see def:TranGra),*
- *S is the starting graph,*
- *Π is the graph grammar validation condition, that verifies the current state of the graph.*

We use Π as the validation condition of the graph grammar because sometimes it is necessary to execute a sequence of transformations and some of the intermediate states may not form a correct graph. The validation condition Π usually is not expressed in the graph's definition, but in theory we always assume that there are no non-terminals in the final graph, so we have at least one condition:

$$\forall v \in V : lab_V(v) \in \Delta \Rightarrow G \text{ is a final graph}$$

The non-terminal nodes are used only in temporary states where a sequence of transformations is used to move the graph from one final state to another one.

At each level, the transformations will work on a subset of V and E, i.e. it is true that:

- for each transformation Ω $V_\Omega \subseteq V$
- for any node $x \in V - S$ exist the transformations Υ such that $x \in V_\Upsilon$.

Definition 3 (Graph Transformation). *A five tuple*

$$(L, R, roots, eval_V, eval_E) \in \Phi_\Omega$$

is a transformation rule when:

- *L, R are the graphs called the left- and right-side of a transformation, respectively; each of them can be a set of weakly-connected graphs.*
- *For a given left-hand-side graph L consisting of k connected components, cont is a set of indices, $\{i_1, i_2, \ldots, i_k\}$, such that i_j is a least index (referred to as a root) in a j-th connected component. The subset roots $\subseteq V_L$ contains vertices with indices belonging to cont.*
- *$eval_X : X_R \rightarrow 2^{Att\_name_X \times Att\_val_X}$ $(X = V, E)$ is the node/edge attribute evaluation function, respectively (see att_X functions in Definition [def:dsg])*
- *the graph transformation mechanism in based on the single push-out mechanism:*
  1. *the morphism $h : L \rightarrow H$ is designated (basing on cont function),*
  2. *nodes of the graph $h(L - R)$ are removed from H,*
  3. *edges with removed nodes are removed too,*
  4. *graph R is added to H (nodes and edges belonging to $L \cap R$ joins the R graph with $H - (L - R)$, and for them the transformations only change their attributes values)*

The $cont, eval_V$ and $eval_E$ functions are the formal representation of an interface between the graph transformations and the expert systems, representing domain-specific knowledge.

Let Ξ be an expert system which both sees the current digital twin graph and represents some knowledge (e.g., it is able to point out that a luminary illuminates a given street based on their geometries), so it launches some transaction that enriches the digital twin information. Formally, it will be defined as follows.

This concept is the formal background of the spatially-triggered graph transformations (STGT) methodology.

Definition 4 (Transformation firing). *For any graph H and any transformation* $\omega = (L, R, match, eval_V, eval_E) \in \Phi_\Omega$ *the expert system fires the transaction when:*

1. Ξ *designates the function* $match : roots \to H,$[4]
2. Ξ *designates* $eval_V, eval_E$ *functions, i.e., defines the values of attributes of the graph R,*
3. *the transformation* ω *is applied to the H graph.*

Let us note that the left side of the production is not a single graph L, but a sequence of graphs, such that each vertex with the lowest index is mapped to a corresponding vertex in the input (transformed) graph.

4   Solution Outline

To illustrate the proposed solution using a practical example, we will guide the reader through the individual steps of graph generation and analysis. Each step of the presented method will be presented in relation to the formal base presented in Sect. 3.

The structure of this section is as follows. First, in Sect. 4.1, a graph representing 100% of the data contained in the OpenStreetMap repositories for the area of interest is generated.

The data originating from OSM forms the background (the "common denominator") for integration of datasets related directly to the *smart city* solution being implemented. In this case, the dataset contains the parameters of camera-based traffic intensity sensors. Section 4.2 describes the step of adding nodes representing these objects to the graph.

However, at this stage, the sensor nodes are not yet semantically assigned to the real-world objects they monitor, i.e. to the streets being observed by each camera. Section 4.3 discusses the application spatially-triggered graph transformations to accomplish this. A GIS analysis tool (formally, the "expert system" as described in Sect. 3) is used to first estimate the area "seen" by each camera, and then to detect the streets which intersect these areas. These detected relationships are persisted as graph edges; this *materialisation* of phenomena detected in the data is one of the key characteristics of STGT.

Finally, as the sensor network is relatively sparse, it seems beneficial to *propagate* this information in order to estimate the traffic levels on unmonitored streets. Section 4.4 shows how the network structure of OpenStreetMap data can be utilised to detect areas in which the certainty of estimate traffic intensity is the lowest, making them ideal candidates for furhter extension of the sensor network.

---

[4] Theoretically, such a designation can be made by the graph transformation mechanism (defined in the point 3), but in practice, it leads us to the NP-complete graph isomorphism problem.

4.1   OpenStreetMap Graph Generation

The first stage involves building a graph, containing 100% of data included in the source OpenStreetMap data.

The OpenStreetMap data model is simple, as it contains only three types of entities:

- *nodes*, which are always points with a latitude/longitude geographic location,
- *ways*, which are ordered sequences of points,
- *relations*, which are used to define relationships between other objects (of any of the three types).

Each object has a unique identifier (integer). The semantics of each object are defined by a set of key-value pairs, called *tags*. The taxonomy of keys and values is defined in OSM documentation[5].

We assume that the OSM data is available through an Overpass API[6], or in a PostgreSQL/PostGIS database created using a tool such as Osmosis[7] or Osmium[8].

The resulting graph will typically be managed by a graph database, such as Neo4J[9] or Apache AGE[10]. These solutions use the Cypher[11] language to model and query the data, therefore some examples will present the Cypher code used for experiments.

Nodes. For each OSM node, a graph node with the label `Node` will be created. All node data is stored as graph attributes:

- the OSM identifier, as the `osm_id` attribute,
- the geometry (geographic location of the point), stored as a Well-Known Text (WKT)[12] value in the `geometry` attribute,
- all other tags and values as attributes.

The Cypher code representing an example node is provided in Listing 1, and its visual representation is in Fig. 1.

Ways. Ways are ordered sequences of nodes. Hence, for each way, a graph node with the label `Way` will be created. The attributes are modelled in the same way as for nodes (see Sect. 4.1), with the distinction that the geometry is usually a WKT linestring.

---

[5] https://wiki.openstreetmap.org/wiki/Map_features.
[6] http://overpass-api.de.
[7] https://wiki.openstreetmap.org/wiki/Osmosis.
[8] https://wiki.openstreetmap.org/wiki/Osmium.
[9] https://neo4j.com.
[10] https://age.apache.org.
[11] https://neo4j.com/developer/cypher/.
[12] https://libgeos.org/specifications/wkt/.

```
CREATE (n:Node {
  osm_id: 6958500807, tourism: 'museum',
  name: 'Muzeum AGH', wikidata: 'Q11786993'
});
```

Listing 1: Cypher representation of an OSM node.

Fig. 1. Graph representation of an OpenStreetMap node.

In the graph, all nodes which are members of a way will have MEMBER_OF relationships towards the way, and the order is indicated by the order property.

Cypher code representing an example node is provided in Listing 2, and its visual representation is in Fig. 2.

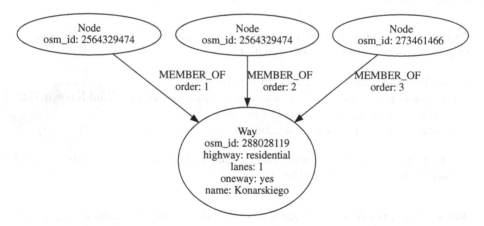

Fig. 2. Graph representation of an OpenStreetMap way; node attributes ommitted for clarity.

Relations. OpenStreetMap *relations* model relationships among other OSM objects. They are ordered lists of nodes, ways and/or other relations. Relation members may have *roles*, such as:

– outer, signifying that a *way* is part of an outer boundary of an area,

```
MERGE (w1:Way {
  osm_id: 288028119, highway: "residential", lanes: 1,
  oneway: "yes", name: "Konarskiego"})
MERGE (n1:Node {osm_id: 2564329474})
MERGE (n2:Node {osm_id: 2720618781})
MERGE (n3:Node {osm_id: 273461466})
MERGE (n1)-[:MEMBER_OF {order: 1}]->(w1)
MERGE (n2)-[:MEMBER_OF {order: 2}]->(w1)
MERGE (n3)-[:MEMBER_OF {order: 3}]->(w1);
```

Listing 2: Cypher representation of an OSM way.

- **subarea**, signifying that an area is part of another, larger area (e.g. a city district is part of the city itself),
- **admin_centre**, signifying that a given point (*node*) is designated as the administrative centre of a city.

Relations can be used to model any relationships, from administrative areas to bus lines. The relation's tags determine its semantics.

We model relations similarly to ways, i.e. the identifier is stored as the osm_id attribute, and all tags as other attributes.

Each relation member will have a MEMBER_OF relationship towards the relation, with its order and role indicated by the order and role properties, respectively.

Due to space limitations and the structural similarity of the graph representation of relations and ways, an example is not provided.

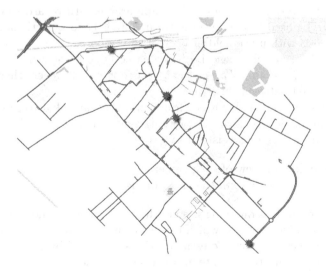

Fig. 3. Locations of 6 camera-based traffic sensors in Siechnice; 3 of these are located in one intersection in the city centre.

4.2 Sensor Modelling

To discuss the issue of sensor modelling, let us consider an example of Siechnice. It is a town in south-western Poland with a population of 9,302 (as of 2022). The local authorities are keen on innovation, and in the years 2019–2022, Siechnice introduced several intelligent solutions in cooperation with AGH University, within the Human Smart Cities project.

There are 6 cameras with vehicle-counting functionality deployed in the city. Their locations have been shown in Fig. 3. Each camera is described by its geographic location, its viewing angle (degrees) and its azimuth (degrees).

The cameras are modelled as nodes with label *Sensor*, add attributes as shown in Fig. 4. Please note that at the moment, sensors.

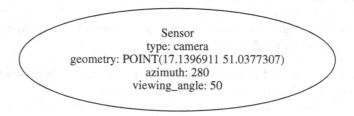

Fig. 4. Graph representation of a camera-based sensor.

4.3 Sensor Assignment

In the presented example, each camera is configurable with regard to detection zones, which can be used to measure the traffic intensity in a particular lane, as long as that lane is within the field of view of the camera.

Please note that, at this stage, both the cameras and the streets are parts of the graph, as *Sensor* and *Way* objects, respectively. However, there are no relationships defined between them yet.

To assign the sensors to the street fragments they are able to monitor, we use the STGT methodology, formally presented in Sect. 3.1.

The proposed procedure consists of two stages:

1. estimating the field of view of each camera,
2. determining the streets within that field of view.

The first stage involves querying the graph for all cameras and, for each one, constructing shapes (geometries) which model the field of view. These are added to the graph as separate nodes. To estimate the area "seen" by each camera, we will first use a procedure to generate an isosceles triangle, where:

– the *apex* is at the location of the camera,

- the length of the *arms* represents a reasonable viewing range of the camera (e.g. 40 m),
- the angle of the *axis* is the azimuth of the camera,
- the angle between the *arms* equals the viewing angle.

Therefore, at the first stage, the graph, containing the *Sensor* nodes, is supplemented with *FieldOfView* nodes, using the transformation presented in Fig. 5. The new node has a `geometry` attribute, with a polygon feature modelling the field of view. As indicated in Sect. 3.1, this would be calculated using an external tool, such as GeoPandas, or a spatial database (e.g. PostGIS).

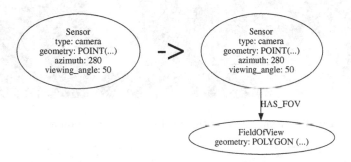

Fig. 5. Graph transformation adding the field of view to a camera-based sensor.

The external tool is used to find the pairs of *FieldOfView* and *Way* nodes with intersecting geometries. In this step, no new nodes are added, but a *COVERS* relationship is added between the two "intersecting" nodes, as shown in Fig. 7. Therefore, the spatial relationships *detected* in the data are now *materialized*, making them suitable for further analysis.

In the last stage, a *SEES* relationship between the *Sensor* node and the *Way* is added, using only the relationships detected and materialised in the previous step, as shown in Fig. 8. Therefore, the relationship between the sensor and the road is materialised, which means that the fact that a given camera is able to measure traffic on a given road.

It should also be noted that this approach makes the process well-documented, replicable and reversible; these characteristics alone constitute added value of the graph-based methodology compared to approaches based solely on GIS tools or spatial databases. However, further benefits are presented in Sect. 4.4 below.

4.4   Traffic Flow Modelling

At this stage, we have assigned the camera sensors to the roads they are able to monitor in a direct manner. However, given the usual sparsity of the sensor system, the traffic intensity in most streets fragments is not monitored, and

Fig. 6. Visualisation of the created field of view on a map.

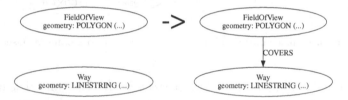

Fig. 7. Materialiasation of the *COVERS* relationship between the camera's field of view and the geometries.

therefore we are unable include them e.g. in the dynamic streetlight dimming system (Fig. 6).

However, the graph-based structure makes it appropriate for network analysis, which may result both in no-cost expansion of the dimming system's coverage and in determination of the candidates for expansion.

Since the steps of this procedure use traditional graph transformation, and due to space limitations, only a brief, textual description is provided:

1. New *Flow* nodes, representing possible traffic flows, are introduced into the graph – two for each two-way street, and one for each one-way street; each *Flow* node has a *ON_STREET* edge towards "its" street (*Way* node), with the `direction` attribute indicating the flow direction (forward or reverse).
2. Possible street-to-street flows are modelled by introducing *CONTIN-UES_INTO* edges between *Flow* nodes; e.g., for an intersection where a vehicle can go straight on, turn left or turn right, there will be 3 edges towards the respective destination flows.

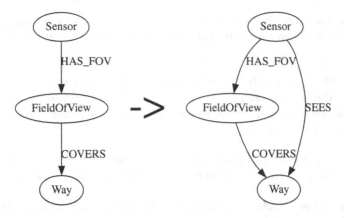

Fig. 8. Materialiasation of the *SEES* relationship between the camera-based sensor and the road.

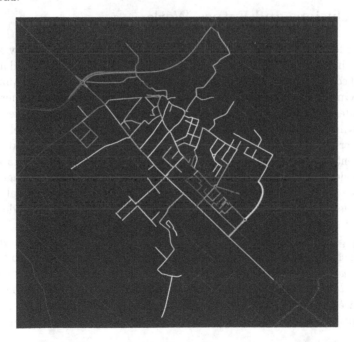

Fig. 9. Results of estimation certainty propagation in the city of Siechnice.

3. A `certainty` attribute with the value of 1 is added to *Flow* nodes belonging to *Way* nodes with *SEES* relationships from *Sensor* nodes.
4. A *certainty propagation* network algorithm is executed, where the values of the `certainty` attribute are propagated to continuation *Flow* nodes, where the value is divided by the number of possible continuations; e.g., for the example presented in point 2 above, if the source *Flow* node's `certainty`

equals 1, the three destination *Flow* nodes will have a `certainty` of $\frac{1}{3}$. This procedure is repeated until there are no modifications of certainty of any *Flow*.

As a result, it is possible to estimate the certainty with which it is possible to *estimate* the traffic intensity in each street. Example results for the city of Siechnice are presented in Fig. 9, where the colour indicates the certainty value. This allows us to distinguish:

- streets with reasonably high traffic estimation certainty – the ones which can be equipped with *virtual sensors* and included in the dynamic dimming system,
- streets with low certainty – which should be the natural candidates for deployment of new sensors.

5   Conclusions and Future Work

The graph formalism is suitable for modelling smart city concepts by gathering knowledge in a robust way which results in a digital twin. Labelled and attributes nodes and edges are suitable for expressing semantic, qualitative and quantitative features. On top of that graph transformations enable simulation, inference and general interpretation of such knowledge providing actual incentives, such as energy savings, increase of comfort, smart control, or development planing.

However there is a procedural gap between actual reality to be modelled and its representation being the digital twin. It is data integration. Establishing a proper model is often based on semantic fuzziness of natural language such as: *near by*, *looks at*, *crosses over*, *leads to* etc., that binds geographical locations of the modelled physical features. It poses a challenge if such a model is to be established automatically from large heterogeneous datasets.

The gap can be sealed with the proposed STGT methodology. It enables graph transformations to be triggered by spatial locations. Thanks to that questions like: *what street does the camera overlook, can traffic flow from street A to street B at this intersection, is this sensor redundant* do not have to be answered by human experts but instead the answers can be inferred at large. It results in an ability to integrate heterogeneous data in a single, coherent, graph-based model.

The proposed approach has been proven within the framework of the pilot projects in Kraków and Siechnice, Poland. It is to be tested at scale as a main tool to assess modernization capacity of large urban areas in 2023.

References

1. Deakin, M., Al Waer, H.: From intelligent to smart cities. Intell. Build. Int. **3**(3), 133–139 (2011). https://doi.org/10.1080/17508975.2011.586673

2. Ernst, S., Kotulski, L.: Estimation of road lighting power efficiency using graph-controlled spatial data interpretation. In: Paszynski, M., Kranzlmüller, D., Krzhizhanovskaya, V.V., Dongarra, J.J., Sloot, P.M.A. (eds.) ICCS 2021. LNCS, vol. 12742, pp. 585–598. Springer, Cham (2021). https://doi.org/10.1007/978-3-030-77961-0_47

3. Ernst, S., Starczewski, J.: How spatial data analysis can make smart lighting smarter. In: Nguyen, N.T., Chittayasothorn, S., Niyato, D., Trawiński, B. (eds.) ACIIDS 2021. LNCS (LNAI), vol. 12672, pp. 272–285. Springer, Cham (2021). https://doi.org/10.1007/978-3-030-73280-6_22

4. European Committee for Standarization: CEN/TR 13201-1: Road lighting – Part 1: Guidelines on selection of lighting classes. Technical report, European Committee for Standarization (2014)

5. Jones, D., Snider, C., Nassehi, A., Yon, J., Hicks, B.: Characterising the digital twin: a systematic literature review. CIRP J. Manuf. Sci. Technol. 29, 36–52 (2020). https://doi.org/10.1016/j.cirpj.2020.02.002. https://www.sciencedirect.com/science/article/pii/S1755581720300110

6. Kotulski, L., Szpyrka, M.: Graph representation of hierarchical Alvis model structure. In: Proceedings of the 2011 International Conference on Foundations of Computer Science FCS 2011 (Part of Worldcomp 2011), Las Vegas, Nevada, USA, pp. 95–101 (2011)

7. Mahoor, M., Salmasi, F.R., Najafabadi, T.A.: A hierarchical smart street lighting system with brute-force energy optimization. IEEE Sens. J. 17(9), 2871–2879 (2017). https://doi.org/10.1109/JSEN.2017.2684240

8. Paiho, S., Tuominen, P., Rökman, J., Ylikerälä, M., Pajula, J., Siikavirta, H.: Opportunities of collected city data for smart cities. IET Smart Cities 4(4), 275–291 (2022). https://doi.org/10.1049/smc2.12044

9. Pardo-Bosch, F., Blanco, A., Sesé, E., Ezcurra, F., Pujadas, P.: Sustainable strategy for the implementation of energy efficient smart public lighting in urban areas: case study in San Sebastian. Sustain. Urban Areas 76, 103454 (2022). https://doi.org/10.1016/j.scs.2021.103454

10. Szpyrka, M., Matyasik, P., Biernacki, J., Biernacka, A., Wypych, M., Kotulski, L.: Hierarchical communication diagrams. Comput. Inform. 35(1), 55–83 (2016)

11. Wojnicki, I., Ernst, S., Kotulski, L.: Economic impact of intelligent dynamic control in urban outdoor lighting. Energies 9(5), 314 (2016). https://doi.org/10.3390/en9050314

12. Wojnicki, I., Ernst, S., Kotulski, L., Sędziwy, A.: Advanced street lighting control. Expert Syst. Appl. 41(4, Part 1), 999–1005 (2014). https://doi.org/10.1016/j.eswa.2013.07.044

13. Wojnicki, I., Kotulski, L.: Improving control efficiency of dynamic street lighting by utilizing the dual graph grammar concept. Energies 11(2), 402 (2018). https://doi.org/10.3390/en11020402

Forecasting Cryptocurrency Prices Using Contextual ES-adRNN with Exogenous Variables

Slawek Smyl[1], Grzegorz Dudek[2(✉)], and Paweł Pełka[2]

[1] Częstochowa, Poland
slawek.smyl@gmail.com

[2] Electrical Engineering Faculty, Czestochowa University of Technology,
Częstochowa, Poland
{grzegorz.dudek,pawel.pelka}@pcz.pl

Abstract. In this paper, we introduce a new approach to multivariate forecasting cryptocurrency prices using a hybrid contextual model combining exponential smoothing (ES) and recurrent neural network (RNN). The model consists of two tracks: the context track and the main track. The context track provides additional information to the main track, extracted from representative series. This information as well as information extracted from exogenous variables is dynamically adjusted to the individual series forecasted by the main track. The RNN stacked architecture with hierarchical dilations, incorporating recently developed attentive dilated recurrent cells, allows the model to capture short and long-term dependencies across time series and dynamically weight input information. The model generates both point daily forecasts and predictive intervals for one-day, one-week and four-week horizons. We apply our model to forecast prices of 15 cryptocurrencies based on 17 input variables and compare its performance with that of comparative models, including both statistical and ML ones.

Keywords: Hybrid forecasting models · Recurrent neural networks · Cryptocurrency price forecasting

1 Introduction

Forecasting cryptocurrency prices is a very difficult task due to several reasons. Firstly, the cryptocurrency market is highly volatile and unstable, with prices fluctuating rapidly and unpredictably [1]. This is because the market is not regulated, and there is no central authority that controls the supply and demand of cryptocurrencies. Unlike traditional assets such as stocks, cryptocurrencies lack a fundamental value. This means that their value is not directly linked to any underlying assets or earnings. Instead, their value is primarily

G.D. thanks prof. W.K. Härdle for his guidance on cryptocurrencies. G.D. and P.P. were partially supported by grant 020/RID/2018/19 "Regional Initiative of Excellence" from the Polish Minister of Science and Higher Education, 2019–23.

J. Mikyška et al. (Eds.): ICCS 2023, LNCS 14073, pp. 450–464, 2023.
https://doi.org/10.1007/978-3-031-35995-8_32

determined by market demand and supply dynamics, making their price highly volatile and subject to sudden fluctuations. Secondly, cryptocurrencies are relatively new, and there is a lack of historical data, making it difficult to identify trends and patterns. Thirdly, the market is influenced by various factors [2,3], including government regulations (related to the legality of cryptocurrencies, taxation policies, and anti-money laundering laws), global economic conditions, and the emergence of new cryptocurrencies, which can make it challenging to predict price movements accurately. Moreover, market sentiment and speculation can have a significant impact on cryptocurrency prices. Cryptocurrencies are still a relatively new and emerging asset class, and as such, they are more susceptible to hype, media coverage, and market sentiment. This can cause price bubbles and crashes, making it challenging to forecast their prices accurately. Finally, the cryptocurrency market operates 24/7, and there is no closing or opening bell, which means that prices can change at any time, making it challenging to keep up with the latest developments and adjust forecasts accordingly.

To capture the influence of external factors on the cryptocurrency price, the forecasting models use different inputs, which can be categorized as follows [3–9]:

1. Economic indicators. Cryptocurrency prices may be influenced by macroeconomic indicators such as GDP, inflation, and interest rates.
2. Trading volume. The volume of cryptocurrency trades can indicate market demand, which can affect price movements.
3. Technical indicators. Technical analysis tools such as moving averages, Bollinger bands, and RSI can help identify trends and patterns in price movements.
4. News sentiment. The sentiment analysis of news articles and social media posts related to cryptocurrencies can provide insight into the market's mood and direction.
5. Network activity. The activity on the blockchain network, such as the number of transactions and the hash rate, can reflect the popularity and adoption of a particular cryptocurrency.
6. Regulatory developments. Government regulations and policies related to cryptocurrencies can significantly impact their prices.

Multivariate forecasting models used for cryptocurrency prices can be categorized into three groups: classical statistical or econometrics models, computational intelligence or machine learning (ML) models, and hybrid models [10]. One popular representative of the classical statistical models is the Vector Autoregression (VAR) model [11,12]. This model uses a system of equations to estimate the relationship between multiple variables and can capture the dynamic interdependence between them. However, it assumes linear relationships between variables, which may not always hold in the highly complex and nonlinear cryptocurrency market. Therefore, it may not be suitable for accurately predicting cryptocurrency prices in all scenarios.

ML models have several advantages over classical statistical models in forecasting cryptocurrency prices. One advantage is their ability to capture nonlinear relationships between variables. Additionally, ML models can handle large amounts of data, can effectively extract relevant features from it, and are often better than statistical models at identifying patterns and trends. They can also adapt to new information and adjust their forecasts accordingly. However, ML can be more computationally intensive and requires more data preprocessing compared to statistical models. Also, as a general rule, they require more data.

Some examples of ML models for forecasting cryptocurrency prices are as follows. A model proposed in [13] utilizes on-chain data as inputs, which are unique records listed on the blockchain that are inherent in cryptocurrencies. To ensure stable prediction performance in unseen price ranges, the model employs change point detection to segment the time series data. The model consists of multiple long short-term memory (LSTM) modules for on-chain variable groups and the attention mechanism. The research described in [14] uses a multivariate prediction approach and compare different types of recurrent neural networks (LSTM, Bi-LSTM, and GRU). Seven input variables are considered: date, open, high, low, close, adjusted close, and volume. Paper [15] aims to develop a prediction algorithm for Bitcoin prices using random forest regression and LSTM. The study also seeks to identify the variables that have an impact on Bitcoin prices. The analysis utilizes 47 explanatory variables, which are categorized into eight groups, including Bitcoin price variables, technical features of Bitcoin, other cryptocurrencies, commodities, market index, foreign exchange, public attention, and dummy variables of the week. In [16], a set of high-dimension features including property and network, trading and market, attention and gold spot price was used for Bitcoin daily price prediction. The authors compared statistical models such as logistic regression and linear discriminant analysis with ML models including random forest, XGBoost, support vector machine and LSTM. A comprehensive comparison of statistical and ML models for cryptocurrency price prediction can be found in [10].

Our study introduces a hybrid model that merges statistical and ML methods, specifically exponential smoothing (ES) and recurrent neural networks (RNNs). This model, referred to as cES-adRNN, consists of two tracks designed to incorporate context information, effectively handle time series of exogenous variables and enhance the accuracy of forecasting. The proposed model offers several advantages over existing ML models:

– Unlike traditional ML models that often require data preprocessing to simplify the forecasting problem, our model can handle raw time series data without any preprocessing. It has built-in mechanisms for preprocessing time series dynamically on-the-fly.
– While classical ML models require additional procedures to select relevant input information, our recurrent cells have an internal attention mechanism that dynamically weighs inputs, making the model more efficient.
– Unlike non-recurrent ML models, RNN, which is a component of our model can model temporal relationships in sequential data. We apply dilated recurrent cells, which are fed by both recent and delayed states, to capture short-term, long-term, and seasonal dynamics. Moreover, our hierarchical RNN

architecture extends the receptive fields in subsequent layers, enabling better modeling of long-term and seasonal relationships.

- Most forecasting ML models produce only point forecasts. Our model is able to generate both point forecasts and predictive intervals in the same time.
- While most forecasting ML models learn on a single time series, our model can learn from multiple similar series, leading to better generalization.

The main contributions of this study lie in the following three aspects:

1. We extend our cES-adRNN model proposed in [17] by introducing exogenous variables. This enables the model to predict time series based not only on the past values of the series (univariate case) but also on other, parallel time series which are correlated with the predicted series.
2. We modify cES-adRNN by introducing additional per-series parameters allowing to capture individual properties of the series of exogenous variables.
3. To validate the efficacy of our proposed approach, we conduct extensive experiments on 15 cryptocurrency datasets comprising 17 variables. Our experimental results show the high performance of the modified cES-adRNN model and its potential to forecast cryptocurrency prices more accurately than comparative models including both statistical and ML models.

The remainder of this paper is organized as follows. In Sect. 2, we present the data and define a forecasting problem. Section 3 presents the proposed forecasting model. Section 4 reports our experimental results. Finally, we conclude this work in Sect. 5.

2 Data and Forecasting Problem

The real-world data was collected from BRC Blockchain Research Center (https://blockchain-research-center.com; accessible for BRC members). The dataset BitInfoCharts provides a wide range of tools for accessing information on the cryptocurrency sphere. It includes market, blockchain operation, and social media data for 16 cryptocurrencies: BTC, ETH, DOGE, ETC, XRP, LTC, BCH, ZEC, BSV, DASH, XMR, BTG, RDD, VTC, BLK, FTC. In this study, we omitted RDD from this list due to the short period of data (from 13 January 2021). The dataset includes 19 variables (time series) for each cryptocurrency:

- active_addresses_per_day
- avg_block_size_per_day
- avg_block_time_per_day
- avg_fee_to_reward
- avg_hashrate_per_day
- avg_mining_difficulty_per_day
- avg_price_per_day
- avg_transaction_fee_per_day
- avg_transaction_value_per_day
- google_trends

- market_capitalization
- med_transaction_fee_per_day
- med_transaction_value_per_day
- mining_profitability
- num_transactions_per_day
- num_tweets_per_day
- num_unique_addresses_per_day
- sent_coins
- top_100_to_all_coins

The data availability and quality differ from coin to coin, due to the issue time, crypto design, and the third party vendors. Due to incompleteness or poor quality, we excluded two variables, `google_trends` and `num_tweets_per_day`.

Our goal is to predict prices of cryptocurrencies (`avg_price_per_day`) based on historical values of all variables. Thus, we predict a sequence of cryptocurrency prices of length h, $\{z_\tau\}_{\tau=M+1}^{M+h}$, given past observations of prices, $\{z_\tau\}_{\tau=1}^{M}$, and N exogenous variables, $\{p_\tau^i\}_{\tau=1}^{M}$, $i = 1, .., N$, where h is the forecast horizon and M is the time series length (training part). In Fig. 1 the forecasted time series of cryptocurrency prices are shown.

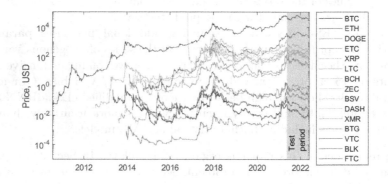

Fig. 1. Prices of cryptocurrencies.

We consider three forecast horizons: 1, 7 and 28 days ahead. For each horizon, a separate model is constructed. We define one-year test period from 1 June 2021 to 31 May 2022. Moving in the test period by one day, the model generates forecasts for the next h days, using past data for training.

3   Model

The model is a modified version of contextually enhanced ES-dRNN with dynamic attention (cES-adRNN) proposed in [17]. It combines exponential smoothing (ES) and recurrent neural network (RNN). The model is composed of two simultaneously trained tracks: context track and main track. The main track learns to generate point forecasts for horizon h and also predictive intervals. The context track learns to generate additional inputs for the main track based on representative time series. The block diagram of the model is shown in Fig. 2.

The model is trained in cross-learning mode (on all cryptocurrency data). Some components of the model are able to capture properties of individual time series, while others learn shared features of all series. Thus the data is exploited hierarchically, utilizing both local and global components to extract and synthesize information at both the individual series and collective dataset levels.

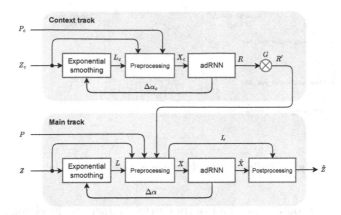

Fig. 2. Block diagram of the forecasting model.

3.1 Main Track

Inputs to the main track are price time series (Z) and exogenous time series (P).

Exponential Smoothing Component. (ES) smooths price time series. It has a basic form given by the formula:

$$l_{t,\tau} = \alpha_t z_\tau + (1 - \alpha_t) l_{t,\tau-1} \tag{1}$$

where $l_{t,\tau}$ is a level component, and $\alpha_t \in [0,1]$ is a smoothing coefficient, both for recursive step t.

A dynamic variant of ES is utilized, wherein the smoothing coefficient is not constant, unlike in the standard version, but instead changes dynamically at each step t to adapt to the current time series properties, as proposed in [18]. The adaptation of the smoothing coefficient is performed using correction $\Delta\alpha_t$, which is learned by RNN, as follows [18]:

$$\alpha_{t+1} = \sigma(I\alpha + \Delta\alpha_t) \tag{2}$$

where $I\alpha$ represents the starting value of α, and the sigmoid function σ is employed to ensure that the coefficient remain within the range from 0 to 1.

Preprocessing Component prepares input and output data for RNN training. Input data includes preprocessed sequences of the price and exogenous variables time series, i.e. sequences covered by the sliding input window of length n, Δ_t^{in}, while output data includes preprocessed sequences of the price series covered by the sliding output window of length h, Δ_t^{out} (see Fig. 3).

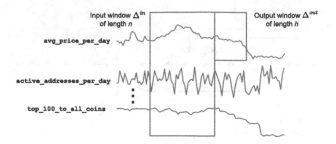

Fig. 3. Sliding windows for generating training data.

To normalize both input and output price sequences, level component (1) extracted by ES is used. Input sequences are preprocessed as follows:

$$x_\tau^{in} = \log \frac{z_\tau}{\hat{l}_t} \tag{3}$$

where $\tau \in \Delta_t^{in}$, and $\hat{l}_t$ is the level component predicted by ES for the last point in Δ_t^{in} in step t.

The use of the log function for squashing in (3) is aimed at preventing outliers from interfering with the learning process.

The output sequences are normalized as follows:

$$x_\tau^{out} = \frac{z_\tau}{\hat{l}_t} \tag{4}$$

where $\tau \in \Delta_t^{out}$.

The preprocessed input and output sequences are represented by vectors: $\mathbf{x}_t^{in} = [x_\tau^{in}]_{\tau \in \Delta_t^{in}} \in \mathbb{R}^n$ and $\mathbf{x}_t^{out} = [x_\tau^{out}]_{\tau \in \Delta_t^{out}} \in \mathbb{R}^h$, respectively. It should be noted that the input and output patterns have a dynamic nature and are modified in each training epoch due to the adaptation of the level component. This can be viewed as the learning of the optimal representation for RNN.

The input sequences of the i-th exogenous variable are normalized as follows:

$$x_\tau^{p_i} = \log_{10} \left(\frac{p_\tau^i}{\bar{p}^i} + 1 \right) \tag{5}$$

where $\tau \in \Delta_t^{in}$, and $\bar{p}^i$ is the average value of the i-th exogenous variable in the training period.

The regressors take broad range of positive numbers and zero. Therefore the ratio $\frac{p_\tau^i}{\bar{p}^i}$ can be very small and occasionally zero. For this reason we add 1 in the formula above. The preprocessed input sequences of exogenous variables are represented by vectors $\mathbf{x}_t^{p_i} = [x_\tau^{p_i}]_{\tau \in \Delta_t^{in}} \in \mathbb{R}^n$, $i = 1, ..., N$.

To construct input patterns for RNN, vectors $\mathbf{x}_t^{in}$ and $\mathbf{x}_t^{p_i}, i = 1, ..., N$ are concatenated and extended by two components: a local level of the price series, $\log_{10}(\hat{l}_t)$, and a modulated context vector produced by the context track, $\mathbf{r}_t'$:

$$\mathbf{x}_t^{in'} = [\mathbf{x}_t^{in}, \mathbf{x}_t^{p_1}, ..., \mathbf{x}_t^{p_N}, \log_{10}(\hat{l}_t), \mathbf{r}_t'] \tag{6}$$

Recurrent Neural Network is trained on training samples $(\mathbf{x}_t^{in'}, \mathbf{x}_t^{out})$ generated continuously by shifting sliding windows Δ_t^{in} and Δ_t^{in} by one day. RNN produces four components: point forecast vector, $\hat{\mathbf{x}}_t^{RNN}$, two quantile vectors defining a predictive interval (PI), $\underline{\hat{\mathbf{x}}}_t^{RNN}$ and $\overline{\hat{\mathbf{x}}}_t^{RNN}$, and correction for the smoothing coefficient, $\Delta\alpha_t$.

RNN architecture is depicted in Fig. 4. Initially, to reduce the dimensionality of the exogenous data, each n-dimensional vector $\mathbf{x}_t^{p_i}$ is embedded into d-dimensional space $(d < n)$ by a linear layer. Note that the embedding is trained as part of the model itself.

After embedding, the exogenous vectors are concatenated and modulated by per-series parameters collected in modulation vectors $\mathbf{p}^{(j)} \in \mathbb{R}^{dN}, j = 1, ..., J$, where J is the size of the main batch. Modulation vectors initialized as ones are used to element-wise multiply the exogenous vector:

$$\mathbf{x}_t^{p'(j)} = \mathbf{x}_t^{p(j)} \otimes \mathbf{p}^{(j)} \tag{7}$$

Modulation vectors $\mathbf{p}^{(j)}$ are updated along with other learnable parameters of the model using the same optimization algorithm (stochastic gradient descent), with the ultimate objective of minimizing the loss function.

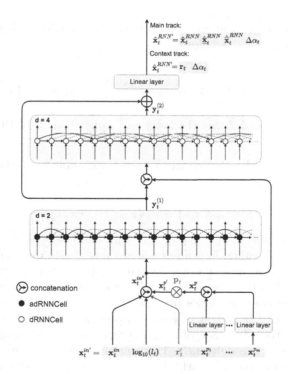

Fig. 4. Architecture of adRNN (elements in red do not exist in the context track). (Color figure online)

RNN utilizes two types of recurrent cells [18]: the dRNNCell, a dilated cell that is fed by both recent and delayed states, and the adRNNCell, a dilated cell with an attention mechanism. The latter combines two dRNNCells, with the first producing an attention vector which components are treated as weights for the inputs to the second, generating an output vector from which the forecast is constructed. Hence, the components of the input vector $\mathbf{x}_t^{in'}$ can be strengthened or weakened by the attention vector depending on their predictive power. Note that the attention vector has a dynamic nature, as it is adapted to the current inputs at time t (see [18] for details).

RNN is composed of two recurrent layers: first one contains adRNNCell dilated 2, while the second one contains dRNNCell dilated 4. The number of layers and dilations were determined through experimentation and may not be the best choice for other forecasting tasks. By stacking multiple layers with hierarchical dilations, more abstract features can be extracted in successive layers and a larger receptive field can be obtained, making it easier to learn the long-term dependencies of different scales. To facilitate the gradient optimization of the model, ResNet-style shortcuts are used between layers. It's worth noting that the input vector $\mathbf{x}_t^{in'}$ is not only inputted into the first layer, but also into the second one by extending the output vector from the first layer.

Postprocessing Component converts outputs of RNN, i.e. point forecasts and PIs, into real values as follows:

$$\hat{z}_\tau = \exp\left(\hat{x}_\tau^{RNN}\right)\hat{l}_t, \quad \hat{\underline{z}}_\tau = \exp\left(\underline{\hat{x}}_\tau^{RNN}\right)\hat{l}_t, \quad \hat{\overline{z}}_\tau = \exp\left(\hat{\overline{x}}_\tau^{RNN}\right)\hat{l}_t \qquad (8)$$

where $\tau \in \Delta_t^{out}$, $\hat{\underline{z}}_\tau$ and $\hat{\overline{z}}_\tau$ are the lower and upper bounds of PI at time τ.

Loss Function enables the model to learn both point forecasts and PIs. It is defined as follows [19]:

$$L = \rho(x, \hat{x}_{q^*}) + \gamma(\rho(x, \hat{x}_{\underline{q}}) + \rho(x, \hat{x}_{\overline{q}})) \qquad (9)$$

where $\rho(x, \hat{x}_q) = (x - \hat{x}_q)(q - \mathbf{1}_{(x<\hat{x}_q)})$ is a pinball loss, $q \in (0,1)$ is a quantile order, x is an actual value, $\hat{x}_q$ is a forecasted value of q-th quantile of x, $q^* = 0.5$ corresponds to the median, $\underline{q} \in (0, q^*)$ and $\overline{q} \in (q^*, 1)$ correspond to the lower and upper bound of PI, respectively, and $\gamma \geq 0$ is a control parameter.

The loss function (9) enables both point forecasts and PIs to be optimized at the same time. The weight of each component can be adjusted by γ. The pinball loss has the advantage of reducing forecast bias by penalizing positive and negative deviations differently. This can be done by adjusting q^* to be less than or greater than 0.5. Similarly the bias in PI can be reduced. This concept is further explained in [20,21].

3.2   Context Track

A context track generates a dynamic context vector, which is based on the history of a representative group of series. This vector is adjusted to the series being forecasted and added as supplementary input to the main track RNN. This extends the per-series input data with information from the context series, thereby enhancing the accuracy of individual series forecasts.

The context track is trained on the selected price time series (Z_c) and associated exogenous time series (P_c). In this study, Bitcoin price series along with its corresponding exogenous data are selected as the context series. This is because Bitcoin is the most popular, widely accepted and has the longest historical data among all cryptocurrencies.

The context track components are: ES, preprocessing, adRNN and modulation (see Fig. 2). ES is the same as for the main track. Preprocessing component normalizes time series sequences in the same way as its main track counterpart. It uses Eqs. (3), (4) and (5) for this. Then, it constructs the input pattern for RNN, which has the form of (6) excluding modulated context vector $\mathbf{r}'_t$.

Context adRNN has the same architecture as the main adRNN except that: (i) inputs do not include $\mathbf{r}'_t$, (ii) exogenous vector is not modulated (no $\mathbf{p}_t$), and (iii) the output is composed of u-component context vector $\mathbf{r}_t$ and correction for ES, $\Delta\alpha_t$.

To adjust the context vector to the individual price series forecasted by the main track, this vector is element-wise multiplied by the u-component modulation vectors, $\mathbf{g}^{(j)}, j = 1, ..., J$, which are learned for each of the J series from the main batch. Modulated context vector is of the form:

$$\mathbf{r}_t^{'(j)} = \mathbf{r}_t \otimes \mathbf{g}^{(j)} \tag{10}$$

Modulation vectors $\mathbf{g}^{(j)}$ as well as vectors $\mathbf{p}^{(j)}$ are updated by the overall optimization procedure and have a static nature, i.e. they does not change while stepping through the batch time steps.

4   Experimental Study

In this section, we apply our proposed model to forecast prices of 15 cryptocurrencies and compare its performance with that of comparative models including both statistical and ML ones. Although cES-adRNN is able to predict both point forecasts and predictive intervals, in this study we do not explore the model's capabilities to create probabilistic forecasts, focusing on point forecasts. The model was implemented in Python using PyTorch. It was run on an eight-core CPU (AMD Ryzen 7 1700, 3.0 GHz, 32 GB RAM).

During the model development and hyperparameter search, the data used was from the period preceding the one-year test period, which spanned from 1 June 2021 to 31 May 2022. Then, using the best hyperparameters, a final training was conducted and the model was tested on the data from the one-year test period. The training procedure generally followed [17], but there a few changes, necessitated by a small number of series:

- batch size was changed once only, from 2 to 4, in epoch 6,
- although the batch size could not be increased much, but a similar effect of obtaining more smooth and higher quality gradient was achieved by increasing number of training steps applied to each batch, starting from 15, and doubling it in epoch 2, tripling to 45 in epoch 3, increasing to 60 in epoch 4, and to 75 in epoch 5,
- additionally, in [17], the epoch was defined as executing 2000–2500 updates; here due to very small dataset size, the overtraining was starting much earlier, forcing us to reduce the number of updates per epoch to 300–500 for horizon of 1, 150–200 for horizon of 7, and just 50–70 for horizon of 28,
- we used the ensemble size of 30.

As the performance metrics, the following measures were used: MAPE - mean absolute percentage error, RMSE - root mean square error, MPE - mean absolute percentage error, and StdPE - standard deviation of percentage error.

The models used for comparison were:

- Naive – naive model: the forecasted price for day i is the same as the price for day $i - h$, where $h = 1, 7$ or 28 is the forecast horizon.
- ARIMA – autoregressive integrated moving average model [22],
- ES – exponential smoothing model [22],
- FNM – fuzzy neighbourhood model [22]
- MLP – perceptron with a single hidden layer and sigmoid nonlinearities [23],
- DeepAR – autoregressive RNN model for probabilistic forecasting [24],
- N-BEATS – deep NN with hierarchical doubly residual topology [25],
- Transformer – transformer NN with attention mechanism [26],
- WaveNet – autoregressive deep NN model combining causal filters with dilated convolutions [27].

We used R implementations of ARIMA and ES (package `forecast`), and GluonTS implementations of MLP, DeepAR, N-BEATS, Transformer and WaveNet [28]. A Naive model and FNM were implemented in Matlab.

Our experimentation involved two paths. The first path was to predict cryptocurrency prices solely based on their historical values, without using exogenous variables. The second path incorporated time series of both prices and exogenous variables as input. Tables 1 and 2 show the results of both experimentation paths. The models with exogenous variables are marked with the "+" symbol (note that some of the models are not able to incorporate exogenous data).

The superiority of our proposed model over other models is clearly demonstrated by the results presented in Tables 1 and 2: our model fed by exogenous variables, cES-adRNN+, produced the most accurate forecasts for all horizons. The differences in MAPE between cES-adRNN variants without and with exogenous inputs were: 3.0% for $h = 1$, 3.8% for $h = 7$, and 7.2% for $h = 28$. Note that in case of other models, introducing exogenous variables not always lead to error decreasing (this is especially evident for $h = 1$). For a horizon of 1, it is challenging for forecasting models to outperform the Naive model. However, our proposed model is the only one that generated errors lower than those of

Table 1. Forecasting metrics for models without exogenous variables (univariate case).

Model	Horizon 1				Horizon 7				Horizon 28			
	MAPE	RMSE	MPE	StdPE	MAPE	RMSE	MPE	StdPE	MAPE	RMSE	MPE	StdPE
Naive	**3.25**	**70.6**	0.36	**4.68**	10.63	256.0	2.86	11.52	22.43	567.3	9.66	20.87
ARIMA	3.38	73.7	**0.22**	4.82	7.92	194.4	1.23	9.72	16.45	418.5	3.99	18.32
ES	3.29	71.1	0.33	4.80	7.53	181.3	1.56	9.84	15.60	386.2	5.27	21.63
FNM	5.48	101.8	0.97	7.49	10.18	249.0	3.35	11.79	20.79	530.9	9.51	23.75
MLP	3.38	75.1	0.86	4.75	7.87	198.7	1.80	9.21	15.97	462.1	**2.05**	15.89
DeepAR	3.30	72.1	0.33	4.73	7.71	191.5	2.41	9.14	16.68	**381.4**	8.73	16.02
N-BEATS	3.27	71.0	0.48	4.69	**7.33**	**179.8**	**0.06**	**8.81**	16.48	382.6	5.79	16.46
Transformer	5.44	157.7	4.68	4.95	8.09	188.5	3.09	9.42	17.95	647.1	3.42	18.22
WaveNet	10.23	188.9	4.84	15.89	8.10	191.3	3.48	9.75	18.18	403.0	10.56	17.94
cES-adRNN	3.29	72.3	0.39	4.75	7.40	184.2	1.58	8.99	**15.05**	391.7	5.40	**15.74**

Table 2. Forecasting metrics for models with exogenous variables (multivariate case).

Model	Horizon 1				Horizon 7				Horizon 28			
	MAPE	RMSE	MPE	StdPE	MAPE	RMSE	MPE	StdPE	MAPE	RMSE	MPE	StdPE
MLP+	3.33	75.4	0.31	4.74	7.74	201.1	1.20	9.06	15.54	430.7	4.64	15.88
DeepAR+	3.31	71.8	0.42	4.72	7.27	181.0	−0.27	**8.59**	15.95	432.8	8.04	15.82
N-BEATS+	3.35	70.7	0.69	4.70	7.35	179.8	0.81	8.85	16.87	375.5	6.24	16.29
Transformer+	5.91	173.6	5.28	4.96	7.54	188.6	0.61	9.12	14.50	468.1	**0.26**	15.35
WaveNet+	4.01	86.4	2.10	6.56	7.86	187.0	3.00	9.48	19.29	572.0	12.98	17.50
cES-adRNN+	**3.19**	**68.9**	**0.21**	**4.58**	**7.12**	**179.6**	0.55	8.67	**13.97**	**374.3**	2.08	**15.11**

Table 3. GWtest metric.

	Naive	ARIMA	ETS	FNM	MLP	DeepAR	N-BEATS	Transformer	WaveNet	cES-adRNN	MLP+	DeepAR+	N-BEATS+	Transformer+	WaveNet+	cES-adRNN+
Horizon 1	45.8	29.8	32.0	15.6	30.2	15.6	31.1	8.4	31.1	0.0	31.6	25.8	39.1	31.1	36.4	**57.3**
Horizon 7	2.2	32.4	53.3	5.3	20.4	27.1	31.6	52.4	40.0	30.7	76.4	44.0	77.3	67.6	78.7	**89.8**
Horizon 28	3.6	44.0	64.4	10.7	43.6	25.3	56.9	72.4	42.7	22.7	57.8	15.6	75.1	69.3	83.1	**95.1**

the Naive model. For longer horizons, the Naive model produced forecasts with much greater errors than those of other models.

The cES-adRNN+ model shows the lowest StdPE values, indicating that its predictions are less dispersed compared to the baseline models. Additionally, the MPE values for cES-adRNN+ are also among the lowest. MPE reflects a forecast bias. While our model can reduce bias by selecting an appropriate quantile order for the loss function, it's important to note that bias reduction may negatively impact forecast error, which is our primary quality measure. As a result, we did not further reduce bias in our model.

To further evaluate the accuracy of the models, we conducted a pairwise one-sided Giacomini-White test for conditional predictive ability for each cryptocurrency and forecast horizon [29]. Table 3 shows the results: GWtest value, i.e. the percentage of cases where a given model outperformed other models in terms of MAPE at a significance level of $\alpha = 0.05$. For instance, a GWtest value of 95.1 was obtained for $h = 28$ by cES-adRNN+, indicating that this model had a significantly lower MAPE than other models in 95.1% of pairwise comparisons. For horizon of 1, our model demonstrates a highest value of GWtest among all models (57.3), but note that the second highest value is for the Naive model (45.8). For longer horizons, GWtest increased for cES-adRNN+ to 89.8 for $h = 7$, and 95.1 for $h = 28$.

Figure 5 presents examples of BTC price forecasts generated by the most accurate models. The top panel displays one-day ahead forecasts for the first three months of the test period, revealing a characteristic inertia or lag in the predictions. The middle panel shows 7-day ahead forecasts for the first three weeks of the test period, while the bottom panel displays 28-day ahead forecasts for the first three months. Notably, the models are unable to accurately predict BTC price fluctuations for longer horizons due to insufficient information in the input variables.

The model training process took several hours on a desktop-class computer, utilizing only CPUs (without GPUs) running in parallel with all available cores. The final forecast was then aggregated after the training. The forecasting process is expected to be much faster, as it does not involve gradient calculations. In a real-life scenario, we would expect daily forecasts based on saved NN weights, which could be generated in less than a minute. The retraining process could be performed less frequently, for example, on a monthly basis.

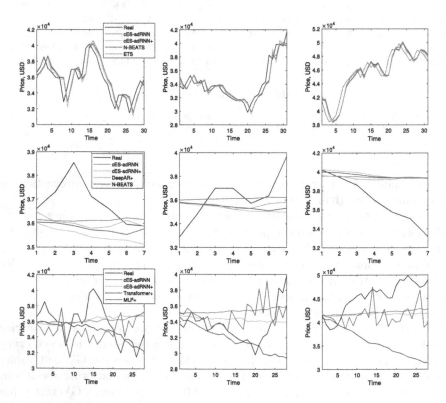

Fig. 5. Examples of forecasts for BTC: top panel - $h = 1$, middle panel - $h = 7$, bottom panel - $h = 28$.

5   Conclusions

Cryptocurrencies are notoriously volatile and their prices are influenced by a range of factors, such as market sentiment, regulatory changes, and technological advancements. Moreover, the lack of discernible patterns in the price data and its erratic fluctuations make accurately predicting cryptocurrency prices a daunting task. Despite these challenges, our study employs state-of-the-art ML model, cES-sdRNN, that incorporates various mechanisms and procedures to improve forecasting efficiency, such as a hybrid architecture, context track, recurrent cells with dilation and attention mechanisms, dynamic ES model, cross-learning, quantile loss function, ensembling, and overfitting prevention mechanisms. Our proposed model, which uses exogenous variables as input, outperformed all comparative models, achieving the highest accuracy across all forecast horizons.

References

1. Giudici, G., Milne, A., Vinogradov, D.: Cryptocurrencies: market analysis and perspectives. J. Ind. Bus. Econ. **47**, 1–18 (2020)
2. Sovbetov, Y.: Factors influencing cryptocurrency prices: evidence from bitcoin, ethereum, dash, litcoin, and monero. J. Econ. Finan. Anal. **2**, 1–27 (2018)
3. Walther, T., Klein, T., Bouri, E.: Exogenous drivers of Bitcoin and Cryptocurrency volatility–a mixed data sampling approach to forecasting. J. Int. Finan. Markets. Inst. Money **63**, 101133 (2019)
4. Gradojevic, N., Kukolj, D., Adcock, R., Djakovic, V.: Forecasting Bitcoin with technical analysis: a not-so-random forest? Int. J. Forecast. **39**, 1–17 (2023)
5. Mudassir, M., Bennbaia, S., Unal, D., Hammoudeh, M.: Time-series forecasting of Bitcoin prices using high-dimensional features: a machine learning approach. Neural Comput. Appl. (2020). https://doi.org/10.1007/s00521-020-05129-6
6. Ahmed, W.M.: Robust drivers of Bitcoin price movements: an extreme bounds analysis. North Am. J. Econ. Finan. **62**, 101728 (2022)
7. Kraaijeveld, O., De Smedt, J.: The predictive power of public Twitter sentiment for forecasting cryptocurrency prices. J. Int. Finan. Markets. Inst. Money **65**, 101188 (2020)
8. Bouri, E., Lau, C.K.M., Lucey, B., Roubaud, D.: Trading volume and the predictability of return and volatility in the cryptocurrency market. Financ. Res. Lett. **29**, 340–346 (2019)
9. Saad, M., et al.: Toward characterizing blockchain-based cryptocurrencies for highly accurate predictions. IEEE Syst. J. **14**, 321–332 (2019)
10. Khedr, A.M., et al.: Cryptocurrency price prediction using traditional statistical and machine-learning techniques: a survey. Intell. Syst. Account. Finan. Manag. **28**, 3–34 (2021)
11. Hotz-Behofsits, C., Huber, F., Zörner, T.O.: Predicting crypto-currencies using sparse non-Gaussian state space models. J. Forecast. **37**, 627–640 (2018)
12. Giudici, P., Abu-Hashish, I.: What determines bitcoin exchange prices? A network VAR approach. Financ. Res. Lett. **28**, 309–318 (2019)
13. Kim, G., Shin, D.-H., Choi, J.G., Lim, S.: A deep learning-based Cryptocurrency price prediction model that uses on-chain data. IEEE Access **10**, 56232–56248 (2022)

14. Hansun, S., Wicaksana, A., Khaliq, A.Q.: Multivariate cryptocurrency prediction: comparative analysis of three recurrent neural networks approaches. J. Big Data **9**, 1–15 (2022)
15. Chen, J.: Analysis of Bitcoin price prediction using machine learning. J. Risk Finan. Manag. **16**, 51 (2023)
16. Chen, Z., Li, C., Sun, W.: Bitcoin price prediction using machine learning: an approach to sample dimension engineering. J. Comput. Appl. Math. **365**, 112395 (2020)
17. Smyl, S., Dudek, G., Pełka, P.: Contextually enhanced ES-dRNN with dynamic attention for short-term load forecasting. arXiv preprint arXiv:2212.09030 (2022)
18. Smyl, S., Dudek, G., Pelka, P.: ES-dRNN with dynamic attention for short-term load forecasting. In: 2022 International Joint Conference on Neural Networks (IJCNN), pp. 1–8. IEEE (2022)
19. Smyl, S., Dudek, G., Pełka, P.: ES-dRNN: a hybrid exponential smoothing and dilated recurrent neural network model for short-term load forecasting. arXiv preprint arXiv:2112.02663 (2021)
20. Smyl, S.: A hybrid method of exponential smoothing and recurrent neural networks for time series forecasting. Int. J. Forecast. **36**, 75–85 (2020)
21. Dudek, G., Pełka, P., Smyl, S.: A hybrid residual dilated LSTM and exponential smoothing model for midterm electric load forecasting. IEEE Trans. Neural Netw. Learn. Syst. **33**, 2879–2891 (2021)
22. Dudek, G.: Pattern similarity-based methods for short-term load forecasting-Part 2: models. Appl. Soft Comput. **36**, 422–441 (2015)
23. Dudek, G.: Neural networks for pattern-based short-term load forecasting: a comparative study. Neurocomputing **205**, 64–74 (2016)
24. Salinas, D., Flunkert, V., Gasthaus, J., Januschowski, T.: DeepAR: probabilistic forecasting with autoregressive recurrent networks. Int. J. Forecast. **36**, 1181–1191 (2020)
25. Oreshkin, B.N., Carpov, D., Chapados, N., Bengio, Y.: N-BEATS: neural basis expansion analysis for interpretable time series forecasting. arXiv preprint arXiv:1905.10437 (2019)
26. Vaswani, A., et al.: Attention is all you need. In: Advances in Neural Information Processing Systems, vol. 30 (2017)
27. Oord, A., et al.: WaveNet: a generative model for raw audio. arXiv preprint arXiv:1609.03499 (2016)
28. Alexandrov, A., et al.: Gluonts: probabilistic and neural time series modeling in python. J. Mach. Learn. Res. **21**, 4629–4634 (2020)
29. Giacomini, R., White, H.: Tests of conditional predictive ability. Econometrica **74**(6), 1545–1578 (2006)

Enhanced Emotion and Sentiment Recognition for Empathetic Dialogue System Using Big Data and Deep Learning Methods

Marek Kozłowski[1]([✉])[ID], Karolina Gabor-Siatkowska[1][ID], Izabela Stefaniak[2][ID], Marcin Sowański[1][ID], and Artur Janicki[1]([✉])[ID]

[1] Warsaw University of Technology, ul. Nowowiejska 15/19, 00-665 Warsaw, Poland
{Marek.Kozlowski,Karolina.Gabor-Siatkowska.dokt,Marcin.Sowanski.dokt,
Artur.Janicki}@pw.edu.pl
[2] Institute of Psychiatry and Neurology, ul. Sobieskiego 9, 02-957 Warsaw, Poland
istefaniak@ipin.edu.pl

Abstract. The article presents the results of work on improving sentiment and emotion recognition for Polish texts using a big data-based
expansion process and larger neural language models. The proposed
recognition method is intended to serve in a therapeutic dialogue system
to analyze sentiment and emotion in human utterances. First, the language model is enhanced, by replacing the BERT neural language model
with RoBERTa. Next, the emotion-based text corpus is enlarged. A novel
process of augmenting an emotion-labeled text corpus using semantically similar data from an unlabeled corpus, inspired by semi-supervised
learning methods, is proposed. The process of using the Common Crawl
web archive to create an enlarged corpus, named CORTEX+pCC, is
presented. An empathetic dialogue system named Terabot, incorporating the elaborated method, is also described. The system is designed to
employ elements of cognitive-behavioral therapy for psychiatric patients.
The improved language model trained on the enlarged CORTEX+pCC
corpus resulted in remarkably improved sentiment and emotion recognition. The average accuracy and F1 scores increased by around 3% and
8% relative, which will allow the dialogue system to operate more appropriately for the emotional state of the patient.

Keywords: Natural language processing · Artificial intelligence · Big
data · Humanities · Text-based emotion recognition · Corpus
augmentation · Healthcare · Psychiatric therapy

1 Introduction

Chatbots and dialogue systems are conversational agents that interact with users
on a specific topic using natural language sentences. They are designed to interact

This research was funded by the Center for Priority Research Area Artificial Intelligence
and Robotics of the Warsaw University of Technology within the Excellence Initiative:
Research University (IDUB) program.

verbally with people by analyzing spoken or written data, so that they can provide an appropriate response. Their popularity is rising in many fields, e.g., education and healthcare. Dialogue systems are created to perform some kind of action, therefore they will try to narrow down the conversation to get all needed information to do so. Chatbots, in contrast, are designed for extended conversations.

Goal-oriented dialogue systems, also called conversational, aim at completing a specific task through conversation with a user. Typically, such a system consists of a few sub-systems: natural language understanding with intent recognition and slot filling [5], dialogue state tracking [35], dialogue management [27], and language generation [24].

Recent development in conversational AI has been led almost exclusively by transformer-based neural language models specialized in dialogue generation. Since the initial release of a Transformer [30], or more specifically the Bidirectional Encoder Representations from Transformers (BERT) model [10], improvements to dialogue generation have been made mostly by increasing the number of network parameters. This strategy has proven to be so effective that in just a couple of years the set of problems in state-of-the-art models have shifted from problems with short, dull and uncontrollable answers to problems of how much understanding of meaning these models have. The most notable example of a conversational AI state-of-the-art model is ChatGPT, which uses GPT-4 [2] model, that is capable of attaining the highest possible score on AP Biology, AP Psychology, and a high score on Medical Knowledge Self-Assessment Program. While some researchers raise concerns about the reliability of the model's responses and its tendency to generate unrealistic content, others argue that ChatGPT has shown promise in the field of psychological therapy.

In recent years, there has been increasing acknowledgment of the important role mental health plays in achieving personal and global goals. The growing number of people with mental disorders forces one to look for new solutions to help patients [33]. One of the ideas is to support or supplement the therapist's work with tools that use new technologies. Some forms of therapy, e.g., cognitive behavioral therapy (CBT) or therapeutic techniques such as exposure in virtual reality settings, can, due to their well-described and structured nature, be applied in tools that use new technologies [11].

Chatbots have been tried for psychological therapies since the ELIZA chatbot [29]. A Woebot chatbot, working for the English language, has turned out to be helpful in therapy for depression [13]. Avatar-based dialogue systems (although human-controlled) have been shown to be successful in therapy for auditory hallucinations [6,26]. The combined use of computer-based therapies may offer many possibilities for the treatment of physical and mental disabilities [12].

The work described in this article is part of the project devoted to the creation of a therapeutic dialogue system for Polish which would be able to realize elements of CBT by conducting empathetic dialogues with a patient. Such a system can support a human therapist, and free up some of a clinician's time;

they can then use their time and skills, for example, to take better care of other patients [4]. In this work we will propose such a dialogue system, named Terabot, and present its main components.

A dialogue system to be used in therapy needs not only to be able to respond according to the user's intent and the topics mentioned, but also to be empathetic. This means that its utterances should correspond with the user's emotional state. For this purpose, text-based sentiment and emotion recognition needs to be employed. In our previous work [36] we created such a module and tested it on a sentiment and emotion-labeled dataset, named CORTEX. In this work we propose big data- and deep learning-based methods to expand our training dataset. We show how they influence the sentiment and emotion recognition models.

2   Terabot – A Therapeutic Dialogue System

We designed a dialogue system, called Terabot, which will help in CBT therapy for psychiatric patients. It is equipped with a voice interface and is able to operate in Polish. Its main purpose is to help patients cope with difficult and overwhelming emotions influencing their life. In our study, we proposed the use of a dialogue agent, whose task will be to conduct exercises helping patients to cope with emotions (anxiety, anger, shame). The tool can be used as a supplement to therapies for many mental disorders, including psychotic disorders and personality problems.

We decided not to choose a neural dialogue system, due to the lack of sufficient training dialogue data. In addition, the lack of controllability of neural systems has been recently widely discussed [1,28,32]. Therefore, due to a very sensitive application context (therapy for a psychiatric patient), we decided to use a goal-oriented system, to retain better control of the system's actions, and to minimize the risk of giving an inappropriate answer to the patient.

A schematic diagram of Terabot is depicted in Fig. 1. First, when a patient talks, the speech is recognized by the Google Web Speech API and transformed into text. Next, the text is further analyzed using the DIET Classifier [3] to identify the intentions and slots, and the sentiment/emotion recognizer, which is the part we focus on in this article. The DIET Classifier is pre-trained with the training data, originating from mock therapeutic sessions. Typical intents recognized by our dialogue system are: *Choose an exercise type, Describe the feeling, Explain the cause of the anxiety.* The slots are filled in with, e.g., an exercise name or recognized patient emotional state.

In the next step a decision is made about an action, resulting from a weighted combination of a rule policy, memoization policy (i.e., based on stories kept in memory) and Transformer Embedding Dialogue (TED) policy [31]. The TED policy takes into account the current dialogue state, i.e., among others, the patient intent(s) and the slot values. The latter include the current emotional state of the patient, recognized by the text-based sentiment/emotion recognition module, as described in this study.

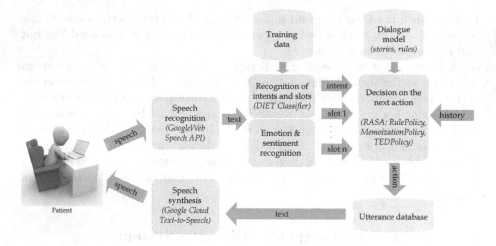

Fig. 1. Block diagram of the proposed therapeutic dialogue system.

If the next system action is an utterance (which is often the case), an appropriate text for the Terabot dialogue turn is found in the utterance database. Then, it is then transformed back to a speech signal by the Google Cloud Text-to-Speech (TTS) service. Both the DIET Classifier and the action decision pipeline are realized using RASA [14], an open-source framework for natural language understanding, dialogue management, and integration[1].

2.1 Sentiment and Emotion Classification

When designing Terabot, we evaluated several approaches to emotion and sentiment recognition, both simpler and more complex ones [36]. We started with baseline models such as multinomial Naïve Bayes and linear support vector machine (SVM) classifiers trained on top of bag-of-words (BoW) representation applied to token bigrams. Another model was based on the fastText algorithm, obtained from pre-trained word embeddings (300-dimensional variant) available in the *fasttext* library, for both English and Polish, and also using token bigrams.

The most complex approach was to fine-tune the pre-trained BERT$_{\text{BASE}}$ models. These models outperformed the simpler methods, reaching around 92% accuracy for sentiment classification and around 75% for emotion classification, for Polish datasets. The F1-score yielded similar values.

3   Datasets

3.1   CORTEX Dataset

When working on the sentiment and emotion classifier for a Polish dialogue system, we faced the problem of lack of a relevant corpus for Polish which would

---

[1] https://rasa.com/products/rasa-platform/.

contain conversational texts. Most of the existing dialogue corpora are either domain-specific and task-oriented or social media-oriented, and, therefore, they are difficult to apply in a therapeutic setting. There are, however, examples of emotionally grounded conversational datasets for English, such as DailyDialog [16] and EmpatheticDialogues [21], which are labeled with emotions at the utterance and dialogue levels, respectively.

The DailyDialog dataset (DD) consists of daily conversations obtained through crawling websites for learners of English. It counts about 13k dialogues, manually labeled with six basic emotions: anger, disgust, fear, happiness, sadness, and surprise. The emotion distribution in the dataset is highly imbalanced, with *happiness* being more than 10 times more frequent than the other emotions.

EmphatheticDialogues (ED) is a novel dataset with about 25k personal dialogues. Each dialogue is grounded in a specific situation where the speaker felt a given emotion, with the listener responding actively. This dataset is larger compared to DD, and experiments show that models built on this dataset are usually perceived by human evaluators as more empathetic, compared to models merely trained on large-scale Internet-crawled opinion-oriented data.

To create an emotion-labeled corpus with conversational data for Polish, we decided to use available resources for English and obtain the desired Polish version via neural machine translation (NMT). Since ED lacked examples of labeled neutral sentences, we added samples of neutral texts from the DD set. In this way, we built a parallel bilingual (English and Polish) corpus of emotional texts, labeled with three sentiment polarity classes and nine emotional classes. It is designed to serve experiments on sentiment and emotion recognition in a conversational context (e.g., dialogue systems or virtual assistants). We named the corpus CORTEX (for CORpus of Translated Emotional teXts) and made it publicly available[2].

During various experiments described in [36] we realized that one of the challenges is a limited number of examples in the training dataset (around 21k instances in total). One of the ways to augment the training dataset was to employ human annotators who would manually find and label new data. We did not opt for this method, as it is tedious, time-consuming, costly, and error-prone. We decided to invest in an automatic approach, which consists in identifying semantically similar sentences in big data: a web archive corpus, Common Crawl, described in the next section.

3.2   Common Crawl Dataset

The exponential growth of the Internet community has resulted in the production of a vast amount of unstructured data, including web pages, blogs, and social media. Such a volume consisting of hundreds of billions of words is unlikely to be analyzed by humans. In this work we applied the *LanguageCrawl* tool [23], which

---

[2] CORTEX is freely available for download at https://github.com/azygadlo/CORTEX.

allows natural language processing (NLP) researchers to easily build a single-language (e.g., Polish) web-scale corpus using the Common Crawl Archive [18, 25] – an open repository of web crawl information containing petabytes of data.

The Common Crawl Archive is built by the non-profit Common Crawl organization founded by Gil Elbaz in California [7]. Its purpose is to enable wider access to web information by manufacturing and to support an open web crawl repository. The data is made available both in raw HTML format (WARC) and text-only format (WET), and is freely accessible to anyone either via free direct download[3] or the commercial Amazon S3 service, for which a deposit is required.

The LanguageCrawl toolkit provides a highly concurrent actor-based architecture for building a local Common Crawl Archive. In [23] the authors presented three use cases: filtering of Polish websites, the construction of n-gram corpora, and the training of a continuous skipgram language model with hierarchical softmax. To create a Polish subset of Common Crawl, called hereinafter the Polish Common Crawl (pCC), the toolkit was applied in the first scenario (filtering websites).

Processing data from the Common Crawl Archive was a colossal task, which was severely limited by the Internet connection bandwidth, the cluster of multi-core servers, and a bottleneck for data fetching. Therefore, it took several months to download enough data to construct a reasonable Polish website corpus.

Originally, the crawl data were stored in the Web Archive (WARC) format[4]. The WARC format retains and processes data from the Common Crawl Archive dump, which can be as large as hundreds of terabytes in size and contains billions of websites, in a more effective and manageable way. The raw crawl data is wrapped around the WARC format, ensuring a straightforward mapping to the crawl action. The HTTP request and response are stored, along with metadata information. In the case of the HTTP response, the HTTP header information is stored. This allows a high number of inquisitive insights to be gathered. The website content collected takes the form of an HTML document.

In our case, this rich format WARC is too complex, therefore we decided to use the WET Response format[5], which gathers plain-text web content instead of HTML. As most NLP tasks require only textual data, and we have access to limited resources, we decided to construct our tool around WET files, containing a minimal amount of metadata.

During our CORTEX expansion process we worked on corpora extracted from a few months of the 2015, 2018 and 2019 Common Crawl Archive, approximately 1PB in size and containing approximately 1.5 trillion web pages. Since we considered only WET files, we fetched around 100 TB of compressed textual data (see the process sketched in Fig. 2). Although the amount of data processed is enormous, the Polish language constitutes only a tiny fraction of it: we estimated it to be approximately 0.2 to 0.3%. Finally, our pipeline retrieved a few billion web pages with some Polish content. We decided to use only those web

---

[3] http://commoncrawl.org/.

[4] https://www.loc.gov/preservation/digital/formats/fdd/fdd000236.shtml.

[5] https://commoncrawl.org/the-data/get-started.

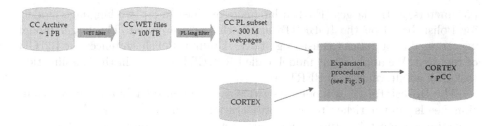

Fig. 2. Process of expanding the CORTEX database from big data perspective.

pages with most of the text written in Polish. This constraint led to 300 million web pages, which we used ultimately for the CORTEX expansion process. The whole filtering process, which employed 15 servers with 48 vCPUs and 256 GiB RAM each, took six months.

4 Experiments

4.1 Improving the Classification Model

When comparing various text-based emotion classification models in [36], the BERT models always yielded the best results. Since text representations generated by BERT have proved to be effective in many diverse NLP problems, e.g., classification, regression, machine translation, question answering, we continued to research these transformer-based language models in this project.

In the past three years, many modifications in the architecture of Transformer neural networks or their training procedures have been proposed. Transformers have the potential to learn longer dependencies, but are limited by a fixed-length context in the setting of language modeling. In [9] the authors proposed a novel neural architecture, Transformer-XL, which enabled learning dependency beyond a fixed length without disrupting temporal coherence. The authors of a Reformer model [15] introduced two techniques to improve the memory and training-time efficiency of large transformers, i.e., replacement of dot-product attention by one that uses locality-sensitive hashing, and applying reversible residual layers instead of the standard residuals, which allows the storing of activations only once in the training process. The XLNet model, presented in [34], was a generalized autoregressive pretraining method that enabled bidirectional context learning by maximizing the expected likelihood over all permutations of the factorization order. In addition, it overcame the limitations of the original BERT training procedure. The authors in [17] proved that BERT was significantly undertrained, and they proposed an optimized and larger neural model called Robustly Optimized BERT (RoBERTa).

In our work we decided to choose and evaluate a Polish version of RoBERTa. Although some other transformer-based language models for Polish are available, none has come even close to the scale, in terms of corpus size and number of parameters, of the largest English-language models. In [8] two language models for Polish based on the RoBERTa approach were proposed. The larger model was trained on a dataset consisting of over 1 billion Polish sentences, or 135 GB of raw text. We applied this model, called RoBERTa$_{LARGE}$, in the classification tasks, as an alternative to BERT$_{BASE}$.

The RoBERTa$_{LARGE}$ model has twice as many encoder blocks, more attention heads, and a richer token representation than BERT$_{BASE}$. As a result, it contains almost 3.5 times more parameters: 355M compared to 110M for BERT$_{BASE}$. We trained both models for 4 epochs with an effective batch size of 24, and, based on the validation metrics obtained after each training epoch, we selected the best models. The results will be presented and discussed in Sect. 5.

4.2   Dataset Expansion Process

In addition to improving the language model, we decided to extend the dataset used for training. Our idea was to find sentences in the pCC plain text corpora which would be semantically similar to sentences in the training part of the CORTEX dataset. Therefore, we sampled pCC to retrieve around 200 million web documents containing at least 10 sentences in a continuous manner. These sentences were considered for the potential expansion of CORTEX.

To resolve the problem of semantic similarity between sentences we used the state-of-the art sentence embeddings framework/library called SentenceTransformers[6] [22]. It is a Python framework for state-of-the-art sentence, text, and image embeddings. It is based on PyTorch and Transformers, and offers a large collection of pre-trained models tuned for various tasks. The framework can be used to compute sentence or text embeddings for more than 100 languages. These embeddings can then be compared, e.g., with cosine similarity to find sentences with a similar meaning.

Transformers were introduced in NLP in the late 2010s, to allow parallel computation and also to reduce the problem of long dependencies. The main advantages of transformers are: a) sentences are processed as a whole rather than word by word; b) self attention: to compute similarity scores between words in a sentence; c) positional embeddings: to encode information about position in the sentence.

The most popular NLP deep learning architecture, BERT, discussed previously in Sect. 4.1, is also based on transformers. BERT established a new state-of-the-art performance in many natural language understanding (NLU) tasks, e.g., in sentence-pair regression tasks like semantic textual similarity (STS). However, it requires that both sentences are fed into the network, which causes a massive computational overhead: finding the most similar pair in a collection of 10,000 sentences requires about 50 million inference computations. The construction of BERT makes it unsuitable for either semantic similarity search or for unsupervised tasks like clustering.

---

[6] https://www.sbert.net.

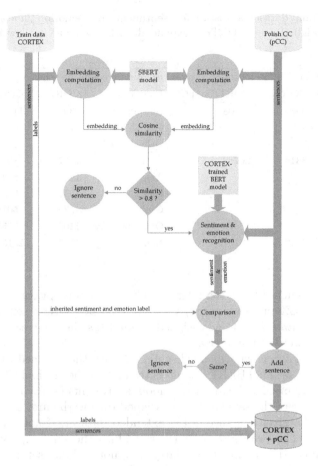

Fig. 3. Scheme of the formation of the improved CORTEX database with the Polish Common Crawl.

In [22], the authors proposed Sentence-BERT (SBERT), a modification of the pre-trained BERT network that uses Siamese and triplet network structures to derive semantically meaningful sentence embeddings that can be compared using cosine similarity. This reduces the effort of finding the most similar pair from 65 h with BERT/RoBERTa to about 5 s with SBERT, while maintaining the accuracy of BERT.

Therefore, we decided to use the SBERT-based approach to compute embeddings of sentences coming from the CORTEX training dataset and pCC, and next we evaluated their semantic similarity using cosine measure between embeddings. More precisely, for each sentence/prompt from the CORTEX training subset we found the top-10 semantically closest ones within pCC. In addition, we decided that their cosine similarity must be over a threshold of 0.8 (the value was set heuristically). This way we expanded the CORTEX prompts with semantically similar ones from pCC, and assigned them corresponding categories

Table 1. Train dataset statistics for sentiment and emotion labels schemas, for CORTEX (original) and CORTEX expanded with CommonCrawl web sites (COR-TEX+pCC).

Sentiment recognition schema			Emotion recognition schema		
Sentiment	CORTEX	CORTEX+pCC	Emotion	CORTEX	CORTEX+pCC
Positive	5985	24608	Happiness	2391	11107
			Confidence	1705	6460
			Other positive	1889	7041
Negative	8314	26768	Sadness	2267	8093
			Anger	1804	5243
			Fear	1565	5759
			Guilt	1576	5018
			Other negative	1102	2655
Neutral	7149	28157	Neutral	1787	28157

(mood, sentiment) of the given reference CORTEX prompt. In other words, for a given CORTEX prompt (*query*) we found the pCC-origin semantically closest sentences (*candidates*) and assigned *candidates* the same categories (mood, sentiment) as were in the *query*.

Next, we analyzed the expanded sentences with the assigned CORTEX categories using the original BERT classifiers trained on the original CORTEX training dataset. We predicted categories (mood, sentiment) for each "expanded" sentence. If they were the same as those assigned during the first expansion phase, we retained the given sentence in the extended dataset. If not, such a sentence was removed. The proposed approach resembles the semi-supervised learning (SSL) technique [19,20], in which a large unannotated dataset is assigned labels based on a classifier trained on a much smaller annotated dataset.

The whole process of CORTEX corpus expansion is shown in Fig. 3. It took four weeks on a single server with four V100 GPUs and 32 GiB RAM memory each. As a result, we created a new extended training dataset, which is almost four times larger than the original CORTEX training dataset, which contained 21,800 records (sentences). After expansion, the training set reports around 79,500 records. This dataset will hereinafter be referred to as COR-TEX+pCC. The detailed statistics of the training set are shown in Table 1.

5   Results and Discussion

To verify the impact of the proposed changes on sentiment and emotion recognition, we compared it against the baseline results presented in [36]. We used the same test set as in [36] and observed precision, recall, F1-score, accuracy (micro F1), macro averaged F1, and support-weighted F1. First, we will present the results achieved by using an improved language model. Next, in addition to the model change, we will show the impact of using an extended training dataset.

5.1 Impact of Improved Language Model

Table 2 shows the comparison of text-based sentiment and emotion recognition for Polish, using the $BERT_{BASE}$ and $RoBERTA_{LARGE}$ neural language models. The training and testing datasets remained unchanged.

Unsurprisingly, for the 3-class scenario, evaluation metrics reached much higher values than for the 9-class scenario, because of the number of degrees of freedom. We reached almost 94% accuracy for sentiment classification and over 80% for emotion classification. The F1-score yielded similar values.

Table 2. Sentiment and emotion classification results for Polish $BERT_{BASE}$ and $RoBERTa_{LARGE}$ models trained on original CORTEX trainset.

Task	Metric	$BERT_{BASE}$	$RoBERTa_{LARGE}$
Sentiment	Accuracy	0.9224	0.9385
(3 class)	Weighted F1	0.9226	0.9385
Emotion	Accuracy	0.7519	0.8040
(9 class)	Weighted F1	0.7515	0.8035

The results confirmed that $RoBERTA_{LARGE}$ yielded better results than $BERT_{BASE}$: the accuracy and F1 of sentiment recognition for Polish increased by more than 1.5% relative. As for emotion recognition, the advantage was even clearer: the accuracy and F1-score (weighted) increased by more than 5% relative.

We found this improvement remarkable, but expected there was still room for improvement using the training dataset expansion. These results are discussed in the next subsection.

5.2 Impact of Extended Training Dataset

We decided to train the $RoBERTA_{LARGE}$ model, which yielded better results in the previous experiment, with the extended training corpus CORTEX+pCC. Next, we tested it using the same test set as in the previous experiment.

We evaluated the results of sentiment and emotion recognition; the results are displayed in Tables 3 and 4, respectively. We conducted many analyses including quality assessment per category, and aggregated measures (micro, macro, weighted), comparing the baseline results ($BERT_{BASE}$ trained on the original CORTEX corpus) with the proposed approach, i.e., the $RoBERTA_{LARGE}$ model trained on the extended CORTEX+pCC training dataset.

We observed a significant quality improvement in favor of the proposed approach. We outperformed the original approach in both tasks: sentiment classification and emotion classification. In sentiment classification we reported accuracy (micro F1) and weighted-averaged F1 around 95%, i.e., 3% relative higher than in the original approach. The best predictions concerned the neutral label (F1 = 98%, slight improvement from 97% in the original approach). The best partial F1 improvement was reported for positive label, from 88% to 92%.

Table 3. Results of sentiment classification for frozen test data for BERT/RoBERTa model trained on two Polish datasets: CORTEX (original) and CORTEX expanded with CommonCrawl web sites (CORTEX+pCC).

Sentiment	Support	Metric	BERT/ CORTEX	RoBERTa/ CORTEX+pCC
Positive	826	Precision	0.88	0.93
		Recall	0.89	0.90
		F1-score	0.88	0.92
Negative	1084	Precision	0.91	0.93
		Recall	0.92	0.95
		F1-score	0.91	0.94
Neutral	955	Precision	0.98	0.98
		Recall	0.95	0.97
		F1-score	0.97	0.98
Average	2865	Accuracy	0.92	0.95
		Macro avg	0.92	0.94
		Weighted avg	0.92	0.95

We also achieved remarkable gains in the emotion-classification problem, which is a more complex problem because of the number of classes. During emotion-detection evaluation we noticed that all aggregated quality measures, i.e., accuracy, macro-averaged F1, and weighted-averaged F1, are about 8% relative higher than the baseline measures.

The three best predictive labels were neutral (F1 = 96%), fear, and sadness (F1 = 84%). The best partial F1 improvement was reported for labels confidence, sadness, and guilt; the quality gains in those categories were around 8–9% relative, in comparison with the previously scored values presented in [36]. In terms of the application of the emotion recognizer in the empathetic dialogue system, such an improvement in detecting such sensitive emotions was highly desired.

Table 4. Results of emotion classification with BERT/RoBERTa model trained on original trainset (CORTEX) and data expanded with CommonCrawl web sites (COR-TEX+pCC).

Emotion	Support	Metric	BERT/ CORTEX	RoBERTa/ CORTEX+pCC
Anger	226		0.70	0.77
Confidence	222		0.75	0.83
Fear	208		0.80	0.84
Guilt	199		0.73	0.82
Happiness	328	F1-score	0.75	0.83
Neutral	238		0.92	0.96
Other negative	160		0.66	0.79
Other positive	276		0.67	0.76
Sadness	291		0.76	0.84
Average	2148	Accuracy	0.75	0.83
		Macro avg	0.75	0.83
		Weighted avg	0.75	0.83

6 Conclusions

In this article we presented a novel intelligent process of enlarging (augmenting) an emotion-labeled text corpus using semantically similar sentences from big data: a massive yet unlabeled corpus – in this case the Common Crawl web archive. The proposed approach was inspired by the SSL approach, known in other areas of machine learning. Similarly to SSL applications, here we also observed that using an *imperfect* sentiment and emotion classifier (taken from [36]) we were able to automatically enlarge the training dataset using unlabeled data, which ultimately has led to constructing a *strongly improved* classifier.

We also proposed replacing the previously used BERT$_{BASE}$ neural language model with a model with richer architecture: RoBERTa$_{LARGE}$. This model, trained with the enlarged CORTEX+pCC corpus, allowed for an improvement of text-based sentiment and emotion recognition for Polish: the average accuracy and F1 increased by around 3% and 8%, respectively. The ultimate accuracy scores reached 95% and 83% for sentiment and emotion recognition, respectively. However, despite such progress we must be aware that the recognition is still not error-free.

Last but not least, in this article we presented an empathetic dialogue system for therapeutic purposes, named Terabot, for which the proposed text-based sentiment and emotion-recognition module is intended. Thanks to the remarkably improved recognition of guilt, happiness, sadness, and other emotions, the dialogue system will work better to understand the emotional state of a patient undergoing therapy.

References

1. Bommasani, R., et al.: On the opportunities and risks of foundation models. CoRR (2021). https://doi.org/10.48550/arXiv.2108.07258
2. Brown, T., et al.: Language models are few-shot learners. In: Advances in Neural Information Processing Systems, vol. 33, pp. 1877–1901 (2020)
3. Bunk, T., Varshneya, D., Vlasov, V., Nichol, A.: Diet: lightweight language understanding for dialogue systems (2020). https://doi.org/10.48550/arXiv.2004.09936
4. Carroll, K., Rounsaville, B.: Computer-assisted therapy in psychiatry: be brave-it's a new world. Curr. Psychiatry Rep. **12**, 426–432 (2010). https://doi.org/10.1007/s11920-010-0146-2
5. Chen, Q., Zhuo, Z., Wang, W.: BERT for joint intent classification and slot filling. arXiv preprint arXiv:1902.10909 (2019)
6. Craig, T.K., et al.: Avatar therapy for auditory verbal hallucinations in people with psychosis: a single-blind, randomised controlled trial. Lancet Psychiatry **5**(1), 31–40 (2018). https://doi.org/10.1016/S2215-0366(17)30427-3
7. Crouse, S., Nagel, S., Elbaz, G., Malamud, C.: Common Crawl Foundation. http://commoncrawl.org (2008)
8. Dadas, S., Perełkiewicz, M., Poświata, R.: Pre-training polish transformer-based language models at scale. In: Rutkowski, L., Scherer, R., Korytkowski, M., Pedrycz, W., Tadeusiewicz, R., Zurada, J.M. (eds.) ICAISC 2020. LNCS (LNAI), vol. 12416, pp. 301–314. Springer, Cham (2020). https://doi.org/10.1007/978-3-030-61534-5_27
9. Dai, Z., Yang, Z., Yang, Y., Carbonell, J., Le, Q.V., Salakhutdinov, R.: Transformer-XL: attentive language models beyond a fixed-length context. arXiv preprint arXiv:1901.02860 (2019)
10. Devlin, J., Chang, M.W., Lee, K., Toutanova, K.: BERT: pre-training of deep bidirectional transformers for language understanding. arXiv preprint arXiv:1810.04805 (2018)
11. Dino, F., Zandie, R., Abdollahi, H., Schoeder, S., Mahoor, M.H.: Delivering cognitive behavioral therapy using a conversational social robot. In: 2019 IEEE/RSJ International Conference on Intelligent Robots and Systems (IROS), pp. 2089–2095 (2019). https://doi.org/10.1109/IROS40897.2019.8968576
12. Fernández-Caballero, A., et al.: Human-avatar symbiosis for the treatment of auditory verbal hallucinations in schizophrenia through virtual/augmented reality and brain-computer interfaces. Front. Neuroinform. **11** (2017). https://doi.org/10.3389/fninf.2017.00064
13. Fitzpatrick, K.K., Darcy, A., Vierhile, M.: Delivering cognitive behavior therapy to young adults with symptoms of depression and anxiety using a fully automated conversational agent (Woebot): a randomized controlled trial. JMIR Ment Health **4**(2), e19 (2017). https://doi.org/10.2196/mental.7785. http://mental.jmir.org/2017/2/e19/
14. Jiao, A.: An intelligent chatbot system based on entity extraction using RASA NLU and neural network. In: Journal of Physics: Conference Series, vol. 1487, no. 1, p. 012014 (2020). https://doi.org/10.1088/1742-6596/1487/1/012014

15. Kitaev, N., Kaiser, Ł., Levskaya, A.: Reformer: the efficient transformer. arXiv preprint arXiv:2001.04451 (2020)
16. Li, Y., Su, H., Shen, X., Li, W., Cao, Z., Niu, S.: DailyDialog: a manually labelled multi-turn dialogue dataset. In: Proceedings of the Eighth International Joint Conference on Natural Language Processing, Taipei, Taiwan (Volume 1: Long Papers), pp. 986–995. Asian Federation of Natural Language Processing (2017). https://www.aclweb.org/anthology/I17-1099
17. Liu, Y., et al.: RoBERTa: a robustly optimized BERT pretraining approach. arXiv preprint arXiv:1907.11692 (2019)
18. Mühleisen, H., Bizer, C.: Web data commons-extracting structured data from two large web corpora. LDOW **937**, 133–145 (2012)
19. Oliver, A., Odena, A., Raffel, C.A., Cubuk, E.D., Goodfellow, I.: Realistic evaluation of deep semi-supervised learning algorithms. In: Advances in Neural Information Processing Systems, vol. 31 (2018)
20. Pudo, M., Szczepanek, N., Lukasiak, B., Janicki, A.: Semi-supervised learning with limited data for automatic speech recognition. In: Proceedings of the IEEE 7th Forum on Research and Technologies for Society and Industry Innovation (RTSI 2022), Paris, France, pp. 136–141 (2022). https://doi.org/10.1109/RTSI55261.2022.9905112
21. Rashkin, H., Smith, E.M., Li, M., Boureau, Y.L.: Towards empathetic open-domain conversation models: a new benchmark and dataset. arXiv preprint arXiv:1811.00207 (2018)
22. Reimers, N., Gurevych, I.: Sentence-BERT: sentence embeddings using Siamese BERT-networks. arXiv preprint arXiv:1908.10084 (2019)
23. Roziewski, S., Kozłowski, M.: LanguageCrawl: a generic tool for building language models upon Common Crawl. Lang. Resour. Eval. **55**(4), 1047–1075 (2021)
24. Sharma, S., He, J., Suleman, K., Schulz, H., Bachman, P.: Natural language generation in dialogue using lexicalized and delexicalized data. In: International Conference on Learning Representations: Workshop (2017)
25. Smith, J.R., Saint-Amand, H., Plamada, M., Koehn, P., Callison-Burch, C., Lopez, A.: Dirt cheap web-scale parallel text from the common crawl. In: ACL (1), pp. 1374–1383 (2013)
26. Stefaniak, I., Sorokosz, K., Janicki, A., Wciórka, J.: Therapy based on avatar-therapist synergy for patients with chronic auditory hallucinations: a pilot study. Schizophr. Res. **211**, 115–117 (2019). https://doi.org/10.1016/j.schres.2019.05.036. https://www.sciencedirect.com/science/article/pii/S0920996419302130
27. Su, P.H., et al.: On-line active reward learning for policy optimisation in spoken dialogue systems. In: Proceedings of the 54th Annual Meeting of the Association for Computational Linguistics (Volume 1: Long Papers), pp. 2431–2441 (2016)
28. Tamkin, A., Brundage, M., Clark, J., Ganguli, D.: Understanding the capabilities, limitations, and societal impact of large language models. CoRR (2021). https://doi.org/10.48550/arXiv.2102.02503
29. Vaidyam, A.N., Wisniewski, H., Halamka, J.D., Kashavan, M.S., Torous, J.B.: Chatbots and conversational agents in mental health: a review of the psychiatric landscape. Can. J. Psychiatry **64**(7), 456–464 (2019)
30. Vaswani, A., et al.: Attention is all you need. In: Advances in Neural Information Processing Systems, vol. 30 (2017)
31. Vlasov, V., Mosig, J.E.M., Nichol, A.: Dialogue transformers (2019). https://doi.org/10.48550/arXiv.1910.00486
32. Weidinger, L., et al.: Ethical and social risks of harm from language models. CoRR (2021). https://doi.org/10.48550/arXiv.2112.04359

33. World Health Organization, et al.: The WHO special initiative for mental health (2019–2023): universal health coverage for mental health. Technical report, World Health Organization (2019)
34. Yang, Z., Dai, Z., Yang, Y., Carbonell, J., Salakhutdinov, R.R., Le, Q.V.: XLNet: generalized autoregressive pretraining for language understanding. In: Advances in Neural Information Processing Systems, vol. 32 (2019)
35. Zhong, V., Xiong, C., Socher, R.: Global-locally self-attentive dialogue state tracker. In: Association for Computational Linguistics (2018)
36. Zygadło, A., Kozłowski, M., Janicki, A.: Text-based emotion recognition in English and Polish for therapeutic chatbot. Appl. Sci. **11**(21), 10146 (2021)

Automatic Delta-Adjustment Method Applied to Missing Not At Random Imputation

Ricardo Cardoso Pereira[1(✉)], Pedro Pereira Rodrigues[2],
Mário A. T. Figueiredo[3,4], and Pedro Henriques Abreu[1]

[1] Centre for Informatics and Systems of the University of Coimbra, Department of Informatics Engineering, University of Coimbra, 3030-290 Coimbra, Portugal
{rdpereira,pha}@dei.uc.pt
[2] Center for Health Technology and Services Research, Faculty of Medicine (MEDCIDS), University of Porto, 4200-319 Porto, Portugal
pprodrigues@med.up.pt
[3] Instituto Superior Técnico, University of Lisbon, 1049-001 Lisbon, Portugal
mario.figueiredo@tecnico.ulisboa.pt
[4] Instituto de Telecomunicações, 1049-001 Lisbon, Portugal

Abstract. Missing data can be described by the absence of values in a dataset, which can be a critical issue in domains such as healthcare. A common solution for this problem is imputation, where the missing values are replaced by estimations. Most imputation methods are suitable for the Missing Completely At Random (MCAR) and Missing At Random (MAR) mechanisms but produce biased results for Missing Not At Random (MNAR) values. An effective approach to mitigate this bias effect is to use the delta-adjustment method. This method assumes the imputation is performed for the MAR mechanism and adjusts the imputed values to become valid under MNAR assumptions by applying a correction factor. Such adjustment is usually defined manually by a domain expert, which often makes this method unfeasible. In this work, we propose an automatic procedure to find an approximate delta adjustment value for every feature of the dataset, which we call Automatic Delta-Adjustment Method. The proposed procedure is validated in an experimental setup comprising 10 datasets of the healthcare domain injected with MNAR values. The results from seven state-of-the-art imputation methods are compared with and without the adjustment, and applying the correction provides a significantly lower imputation error for all methods.

Keywords: Missing Data · Imputation · Delta-Adjustment · Missing Not At Random

1 Introduction

Missing data is a common problem in real-world data. It can be described by the absence of values in one or more features of a dataset. However, the missing

© The Author(s), under exclusive license to Springer Nature Switzerland AG 2023
J. Mikyška et al. (Eds.): ICCS 2023, LNCS 14073, pp. 481–493, 2023.
https://doi.org/10.1007/978-3-031-35995-8_34

values can assume different characteristics that are directly related with the missingness causes. There are three different mechanisms that categorize these causes [6,19]:

- Missing Completely At Random (MCAR), which describes missing values that are the result of a purely random event, meaning the missingness causes are unrelated to any data. An example would be if someone randomly misses a question in a form;
- Missing At Random (MAR), which states that the missing values are related to part of the observed data (e.g., a specific range of a feature). For example, in a medical report, female people would have the results for prostate-related exams missing;
- Missing Not At Random (MNAR), which refers to missing values that are related to themselves or to other unobserved data, and, therefore, the missingness causes are unknown. For example, in a life insurance questionnaire, people that smoke a lot may not want to disclose how many cigarettes they smoke per day.

The existence of missing values impacts negatively any task performed with the data, such as predictive and statistical inference. Therefore, this issue is usually handled in a pre-processing stage to avoid being propagated. The most common approach to address missing data is imputation, which provides plausible estimates to replace the missing values. Such estimations can be performed through simple statistical strategies or even complex machine learning models [11]. However, in general terms, the imputation approaches tend to only be suitable for the MCAR and MAR mechanisms. Since the models base their estimations on the available data, the imputation tends to be biased when performed under MNAR assumptions because the existent data is not enough to properly model its missingness causes [10]. Nevertheless, there are a few approaches to reduce this bias, one being the delta-adjustment method [20]. Assuming that the imputation is performed with a model designed for the MAR mechanism, the missing values are likely to be either underestimated or overestimated. In other words, they are probably shifted towards a lower or higher domain. The delta-adjustment method provides a simple and transparent solution to correct this shift: add or multiply the imputed values by a correction factor. However, this factor must be manually defined by domain experts, which makes the delta-adjustment method often unfeasible to be applied considering the complexity and cost of consulting these experts. Moreover, the lack of scientific rigor in this process may also compromise the generalization of the results.

In this work we introduce an automatic procedure to estimate the approximate delta adjustment values for every feature of the dataset, which we call Automatic Delta-Adjustment Method (ADAM). The procedure explores the distance between the biased imputation and other estimations that are sampled from Gaussian distributions with extreme means that stretch the range of considered values. We conducted an experimental setup with seven state-of-the-art imputation methods, comprising 10 datasets of the healthcare domain that were injected with missing values under MNAR. We compared the results of the imputation with and without the adjustments provided by ADAM trough the Mean Absolute Error (MAE), and validated the results trough the Wilcoxon signed-rank test with a significance level of 5%. The use of ADAM's adjustment provided significant improvements for all imputation methods and missing rates.

The remainder of the paper is organized in the following way: Sect. 2 presents the related work regarding how to address the MNAR mechanism; Sect. 3 describes in detail the proposed ADAM method; Sect. 4 presents the design of the experimental setup; Sect. 5 displays the analysis of the obtained results; and Sect. 6 states the final conclusions and future directions.

2   Related Work

Any imputation method will likely produce biased results for the MNAR mechanism since the missing values depend on unknown data. Incidentally, most features suffering from MNAR values have their distribution shifted when only the available data is considered. This shift will be propagated to the imputation model, which leads to the biased estimations. The impact of such bias can be measured by performing sensitivity analysis, where the parameters of the imputation models are varied in order to understand the magnitude of the bias and if the estimations can be admitted as valid [2]. Such strategy is not a solution for the problem, but allows for a more informed decision about whether the imputation models are usable or not. Nevertheless, performing a correct sensitivity analysis often requires previous knowledge about the features' distribution and domain, information that is usually obtained from domain experts. Consulting such experts is a complex and expensive task, which leads to this type of study often being unfeasible.

For these reasons, the more common way to address MNAR is to use imputation methods suitable for MCAR and MAR, and apply procedures that make the estimations more resilient to the MNAR assumptions. A common strategy for this purpose is multiple imputation, where the missing values are imputed multiple times while varying specific parameters (usually in an automated way) to mitigate the uncertainty. The multiple estimations are then analyzed and aggregated into a final result. A state-of-the-art method which is based on this strategy is the Multiple Imputation by Chained Equations (MICE) [26]. Another useful approach to achieve estimations valid under MNAR is the delta-adjustment method. As previously described, this method provides a simple and understandable way to correct the distribution shift, which is to add or multiply the imputed values by a correction factor [20]. This approach, although very effective, relies once again on domain experts since they need to define the correction factors manually for each feature. Consequently, the method is often unfeasible to be used for the same reasons reported for the sensitivity analysis. More recently, neural network-based models have also been successfully used to perform imputation under MNAR, in particular denoising and variational autoencoders [16]. Such models are very resilient to noise and can accommodate the MNAR characteristics better than other imputation methods [7,8,15]. Nevertheless, there are imputation models designed for MCAR and MAR that are also often included in experimental baselines of MNAR studies, particularly the k-Nearest Neighbors (kNN), matrix completion methods, and the common mean/mode imputation [3,17]. However, such methods tend to produce poor results considering the existent bias.

Focusing on the delta-adjustment method, the following works address imputation under MNAR assumptions using it. Carreras et al. [5] conducted a sensitivity analysis study to understand the impact of assuming and treating missing data as MAR or MNAR in end-of-life care studies. The MICE method was used for both mechanisms, but for MNAR it was integrated with the delta-adjustment method. Four different adjustment values were considered, which were defined as the equispaced values between zero and half of the interquartile range of the features. The experiments were conducted with data from the ACTION study, a randomized controlled trial testing advance care planning in patients with advanced lung or colorectal cancer. The authors concluded that the imputation assuming MAR reflected that the missing values were related to poorer health conditions. These correlations changed when the MNAR mechanism was assumed, which shows that the obtained conclusions are sensible to the violation of the MAR assumptions.

Leacy et al. [9] performed a similar sensitivity analysis study to understand how the departure from MAR to MNAR influenced the tasks of estimating the prevalence of a partially observed outcome and performing parametric causal mediation analyses with a partially observed mediator. The study used data from a tuberculosis (TB) and human immunodeficiency virus (HIV) prevalence survey that was conducted as part of the Zambia-South Africa TB and AIDS Reduction Study, between 2006 and 2010. The shift from MAR to MNAR was once again performed with the delta-adjustment method integrated in a multiple imputation procedure. Three adjustment values were manually chosen based on the experts opinion and on data from 3 consecutive annual rounds of HIV counseling and testing in the Karonga District of Malawi (2007 to 2010). Each of these values represented different magnitudes of departure from the MAR assumptions. The authors concluded that the estimation of the overall HIV prevalence was considerably different when assuming the MAR or MNAR mechanisms, particularly for strong departures between them.

Rezvan et al. [18] also conducted a sensitivity analysis study where the missing values imputed under MAR with multiple imputation were shift to MNAR with the delta-adjustment method. The data used in the experiments was from the Longitudinal Study of Australian Children, and the goal was to estimate the association between exposure to maternal emotional distress at the age of four/five years and total difficulties at the age of eight/nine years. The adjustment values were defined with the help of domain experts through an elicitation process that allows for the formulation of the expert's feedback into a probability distribution. The authors concluded that there are significant increases in the magnitude of the association between maternal distress and total difficulties when the MNAR assumptions are assumed with a large departure from MAR.

Tan et al. [23] proposed a review study about the use of controlled multiple imputation in randomized controlled trials where missing data exists. The analysis considered the trials in phases II, III and IV published in The Lancet and New England Journal of Medicine between January 2014 and December 2019, covering primary and sensitivity analysis studies. The findings show that 56% of the controlled multiple imputation was performed with the delta-adjustment method. Nevertheless, most of the works report the used delta values but do not provide justifications to why the experts decided towards those values.

In conclusion, the delta-adjustment method has been widely used to shift missing values under MAR assumptions to the MNAR mechanism, reducing greatly the departure between the two. Its use has been particularly relevant in the healthcare domain, where the MNAR mechanism appears abundantly. However, the definition of the delta values is always performed by domain experts, which is the problem that we are addressing here. To the best of our knowledge, the automatic estimation of the approximate delta values is being introduced for the first time in our work.

3   Automatic Delta-Adjustment Method

In this work an automatic procedure that is capable of estimating approximate delta adjustment values is proposed. Considering the MNAR characteristics, it is impossible to find an optimal delta value since it would depend on the missing values themselves. Therefore, we rely on statistics to find an approximate value that will bring the estimates made under MAR assumptions valid under MNAR. We called our approach Automatic Delta-Adjustment Method (ADAM).

ADAM's goal is to find a factor comprised within $[0, 1]$ for each feature of the dataset. This factor will be multiplied by the imputed values in order to adjust them. The proposed procedure to estimate the factor for a specific feature X_i is presented in Algorithm 1. The procedure assumes that the missing values correspond to the smaller values of the feature (the opposite scenario is addressed later), and comprises these main steps:

1. The mean of the available values within the first quartile is calculated (μ_{Q1}). This value is representative of the lower tail of the feature, and it is used instead of the minimum because it better represents the group of smaller values and it is resilient to extreme factors. Additionally, the standard deviation of all the available values in the feature is also calculated (σ_{all});
2. The missing values are imputed three independent times by sampling from a Gaussian distribution where the standard deviation is one and the mean varies μ_{Q1} minus σ_{all} according to the empirical rule: $\mu_{Q1} - \sigma_{all}$, $\mu_{Q1} - 2\sigma_{all}$, and $\mu_{Q1} - 3\sigma_{all}$. Such variation is used to define a reasonable range for the missing values, since the imputed values with MAR models are likely to be shifted towards a higher domain. Moreover, when later calculating the adjustment factor, these three imputations are weighted so that the more extreme values (i.e., further away from μ_{Q1}) have a decreased impact on the calculation.

Algorithm 1. Pseudocode for the ADAM procedure. The missing values are on the lower tail of the feature X_i. The algorithm receives as input the feature data containing missing values ($X_i\_missing$) and already imputed ($X_i\_imputed$), and it returns the adjusted data for the feature ($X_i\_adjusted$).

Input: $X_i\_missing$, $X_i\_imputed$
Output: $X_i\_adjusted$
1: μ_{Q1} = mean($Q1$ values from $X_i\_missing$)
2: σ_{all} = standard_deviation($X_i\_missing$)
3: **for** each j in $\{1, 2, 3\}$ **do**
4:     $gaussian\_imputed_j$ = missing data $\sim \mathcal{N}(\mu_{Q1} - j * \sigma_{all}, 1)$
5: **end for**
6: $scalar_0$ = PCA (1 comp.) applied to $X_i\_imputed$
7: **for** each j in $\{1, 2, 3\}$ **do**
8:     $scalar_j$ = PCA (1 comp.) applied to $gaussian\_imputed_j$
9:     $dist_j$ = euclidean_distance($scalar_0$, $scalar_j$)
10: **end for**
11: Normalize $dist_j, \forall j \in \{1, 2, 3\}$ so that $\sum_1^3 dist_j = 1$
12: $factor_i = (3dist_3 + 2dist_2 + dist_1) / 6$
13: $X_i\_adjusted = X_i\_imputed - factor_i * X_i\_imputed$
14: **return** $X_i\_adjusted$

---

3. The Principal Component Analysis (PCA)[1] method is used to condense the data from the features into a single representative numeric value. This transformation is applied individually to the dataset imputed with the MAR model and the datasets imputed in the previous step, which leads to four different scalars by feature. To achieve these results the datasets must be transposed before the transformation since we are condensing the data from the features and not the features themselves. The obtained values for each feature are then used to estimate how far away are the imputations from the previous step when compared to the original imputation with the MAR model. For this purpose, we calculate the euclidean distance between the value representing the imputation with the MAR model and each one of the remaining scalars;
4. To calculate the factor, the distances from the previous step are normalized so that their sum is equal to one (which is necessary to achieve a factor within $[0, 1]$), and a weighted mean is calculated so that the imputations performed according to the empirical rule have an impact in the factor proportional to their means.

After performing the previous step, we have an independent factor ($factor_i$) within $[0, 1]$ for each feature (X_i) of the dataset. To adjust the values imputed with the MAR model, the following equation is applied to each of the D features:

$$X_i = X_i - factor_i * X_i, \forall i \in \{1, ..., D\} \tag{1}$$

---

[1] PCA is a feature extraction technique often used for dimensionality reduction. It computes the principal components of the data and returns the first n, which is a user-defined parameter [1].

As previously stated, the described procedure assumes that the missing values are the smaller values of the feature. However, the opposite scenario where the missing data corresponds to the larger values is also valid. To address this scenario, the procedure suffers two specific changes:

– In step 1), the last quartile is instead calculated (μ_{Q4}) since it now represents the higher tail of the feature.
– In step 2), the mean is varied on the opposite direction ($\mu_{Q4}+\sigma_{all}$, $\mu_{Q4}+2\sigma_{all}$, and $\mu_{Q4}+3\sigma_{all}$), since the imputed values with MAR models are now shifted towards a lower domain.

Finally, the adjustment operation also changes since the imputed values are now being shifted upwards:

$$X_i = X_i + factor_i * X_i, \forall i \in \{1, ..., D\} \tag{2}$$

The identification of the feature's missing tail (i.e., if the missing values correspond to the smaller or larger values of the feature) should be made manually through exploratory data analysis or even by using domain knowledge.

4   Experimental Setup

To evaluate if ADAM is effective in adjusting the imputed values, an experiment was conducted to compare the imputation results before and after applying it. The experimental setup comprised the following seven state-of-the-art imputation methods:

– Mean imputation, where the mean of the available values of each feature are used to impute;
– Multiple Imputation by Chained Equations (MICE), which is a multiple imputation-based approach where several Bayesian ridge regressions are fitted in a round-robin procedure with 100 iterations. Each regression defines as the dependent variable one of the features containing missing values, and uses the remaining as the independent variables [4];
– k-Nearest Neighbors (kNN) imputation with $k = 5$, which selects the k nearest neighbors of the instance being imputed by calculating the Euclidean distance between the available values, and uses the mean of these k neighbors to impute the features with missing data [21];
– SoftImpute, which is a matrix completion iterative method based on nuclear-norm regularization, that estimates the missing values through soft-threshold singular value decomposition [12]. A maximum of 100 iterations was considered;
– Denoising Autoencoder (DAE), which is an autoencoder trained with data containing additional noise, which in this context is the existence of missing values. Incidentally, an autoencoder is a special type of neural network that tries to reproduce the input data at the output layer, usually by learning a

compressed representation of that data [16, 25]. The architecture of the networks was defined with the following hyperparameters: a single hidden layer with a number units equal to half of the input dimension; ReLU as the activation function; batches of 64 instances; a maximum of 200 epochs; Adam as the optimization algorithm; Mean Squared Error as the loss function; a learning rate of 0.001; a dropout layer with a rate of 25% for regularization; early stop if the validation loss has no improvements over 100 epochs; a reduction of the learning rate by 80% if there are no improvements over 100 epochs; and Sigmoid as the activation function for the output layer, so that the data is normalized within [0, 1];

– Variational Autoencoder (VAE), which is a generative variant of the autoenconder that learns the multidimensional parameters of a Gaussian distribution (i.e., mean and standard deviation), and by sampling from these is able to generate new data with similar characteristics [13, 15, 16]. The architecture of the networks was defined with the same hyperparameters as the DAE, adding the two layers needed to represent the Gaussian parameters.

– Generative Adversarial Imputation Nets (GAIN), which is a direct application of the well-known generative adversarial networks to the problem of missing data [28]. The generator network performs the imputation, while the discriminator tries to distinguish between original and imputed data. The networks were parametrized with the hyperparameters reported by the authors of the method.

The architecture and hyperparameters used for the seven imputation methods were defined through a grid search procedure. This is a common strategy that aims to obtain optimal hyperparameters that conform to common use cases. Regarding the implementation of these algorithms, the autoencoders were implemented with the Keras library, the SoftImpute was coded from scratch, the GAIN code was obtained from the original authors[2], and the remaining methods were directly used from the Scikit-learn library. Furthermore, the implementation of ADAM is available on GitHub[3].

The experiment considered 10 public datasets from the healthcare domain, covering clinical research based on routinely collected data for different pathologies. This domain was chosen since it frequently suffers from missing data under the MNAR mechanism [14]. All datasets are available at the UC Irvine repository[4], and they cover different ranges of instances and features, as Table 1 shows.

In order to have a controlled experiment, the datasets were all complete (i.e., without missing data) and the missing values were artificially generated according to the MNAR mechanism. The used strategy removed the smaller values [24] or larger values [27] of the orderable features upon a certain missing rate (which excludes nominal categorical features since they are non-orderable). Such strategy was applied in a multivariate fashion where the missing rate is defined for the entire dataset and several features are injected with missing

---

[2] https://github.com/jsyoon0823/GAIN.

[3] https://github.com/ricardodcpereira/ADAM.

[4] https://archive.ics.uci.edu/ml.

Table 1. Datasets used in the experimental setup.

Dataset	# Instances	# Features	
		Continuous	Categorical
wisconsin	569	31	0
ctg	2126	21	2
pima	768	9	0
liver	583	10	1
diabetic-retinopathy	1151	16	4
parkinsons	195	23	0
bc-coimbra	116	10	0
thoracic-surgery	470	14	3
spine	310	13	0
mammographic-masses	830	2	4

values simultaneously [22]. Consequently, each feature has a different number of missing values, which grouped together sum up to the desired global missing rate. The imputation is performed for all features at once, and the results are assessed through the Mean Absolute Error (MAE) calculated between the ground truth (i.e., original data) and the imputed values.

All datasets were normalized within $[0, 1]$ and split into train and test sets with 70% and 30% of the instances, respectively. The normalizer uses the minimum and maximum values from the train set and it is then applied to both sets. This strategy keeps the test data isolated from the training data, preventing bias in the test set. However, when dealing with high missing rates (usually above 50%), it is possible for the test set to contain unseen values, which makes its normalization boundaries go slightly beyond the aimed $[0, 1]$ domain. For the neural network-based methods, 20% of the train set was used for validation. The non-orderable features (e.g., categorical nominal) were transformed through one-hot encoding (i.e., dummy coding). The missing values were injected independently in each of the described sets in order to ensure that all of them had equal missing rates and MNAR assumptions. Five different missing rates were considered (5%, 10%, 20%, 40%, and 60%) in order to cover different levels of missingness. For the neural network-based methods the missing values were pre-imputed with the mean imputation.

To mitigate bias and stochastic behaviors the experiment was executed 30 independent times, with the data being randomly split into the train and test sets in each run. The results here presented are the mean of these 30 runs. Each run was executed in a computer with the following specifications: Windows 11, CPU AMD Ryzen 5600X, 16 GB RAM, and GPU NVIDIA GeForce GTX 1060 6 GB. The time complexity of applying ADAM for each of the imputation methods was not directly measured, but the impact was not significant.

Table 2. Mean Absolute Error results of the imputation methods with and without the adjustment provided by ADAM. The left tail of the features was missing (i.e., smaller values).

Imp.	ADAM	Missing Rate				
		5%	10%	20%	40%	60%
AE	No	0.236 ± 0.09	0.238 ± 0.09	0.256 ± 0.10	0.322 ± 0.15	0.515 ± 0.34
	Yes	**0.192 ± 0.07**	**0.195 ± 0.06**	**0.212 ± 0.07**	**0.274 ± 0.12**	**0.464 ± 0.31**
GAIN	No	0.254 ± 0.08	0.252 ± 0.08	0.266 ± 0.10	0.324 ± 0.15	0.519 ± 0.34
	Yes	**0.206 ± 0.06**	**0.208 ± 0.06**	**0.220 ± 0.07**	**0.277 ± 0.12**	**0.469 ± 0.31**
MICE	No	0.184 ± 0.09	0.186 ± 0.09	0.204 ± 0.10	0.280 ± 0.14	0.475 ± 0.34
	Yes	**0.154 ± 0.07**	**0.157 ± 0.07**	**0.174 ± 0.07**	**0.245 ± 0.12**	**0.436 ± 0.31**
Mean	No	0.269 ± 0.08	0.264 ± 0.08	0.278 ± 0.09	0.334 ± 0.14	0.523 ± 0.33
	Yes	**0.217 ± 0.05**	**0.216 ± 0.06**	**0.229 ± 0.07**	**0.284 ± 0.11**	**0.472 ± 0.30**
SoftImp	No	0.178 ± 0.07	0.184 ± 0.07	0.197 ± 0.08	0.255 ± 0.11	0.429 ± 0.29
	Yes	**0.152 ± 0.06**	**0.160 ± 0.06**	**0.174 ± 0.06**	**0.233 ± 0.10**	**0.412 ± 0.28**
VAE	No	0.285 ± 0.11	0.279 ± 0.10	0.294 ± 0.12	0.348 ± 0.16	0.533 ± 0.35
	Yes	**0.227 ± 0.07**	**0.224 ± 0.07**	**0.238 ± 0.08**	**0.292 ± 0.12**	**0.477 ± 0.32**
kNN	No	0.185 ± 0.09	0.194 ± 0.09	0.219 ± 0.10	0.292 ± 0.14	0.487 ± 0.34
	Yes	**0.153 ± 0.07**	**0.162 ± 0.07**	**0.184 ± 0.07**	**0.254 ± 0.11**	**0.446 ± 0.31**

5   Results

The obtained results for the adjustment of imputed values provided by ADAM are presented in Tables 2 and 3. For Table 2 the left tail was missing (i.e., smaller values of the features), and for Table 3 the right tail was removed (i.e., larger values of the features).

In an overall analysis, all methods achieved smaller imputation errors after the imputed values were adjusted through ADAM. This behavior is consistent for all missing rates and for both MNAR strategies with the smaller and larger values being removed, with the global error improvement being 11%. The enhancement also appears to be stable among the different levels of missingness, peaking at 18% for the 5% missing rate, which shows ADAM's resilience to variations on this factor.

To understand if the obtained results were statistically significant we applied the Wilcoxon signed-rank test with a significance level of 5%. This test was chosen because the normality assumptions were not met, and we have paired MAE values for each imputation method (before and after applying ADAM). The test was applied independently for each missing rate, and the one-sided alternative was used since we were only interested in evaluating if the MAE values after applying ADAM are significantly lower. The obtained p-values showed that the results are statistically significant with $p < 0.001$ for all settings, which corroborates the good performance obtained by ADAM.

Table 3. Mean Absolute Error results of the imputation methods with and without the adjustment provided by ADAM. The right tail of the features was missing (i.e., larger values).

Imp.	ADAM	Missing Rate				
		5%	10%	20%	40%	60%
AE	No	0.710 ± 0.27	0.847 ± 0.52	1.417 ± 1.91	2.770 ± 3.37	4.011 ± 4.81
	Yes	**0.610 ± 0.26**	**0.744 ± 0.51**	**1.306 ± 1.91**	**2.655 ± 3.36**	**3.894 ± 4.80**
GAIN	No	0.725 ± 0.27	0.849 ± 0.52	1.410 ± 1.91	2.769 ± 3.37	4.036 ± 4.81
	Yes	**0.628 ± 0.26**	**0.747 ± 0.51**	**1.301 ± 1.91**	**2.654 ± 3.36**	**3.924 ± 4.80**
MICE	No	0.579 ± 0.23	0.716 ± 0.47	1.291 ± 1.87	2.678 ± 3.35	3.943 ± 4.80
	Yes	**0.496 ± 0.22**	**0.624 ± 0.46**	**1.183 ± 1.85**	**2.557 ± 3.34**	**3.813 ± 4.79**
Mean	No	0.774 ± 0.28	0.896 ± 0.53	1.448 ± 1.92	2.784 ± 3.36	4.019 ± 4.81
	Yes	**0.689 ± 0.27**	**0.804 ± 0.52**	**1.346 ± 1.91**	**2.671 ± 3.35**	**3.902 ± 4.80**
SoftImp	No	0.676 ± 0.27	0.826 ± 0.51	1.422 ± 1.90	2.834 ± 3.39	4.187 ± 4.84
	Yes	**0.579 ± 0.26**	**0.727 ± 0.50**	**1.319 ± 1.90**	**2.738 ± 3.38**	**4.119 ± 4.84**
VAE	No	0.758 ± 0.29	0.884 ± 0.54	1.441 ± 1.92	2.781 ± 3.37	4.018 ± 4.81
	Yes	**0.671 ± 0.28**	**0.790 ± 0.53**	**1.338 ± 1.92**	**2.669 ± 3.36**	**3.903 ± 4.80**
kNN	No	0.618 ± 0.24	0.761 ± 0.49	1.342 ± 1.89	2.722 ± 3.36	3.976 ± 4.81
	Yes	**0.509 ± 0.23**	**0.650 ± 0.48**	**1.224 ± 1.88**	**2.600 ± 3.35**	**3.852 ± 4.80**

6   Conclusions

In this work we proposed a procedure called Automatic Delta-Adjustment Method (ADAM) to automatically estimate the delta-adjustment values. We compared the results obtained by seven state-of-the-art imputation methods with and without the adjustments provided by ADAM. The experimental setup comprised 10 datasets from the healthcare context that were injected with missing values under MNAR in a multivariate fashion. We concluded that the adjustment performed by ADAM led to error improvements in all imputations methods and missing rates, achieving a global enhancement of 11%.

Motivated by the results achieved with ADAM, future work will be focused on integrating it with an auxiliary procedure to automatically identify the features' missing tails. A possible direction is to model this task as a binary classification problem and use machine learning to solve it. Furthermore, we want to incorporate information from other datasets in the adjustments calculations, so that external data can be used to help reduce bias in MNAR settings. Finally, we also want to compare ADAM results to imputed data that was manually adjusted by domain experts.

Acknowledgements. This work is supported in part by the FCT - Foundation for Science and Technology, I.P., Research Grant SFRH/BD/149018/2019. This work is also funded by the FCT - Foundation for Science and Technology, I.P./MCTES through national funds (PIDDAC), within the scope of CISUC R&D Unit - UIDB/00326/2020 or project code UIDP/00326/2020.

References

1. Abdi, H., Williams, L.J.: Principal component analysis. Wiley Interdisc. Rev.: Comput. Stat. **2**(4), 433–459 (2010)
2. Austin, P.C., White, I.R., Lee, D.S., van Buuren, S.: Missing data in clinical research: a tutorial on multiple imputation. Can. J. Cardiol. **37**(9), 1322–1331 (2020)
3. Beaulieu-Jones, B.K., Lavage, D.R., Snyder, J.W., Moore, J.H., Pendergrass, S.A., Bauer, C.R.: Characterizing and managing missing structured data in electronic health records: data analysis. JMIR Med. Inf. **6**(1), e11 (2018)
4. Van Buuren, S., Groothuis-Oudshoorn, K.: mice: multivariate imputation by chained equations in R. J. Stat. Softw. **45**, 1–68 (2010)
5. Carreras, G., et al.: Missing not at random in end of life care studies: multiple imputation and sensitivity analysis on data from the action study. BMC Med. Res. Methodol. **21**(1), 1–12 (2021)
6. García-Laencina, P.J., Sancho-Gómez, J.L., Figueiras-Vidal, A.R.: Pattern classification with missing data: a review. Neural Comput. Appl. **19**(2), 263–282 (2010)
7. Gondara, L., Wang, K.: Recovering loss to followup information using denoising autoencoders. In: 2017 IEEE International Conference on Big Data (Big Data), pp. 1936–1945 (2017)
8. Gondara, L., Wang, K.: MIDA: multiple imputation using denoising autoencoders. In: Phung, D., Tseng, V.S., Webb, G.I., Ho, B., Ganji, M., Rashidi, L. (eds.) PAKDD 2018. LNCS (LNAI), vol. 10939, pp. 260–272. Springer, Cham (2018). https://doi.org/10.1007/978-3-319-93040-4_21
9. Leacy, F.P., Floyd, S., Yates, T.A., White, I.R.: Analyses of sensitivity to the missing-at-random assumption using multiple imputation with delta adjustment: application to a tuberculosis/HIV prevalence survey with incomplete HIV-status data. Am. J. Epidemiol. **185**(4), 304–315 (2017)
10. Leurent, B., Gomes, M., Faria, R., Morris, S., Grieve, R., Carpenter, J.R.: Sensitivity analysis for not-at-random missing data in trial-based cost-effectiveness analysis: a tutorial. Pharmacoeconomics **36**(8), 889–901 (2018)
11. Little, R.J., Rubin, D.B.: Statistical Analysis with Missing Data, vol. 793. John Wiley & Sons, New York (2019)
12. Mazumder, R., Hastie, T., Tibshirani, R.: Spectral regularization algorithms for learning large incomplete matrices. J. Mach. Learn. Res. **11**, 2287–2322 (2010)
13. McCoy, J.T., Kroon, S., Auret, L.: Variational autoencoders for missing data imputation with application to a simulated milling circuit. IFAC-PapersOnLine **51**(21), 141–146 (2018)
14. Peek, N., Rodrigues, P.P.: Three controversies in health data science. Int. J. Data Sci. Anal. **6**(3), 261–269 (2018). https://doi.org/10.1007/s41060-018-0109-y
15. Pereira, R.C., Abreu, P.H., Rodrigues, P.P.: Partial multiple imputation with variational autoencoders: tackling not at randomness in healthcare data. IEEE J. Biomed. Health Inf. **26**(8), 4218–4227 (2022)
16. Pereira, R.C., Santos, M.S., Rodrigues, P.P., Abreu, P.H.: Reviewing autoencoders for missing data imputation: technical trends, applications and outcomes. J. Artif. Intell. Res. **69**, 1255–1285 (2020)
17. Qiu, Y.L., Zheng, H., Gevaert, O.: Genomic data imputation with variational autoencoders. GigaScience **9**(8) (2020)
18. Rezvan, P.H., Lee, K.J., Simpson, J.A.: Sensitivity analysis within multiple imputation framework using delta-adjustment: application to longitudinal study of australian children. Longitudinal Life Course Stud. **9**(3), 259–278 (2018)

19. Rubin, D.B.: Inference and missing data. Biometrika **63**(3), 581–592 (1976)
20. Rubin, D.B.: Multiple Imputation for Nonresponse in Surveys, vol. 81. John Wiley & Sons, New York (2004)
21. Santos, M.S., Abreu, P.H., García-Laencina, P.J., Simão, A., Carvalho, A.: A new cluster-based oversampling method for improving survival prediction of hepatocellular carcinoma patients. J. Biomed. Inf. **58**, 49–59 (2015)
22. Santos, M.S., Pereira, R.C., Costa, A.F., Soares, J.P., Santos, J., Abreu, P.H.: Generating synthetic missing data: a review by missing mechanism. IEEE Access **7**, 11651–11667 (2019)
23. Tan, P.T., Cro, S., Van Vogt, E., Szigeti, M., Cornelius, V.R.: A review of the use of controlled multiple imputation in randomised controlled trials with missing outcome data. BMC Med. Res. Methodol. **21**(1), 1–17 (2021)
24. Twala, B.: An empirical comparison of techniques for handling incomplete data using decision trees. Appl. Artif. Intell. **23**(5), 373–405 (2009)
25. Vincent, P., Larochelle, H., Bengio, Y., Manzagol, P.A.: Extracting and composing robust features with denoising autoencoders. In: Proceedings of the 25th International Conference on Machine learning, pp. 1096–1103 (2008)
26. White, I.R., Royston, P., Wood, A.M.: Multiple imputation using chained equations: issues and guidance for practice. Stat. Med. **30**(4), 377–399 (2011)
27. Xia, J., et al.: Adjusted weight voting algorithm for random forests in handling missing values. Pattern Recogn. **69**, 52–60 (2017)
28. Yoon, J., Jordon, J., Schaar, M.: Gain: missing data imputation using generative adversarial nets. In: International Conference on Machine Learning, pp. 5689–5698. PMLR (2018)

Data Integration Landscapes: The Case for Non-optimal Solutions in Network Diffusion Models

James Nevin[✉][iD], Paul Groth[iD], and Michael Lees[iD]

University of Amsterdam, Amsterdam, The Netherlands
{j.g.nevin,p.t.groth,m.h.lees}@uva.nl

Abstract. The successful application of computational models presupposes access to accurate, relevant, and representative datasets. The growth of public data, and the increasing practice of data sharing and reuse, emphasises the importance of data provenance and increases the need for modellers to understand how data processing decisions might impact model output. One key step in the data processing pipeline is that of data integration and entity resolution, where entities are matched across disparate datasets. In this paper, we present a new formulation of data integration in complex networks that incorporates integration uncertainty. We define an approach for understanding how different data integration setups can impact the results of network diffusion models under this uncertainty, allowing one to systematically characterise potential model outputs in order to create an output distribution that provides a more comprehensive picture.

Keywords: Complex networks · Data integration · Entity resolution · Network diffusion models

1 Introduction

Since all computational models offer simplified views of reality, there inevitably arises uncertainty in the results that they produce [37]. This uncertainty can be either epistemic, driven by lack of knowledge of the system, or aleatory, driven by the inherent randomness in the system of study. Understanding and quantifying epistemic uncertainty has been a significant field of research in computational modelling [14,36]. While this uncertainty can be caused by many factors, the majority of the current literature has addressed parameter [35], structural [32], measurement [9], and interpolation [19] uncertainty.

The growth of public data, and the increasing practice of data sharing and reusing data from multiple sources, highlight the need for modellers to understand how data processing decisions can impact model uncertainty. These data processing pipelines are complex due to both the number of data sources but also due to the large space of possible choices in algorithms and decisions that need to be made [8]. While the importance of data quality in computational models

J. Mikyška et al. (Eds.): ICCS 2023, LNCS 14073, pp. 494–508, 2023.
https://doi.org/10.1007/978-3-031-35995-8_35

is well-understood [17,31], *to date there has been only limited investigation into quantifying the effects that data handling decisions can have on computational models.*

This paper thus presents a deeper investigation into this problem focused on data integration. Specifically, it defines a formalism for systematically assessing these effects for diffusion models on complex networks, and we validate this approach with an experiment on a benchmark dataset. In the next section, we review the related work more deeply and more fully articulate the problem.

2   Background and Related Work

One of the key areas in data handling is data integration, the combining of multiple distinct datasets from different sources into a single, unified view [10]. A critical aspect of data integration is entity resolution, in which real-world entities imperfectly recorded in datasets are identified and consolidated. Entity resolution is a well-studied problem [4,13], and has been tackled from both a theoretical [11] and practical [18] standpoint.

Since its inception, the entity resolution problem has been defined as an optimisation issue [11], where a single, 'best' resolved set of entities is identified and used for creating the unified dataset. In this case, 'best' usually means the resolved entity set that most closely resembles the unobserved reality (according to some arbitrary measure) that generated the unresolved data. Even with more advanced techniques like deep learning and parallelisation being employed, the problem is still seen as one of optimising to a single entity set [8,22]. Only recently has the entity resolution problem begun to be considered based on downstream intent for the resolved data [12].

However, in almost all cases there is a level of uncertainty in the entity resolution problem. While a solution may be found that is indeed optimal based on the specific optimisation problem definition, there is no guarantee that this definition serves as a perfect base to recreate the true unobserved reality. In fact, usually insufficient data have been collected to identify perfectly this true reality. Hence, different optimisation problem definitions can produce different sets of entities, and it can be challenging to decide which to use to resolve a dataset. This is often not accounted for in the literature, where instead only a single optimisation problem and subsequent resolved dataset are considered.

One case where this can cause problems is in the modelling of diffusion processes on complex networks. Many diffusion models are known to exhibit critical threshold/phase transition behaviour [23,29], where small changes in model parameters or network topology can have a drastic effect on model output. These models have application in a wide range of important societal problems, e.g., epidemic spreading [15], opinion dynamics [1], and polarisation [30]. When data are used to create complex networks, the set of entities used can create sufficiently large changes in network topology to cross over threshold levels for diffusion models [28].

This sensitivity has already been demonstrated with different data integration setups, which includes the entity resolution step [28]. But, as of yet, there is no formal framework for describing the levels of uncertainty in different data integration setups and their effect on diffusion models. With the, often observed, critical threshold behaviour of these diffusion models, it is also important to characterise model behaviour in 'low probability' data integration setups, as the resolved graph may exhibit drastically different topology and thus diffusion model output. When considering aggregate measures, there is a need to capture a distribution of model outputs over different data integration setups, rather than a single, best setup.

Hence, we argue for a need for a *data integration landscape* in complex network analysis. Rather than performing a single data integration, multiple different setups should be defined, tested, and weighted based on confidence in their accuracy using flexible entity resolution approaches. Diffusion models can be tested on the resultant resolved networks and their associated weighting, giving a distribution of model outputs that captures the uncertainty arising from the imperfect data.

3   Data Integration Landscapes

We now define formally the data integration process such that multiple possible integrations are possible, which together create a data integration landscape.

There exists a true graph/complex network, G_T, which is made up of a set of true vertices, V_T, and edges, E_T, where the vertices represent different real-world entities and the edges relationships between these entities. Given a network model, M, whose behaviour on G_T one wishes to understand, a landscape of potential integrated graphs is created on which to estimate M's behaviour on G_T.

Assume that G_T is not perfectly observable; instead, there is a set of observed graphs, $G_1, G_2, ..., G_n$, each of which offer some imperfect or incomplete view of G_T. Let the union of all observed graphs be

$$\mathcal{G} := \bigcup_{i=1}^{n} G_i. \tag{1}$$

Let $V(G)$ be the set of vertices of some graph G. We define $N := |V(\mathcal{G})|$.

3.1   Entity Resolution

Two vertices v and w are considered equal ($v = w$) if they correspond to the same real-world entity. We assume the following:

1. $\forall v \in V_T, \exists w \in V(\mathcal{G})$ s.t. $v = w$
2. $\forall w \in V(\mathcal{G}), \exists! v \in V_T$ s.t. $v = w$

i.e. every vertex in the true graph is present in at least one of the observed graphs, and every vertex in the observed graphs corresponds to exactly one vertex in the true graph.

Consider a partitioning of the vertices in $\mathcal{G}$, $\{P_1, P_2, ..., P_k\}$. Each P_i is a subset of vertices from $V(\mathcal{G})$, with $P_i \cap P_j = \emptyset$ for $i \neq j$ and $\bigcup_{i=1}^{k} P_i = V(\mathcal{G})$. We call such a partitioning a *state-of-the-world*, which we generally denote with S. A state-of-the-world describes a potential reality where all vertices in the same partition are equal i.e. they represent the same real-world entity. For an assumed state-of-the-world, each $v \in V_T$ corresponds to exactly one partition. We note that there exists exactly one *true* state-of-the-world, S_T, which perfectly captures the real-world equality relationships between vertices in V_T and $V(\mathcal{G})$.

The number of potential states-of-the-world is equal to the total number of ways of partitioning the vertices in $\mathcal{G}$. With $|V(\mathcal{G})| = N$, this is B_N, the Nth Bell number:

$$B_N = \sum_{k=0}^{N} \left\{ {N \atop k} \right\},\tag{2}$$

where $\left\{ {N \atop k} \right\}$ is the Stirling number of the second kind given by

$$\left\{ {N \atop k} \right\} = \frac{1}{k!} \sum_{i=0}^{k} (-1)^i \binom{k}{i} (k-i)^N.\tag{3}$$

We now assume that we have some (partial) information/data about the observed vertices, which we call vertex attributes. This could include things like names, geographical location, email addresses, etc. This information allows us to identify different potential states-of-the-world as more or less likely. For example, a state-of-the-world where all vertices in a partition have the same last name might be more likely than one with different last names in a given partition. Indeed, based on one's understanding of the vertices' attributes, one may decide that a large number of the potential states-of-the-world are impossible.

Let $\mathcal{S} = \{S_1, S_2, ..., S_m\}$ be the set of all potential states-of-the-world that are not impossible. Let $\mathbb{P}(S_i = S_T)$ be the probability that S_i is the true state-of-the-world. We then have:

1. $\mathbb{P}(S_i = S_T \cap S_j = S_T) = 0$ for $i \neq j$
2. $\sum_{i=1}^{m} \mathbb{P}(S_i = S_T) = 1$

The defining of probabilities of states-of-the-world is use-case specific. With perfectly detailed data, free of errors and fully documented, it may be possible to assign a probability of 1 to a specific state-of-the-world. With no data, all states may be considered equally likely. Reality will likely fall between these two extremes, and there is unlikely to be a perfect or consistent approach to assigning probabilities for all domains.

3.2   Graph Resolution

We now introduce (graph) *resolution* functions, which complete the data integration process started by entity resolution and, in conjunction with the associated probabilities, define a data integration landscape. Resolution functions are functions on $\mathcal{G} \times S$ that return a single, *resolved* graph. For resolution function R and state-of-the-world $S = \{P_1, P_2, ..., P_k\}$, $R(\mathcal{G}, S)$ returns a graph with k vertices, each of which correspond to a different partition in S.

The purpose of resolution functions is, under an assumed state-of-the-world, to recreate the unobserved G_T from the observed $\mathcal{G}$. The partitions in the state-of-the-world, S, dictate the vertices in $R(\mathcal{G}, S)$; the resolution function R determines which of these vertices are connected and what their resolved attributes are.

As a simple example, suppose we have two partitions, P_i and P_j, in our assumed state-of-the-world. These will then correspond to two vertices in our resolved graph. One way our resolution function could determine whether to connect these two vertices would be to consider the connectedness of the vertices in P_i and P_j in $\mathcal{G}$: if over 50% of the vertices in P_i have an edge to a vertex in P_j, connect the vertices in the resolved graph, for example. The attributes of the vertices in the resolved graph could be decided using means of the attributes of the vertices in P_i and P_j, as one possible approach.

The graphs created using a resolution function, in conjunction with the probabilities of the states-of-the-world that generate them, allow one to identify a distribution of potential true graphs. Let the set of graphs generated through resolution function R and states-of-the-world S be $\mathcal{G}_{R,S}$. Then for $G \in \mathcal{G}_{R,S}$:

$$\mathbb{P}(G = G_T) = \sum_{S \in \mathcal{S}: R(\mathcal{G}, S) = G} \mathbb{P}(S = S_T), \tag{4}$$

where it is assumed that, given the true state-of-the-world, a resolution function returns G_T (Sect. 3.3 offers more details). In most cases, only one state-of-the-world will generate a given resolved graph, but this is not strict.

Using this distribution of potential true graphs, one can better understand the potential behaviour of the network model, M, on G_T. For some model metric, X:

$$\mathbb{P}(X = x) = \sum_{G \in \mathcal{G}_{R,S}} \mathbb{P}(X = x | G = G_T)\mathbb{P}(G = G_T). \tag{5}$$

The $\mathbb{P}(X = x | G = G_T)$ values can be estimated using various model analysis techniques, such as Monte Carlo simulation, and the approach will likely be model- (and possibly graph-) specific.

The approach described in Eq. 5 can be extended to calculate various other values of interest, such as the risk of exceeding a certain threshold:

$$\mathbb{P}(X > t) = \sum_{G \in \mathcal{G}_{R,S}} \mathbb{P}(X > t | G = G_T)\mathbb{P}(G = G_T), \tag{6}$$

or identifying in which states-of-the-world risk is high:

$$\{S : \mathbb{P}(X > t | R(\mathcal{G}, S) = G_T) > r\}. \tag{7}$$

Again, many of these intermediate values/probabilities will need to be estimated, but there exists a wealth of research into such topics [3,25].

3.3 Multiple Resolution Functions

We can extend the formulation above to include the scenario where there is a set of potential resolution functions, $\mathcal{R} = \{R_1, R_2, ..., R_k\}$. Furthermore, we introduce a series of simplifying assumptions that lead to a straightforward probability assignment.

A resolution function is called *true* if it perfectly maps the true state-of-the-world to G_T i.e. $R(\mathcal{G}, S_T) = G_T$. Note that there is more than one possible true resolution function.

The resolution functions in $\mathcal{R}$ may or may not be true. This set of potential resolution functions and the potential states-of-the-world define a set of possible resolved graphs.

$$G_{\mathcal{R},\mathcal{S}} = \{R(\mathcal{G}, S) : R \in \mathcal{R} \wedge S \in \mathcal{S}\}. \tag{8}$$

As before, we would like to assign a probability to each of these potential resolved graphs being equal to the true graph. For $G \in G_{\mathcal{R},\mathcal{S}}$:

$$\mathbb{P}(G = G_T) = \sum_{S \in \mathcal{S}} \mathbb{P}(G = G_T | S = S_T)\mathbb{P}(S = S_T), \tag{9}$$

since the events $S = S_T$ are mutually exclusive and exhaustive. If we assume that the resolution functions in $\mathcal{R}$ are deterministic, we have:

$$\sum_{S \in \mathcal{S}} \mathbb{P}(G = G_T | S = S_T)\mathbb{P}(S = S_T) =$$

$$\sum_{S \in \mathcal{S}:\exists R \in \mathcal{R} \text{ s.t. } R(\mathcal{G},S)=G} \mathbb{P}(G = G_T | S = S_T)\mathbb{P}(S = S_T)+$$

$$\sum_{S \in \mathcal{S}:\nexists R \in \mathcal{R} \text{ s.t. } R(\mathcal{G},S)=C} \mathbb{P}(G = G_T | S = S_T)\mathbb{P}(S = S_T), \tag{10}$$

as then each state-of-the-world will map to exactly one resolved graph with each potential resolution function.

We now introduce an assumption in order to simplify Eq. 10: exactly one resolution function in $\mathcal{R}$ is true.

If it is possible for none of the resolution functions to be true, then there is a non-zero probability of $G_T \notin \mathcal{G}_{\mathcal{R},\mathcal{S}}$. In this case, we cannot say anything about network model behaviour on G_T. Thus, we assume at least one resolution function is true.

If two or more resolution functions are true, then the true state-of-the-world will map to the same graph with these true resolution functions. Thus, the likelihood of resolution functions being true becomes a joint likelihood with the potential states-of-the-world being true. While this can be handled, it does not offer any simplification to our formula. Hence, we assume exactly one resolution function in $\mathcal{R}$ to be true.

In light of this, we impose a further restriction: no state-of-the-world can map to the same graph with different resolution functions. If a state-of-the-world maps to the same graph with two resolution functions, either the state-of-the-world is not true or neither of the resolution functions are true. Both of these disrupt our probability assignment and calculation, so we prefer to avoid them. This restriction is not too prohibitive; it mainly requires that the resolution functions in $\mathcal{R}$ are sufficiently different.

Equation (10) can now be simplified to be summed over pairs of states-of-the-world and resolution functions in the first sum,

$$\sum_{S \in \mathcal{S}} \mathbb{P}(G = G_T | S = S_T) \mathbb{P}(S = S_T) =$$

$$\sum_{S \in \mathcal{S}, R \in \mathcal{R}:R(\mathcal{G},S)=G} \mathbb{P}(G = G_T | S = S_T) \mathbb{P}(S = S_T) +$$

$$\sum_{S \in \mathcal{S}: \nexists R \in \mathcal{R} \text{ s.t. } R(\mathcal{G},S)=G} \mathbb{P}(G = G_T | S = S_T) \mathbb{P}(S = S_T). \quad (11)$$

Note that, for terms in the final sum in Eq. 11, $\mathbb{P}(G = G_T | S = S_T) = 0$. Given that S is the true state-of-the-world and there is a true resolution function in $\mathcal{R}$, one of the graphs S maps to must be the true graph. However, the sum is over the states-of-the-world which do not map to G. Hence, $G \neq G_T$.

Furthermore, one can replace G in the first sum with the appropriate state-of-the-world and resolution function.

$$\sum_{S \in \mathcal{S}} \mathbb{P}(G = G_T | S = S_T) \mathbb{P}(S = S_T) =$$

$$\sum_{S \in \mathcal{S}, R \in \mathcal{R}:R(\mathcal{G},S)=G} \mathbb{P}(R(\mathcal{G}, S) = G_T | S = S_T) \mathbb{P}(S = S_T) =$$

$$\sum_{S \in \mathcal{S}, R \in \mathcal{R}:R(\mathcal{G},S)=G} \mathbb{P}(R(\mathcal{G}, S_T) = G_T) \mathbb{P}(S = S_T), \quad (12)$$

where in the last line we have used the condition $S = S_T$ to make a substitution. This formulation now has the product of two probabilities: the first is the probability that the resolution function maps the true state-of-the-world to the true graph (i.e. the probability that the resolution function is true); the second is the probability that the state-of-the-world is equal to the true state-of-the-world. Hence,

$$\mathbb{P}(G = G_T) = \sum_{S \in \mathcal{S}, R \in \mathcal{R}:R(\mathcal{G},S)=G} \mathbb{P}(R \text{ is true}) \mathbb{P}(S = S_T). \quad (13)$$

Using this equation, the probabilities of resolved graphs being equal to the true graph can be assigned by independently assigning probabilities to the different states-of-the-world being equal to the true state-of-the-world and to the resolution functions being true.

A given network model can then be analysed on these graphs in the same way as detailed in the previous subsection.

4   Probability Assignment

The key challenge to applying the data integration landscape formulation outlined above is the assignment of probabilities to states-of-the-world, which we
discuss in detail in this section. How this challenge is addressed practically will
depend on the level of uncertainty in the data. At one extreme where there
is zero uncertainty, a probability of 1 is assigned to one state-of-the-world. In
the other extreme of maximum uncertainty, equal probability is assigned to all
states-of-the-world.

The certainty in the data, and thus assignment of probabilities to states-of-
the-world, should be dependent on the similarity between vertex attributes, and
the distribution thereof. The quantification of similarity between entities is a
well-studied field in data science [4, 8, 13], and most entity resolution approaches
rely on this quantification in some way. Even highly sophisticated, state-of-the-
art approaches are typically represented with a final result indicating pairwise
similarities between entities [8]. With this in mind, we use pairwise similarities,
or, equivalently, distances between vertices to measure uncertainty in the data
and thus assign probabilities.

Consider a scenario in which vertices can be divided into two groups, where
all vertices in a group are highly similar to each other, and all vertices between
groups are highly dissimilar. A state-of-the-world that partitions the vertices
into these two groups should have a high probability. We would thus consider
this data to have low uncertainty – we can easily identify a unique or small
set of reasonable states-of-the-world based on the similarities between vertices.
As the distinction between similar and dissimilar vertices becomes less clear, it
becomes more difficult to narrow down the set of states-of-the-world that should
be considered, and thus our uncertainty grows.

Hence, the overall level of uncertainty in the data can be broadly captured
in the distribution of pairwise similarities/distances. In data with low uncertainty, we can easily distinguish between vertices that are equal and vertices
that are unequal: the distances between equal vertices will be significantly lower
than the distances between unequal vertices. In data with high uncertainty, we
cannot easily make this distinction: the distances between equal vertices will be
more similar to the distances between unequal vertices, making them harder to
distinguish.

Figure 1 shows illustrative **distance distributions** for different levels of certainty in the data. Figure 1a shows low uncertainty data. In this case, the distances between matched/equal vertices (blue) are exclusively smaller than the
distances between unmatched/unequal vertices (orange). This clear distinction
means that testing multiple entity resolutions will not be useful, since the likelihood of one state-of-the-world will be significantly higher than all others.

Figure 1b shows distance distributions for data with moderate uncertainty.
In such cases, there is generally a clear distinction in the distances between
matched and unmatched vertices, but there is also a sizeable overlap. Working
in this region of overlap will allow us to capture the various different possibilities,
and we will have clearly defined low and high probability states-of-the-world.

Finally, Fig. 1c shows distances in a high uncertainty dataset where there is significant overlap in the distances seen between matched and unmatched vertices. In this scenario, the application of a data integration landscape will be useful, but it will be more difficult to distinguish between high and low probability states-of-the-world, and there is a high chance of matching unmatched vertices and not matching matched vertices.

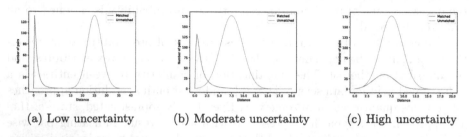

     (a) Low uncertainty   (b) Moderate uncertainty   (c) High uncertainty

Fig. 1. Illustrative distribution of distances in: (a) low uncertainty data where matched distances are much lower than unmatched distances; (b) moderate uncertainty data where there is some overlap in the matched and unmatched distances; (c) high uncertainty data where there is close to no distinction between distances.

In all of the illustrative examples above, we distinguish between the distance distributions of the matched and unmatched vertices. In practice, this distinction will not be possible, and only the joint distance distribution is seen. However, the level of distinction in multiple peaks in our joint distribution will indicate the level of uncertainty in the data. Another factor to consider is the number of matched versus unmatched vertices. Consider a dataset with 1000 total vertices, of which 750 are original and 250 are duplicates (where each duplicate corresponds to a different original vertex). In such a scenario, there will be 250 distances relating to matched vertices and 499250 unmatched distances. Hence, the peak in matched distances will be orders of magnitude lower than the peak of unmatched vertices, and thus more difficult to distinguish in the joint distance distribution. The joint distribution of distances indicates what possible states-of-the-world should be considered.

5   Experiment

In order to validate the data integration landscape formulation, we offer an instantiation of a landscape using standard entity resolution approaches, and apply it to a benchmark dataset with a standard network diffusion model [27].

5.1   Instantiation

Instantiating a data integration landscape as defined in Sect. 3 requires choice of both resolution function(s) and assignment of probabilities to the states-of-the-world. For this paper, we focus primarily on the probability assignment, and use an agglomerative clustering approach for entity resolution. We use a single graph resolution function. We provide a fully automated, naive approach for selecting states-of-the-world and assigning them probabilities.

Agglomerative Clustering Approach. As network models can be sensitive to changes in graph topology, it is important to select states-of-the-world with varying numbers of partitions and thus significantly different resolved graphs, in order to create a full, robust picture.

For the entity resolution step of instantiating the data integration landscape, our approach to initially identify states-of-the-world is based on classical hierarchical clustering [26]. Hierarchical clustering iteratively combines or splits clusters (equivalent to partitions) in a dataset based on some measure of distance/dissimilarity, and similar approaches have been previously applied for entity resolution on graphs [5].

Let $v_1, v_2, ..., v_N$ be the vertices in $\mathcal{G}$. We define $d(v_i, v_j)$ as the *distance* between vertex i and vertex j based on their attributes. This distance can be defined using various classic entity resolution approaches, such as string similarity, numerical difference, etc. Furthermore, we define the distance between two clusters, C_i and C_j, as $D(C_i, C_j)$. This distance can also be determined in numerous ways, such as complete-linkage clustering.

Using these definitions of vertex and cluster distances, we create entity clusterings using an agglomerative approach. The initial clustering has N clusters, each equal to a vertex in $\mathcal{G}$. This defines the first state-of-the-world, $S_N := \{\{v_1\}, \{v_2\}, \{v_3\}, ..., \{v_N\}\}$. States-of-the-world are then defined recursively as per Algorithm 1. Following this algorithm and labelling convention, we have a set of states-of-the-world, $S_N, S_{N-1}, ..., S_1$, where each state S_i has i partitions.

Algorithm 1. Agglomerative Clustering

1. Let the current clustering be $S_i := \{C_1, C_2, ..., C_i\}$
2. Find the closest clusters $C_x, C_y = \arg\min_{C_x, C_y \in S_i} D(C_x, C_y)$ where we order the clusters such that $x < y$
3. Define the next clustering $S_{i-1} := \{C_1, C_2, ..., C_{x-1}, C_{x+1}, ..., C_{y-1}, C_{y+1}, ..., C_i, C_x \cup C_y\}$

This set of states-of-the-world contains all states to which we assign a non-zero probability. For consistency, we assign probabilities proportionally to the distance metrics d and D used in selecting the states-of-the-world. This probability aims to capture how 'good' a particular clustering is i.e. how small distances are within clusters, and how large they are between clusters. We calculate the mean silhouette score of each state-of-the-world, where we reuse the pairwise distances calculated previously. This score is a measure of the quality of a clustering [34]. These mean silhouette scores are then scaled to create a probability mass function over the states-of-the-world.

The resolution function used is as follows: for two partitions P_1 and P_2 in a given state-of-the-world, the vertices in the resolved graph corresponding to these two partitions will be connected if at least one vertex in P_1 is connected to at least one vertex in P_2 in the union graph. Vertex attributes are not relevant to the downstream model, and are thus not resolved.

5.2   Dataset

We apply the above instantiation to a semi-synthetic dataset. The tabular data used is the FEBRL dataset [6], which contains 1000 records. Each record includes information like given name, surname, address, and more. There are 500 error-free records and 500 duplicate records, which contain errors based on real world studies [7]. We drop all rows with missing values, leaving 723 records. We do not explicitly identify the number of duplicate points. We create a Barabási-Albert graph [2] with 723 vertices. Vertices are created with 40 links to existing vertices. Each of the records is assigned to a vertex in the graph, thus creating the dataset.

We define vertex and cluster distances as follows: for two vertices, v_i and v_j, we calculate the commonly-used Levenstein string edit distance [18,20] between their given names, surnames, and street address – $d(v_i, v_j)$ is then equal to the average of these three string distances. We define the distance between two clusters, C_i and C_j, using complete-linkage i.e. $D(C_i, C_j) = \max_{v \in C_i, w \in C_j} d(v, w)$. Using complete-linkage means that we do not need to perform new cluster distance calculations over iterations; instead, we repeatedly reuse the distances between vertices, selecting the maximum between-vertex distances.

As the mean silhouette scores show inconsistent behaviour with very large or small numbers of clusters, we select only the states-of-the-world with between 100 and 700 partitions to assign a non-zero probability to. The mean silhouette scores for these states are scaled so that they sum to 1, giving our assigned probabilities.

5.3   Model

We run a Susceptible-Infected-Recovered (SIR) model on the resolved graphs [15]. This model simulates a spreading process through a population, where spreading is done over edges in the graph. Using an infection rate of 0.005, a recovery rate of 0.2, and an initial population infection size of 0.05, we run 1000 simulations of a spread on each resolved graph [16].

The model metric we use is the fraction of the population that becomes infected after convergence. We compare the behaviour of the model and this metric on the different resolved graphs, and contrast this with the behaviour on the most likely graph (the graph with the highest mean silhouette score). We denote this most likely graph by G_L. We thus can contrast the expected behaviour of the model when taking the different potential states-of-the-world into account, and the behaviour when only using the 'best' or most likely state.

We can calculate the expected proportion of the population becoming infected across all graphs by:

$$\mathbb{E}[X] = \sum_{G \in \mathcal{G}} \mathbb{E}[X|G = G_T]\mathbb{P}[G = G_T], \tag{14}$$

where X is the random variable representing the proportion of the population infected, $\mathcal{G}$ is the set of resolved graphs, and $\mathbb{P}[G = G_T]$ is the mean scaled silhouette score associated with resolved graph G. $\mathbb{E}[X|G = G_T]$ is estimated by taking the sample mean of the proportion of the population infected over the 1000 simulations on resolved graph G. Note that $\mathbb{E}[X|G = G_T]$ is a random variable over the events $G = G_T$.

5.4   Results

Applying Eq. 14 gives an expected population infected of 0.850. When comparing to the most likely graph, $\mathbb{E}[X|G_L = G_T] = 0.847$. Hence, the overall expected behaviour of the model on the most likely graph does not drastically differ from the expected behaviour across all resolved graphs.

We now consider the distribution of $\mathbb{E}[X|G = G_T]$ over the different graphs. As the events $G = G_T$ are discrete, we illustrate the cumulative distribution function of $\mathbb{E}[X|G = G_T]$. Figure 2 shows this cumulative distribution. For some value x, the cumulative probability is calculated as $\sum_{G:\mathbb{E}[X|G=G_T] \leq x} \mathbb{P}[G = G_T]$.

Also plotted are quantiles and the expectation of X on G_L. The solid line shows $\mathbb{E}[X|G_L = G_T]$, while the dashed lines show values a and b with $\mathbb{P}[X < a|G_L = G_T] = 0.025$ and $\mathbb{P}[X > b|G_L = G_T] = 0.025$, with $a < b$ (the 2.5^{th} and 97.5^{th} percentiles). These two values a and b are also estimated based on the simulations run on graph G_L.

From Fig. 2, we draw the following conclusions. Despite the fact that $\mathbb{E}[X] \approx \mathbb{E}[X|G_L = G_T]$, these values are only at the 40th percentile of $\mathbb{E}[X|G = G_T]$ i.e. with probability 0.6, $\mathbb{E}[X|G = G_T] > \mathbb{E}[X|G_L = G_T]$.

Given $G_L = G_T$, we have that $\mathbb{P}[X > 0.89] \approx 0.025$. However, $\mathbb{P}[\mathbb{E}[X|G = G_T] > 0.89] = 0.2$. In other words, there are states-of-the-world with a joint probability of 0.2 in which the expected proportion of the population infected exceeds the 97.5^{th} percentile of the distribution of the population infected on the most likely graph. Likewise, there are states with a joint probability of approximately 0.175 in which the expected value is below the 2.5^{th} percentile of the distribution on the most likely graph. So there is a total probability of 0.375 that the expected infection rate lies within the extreme 5% of the distribution on the most likely graph.

These extreme differences can be practically significant. A difference in expected population infected of 5% can mean the difference between hospitals being able to handle patients, or an idea driving a change in law.

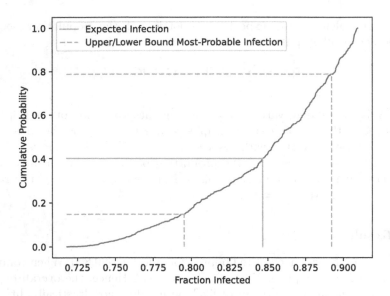

Fig. 2. Cumulative probability of fraction of population infected over different resolved graphs (blue curve); mean and percentiles of fraction of population infected for most likely state-of-the-world (yellow lines). (Color figure online)

6 Conclusions

The ability to combine multiple datasets provides unique possibilities for computational science [21,33]; however, when doing so it becomes important to consider how these combinations can impact downstream results. Traditional data integration focuses on providing one single, best resolved dataset. However, for computational models, this potentially misses a range of (sometimes extreme) model behaviour that is crucial for drawing well-rounded and considered conclusions.

With this in mind, we have proposed *data integration landscapes*, a formulation for describing a set of data integrations and their likelihood on complex networks. Furthermore, we offer a simple, but effective, practical approach for instantiating such a landscape. Experimentally, we show the extra information in diffusion model output that can be gained through implementing said approach.

There are a number of potential avenues for furthering this methodology. In our experiment, we only considered the case of one resolution function. In some cases this may not be sufficient. For example, one of the roles of the resolution function is link prediction, which is a difficult task [24]. Thus, it is likely that using different resolution functions could be crucially important for creating data integration landscapes in different, more complex use cases.

There is also room for more complex approaches to the state-of-the-world probability assignment. Our working example employed simple string distance metrics and complete linkage to define partition goodness. More powerful approaches like supervised learning and latent-representation can also be applied, as long as the likelihood assignment can function as a probability.

Finally, while we have focused primarily on diffusion models on complex networks, this formulation could possibly be extended and applied to different data-dependent computational models.

References

1. Anderson, B.D., Ye, M.: Recent advances in the modelling and analysis of opinion dynamics on influence networks. Int. J. Autom. Comput. **16**(2), 129–149 (2019)
2. Barabási, A.L., Albert, R.: Emergence of scaling in random networks. Science **286**(5439), 509–512 (1999)
3. Barbu, A., Zhu, S.C.: Monte Carlo Methods, vol. 35. Springer, Singapore (2020). https://doi.org/10.1007/978-981-13-2971-5
4. Benjelloun, O., Garcia-Molina, H., Menestrina, D., Su, Q., Whang, S.E., Widom, J.: Swoosh: a generic approach to entity resolution. VLDB J. **18**(1), 255–276 (2009)
5. Bhattacharya, I., Getoor, L.: Entity resolution in graphs. Min. Graph Data **311** (2006)
6. Christen, P.: Febrl- an open source data cleaning, deduplication and record linkage system with a graphical user interface. In: Proceedings of the 14th ACM SIGKDD International Conference on Knowledge Discovery and Data Mining, pp. 1065–1068 (2008)
7. Christen, P., Pudjijono, A.: Accurate synthetic generation of realistic personal information. In: Theeramunkong, T., Kijsirikul, B., Cercone, N., Ho, T.-B. (eds.) PAKDD 2009. LNCS (LNAI), vol. 5476, pp. 507–514. Springer, Heidelberg (2009). https://doi.org/10.1007/978-3-642-01307-2_47
8. Christophides, V., Efthymiou, V., Palpanas, T., Papadakis, G., Stefanidis, K.: An overview of end-to-end entity resolution for big data. ACM Comput. Surv. (CSUR) **53**(6), 1–42 (2020)
9. Dieck, R.H.: Measurement Uncertainty: Methods and Applications. ISA (2007)
10. Dong, X.L., Srivastava, D.: Big data integration. Synth. Lect. Data Manag. **7**(1), 1–198 (2015)
11. Fellegi, I.P., Sunter, A.B.: A theory for record linkage. J. Am. Stat. Assoc. **64**(328), 1183–1210 (1969)
12. Genossar, B., Shraga, R., Gal, A.: FlexER: flexible entity resolution for multiple intents. arXiv preprint arXiv:2209.07569 (2022)
13. Getoor, L., Machanavajjhala, A.: Entity resolution: theory, practice & open challenges. Proc. VLDB Endow. **5**(12), 2018–2019 (2012)
14. Goodwin, G.C., Ninness, B., Salgado, M.E.: Quantification of uncertainty in estimation. In: 1990 American Control Conference, pp. 2400–2405. IEEE (1990)
15. Kermack, W.O., McKendrick, A.G.: A contribution to the mathematical theory of epidemics. Proc. R. Soc. Lond. Ser. A Containing Papers of a Mathematical and Physical Character **115**(772), 700–721 (1927)
16. Kiss, I.Z., Miller, J.C., Simon, P.L., et al.: Mathematics of Epidemics on Networks, vol. 598, p. 31. Springer, Cham (2017)
17. Kolossa, A., Kopp, B.: Data quality over data quantity in computational cognitive neuroscience. Neuroimage **172**, 775–785 (2018)
18. Köpcke, H., Thor, A., Rahm, E.: Evaluation of entity resolution approaches on real-world match problems. Proc. VLDB Endow. **3**(1–2), 484–493 (2010)

19. Lepot, M., Aubin, J.B., Clemens, F.H.: Interpolation in time series: an introductive overview of existing methods, their performance criteria and uncertainty assessment. Water **9**(10), 796 (2017)
20. Levenshtein, V.I., et al.: Binary codes capable of correcting deletions, insertions, and reversals. In: Soviet Physics Doklady, vol. 10, pp. 707–710. Soviet Union (1966)
21. Ley, C., Bordas, S.P.: What makes data science different? A discussion involving statistics 2.0 and computational sciences. Int. J. Data Sci. Anal. **6**, 167–175 (2018)
22. Li, Y., Li, J., Suhara, Y., Wang, J., Hirota, W., Tan, W.C.: Deep entity matching: challenges and opportunities. J. Data Inf. Qual. (JDIQ) **13**(1), 1–17 (2021)
23. López-Pintado, D.: Diffusion in complex social networks. Games Econom. Behav. **62**(2), 573–590 (2008)
24. Lü, L., Zhou, T.: Link prediction in complex networks: a survey. Physica A **390**(6), 1150–1170 (2011)
25. Metropolis, N., Ulam, S.: The Monte Carlo method. J. Am. Stat. Assoc. **44**(247), 335–341 (1949)
26. Murtagh, F., Contreras, P.: Algorithms for hierarchical clustering: an overview. Wiley Interdisc. Rev.: Data Min. Knowl. Discov. **2**(1), 86–97 (2012)
27. Nevin, J.: Data Integration Landscape Naive Implementation. University of Amsterdam, V1 (2023). https://doi.org/10.17632/9jdzy6jr82.1
28. Nevin, J., Lees, M., Groth, P.: The non-linear impact of data handling on network diffusion models. Patterns **2**(12), 100397 (2021)
29. Radosz, W., Doniec, M.: Three-state opinion Q-voter model with bounded confidence. In: Paszynski, M., Kranzlmüller, D., Krzhizhanovskaya, V.V., Dongarra, J.J., Sloot, P.M.A. (eds.) ICCS 2021. LNCS, vol. 12744, pp. 295–301. Springer, Cham (2021). https://doi.org/10.1007/978-3-030-77967-2_24
30. Rainer, H., Krause, U.: Opinion dynamics and bounded confidence: models, analysis and simulation (2002)
31. Rice, E., Holloway, I.W., Barman-Adhikari, A., Fuentes, D., Brown, C.H., Palinkas, L.A.: A mixed methods approach to network data collection. Field Methods **26**(3), 252–268 (2014)
32. Roy, C.J., Oberkampf, W.L.: A comprehensive framework for verification, validation, and uncertainty quantification in scientific computing. Comput. Methods Appl. Mech. Eng. **200**(25–28), 2131–2144 (2011)
33. Rude, U., Willcox, K., McInnes, L.C., Sterck, H.D.: Research and education in computational science and engineering. SIAM Rev. **60**(3), 707–754 (2018)
34. Shahapure, K.R., Nicholas, C.: Cluster quality analysis using silhouette score. In: 2020 IEEE 7th International Conference on Data Science and Advanced Analytics (DSAA), pp. 747–748. IEEE (2020)
35. Smith, R.C.: Uncertainty Quantification: Theory, Implementation, and Applications, vol. 12. SIAM (2013)
36. Sullivan, T.J.: Introduction to Uncertainty Quantification, vol. 63. Springer, Cham (2015)
37. Wit, E., van den Heuvel, E., Romeijn, J.W.: 'All models are wrong...': an introduction to model uncertainty. Statistica Neerlandica **66**(3), 217–236 (2012)

Excessive Internet Use in the Organizational Context: A Proposition of the New Instrument to Measure Cyberloafing at Work

Beata Bajcar 🆔 and Jolanta Babiak(✉) 🆔

Wrocław University of Science and Technology, Wrocław, Poland
{beata.bajcar,jolanta.babiak}@pwr.edu.pl

Abstract. Cyberloafing considered as non-work-related excessive Internet use at work is embedded in everyday's work across organizations. Despite growing concerns about the waste of energy, time, money, and corporate data security caused by cyberloafing, there is still debate about the impact of cyberloafing on key work-related factors of job demands, work performance, satisfaction, and stress. The existing measures of cyberloafing in organizations seem obsolete, so the need to create a tool tailored to changing repertoire of cyberactivities has become warranted. Therefore, we developed and empirically verified a new Cyberloafing Scale, the CBLS-15, to measure four dimensions and a total score of this phenomenon. The CBLS-15 scale includes 15 items grouped into four dimensions 1) *Information browsing* (IB), 2) *Social networking* (SN), 3) *Personal matters* (PM), and 4) *Gambling/Adult content* (GA). In support of the external validity of the CBS-15, we found positive associations of cyberloafing with workload, cognitive demands, role conflict, and stress, and negative relationship with work satisfaction and work performance. The CBLS-15 can be useful for researchers and practitioners as a diagnostic tool. Our results are a valuable contribution to the literature on cyberloafing in modern organizations, providing important insights into how work-related factors may influence non-work Internet use at work.

Keywords: Cyberloafing · Excessive Internet Use · Job Demands · Work Performance · Work Satisfaction · Stress

1 Introduction

The use of Information and Communication Technology for processing data and communicating is currently widespread among employees of any organization in highly digitized workplaces. Computer-human interaction is an everyday experience in most personal and social functions. In some cases, cyber-activity or presence in virtual reality is necessary due to the type of performed work. While organizations have a general expectation that their employees are proficient in the use of modern technology, employees may also be using it for non-work purposes. This behavior, called cyberloafing, may be understood as employees' use of ICT technologies, including advanced devices (e.g., computers, smartphones, nowadays also smartwatches), and the Internet to access various websites, including social media, during work hours for personal or non-work reasons [1, 2].

© The Author(s), under exclusive license to Springer Nature Switzerland AG 2023
J. Mikyška et al. (Eds.): ICCS 2023, LNCS 14073, pp. 509–523, 2023.
https://doi.org/10.1007/978-3-031-35995-8_36

Cyberloafing, interchangeably called cyberslacking is said to have a dual or even complex nature [3, 4]. It can contribute to negative effects, such as frequent distractions at work [5], poorer performance [6], and time-wasting [7, 8]. However, other evidence suggests that cyberloafing, by providing a little breather from everyday work pressure, has a positive impact on employee well-being in terms of relief from work stress [9]. Differences in research findings may stem from various sources that influence employees' coping behavior, which nowadays often has a form of cyberloafing [10].

1.1 Dimensionality of Cyberloafing

Without deciding on the positive or negative impact of cyberloafing for organizations or employees, one of the key issues is the ability to assess cyberloafing with an up-to-date tool, adapted to current organizational conditions. There has been a considerable scientific effort to develop tools to measure some of the non-work-related online activities including the use of e-mail for personal purposes, surfing information websites, participating in social media platforms, blogging, watching online videos, gambling, and others. These activities were generally reflected in subdimensions of the self-reported questionnaires [1, 5, 11–22] (see Table 1). However, there were few, if any, replication studies that supported the validity of existing scales, thus rendering their current usefulness limited.

Table 1. Dimensions used in cyberloafing measures.

Authors	Scale dimensions
Lim [1], Lim, Teo [11]	1) browsing activities; 2) emailing activities
Mahatanankoon et al. [12]	1) e-commerce; 2) information research; 3) interpersonal communication
Anandarajan et al. [13]	1) disruptive; 2) recreational; 3) personal learning; 4) ambiguous
Blau et al. [14]	1) non-work-related e-mail; 2) browsing-related; 3) interactive
Mastrangelo et al. [5]	1) non-productive; 2) counterproductive
Blanchard, Henle [15]	1) major activities; 2) minor activities
Coker [16]	1) workplace Internet leisure
Anandarajan et al. [17]	1) hedonic; 2) self-development; 3) citizenship; 4) work/family behavior
Vitak et al. [18]	1) cyberslacking variety; 2) cyberslacking communication frequency
Aghaz, Sheikh [19]	1) activities: social, informational, leisure virtual; 2) behaviors: learning, recovery, deviant, addiction
Akbulut et al. [20], Koay[21], Şahin [22]	1) sharing; 2) shopping; 3) real-time updating; 4) accessing online content; 5) gambling

The existing tools have some noticeable limitations. One of the early scales developed by Lim [1] measures general Internet browsing and e-mailing but omitted other diversified Internet activities. Subsequently developed a 9-item scale by Vitak et al. [18] is quite short and does not capture important categories of Internet activities, such as online auctions, booking vacations, or visiting adult websites. Similar shortcomings apply to a questionnaire developed by Mahatanankoon et al. [12] where recent trends in cyberbehavior, such as social media usage, blogging, etc., were lacking. Other questionnaires

[12, 15, 16, 19] are relatively long and quite onerous to complete. For example, Akbulut et al. [20], Anandarajan et al. [17, 19], and Mastrangelo et al. [5], created comprehensive albeit extra-long instruments containing 30 to 41 items. Cocker [16] developed a 17-item unidimensional scale, which measures the frequency and duration of key areas of cyberloafing. However, this study was conducted in a relatively small sample (n = 268), in which 74% of the participants were women. A study by Blau et al. [14] was also conducted on a rather small sample size and it relied on an older Lim's scale with 6 new items developed by authors. As many scholars noted, older cyberloafing scales are obsolete and do not correspond to modern online activities and cyberloafing behavior [3, 23]. Thus, it seems justified to construct a new up-to date cyberloafing measurement tool, accounting for currently dominating types of online activity. Therefore, we proposed a new Cyberloafing Scale-15 (CBLS-15), usefull for researchers to measure cyberloafing in different organizational context.

1.2 Antecedents and Consequences of Cyberloafing

There is a great deal of research on what motivates employees to engage in cyberloafing. Some of the key organizational predictors include employee workload [24–26], role conflict, or role ambiguity [24, 26, 27]. According to job demands-resources theory [28], employee workload (i.e., one's perceived volume of work), cognitive demands (i.e., work difficulty and complexity; intellectual strain), and role conflict (incompatibility in terms of responsibility) are considered major organizational stressors that can hinder performance. In this context, studies revealed that workload and cognitive demands [4] are significant predictors of high levels of stress at work and excessive Internet use [26, 29]. On the contrary other findings have shown that high job demands can reduce cyberloafing [30] or that low levels of workload encourage cyberloafing [10, 31]. Some researchers suggest that harmful job demands contribute to employees' experience of stress and emotional exhaustion at work and can lead to cyberloafing. Employees may choose to cyberloaf as a way to calm down, de-stress, recover and replenish personal resources. According to Hobfoll's Conservation of Resources theory [32], when employees cannot cope with demands at work due to limited resources, such as energy, concentration, and time, then they experience stress. Consequently, they are more likely to engage in cyberloafing to escape the strain, find relief and maintain respite and positive affect. In this way, they prevent further loss of resources and possibly regain the ones lost [33–35]. So, cyberloafing can be considered a stress-reducing mechanism when workload becomes distressing [4]. However, this strategy may prove to be short-lived [36]. Cyberloafing itself can become a source of stress and frustration due to work distraction, loss of attentional resources, and untimely completion of tasks [7, 36]. This can create a downward spiral of further resource loss. It is still not clear if cyberloafing is a resource gain or is it rather a drain of resources. Latest reviews of studies showed that among external antecedents of cyberloafing were work-related, supervision-related, and organization-related factors [3, 37] (e.g., tenure, organizational position, income, leadership, meaningfulness of work, autonomy, Internet control policy, organizational norms).

It has become clear that cyberloafing is an expensive problem. It can cost organizations as much as $85 billion a year [38]. This is most likely due to the disturbed use of the company's resources e.g., time, computers, and smartphones, as well individual resources, e.g., focus, concentration, or emotions. There are other negative consequences of cyberloafing that organizations already recognized, such as online gambling or increasing the risk of disclosing insight data to third parties [4]. Cyberloafing thus appears as one of counterproductive work behaviors [1, 2, 11, 20], related to withdrawing behaviors [2, 37]. Recent meta-analysis supported positive relationship between cyberloafing and counterproductive work behavior [39].

Some scholars observed a negative relationship between cyberloafing and work performance [30, 40], productivity [41], and effectiveness [12, 42]. Other findings however suggest a positive effect of cyberloafing on work performance as well as employee creativity, acquisition of new skills, employee interest in work, regain of attention span, feeling of enthusiasm and satisfaction, and in consequence higher productivity at work [16, 43–45]. Coker [16] found that work productivity is positively related to leisure Internet browsing at work when it does not exceed around 12 percent of work time. Furthermore, the relationship between cyberloafing and work performance can take on an inverted U-shaped relationship [46], especially when cyberloafing occurs in less frequent and shorter episodes [16]. As research shows, cyberloafing can produce both detrimental or beneficial outcomes [3, 23]. The relationship between cyberloafing and work performance requires further investigation.

No less intriguing is the relationship between employee work satisfaction and cyberloafing. Satisfied employees are most of the time motivated and their performance is high [47]. But when it comes to cyberloafing as a work behavior, the picture is not that clear anymore [3, 37]. Cyberloafing can act as a moderator in the relationship between physical workplace aggression and satisfaction [9] and serve as a stress coping against workplace strain, thereby increasing work satisfaction. Mohammad et al. [43] revealed that cyberloafing positively predicts work satisfaction which subsequently leads to higher work productivity. Similarly, other authors emphasized a positive relationship between cyberloafing and work satisfaction [43, 48–50], or non-significant relationships between cyberloafing and work satisfaction [5, 10, 12, 18, 39, 51]. There may be many reasons why cyberloafing enhances work satisfaction, but one thing is clear; people like to spend time on the Internet while at work. Still fairly few findings revealed a negative association between cyberloafing and work satisfaction [52, 53]. Presumably, dissatisfied employees use the Internet to distract themselves from negative work-related emotions. In this situation, cyberloafing acts as a substitute for working, filling up time that should be spent on completing organizational tasks.

To summarise, there is no consensus among researchers and practitioners about the detrimental or beneficial impact that cyberloafing may have on employees' well-being and performance at work [54]. It is therefore very important to investigate the relationship between cyberloafing and basic work-related factors that may either contribute to or result from cyberloafing. In this study, we therefore seek to gain a deeper understanding on the dimensionality of cyberloafing and the relationship between cyberloafing and work-related characteristics.

Based on the evaluation of a wide range of contemporary cyberloafing behaviors in the workplace, our main goal was to develop a valid and reliable scale to measure cyberloafing (CBLS-15) and investigate its psychometric properties. To test the external validity of the CBLS-15, we have also examined the relationship between cyberloafing and work-related factors categorized as predictors, such as workload, cognitive demands, role conflict, and consequences, such as work satisfaction, stress, and work performance.

2   Method

2.1   Participants and Procedure

Data were collected from three independent samples via online survey using the snowball procedure. The total sample consisted of 1824 participants from Poland (846 males), working in various industries and professions, such as salespersons, IT professionals, educators, medical staff, lawyers, administrative, service workers, etc.). Working with open access to the Internet was an inclusion criterion. Sample 1 consisted of 853 participants (385 males), aged from 18 to 79 years ($M = 35.0$, $SD = 11.4$). Sample 2 comprised 703 employees (334 males), aged between 19 and 77 years ($M = 35.2$, $SD = 11.3$). An average work tenure was 12.3 ($SD = 10.0$) years in sample 1 and 12.4 ($SD = 9.8$) years in sample 2. In sample 3 there took part 268 employees (127 males), aged from 18 to 74 years ($M = 34.0$, $SD = 12.1$), and with average work tenure of 12.2 years ($SD = 11.5$). Participation in the study was anonymous and voluntary. Respondents provided informed consent before they completed all self-reported measures. All study procedures have been approved by the Ethics Committee of Scientific Research at Wrocław University of Science and Technology.

2.2   Measures

Cyberloafing. To measure cyberloafing behavior we used *Cyberloafing Scale* (CBLS-15), which we developed and empirically verified. The final version of the instrument consisted of four dimensions representing various non-work-related online activities, in which employees engage at work: (1) *Information browsing* (IB) reflecting browsing content related to information and interests; (2) *Social networking* (SN) – active participation in social media; (3) *Personal matters* (PM) – handling e-mails and personal matters; (4) *Gambling/Adult content* (GA) – visiting adult or gambling sites. The cyberloafing level was also considered as a total score (CBL). All 15 items were rated on a 5-point scale (1 – never; 5 – very often).

Job Demands. We measured three workplace characteristics: (1) workload, (2) cognitive demands, and (3) role conflict, using three 4-item subscales (quantitative demands, cognitive demands, role conflict) of the Copenhagen Psychosocial Questionnaire (COPSOQ II) [55]. Workload (measured by the quantitative demands subscale) relates to a high volume of work and demands; cognitive demands relate to feeling overwhelmed and cognitively loaded, and role conflict refers to the degree of incompatibility between the requirements and objectives set by management and the employee's feeling that certain work activities are performed unnecessarily. Workload and cognitive demands were rated on a 5-point scale (1 – never; 5 – always). Role conflict was rated on a 5-point scale (1 – to a very small extent; 5 – to a very large extent). The reliability of the measures in this study was satisfactory (Cronbach $\alpha = .70 - .81$).

Stress at Work. Stress at work was measured using 4 items of the *stress* subscale of the COPSOQ II [55], rated on the scale (1 – at all to 5 - all the times). Cronbach's α of this measure was at the level of .77.

Work Performance and Work Satisfaction. We measured work performance and work satisfaction, using one item per each variable, i.e. "How satisfied are you with your work?", "How effective are you at your work?". Participants responded on a 5-point scale (1 - to a very small extent; 5 - to a very large extent).

3   Results

In our study, we verified the factorial validity and reliability of a new CBLS-15 instrument to measure cyberloafing, individual differences, and relationships of cyberloafing dimensions with work-related characteristics. All statistical analyses were performed using the IBM SPSS-22 and IBM AMOS-22 software.

3.1   Factor Structure and Reliability of the CBLS-15 Measure

First, in sample 1, we performed the exploratory factor analysis on the initial set of 27 items (see Table 2). Based on the principal component method, Cattell criterion extraction, and Promax rotation, a 4-factor solution was obtained, which explained 60% of the variance. All factor loadings ranged between .41 and .87. However, for the sake of parsimony, we included 4 items with the highest loadings in each factor for further analysis (above .50) (see Table 2).

The first factor explained a relatively high level of variance (39%), indicating the potential unidimensionality of the cyberloafing. We thus decided to test the 1-factor model. Therefore, in the confirmatory factor analysis (using the AMOS 27.0 software) we verified both, 1-factor, and 4-factor models in samples 1 and 2 (Table 3). Following Byrne's recommendations [56], all goodness of fit indices indicated a very good fit of the measurement models; χ^2/df index was lower than 5, the root mean square error of approximation (RMSEA) and standardized root mean square residual (SRMR) did not exceed .08; the adjusted goodness of fit index (AGFI), comparative fit index (CFI), and Tucker-Lewis fit index (TLI) values were higher than .90.

The results of these analyses showed that 4-factor, as well as 1-factor solutions, were well fitted to the data (Table 3). There were no substantive changes in the model's goodness of fit indices between samples 1 and 2 (ΔCFI < 0.010 and ΔRMSEA < 0.015), indicating the invariance of the tested factor solutions. This may suggest that the same factor model could be applied across groups, and that the same item parameters (i.e., factor loadings and intercepts) could be constrained to the same value between samples (Table 3). However, because both, 1- and 4-factor models had an acceptable fit [56], we included the four subscales of the CBLS-15 and a total score in further analyses.

Next, we calculated the reliability indicators for the CBLS-15 in terms of internal consistency, discriminatory power, and temporal stability (Table 4). In samples 1 and 2, the internal consistency (Cronbach's α) was at an excellent or high level, except for the GA subscale (a moderate level). The discriminatory power was represented by three coefficients (mean inter-item correlation, item-total correlation, and intra-class

Table 2. Factor loadings of the CBLS-15 items in the four-factor model.

Item content	Sample1 $N = 853$				Sample 2 $N = 703$			
	1	2	3	4	1	2	3	4
1. I browse pages with offers for purchases, even if I'm not looking for anything specific	.87				.63			
7. I search for various sales or deals online	.86				.74			
6. I check the latest news	.79				.55			
13. I browse sites of interest, e.g., health, culinary, travel	.75				.84			
15. I spend time on social media (Facebook, Instagram, TikTok, etc.) without specific intent		.82				.88		
11. I chat with my family and friends using various chat apps (e.g., Messenger, Snapchat, Instagram)		.73				.64		
2. I respond (e.g., "like") to posts or tweets, immediately after receiving a notification		.70				.72		
3. I listen to music/podcasts/streams/lives on the Internet		.50				.56		
4. I shop online			.87				.62	
12. I handle various matters via the Internet (e.g., online banking, medical appointments, making reservations, etc.)			.74				.73	
10. I send and receive personal e-mails			.68				.56	
14. I download programs and apps for my use			.55				.67	
9. I visit betting sites (e.g., sports, lotteries, casinos)				.77				.53
5. I play games online				.63				.63
8. I view adult content online				.62				.52

Table 3. Confirmatory factor analysis of the CBLS-15 measurement models.

Model	χ^2/df	RMSEA	SRMR	AGFI	TLI	CFI	ΔCFI	ΔRMSEA
Sample 1 (N = 853)								
1-factor model	4.00	.059	.036	.92	.96	.97		
4-factor model	3.02	.049	.037	.95	.96	.98	.013	.010
Sample 2 (N = 703)								
1-factor model	3.23	.056	.036	.93	.96	.97		
4-factor model	2.34	.044	.028	.95	.97	.98	.010	.013

correlation) and achieved also satisfactory level. The mean inter-item correlations ranged from .35 to .60, the intra-class correlation (ICC) values ranged from .84 to .93, whereas the item-total correlations were between .30 and .74.

To verify the temporal stability of the CBLS-15 measure, the test-retest procedure was performed. In sample 3 ($n = 268$), a double-measures of cyberloafing were administered at a 4-week interval. The correlations between the first and second measure of cyberloafing subscales and total scores were significant and high (see Table 4), indicating a satisfactory level of stability of the cyberloafing measure over time. As depicted in Table 5, moderate to high correlations of the four CBLS-15 factors with cyberloafing

Table 4. Reliability parameters of the CBLS-15 dimensions.

Cyberloafing dimension		Sample 1[a]				Sample 2[b]				Sample 3[c]
		α	Mean IIC	ICC	Itemtotal	α	Mean IIC	ICC	Item-total	r_{tt}
1. Information browsing	IB	.86	.60	.86	.61–.74	.82	.52	.82	.56–.72	.63[**]
2. Social networking	SN	.73	.42	.73	.33–.63	.75	.44	.74	.37–.69	.67[**]
3. Personal matters	PM	.81	.51	.81	.50–.72	.79	.47	.81	.51–.63	.66[**]
4. Gambling/Adult content	GA	.68	.44	.65	.46–.58	.64	.40	.64	41–.50	.69[**]
5. Cyberloafing total	CBL	.90	.36	.90	.30–.72	.89	.35	.89	.32–.72	.66[**]

Note. [a] $N = 853$, [b] $N = 703$, [c] $N = 268$. α – Cronbach's reliability coefficient, mean IIC – mean inter-item correlations; ICC – intra-class correlation, ITC – item-total correlation, r_{tt} - test-retest correlation

total score were revealed in samples 1 and 2. The intercorrelations among CBLS-15 subscales were small to moderate in samples 1 and 2.

Table 5. Means, standard deviations, and correlations between CBLS-15 subscales.

Cyberloafing dimension		Sample 1[a]		Sample 2[b]						
		M	SD	M	SD	1	2	3	4	5
1. Information browsing	IB	2.59	1.20	2.42	.97	1	.59[**]	.74[**]	.39[**]	.88[**]
2. Social networking	SN	2.94	1.21	2.74	.13	.60[**]	1	.61[**]	.31[**]	.84[**]
3. Personal matters	PM	2.36	1.07	2.38	.91	.70[**]	.63[**]	1	.36[**]	.87[**]
4. Gambling/Adult content	GA	1.25	.60	1.36	.64	.38[**]	.34[**]	.37[**]	1	.54[**]
5. Cyberloafing total	CBL	2.35	.86	2.28	.76	.88[**]	.85[**]	.87[**]	.52[**]	1

Note. [*]$p < .05$, [**]$p < .01$. Correlations below the diagonal refer to sample 1 and above the diagonal to sample 2

3.2 Cyberloafing and Work-Related Characteristics

A one-way analysis of variance (ANOVA) revealed no significant differences between women and men (Table 6) on the IB, SN, and PM dimensions, but significant on the GA subscale and for the CBLS total score. This means that men displayed higher level of general cyberloafing and gambling than women did.

Additionally, there were significant differences between cyberloafing dimensions and the total score in dependence on Internet access restrictions (Table 6). Employees with low restrictions on accessing the Internet were more likely to cyberloaf (specifically, concerning the activities of: IB, SN, PM, and CBL total score) than employees with high restrictions on accessing the Internet. Only the GA subscale was independent of restricted Internet access in the organization.

Having inspected the intercorrelation matrix among cyberloafing we noticed that except for the PM dimension, all other dimensions of cyberloafing (as a total score and the four factors) correlated significantly with sociodemographic factors. As presented in Table 7, the level of correlations ranges from low to moderate. Three cyberloafing dimensions (IB, PM, GA) correlated significantly but weakly with age and job tenure

Table 6. Anova results for CBLS-15 dimensions according to gender and restricted Internet access.

Cyberloafing dimension	Women		Men		F	Partial η^2	Internet not restricted		Internet restricted		F	Partial η^2
	M	SD	M	SD			M	SD	M	SD		
1. Information browsing	2.39	1.02	2.46	.92	.80	.001	2.43	.88	2.25	.91	5.47*	.008
2. Social networking	2.70	1.17	2.78	1.09	.84	.001	2.73	.95	2.39	.99	17.46***	.024
3. Personal matters	2.33	.93	2.44	.88	2.81	.004	2.39	.82	2.21	.82	6.27*	.009
4. Gambling/Adult content	1.22	.52	1.51	.71	40.52***	.055	1.33	.56	1.37	.67	.49*	.001
5. Cyberloafing total	2.22	.79	2.35	.73	5.01*	.007	2.28	.66	2.10	.73	9.36**	.013

Note. $^*p < .05$, $^{**}p < .01$, $^{***}p < .001$

Table 7. Means, standard deviations, intercorrelations between study variables.

Cyberloafing dimension	Age	Work tenure	Work load	Cognitive demands	Role conflict	Stress	Work satisfaction	Work performance
1. Information browsing (IB)	−.20**	−.14**	.16**	.14**	.15**	.14**	−.18**	−.10**
2. Social networking (SN)	−.44**	−.40**	.17**	.09**	.16**	.11*	−.17**	−.10**
3. Personal matters (PM)	−.19**	−.14**	.19**	.07*	.12**	.12**	−.15**	−.03
4. Gambling/Adult content (GA)	−.18**	−.19**	.09**	.12**	.15**	.09**	−.13**	−.14**
5. Cyberloafing total (CBL)	−.32**	−.32**	.17**	.13**	.17**	.14**	−.20**	−.11**

Note. $N = 703$. $^*p < .05$, $^{**}p < .01$. IB - Information searching and browsing, SN - Social Networking Site, PM - Personal matters, GA - Gambling/Adult content, CBL - Cyberloafing total.

but SN dimension and total score correlated negatively and moderately, highlighting that younger employees and employees with less work experience use the Internet extensively (especially in terms of SN). All dimensions and the total score of cyberloafing correlated significantly and weakly with high workload, high cognitive demands, and high role conflict. Cyberloafing as a total score was significantly negatively related to work-related outcomes i.e., work satisfaction, stress, and slightly to work performance. All CBLS-15 dimensions and the total score were weakly negatively correlated with work performance, except for PM dimension. Slightly higher correlation values were between cyberloafing dimensions and the total score and work satisfaction.

4 Discussion

In this paper we developed a new Cyberloafing Scale (CBLS-15) that complements previous questionnaires [1, 12–18, 20]. We have identified and validated a one-factor and a four-factor structure of the CBLS-15. The four factors reflect multidimensional nature of cyberloafing and represent specific online activity, such as information browsing, social networking, personal matters, and gambling/adult content. The one-factor model denotes a general cyberloafing measure as an aggregate score. Our findings also revealed significant positive relationships between cyberloafing (i.e., total score and the four dimensions) and workload, cognitive demands at work, role conflict, as well as stress, supporting the external validity of the CBLS-15 measure. These results are consistent with previous studies showing that excessive workload [24–26], cognitive difficulty at work [26, 29], and experiencing role conflict [24, 26, 27] can predict cyberloafing. Work demands can be stressful and can lead to behaviors associated with withdrawal or escape from work threats and engaging in more pleasant activities, such as cyberloafing. Similarly to previous studies [24, 26, 29], we have found that cyberloafing was related to high work stress. Probably, employees who experience stress at work may intensively engage in non-work Internet use to avoid or reduce negative emotions and facilitate mental recovery in the workplace [9, 19, 24, 26, 36].

Our findings indicated that all dimensions and a total score of CBLS-15 were negatively related to work outcomes such as work satisfaction and performance. This is consistent with previous studies on the link between low satisfaction and non-work use of the Internet [52, 53]. However, it does not correspond with other findings that showed positive associations between these variables [43, 48–50], or non-significant relationships between cyberloafing and work satisfaction [5, 10, 12, 18, 39, 51]. Still, our results suggest that employees are more likely to use the Internet for various purposes such as mindless browsing, online shopping, engaging in communication on social media, or even gambling, when work satisfaction is low.

Employees' work performance, an important work outcome, was negatively related to cyberloafing total score and IB, SN, and GA dimensions. This effect corroborates other results showing that cyberloafing negatively relates to work performance [6, 12, 40–42]. So, cyberloafing appears to be a modern manifestation of counterproductive withdrawal behavior at work [1, 2, 11, 20]. This can be a potentially worrying issue, as the availability of Internet access is virtually unlimited if employees are not supervised or if they do not adhere to social norms of civic behavior at work. However, the relationship between cyberloafing and work performance is not fully understood as yet [3, 23, 46, 57]; it is probably moderated by different conditions such as cognitive load, level of arousal, timing, and frequency of cyberloafing [57]. Although some scientific work has emphasized the benefits of cyberloafing for the well-being of employees, such as relaxation, a pleasant break, and a stress-reducing technique [9, 24, 35] it can still drain personal resources and reduce task performance [23, 34, 36].

Our results also showed that in the absence of restrictions on the Internet use, cyberloafing behavior is significantly higher than when restrictions are imposed. The greatest difference related to the use of social media. This is a sign of the times, as social networking can be considered one of the most favorite activity for 95.2% of Internet users [58]. Finally, similar to previous studies [3, 37], both age and job tenure positively related to cyberloafing. This result indicated that the younger generation of employees is accustomed to using technology much more often than older people and they are highly computer-savvy. This is a great human potential that should be managed with equal skill.

4.1 Practical Implications

The following practical implications refer to work-related factors that have been found significantly related to cyberloafing in this study. This research provides useful insights for managers to better understand cyberloafing and its relationship with some performance criteria. When confronted with excessive work demands employees may become distressed and resort to cyberloafing as a coping mechanism. Conversely, when their job demands are balanced, they tend to feel more satisfied and calmer. Therefore, it is essential that managers recognize their employees' abilities, assign tasks appropriately, and monitor their progress regularly. Sometimes employees' cyberloaf because of their habit of regularly checking the Internet for various purposes. At the same time, they may be unaware that such uncontrolled distractions extend time to complete tasks, which potentially produces frustration [34]. Providing specific training to help employees understand that cyberloafing can have detrimental effect on their level of effectiveness and productivity could prove to be resourceful. A major cause of stress and frustration, often unrecognized by management, is role conflict, or a situation where an employee cannot figure out what is expected of them. Managers shall ensure that tasks are clearly communicated, do not overlap, and that employees understand their responsibilities [3]. After all, work performance is a function of well-designed, appropriately delegated, and well-understood tasks, all of which impact work satisfaction. Additionally, managers shall demonstrate responsible Internet use themselves and set the standard for fairness and appropriate workplace conduct. Internet use cannot be avoided or fully prohibited, but company policy shall make it explicit to what extent Internet use is or is not acceptable. Communicating expectations in this regard will help employees follow company regulations. Lastly, implementing and using monitoring tools (e.g., dedicated software that tracks Internet usage and marks inappropriate activities) may be needed. This could enable management to identify problem areas that shall be addressed.

4.2 Limitations and Future Directions

Our study has some limitations that can be resolved in future work. First, this study is cross-sectional and all data were self-reported. Thus, we cannot draw any conclusions on a causative effect between variables. To reduce the bias of false responses and social desirability or memory recall issues, future research on the CBLS-15 validity shall include objective measures, like computer usage logs or supervisor evaluations. Second, this study was conducted in Poland, thus the CBLS-15 shall be tested in other countries to increase the generalizability of the findings or to identify specific cyberloafing behaviors in other cultural contexts. Finally, future studies can extend our findings by verifying the validity of the CBLS-15 and using other individual predictors of cyberloafing, such as temperament, personality, personal values, ethics, psychological detachment, the need for recovery, and trust in management. In addition, a longitudinal design and a more advanced analytical approach could establish more complex or causal relationships between variables.

5 Conclusions

To conclude, in this study we proposed a new Cyberloafing Scale-15 (CBLS-15) to measure currently prevailing aspects of cyberloafing in the organizational context. The CBLS-15 can be useful for researchers and for practitioners as a diagnostic tool. Our findings provide important insights into how work-related factors can affect non-work-related Internet use in the workplace, making valuable contribution to the literature on cyberloafing in modern organizations. The strength of our study also lies in its diverse participant pool and gender balance, increasing the external validity of the findings. The study highlights the importance of assessing and monitoring various cyberloafing behaviors in the organization to better cope and prevent problematic cyberbehaviors at work. Our study is an initial step in the empirical verification of a new measure CBCS-15, which will be continued in the further studies.

References

1. Lim, V.K.G.: The IT way of loafing on the job: cyberloafing, neutralizing and organizational justice. J. Organ. Behav. **23**(5), 675–694 (2002)
2. Askew, K., Buckner, J.E., Taing, M.U., Ilie, A., Bauer, J.A., Coovert, M.D.: Explaining cyberloafing: the role of the theory of planned behavior. Comput. Hum. Behav. **36**, 510–519 (2014)
3. Tandon, A., Kaur, P., Ruparel, N., Islam, J.U., Dhir, A.: Cyberloafing and cyberslacking in the workplace: systematic literature review of past achievements and future promises. Internet Res. **32**(1), 55–89 (2022)
4. Lim, V.K.G., Teo, T.S.H.: Cyberloafing: a review and research agenda. Appl. Psychol. (2022)
5. Mastrangelo, P.M., Everton, W., Jolton, J.A.: Personal use of work computers: distraction versus destruction. Cyberpsychol. Behav. **9**(6), 730–741 (2006)
6. Andreassen, C.S., Torsheim, T., Pallesen, S.: Use of online social network sites for personal purposes at work: does it impair self-reported performance? Compr. Psychol. **3**(1), Article 18 (2014)
7. Kim, S., Christensen, A.L.: The dark and bright sides of personal use of technology at work: a job demands–resources model. Hum. Resour. Dev. Rev. **16**(4), 425–447 (2017)
8. Conner, C.: Wasting time at work: the epidemic continues. Forbes, 31 July 2015. https://www.forbes.com
9. Andel, S.A., Kessler, S.R., Pindek, S., Kleinman, G., Spector, P.E.: Is cyberloafing more complex than we originally thought? Cyberloafing as a coping response to workplace aggression exposure. Comput. Hum. Behav. **101**, 124–130 (2019)
10. Giordano, C., Mercado, B.K.: Cyberloafing: Investigating the importance and implications of new and known predictors. Collabra Psychol. **9**(1), 1–18 (2023)
11. Lim, V.K., Teo, T.S.: Prevalence, perceived seriousness, justification and regulation of cyberloafing in Singapore. Inf. Manag. **42**(8), 1081–1093 (2005)
12. Mahatanankoon, P., Anandarajan, M., Igbaria, M.: Development of a measure of personal web usage in the workplace. Cyberpsychol. Behav. **7**(1), 93–104 (2004)
13. Anandarajan, M., Devine, P., Simmers, C.A.: A multidimensional sealing approach to personal web usage in the workplace. In: Anandarajan, M., Simmers, C.A. (eds.) Personal Web Usage in the Workplace: A Guide to Effective Human Resources Management, pp. 61–79. Information Science Publishing, Hershey, PA (2004)

14. Blau, G., Yang, Y., Ward-Cook, K.: Testing a measure of cyberloafing. J. Allied Health **35**(1), 9–17 (2006)
15. Blanchard, A.L., Henle, C.A.: Correlates of different forms of cyberloafing: the role of norms and external locus of control. Comput. Hum. Behav. **24**(3), 1067–1084 (2008)
16. Coker, B.L.: Freedom to surf: the positive effects of workplace Internet leisure browsing. N. Technol. Work. Employ. **26**(3), 238–247 (2011)
17. Anandarajan, M., Simmers, C.A., D'Ovidio, R.: Exploring the underlying structure of personal web usage in the workplace. Cyberpsychol. Behav. Soc. Netw. **14**(10), 577–583 (2011)
18. Vitak, J., Crouse, J., LaRose, R.: Personal Internet use at work: understanding cyberslacking. Comput. Hum. Behav. **27**(5), 1751–1759 (2011)
19. Aghaz, A., Sheikh, A.: Cyberloafing and job burnout: an investigation in the knowledge-intensive sector. Comput. Hum. Behav. **62**, 51–60 (2016)
20. Akbulut, Y., Dursun, Ö.Ö., Dönmez, O., Şahin, Y.L.: In search of a measure to investigate cyberloafing in educational settings. Comput. Hum. Behav. **55**, 616–625 (2016)
21. Koay, K.Y.: Assessing cyberloafing behaviour among university students: a validation of the cyberloafing scale. Pertanika J. Soc. Sci. Humanit. **26**(1), 409–424 (2018)
22. Şahin, M.D.: Effect of item order on certain psychometric properties: a demonstration on a cyberloafing scale. Front. Psychol. **12**, 590545 (2021)
23. Koay, K.-Y., Soh, P.C.-H.: Does cyberloafing really harm employees' work performance?: An overview. In: Xu, J., Cooke, F.L., Gen, M., Ahmed, S.E. (eds.) Proceedings of the Twelfth International Conference on Management Science and Engineering Management. Lecture Notes on Multidisciplinary Industrial Engineering, pp. 901–912. Springer International Publishing, Cham (2018)
24. Henle, C.A., Blanchard, A.L.: The interaction of work stressors and organizational sanctions on cyberloafing. J. Manag. Issues **20**(3), 383–400 (2008)
25. Hensel, P.G., Kacprzak, A.: Job overload, organizational commitment, and motivation as antecedents of cyberloafing: evidence from employee monitoring software. Eur. Manag. Rev. **17**(4), 931–942 (2020)
26. Elrehail, H., Rehman, S.U., Chaudhry, N.I., Alzghoul, A.: Nexus among cyberloafing behavior, job demands and job resources: a mediated-moderated model. Educ. Inf. Technol. **26**(4), 4731–4749 (2021)
27. Varghese, L., Barber, L.K.: A preliminary study exploring moderating effects of role stressors on the relationship between big five personality traits and workplace cyberloafing. Cyberpsychology **11**(4), Article 4 (2017)
28. Bakker, A.B., Demerouti, E.: The job demands-resources model: state of the art. J. Manag. Psychol. **22**(3), 309–328 (2007)
29. Bajcar, B., Babiak, J.: Job characteristics and cyberloafing among Polish IT professionals: mediating role of work stress. In: Soliman, K.S. (ed.) Proceedings of the 36th International Business Information Management Association Conference, pp. 6565–6578. (IBIMA), King of Prussia, PA (2020)
30. Andreassen, C.S., Torsheim, T., Pallesen, S.: Predictors of use of social network sites at work - a specific type of cyberloafing. J. Comput.-Mediat. Commun. **19**(4), 906–921 (2014)
31. Pindek, S., Krajcevska, A., Spector, P.E.: Cyberloafing as a coping mechanism: dealing with workplace boredom. Comput. Hum. Behav. **86**, 147–152 (2018)
32. Hobfoll, S.E.: Conservation of resources. A new attempt at conceptualizing stress. Am. Psychol. **44**(3), 513–524 (1989)
33. Lim, P.K., Koay, K.Y., Chong, W.Y.: The effects of abusive supervision, emotional exhaustion and organizational commitment on cyberloafing: a moderated-mediation examination. Internet Res. **31**(2), 497–518 (2021)

34. Lim, V.K., Chen, D.J.: Cyberloafing at the workplace: gain or drain on work? Behav. Inf. Technol. **31**(4), 343–353 (2012)
35. Zhu, J., Wei, H., Li, H., Osburn, H.: The paradoxical effect of responsible leadership on employee cyberloafing: a moderated mediation model. Hum. Resour. Dev. Q. **32**(4), 597–624 (2021)
36. Koay, K.Y., Soh, P.C.-H., Chew, K.W.: Antecedents and consequences of cyberloafing: evidence from the Malaysian ICT industry. First Monday **22**(3), 1–16 (2017)
37. Weissenfeld, K., Abramova, O., Krasnova, H. (eds.): Antecedents for cyberloafing – a literature review. In: 14th International Conference on Wirtschaftsinformatik, pp. 1687–1701, Siegen (2019)
38. Zakrzewski, C.: The key to getting workers to stop wasting time online. Wall Street Journal (2016). http://www.wsj.com/articles/the-key-to-getting-workers-to-stop-wasting-time-online-1457921545
39. Mercado, B.K., Giordano, C., Dilchert, S.: A meta-analytic investigation of cyberloafing. Career Dev. Int. **22**(5), 546–564 (2017)
40. Askew, K.: The relationship between cyberloafing and task performance and an examination of the theory of planned behavior as a model of cyberloafing. Unpublished doctoral dissertation, University of South Florida, Tampa (2012)
41. Bock, G.-W., Ho, S.L.: Non-work related computing (NWRC). Commun. ACM **52**(4), 124–128 (2009)
42. Ramayah, T.: Personal web usage and work inefficiency. Bus. Strategy Ser. **11**(5), 295–301 (2010)
43. Mohammad, J., Quoquab, F., Halimah, S., Thurasamy, R.: Workplace Internet leisure and employees' productivity. Internet Res. **29**(4), 725–748 (2019)
44. Coker, B.L.S.: Workplace Internet leisure browsing. Hum. Perform. **26**(2), 114–125 (2013)
45. Sao, R., Chandak, S., Patel, B., Bhadade, P.: Cyberloafing: effects on employee job performance and behavior. Int. J. Recent Technol. Eng. **8**(5), 1509–1515 (2020)
46. She, Z., Li, Q.: When too little or too much hurts: evidence for a curvilinear relationship between cyberloafing and task performance in public organizations. J. Bus. Ethics **183**(4), 1141–1158 (2023)
47. Judge, T.A., Thoresen, C.J., Bono, J.E., Patton, G.K.: The job satisfaction-job performance relationship: a qualitative and quantitative review. Psychol. Bull. **127**(3), 376–407 (2001)
48. Stanton, J.M.: Company profile of the frequent internet user. Commun. ACM **45**(1), 55–59 (2002)
49. Farivar, F., Richardson, J.: Workplace digitalisation and work-nonwork satisfaction: the role of spillover social media. Behav. Inf. Technol. **40**(8), 747–758 (2021)
50. Messarra, L.C., Karkoulian, S., McCarthy, R.: To restrict or not to restrict personal internet usage on the job. Educ. Bus. Soc. Contemp. Middle Eastern Issues **4**(4), 253–266 (2011)
51. Garrett, R.K., Danziger, J.N.: Disaffection or expected outcomes: understanding personal Internet use during work. J. Comput. Mediat. Commun. **13**(4), 937–958 (2008)
52. O'Neill, T.A., Hambley, L.A., Bercovich, A.: Prediction of cyberslacking when employees are working away from the office. Comput. Hum. Behav. **34**, 291–298 (2014)
53. Galletta, D., Polak, P.: An empirical investigation of antecedents of Internet abuse in the workplace. SIGHCI 2003 Proc. **14**, 47–51 (2003)
54. Oravec, J.C.: Cyberloafing and constructive recreation. In: Khosrow-Pour, M. (ed.) Encyclopedia of Information Science and Technology, pp. 4316–4325. IGI Global, Hershey, PA (2018)
55. Pejtersen, J.H., Kristensen, T.S., Borg, V., Bjorner, J.B.: The second version of the Copenhagen psychosocial questionnaire. Scand. J. Public Health **38**(3 Suppl), 8–24 (2010)
56. Byrne, B.M.: Structural Equation Modeling with AMOS: Basic Concepts, Applications, and Programming. Routledge, New York (2016)

57. Jiang, H., Siponen, M., Tsohou, A.: Personal use of technology at work: a literature review and a theoretical model for understanding how it affects employee job performance. Eur. J. Inf. Syst. **32**(2), 331–345 (2023)

58. Global internet habits: Phone usage and social media still most popular. Brussels Times, 10 September 2022. https://www.brusselstimes.com

FAIR-FATE: Fair Federated Learning with Momentum

Teresa Salazar[1]([envelope]) [iD], Miguel Fernandes[1] [iD], Helder Araújo[2] [iD],
and Pedro Henriques Abreu[1] [iD]

[1] Centre for Informatics and Systems, Department of Informatics Engineering of the
University of Coimbra, University of Coimbra, Coimbra, Portugal
{tmsalazar,mfernandes,pha}@dei.uc.pt
[2] Institute of Systems and Robotics, Department of Electrical and Computer
Engineering of the University of Coimbra, University of Coimbra, Coimbra, Portugal
helder@isr.uc.pt

Abstract. While fairness-aware machine learning algorithms have been receiving increasing attention, the focus has been on centralized machine learning, leaving decentralized methods underexplored. Federated Learning is a decentralized form of machine learning where clients train local models with a server aggregating them to obtain a shared global model. Data heterogeneity amongst clients is a common characteristic of Federated Learning, which may induce or exacerbate discrimination of unprivileged groups defined by sensitive attributes such as race or gender. In this work we propose FAIR-FATE: a novel FAIR FederATEd Learning algorithm that aims to achieve group fairness while maintaining high utility via a fairness-aware aggregation method that computes the global model by taking into account the fairness of the clients. To achieve that, the global model update is computed by estimating a fair model update using a Momentum term that helps to overcome the oscillations of non-fair gradients. To the best of our knowledge, this is the first approach in machine learning that aims to achieve fairness using a fair Momentum estimate. Experimental results on real-world datasets demonstrate that FAIR-FATE outperforms state-of-the-art fair Federated Learning algorithms under different levels of data heterogeneity.

Keywords: Fairness · Federated Learning · Machine Learning · Momentum

1 Introduction

With the widespread use of machine learning algorithms to make decisions which impact people's lives, the area of fairness-aware machine learning has been receiving increasing attention. Fairness-aware machine learning algorithms ensure that predictions do not prejudice unprivileged groups of the population with respect to sensitive attributes such as race or gender [5]. However, the focus has been on centralized machine learning, with decentralized methods receiving little attention.

© The Author(s), under exclusive license to Springer Nature Switzerland AG 2023
J. Mikyška et al. (Eds.): ICCS 2023, LNCS 14073, pp. 524–538, 2023.
https://doi.org/10.1007/978-3-031-35995-8_37

Federated Learning is an emerging decentralized technology that creates machine learning models using data distributed across multiple clients [15]. In Federated Learning, the training process is divided into multiple clients which train individual local models on local datasets without the need to share private data. At each communication round, each client shares its local model updates with the server that uses them to create a shared global model.

While centralized machine learning batches can typically be assumed to be IID (independent and identically distributed), this assumption is unlikely to be true in Federated Learning settings [11]. Since in Federated Learning each client has its own private dataset, it may not be representative of all the demographic groups and thus can lead to discrimination of unprivileged groups defined by sensitive attributes. To solve the issue of heterogeneous client distributions, Federated Learning solutions based on Momentum gradient descent have been proposed [8,14], but none of these has focused on fairness in machine learning. Moreover, typical centralized fair machine learning solutions require centralized access to the sensitive attributes information, and, consequently, cannot be directly applied to Federated Learning without violating privacy constraints. Because of these reasons, finding Federated Learning algorithms which facilitate collaboration amongst clients to build fair machine learning models while preserving their data privacy is a great challenge.

Motivated by the importance and challenges of developing fair Federated Learning algorithms, we propose FAIR-FATE: a novel fairness-aware Federated Learning algorithm that aims to achieve group fairness while maintaining classification performance. FAIR-FATE uses a new fair aggregation method that computes the global model by taking into account the fairness of the clients. In order to achieve this, the server has a validation set that is used at each communication round to measure the fairness of the clients and the current global model. Afterwards, the fair Momentum update is computed using a fraction of the previous fair update, as well as the average of the clients' updates that have higher fairness than the current global model, giving higher importance to fairer models. Since Momentum-based solutions have been used to help overcoming oscillations of noisy gradients, we hypothesize that the use of a fairness-aware Momentum term can overcome the oscillations of non-fair gradients. Therefore, this work aims to address the following research question:

Can a fairness-aware Momentum term be used to overcome the oscillations of non-fair gradients in Federated Learning?

Several experiments are conducted on real world datasets [2,3,13,19] with a range of non-identical data distributions at each client, considering the different sensitive attributes and target classes in each dataset. Experimental results demonstrate that FAIR-FATE outperforms state-of-the-art fair Federated Learning algorithms on fairness without neglecting model performance under different data distributions.

2   Related Work

Fairness-aware machine learning has been gaining increasing attention. Fairness-aware algorithms have been categorized into three different groups, according to the stage in which they are performed: pre-processing, in-processing and post-processing [5,18]. The algorithm proposed in this work falls into the in-processing category. Fairness metrics can be divided into two main groups: group fairness and individual fairness [5]. Group fairness states that privileged and unprivileged groups should have equal probability of outcomes. On the other hand, individual fairness aims to give similar predictions to similar individuals. In this work, we focus on optimizing group fairness.

Few efforts have been made in the area of fairness-aware Federated Learning. Abay et al. [1] propose to apply three existing fair machine learning techniques to the Federated Learning setting. However, studies have shown that local debiasing strategies are ineffective in the presence of data heterogeneity [12]. Kanaparthy et al. [12] propose simple strategies for aggregating the local models based on different heuristics such as using the model which provides the least fairness loss for a given fairness notion or using the model with the highest ratio of accuracy with fairness loss. Ezzeldin et al. [7] propose a method to achieve global group fairness via adjusting the weights of different clients based on their local fairness metric. However, this work assumes that there is a local debiasing method at each client. Mehrabi et al. [16] propose "FedVal" that is a simple aggregation strategy which performs a weighted averaged of the clients models based on scores according to fairness or performance measures on a validation set.

To the best of our knowledge, we have yet to see a fairness-aware machine learning approach that uses Momentum to estimate a fair update of the global model. Furthermore, FAIR-FATE is flexible since it does not rely on any fairness local strategy at each client, contrary to the works in [1,7]. Finally, very few works consider multiple data heterogeneity scenarios (non-IID), which is a characteristic of Federated Learning settings.

3   Problem Statement

We first define the following notations used throughout the paper. We assume there exist K clients in the Federated Learning setting, where each client has its private local dataset $D_k = \{\mathbb{X}, Y\}, k \in K$. Each dataset D_k, contains n_k instances where the total number of instances is defined as $\sum_{k=1}^{K} n_k$. We further assume a binary classification setting, where $\mathbb{X} \in \mathcal{X}$ is the input space, $Y \in \mathcal{Y} = \{0, 1\}$ is the output space, $\hat{Y} \in \{0, 1\}$ represents the predicted class, and $S \in \{0, 1\}$ is the binary sensitive attribute of $\mathbb{X}$.

3.1   Federated Learning and Momentum

The objective of Federated Learning is to train a global model, θ, which resides on the server without access to the clients' private data. In Federated Learning, multiple clients collaborate to train a model that minimizes a weighted average of the loss across all clients. The objective of this Federated Learning framework can be written as [15]:

$$\min_\theta f(\theta) = \sum_{k \in K} \frac{n_k}{n} G_k(\theta), \qquad G_k(\theta) = \frac{1}{n_k} \sum_{i=1}^{n_k} f_i(\theta), \qquad (1)$$

where $G_k(\theta)$ is the local objective for the client k and $f_i(\theta)$ is the loss of the datapoint i of client k [15].

The first work on Federated Learning was proposed in [15], where the authors present an algorithm named Federated Averaging (FedAvg) that consists of a way of generating the global model by periodically averaging the clients' locally trained models.

Since vanilla FedAvg can be slow to converge [17], Federated Learning solutions based on Momentum gradient descent have been proposed [8,14]. Momentum gradient descent is a technique where an exponentially weighted averaged of the gradients is used to help overcoming oscillations of noisy gradients. By providing a better estimate of the gradients, it is faster than stochastic gradient descent. Moreover, decaying Momentum rules have been previously applied to Federated Learning [8] with the objective of improving training and being less sensitive to parameter tuning by decaying the total contribution of a gradient to future updates.

3.2 Fairness Metrics

We consider the following group fairness metrics:

Definition 1 (Statistical Parity (SP)). *A predictor has Statistical Parity [20] if the proportion of individuals that were predicted to receive a positive output is equal for the two groups:*

$$SP = \frac{P[\hat{Y} = 1 \mid S = 0]}{P[\hat{Y} = 1 \mid S = 1]} \qquad (2)$$

Definition 2 (Equality of Opportunity (EO)). *A classifier satisfies Equality of Opportunity (EO) if the True Positive Rate is the same across different groups [9]:*

$$EO = \frac{P[\hat{Y} = 1 \mid S = 0, Y = 1]}{P[\hat{Y} = 1 \mid S = 1, Y = 1]} \qquad (3)$$

Definition 3 (Equalized Odds (EQO)). *A classifier satisfies Equalized Odds (EQO) if the True Positive Rate and False Positive Rate is the same across different groups [9]:*

$$EQO = \frac{\frac{P[\hat{Y}=1|S=0,Y=0]}{P[\hat{Y}=1|S=1,Y=0]} + \frac{P[\hat{Y}=1|S=0,Y=1]}{P[\hat{Y}=1|S=1,Y=1]}}{2} \qquad (4)$$

All of the metrics presented have an ideal value of 1. If a ratio is bigger than 1, $S = 0$ becomes $S = 1$ and $S = 1$ becomes $S = 0$.

3.3 Objective

The goal of our problem is to minimize the loss function $f(\theta)$ of the global model as well as to promote fairness without knowledge of the private data of the clients participating in the federation. We formulate a multi-objective optimization problem with two objectives: one for minimizing the model error, and one for promoting fairness:

$$\min_{\theta} \quad \underbrace{(1 - \lambda_t) \sum_{k \in K_t} \frac{n_k}{n} G_k(\theta)}_{\text{model error objective}} \; + \lambda_t \underbrace{\sum_{k \in K_{tFair}} \frac{F_{\theta_{t+1}^k}}{F_{total}} G_k(\theta)}_{\text{fairness promotion objective}} \tag{5}$$

The model error objective is represented by the first term in the optimization problem, which involves minimizing the sum of the error functions over the clients participating in the federation round, K_t. This objective ensures that the resulting model performs well on the entire dataset. The fairness promotion objective is represented by the second term in the optimization problem, which involves minimizing the sum of the error functions over the clients of K_t that have higher fairness values than the global model at the server, K_{tFair}, giving higher importance to clients which have higher values of fairness, $F \in \{SP, EO, EQO\}$. Hence, our objective does not only minimize the global loss to make the global model accurate, but also promotes fairness so that the final global model is both fair and accurate. To balance the trade-off between these two objectives we use λ_t which is calculated at each federation round, t. The next section presents further details on how FAIR-FATE implements this objective.

4 FAIR-FATE: Fair Federated Learning with Momentum

FAIR-FATE is a novel fair Federated Learning algorithm developed in this work which consists of a new approach of aggregating the clients' local models in a fair manner, and, simultaneously, using a fairness-aware Momentum term that has the objective of overcoming the oscillations of non-fair gradients.

In this approach, the server has access to a small validation set, similar to the works in [4,16], which discuss different ways of obtaining this set.[1] Given an update from a client, the server can use the validation set to verify whether the client's update would improve the global model's fairness.

---

[1] For the sake of simplicity, we chose to randomly select a portion of the dataset as the validation set. To ensure the results are based on a representative sample, we conducted the experiments 10 times with a different validation set randomly selected each time.

The algorithm starts by initializing a global model θ_t. Then, at each round, the server sends its current model, θ_t, to a random subset of clients, K_t of size $|K_t|$, $|K_t| < |K|$. The chosen clients then train their local models using their private data, producing an updated model θ_{t+1}^k.

Two Model Updates: α_N and α_F. To simultaneously achieve the two objectives described in Sect. 3.3, the server calculates two model updates: α_N, corresponding to the model error objective, and α_F, corresponding to the fairness promotion objective. α_N is calculated by averaging the local updates of each client, given by:

$$\alpha_N = \sum_{k \in K_t} \frac{n_k}{n} (\theta_{t+1}^k - \theta_t), \tag{6}$$

where θ_t is the global model at communication round t, and θ_{t+1}^k is the local model of client k at communication round $t + 1$. On the other hand, α_F is a weighted average of the clients' updates which have higher fairness than the current global model at the server on the validation set, given by:

$$\alpha_F = \sum_{k \in K_{tFair}} \frac{F_{\theta_{t+1}^k}}{F_{total}} (\theta_{t+1}^k - \theta_t), \text{ where } F_{total} = \sum_{k \in K_{tFair}} F_{\theta_{t+1}^k}, \tag{7}$$

As such, the weighted average gives higher importance to models with higher fairness values on the validation set.

Decaying Momentum. Afterwards we use Momentum, and, more specifically, decaying Momentum [6,8], calculating β_t as follows:

$$\beta_t = \beta_0 \times \frac{(1 - \frac{t}{T})}{(1 - \beta_0) + \beta_0(1 - \frac{t}{T})}, \tag{8}$$

where β_0 is the initial Momentum parameter. The fraction $(1 - \frac{t}{T})$ refers to the proportion of iterations remaining. The Momentum update, v_{t+1}, is calculated by summing a fraction of the previous update, v_t, and the global model fair update, α_F, as follows:

$$v_{t+1} = \beta_t v_t + (1 - \beta_t)\alpha_F, \tag{9}$$

where β_t controls the amount of the global fair update against the previous update.

The use of decaying Momentum is motivated by two main reasons. Firstly, when α_F is calculated using the local updates of the clients which have higher fairness than the server, it can be affected by the variability and noise of the data of each individual client. This can result in oscillations and instability in the overall training process. By introducing Momentum, we can smooth out these oscillations and stabilize the overall training process, leading to better convergence of the overall optimization problem. Secondly, using decaying Momentum can take advantage of momentum for speed-up in the early phases of training, but decay Momentum throughout training [6,8]. This can prevent neural network weights from growing excessively fast, which has been found to be crucial for achieving optimal performance [10].

Exponential Growth. Finally, the global model, θ_{t+1}, is updated by summing the previous model, θ_t, with the fair Momentum update, v_{t+1}, and the normal update, α_N, as follows:

$$\theta_{t+1} = \theta_t + \lambda_t v_{t+1} + (1 - \lambda_t)\alpha_N, \tag{10}$$

where λ_t has the objective of controlling the amount of estimated fairness against the actual model update. It follows an exponential growth calculated as follows:

$$\lambda_t = \lambda_0(1 + \rho)^t, \qquad \lambda_t \leq MAX, \tag{11}$$

where ρ, λ_0 and MAX are hyperparameters that can be tuned which represent the growth rate, the initial amount, and the upper bound of λ_t, respectively.

By exponentially increasing λ, FAIR-FATE allows that in the early stages of training, the model will prioritize the optimization of the main objective function using α_N. In the latter stages, when it is expected that the model has a good performance and there is a better Momentum estimate, FAIR-FATE will have a higher value of λ and, as a consequence, it will prioritize the global model's fair update, α_F. Regarding the MAX hyperparameter, a higher value will prioritize fair model updates, while a lower value will take more the model performance into consideration. Hence, as it will be observed in the experimental section, the MAX hyperparameter can be used to control the fairness and model performance trade-off. Algorithm 1 presents the FAIR-FATE algorithm.

5   Experiments

In this section, a comprehensive set of experiments to evaluate FAIR-FATE is conducted in a Federated Learning scenario.

5.1   Datasets

FAIR-FATE is evaluated on four real-world datasets which have been widely used in the literature [5] to access the performance of fairness-aware algorithms: the COMPAS [2], the Adult [3], the Law School [19], and the Dutch [13] datasets. The COMPAS dataset's prediction task is to calculate the recidivism outcome, indicating whether individuals will be rearrested within two years after the first arrest, with race as the sensitive attribute. The Adult dataset's prediction task is to determine whether a person's income exceeds $50K/yr or not, with gender as the sensitive attribute. The Law School dataset's prediction task is to predict whether a candidate would pass the bar exam, with race as the sensitive attribute. Finally, the Dutch Census dataset's prediction task is to determine a person's occupation which can be categorized as high level (prestigious) or low level profession, with gender as the sensitive attribute.

Algorithm 1. FAIR-FATE

1: Initialize global model: θ_t, number of rounds: T, growth rate: ρ, upper bound of λ_t: MAX, Momentum parameter: β_t, initial Momentum parameter: β_0, initial amount: λ_0, initial Momentum update: $v_0 = 0$.

2: **for** each round $t = 1,2,..., T$ **do**

3:     Select K_t = random subset of clients K

4:     **for** each client $k \in K_t$ **do**

5:         Send θ_t and receive the locally trained model, θ_{t+1}^k

6:     **end for**

7:     $K_{tFair} \in K_t$,   where $F_{\theta_{t+1}^k} >= F_{\theta_t}$

8:     $F_{total} = \sum_{k \in K_{tFair}} F_{\theta_{t+1}^k}$

9:     $\alpha_F = \sum_{k \in K_{tFair}} \frac{F_{\theta_{t+1}^k}}{F_{total}}(\theta_{t+1}^k - \theta_t)$

10:     $\alpha_N = \sum_{k \in K_t} \frac{n_k}{n}(\theta_{t+1}^k - \theta_t)$

11:     $\beta_t = \beta_0 \times \frac{(1 - \frac{t}{T})}{(1-\beta_0)+\beta_0(1-\frac{t}{T})}$

12:     $v_{t+1} = \beta_t v_t + (1 - \beta_t)\alpha_F$

13:     $\lambda_t = \lambda_0(1 + \rho)^t$, $\lambda_t \leq MAX$

14:     $\theta_{t+1} = \theta_t + \lambda_t v_{t+1} + (1 - \lambda_t)\alpha_N$

15: **end for**

5.2   Non-identical Client Data

Because of its decentralized nature, Federated Learning exacerbates the problem of bias since clients' data distributions can be very heterogeneous. This means that there may exist a client that contains a very high representation of datapoints belonging to specific sensitive group with a certain outcome (e.g. unprivileged group with positive outcome - $S = 0, Y = 1$) and another client that has a very low representation of the same group.

Following the works in [7,11], we study the algorithms under different sensitive attribute distributions across clients. We use a non-IID synthesis method based on the Dirichlet distribution proposed in [11] and apply it in configurable sensitive attribute distribution such as the work in [7] but including the target distributions as well. More specifically, for each sensitive attribute value s and target value y, we sample $p_{s,y} \sim Dir(\sigma)$ and allocate a portion $p_{s,y,k}$ of the datapoints with $S = s$ and $Y = y$ to client k. The parameter σ controls the heterogeneity of the distributions in each client, where $\sigma \longrightarrow \infty$ results in IID distributions [7].

Figure 1 presents different examples of heterogeneous data distributions on the COMPAS dataset using race as the sensitive feature for 10 clients with $\sigma = 0.5$ and $\sigma \to \infty$. On one hand it can be observed that for $\sigma = 0.5$, different clients have very different representations of privileged and unprivileged groups and positive and negative outcomes. For example, while client 10 has about 40% of the Caucasians that won't recidivate, client 9 has only about 10%. Moreover, it can be observed that different clients have different number of datapoints.

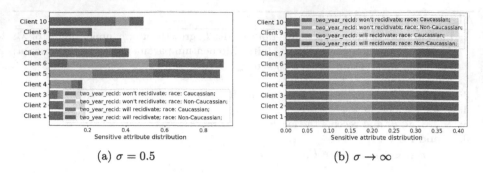

Fig. 1. Examples of heterogeneous data distributions on the COMPAS dataset using race as the sensitive feature for 10 clients with $\sigma = 0.5$ and $\sigma \to \infty$

On the other hand, for $\sigma \to \infty$, the representations of the different groups for different clients are the same.

5.3   Implementation and Setup Details

We train a fully connected feed-forward neural network for binary classification on the described datasets with one hidden layer with 10 $tanh$ units. On the local clients models, mini-batch stochastic gradient descent and the standard binary cross-entropy loss are used.

The data instances are split into 60% for the training set, 20% for the server's validation set and 20% for the testing set. The training data is divided into private training datasets, one for each client. The private training datasets owned by the clients are used to train the local models. The validation set is used to calculate the fairness of the models for the fair Momentum update as explained in the previous sections. The testing dataset is used to evaluate the model performance and fairness of the proposed approach and the baselines. The presented results are based on the average values over 10 runs.

We set $\lambda_0 = \{0.1, 0.5\}$, $\rho = \{0.04, 0.05\}$, $MAX = \{0.8, 0.9, 1.0\}$, $\beta_0 = \{0.8, 0.9, 0.99\}$, $F = \{SP, EO, EQO\}$. Additionally, we set $\sigma = \{0.5, 1.0\}$, $T = 100$, $E = 10$ (local epochs), $B = 10$ (local batches), $\eta_l = 0.01$ (local learning rate). Based on the number of instances of the COMPAS, the Adult, the Law School and the Dutch datasets, we set $|K|$ to 10, 15, 12 and 20, respectively, and $|K_t|$ to about a third of $|K|$: 3, 5, 4, 6, respectively.[2]

---

[2] Source code can be found at: https://github.com/teresalazar13/FAIR-FATE.

The following algorithms are used as the baselines:

- FedAvg - the classical Federated Averaging algorithm presented in [15];
- FedMom - the Federated Averaging with Momentum presented in [14];
- FedDemon - the Federated Averaging with Momentum Decay presented [8];
- FedAvg+LR - the fair Federating Learning approach proposed in [1] where each client uses a local reweighing technique;
- FedAvg+GR - the fair Federating Learning approach proposed in [1] where each client uses a global reweighing technique;
- FedVal - the fair Federating Learning approach proposed in [16] where the server performs a weighted average based on the validation scores.

5.4   Experimental Results

Results of Last Communication Round. Tables 1 and 2 present the corresponding results of the last communication round for the COMPAS, the Adult, the Law School and the Dutch datasets[3]. Regarding the COMPAS dataset, FAIR-FATE was able to surpass the baselines 78% of the times. In particular, for $\sigma = 0.5$, FAIR-FATE presented substantial improvements: 7% for SP, 11% for EO, and 16% for EQO. Concerning the Adult dataset, FAIR-FATE was always able to surpass the baselines with an average improvement of fairness of 10%. In particular, for $\sigma = 0.5$, FAIR-FATE presented substantial improvements: 8% for SP, 7% for EO, and 19% for EQO. Regarding the Law School and the Dutch datasets, FAIR-FATE surpassed the baselines 78% and 88% of the times, respectively. Overall, the baselines only surpassed FAIR-FATE for random train-test splits (RND), which do not represent typical Federated Learning settings.

Stability and Convergence. Figure 2 presents the results over communication rounds under different sensitive attribute distributions on the Adult dataset. It can be observed that FAIR-FATE is able to surpass the baselines in terms of fairness, especially in heterogeneous configurations ($\sigma = 0.5$ and $\sigma = 1.0$), without neglecting the classification performance. Moreover, the stability of the algorithms decreases in the presence of heterogeneous data. However, FAIR-FATE is able to overcome the oscillations of non-fair gradients using a fairness-aware Momentum term and, consequently, the fairness performance does not drop after some communication rounds.

---

[3] Since there are variations in the results due to different hyperparameters, we choose the outcome that achieves a good balance between accuracy and fairness, but prioritizes fairness by allowing a small sacrifice in accuracy if necessary.

Table 1. Results for the COMPAS and Adult datasets under different σ heterogeneity levels and random (RND) splits. FedAvg (1), FedMom (2), FedDemon (3), FedAvg+LR (4), FedAvg+GR (5), FedVal ($F = SP$), (6) FedVal, ($F = EO$) (7), FedVal ($F = EQO$) (8), FAIR-FATE ($F = SP$) **(9)**, FAIR-FATE ($F = EO$) **(10)**, FAIR-FATE ($F = EQO$) **(11)**

	Alg.	COMPAS (race)			Adult (gender)		
		σ		RND	σ		RND
		0.5	1.0		0.5	1.0	
ACC	(1)	0.60+-0.04	0.64+-0.03	0.68+-0.01	0.79+-0.05	0.83+-0.01	0.84+-0.01
	(2)	0.62+-0.05	0.63+-0.04	0.69+-0.01	0.83+-0.02	0.82+-0.01	0.82+-0.01
	(3)	0.68+-0.01	0.68+-0.01	0.68+-0.01	0.84+-0.01	0.84+-0.01	0.84+-0.01
	(4)	0.62+-0.04	0.65+-0.02	0.68+-0.01	0.81+-0.03	0.82+-0.01	0.84+-0.01
	(5)	0.62+-0.04	0.65+-0.03	0.68+-0.01	0.80+-0.03	0.82+-0.01	0.84+-0.01
	(6)	0.61+-0.04	0.64+-0.03	0.68+-0.01	0.75+-0.10	0.83+-0.01	0.84+-0.01
	(7)	0.62+-0.05	0.66+-0.03	0.68+-0.01	0.78+-0.06	0.82+-0.03	0.84+-0.01
	(8)	0.62+-0.04	0.66+-0.03	0.68+-0.01	0.76+-0.07	0.82+-0.03	0.84+-0.01
	(9)	0.57+-0.02	0.61+-0.04	0.66+-0.01	0.74+-0.14	0.80+-0.06	0.80+-0.02
	(10)	0.63+-0.04	0.65+-0.02	0.67+-0.01	0.74+-0.09	0.78+-0.06	0.83+-0.02
	(11)	0.63+-0.04	0.65+-0.03	0.67+-0.01	0.75+-0.09	0.79+-0.05	0.77+-0.01
SP	(1)	0.59+-0.34	0.86+-0.17	0.83+-0.04	0.41+-0.30	0.27+-0.10	0.31+-0.02
	(2)	0.75+-0.23	0.83+-0.20	0.82+-0.03	0.39+-0.16	0.63+-0.24	0.46+-0.20
	(3)	0.83+-0.04	0.83+-0.04	0.83+-0.04	0.31+-0.03	0.31+-0.03	0.31+-0.03
	(4)	0.93+-0.10	0.91+-0.08	0.94+-0.02	0.69+-0.19	0.70+-0.10	0.67+-0.05
	(5)	0.76+-0.28	0.87+-0.16	0.94+-0.02	0.66+-0.27	0.64+-0.20	0.67+-0.05
	(6)	0.87+-0.17	0.88+-0.13	0.83+-0.04	0.71+-0.20	0.52+-0.18	0.32+-0.02
	(9)	**1.00+-0.01**	**0.98+-0.01**	**0.95+-0.03**	**0.79+-0.16**	**0.81+-0.11**	**0.79+-0.13**
		+7%	**+7%**	**+1%**	**+8%**	**+11%**	**+8%**
EO	(1)	0.46+-0.34	0.81+-0.19	0.85+-0.05	0.51+-0.27	0.68+-0.21	0.78+-0.04
	(2)	0.74+-0.24	0.73+-0.34	0.83+-0.05	0.66+-0.26	0.65+-0.17	0.77+-0.17
	(3)	0.84+-0.05	0.84+-0.05	0.84+-0.05	0.78+-0.04	0.78+-0.04	0.78+-0.04
	(4)	0.77+-0.31	0.91+-0.08	**0.98+-0.03**	0.66+-0.21	0.69+-0.15	0.76+-0.03
	(5)	0.54+-0.38	0.84+-0.21	**0.97+-0.03**	0.55+-0.28	0.71+-0.14	0.75+-0.04
	(7)	0.58+-0.34	0.83+-0.22	0.85+-0.05	0.72+-0.15	0.80+-0.12	0.79+-0.04
	(10)	**0.95+-0.06**	**0.96+-0.02**	0.95+-0.03	**0.85+-0.15**	**0.95+-0.02**	**0.90+-0.07**
		+11%	**+5%**	**-3%**	**+7%**	**+15%**	**+11%**
EQO	(1)	0.39+-0.32	0.77+-0.19	0.76+-0.06	0.36+-0.21	0.42+-0.15	0.48+-0.03
	(2)	0.65+-0.26	0.71+-0.26	0.73+-0.05	0.47+-0.24	0.53+-0.21	0.60+-0.19
	(3)	0.75+-0.06	0.75+-0.06	0.75+-0.06	0.48+-0.04	0.48+-0.04	0.48+-0.04
	(4)	0.78+-0.29	0.87+-0.11	**0.93+-0.03**	0.58+-0.23	0.75+-0.09	0.80+-0.05
	(5)	0.59+-0.26	0.82+-0.21	**0.93+-0.04**	0.46+-0.30	0.69+-0.19	0.80+-0.05
	(8)	0.62+-0.29	0.83+-0.17	0.76+-0.06	0.59+-0.21	0.56+-0.12	0.49+-0.03
	(11)	**0.94+-0.05**	**0.93+-0.03**	0.92+-0.04	**0.78+-0.10**	**0.79+-0.11**	**0.82+-0.10**
		+16%	**+6%**	**-1%**	**+19%**	**+4%**	**+2%**

Table 2. Results for the Law School and Dutch Census datasets under different σ heterogeneity levels and random (RND) splits. FedAvg (1), FedMom (2), FedDemon (3), FedAvg+LR (4), FedAvg+GR (5), FedVal ($F = SP$), (6) FedVal, ($F = EO$) (7), FedVal ($F = EQO$) (8), FAIR-FATE ($F = SP$) (9), FAIR-FATE ($F = EO$) (10), FAIR-FATE ($F = EQO$) (11)

	Alg.	Law School (race)			Dutch Census (gender)		
		σ		RND	σ		RND
		0.5	1.0		0.5	1.0	
ACC	(1)	0.89+-0.01	0.89+-0.02	0.90+-0.01	0.81+-0.02	0.82+-0.01	0.83+-0.00
	(2)	0.89+-0.01	0.90+-0.01	0.90+-0.01	0.81+-0.01	0.82+-0.01	0.83+-0.00
	(3)	0.90+-0.01	0.90+-0.01	0.90+-0.01	0.82+-0.01	0.82+-0.01	0.82+-0.01
	(4)	0.89+-0.00	0.88+-0.03	0.89+-0.01	0.81+-0.03	0.82+-0.01	0.82+-0.03
	(5)	0.90+-0.01	0.88+-0.01	0.89+-0.01	0.81+-0.02	0.82+-0.01	0.82+-0.00
	(6)	0.88+-0.03	0.89+-0.01	0.90+-0.01	0.78+-0.04	0.82+-0.01	0.83+-0.00
	(7)	0.80+-0.12	0.88+-0.02	0.90+-0.00	0.81+-0.02	0.82+-0.01	0.83+-0.00
	(8)	0.85+-0.07	0.89+-0.02	0.99+-0.00	0.80+-0.03	0.82+-0.01	0.83+-0.00
	(9)	0.89+-0.00	0.89+-0.01	0.89+-0.01	0.75+-0.07	0.78+-0.04	0.81+-0.01
	(10)	0.90+-0.01	0.89+-0.03	0.90+-0.00	0.77+-0.07	0.77+-0.08	0.82+-0.00
	(11)	0.90+-0.01	0.90+-0.01	0.90+-0.01	0.75+-0.08	0.76+-0.06	0.81+-0.01
SP	(1)	0.87+-0.13	0.88+-0.09	0.89+-0.02	0.60+-0.12	0.61+-0.09	0.60+-0.02
	(2)	0.91+-0.12	0.91+-0.07	0.91+-0.04	0.62+-0.09	0.59+-0.10	0.60+-0.01
	(3)	0.91+-0.03	0.91+-0.03	0.91+-0.03	0.60+-0.01	0.60+-0.01	0.60+-0.01
	(4)	0.96+-0.04	0.95+-0.06	0.98+-0.01	0.72+-0.04	0.72+-0.04	0.74+-0.02
	(5)	0.95+-0.05	0.97+-0.03	0.98+-0.01	0.72+-0.10	0.73+-0.08	0.74+-0.02
	(6)	0.89+-0.11	0.92+-0.05	0.89+-0.02	0.67+-0.13	0.70+-0.09	0.60+-0.01
	(9)	**1.00+-0.00**	**0.99+-0.01**	**0.99+-0.01**	**0.78+-0.12**	**0.84+-0.10**	0.74+-0.03
		+4%	**+2%**	**+1%**	**+6%**	**+12%**	0%
EO	(1)	0.57+-0.41	0.79+-0.30	0.91+-0.02	0.83+-0.08	0.89+-0.07	0.90+-0.02
	(2)	0.52+-0.47	0.73+-0.39	0.92+-0.03	0.86+-0.05	0.88+-0.13	0.90+-0.02
	(3)	0.92+-0.03	0.92+-0.03	0.92+-0.03	0.91+-0.02	0.91+-0.02	0.91+-0.02
	(4)	0.67+-0.46	0.76+-0.41	**0.99+-0.01**	0.93+-0.05	0.91+-0.04	0.91+-0.02
	(5)	0.86+-0.30	0.77+-0.41	**0.99+-0.01**	0.90+-0.07	0.90+-0.05	0.91+-0.02
	(7)	0.62+-0.29	0.88+-0.12	0.91+-0.03	0.86+-0.05	0.90+-0.05	0.90+-0.02
	(10)	**0.96+-0.04**	**0.98+-0.01**	0.96+-0.01	**0.94+-0.04**	**0.99+-0.01**	**0.97+-0.03**
		+4%	**+6%**	**-3%**	**+1%**	**+8%**	**+6%**
EQO	(1)	0.50+-0.32	0.67+-0.29	0.75+-0.05	0.59+-0.17	0.62+-0.15	0.58+-0.02
	(2)	0.49+-0.39	0.62+-0.34	0.79+-0.08	0.63+-0.16	0.60+-0.14	0.58+-0.02
	(3)	0.78+-0.05	0.78+-0.05	0.78+-0.05	0.59+-0.02	0.59+-0.02	0.59+-0.02
	(4)	0.64+-0.44	0.74+-0.40	**0.97+-0.02**	0.78+-0.05	0.80+-0.06	0.82+-0.03
	(5)	0.83+-0.16	0.82+-0.19	**0.97+-0.02**	0.76+-0.14	0.78+-0.12	0.82+-0.03
	(8)	0.64+-0.25	0.78+-0.12	0.74+-0.04	0.62+-0.13	0.73+-0.14	0.82+-0.03
	(11)	**0.92+-0.06**	**0.95+-0.04**	0.90+-0.04	**0.81+-0.06**	**0.89+-0.08**	**0.83+-0.04**
		+9%	**+13%**	**-7%**	**+3%**	**+9%**	**+1%**

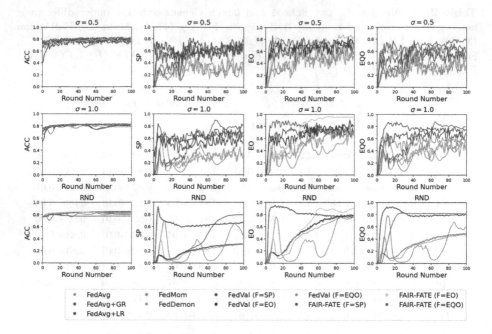

Fig. 2. Performance over communication rounds on the `Adult` dataset for the different algorithms under different σ heterogeneity levels and random (RND) splits.

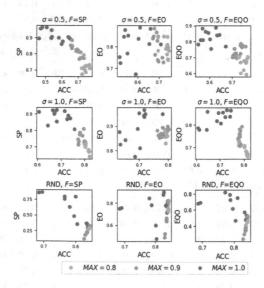

Fig. 3. FAIR-FATE's fairness-accuracy trade-off on the `Adult` dataset considering MAX under different σ heterogeneity levels and random (RND) splits.

Fairness vs Model Performance Trade-Off Using MAX. We further investigate the fairness and model performance trade-off of FAIR-FATE, controlled by the MAX hyperparameter. Figure 3 present FAIR-FATE's fairness-accuracy trade-offs considering the MAX hyperparameter on the `Adult` dataset. It can be observed that generally MAX can be used to select different trade-offs, with higher values of MAX resulting in higher values of fairness and lower values of accuracy, and lower values of MAX resulting in lower values of fairness and higher values of accuracy. On a broad level, it can be observed that the results produce a shape similar to a pareto front, especially for $F = SP$. However, for $F = EO$ and $F = EQO$ this phenomenon is not as noticeable. This might be justified by the fact that these metrics consider the actual predictions ($\hat{Y}$), and, consequently, are more aligned with the performance of the algorithm.

Effect of ρ, λ_0 and β_0. Experimental results suggest that the values of β_0, λ_0, and ρ should be chosen based on experimentation. Lower values of ρ and λ_0 could be preferable in cases where there is significant heterogeneity of the data with respect to the sensitive attributes. Those lower values allow the global model to adjust to the diversity in the data more slowly and capture the complexity of the data. Regarding the β_0 hyperparameter, it is more difficult to analyse its impact. As a result, we recommend tuning these hyperparameters to achieve the best results.

6   Conclusion and Future Work

We proposed FAIR-FATE: a fairness-aware Federated Learning algorithm that uses a novel aggregation technique that estimates a fair Momentum update based on the fairness of each client participating in the federation. In addition, FAIR-FATE does not need clients' restricted information, respecting the data privacy constraints associated with Federated Learning. Furthermore, FAIR-FATE is a flexible algorithm that does not rely on any local fairness strategy at each client.

To the best of our knowledge, this is the first approach in machine learning that aims to achieve fairness using a fair Momentum estimate. Experimental results on real-world datasets demonstrate that FAIR-FATE can effectively increase the fairness results under different data distributions compared to state-of-the-art fair Federated Learning approaches.

Interesting directions for future work are: testing FAIR-FATE using multiple fairness metrics and sensitive attributes; extending the work to various application scenarios such as clustering; focusing on individual fairness notions.

Acknowledgments. This work is funded by the FCT - Foundation for Science and Technology, I.P./MCTES through national funds (PIDDAC), within the scope of CISUC R&D Unit - UIDB/00326/2020 or project code UIDP/00326/2020. This work was supported in part by the Portuguese Foundation for Science and Technology (FCT) Research Grants 2021.05763.BD.

References

1. Abay, A., Zhou, Y., Baracaldo, N., Rajamoni, S., Chuba, E., Ludwig, H.: Mitigating bias in federated learning. arXiv preprint arXiv:2012.02447 (2020)
2. Angwin, J., Larson, J., Mattu, S., Kirchner, L.: Machine Bias (2016). https://www.propublica.org/article/machine-bias-risk-assessments-in-criminal-sentencing
3. Asuncion, A., Newman, D.: Adult data set. In: UCI Machine Learning Repository (2017). https://archive.ics.uci.edu/ml/datasets/adult
4. Bhagoji, A.N., Chakraborty, S., Mittal, P., Calo, S.: Analyzing federated learning through an adversarial lens. In: Chaudhuri, K., Salakhutdinov, R. (eds.) Proceedings of the 36th International Conference on Machine Learning. Proceedings of Machine Learning Research, vol. 97, pp. 634–643. PMLR (2019)
5. Caton, S., Haas, C.: Fairness in machine learning: a survey. arXiv preprint arXiv:2010.04053 (2020)
6. Chen, J., Wolfe, C., Li, Z., Kyrillidis, A.: Demon: improved neural network training with momentum decay. In: 2022 IEEE International Conference on Acoustics, Speech and Signal Processing (ICASSP), ICASSP 2022, pp. 3958–3962. IEEE (2022)
7. Ezzeldin, Y.H., Yan, S., He, C., Ferrara, E., Avestimehr, S.: FairFed: enabling group fairness in federated learning. arXiv preprint arXiv:2110.00857 (2021)
8. Fernandes, M., Silva, C., Arrais, J.P., Cardoso, A., Ribeiro, B.: Decay momentum for improving federated learning. In: ESANN 2021, pp. 17–22 (2021). https://doi.org/10.14428/esann/2021.ES2021-106
9. Hardt, M., Price, E., Srebro, N.: Equality of opportunity in supervised learning. In: Advances in Neural Information Processing Systems, vol. 29 (2016)
10. Heo, B., et al.: AdamP: slowing down the slowdown for momentum optimizers on scale-invariant weights. arXiv preprint arXiv:2006.08217 (2020)
11. Hsu, T.M.H., Qi, H., Brown, M.: Measuring the effects of non-identical data distribution for federated visual classification. arXiv preprint arXiv:1909.06335 (2019)
12. Kanaparthy, S., Padala, M., Damle, S., Gujar, S.: Fair federated learning for heterogeneous face data. arXiv preprint arXiv:2109.02351 (2021)
13. Van der Laan, P.: The 2001 census in the Netherlands. In: Conference the Census of Population (2000)
14. Liu, W., Chen, L., Chen, Y., Zhang, W.: Accelerating federated learning via momentum gradient descent. IEEE Trans. Parallel Distrib. Syst. 1 (2020). https://doi.org/10.1109/TPDS.2020.2975189
15. McMahan, H., Moore, E., Ramage, D., Hampson, S., Arcass, B.A.: Communication-efficient learning of deep networks from decentralized data. In: AAAI Fall Symposium. Google Inc. (2017)
16. Mehrabi, N., de Lichy, C., McKay, J., He, C., Campbell, W.: Towards multi-objective statistically fair federated learning. arXiv preprint arXiv:2201.09917 (2022)
17. Qian, N.: On the momentum term in gradient descent learning algorithms. Neural Netw. 12(1), 145–151 (1999)
18. Salazar, T., Santos, M.S., Araujo, H., Abreu, P.H.: FAWOS: fairness-aware oversampling algorithm based on distributions of sensitive attributes. IEEE Access 9, 81370–81379 (2021). https://doi.org/10.1109/ACCESS.2021.3084121
19. Wightman, L.: LSAC national longitudinal bar passage study. In: LSAC Research Report Series (1998)
20. Zafar, M.B., Valera, I., Rogriguez, M.G., Gummadi, K.P.: Fairness constraints: mechanisms for fair classification. In: Artificial Intelligence and Statistics, pp. 962–970. PMLR (2017)

Multi-agent Cellular Automaton Model for Traffic Flow Considering the Heterogeneity of Human Delay and Accelerations

Krzysztof Małecki[1]([✉]) [ID], Patryk Górka[1], and Maria Gokieli[2] [ID]

[1] Faculty of Computer Science and Information Technology, West Pomeranian University of Technology, Żołnierska 52 Str., 71-210 Szczecin, Poland
{kmalecki,gp46518}@zut.edu.pl

[2] Faculty of Mathematics and Natural Sciences - School of Exact Sciences, Cardinal Stefan Wyszyński University, Wóycickiego 1/3, 01-938 Warsaw, Poland
m.gokieli@uksw.edu.pl

Abstract. We propose a multi-agent cellular automata model for analysing the traffic flow with various types of agents (drivers). Agents may differ by their vehicles' acceleration/deceleration values and the delay value of their decision-making. We propose a model in which the main parameters are chosen to reflect different types of driving. Based on valuable previous works, accurate data for possible acceleration/deceleration are used. Additionally, to accurately reflect the cars' dimensions and their limited movement in a traffic jam, a small-cell cellular automaton is used, where a set of cells represents one car. We present the results of a numerical simulation showing the influence of the main factors of the driving type on the traffic flow. Research shows that aggressive braking has a greater negative impact on traffic flow than aggressive acceleration.

Keywords: Cellular Automata (CA) · Traffic Flow · Agent-Based Modeling (ABM)

1 Introduction

There are many different models and analyses of car traffic and its effectiveness [1–4]. Regarding the microscopic approach, one can point out classical models, which include, on the one hand, time–continuous car–following model like the Intelligent Driver Model [5], and on the other hand, discrete traffic models: Nagel-Schreckenberg [6] and Chopard-Luthi-Queloz [7], where the cellular automata (CA) paradigm was applied. Although these models are relatively simple, they allow for analysis of numerous dependencies [8–15], etc. One can also identify a trend where drivers/cars are represented as agents with various abilities and driving styles [16–18]. In [17], the authors used agent-based modelling to point to on-street parking problems related to drivers' behaviours. While in [18], the authors used small–cell CA and multi-agent system to indicate the essential aspect of road traffic, which is a corridor of life.

© The Author(s), under exclusive license to Springer Nature Switzerland AG 2023
J. Mikyška et al. (Eds.): ICCS 2023, LNCS 14073, pp. 539–552, 2023.
https://doi.org/10.1007/978-3-031-35995-8_38

Among others published research, a particular number of road traffic models are oriented on the impact of vehicles' acceleration and deceleration on traffic flow. The first works assumed that the value of acceleration and deceleration is constant for all vehicles. The value 1 was suggested, understood as acceleration by one CA cell or braking by one CA cell—without distinguishing the size of the cell [8–10]. In [19], the authors performed measurements for several selected vehicles and determined the actual values of acceleration and braking. The average acceleration value was 3.22 m/s^2, and deceleration -8.49 m/s^2. In [20], the authors proposed a new CA model which avoids unrealistic deceleration behaviour found in most previous CA models. They introduced a parameter Δv, according to which vehicles change their speed. It was then possible to model the traffic flow with different characteristics of acceleration and braking values, i.e. vehicles belonging to different groups (passenger cars, trucks, etc.). In the next article [21], a new simple model was proposed, in which the braking possibilities were limited (as opposed to the Nagel-Schreckenberg model, where they were unlimited [6]). Additionally, the possibility of accelerating vehicles with a certain probability was introduced (compared to the aforementioned model [6], in which only braking was based on a fixed random value). Hence, the presented fundamental diagrams had more complex than those shown in earlier works. The extension of [20] was presented in [22]. The authors gave the simulation results on a two–lane system with periodic conditions and two types of vehicles with different lengths and different limited velocities. The introduced model reproduces a realistic density dependence of the number of lane changes observed in reality and the different traffic states.

In this article, we propose a multi–agent CA model to study the impact of drivers' behaviour and their delay ratio on traffic flow. In order to obtain the possibility of indicating undesirable behaviours, we develop a CA model that allows vehicles (agents of a multi–agent system) to move with various parameters of acceleration/deceleration and a reaction distance, which is the additional gap to the preceding car that the driver would like to leave. This idea is a novelty of this work. Our model can reproduce some common characteristics of real traffic, such as the variability of acceleration and deceleration value, the density of the traffic flow, and various drivers' reactions. We hope it can help to carry out research on major negative factors for the road traffic flow.

2   Model

This work is based on a single–lane CA model for traffic flow, called LAI model [20] and its extension [22]. The basic model consists of a number of vehicles moving in one direction on a one-dimensional lattice, consisting of a number of cells arranged in a ring topology (Fig. 1). Hence, the number of vehicles is fixed. This allows for fixing an accurate density of traffic. Each cell is either empty or occupied by just one vehicle (or part of one vehicle) travelling with velocity v, which takes integer values ranging from 0 to v_{max}. It is assumed that vehicles move from left to right. The system evolves in time steps Δt, which is taken to be 1 s. In each time step, the agents decide on their velocity and then move accordingly. Each decision on the velocity is related to: a) the distance to the neighbouring vehicle and its velocity that the agent is assumed to estimate, b) the preferences of the driver, c) a small random factor.

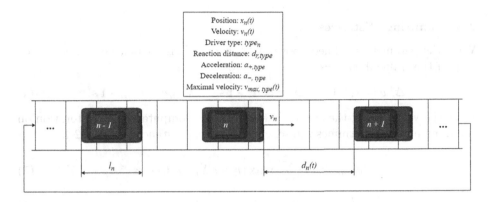

Fig. 1. The general approach in this study.

2.1 Parameters

In our multi-agent CA model, it is possible to have several types of agents. That is, several parameters that we introduce so as to model the traffic flow are based on the driver's (agent's) characteristics. We thus first introduce the driver's type. We consider five driver types in the present study.

$$type \in \{I, II, III, IV, V\}. \tag{1}$$

We then associate the following parameters to this driver's type:

$$type \mapsto (d_r, a_+, a_-, v_{max}) \tag{2}$$

where v_{max} is the maximum velocity that can be attained by the driver, $a+$ and a_- are the acceleration and deceleration values used by the driver, and d_r is a coefficient which governs what we call the reaction distance, i.e. a distance needed by the driver to take action, which results in an additional gap between the agent and its predecessor. We assume that d_r is the reaction distance when the speed is maximal, and then it is equal to the gap. For smaller velocities, the gap is defined by (7) below. This coefficient and the notion of reaction distance (or gap) are new in our study.

We also have common parameters, independent of the driver's type, which is the maximal braking deceleration

$$a_{min} = -5 \, m/s^2, \tag{3}$$

and a factor $R_s \in (0,1)$, which is the probability of a random slowing down, as present in [6] and in most later works.

The values of (d_r, a_+, a_-, v_{max}) used for different drivers in the present paper are given in Table 1. In what follows, we may omit their dependence on the driver's type in the notation.

2.2   Limiting Distances

With the parameters defined above, we now define the velocity changes in each step of the traffic dynamics:

$$\Delta^+ v = a_+ \cdot 1\,s, \quad \Delta^- v = a_- \cdot 1\,s, \quad \Delta_{min} v = a_{min} \cdot 1\,s. \tag{4}$$

Also, the velocity v of the agent — which will be computed in each iteration on the dynamics — determines the so-called limiting distances, as in [22]):

$$d_{brake}(v) = \sum_{i=1}^{\infty} \max\left(v + i\Delta_{min}v, 0\right) \cdot 1\,s \tag{5}$$

$$d_{safe}(v) = v \cdot 1\,s + d_{brake}(v). \tag{6}$$

Here, $d_{brake}(v)$ is the braking distance, i.e. the distance needed to reduce the velocity to 0, when using the maximal deceleration a_{min}. The safe distance $d_{safe}(v)$ is the distance covered in $1\,s$ with velocity v plus the braking distance. Note that the sum in (5) is actually finite and that the formula coincides with [22, (5)]. Also, our (6) coincides with [22, (4)] (denoted by d_f there).

Our new reaction distance (or gap) g is defined as follows: and for smaller velocities, decreases linearly:

$$(v, type) \mapsto g = d_r \frac{v}{v_{max}} \tag{7}$$

i.e., the gap is equal to d_r when the velocity v is maximal, and decreases linearly to zero for smaller velocities. This reaction distance may also be interpreted as the speed of the driver's reaction.

As for the parameters, we may omit the dependence of g on the driver's type, and write just $g(v)$, by the sake of simplicity of notation.

2.3   Dynamics

Our system consists of N vehicles evolving in time steps $t = 0, 1, \ldots$. In each time moment t and for each vehicle n we store:

$$(x_n(t), d_n(t), v_n(t), l_n, type_n), \tag{8}$$

where: $x_n(t)$ is the vehicle's n position, $d_n(t) = x_{n+1}(t) - l_{n+1} - x_n(t)$ is its distance to the preceding one, $v_n(t)$ its velocity at time t. The static variables $l_n, type_n$ are the length and type of the vehicle n. Let us describe the evolution from time step t to $t+1$.

Dynamic Safe Distances. We first define, as a function of $v_n(t)$, the following additional distances, which are going to be crucial for the agent's decision on its speed in the next time moment:

$$(v_n(t), type) \mapsto \begin{cases} d_{-,n}(t) = d_{safe}(v_n(t) + \Delta^- v) + g(v_n) \\ d_{0,n}(t) = d_{safe}(v_n(t)) + g(v_n) \\ d_{+,n}(t) = d_{safe}(v_n(t) + \Delta^+ v) + g(v_n) \end{cases}, \quad (9)$$

where d_{safe} and g are defined in (6)–(7). Thus, $d_{0,n}(t)$ is the sum of the safe distance for the actual velocity, and of the reaction gap for the same velocity; $d_{-,n}(t)$ is the sum of a distance which is safe if the agent decreases its velocity, and of the reaction gap for the actual velocity; $d_{+,n}(t)$ is the sum of a distance which is safe if the agent increases its velocity, and of the reaction gap for the actual velocity. One may also consider a model with the same velocities in both components of the sum; however, this would demand a slight modification of the gap definition (7). It is a matter of choice and is not crucial for the present study.

In the same way, we define the dynamic distances, which take additionally into account the velocity of the preceding vehicle, that the driver of the vehicle n is assumed to estimate (see [22, (1)–(3)]):

$$D_{*,n}(t) = \max(d_{*,n}(v_n(t)) - d_{brake}(v_{n+1}(t)), 0) \quad (10)$$

where $* \in \{-, 0, +\}$.

Randomness. In each time step, we take a random number $r(t) \in [0,1]$.

The Decision on the Velocity Change. The determination of $v_n(t+1)$, follows the main lines of [22, (S3)]. However, we do not use the random factor for acceleration, which the authors introduce there, as we do not want to overload the model with randomness and instead determine the influence of various drivers' behaviour on the dynamics.

$$v_n(t+1) = \begin{cases} \min(v_n(t) + \Delta^+ v; v_{max}) & \text{if } d_n(t) \geq D_{+,n}(t) \\ & \text{(accelaration)} \\ \\ v_n(t) & \text{if } (d_n(t) \in [D_{0,n}, D_{+,n}(t)) \text{ or } v_n(t) = v_{max}) \\ & \text{and } r(t) > R_s \\ & \text{(random keeping the same speed)} \\ \\ \max(v_n(t) + \Delta^- v; 0) & \text{if } (d_n(t) \in [D_{0,n}, D_{+,n}(t)) \text{ or } v_n(t) = v_{max}) \\ & \text{and } r(t) \leq R_s \\ & \text{(random slowing down)} \\ \\ \max(v_n(t) + \Delta^- v; 0) & \text{if } d_n(t) \in [D_{-,n}, D_{0,n}(t)) \\ & \text{(braking)} \\ \\ \max(v_n(t) + \Delta_{min} v; 0) & \text{if } d_n(t) < D_{-,n} \\ & \text{(emergency braking)} \end{cases} \quad (11)$$

Movement. Once the decision on the speed is taken, the vehicles move accordingly; their velocities and the distances between them are updated:

$$x_n(t+1) = x_n(t) + v_n(t+1), \tag{12}$$

$$d_n(t+1) = x_{n+1}(t+1) - l_{n+1} - x_n(t+1). \tag{13}$$

2.4 Agents Behaviour

We give in Table 1 the five types of agents, their values of acceleration and deceleration and the gap factor parameter d_r, which determines the gap to the previous car according to (7).

Table 1. Parameter settings for different agent types.

Agent type	Gap (d_r)	Acceleration (a_+)	Deceleration (a_-)
I	(0;5;10)	slow (1)	slow (−1)
II	(0;5;10)	moderate (2)	moderate (−2)
III	(0;5;10)	aggressive (4)	aggressive (−4)
IV	(0;5;10)	aggressive (4)	slow (−1)
V	(0;5;10)	slow (1)	aggressive (−4)

Agents of type I and II are the most standard, considered in most of the cited research. Their specificity—the value of acceleration and deceleration, and the gap (here $d_r = 0$)—is taken from other publications: [19, 20, 23]. Agents of types III, IV and V correspond to the aggressive driving style. Agent III is totally aggressive; the acceleration and deceleration values are almost the highest as for real data for various cars' descriptions [19]. The other agents (type IV and V) are defined in view of determining which one of the factors—the value of acceleration or braking—is more significant in maintaining traffic flow. Agent of type IV is aggressive in the context of acceleration, while Agent of type V is aggressive in the context of braking.

3   Numerical Results

Numerical tests were performed for one type of vehicle with a length of 8 CA cells, which is about 5 m. The length of each cell is set to 0.625 m. A small CA cell and mapping a car on a few CA cells allow for smaller vehicle movements, which is essential in the case of such research, in which small shifts make a big difference. The system size is assumed to be 3200 segments, corresponding to an actual road length of around 2 km. One time step approximately equals 1 s in real-time. Thus, the maximum velocity in the model is $v_{max} = 24$, which corresponds to

54 km/h in urban traffic. The above values (CA cell size, 1 s of time step) allowed us to obtain the integer values of real velocity (i.e. $v = 4$ corresponds to 9 km/h, $v = 8$–18 km/h, etc.). All charts with fundamental diagrams show the results of the arithmetic mean of 100 simulations for each density. Each simulation lasted 2,200 iterations. To analyze the results, the first 200 time steps of the simulation are discarded to let short-lived states die out and reach the system's steady state.

We used a PC with AMD Ryzen 9 (5950X), 3.40 GHz, 32GB DDR4 3200 MHz CL16 dual-channel, working under MS Windows 11 (64-bits) for all simulations.

The first study shows the impact of drivers' behaviours on the traffic flow for left, right and both lanes, respectively (Fig. 2). In this study, the parameter d_r is constant and equals 0 for each agent type. Each line presents a traffic flow for a 100% of the certain type of agent (Table 1) and road occupation. As the chart presents, each form of aggressiveness reduces the flow of vehicles. Moreover, the non-trivial conclusion is that aggressiveness in braking is more destructive to the traffic flow than aggressiveness in acceleration. Agents type III and V, with the highest deceleration value, make traffic flow the least efficient, while agents type IV slightly reduce it. In addition, it is worth noting the change in the shape of the curve representing the road occupancy for agents type V. Its slope changes sharply in the range of road occupancy from 40% to 70%. This change requires further analysis in future studies.

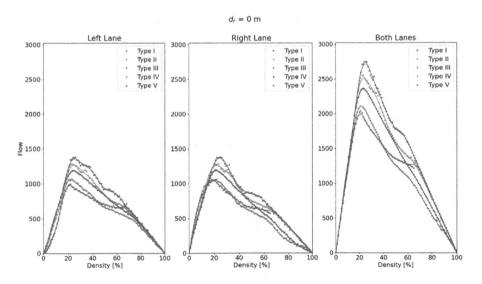

Fig. 2. Fundamental diagram of the presented model for various types of drivers, and $d_r = 0$.

The space-time plots were additionally generated for this simulation study (Fig. 3). We mark in red the moment when the lane changes occur. In this way, the specificity of the behaviour of individual agents is revealed. Agents III and V

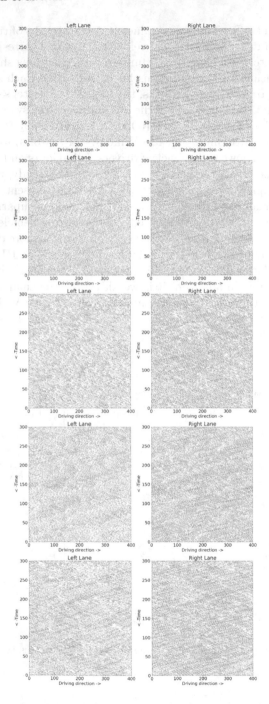

Fig. 3. The space-time plots of the present model, where $d_r = 0$. Each line presents a diagram for the left and right lanes. The next lines present the results for agents I-V (from the top). The vehicles move from the left to the right, and the vertical direction (down) of each graph means the next iterations.

show up a higher frequency of changing lanes, which results in numerous changes in the speed of other vehicles and consequently reduces the total traffic flow.

Furthermore, Fig. 2 reflects the most common approach when the reaction distance (the gap between the vehicles) is 0: $d_r = 0$. Our aim is to prepare a CA model to study the effect of the extra gap to the previous car on traffic flow. For that reason, based on the CA model extension, the next simulations have been done, and the results are presented as fundamental diagrams (Figs. 4 and 5). The next simulation study was based on all the above–defined agent types, but all of them use the privilege to have a distance to the previous car set to 5 ($d_r = 5$). This value is reduced as the vehicle's speed decreases, as it was described in Sect. 2.1.

The possibility of an additional distance to the vehicle ahead revealed a smaller range of maximum traffic flow value in relation to individual agents, about 25% (Fig. 4) relative to 37% (Fig. 2). In addition, the flow characteristics of traffic have changed. Increasing the gap caused disturbances in the trends in the area of road occupation oscillating in the range of about 65% to 95%. In the first study, agents type III generated the lowest ratio of traffic flow, while in the second simulation, agents type IV generated the lowest throughput.

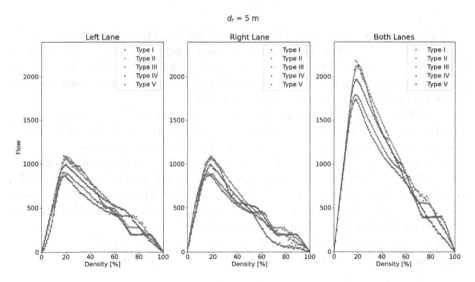

Fig. 4. Fundamental diagram of the presented model for various types of drivers, and $d_r = 5$.

In the next study, we increased the value of parameter d_r, which now equals 10. The results are presented in Fig. 5. Traffic flow decreased by about 48% relative to agent type I and about 33% relative to agent type V in the study with parameter $d_r = 0$. While the relationship between this study and the study with $d_r = 5$ is a decrease of about 19% and about 13%, relative to agents type I and V.

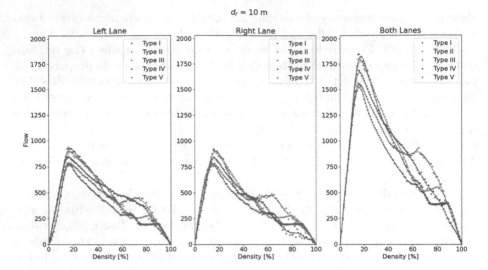

Fig. 5. Fundamental diagram of the presented model for various types of drivers, and $d_r = 10$.

The space-time plots were also generated for this simulation study (Fig. 6). Again, agents III and V have increased lane change variability. However, it is smaller than when the parameter $d_r = 0$, which means that the value of this parameter is important for the overall traffic flow. The increased distance between vehicles has an impact on traffic stability.

In addition, when the distance between the vehicles was increased, there were disturbances in traffic flow for agents type I, II, III and V, in the range of the road occupancy from about 60% to about 85%. This applies to disturbances on each lane separately and on both lanes together. The results show a situation where slow acceleration and braking generate an advantage in the form of a fluent going of the whole stream of vehicles, even though for the road occupancy above 60% traffic flow is lower for agents type III, and for the road occupancy from about 80% for other agent types, for $d_r = 10$ (Fig. 5).

So as to present the difference between agent types more clearly, fundamental diagrams for various values of d_r parameter are given in Figs. 7, 8.

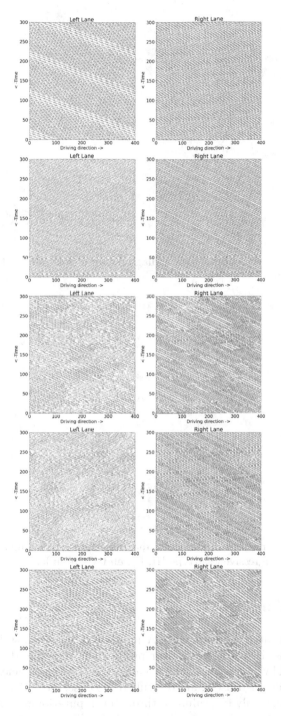

Fig. 6. The space-time plots of the present model, where $d_r = 10$. Each line presents a diagram for the left and right lanes. The next lines present the results for agents I-V (from the top).

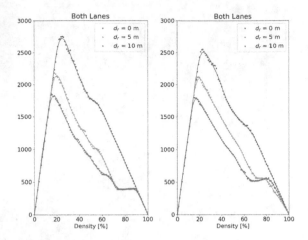

Fig. 7. The comparison of fundamental diagrams for various values of d_r for agent type I and II (from left).

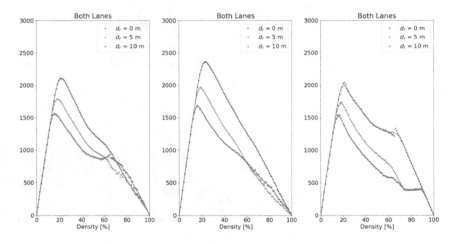

Fig. 8. The comparison of fundamental diagrams for various values of d_r for agents with any kind of aggressiveness (type III, IV, and V - from left).

4   Conclusion

This paper presents a multi-agent CA model to evaluate how heterogeneity in road traffic caused by different driver profiles affects traffic dynamics. Drivers' profiles, defined by varying acceleration and decelerations policies and the redundant gap to other road users, are difficult to be observed in standard road measurements and can only be evaluated through computational simulations. Therefore, five types of drivers were defined, and the differences between them were shown based on fundamental and space-time diagrams. Aggressive braking has been shown to be more destructive to traffic flow than aggressive acceleration. In addition, the model uses the representation of a single vehicle as a set of several CA cells, which allows for a more realistic representation of urban speeds and parameters considered in this article (acceleration and deceleration values and the distance between vehicles).

Finally, the proposed model preserves computational simplicity, and its rules make it possible to use parallel computing. This critical feature of the CA models is therefore held.

The plan for future research is to examine the effect of the distance to the preceding vehicle in relation to the agent types. This parameter turned out to be an important factor causing a change in the characteristics of traffic flow, especially when road occupancy increases. This knowledge is essential from the point of view of designers of algorithms controlling autonomous vehicles.

Acknowledgements. This research was supported by ZUT Highfliers School (Szkoła Orłów ZUT) project coordinated by Assoc. Prof. Piotr Sulikowski within the framework of the program of the Minister of Education and Science (Grant No. MNiSW/2019/391/DIR/KH, POWR.03.01.00-00-P015/18), co-financed by the European Social Fund, the amount of financing PLN 2,634,975.00.

References

1. Macioszek, E.: Analysis of driver behaviour at roundabouts in Tokyo and the Tokyo surroundings. In: Macioszek, E., Sierpiński, G. (eds.) TSTP 2019. AISC, vol. 1083, pp. 216–227. Springer, Cham (2020). https://doi.org/10.1007/978-3-030-34069-8_17

2. Sierpiński, G.: Revision of the modal split of traffic model. In: Mikulski, J. (ed.) TST 2013. CCIS, vol. 395, pp. 338–345. Springer, Heidelberg (2013). https://doi.org/10.1007/978-3-642-41647-7_41

3. Macioszek, E.: Models of critical gaps and follow-up headways for turbo roundabouts. In: Macioszek, E., Akçelik, R., Sierpiński, G. (eds.) TSTP 2018. LNNS, vol. 52, pp. 124–134. Springer, Cham (2019). https://doi.org/10.1007/978-3-319-98618-0_11

4. Macioszek, E., Sierpiński, G., Czapkowski, L.: Methods of modeling the bicycle traffic flows on the roundabouts. In: Mikulski, J. (ed.) TST 2010. CCIS, vol. 104, pp. 115–124. Springer, Heidelberg (2010). https://doi.org/10.1007/978-3-642-16472-9_12

5. Treiber, M., Hennecke, A., Helbing, D.: Congested traffic states in empirical observations and microscopic simulations. Phys. Rev. E **62**, 1805–1824 (2000)

6. Nagel, K., Schreckenberg, M.: A cellular automaton model for freeway traffic. J. Phys. I France **2**, 2221–2229 (1992)

7. Chopard, B., Luthi, P.O., Queloz, P.A.: Cellular automata model of car traffic in a two-dimensional street network. J. Phys. A: Math. Gen. **29**, 2325–2336 (1996)

8. Schadschneider, A., Schreckenberg, M.: Traffic flow models with 'slow-to-start'rules. Ann. der Phys. **509**, 541–551 (1997)

9. Nagel, K., Wolf, D.E., Wagner, P., Simon, P.: Two-lane traffic rules for cellular automata: a systematic approach. Phys. Rev. E **58**, 1425 (1998)

10. Rickert, M., Nagel, K., Schreckenberg, M., Latour, A.: Two lane traffic simulations using cellular automata. Phys. A: Stat. Mech. Appl. **231**, 534–550 (1996)
11. Liu, M., Shi, J.: A cellular automata traffic flow model combined with a BP neural network based microscopic lane changing decision model. J. Intell. Transp. Syst. **23**, 309–318 (2019)
12. Małecki, K., Gabryś, M.: The computer simulation of cellular automata traffic model with the consideration of vehicle-to-infrastructure communication technology. SIMULATION **96**, 911–923 (2020)
13. Wójtowicz, J., Wadowski, I., Dyrda, B., Gwizdałła, T.M.: Traffic on small grids and the ramp problem. In: Mauri, G., El Yacoubi, S., Dennunzio, A., Nishinari, K., Manzoni, L. (eds.) ACRI 2018. LNCS, vol. 11115, pp. 196–206. Springer, Cham (2018). https://doi.org/10.1007/978-3-319-99813-8_18
14. Małecki, K.: The use of heterogeneous cellular automata to study the capacity of the roundabout. In: Rutkowski, L., Korytkowski, M., Scherer, R., Tadeusiewicz, R., Zadeh, L.A., Zurada, J.M. (eds.) Artificial Intelligence and Soft Computing, pp. 308–317. Springer International Publishing, Cham (2017)
15. Iwan, S., Małecki, K.: Utilization of cellular automata for analysis of the efficiency of urban freight transport measures based on loading/unloading bays example. Transp. Res. Procedia **25**, 1021–1035 (2017)
16. Chmielewska, M., Kotlarz, M., Was, J.: Computer simulation of traffic flow based on cellular automata and multi-agent system. In: Wyrzykowski, R., Deelman, E., Dongarra, J., Karczewski, K., Kitowski, J., Wiatr, K. (eds.) Parallel Processing and Applied Mathematics, pp. 517–527. Springer International Publishing, Cham (2016)
17. Małecki, K.: A computer simulation of traffic flow with on-street parking and drivers' behaviour based on cellular automata and a multi-agent system. J. Comput. Sci. **28**, 32–42 (2018)
18. Małecki, K., Kamiński, M., Wąs, J.: A multi-cell cellular automata model of traffic flow with emergency vehicles: effect of a corridor of life and drivers' behaviour. J. Comput. Sci. **61**, 101628 (2022)
19. Shang, H.Y., Peng, Y.: A new three-step cellular automaton model considering a realistic driving decision. J. Stat. Mech: Theory Exp. **2012**, P10001 (2012)
20. Lárraga, M.E., Alvarez-Icaza, L.: Cellular automaton model for traffic flow based on safe driving policies and human reactions. Phys. A: Stat. Mech. Appl. **389**, 5425–5438 (2010)
21. Chmura, T., Herz, B., Knorr, F., Pitz, T., Schreckenberg, M.: A simple stochastic cellular automaton for synchronized traffic flow. Phys. A: Stat. Mech. Appl. **405**, 332–337 (2014)
22. GUZMÁN, H., Larraga, M.E., Alvarez-Icaza, L., et al.: A two lanes cellular automata model for traffic flow considering realistic driving decisions. J. Cell. Automata **10** (2015)
23. Li, Z.H., Zheng, S.T., Jiang, R., Tian, J.F., Zhu, K.X., Di Pace, R.: Empirical and simulation study on traffic oscillation characteristic using floating car data. Phys. A: Stat. Mech. Appl. **605**, 127973 (2022)

A Contrastive Self-distillation BERT with Kernel Alignment-Based Inference

Yangyan Xu[1,2], Fangfang Yuan[1(✉)], Cong Cao[1(✉)], Majing Su[3], Yuhai Lu[1], and Yanbing Liu[1,2]

[1] Institute of Information Engineering, Chinese Academy of Sciences, Beijing, China
{xuyangyan,yuanfangfang,caocong,luyuhai,liuyanbing}@iie.ac.cn
[2] School of Cyber Security, University of Chinese Academy of Sciences, Beijing, China
[3] The 6th Research Institute of China Electronic Corporations, Beijing, China
sumj@ncse.com.cn

Abstract. Early exit, as an effective method to accelerate pre-trained language models, has recently attracted much attention in the field of natural language processing. However, existing early exit methods are only suitable for low acceleration ratios due to two reasons: (1) The shallow classifiers in the model lack semantic information. (2) Exit decisions in the intermediate layers are unreliable. To address the above issues, we propose a Contrastive self-distillation BERT with kernel alignment-based inference (CsdBERT), which aims to let shallow classifiers learn deep semantic knowledge to make comprehensive predictions. Specifically, we classify the early exit classifiers into teachers and students based on classification loss to distinguish the representation ability of the classifiers. Firstly, we present a contrastive learning approach between teacher and student classifiers to maintain the consistency of class similarity between them. Then, we introduce a self-distillation strategy between these two kinds of classifiers to solidify learned knowledge and accumulate new knowledge. Finally, we design a kernel alignment-based exit mechanism to identify samples of different difficulty for accelerating BERT inference. Experimental results on the GLUE and ELUE benchmarks show that CsdBERT not only achieves state-of-the-art performance, but also maintains 95% performance at 4× speed.

Keywords: Early exit · Contrastive learning · Self-distillation · Centered kernel alignment

1 Introduction

Pre-trained language models (PLMs) have become the most promising models in the field of natural language processing (NLP), such as BERT [1], GPT [2], XLNet [3], RoBERTa [4], ALBERT [5], etc., which bring significant improvements to NLP tasks. Despite the great success of PLMs, they incur computational consumption, which leads to very slow inference and high latency. To cope with these issues, static model compression techniques are used to accelerate PLMs inference, such as knowledge distillation [6], quantization [7], and pruning [8], etc. However, this class of static compression methods aims to obtain a compact model, resulting in a dramatic performance degradation. Conversely, dynamic early exit [9] has proven to be an effective way to reduce

© The Author(s), under exclusive license to Springer Nature Switzerland AG 2023
J. Mikyška et al. (Eds.): ICCS 2023, LNCS 14073, pp. 553–565, 2023.
https://doi.org/10.1007/978-3-031-35995-8_39

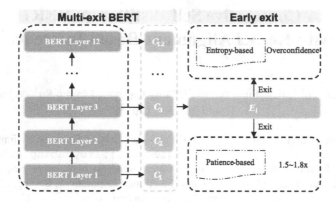

Fig. 1. On the left is an example of a multi-exit BERT, and on the right are the two mainstream exit mechanisms (where c_1 to c_{12} denote early exit classifiers and E_i denotes exit at some layer i).

computation and latency. The main idea of dynamic early exit models [9–13] is to add additional classifiers to the different layers of PLMs. Dynamic early exit PLMs mainly involve training these classifiers in the fine-tuning stage, thus exiting early in the inference stage. Hence, it is important to design an effective exit mechanism for the inference stage.

Existing exit methods can be categorized into entropy-based and patience-based methods, as shown in Fig. 1. The entropy-based methods [10, 11, 13, 14] aim to compute the entropy of the predicted probability distribution as an estimate of the confidence of exiting classifiers. These methods are overconfident in some classifiers, making them unreliable indicators of confidence, and the ability of the low layers may not match their high confidence scores. The patience-based methods [9, 12, 15] rely on the cross-layer consistency prediction to make an exit decision. Unfortunately, these methods using the patience index only provide very limited acceleration ratios and cannot be applied to complex real-world scenarios.

To solve the above problems, we propose CsdBERT: a Contrastive self-distillation BERT with kernel alignment-based inference to reduce computational cost and inference latency. Our CsdBERT contains a contrastive learning approach and a self-distillation strategy in the fine-tuning phase, and a kernel alignment-based exit decision in the inference phase. Specifically, we first rank the early exit classifiers by classification loss, which can effectively distinguish strong teacher classifiers from weak student classifiers. For the contrastive learning approach, teachers and students are obtained from the encoder outputs of the early exit classifiers, and the ensemble teacher is obtained by averaging multiple teachers to allow the student classifiers to learn the contrastive training signal of the ensemble teacher at the encoder level. For the self-distillation strategy, the teachers composed of soft-labels become an expert teacher through the weight of difficulty perception, so that students can learn the rich knowledge of the expert teacher and reduce forgetting. Finally, we design a kernel alignment-based exit mechanism to calibrate the predictions of the model based on the instance difficulty, which satisfies the requirements of acceleration ratios in different scenarios.

Extensive experiments are carried out on the GLUE and ELUE benchmarks, and the results show that our proposed method not only outperforms the state-of-the-art methods, but also achieves the highest acceleration ratio.

2   Related Work

Large-scale pre-trained language models based on the Transformer architecture show excellent performance in the field of NLP. However, such models have a large number of parameters, resulting in large memory requirements and computational consumption during inference. To deal with this problem, studies [16–21] on improving the efficiency of over-parameterized models has gradually emerged.

Static Compression. Knowledge distillation [22], as a model compression technique, compacts the model structure to a smaller model, and keeps static for all instances in the inference process. In the pre-training stage, [16] reduces the size of the BERT model by knowledge distillation, which makes the training cost of the model less and the inference time shorter. [17] proposes deep self-attention distillation, and student models are trained by deep imitation of self-attention module, which shows that using the knowledge of the last Transformer layer can alleviate the difficulties of layer mapping between teacher and student models. [18] proposes progressive module replacing, which provides a new perspective for model compression without additional loss functions. [23] demonstrates that the teacher network can learn to better transfer knowledge to the student network by distilling the feedback on the student network's performance in a meta-learning framework. However, these static compression methods have to distill the model from scratch to meet different speedup requirements and treat instances requiring different computational costs indiscriminately.

Dynamic Early Exit. To meet different speedup constraints, instance adaptation methods [14,24] have been proposed to adjust the number of executed layers for different instances. Among them, dynamic early exit is an efficient way to adaptively speed up inference, which is first used in computer vision tasks [25–27]. [28] makes full use of the idea of early exit, and proposes calibrated confidence scores to make early exit decisions, which are applied to NLP tasks to show the effectiveness of the method. [10] proposes a two-stage training method. In the first stage, the classifier of the last Transformer layer is fine-tuned; In the second stage, the parameters fine-tuned in the first stage are frozen, and then the remaining eleven classifiers are updated. To make predictions with fewer Transformer layers, [9] proposes a joint training method to train each early exit classifier, and make the model stop inference dynamically through cross-layer consistent prediction. In addition to the joint training and two-stage training methods, [29] proposes an alternating training method, which combines the advantages of the first two training methods and extends the idea of early exit to regression tasks. [11] proposes a speed-adjustable BERT model, which improves the inference time of NLP model through sample-wise adaptation and self-distillation mechanism. [30] analyzes the working mechanism of dynamic early exit and shows that dynamic selection of appropriately sized models in a cascading manner can provide a comprehensive representation for prediction. To improve the performance of early exit classifiers, [15] reveals that each early exit classifier can learn from each other, and improves the optimization results through a cross-level optimization algorithm. [12] uses mutual learning and gradient alignment for knowledge distillation, which shows that deep classifiers can also learn from shallow classifiers, and the conflict between cross-entropy loss and distillation loss can be eliminated by gradient alignment. [14] proposes a unified horizontal and vertical multi-perspective early exit framework to achieve a trade-off between efficiency and performance. However, most of the research on this subject has not fully exploited semantic information of high-layer classifiers, and the judgment of exit decisions is not accurate enough.

3   Method

Our CsdBERT aims at learning deep semantic knowledge to realize efficient inference. The main framework of CsdBERT is shown in Fig. 2. Firstly, we train early exit classifiers by classification loss, and then we present a contrastive learning design and elaborate a self-distillation strategy. Finally, we design a kernel alignment-based exit mechanism.

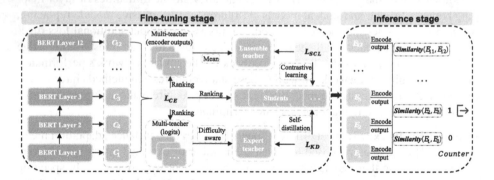

Fig. 2. The framework of our proposed CsdBERT. There are three main loss functions in the fine-tuning stage: classification loss $\mathcal{L}_{CE}$, contrastive loss $\mathcal{L}_{SCL}$, and distillation loss $\mathcal{L}_{KD}$. *Centered kernel alignment* is used as a similarity index to measure the output of the encoder in the exit layer, and the prediction can be accurately output in advance by counting.

3.1   Early Exit

The base version of the BERT model typically contains twelve layers of transformers. The input sample is first tokenized as a sequence of subwords, i.e., $X = [x_1, x_2, ..., x_K]$, with the corresponding ground truth label y, where K is the length of the sequence. Firstly, a special token $[CLS]$ is added to the head of the sequence so that each layer's corresponding hidden state $h_{[CLS]}^{(l)}$ is encoded through a multilayer encoding process that includes all representative information about all tokens. For each layer, the encoding process is defined as follows:

$$\left[h_{[CLS]}^{(l)}, h_1^{(l)}, \ldots, h_N^{(l)}\right] = f_\theta^{(l)}\left(\left[h_{[CLS]}^{(l-1)}, h_1^{(l-1)}, \ldots, h_N^{(l-1)}\right]; \theta\right), \quad (1)$$

where $f_\theta^{(l)}$ is the Transformer encoder at the l_{th} layer of the θ parameterization. $h_{[CLS]}^{(l)}$ is considered to be the features that train the early exit classifier $c^{(l)}$.

The predicted probability distribution $\hat{y}^{(l)}$ of the early exit classifier for the ground-truth label is computed as follows:

$$c^{(l)} = \tanh\left(W_c^{(l)} h_{[CLS]}^{(l)} + b_c^{(l)}\right), \quad (2)$$

$$\hat{y}_c^{(l)} = \text{softmax}\left(c^{(l)}\right), \quad (3)$$

where tanh is the activation function, $W_c^{(l)}$ is the weight, and $b_c^{(l)}$ is the bias. The loss function for the target task is a categorical cross-entropy, which is defined as

$$\mathcal{L}_{CE}^{(l)} = -\sum_{l=1}^{L} \mathbb{I}(y) \circ \log\left(\hat{y}_c^{(l)}\right), \quad (4)$$

where y and $\hat{y}_c^{(l)}$ denote the ground truth and probability distribution of the l_{th} layer. $\mathbb{I}(y)$ represents a one-hot vector with the y_{th} component being one. $\circ$ denotes the element-wise multiplication operation.

3.2 Classification Loss

To train early exit classifiers of BERT, additional classifiers are added. For each classifier $c^{(l)}$ ($l = [1, 2, ..., L]$), its cross-entropy loss $\mathcal{L}_l^{CE}$ is first calculated. Then, we use the summed loss to facilitate the joint training of these classifiers:

$$\mathcal{L}_{CE} = \sum_{l=1}^{L} \mathcal{L}_l^{CE}, \quad (5)$$

where L is the number of early exit classifiers.

Based on the cross-entropy loss, we sort the encoder outputs corresponding to all classifiers, and select M students to be the set S (where $S = [S_1, S_2, ..., S_M]$) with larger losses and the remaining ones as teachers $T^{(encoder)}$. The ensemble teacher T_{ens} is obtained by averaging over multiple teachers.

3.3 Contrastive Loss

To alleviate the inherent semantic bias of the student classifiers in the training process, we introduce self-supervised contrastive loss [31], which explores the merit of the ensemble teacher classifier. The contrastive learning between the student classifiers and the ensemble teacher classifier can effectively improve the ability of shallow classifiers to extract semantic information. Our goal is to encourage the same samples to be pulled closer and the other samples be pushed away between the ensemble teacher and the student classifiers. The specific form is as follows:

$$l_{SCL}^{(x_i, x_j)} = \frac{e^{\text{sim}(T_{ens}(x_i), S_m(x_j))/\tau}}{\sum_{k=1, k \neq i}^{2Q} e^{\text{sim}(T_{ens}(x_i), S_m(x_k))/\tau}}, \tag{6}$$

where the sample x_i is from the ensemble teacher classifier T_{ens} and the sample x_j is from the student classifier S_m ($m = [1, 2, ..., M]$, where M is the number of students). Q is the batch size. For the ensemble teacher T_{ens} and student S_m, the similarity between them is computed, where $\text{sim}(., .)$ is the cosine similarity (the dot product of the input samples). τ denotes the contrastive temperature [32]. As suggested by [31], we also use the normalised softmax in Eq. (6), instead of using the cosine similarity measure directly. The contrastive loss takes the form of

$$\mathcal{L}_{SCL}^{(m)} = -\frac{1}{Q} \sum_{j=1}^{Q} \log l_{SCL}^{(x_i, x_j)}, \tag{7}$$

where the contrastive loss is defined as the arithmetic mean of the cross-entropy of the normalised similarity $l_{SCL}^{(x_i, x_j)}$. Then, the final contrastive loss is

$$\mathcal{L}_{SCL} = \frac{\sum_{m=1}^{M} m \cdot \mathcal{L}_{SCL}^{(m)}}{\sum_{m=1}^{M} m}. \tag{8}$$

3.4   Distillation Loss

Apart from maintaining the consistency of class similarities, distilling semantic knowledge from the teacher classifiers is also essential to alleviate the catastrophic forgetting. To this end, we divide soft-labels *logits* of classifiers into students S and teachers $T^{(logits)}$ (where $S = [S_1, S_2, ..., S_M]$, $T^{(logits)} = [T_1^{(logits)}, T_2^{(logits)}, ..., T_N^{(logits)}]$, M is the total number of students and N is the total number of teachers). Firstly, we design the difficulty function:

$$Dif^{(n)} = \frac{\sum_{i=1}^{Y} p_t^n(i) \log p_t^n(i)}{\log \frac{1}{Y}}, \tag{9}$$

where Y is the number of labeled classes, $n = [1, 2, ..., N]$. $p_t^n(i)$ is the distribution of output probability of the teacher classifier. Then, we multiply the corresponding teacher by $Dif^{(n)}$ as a weight to become an expert teacher T_{exp}:

$$T_{exp} = \sum_{n=1}^{N} Dif^{(n)} \cdot T_n^{(logits)}, \tag{10}$$

The distillation is performed as

$$\mathcal{L}_{KD} = \sum_{m=1}^{M} \mathcal{L}_{S_m \to T_{exp}}^{KD}, \tag{11}$$

where $\mathcal{L}^{KD}$ is the distillation loss [11]. S_m denotes the student ($m = [1, 2, ..., M]$). Since the model structure of the expert teacher T_{exp} and the student S_m is the same, self-distillation is carried out between them.

3.5 Total Loss

In the fine-tuning stage, we train early exit classifiers by the above three losses. The total loss function of CsdBERT is summarized as

$$\mathcal{L}_{Total} = \mathcal{L}_{CE} + \alpha\mathcal{L}_{SCL} + \beta\mathcal{L}_{KD}, \tag{12}$$

where α and β are hyper-parameters to balance the effect between contrastive loss and distillation loss.

3.6 Centered Kernel Alignment

To achieve a high acceleration ratio, we use an early exit based on centered kernel alignment (CKA) [33] in the inference stage. We first employ the encoder output E_i of each exit layer as input. Then, we use CKA as the similarity index. The similarity of the encoder outputs is computed two-by-two in turn, which is denoted as

$$\text{CKA}(E_i, E_{i+1}) = \frac{\text{HSIC}(E_i, E_{i+1})}{\sqrt{\text{HSIC}(E_i, E_i)\,\text{HSIC}(E_{i+1}, E_{i+1})}}, \tag{13}$$

where $E = [E_1, E_2, ..., E_L]$, L is the number of early exit classifiers. HSIC is

$$\text{HSIC}(E_i, E_{i+1}) = \frac{1}{(n-1)^2}\,\text{tr}(E_i H E_{i+1} H), \tag{14}$$

where H denotes the centering matrix, $H_n = I_n - \frac{1}{n}\mathbf{1}\mathbf{1}^{\mathrm{T}}$. n is the size of the E_i row vector. We count once when $Similarity(i+1)$ minus $Similarity(i)$ is less than θ ($Similarity(i) = CKA(E_i, E_{i+1})$), corresponding to an *inference threshold* of 1. We count twice, then *inference threshold* is 2, and so on.

4   Experiments

Datasets. We conduct experiments on the GLUE [34] and ELUE [35] benchmarks, as shown in Table 1 and Table 2, respectively. On the GLUE benchmark, we test on Recognizing Textual Entailment (RTE), Question Natural Language Inference (QNLI), and Multi-Genre Natural Language Inference Matched (MNLI) for the Natural Language Inference (NLI) task; Quora Question Pairs (QQP) for Paraphrase Similarity Matching. On the ELUE benchmark, we test on Microsoft Research Paraphrase Matching (MRPC) for Paraphrase Similarity Matching; Stanford Sentiment Treebank (SST-2) and IMDb for Sentiment Classification; SciTail and SNLI for the NLI task.

Baselines. We compare our method with two types of baselines: (1) **Compressed models**. We choose BERT-6L, BERT-of-Theseus [18], LayerDrop [36], DistilBERT [16],

Table 1. The GLUE benchmark.

Tasks	Datasets	#Train	#Dev	#Test
Natural Language Inference	RTE	2,490	277	3,000
Natural Language Inference	QNLI	104,743	5,463	5,463
Paraphrase Similarity Matching	QQP	363,849	40,430	390,965
Natural Language Inference	MNLI	392,702	9,815	9,796

Table 2. The ELUE benchmark.

Tasks	Datasets	#Train	#Dev	#Test
Paraphrase Similarity Matching	MRPC	3,668	408	1,725
Sentiment Classification	SST-2	8,544	1,101	2,208
Sentiment Classification	IMDb	20,000	5,000	25,000
Natural Language Inference	SciTail	23,596	1,304	2,126
Natural Language Inference	SNLI	549,367	9,842	9,824

and BERT-PKD [37] as baselines. (2) **Early exit models**. We choose state-of-the-art early exit models, including PABEE [9], DeeBERT [10], FastBERT [11], GAML-BERT [12], and CascadeBERT [30].

Evaluation Metrics. For QQP and MRPC, we report the unweighted average of accuracy and F1 score. For the other datasets, we simply adopt accuracy as the metric. We use the above metrics to evaluate our CsdBERT and baselines in all experiments, and the detailed information of metrics are shown in the literature [34,35].

Table 3. Results(%) on the GLUE benchmark.

Models	#Param	Speed-up	RTE	QNLI	QQP	MNLI	Average
BERT-base	109M	1.00×	67.1	86.8	88.7	83.5	81.5
BERT-6L	66M	1.96×	60.3	80.2	84.8	77.1	75.6
BERT-of-Theseus	66M	1.96×	63.6	82.4	86.5	80.7	78.3
LayerDrop	66M	1.96×	62.8	81.7	86.2	80.1	77.7
DistilBERT	66M	1.96×	63.1	82.2	86.4	79.8	77.9
BERT-PKD	66M	1.96×	63.4	82.5	86.8	80.6	78.3
CascadeBERT	66M	1.96×	64.6	89.4	71.2	83.0	77.1
GAML-BERT	109M	1.96×	66.8	85.1	89.1	83.2	81.1
PABEE	109M	1.89×	64.5	83.1	87.5	81.5	79.2
DeeBERT	109M	1.88×	63.9	82.5	87.3	81.2	78.7
FastBERT	109M	1.93×	64.9	83.6	88.1	82.1	79.7
CsdBERT(Ours)	109M	1.96×	**73.3**	**92.1**	**90.0**	**85.3**	**85.2**

Implementation Details. Following [35], the batch size is set to 32 and the learning rate is set to 2e−5. The contrastive temperature τ is set to 0.5 and θ is set to 0.06.

We fine-tune 5 epochs using Adam optimizer, and take the distillation temperature to be 4. Furthermore, we set α equal to 0.01 and β equal to 1. The implementation of CsdBERT is based on the HuggingFace Transformers Library [38].

Table 4. Results(%) on the ELUE benchmark.

Models	#Param	Speed-up	MRPC	SST-2	IMDb	SciTail	SNLI	Average
ElasticBERT-base	109M	1.00×	87.9	88.6	93.9	93.8	91.3	91.1
ElasticBERT-entropy	109M	1.88×	87.5	88.3	88.2	94.5	90.0	89.7
ElasticBERT-patience	109M	1.89×	86.7	88.7	88.0	93.9	90.1	89.5
CsdBERT(Ours)	109M	1.96×	**91.8**	**93.1**	**88.7**	**95.9**	**91.0**	**92.1**

4.1 Experimental Results on GLUE and ELUE

Comparison with State-of-the-Art Methods. To verify the effectiveness of CsdBERT, we first evaluate CsdBERT and our baselines on the GLUE benchmark with BERT as the backbone model. From Table 3, we can see that CsdBERT outperforms the state-of-the-art methods with the same speedup ratio. For instance, the accuracy of our method on RTE is 73.3%, which improves 6.5% compared with the best result of GAML-BERT. This is because our method not only enhances the ability of classifiers to recognize semantic knowledge, but also determines the difficulty of different samples to exit precisely in advance.

Further Comparison. In order to further explore the efficiency of our method, we conduct experiments on the ELUE benchmark with ElasticBERT [35] as the backbone model. From Table 4, we can see that CsdBERT outperforms comparative models. The reason is that we alleviate semantic bias between the strong classifier (the ensemble teacher classifier made by combining multiple teachers) and the weak classifiers (student classifiers) by contrastive learning. Then, through self-distillation, our weak classifiers can learn rich semantic knowledge from the strong classifier.

Table 5. Ablation study of CsdBERT.

Models	RTE	MRPC	SST-2
CsdBERT	**73.3**	**91.8**	**93.1**
w/o $\mathcal{L}_{SCL}$	70.4	90.3	87.4
w/o $\mathcal{L}_{KD}$	69.7	90.0	88.3

4.2 Ablation Study

We conduct ablation study to validate the effectiveness of the essential components of CsdBERT, and the results are shown in Table 5.

(1) w/o $\mathcal{L}_{SCL}$. The accuracy corresponding to all the three tasks shows a relatively large decrease in the absence of contrastive loss. This is because contrastive loss ensures the student classifiers to get rich contrastive training signals from the ensemble teacher classifier.

(2) w/o $\mathcal{L}_{KD}$. If distillation loss is removed, the performance of classifiers in each layer is degraded. This indicates that the student classifiers can learn new knowledge of the expert teacher classifier.

4.3 Parameter Analysis

Total Number of Teachers N. As shown in Fig. 3 (a), the total number of teachers has a different impact on performance. In order to select the optimal number of teachers, we conducted experiments on MRPC at $2\times$ speed. We analyse the effect of N by varying its value from 3 to 8. With the increase of teachers, the score first increases and then declines dramatically. As the score is highest when $N = 6$, we use $N = 6$ in all experiments reported.

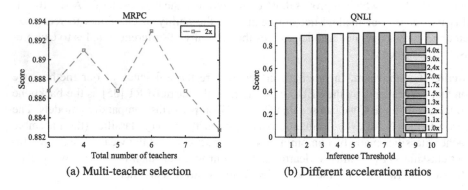

(a) Multi-teacher selection   (b) Different acceleration ratios

Fig. 3. Teacher selection and acceleration ratio. For the MRPC task, *Score* is the average of accuracy and F1 score. For the QNLI task, *Score* represents accuracy.

Inference Threshold R. On QNLI, we show the speedup ratios corresponding to different values of inference thresholds. Concretely, compared to BERT's accuracy on QNLI, CsdBERT still maintains 95% performance at $4\times$ speed. As illustrated in Fig. 3 (b), when R is 1, the speedup ratio is the highest.

5 Conclusion

In this paper, we propose CsdBERT for accelerating BERT inference. We first keep the consistency of class similarity between strong ensemble teacher and weak student classifiers via contrastive learning. Then, our student classifiers learn richer knowledge of the expert teacher via self-distillation. Finally, our designed kernel alignment-based mechanism can reflect the real difficulty of each instance and output more reliable predictions. Experimental results on the GLUE and ELUE benchmarks show that Csd-BERT outperforms the state-of-the-art methods, and maintains 95% performance at 4× speed. In the future, we will explore CsdBERT deployed to complex real-world scenarios.

Acknowledgement. This work is supported by Key Research and Development Program Projects of Xinjiang (No. 2022B03010-2), Strategic Priority Research Program of the Chinese Academy of Sciences (No. XDC02030400).

References

1. Devlin, J., Chang, M.W., Lee, K., Toutanova, K.: BERT: pre-training of deep bidirectional transformers for language understanding. In: Proceedings of the 2019 Conference of the North American Chapter of the Association for Computational Linguistics: Human Language Technologies, NAACL-HLT, pp. 4171–4186 (2019)
2. Radford, A., Narasimhan, K., Salimans, T., Sutskever, I.: Improving language understanding by generative pre-training. OpenAI (2018)
3. Yang, Z., Dai, Z., Yang, Y., Carbonell, J.G., Salakhutdinov, R., Le, Q.V.: Xlnet: generalized autoregressive pretraining for language understanding. In: Advances in Neural Information Processing Systems 32: Annual Conference on Neural Information Processing Systems 2019, NeurIPS, pp. 5754–5764 (2019)
4. Liu, Y., et al.: Roberta: a robustly optimized BERT pretraining approach. arXiv preprint arXiv:1907.11692 (2019)
5. Lan, Z., Chen, M., Goodman, S., Gimpel, K., Sharma, P., Soricut, R.: ALBERT: a lite BERT for self-supervised learning of language representations. In: International Conference on Learning Representations, ICLR (2020)
6. Hinton, G., Vinyals, O., Dean, J.: Distilling the knowledge in a neural network. arXiv preprint arXiv:1503.02531 (2015)
7. Tang, H., Zhang, X., Liu, K., Zhu, J., Kang, Z.: MKQ-BERT: quantized BERT with 4-bits weights and activations. arXiv preprint arXiv:2203.13483 (2022)
8. Lassance, C., Maachou, M., Park, J., Clinchant, S.: Learned token pruning in contextualized late interaction over BERT (colbert). In: SIGIR 2022: The 45th International ACM SIGIR Conference on Research and Development in Information Retrieval, pp. 2232–2236 (2022)
9. Zhou, W., Xu, C., Ge, T., McAuley, J.J., Xu, K., Wei, F.: BERT loses patience: fast and robust inference with early exit. In: Advances in Neural Information Processing Systems 33: Annual Conference on Neural Information Processing Systems 2020, NeurIPS (2020)
10. Xin, J., Tang, R., Lee, J., Yu, Y., Lin, J.: DeeBERT: dynamic early exiting for accelerating BERT inference. In: Proceedings of the 58th Annual Meeting of the Association for Computational Linguistics, ACL, pp. 2246–2251 (2020)
11. Liu, W., Zhou, P., Wang, Z., Zhao, Z., Deng, H., Ju, Q.: Fastbert: a self-distilling bert with adaptive inference time. In: Proceedings of the 58th Annual Meeting of the Association for Computational Linguistics, ACL, pp. 6035–6044 (2020)

12. Zhu, W., Wang, X., Ni, Y., Xie, G.: GAML-BERT: improving BERT early exiting by gradient aligned mutual learning. In: Proceedings of the 2021 Conference on Empirical Methods in Natural Language Processing, EMNLP, pp. 3033–3044 (2021)

13. Kong, J., Wang, J., Zhang, X.: Accelerating pretrained language model inference using weighted ensemble self-distillation. In: Wang, L., Feng, Y., Hong, Yu., He, R. (eds.) NLPCC 2021. LNCS (LNAI), vol. 13028, pp. 224–235. Springer, Cham (2021). https://doi.org/10.1007/978-3-030-88480-2_18

14. Kong, J., Wang, J., Yu, L., Zhang, X.: Accelerating inference for pretrained language models by unified multi-perspective early exiting. In: Proceedings of the 29th International Conference on Computational Linguistics, COLING, pp. 4677–4686 (2022)

15. Zhu, W.: Leebert: learned early exit for BERT with cross-level optimization. In: Proceedings of the 59th Annual Meeting of the Association for Computational Linguistics and the 11th International Joint Conference on Natural Language Processing, ACL/IJCNLP, pp. 2968–2980 (2021)

16. Sanh, V., Debut, L., Chaumond, J., Wolf, T.: Distilbert, a distilled version of bert: smaller, faster, cheaper and lighter. arXiv preprint arXiv:1910.01108 (2019)

17. Wang, W., Wei, F., Dong, L., Bao, H., Yang, N., Zhou, M.: Minilm: deep self-attention distillation for task-agnostic compression of pre-trained transformers. In: Advances in Neural Information Processing Systems 33: Annual Conference on Neural Information Processing Systems 2020, NeurIPS (2020)

18. Xu, C., Zhou, W., Ge, T., Wei, F., Zhou, M.: BERT-of-theseus: compressing BERT by progressive module replacing. In: Proceedings of the 2020 Conference on Empirical Methods in Natural Language Processing, EMNLP, pp. 7859–7869 (2020)

19. Ryu, M., Lee, G., Lee, K.: Knowledge distillation for BERT unsupervised domain adaptation. Knowl. Inf. Syst. **64**(11), 3113–3128 (2022)

20. Yao, S., Tan, J., Chen, X., Zhang, J., Zeng, X., Yang, K.: ReprBERT: distilling BERT to an efficient representation-based relevance model for e-commerce. In: KDD 2022: The 28th ACM SIGKDD Conference on Knowledge Discovery and Data Mining, pp. 4363–4371 (2022)

21. Gao, Y., Zhang, B., Qi, X., So, H.K.: DPACS: hardware accelerated dynamic neural network pruning through algorithm-architecture co-design. In: Proceedings of the 28th ACM International Conference on Architectural Support for Programming Languages and Operating Systems, vol. 2, ASPLOS, pp. 237–251 (2023)

22. Yuan, F., et al.: Reinforced multi-teacher selection for knowledge distillation. In: Thirty-Fifth AAAI Conference on Artificial Intelligence, AAAI, pp. 14284–14291 (2021)

23. Zhou, W., Xu, C., McAuley, J.J.: BERT learns to teach: knowledge distillation with meta learning. In: Proceedings of the 60th Annual Meeting of the Association for Computational Linguistics, ACL, pp. 7037–7049 (2022)

24. Zhang, Z., Zhu, W., Zhang, J., Wang, P., Jin, R., Chung, T.: PCEE-BERT: accelerating BERT inference via patient and confident early exiting. In: Findings of the Association for Computational Linguistics: NAACL, pp. 327–338 (2022)

25. Teerapittayanon, S., McDanel, B., Kung, H.T.: Branchynet: fast inference via early exiting from deep neural networks. In: 2016 23rd International Conference on Pattern Recognition (ICPR), pp. 2464–2469. IEEE (2016)

26. Kaya, Y., Hong, S., Dumitras, T.: Shallow-deep networks: understanding and mitigating network overthinking. In: Proceedings of the 36th International Conference on Machine Learning, ICML, vol. 97, pp. 3301–3310 (2019)

27. Elbayad, M., Gu, J., Grave, E., Auli, M.: Depth-adaptive transformer. arXiv preprint arXiv:1910.10073 (2019)

28. Schwartz, R., Stanovsky, G., Swayamdipta, S., Dodge, J., Smith, N.A.: The right tool for the job: matching model and instance complexities. arXiv preprint arXiv:2004.07453 (2020)

29. Xin, J., Tang, R., Yu, Y., Lin, J.: BERxiT: early exiting for BERT with better fine-tuning and extension to regression. In: Proceedings of the 16th Conference of the European Chapter of the Association for Computational Linguistics: Main Volume, EACL, pp. 91–104 (2021)
30. Li, L., et al.: Cascadebert: accelerating inference of pre-trained language models via calibrated complete models cascade. In: Findings of the Association for Computational Linguistics: EMNLP, pp. 475–486 (2021)
31. Atito, S., Awais, M., Kittler, J.: Sit: Self-supervised vision transformer. arXiv preprint arXiv:2104.03602 (2021)
32. Chen, T., Kornblith, S., Norouzi, M., Hinton, G.E.: A simple framework for contrastive learning of visual representations. In: Proceedings of the 37th International Conference on Machine Learning, ICML, vol. 119, pp. 1597–1607 (2020)
33. Kornblith, S., Norouzi, M., Lee, H., Hinton, G.E.: Similarity of neural network representations revisited. In: Proceedings of the 36th International Conference on Machine Learning, ICML, vol. 97, pp. 3519–3529 (2019)
34. Wang, A., Singh, A., Michael, J., Hill, F., Levy, O., Bowman, S.R.: Glue: a multi-task benchmark and analysis platform for natural language understanding. arXiv preprint arXiv:1804.07461 (2018)
35. Liu, X., et al.: Towards efficient NLP: a standard evaluation and a strong baseline. arXiv preprint arXiv:2110.07038 (2021)
36. Fan, A., Grave, E., Joulin, A.: Reducing transformer depth on demand with structured dropout. In: International Conference on Learning Representations, ICLR (2019)
37. Sun, S., Cheng, Y., Gan, Z., Liu, J.: Patient knowledge distillation for BERT model compression. In: Proceedings of the 2019 Conference on Empirical Methods in Natural Language Processing and the 9th International Joint Conference on Natural Language Processing, EMNLP-IJCNLP, pp. 4322–4331 (2019)
38. Wolf, T., Debut, L., Sanh, V., Chaumond, J., Delangue, C., Moi, A., et al.: Transformers: state-of-the-art natural language processing. In: Proceedings of the 2020 Conference on Empirical Methods in Natural Language Processing, EMNLP, pp. 38–45 (2020)

Turning Flight Simulation with Fluid-Rigid Body Interaction for Flying Car with Contra-Rotating Propellers

Ayato Takii[1,2]([✉]), Ritsuka Gomi[2], Masashi Yamakawa[2], and Makoto Tsubokura[1,3]

[1] RIKEN Center for Computational Science, 7-1-26 Minatojima-minami-machi, Chuo-ku,
Kobe, Hyogo 650-0047, Japan
`ayato.takii@riken.jp`
[2] Kyoto Institute of Technology, Matsugasaki, Sakyo-ku, Kyoto 606-8585, Japan
[3] Graduate School of System Informatics, Kobe University, 1-1 Rokkodai, Nada-ku, Kobe,
Hyogo 657-8501, Japan

Abstract. Toward realization of the Digital Flight for the next-generation vehicle, numerical simulation of turning flights of a flying car was performed with consideration of fluid-rigid body interaction. The vehicle use in the paper is electric vertical takeoff and landing (eVTOL) octorotor type with four contra-rotating propeller units which successfully performed manned flight test. The simulation is conducted such that the flying car mimicks its real-world counterpart by generating force solely through the rotation of its eight propellers. The moving computational domain method was adopted to realize the free movement of the vehicle in three-dimensional space. The whole computational grid is divided into eight domains to reproduce rotation of propellers. The propeller rotations are achieved without any simplification by applying the multi-axis sliding mesh approach. Moreover, to fly the flying car as intended, an attitude of the body in the flight is controlled properly to by the proportional-integral-derivative (PID) controller. As a result, the vehicle flies under only the lift generated by the propeller and the turning flights of a flying car with coaxial propellers are reproduced. In addition, this simulation results are different from analytical results based on simplified aerodynamic force. Therefore, it is suggested that this method is the effective one for numerical flight tests of a multi-rotor aircraft on a computer.

Keywords: CFD · Coupled Simulation · Moving Boundary

1 Introduction

Currently, there is an urgent necessity to reach net-zero carbon emissions to prevent environmental issues. Most countries are planning to achieve that by 2050–2070, and a promising movement for carbon neutrality is taking shape in the world [1]. One of strategies to reach a carbon neutrality is the streamlining and the electrification of the transportation. In recent years, small unmanned aerial vehicles called drones have become popular. As the next stage, the realization of aircrafts that can be easily used by individuals

© The Author(s), under exclusive license to Springer Nature Switzerland AG 2023
J. Mikyška et al. (Eds.): ICCS 2023, LNCS 14073, pp. 566–577, 2023.
https://doi.org/10.1007/978-3-031-35995-8_40

is expected, and these are expected as countermeasures against environmental problems and urban traffic problems. It's also reported that those flying cars have less greenhouse gas emissions than gasoline vehicles for long-distance transportation [2]. The different type of flying cars, especially electric vertical takeoff and landing (eVTOL) vehicles, are being developed around the world as next-generation mobility [3]. Practical use of eVTOL vehicles, as the new type of mobility, requires numerous tests to confirm safety, and cost a great deal of development time. Until now, aircraft development sites have adopted a method of repeating prototyping and testing, which has resulted in enormous amounts of time and money being spent. For eVTOL aerodynamic performance evaluation, various studies have been conducted. Experimental studies of eVTOL aircraft conducted to measure lift and drag forces through wind tunnel test [4] and to visualize the flow field around rotating rotors using particle image velocimetry [5]. Numerical simulations were employed to study the aerodynamic interaction of coaxial propellers [6]. The simulations considering the entire aircraft were done by modeling propeller lift [7] or by computing actual propeller rotation [8]. However, these studies were in the way of numerical wind tunnel with fixed body in Euler coordinate system and mainly focused on steady state. This means that most of previous studies focuses only on a few aspects of eVTOL aerodynamic performance. To design a safe airframe as intended, it is necessary to analyze the behavior of the airframe more comprehensively and practically.

The aim of realistic numerical flight simulation is to achieve Digital Flight [9], which facilitates the simulation and analysis of various types of flights using only computers, including take-off, landing, and emergency scenarios. This approach simplifies the analysis of the high-risk flight conditions and enables the execution of large-scale experiments that are difficult to carry out physically and socially. Consequently, Digital Flight leads to safer airframe designs at a lower cost. Currently, although the progress of computers' development doesn't catch up with the realization of complete Digital Flight, we have conducted some realistic flight simulations considering fluid-aircraft interaction [10, 11]. In present work, this method is newly applied to the eVTOL flying car. Few reports are available on the numerical flight simulation of eVTOL flying. The numerical simulation has a difficulty with a dynamically moving grid as well as complexity of feedback control for multi-rotor. Numerical simulation of turning flights of aircraft is challenging due to the complex attitude variation, acceleration, and long-distance travel involved. Although turning flight is frequently used in practice, it remains difficult to simulate. This study overcomes these challenges by applying Digital Flight to an eVTOL flying car, and performing a turning flight simulation.

As the model of the eVTOL flying car, we adopt an octorotor type with four contra-rotating propeller units which successfully performed manned flight test. This type is one of the choices for practical use due to the compactness of the body and the safety to keep the vehicle maneuverable in the event of propeller trouble, which is attributed to propeller redundancy. To reproduce realistic flight, the coupled computation between airflow and flying car is performed regarding the eVTOL flying car as a rigid body; that is, the airflow is driven by rotations of propellers, and the flying car flies only by the fluid forces applied to that body. Unsteady motion of the flying car is complex moving boundary problem in computational fluid dynamics. Although there are some approaches for solving moving boundary problem such as the arbitrary Lagrangian-Eulerian (ALE) [13, 14] and the immersed boundary method (IBM) [15, 16], we adopt the moving-grid finite-volume method (MGFV) method [17] which satisfies the geometric conservation law (GCL) automatically and derives the accurate fluid force with unstructured body fitted grid. The MGFV method allows deformation of the computational grid points by discretizing the governing equations in the space-time unified four-dimensional control volume. In addition, we use the moving computational domain (MCD) method [18] to enable a flying car to navigate a wide range of space compared with its own size. In this method, the whole computational grid moves accordingly to the motion of the object so that the flight of the flying car is not constrained by the size of the computational grid. Futhermore, we directly compute propeller rotations by combining with the multi-axis sliding mesh approach [19]. To fly the car as intended, the eight propellers must be controlled properly so that the attitude is stable. The proportional-integral-derivative control is used as the feedback system. This study demonstrates numerical flight simulation considering fluid-rigid body interactions for eVTOL flying car with contra-rotating propellers.

2 Numerical Approach

2.1 Governing Equations

In this study, a flying car flies a long distance consuming a tremendous wall-clock time to complete the numerical simulation. Therefore, this research uses the Euler equation as governing equation of compressible flow to reduce the computation time after considering that this simulation has high Reynolds number and low influence of viscosity. The flow equations and ideal gas law are given as follows:

$$\frac{\partial q}{\partial t} + \frac{\partial E}{\partial x} + \frac{\partial F}{\partial y} + \frac{\partial G}{\partial z} = 0, \tag{1}$$

$$q = \begin{bmatrix} \rho \\ \rho u \\ \rho v \\ \rho w \\ e \end{bmatrix}, E = \begin{bmatrix} \rho u \\ \rho u^2 + p \\ \rho uv \\ \rho uw \\ u(e+p) \end{bmatrix}, F = \begin{bmatrix} \rho v \\ \rho uv \\ \rho v^2 + p \\ \rho vw \\ v(e+p) \end{bmatrix}, G = \begin{bmatrix} \rho w \\ \rho uw \\ \rho vw \\ \rho w^2 + p \\ w(e+p) \end{bmatrix}, \tag{2}$$

$$p = (\gamma - 1)\left[e - \frac{1}{2}\rho\left(u^2 + v^2 + w^2\right) \right], \tag{3}$$

where q is the vector of conserved variables, E, F and G are the convective flux vectors, ρ is the density, u, v and w are the corresponding velocity components in the x, y and z-directions, respectively, p is the pressure and e is the total energy per unit volume. The ratio of specific heats γ is 1.4 in this study. Each variable is nondimensionalized as follows:

$$x = \frac{\tilde{x}}{\tilde{L}}, y = \frac{\tilde{y}}{\tilde{L}}, z = \frac{\tilde{z}}{\tilde{L}}, t = \frac{\tilde{t}}{\tilde{L}/\tilde{U}},$$

$$\tilde{p} = \frac{\tilde{p}}{\tilde{\rho}\tilde{U}^2}, u = \frac{\tilde{u}}{\tilde{U}}, v = \frac{\tilde{v}}{\tilde{U}}, w = \frac{\tilde{w}}{\tilde{U}}. \tag{4}$$

where ($\sim$) represents a dimensional quantity; $\tilde{L}(4\,\text{m})$ is the characteristic length, $\tilde{U}$ (340 m/s) is the characteristic velocity, $\tilde{\rho}$ (1.2 kg/m^3) is the density of the air.

The computational grid utilizes the body-fitted unstructured mesh with collocated arrangement; the values of the physical parameters are defined at cell centers. The governing equations are discretized by the MGFV method. This is a dynamic mesh method which is automatically satisfied by discretization performed using a four-dimensional control volume in the space and time unified domain, which enables deformation of the computational grid points. The convective flux vectors on faces of the control volume are evaluated by the Roe flux difference splitting scheme [20]. The conserved variables in the cell are reconstructed by the monotonic upstream-centered scheme for conservation laws (MUSCL) approach with Hishida's van Leer-like limiter [21]. To solve discretized equations, the dual time stepping is adopted. The two-stage Runge-Kutta method [22] is used for the pseudo time marching.

2.2 Moving Computational Domain Approach

The flying car moves in much wider space than its computational grid size. In conventional approach using a fixed computational domain, the flying car can't travel infinite distance due to the grid size limitation. To deal with such simulation, we use the moving computational domain approach based on the MGFV method. In this MCD method, the whole computational domain including a body inside moves rigidly in the world space (see Fig. 1). Thus, there is no limitation of the flight distance. The rotation of the propellers was achieved through a combination of the multi-axis sliding mesh approach [19] and the approach of moving the entire computational domain.

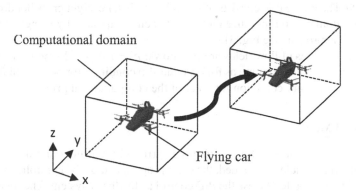

Fig. 1. Conceptual diagram of the moving computational domain method. The whole computational domain including a body inside moves rigidly in the world space.

2.3 Coupled Simulation with Rigid Body

In this study, the flight simulation of the flying car is conducted with fluid-rigid body interaction. A fluid flow around the flying car is driven by rotating propellers of that; the attitude and position of the car are determined by the fluid force of the air. The rigid body motion is determined by the Newton's equation of motion and the Euler's as follows:

$$m\frac{d^2r}{dt^2} = F,$$ (5)

$$I\frac{d\omega}{dt} + \omega \times I\omega = T,$$ (6)

where m is the mass, r is the position, F is the force, I is the inertia tensor, ω is the angular velocity and T is the torque. The force and torque applying to the flying car are calculated by integrating the pressure of the flow fluid on surface of the body. This simulation is weak coupling where the translation and rotation of the flying car are performed once every time step by using the force and the torque from the previous time step.

3 Flight Simulation of eVTOL Flying Car

3.1 Computational Model

The model of a flying car is based on SD-03, which was developed by SkyDrive in Japan [12]. The model has totally 8 propellers in total consisting of four contra-rotating propeller units. The flying car's mass is 400 kg, the overall length of the vehicle $\tilde{L}$, which is used as the characteristic length, is 4 m. The inertia tensor to solve the motion equations is calculated by integrating in the volume domain assuming that the density is uniform. The computational grid was created by MEGG3D [23, 24] shown as Fig. 2. The shape of the grid is a sphere with diameter of 30 $\tilde{L}$ and total number of grid cells is ~3,000,000. The whole grid consists of several domains, i.e. eight propeller domains and the fuselage domain, to reproduce motions of each component by using the multi-axis sliding mesh approach (see Fig. 3).

The initial values of flow fields are as follows: $\rho = 1.0, p = 1.0/\gamma, (u, v, w) = (0, 0, 0)$. The boundary conditions are as follows: the slip condition for body of the flying car, the Riemann boundary condition for outer of the computational grid.

3.2 Control Model

The flying car is subject to forces from the surrounding airflow and moves freely in space. To fly the vehicle as intended, the eight propellers must be controlled properly so that the attitude is stable. We use the PID control as feedback system. The rotation of the propellers is determined by the superposition of four operation variables: the throttle to control the altitude, the aileron to control the roll, the rudder to control the yaw, and the elevator to control the pitch. To get the target attitude, the operating value of the rotation of the propellers $h(t)$ is determined by the following equations:

$$h(t) = K_p e(t) + K_i \int_0^t e(\tau)d\tau + \frac{K_d de(t)}{dt},$$ (7)

$$e(t) = r_{target}(t) - r_{current}(t),$$ (8)

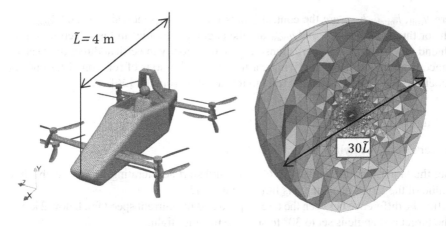

Fig. 2. Surface mesh of the flying car and the computational grid.

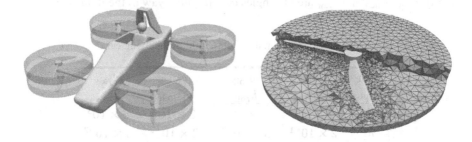

Fig. 3. Domain division for the sliding mesh approach and grid of a propeller domain.

where $e(t)$ is the error, $r_{\text{target}}(t)$ is the target value and $r_{\text{current}}(t)$ is the current value. K_{p}, K_{i} and K_{d} are the proportional, integral and differential gain. These gains are determined by tuning the linear single-input single-output (SISO) transfer function model while checking their step responses. Here, the kinematic model of the flying car is linearized by assuming that the changes in the propeller speed are enough small in the equations of the motion. By applying the mixing process [25], this model can be considered as a SISO system without coupling for target values. The transfer functions in the Laplace domain are regarded as follows:

$$\frac{\omega_{\text{roll}}(s)}{h_{\text{roll}}} = \frac{DK_T}{I_{xx}s}, \tag{9}$$

$$\frac{\omega_{\text{pitch}}(s)}{h_{\text{pitch}}} = \frac{LK_T}{I_{zz}s}, \tag{10}$$

$$\frac{\omega_{\text{yaw}}(s)}{h_{\text{yaw}}} = \frac{K_A}{I_{yy}s}, \tag{11}$$

$$\frac{v_y(s)}{h_{\text{throttle}}} = -\frac{K_T}{ms}, \tag{12}$$

where h_{roll}, h_{pitch}, h_{yaw} are the control inputs for aircraft principal axes and h_{throttle} is input for thrust. ω_{roll}, ω_{pitch} and ω_{yaw} are the angular velocities in the direction of roll, pitch and yaw axes. I is the inertia tensor, v_y is the velocity in vertical direction, D is the distance from the center of the flying car to the propeller axis of rotation. d K_T and K_A are coefficients obtained from the lift and torque of a propeller [26].

3.3 Flight Condition

In present work, the flying car flies as follows.

1. Set the target forward velocity to 60 km/h and start accelerating. Here, the absolute value of the target pitch angle is limited within 15°.
2. After the difference between the target speed and the current speed fell below 2 km/h, the target roll angle is set to 30° to start the turning flight.
3. After the aircraft turns 90°, stop turning flight and let the aircraft go straight.

The gains for each controlled values are shown in Table 1, where v_x is forward velocity, θ_{roll}, θ_{pitch} and θ_{yaw} are the angles of roll, pitch, yaw of the flying car.

Table 1. Control gain in turning flight of the flying car.

Input	v_y	ω_{roll}	ω_{pitch}	ω_{yaw}
Output	h_{throttle}	h_{roll}	h_{pitch}	h_{yaw}
K_p	2×10^4	3×10^2	2×10^3	5×10^3
K_i	2×10^{-1}	3×10^{-3}	2×10^{-2}	5×10^{-2}
K_d	1×10^4	1×10^2	8×10^2	8×10^3

Input	y	v_x	θ_{roll}	θ_{pitch}	θ_{yaw}
Output	v_y	θ_{pitch}	ω_{roll}	ω_{pitch}	ω_{yaw}
K_p	4.5×10^3	2×10^3	7.4	7.4	1.5
K_i	0	1×10^1	0	0	0
K_d	0	2×10^4	0	0	0

4 Result and Discussion

As the result of coupling simulation between fluid and rigid body with the PID control, the flying car flied stably with turning around. The trajectory of turning flight at 60 km/h is shown in Fig. 4. The size of square grid of the figure is 10 m. The car flied ~140 m in the x-direction from start of acceleration to finishing of 90° turning flight. This cruising distance is more than double size of the computational grid. Next, the time variations of flight attitude and forward velocity are shown in Fig. 5(a). The dotted line in the figure indicates the target forward velocity of 60 km/h. After the current velocity approached to the target velocity at 9 s, the velocity of the car is adjusted by increasing or decreasing its pitch angle. In the turning, the roll angle almost fitted the target value 30° although an overshoot occurred at the end of the turn. The time variations of the handling values for maneuvering attitude and the altitude of the flying car are shown in Fig. 5(b). The handling values changed rapidly at the start and end of turning. At time is 14 s, the flying car failed to keep constant altitude under the influence of a huge rudder input occurred at the end of the turn.

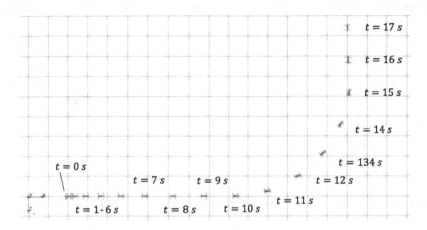

Fig. 4. The trajectory of the flying car in turning flight at 60 km/h.

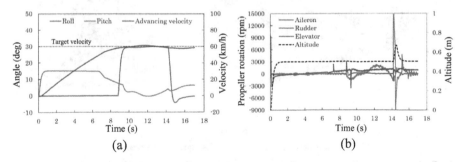

Fig. 5. Time variations of (a) attitude and forward velocity, and (b) handling values for maneuvering the attitude and altitude of the flying car.

Next, this result is compared with a simple theoretical solution. Assuming that the flight altitude and forward velocity are constant, the theoretical turning radius is calculated as follows:

$$r = \frac{v^2}{g\tan\theta},$$

(13)

where v is the forward velocity of the car, g is the gravitational acceleration and θ is the roll angle. By this equation, the theoretical turning radius with the forward velocity of 60 km/h and the roll angle of 30° is 49.0 m. The turning radius in the numerical simulation is calculated by fitting the trajectory data to the circle equation with the least squares method. Used trajectory is the section of 30° roll angle in the x-z plane. A circle of the obtained radius and the trajectory of the flying car are shown in Fig. 6. The turning radius obtained by fitting is 59.8 m. This is 22% larger than the theoretical turning radius. This may be affected by non-constant centripetal force (~10% fluctuation) during turning, as shown in Fig. 6(b). This result clarified that numerical simulation considering the aerodynamics is necessary to predict the turning motion of the flying car accurately.

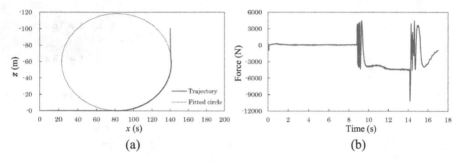

(a)   (b)

Fig. 6. (a) trajectory of the flying car and fitted circle. (b) Time variation of centripetal force.

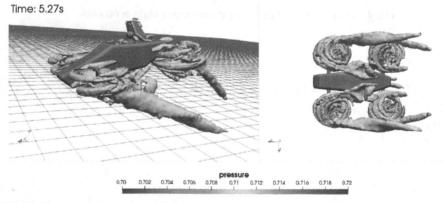

Time: 5.27s

pressure

Fig. 7. Pressure contour on the flying car and isosurface of Q-criterion ($Q = 0.1$) during acceleration.

Figure 7, 8 and 9 show pressure contour on the flying car and isosurface of Q-criterion ($Q = 0.1$) of the flow field around the body: during acceleration, at the start of the turn and during turning, respectively. The left is the perspective view, and the right is the bottom view of the body. In every flight stage, the complex vortices are generated by multiple propellers. While turning, the vortices from the front propellers are dragged backward and reach rear propellers. The vortex shape is more complicated at the transition stage of motion. This numerical simulation revealed such unsteadiness in turning flight.

Time: 9.41s

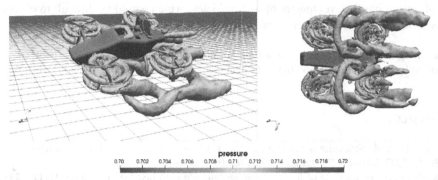

pressure
0.70   0.702   0.704   0.706   0.708   0.71   0.712   0.714   0.716   0.72

Fig. 8. Pressure contour on the flying car and isosurface of Q-criterion ($Q = 0.1$) at start of turning

Time: 11.94s

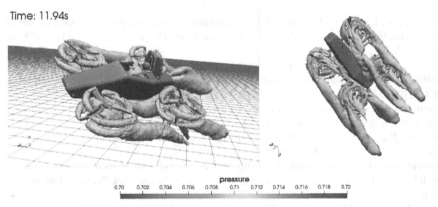

pressure
0.70   0.702   0.704   0.706   0.708   0.71   0.712   0.714   0.716   0.718   0.72

Fig. 9. Pressure contour on the flying car and isosurface of Q-criterion ($Q = 0.1$) during turning.

5 Conclusions

Turning flight simulation was computed for the eVTOL flying car with contra-rotating, considering the fluid and rigid body interaction. The MCD method based on the unstructured MGFV method was used to enable the flying car to move in the infinite world space without limitation of its computational size. Moreover, the multi-axis sliding mesh approach was adopted to rotate eight propellers of the flying car. To fly the flying car as intended, the PID controller is used to control the number of rotations of eight propellers so that the flight attitude such as pitch, roll and yaw were controlled properly. As a result of the coupled computation between the flow field and the rigid body using above approach, the flying car obtained the lift by rotating its propellers and flew in turning. Next, the flight trajectory was compared with the theoretical turning radius. The turning radius obtained from this numerical simulation was 22% larger than the theoretical one, which implies numerical simulation considering aerodynamics is important to predict the flight of the flying car. Finally, this realistic flight simulation allowed for the visualization of complex flow fields around the flying car and has the potential to provide valuable information for the future development of such vehicles, which would be difficult to obtain through experiments and conventional stational simulations.

Acknowledgments. This paper is based on results obtained from a project, JPNP14004, subsidized by the New Energy and Industrial Technology Development Organization (NEDO).

References

1. Chen, L., et al.: Strategies to achieve a carbon neutral society: a review. Environ. Chem. Lett. **20**(4), 2277–2310 (2022). https://doi.org/10.1007/s10311-022-01435-8
2. Kasliwal, A., et al.: Role of flying cars in sustainable mobility. Nat. Commun. **10**(1), 1555 (2019)
3. Bacchini, A., Cestino, E.: Electric VTOL configurations comparison. Aerospace **6**(3), 26 (2019)
4. Pardede, W.M., Adhitya, M.: Take off and landing performance analysis for a flying car model using wind tunnel test method. In: AIP Conference Proceedings, vol. 2227, no. 1, p. 020030. AIP Publishing LLC (2020)
5. Zanotti, A.: Experimental study of the aerodynamic interaction between side-by-side propellers in evtol airplane mode through stereoscopic particle image velocimetry. Aerospace **8**(9), 239 (2021)
6. Cornelius, J.K., Schmitz, S., Kinzel, M.P.: Efficient computational fluid dynamics approach for coaxial rotor simulations in hover. J. Aircr. **58**(1), 197–202 (2021)
7. Alvarez, E.J., Ning, A.: High-fidelity modeling of multirotor aerodynamic interactions for aircraft design. AIAA J. **58**(10), 4385–4400 (2020)
8. Zhu, H., Nie, H., Zhang, L., Wei, X., Zhang, M.: Design and assessment of octocopter drones with improved aerodynamic efficiency and performance. Aerosp. Sci. Technol. **106**, 106206 (2020)
9. Salas, M.D.: Digital flight: the last CFD aeronautical grand challenge. J. Sci. Comput. **28**(2–3), 479–505 (2006)
10. Takii, A., Yamakawa, M., Asao, S., Tajiri, K.: Six degrees of freedom numerical simulation of tilt-rotor plane. In: Rodrigues, J.M.F., et al. (eds.) ICCS 2019. LNCS, vol. 11536, pp. 506–519. Springer, Cham (2019). https://doi.org/10.1007/978-3-030-22734-0_37
11. Takii, A., Yamakawa, M., Asao, S.: Descending flight simulation of tiltrotor aircraft at different descent rates. In: Krzhizhanovskaya, V.V., et al. (eds.) ICCS 2020. LNCS, vol. 12143, pp. 178–190. Springer, Cham (2020). https://doi.org/10.1007/978-3-030-50436-6_13
12. SkyDrive Inc. (2022). https://en.skydrive2020.com/. Accessed 1 Feb 2023
13. Belytschko, T.: Fluid-structure interaction. Comput Struct **12**(4), 459–469 (1980)
14. Hu, H.H., Patankar, N.A., Zhu, M.: Direct numerical simulations of fluid–solid systems using the arbitrary Lagrangian-Eulerian technique. J. Comput. Phys. **169**(2), 427–462 (2001)
15. Peskin, C.S.: Flow patterns around heart valves: a numerical method. J. Comput. Phys. **10**(2), 252–271 (1972)
16. Peskin, C.S.: The immersed boundary method. Acta Numer **11**, 479–517 (2002)
17. Matsuno, K., Mihara, K., Satofuka, N.: A moving-mesh finite-volume scheme for compressible flows. In: Satofuka, N. (ed.) Computational Fluid Dynamics 2000, pp. 705–710. Springer, Heidelberg (2001). https://doi.org/10.1007/978-3-642-56535-9_107
18. Watanabe, K., Matsuno, K.: Moving computational domain method and its application to flow around a high-speed car passing through a hairpin curve. J. Comput. Sci. Technol. **3**(2), 449–459 (2009)

19. Takii, A., Yamakawa, M., Asao, S., Tajiri, K.: Six degrees of freedom flight simulation of tilt-rotor aircraft with nacelle conversion. J. Comput. Sci. **44**, 101164 (2020)
20. Roe, P.L.: Approximate Riemann solvers, parameter vectors, and difference schemes. J. Comput. Phys. **43**(2), 357–372 (1981)
21. Hishida, M., Hashimoto, A., Murakami, K., Aoyama, T.: A new slope limiter for fast unstructured CFD solver FaSTAR. In: Proceedings of 42nd Fluid Dynamics Conference/Aerospace Numerical Simulation Symposium, pp. 1–10 (2010)
22. Wambecq, A.: Rational Runge-Kutta methods for solving systems of ordinary differential equations. Computing **20**(4), 333–342 (1978)
23. Ito, Y., Nakahashi, K.: Surface triangulation for polygonal models based on CAD data. Int. J. Numer. Meth. Fluids **39**(1), 75–96 (2002)
24. Ito, Y.: Challenges in unstructured mesh generation for practical and efficient computational fluid dynamics simulations. Comput. Fluids **85**(1), 47–52 (2013)
25. Marks, A., Whidborne, J.F., Yamamoto, I.: Control allocation for fault tolerant control of a VTOL octorotor. In: Proceedings of 2012 UKACC International Conference on Control, pp. 357–362. IEEE (2012)
26. Gomi, R., Takii, A., Yamakawa, M., Asao, S., Takeuchi, S., Nishimura, M.: Flight simulation from takeoff to yawing of eVTOL airplane with coaxial propellers by fluid-rigid body interaction. Adv. Aerodyn. **5**(1), 1–15 (2023)

Combining Outlierness Scores and Feature Extraction Techniques for Improvement of OoD and Adversarial Attacks Detection in DNNs

Tomasz Walkowiak⬭, Kamil Szyc⬭, and Henryk Maciejewski(✉)⬭

Wroclaw University of Science and Technology, Wrocław, Poland
{tomasz.walkowiak,kamil.szyc,henryk.maciejewski}@pwr.edu.pl

Abstract. Out-of-distribution (OoD) detection is one of the challenges for deep networks used for image recognition. Although recent works have proposed several state-of-the-art methods of OoD detection, no clear recommendation exists as to which of the methods is inherently best. Our studies and recent results suggest that there is no universally best OoD detector, as performance depends on the in-distribution (ID) and OoD benchmark datasets. This leaves ML practitioners with an unsolvable problem - which OoD methods should be used in real-life applications where limited knowledge is available on the structure of ID and OoD data. To address this problem, we propose a novel, ensemble-based OoD detector that combines outlierness scores from different categories: prediction score-based, (Mahalanobis) distance-based, and density-based. We showed that our method consistently outperforms individual SoTA algorithms in the task of (i) the detection of OoD samples and (ii) the detection of adversarial examples generated by a variety of attacks (including CW, DeepFool, FGSM, OnePixel, etc.). Adversarial attacks commonly rely on the specific technique of CNN feature extraction (GAP - global average pooling). We found that detecting adversarial examples as OoD significantly improves if we also ensemble over different feature extraction methods(such as GAP, cross-dimensional weighting (CroW), and layer-concatenated GAP). Our method can be readily applied with popular DNN architectures and does not require additional representation retraining for OoD detection (All results are fully reproducible, the source code is available at https://github.com/twalkowiak/WNN-OOD).

Keywords: Out-of-distribution detection · adversarial attack · CNN

1 Introduction

Applying deep neural networks in safety-critical systems requires that the models reliably recognize out-of-distribution (OoD) examples. Although many state-of-the-art methods of OoD detection have been proposed in recent works, no consensus has been worked out as to which of the methods is inherently best.

© The Author(s), under exclusive license to Springer Nature Switzerland AG 2023
J. Mikyška et al. (Eds.): ICCS 2023, LNCS 14073, pp. 578–592, 2023.
https://doi.org/10.1007/978-3-031-35995-8_41

The methods differ in how the 'outlierness' of the samples is obtained and can be categorized as: classifier prediction score-based [8,10,16], distance-based [14,21], or density-based [3]. A recent comprehensive study [28] suggests that there is no universally best OoD detector, as performance depends on the in-distribution (ID) and OoD benchmark data sets. Our studies summarized in Table 1 confirm this observation. We compare the performance of the methods: distance-based (Mahalanobis), prediction score-based (MSP), and density-based (LOF) and conclude that each of the methods outperforms other methods on some benchmarks. This poses an unsolvable problem - which OoD detection methods should we implement in a real-life deployment under limited knowledge of the structure of ID and OoD. We show that we can avoid this decision by using an ensemble of OoD detectors from different categories (prediction score-based, distance-based, and density-based) - as this combined OoD detector consistently outperforms popular detectors. We also find that ensembling over different representations (feature extraction procedures) can further improve OoD detection. We considered the following feature extractors: global average pooling (GAP), cross-dimensional weighting (CroW), layer-concatenated GAP (lcGAP), Global Maximum Pooling (GMP), and Selective Convolutional Descriptor Aggregation (SCDA).

Our contributions are the following. (i) We proposed an ensemble approach to OoD detection by (a) combining prediction score-based, distance-based, and density-based OoD detectors and (b) by combining OoD scores obtained under different feature extraction methods (GAP, CroW, lcGAP, GMP, SCDA). (ii) We showed on a comprehensive set of ID and OoD benchmarks that this approach consistently improves OoD detection compared with the individual SoTA algorithm. This holds for different DNN models (VGG16, ResNet, WideResNet, DenseNet, ShuffleNetV2, and MobileNetV2). (iii) We also showed that the proposed method detects adversarial examples generated by a wide range of adversarial attacks as OoD. (iv) We performed a sensitivity analysis of the ensemble method to quantify the contribution of the individual ensemble members to the performance of the proposed OoD detector.

2   Related Work

OoD Detection based on Standard Representations. Many OoD methods rely on representations trained to discriminate between classes given in the training data set. These methods differ in the technique used to estimate outlierness scores to detect OoD examples. Popular methods include: the OpenMax score proposed by [2], the MSP (maximum softmax probability) score [8], the ODIN score [16], generalized ODIN [10], the Mahalanobis distance-based score [14], or density-related scores such as kNN-based [26], or LOF-based [3,30]. These methods allow for post hoc detection of OoD inputs without requiring dedicated training of DNNs for OoD detection. Our proposed method belongs to this line of research.

Table 1. Comparison of different OoD detection methods for CIFAR-10 as ID. There is no one universal OoD method for all benchmarks. A different strategy is better for different CNN models and pairs of ID and OoD. The tested OoD methods operate according to different principles: distance-based (Mahalanobis), prediction score-based (MSP), and density-based (LOF). Our results are in line with recent works [28]. This poses an unsolvable problem: which OoD detection method we should implement in a real-life deployment with limited knowledge of ID and OoD.

Model	OoD	AUROC	DATACC	TNR at TPR 95%
		Mahalanobis/MSP/LOF		
ResNet	SVHN	92.6/92.2/**95.8**	85.2/87.2/**89.4**	61.1/43.6/**79.5**
	CIFAR-100	82.0/**87.2**/82.5	74.6/**80.8**/75.4	29.3/34.6/**35.5**
WideResNet	SVHN	**97.8**/93.0/95.3	**92.9**/88.2/90.2	**88.0**/60.0/69.7
	CIFAR-100	**90.3**/86.8/89.0	**83.2**/81.6/81.9	50.4/44.9/**50.8**
MobileNetV2	SVHN	85.4/**86.3**/76.1	80.6/**81.7**/69.8	20.6/**31.4**/20.9
	CIFAR-100	**86.9**/82.3/85.0	**80.4**/78.1/77.3	36.9/33.0/**40.9**

Table 2. Comparison of different feature extraction detection methods using Mahalanobis and LOF as OoD methods and CIFAR-10 as ID. We propose to use different feature extraction techniques in OoD detection problems. It may significantly improve achieved results compared to the default GAP. Various approaches focused on different components (e.g., on edges, patterns, or whole objects), so different features can effectively separate data for other pairs of ID and OoD. So, as with choosing an OoD method, there is no universal feature extraction strategy.

Model	OoD	Method	AUROC	DATACC	TNR at TPR 95%
			CroW/GAP/lcGAP		
ResNet	SVHN	Mah	94.0/92.6/**94.6**	86.4/85.2/**87.8**	67.7/61.1/**70.5**
		LOF	**96.5**/95.8/96.5	**90.3**/89.4/**90.3**	**83.0**/79.5/81.0
	CIFAR-100	Mah	**83.2**/82.0/80.5	**75.8**/74.6/73.0	**32.1**/29.3/28.8
		LOF	83.1/82.5/**83.4**	**75.9**/75.4/75.7	37.1/35.5/**37.5**
WideResNet	SVHN	Mah	**98.0**/97.8/97.0	**93.1**/92.9/91.1	**88.8**/88.0/85.3
		LOF	95.4/95.3/**98.2**	90.3/90.2/**93.7**	70.9/69.7/**90.0**
	CIFAR-100	Mah	**90.5**/90.3/84.8	**83.5**/83.2/77.3	**51.5**/50.4/40.3
		LOF	89.0/89.0/**89.4**	81.8/81.9/**82.2**	51.5/50.8/**52.9**
MobileNetV2	SVHN	Mah	84.9/85.4/**86.2**	80.2/80.6/**81.6**	19.5/20.6/**20.9**
		LOF	76.6/76.1/**92.7**	70.0/69.8/**87.6**	20.9/20.9/**44.6**
	CIFAR-100	Mah	86.9/86.9/**87.0**	80.3/80.4/**80.5**	**37.2**/36.9/37.0
		LOF	85.0/85.0/**86.8**	77.3/77.3/**79.2**	40.4/40.9/**43.8**

OoD Detection for Classification with Additional Training. The methods in this group include outlier-exposed techniques, for example, [9], where models are trained to produce uniformly distributed predictions for the OoD data available during training. [5] propose to train with outliers synthesized in the feature space (rather than the image). Another line of methods attempts to improve representations for OoD detection using contrastive learning and self-supervised techniques [23,27,32]. These methods learn representations in which transformed (augmented) versions of an image are pulled closer to each other while pushing other images further away. This modified feature space often leads to better OoD detection.

3   Method

3.1   Different Categories of OoD Detection Methods

Here we briefly present the OoD detectors used in the proposed ensemble procedure: predictive score-based (MSP), distance-based (Mahalanobis), and density-based (LOF).

Maximum Softmax Prediction (MSP) [8] quantifies outlierness based on the prediction score from the neural network. More specifically, the confidence score for OoD detection is based on the maximum output of the softmax layer of the DNN.

OoD detection based on the Mahalanobis distance is based on the estimation of the multivariate Gaussian (MVN) distribution as a model of the class-conditional posterior distribution. Given the ID training dataset $X_c \subset R^d$ for class $c \in C = \{1, 2, \ldots, m\}$, we estimate the MVN model for class $\mathcal{N}(\mu_c, \Sigma_c)$ with the mean vector μ_c and the covariance matrix Σ_c estimated from X_c (some methods estimate Σ_c from $\bigcup_{c \in C} X_c$, see, e.g., [14]). The confidence score for the detection of OoD of a test sample u is then obtained as the shortest negative Mahalanobis distance for all known classes: $cs_{Mah}(u) = -\min_c \sqrt{(u - \mu_c)^\top \Sigma_c^{-1}(u - \mu_c)}$.

The Local Outlier Factor (LOF) [3] obtains OoD scores based on nonparametric estimates of density. More specifically, it calculates the local reachability density $LRD_k(u, X)$ of a sample u with respect to the known dataset X. LRD is the ratio of an average reachability distance between a given point, its k neighbors, and their neighbors. K-neighbors $(N_k(u, X))$ includes a set of points that lie in the circle of radius k-distance, where k-distance is the distance between the point, and it is the farthest k^{th} nearest neighbor $(\|N_k(u, X)\| >= k)$. The confidence score for the detection of OoD is then the inverse of LOF defined in [3], i.e. $cs_{LOF}(u, X) = -\frac{\sum_{x \in N_k(u, X_i)} LRD_k(x, X)}{\|N_k(u, X)\| LRD_k(u, X)}$.

3.2   Feature Extraction

The standard method for CNN feature extraction is Global Average Pooling (GAP), proposed by [17]. This approach is widely used in networks designed for image classification since it is robust to spatial translations of the input data. However, many works, especially on image retrieval, propose different feature extraction strategies, focusing on local (object details) or global (whole object) descriptions and low-level (e.g., shapes, textures) or high-level (whole image meaning) features. Different feature extractors focus on specific image components, so various features may prove helpful for other pairs of ID and OoD datasets to detect OoD samples effectively.

Therefore, we also propose to analyze other feature extraction methods for OoD detection problems, such as CroW [12], GAP, lcGAP (our, inspired by [15]), GMP, and SCDA [31]. Most methods work based on the selected convolutional layer (usually the last) T with shape (w, h, c), where w refers to the width, h to height, and c to channels. A feature vector V (with length c) is calculated as follows.

For GAP as $\frac{1}{wh} \sum_{i=0}^{w} \sum_{j=0}^{h} T_{i,j,k}$ and $max_{i,j}^{w,h} T_{i,j,k}$ for GMP (Global Maximum Pooling) for each channel k independently.

For CroW (cross-dimensional weighting), T is first weighted channel-wise by weight vectors β_k and then location-wise by a weight matrix α, that is, we define weighted T as $T'_{i,j,k} = \alpha_{ij} \beta_k T_{i,j,k}$. Next, sum-pooled is performed to aggregate T' features.

SCDA (Selective Convolutional Descriptor Aggregation) is based on an activation feature map. First, the aggregation map $A_{i,j}$ is obtained as $\sum_{k=1}^{c} T_{i,j,k}$. For the aggregation map A, there are w, h summed activation responses corresponding to positions w, h. Next, the mask map M of the same size as A is obtained as $M_{i,j}^{w,h} = 1$ if $A_{i,j} > \bar{a}; 0$ otherwise, where $\bar{a}$ is the mean value of all positions in A. The final feature vector is selected based on the largest connected component of the mask map M. The layer-concatenated GAP (lcGAP) method (our original proposition) uses concatenated features from different convolutional layers after each block using GAP.

These methods have different characteristics; most are global descriptions with high-level features. The GAP method is more robust to scale changes because the GMP response of the feature map does not change rapidly with scale change. CroW uses a weighting and aggregation scheme to transform convolutional image features into compact global image features. SCDA focuses on local descriptions. lcGAP considers low-level features.

We show that the feature extraction strategy can affect the defense against adversarial attacks. These attacks aim to cheat the CNN classifier that commonly relies on GAP features. Different feature extraction methods can make features more resilient and less vulnerable to attacks. See Sect. 4.2 for more information on attacks.

3.3   Proposed Ensemble OoD Detector

The results in Tables 1 and 2 suggest no best method among feature extractors and OoD detectors exists. Therefore, we consider combining confidence scores from all available sources using an MLP regressor. The idea of using an OoD ensemble was inspired by [14] (although they used linear regression) and, in general, by the well-known concept of stack generalization ensemble [29].

The pseudocode of the proposed ensemble OoD detector is shown in the Algorithm 1.

Algorithm 1: WNN ensemble OoD detector. WNN.fit method trains the detector, WNN.score computes the confidence score for OoD detection based on the ensemble of OoD detection and feature extraction methods. See Section 3.3 for further details.

Inputs   : data_IN_train, data_IN_test, data_OOD
Outputs: final_confidence_scores
WNN - ensemble OoD detector

  scores = []
  foreach *feature_method in [feature extraction methods]* **do**
      features_IN_train = feature_method.extract_from(data_IN_train)
      features_IN_test = feature_method.extract_from(data_IN_test)
      features_OOD = feature_method.extract(data_OOD)
      foreach *ood_method in [OoD detection methods]* **do**
          ood_method.fit(features_IN_train)
          score_IN_test = ood_method.score_and_norm(features_IN_test)
          score_OOD = ood_method.score_and_norm(features_OOD)
          scores.append([score_IN_test, score_OOD])

  calibration_scores, test_scores = split_data(scores)
  WNN = new MLP_Model()
  WNN.fit(calibration_scores)
  final_confidence_scores = WNN.score(test_scores)

Given the training data, we extract CNN features using all the methods discussed in Sect. 3.2. We then build OoD detectors for each OoD type (see Sect. 3.1) and each feature extractor. This results in a set of 15 different OoD detectors. The idea of the proposed method is to combine the confidence scores returned by these OoD detectors into the final ensemble-based score.

The confidence score ranges for each OoD method are very different. Mahalanobis gives values below 0 (lower limits depend on the data), MSP values range from 0 to 1 (inclusive), whereas LOF shows values below or equal to –1. Therefore, we scale confidence scores to the range 0–1 using the estimated cumulative density on the validation set of inliers (estimation is done using step functions).

Next, we train the MLP that combines the weighted confidence scores from individual OoD detectors. Similarly to [14,33], we use a calibration set consisting of images from in- and out-of-distribution datasets. We use a fully connected NN consisting of one hidden layer (100 neurons) and a single output. We use ReLU activations for the hidden layer and linear activation in the final layer. We employ Adam optimizer during backpropagation. The output of this network serves as the final confidence score. The method is named WNN from the weighted neural network.

4   Experiments and Results

4.1   OoD Detection Problem

We run OoD detection problems using widely used benchmark datasets and CNNs models such as VGG16 [24], ResNet [7], WideResNet, DenseNet [11], ShuffleNetV2 [18] and MobileNetV2 [22]. All models were trained on CIFAR-10. To evaluate the OoD methods, we used CIFAR-10 as ID data and CIFAR-100 or SVHN as OoD data. We kept a 1:1 proportion of known and unknown samples (10,000:10,000). We used the standard metrics in the results: Area Under Receiver Operating Characteristic curve (AUROC), detection accuracy (DTACC) that defines the ratio of correct classification, and True Negative Rate at 95% True Positive Rate (TNR at TPR 95%). The higher the values of all metrics, the better the detection of OoD. As the measures are strongly correlated, for some results, we present only AUROC. All our tests can be replicated using the standard workstations for deep learning; we used Nvidia GeForce GTX 1080/2080 Ti.

No One Universal OoD Method. In the first experiment, we show that there is no universally best OoD method for all benchmarks. In Table 1, we compare the OoD methods described in Sect. 3.2 and show that for different pairs of IN and OoD datasets, different OoD detectors are the best.

For example, for WideResNet, the parametric Mahalanobis method is the best for all benchmarks. For ResNet, depending on the OoD data and chosen metrics, the best performance is achieved for the density-based LOF or the logits-based MSP. These results are consistent with the recent literature [28].

No One Universal Feature Extraction Technique. We propose using different feature extraction strategies (see Sect. 3.2) in the OoD detection problem rather than relying on a single method. In Table 2, we show that for different pairs of IN and OoD samples, different strategies work better, as different image features are needed to separate these image categories. For example, WideResNet using LOF and SVHN performed best for lcGAP for AUROC (2.9 p.p. better than GAP), and ResNet using Mahalanobis and CIFAR-100 performed best for CroW for AUROC (1.2 p.p. better than GAP).

Proposed Method Results. In Table 3, we report the performance of WNN - our ensemble method introduced in Sect. 3.3. We compare our method with individual OoD detectors used in the ensemble (LOF, Mahalanobis, MSP) and with two popular SoTA OoD detectors: KNN [26] (a density-based non-parametric method), and UF [14] (Mahalanobis distance-based, with a common covariance matrix and feature ensemble from different layers of a CNN). Since the WNN results depend on the distribution of the validation data (20% random samples from the original data were used to build the WNN) and the MLP training process, we repeated the experiments ten times and showed the mean and std results. Due to a random split on validation and test data, the results for all other methods (including those not requiring validation) are also presented as mean and standard deviation. The proposed WNN method gives the best results for most benchmarks. To verify the claim that adding other feature extraction strategies than GAP improves OoD detection, we also considered the results for WNN^{GAP}, the proposed ensemble that works only on GAP features. We find that WNN consistently outperforms WNN^{GAP}, which is often the second best.

Experiments on Other OoD Detection Benchmark Datasets. We also verified the performance of the proposed method on other OoD datasets (i.e., Tiny ImageNet, ImageNet-O, and Textures) and also used CIFAR-100 as in-distribution (with SVHN and CIFAR-10 as OoD).[1] We found the same conclusions as reported in the previous section. The results on CIFAR-100 as in-distribution show that WNN outperforms the MSP, Mahalanobis, LOF, and kNN outlier detectors in all test cases (by ca. 4 pp over the second best). Taking different OoD detectors and different feature generators into an ensemble leads to a significant improvement over individual non-ensemble OoD detectors. Thus, the practitioner using the proposed WNN is freed from the need to select the best/most appropriate OoD detector for a given study (which is difficult to do in practice).

4.2 Attacks

We used our method as a defensive approach against adversarial attacks. The goal is to use adversarial examples as OoD samples similar to [14]. These attacks modify the input image by adding unique noise, making the network fooled. Noise usually does not change the human classification assessment. Although there are numerous dedicated defense methods [34], our approach allows for initial protection against many kinds of adversarial attacks.

We tested various adversarial attack methods: CW, DeepFool, FGSM, One-Pixel, PGD, and Square. One of the fundamental methods is FGSM [6] (Fast Gradient Sign Method), where a small perturbation is added to maximize the loss function. The PGD [19] (Projected Gradient Descent) is an extension of the FGSM by repeating the addition of those perturbations multiple times. The Deep Fool [20] method is another iterative approach. This method calculates the distributions based on the willingness to move the input across the decision boundaries with minimal changes. The first-order approximation of Taylor's expansion is used on a linear model to find these distributions.

CW [4] focused on solving the optimization problem of finding the minimized distance between two images (standard and attacked) so that the classification result will differ for both examples. The problem is not trivial due to the highly non-linear nature of deep models, so the authors propose defining the objective function instead, which is much more likely to optimize using popular optimizers. Square [1] is based on a randomized search scheme in which we select localized

---

[1] The detailed results are available https://github.com/twalkowiak/WNN-OOD.

Table 3. OoD detection results for CIFAR-10 (as inliers) versus SVHN/CIFAR-100 (as outliers). Mahalanobis, MSP[8] and LOF[3] are working on features extracted from the last layer of CNN using the classic GAP method. WNN (our method) combines 15 OoD detectors (obtained by three OoD methods and five feature extractors: GAP, CroW, lcGAP, SCDA, and GMP). WNN^{GAP} represents the combination limited to GAP-based features only (i.e., it is a limited version of WNN). We can see that adding information from additional feature extractors increases the OoD detection. The results show the mean and std values of the metrics achieved.

	Method	AUROC	DATACC	TNR	AUROC	DATACC	TNR
		SVHN			CIFAR-100		
VGG16	Mah	90.0±0.09	84.5±0.13	37.8±0.35	87.4±0.08	80.6±0.16	39.5±0.39
	MSP	89.5±0.08	85.1±0.08	30.1±0.38	86.4±0.10	80.1±0.09	32.5±0.48
	LOF	87.2±0.12	80.3±0.10	37.7±0.88	83.6±0.13	76.8±0.12	36.2±0.76
	WNN^{GAP}	90.4±0.11	85.5±0.08	34.4±0.82	87.6±0.13	80.6±0.16	37.8±0.84
	WNN(our)	97.5±0.10	92.2±0.18	86.2±0.92	**88.3±0.13**	**81.2±0.09**	41.3±1.18
	KNN	91.3±0.15	86.4±0.13	39.3±0.88	87.5±0.15	80.5±0.16	38.4±0.35
	UF	**98.9±0.10**	**95.2±0.19**	**95.2±0.35**	86.8±0.70	79.6±0.87	**44.3±1.02**
ResNet	Mah	92.7±0.06	85.3±0.04	61.3±0.32	82.1±0.09	74.7±0.10	29.3±0.39
	MSP	92.3±0.10	87.2±0.10	44.0±0.63	87.2±0.07	80.8±0.11	34.6±0.59
	LOF	95.8±0.05	89.4±0.10	79.8±0.41	82.6±0.12	75.5±0.11	36.5±0.88
	WNN^{GAP}	96.3±0.07	90.5±0.11	79.8±0.48	88.3±0.16	81.1±0.13	40.6±1.06
	WNN(our)	**99.3±0.03**	**95.8±0.10**	**96.2±0.12**	**89.3±0.14**	**82.0±0.25**	**45.1±0.75**
	KNN	96.4±0.05	90.4±0.10	78.9±0.52	87.5±0.08	80.2±0.09	41.5±0.52
	UF	97.6±0.31	92.4±0.46	88.5±1.61	75.7±0.49	68.9±0.41	22.6±0.93
WideResNet	Mah	97.9±0.03	93.0±0.07	88.0±0.30	90.3±0.11	83.1±0.12	50.6±0.56
	MSP	93.0±0.12	88.2±0.13	59.6±0.56	86.8±0.12	81.6±0.12	45.1±0.65
	LOF	95.3±0.05	90.2±0.07	70.0±0.53	89.0±0.12	82.0±0.13	50.9±0.80
	WNN^{GAP}	97.4±0.10	92.7±0.13	85.4±1.26	90.4±0.11	83.4±0.17	51.6±0.55
	WNN(our)	**99.2±0.02**	**95.4±0.11**	**95.8±0.23**	**91.1±0.08**	**83.9±0.14**	**53.9±0.62**
	KNN	96.9±0.06	91.8±0.12	80.0±0.51	90.9±0.10	83.8±0.10	52.7±0.49
	UF	97.4±0.18	92.3±0.42	89.4±1.06	67.9±1.82	63.1±1.48	17.5±0.48
DenseNet	Mah	99.0±0.03	95.1±0.06	94.7±0.17	91.1±0.09	84.3±0.11	57.6±0.43
	MSP	82.5±0.16	79.4±0.13	46.1±0.63	88.5±0.12	82.7±0.12	47.7±1.19
	LOF	97.7±0.03	93.0±0.08	86.3±0.19	89.8±0.10	83.1±0.15	54.5±0.49
	WNN^{GAP}	98.8±0.05	94.8±0.09	93.7±0.26	91.4±0.08	84.6±0.07	56.6±0.57
	WNN(our)	99.2±0.03	95.8±0.05	96.5±0.09	**92.0±0.13**	**85.1±0.11**	**59.7±0.78**
	KNN	98.4±0.04	94.0±0.09	90.8±0.36	90.9±0.07	84.3±0.07	55.5±0.67
	UF	**99.3±0.12**	**96.2±0.40**	**97.0±0.67**	81.2±1.96	74.8±1.65	38.6±2.88
ShuffleNetV2	Mah	93.3±0.11	88.2±0.13	52.2±1.02	87.2±0.08	80.4±0.11	37.7±0.40
	MSP	90.4±0.12	85.8±0.14	41.5±1.17	82.0±0.10	78.3±0.09	29.6±0.73
	LOF	89.9±0.12	83.5±0.16	39.3±0.69	84.7±0.13	77.6±0.12	36.4±0.26
	WNN^{GAP}	93.3±0.15	88.1±0.15	52.9±1.19	87.2±0.07	80.3±0.09	37.8±0.44
	WNN(our)	**97.2±0.07**	**91.0±0.16**	**82.2±0.66**	**87.6±0.13**	**80.7±0.17**	**39.0±0.87**
	KNN	94.5±0.06	89.2±0.08	60.9±0.60	87.5±0.09	80.5±0.13	38.6±0.45
	UF	98.0±0.09	93.3±0.14	90.0±0.69	86.8±0.28	79.2±0.19	39.7±1.33
MobileNetV2	Mah	85.4±0.11	80.6±0.08	20.7±0.30	87.0±0.11	80.4±0.14	37.2±0.53
	MSP	86.3±0.12	81.7±0.12	31.6±0.57	82.2±0.12	78.1±0.13	33.1±0.50
	LOF	76.0±0.15	69.8±0.14	20.8±0.30	85.0±0.12	77.3±0.15	41.1±0.60
	WNN^{GAP}	89.0±0.16	83.4±0.09	36.9±1.93	87.1±0.09	80.3±0.11	39.1±0.61
	WNN(our)	97.6±0.11	92.0±0.16	86.1±0.93	**87.9±0.11**	**80.9±0.13**	**42.5±0.71**
	KNN	86.7±0.11	81.5±0.11	24.9±0.42	87.5±0.10	80.5±0.13	39.3±0.52
	UF	**98.2±0.14**	**93.3±0.31**	**91.0±0.67**	85.6±0.26	78.6±0.20	35.8±3.20

square-shaped updates at random positions. The OnePixel [25] is highly interesting due to the change of only one pixel to fool the network. It uses a differential evolution algorithm to find which pixel should be changed. Candidate solutions contain information with x,y coordinates, and RGB values. During each epoch, the population is randomly modified by a minor factor, and the algorithm works until one of the candidates is an adversarial attack.

In our experiments, we generated 1000 examples for each attack method using the TorchAttacks [13] library. We treated the attacks as OoD data by sampling the same number of images for the test-known subset. Similarly to previous experiments, we used 20% random samples to build the WNN and the regressor in UF. We repeated the experiments ten times (with different test image subsets), presenting the results as mean and standard deviation. We kept high confidence in attack examples, usually ensuring that the closed-set classifier's output would be above 95% certainty for the wrong class.

We have shown the results of our experiments in Table 4. Although we present results for the ResNet model, the results and conclusions obtained for other models are consistent. Popular methods like Mahalanobis, MSP, and LOF are not suitable for defense against adversarial attacks. UF works well only for FGSM and PGD, which are the same family of attacks. KNN works only for PGD in practice. Our method is the best approach for all tested attacks, significantly outperforming other methods like CW (7.2 p.p. better than second-best), OnePixel (6.6 p.p. better), or DeepFool (4.1 p.p. better). See that our proposed method is very stable.

The adversarial attack aims to fool CNN's classifier part based on the GAP features. Using different feature extraction strategies should improve results in defense against adversarial examples. We can compare the two methods WNN^{GAP} and WNN (proposed), where the first uses only GAP features, and the second uses all five feature extraction techniques. We can see that for all attacks, adding new features improves results, with significant improvement for some attacks, like FGSM (16.7 p.p. better than with only GAP), OnePixel (6.7 p.p. better), or CW (6.1 p.p. better).

4.3   Sensitivity Analysis

The proposed WNN method combines different sources of information (different feature extractors and Ood detectors). The question of how these sources affect the final detection performance arises. Comparing the results labeled WNN^{GAP} (WNN only on GAP features) with results labeled WNN (all feature extractors) shown in Tables 3 and 4, we conclude that extending the feature extractors beyond GAP increases OoD performance. The impact of the OoD methods used in the ensemble is shown in Table 5. It clearly shows that more methods (Mahalanobis, MSP, and LOF) are better (than only Mahalanobis or Mahalanobis and MSP).

Table 4. Comparison of AUROC (%) for ResNet trained on CIFAR-10 and different types of attacks: CW [4], DeepFool [20], FGSM [6], OnePixel [25], PGD [19], Square [1]. Our method is the best approach for all tested attacks, outperforming others. Moreover, we showed that extending the WNN by other feature extraction techniques than GAP methods significantly improves the ability of defenses.

Attack	Mah	MSP	LOF	WNN^{GAP}	WNN	UF	KNN
CW	65.1±0.83	80.8±1.10	63.5±1.44	81.9±1.68	**88.0±3.68**	59.8±2.03	70.3±0.96
DeepFool	63.6±0.92	79.8±0.88	60.2±1.10	79.6±0.82	**83.9±2.89**	52.7±2.36	67.3±0.61
FGSM	73.4±0.80	68.7±1.50	70.7±1.00	76.3±1.17	**93.0±0.47**	92.2±0.44	71.7±1.23
OnePixel	61.2±1.48	75.5±1.63	55.4±1.16	75.4±1.21	**82.1±2.43**	53.9±1.42	65.2±1.04
PGD	86.2±0.53	0.8±0.05	98.2±0.12	99.3±0.13	**99.7±0.08**	97.1±0.21	96.7±0.20
Square	78.7±1.08	80.2±0.92	76.7±0.64	87.1±1.40	**90.2±1.67**	75.1±1.54	80.2±0.86

We also checked the effect of the size of the validation set on the WNN results. To test this, we analyzed the proportions of the original test data used for validation in the CIFAR10 vs. CIFAR-100 detection task for Wide Resnet. We show the results in Fig. 1, along with the performance of UF (as another ensemble technique). The validation size affects the performance (AUROC), but WNN still performs better than the MSP method for a small number, such as 20 outlier images (a factor of 0.002) used to build the ensemble. WNN gives significantly lower confidence intervals, even though it uses a larger ensemble model (i.e., MLP regressor) than UF (linear regressor). It is caused by the fact that UF uses both ensemble and the input pre-processing[16] magnitude meta-parameter set up in the validation set. Moreover, the effective data size used to set up the regressor is reduced by half.

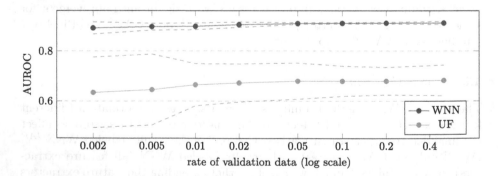

Fig. 1. Effect of validation size on OoD detection for WNN and UF methods. The experiments were performed on Wide ResNet with CIFAR-100 used as outliers. The ratio ranges from 0.002 to 0.5 of the original inliers and outliers images (10,000 in each set). The experiments (splitting the dataset and tuning the ensemble parameters) were repeated 50 times, so the results were random. The solid line shows the mean value of AUROC, whereas the dashed line shows three times the std.

4.4 Limitations

We see three limitations of the proposed method: the need for validation samples to tune the ensemble model, the difficulty in choosing the final methods used in the WNN, and the required calculation of many OoD confidence scores, some of which can be computationally heavy.

The first limitation is typical for a large group of OoD methods (such as [9,14,16]). The problem is that, in most practical scenarios, either OoD samples are unavailable or cover a small fraction of the OoD sample space. The results in Fig. 1 suggest that WNN performance decreases when the amount of OoD available for training decreases, but for just 20 OoD images, the WNN performance is still better than the baseline (MSP). Furthermore, the ability of validation set-based techniques to detect OoD decreases when the OoD samples are from a different OoD domain from the one used to tune the hyperparameters.

Table 5. The impact of the number of OoD methods used in an ensemble. Results present AUCROC for an ensemble over the feature extraction methods and just the Mahalanobis method (WNN^{Mah}), Mahalanobis plus MSP (WNN^{MahMSP}), and all methods (proposed WNN). A steady increase in performance could be observed by adding more OoD methods. The last column shows the best results for other (non-WNN) methods, i.e., the best among Mahalanobis, MSP, LOF, UF, and KNN. We notice that WNN without LOF (the method with large computational complexity is only slightly worse than the proposed WNN.

Model	OoD	WNN^{Mah}	WNN^{MahMSP}	WNN	other best
VGG16	SVHN	90.1±0.16	90.6±0.22	97.5±0.10	98.9±0.10 (UF)
	CIFAR-100	87.4±0.07	87.7±0.13	88.4±0.12	87.5±0.15 (KNN)
ResNet	SVHN	98.9±0.05	99.2±0.03	99.2±0.02	97.6±0.31 (UF)
	CIFAR-100	85.7±0.13	89.2±0.15	89.4±0.11	87.5±0.08 (KNN)
WideResNet	SVHN	99.2±0.03	99.2±0.03	99.2±0.01	97.9±0.03 (Mah)
	CIFAR-100	90.9±0.11	91.0±0.10	91.1±0.12	90.9±0.10 (KNN)
DenseNet	SVHN	99.1±0.05	99.2±0.04	99.2±0.03	99.3±0.12 (UF)
	CIFAR-100	91.4±0.06	91.8±0.09	92.0±0.10	91.1±0.09 (Mah)
ShuffleNetV2	SVHN	96.8±0.07	96.8±0.11	97.2±0.11	98.0±0.09 (UF)
	CIFAR-100	87.3±0.12	87.2±0.10	87.6±0.09	87.5±0.09 (KNN)
MobileNetV2	SVHN	93.6±0.11	94.0±0.08	97.6±0.08	98.2±0.14 (UF)
	CIFAR-100	87.1±0.13	87.2±0.09	87.8±0.13	87.5±0.10 (KNN)

Our WNN method uses OoD detectors obtained by three OoD detection (LOF, Mahalanobis, and MSP) and five feature extraction (CroW, GAP, GMP, lcGAP, and SCDA) methods. We chose the above methods because of the differentiated nature of the operating principle (see details in Sects. 3.1 and 3.2). However, the proposed solution can be used as a framework, combining confidence scores from any OoD detectors. Finding the optimal set of OoD methods for WNN can be a limitation due to the number of methods available in the literature. We suggest using the most diversified OoD detection methods.

The last limitation is the required calculation of many OoD confidence scores. Some of them might be computationally heavy. For instance, in our proposed set of methods WNN uses, LOF is computationally intensive compared to Mahalanobis or MSP due to the required calculation of the nearest-neighbor aspects. For data of dimensions d and train size N, the LOF model's complexity is $O(d * N^2)$, and OoD detection is $O(d * N)$. However, it can be speed up (for a certain distance metric) by using the k-d tree (as implemented in scikit-learn) or R*-tree (as implemented in the ELKI framework), giving $O(d * NlogN)$ for model building and $O(d * logN)$ for detection. If the reduced time complexity is unacceptable, one can build the WNN only on Mahalanobis and MSP. The results presented in Table 5 (column WNN^{MahMSP}) show that this solution is slightly worse than the proposed WNN but still better than other OoD detectors tested.

5   Conclusion

In this work, we proposed an ensemble procedure for OoD detection that combines OoD detectors based on prediction scores (MSP), distance-based (Mahalanobis), and density-based (LOF). Using several benchmarks, we showed that this procedure outperforms each OoD algorithm used in the ensemble. These results are consistent for different DNN architectures.

We also showed that for OoD detection, different feature extraction strategies are worth considering, as this allows us to broaden the representation of objects. Different DNN feature extraction strategies focus on local (object details) or global (whole object) characteristics and either low-level (e.g., shapes, textures) or high-level (entire image meaning) features. We showed that the ensemble OoD detector that combines different feature extractors (we used GAP, CroW, lcGAP, GMP, and SCDA) leads to further improvement in OoD detection. We also found that the proposed method is efficient in recognizing adversarial examples. Moreover, our method more reliably identifies adversarial examples as OoD than individual SoTA OoD detectors for a wide range of adversarial attacks.

Finally, our method can be used as a generic framework that combines different outlierness scores (ensemble over OoD detectors of different natures) and different representations (ensemble over feature extractors). The essential contribution is (a) to show that OoD detection in DNNs consistently improves if based on ensembled information and (b) to propose the practical technique to ensemble over OoD detectors and feature extractors. Incorporating other OoD detectors and feature extractors into our framework may lead to further improvement - we leave this as future work.

Our method can facilitate the application of state-of-the-art, pre-trained DNN models to real-world, safety-critical image and text recognition systems where efficient OoD detection is mandatory.

References

1. Andriushchenko, M., Croce, F., Flammarion, N., Hein, M.: Square attack: a query-efficient black-box adversarial attack via random search. In: Vedaldi, A., Bischof, H., Brox, T., Frahm, J.-M. (eds.) ECCV 2020. LNCS, vol. 12368, pp. 484–501. Springer, Cham (2020). https://doi.org/10.1007/978-3-030-58592-1_29
2. Bendale, A., Boult, T.E.: Towards open set deep networks. In: Proceedings of the IEEE Conference on Computer Vision and Pattern Recognition, pp. 1563–1572 (2016)
3. Breunig, M.M., Kriegel, H.P., Ng, R.T., Sander, J.: LOF: identifying density-based local outliers. SIGMOD Rec. **29**(2), 93–104 (2000). https://doi.org/10.1145/335191.335388
4. Carlini, N., Wagner, D.: Towards evaluating the robustness of neural networks. In: 2017 IEEE Symposium on Security and Privacy (SP), pp. 39–57. IEEE (2017)
5. Du, X., Wang, Z., Cai, M., Li, S.: Towards unknown-aware learning with virtual outlier synthesis. In: Proceedings of the International Conference on Learning Representations (2022)
6. Goodfellow, I.J., Shlens, J., Szegedy, C.: Explaining and harnessing adversarial examples. arXiv preprint arXiv:1412.6572 (2014)
7. He, K., Zhang, X., Ren, S., Sun, J.: Deep residual learning for image recognition. In: Proceedings of the IEEE Conference on Computer Vision and Pattern Recognition, pp. 770–778 (2016)
8. Hendrycks, D., Gimpel, K.: A baseline for detecting misclassified and out-of-distribution examples in neural networks. arXiv preprint arXiv:1610.02136 (2016)
9. Hendrycks, D., Mazeika, M., Dietterich, T.: Deep anomaly detection with outlier exposure. In: Proceedings of the International Conference on Learning Representations (2019)
10. Hsu, Y.C., Shen, Y., Jin, H., Kira, Z.: Generalized odin: detecting out-of-distribution image without learning from out-of-distribution data. In: Proceedings of the IEEE/CVF Conference on Computer Vision and Pattern Recognition, pp. 10951–10960 (2020)
11. Huang, G., Liu, Z., Van Der Maaten, L., Weinberger, K.Q.: Densely connected convolutional networks. In: Proceedings of the IEEE Conference on Computer Vision and Pattern Recognition, pp. 4700–4708 (2017)
12. Kalantidis, Y., Mellina, C., Osindero, S.: Cross-dimensional weighting for aggregated deep convolutional features. In: Hua, G., Jégou, H. (eds.) ECCV 2016. LNCS, vol. 9913, pp. 685–701. Springer, Cham (2016). https://doi.org/10.1007/978-3-319-46604-0_48
13. Kim, H.: Torchattacks: a pytorch repository for adversarial attacks. arXiv preprint arXiv:2010.01950 (2020)
14. Lee, K., Lee, K., Lee, H., Shin, J.: A simple unified framework for detecting out-of-distribution samples and adversarial attacks. In: Proceedings of the 32nd International Conference on Neural Information Processing Systems, pp. 7167–7177. NIPS'18, Curran Associates Inc., Red Hook, NY, USA (2018)
15. Li, Y., Xu, Y., Wang, J., Miao, Z., Zhang, Y.: MS-RMAC: multiscale regional maximum activation of convolutions for image retrieval. IEEE Signal Process. Lett. **24**(5), 609–613 (2017)
16. Liang, S., Li, Y., Srikant, R.: Enhancing the reliability of out-of-distribution image detection in neural networks. In: International Conference on Learning Representations (ICLR) (2018)

17. Lin, M., Chen, Q., Yan, S.: Network in network. arXiv preprint arXiv:1312.4400 (2013)
18. Ma, N., Zhang, X., Zheng, H.-T., Sun, J.: ShuffleNet V2: practical guidelines for efficient CNN architecture design. In: Ferrari, V., Hebert, M., Sminchisescu, C., Weiss, Y. (eds.) Computer Vision – ECCV 2018. LNCS, vol. 11218, pp. 122–138. Springer, Cham (2018). https://doi.org/10.1007/978-3-030-01264-9_8
19. Madry, A., Makelov, A., Schmidt, L., Tsipras, D., Vladu, A.: Towards deep learning models resistant to adversarial attacks. arXiv preprint arXiv:1706.06083 (2017)
20. Moosavi-Dezfooli, S.M., Fawzi, A., Frossard, P.: Deepfool: a simple and accurate method to fool deep neural networks. In: Proceedings of the IEEE Conference on Computer Vision and Pattern Recognition, pp. 2574–2582 (2016)
21. Ren, J., Fort, S., Liu, J., Roy, A.G., Padhy, S., Lakshminarayanan, B.: A simple fix to mahalanobis distance for improving near-ood detection. arXiv preprint arXiv:2106.09022 (2021)
22. Sandler, M., Howard, A., Zhu, M., Zhmoginov, A., Chen, L.C.: MobileNetV 2: Inverted residuals and linear bottlenecks. In: Proceedings of the IEEE Conference on Computer Vision and Pattern Recognition, pp. 4510–4520 (2018)
23. Sehwag, V., Chiang, M., Mittal, P.: SSD: a unified framework for self-supervised outlier detection. In: International Conference on Learning Representations (2021)
24. Simonyan, K., Zisserman, A.: Very deep convolutional networks for large-scale image recognition. arXiv preprint arXiv:1409.1556 (2014)
25. Su, J., Vargas, D.V., Sakurai, K.: One pixel attack for fooling deep neural networks. IEEE Trans. Evol. Comput. 23(5), 828–841 (2019)
26. Sun, Y., Ming, Y., Zhu, X., Li, Y.: Out-of-distribution detection with deep nearest neighbors. arXiv preprint arXiv:2204.06507 (2022)
27. Tack, J., Mo, S., Jeong, J., Shin, J.: CSI: novelty detection via contrastive learning on distributionally shifted instances. In: Larochelle, H., Ranzato, M., Hadsell, R., Balcan, M., Lin, H. (eds.) Advances in Neural Information Processing Systems, vol. 33, pp. 11839–11852. Curran Associates, Inc. (2020)
28. Tajwar, F., Kumar, A., Xie, S.M., Liang, P.: No true state-of-the-art? OOD detection methods are inconsistent across datasets. arXiv preprint arXiv:2109.05554 (2021)
29. Ting, K.M., Witten, I.H.: Issues in stacked generalization. J. Artif. Int. Res. 10(1), 271–289 (1999)
30. Walkowiak, T., Datko, S., Maciejewski, H.: Utilizing local outlier factor for open-set classification in high-dimensional data - case study applied for text documents. In: Bi, Y., Bhatia, R., Kapoor, S. (eds.) IntelliSys 2019. AISC, vol. 1037, pp. 408–418. Springer, Cham (2020). https://doi.org/10.1007/978-3-030-29516-5_33
31. Wei, X.S., Luo, J.H., Wu, J., Zhou, Z.H.: Selective convolutional descriptor aggregation for fine-grained image retrieval. IEEE Trans. Image Process. 26(6), 2868–2881 (2017)
32. Winkens, J., et al.: Contrastive training for improved out-of-distribution detection. arXiv preprint arXiv:2007.05566 (2020)
33. Xingjun, M., et al.: Characterizing adversarial subspaces using local intrinsic dimensionality. In: ICLR (2018)
34. Yuan, X., He, P., Zhu, Q., Li, X.: Adversarial examples: attacks and defenses for deep learning. IEEE Trans. Neural Netw. Learn. Syst. 30(9), 2805–2824 (2019)

From Online Behaviours to Images: A Novel Approach to Social Bot Detection

Edoardo Di Paolo[1]([⊠])(iD), Marinella Petrocchi[2](iD), and Angelo Spognardi[1](iD)

[1] Computer Science Department, Sapienza University of Rome, Rome, Italy
{dipaolo,spognardi}@di.uniroma1.it
[2] IIT-CNR, Pisa, Italy
marinella.petrocchi@iit.cnr.it

Abstract. Online Social Networks have revolutionized how we consume and share information, but they have also led to a proliferation of content not always reliable and accurate. One particular type of social accounts is known to promote unreputable content, hyperpartisan, and propagandistic information. They are automated accounts, commonly called bots. Focusing on Twitter accounts, we propose a novel approach to bot detection: we first propose a new algorithm that transforms the sequence of actions that an account performs into an image; then, we leverage the strength of Convolutional Neural Networks to proceed with image classification. We compare our performances with state-of-the-art results for bot detection on genuine accounts/bot accounts datasets well known in the literature. The results confirm the effectiveness of the proposal, because the detection capability is on par with the state of the art, if not better in some cases.

1 Introduction

With the advent of the internet and Online Social Networks (OSNs), production and fruition of information feature less mediated procedures, where content and quality do not always go through a rigorous editorial process [5,15,39]. Thus, although OSNs make our lives easier by giving us immediate access to information and allowing us to exchange opinions about anything, the danger of being exposed to false or misleading news is high [18,27,35]. The promotion of disinformation on OSNs has often been juxtaposed with the existence of automated accounts known as bots. As an example, Shao et al., in [33], have highlighted the role of Twitter bots, showing how bots were primarily responsible for the early spread of disinformation, interacting with influential accounts through mentions and replies.

This work was partially supported by project SERICS (PE00000014) under the NRRP MUR program funded by the EU - NGEU; by the Integrated Activity Project TOFFEe (TOols for Fighting FakEs) https://toffee.imtlucca.it/; by the IIT-CNR funded Project re-DESIRE (DissEmination of ScIentific REsults 2.0); by 'Prebunking: predicting and mitigating coordinated inauthentic behaviors in social media' project, funded by Sapienza University of Rome.

J. Mikyška et al. (Eds.): ICCS 2023, LNCS 14073, pp. 593–607, 2023.
https://doi.org/10.1007/978-3-031-35995-8_42

The struggle between bots hunters and bots creators has been going on for many years now [6], and the actions of these automated accounts with malicious intent have influenced even the purchase of Twitter itself -just remember the $44 billion deal that went up in smoke precisely because of concerns about the unquantified presence of bots on the platform[1].

In this study, we provide a novel approach to bot detection, by leveraging the remarkable advancements in the field of image recognition [22,25]. To the best of the authors' knowledge, no methodology or tool so far defined for bot detection leverage image recognition. In particular, we will exploit the potential of Convolutional Neural Networks (CNNs) to classify Twitter accounts as bots or not.

Based on the premise that automated accounts are often programmed to carry out spam and/or disinformation campaigns, numerous works in the literature have proposed detection approaches that leverage coordination and synchronism features of accounts, see, e.g. [4,9,47]. The intuition is that the online activities of a group of automated accounts -all devoted carrying out a certain strategy- are more similar to each other than those of genuine accounts. This is the leit motif from which the concept of *Digital DNA*, originally introduced by Cresci et al. in [11], and the detection technique known as *Social Fingerprinting* [8] came to life. The digital DNA of an online account represents the sequence of actions of that account and each action is associated to a symbol from a pre-defined alphabet. By associating each symbol with a color, it is possible to transform the sequence into an image, where each pixel represents the color of the corresponding symbol. The assumption that leads us to exploit image classification to perform bot detection is that images of bot accounts are similar to each other, and different from those of genuine accounts, given the different behavior of the two categories of accounts.

We thus propose an algorithm to transform sequences of digital DNA into images and we run pre-trained CNNs, such as VGG16, ResNet50, and WideResNet50 [20,34] over the generated images. DNA sequences are extracted from Twitter accounts, both genuine and bot, of public datasets well-known in the bot detection literature. Where accounts' timelines are too short to produce good quality images, we have enhanced the latter, even turning features of the accounts other than Digital DNA into part of it.

Main Contributions: The main contributions of this work are as follows:

- Definition and implementation of a new approach to bot detection, based on image recognition;
- Validation of the approach by comparing our performance with state-of-the-art performances on publicly-released datasets: bot detection via image recognition achieves the same performances as obtained in the literature, when not better.

---

[1] Elon Musk terminates $44B Twitter deal. Online: https://nypost.com/2022/07/08/elon-musk-terminates-44-billion-twitter-deal/ August 8, 2022.

We argue that the investigations here presented make the transition from sequence of actions to sequence of pixels for bots detection look promising. Of course, the literature is filled with more than good work in the field. Still, we find the approach itself interesting because it leverages well-established image recognition techniques. Thus, a way forward for further experimentation.

2   Related Work

Bot detection is a topic that began to be studied more than 10 years ago, when social networks became increasingly popular, and interests in spamming, spreading propaganda content, and increasing one's notoriety on the platforms grew tremendously. Different techniques have followed over the years, from classifying via traditional machine learning exploiting features in the accounts profile – the very first attempts in this direction are the papers by Mustafaraj and Metaxas [28] and Yardi et al. [45], both dated 2010 –, to using deep learning techniques. We therefore feel it is appropriate to list some of the work on the subject, without however intending to propose an exhaustive list.

Traditional Machine Learning Approaches. Botometer is probably the most well-known tools in the literature for bot unveiling [43]; it is based on a supervised machine learning approach employing Random Forest classifiers. The last version, Botometer v4, has been recently shown to perform well for detecting both single-acting bots and coordinated campaigns [32]. v4 provides a useful lite version, BotometerLite[2], which does not interface with Twitter, but simply takes the tweet, retrieves the author, and does the necessary follow-up analysis. This light version only needs the information in the user profile to perform bot detection and hence can also process historical data published by accounts that are no longer active.

Over the years, there has been no limit in engineering accounts' features, to feed them to machine learning classifiers, e.g., the length of usernames, the reposting rate, some temporal patterns and the similarity of message contents based on Levenshtein distance [12], just to name a few.

Deep Learning Approaches. Hayawi et al. in [19] propose DeeProBot, where only some of the user profile features (the username's length and the number of followers) are exploited in order to classify single accounts with an LSTM (Long Short-Term Memory) network. Najari et al. in [29] use a GAN (Generative Adversarial Network) associated with a LSTM network. GANs generate bot samples to obtain more information about their behavior. RoSGAS (Reinforced and Self-supervised GNN Architecture Search) [44] is based on multi-agent deep reinforcement learning.

In [23], the authors propose a deep neural network based on a LSTM architecture processing the texts of tweets, thus exploiting NLP techniques. Their intuition is that bot accounts produce similar contents; therefore, analyzing texts

---

[2] https://cnets.indiana.edu/blog/2020/09/01/botometer-v4/.

can help the classification. Authors of [40] propose a text-based approach using a bidirectional LSTM. Work in [41] presents a framework with deep neural networks and active learning to detect bots on Sina Weibo.

All of the cited works have been tested on publicly released bot datasets and achieve very good performances (greater than 0.9), considering standard classification metrics such as accuracy, precision, recall and Area Under the Curve.

Graph-based Approaches. Detection techniques also take into account graph neural networks, where the social network is seen as a graph, where users are the nodes and the edge between two nodes represents a relationship between users, such as, e.g., a followship or retweet relationship. Thus, features derived from the social graph were considered along with profile and timeline fatures to train new models. An example is the work by Alhosseini *et al.* [1] which achieves very high performances, still on publicly released datasets [42].

Behavioral Analysis. Approximately from 2014, a number of research teams, independently, proposed new approaches for detecting coordinated behavior of automated malicious accounts, see, e.g., [7,47]. That line of research does not consider individual account properties, but rather properties in common with a group of accounts, like detection of loosely synchronized actions [4]. It is precisely in the context of the analysis of synchronism and coordination of the account behaviours that the idea of associating symbols with account actions arose, so that the timeline is represented as a string, so called Digital DNA [11]. The concept of Digital DNA is fundamental in the present work and will be introduced in the next section. Recently, Digital DNA has been re-analysed by Gilmary et al. in [16], where they measure the entropy of the DNA sequences. Even in this case, the detection performances result in very high values.

This brief roundup of work might lead one to think that bot detection is a solved task. Unfortunately, bots evolve faster than detection methods [8,14], the latter are not suitable for detecting all kinds of bots [26], and existing datasets for doing training are inherently built according to peculiar accounts characteristics [30,38].

We therefore conclude this section by pointing out how, perhaps, the task can never be solved in its entirety [6,31], and that, since we still have room for investigation, relying on image detection and state-of-the-art tools in this regard seems to us to be a good track to take.

3   Useful Notions

3.1   Digital DNA

The biological DNA contains the genetic information of a living being and is represented by a sequence which uses four characters representing the four nucleotide bases: A (*adenine*), C (*cytosine*), G (*guanine*) and T (*thymine*). Digital DNA is the counterpart of biological DNA and it encodes the behaviour of an online account. In particular, it is a sequence consisting of L characters from a predefined alphabet $\mathbb{B}$:

$$\mathbb{B} = \{ \sigma_1, \sigma_2, \sigma_3, ..., \sigma_N \} \tag{1}$$

In Eq. 1 each σ is a symbol of the alphabet and a digital DNA sequence will be defined as follow:

$$s = (\sigma_1, \sigma_2, ..., \sigma_n), \ \sigma_i \in \mathbb{B} \ \forall \ i = 1, ..., n. \tag{2}$$

Each symbol in the sequence denotes a type of action. In the case of Twitter, a basic alphabet is formed by the 3 actions representing the types of tweets:

$$\mathbb{B} = \left\{ \begin{array}{l} A = \text{tweet}, \\ C = \text{reply}, \\ T = \text{retweet} \end{array} \right\} \tag{3}$$

According to the type of tweets, it is thus possible to encode the account timeline, which could be, e.g., the following $s = ACCCTAAACCCCCTT$. Strings of digital DNA were compared to each other in [9]: the longer the *longest common substring* of a group of accounts, the more likely that group is made up of accounts programmed to complete a similar task, i.e., the more likely those accounts are automated accounts.

3.2   Convolutional Neural Networks

Given the recent and noteworthy [17,22,25,36] advancements in the field of Convolutional Neural Networks (CNNs), we asked ourselves whether these networks could be used to classify Twitter accounts into bot/human.

CNNs are typically composed of three layers: convolutional layers, pooling layers, and fully connected layers. The convolutional layer is the fundamental component of a CNN and it requires most of the computation. The input is an image and the dimension of the input image changes depending on whether it is grayscale or colored. Combined with the convolutional layer, there is the "filter" which is a matrix of small size. From the convolution operation, we have in output a "filtered" image which is a sequence of dot products between the input pixels and the filter. Afterward, an activation function can be applied to the output. It is also possible to optimize the convolution layer through some hyperparameters, such as the "stride" and the "padding". The former represents the amount of movement of the filter over the input, the latter is the process of padding the border of the input. The second type of layer is the "pooling" layer. A pooling layer is used to downsample the given input. There are two main types of pooling: max-pooling and average-pooling. The third type of layer is the "Fully-Connected" (FC) layer, also known as a "dense" layer. The neurons in a FC-layer receive input from all the neurons in the previous layer.

4   From Digital DNA to Images

To the best of our knowledge, no approaches in the literature take advantage of image classification to classify social bots. The aim is transforming each account's DNA sequence into an image. Given the similarity in the sequences of bots' actions with respect to those of genuine accounts, the intuition is that an image classifier might work well in the bot detection task.

The literature offers some DNA-to-image conversion algorithms [21,24]. We tried experimenting with these conversion algorithms, but we did not get significant results since these algorithms are for real DNA strings. Thus, we decided to propose an ad-hoc conversion algorithm, which transforms a digital DNA sequence into a bidimensional object.

Algorithm 1 shows the pseudocode for passing from Digital DNA to an image. Since CNNs expect images of the same size, we first consider the string of maximum length and check whether the length is a perfect square. If not, we consider the perfect square closest to and strictly largest than the maximum length. By doing so, it is possible to transform all the strings to images of equal size[3].

After arbitrarily deciding a RGB color to assign to each symbol in the alphabet, the image is colored pixel by pixel based on the coors assigned to the correspondent symbol. The coloring is done as long as the length of the input string is not exceeded; therefore, if the sequence is not the one with the maximum length, this will result in a black part of the image. All images created are in grayscale; we tried also with colored images, but there was no significant improvement in the final results.

5   Datasets

This section introduces the datasets on which we tested our detection technique. The datasets are all publicly available.

5.1   Cresci-2017

Firstly introduced in [8], this dataset consist of bots and genuine accounts. The kind of bots are various, like bots engaged in online political discussions, bots promoting specific hashtags, and bots advertising Amazon products. In our study, we evaluated 991 bots and 1,083 genuine accounts for a total of 2,074 samples.

The first step of the procedure is the generation of the DNA sequences for each account in the dataset. We rely on the alphabet in Sect. 3.1 (Eq. 3), which considers three symbols, associated to three basic activities on the Twitter platform: Tweet, Retweet, Reply. After the generation of the DNA strings, we apply the algorithm in Algorithm 1 to generate the images.

---

[3] As an example, strings as long as 10000 characters are represented by images of size 100×100.

Algorithm 1. From Digital DNA to image: Pseudocode

Input: List of DNA sequences
Output: DNA images

1: $n \leftarrow$ Length of the longest DNA sequence
2: **if** n is a perfect square **then**
3:     $L \leftarrow \sqrt{n}$
4: **else**
5:     $L \leftarrow$ `get_closest_square_number(n)`
6: **end if**
7: $P \leftarrow$ dict with symbols and colors
8: **for each DNA sequence do**
9:     $I \leftarrow$ `create_image(width=L, height=L)`
10:    **for** `row in range(L)` **do**
11:        **for** `col in range(L)` **do**
12:            $k \leftarrow$ `(row * L) + col`
13:            **if** $k < n$ **then**
14:                $I[row,\ col] \leftarrow P[DNA[k]]$
15:            **end if**
16:        **end for**
17:    **end for**
18: **end for**

(a)   (b)

Fig. 1. Representation as images of a genuine (left) and bot (right) account belonging to Cresci-2017.

Figure 1 show two images, representing a genuine and a bot account, *resp.* belonging to the Cresci-2017 dataset. Some noise in Fig. 1a distinguishes this account from that of the bot (Fig. 1b). Intuitively, a CNN is able to pick up these differences and, thus, classify the accounts in the correct way.

5.2   Cresci-Stock 2018

First introduced by Cresci et al. in [10], this dataset consists of both genuine and automated accounts tweeting so-called 'cashtags', i.e., specific Twitter hashtags that refer to listed companies. Part of the automated accounts has been found to act in a coordinated fashion, in particular by mass retweeting cashtags of low

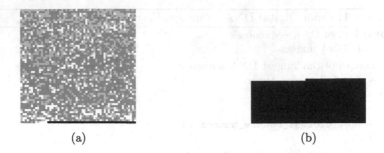

(a)   (b)

Fig. 2. Human and bot accounts on Cresci-Stock-2018 dataset.

capitalization companies. In our study, we used $6,842$ bots and $5,882$ genuine users, for a total of 12.724 labeled samples.

In Fig. 2, it is possible to see the noise which distinguishes a genuine account (Fig. 2a) from a bot account (Fig. 2b).

In this case, the image representing the bot is almost completely white, due to the homogeneity of the actions it performs on the social network and due to the choice of the colors used for the different pixels in creating the images.

5.3 TwiBot20

Firstly introduced in [13], TwitBot20 is a very large dataset with almost 230k Twitter accounts. Of these, the authors provide a total of $11,746$ labeled samples, $6,561$ bots and $5,185$ genuine accounts. The dataset covers diversified bots and genuine users 'to better represent the real-world Twittersphere' and, at the time of the publication of the article presenting it, it represented 'the largest Twitter bot detection benchmark'.

In TwiBot20, bot sequences of activities are very similar to those of genuine accounts. This of course makes the images generated by the sequences similar to each other and there is limited information to highlight specific behavioral patterns. Furthermore, TwiBot20 features accounts with a maximum of 200 tweets per user; therefore, the images are quite small (15×15). We attempted to enlarge the images, but the results were not good. Details are in Sect. 6.

6 Experiments and Results

For the experiments, we use PyTorch Lightning[4] to produce readable and reproducible code and it allows to spend less time on engineering the code. We also adopt WanDB [3] to keep track of metrics. Regarding the loss function, we consider the cross entropy (Eq. 4).

$$Loss = -(y \log(p) + (1 - y) \log(1 - p)) \tag{4}$$

---

[4] https://github.com/PyTorchLightning/pytorch-lightning.

The upperbound to the number of epochs is set to 50. However, we use `EarlyStopping` to monitor the accuracy (or loss) on the validation set: if it the accuracy does to increase (*resp.*, the loss does not decrease) for a predetermined number of epochs, the training stops. Each dataset tested is randomly splitted into training, testing and validation. The only exception is TwiBot20, where the authors of the dataset give this split. We evaluate the classification performances in terms of well-known, standard metrics, such precision, recall, F1 (the harmonic mean of precision and recall), and Matthew Correlation Coefficient (MCC) (i.e., the estimator of the correlation between the predicted class and the real class of the samples). The results achieved in this study are noticeable since they, in some cases, improve the state of the art. In general, we tried several pre-trained models, but the best results were achieved by networks based on the ResNet50 model. During the training phase, we carefully monitored the loss so as to be sure that there were no overfitting problems, and, thus, the model learned to classify correctly.

Comparison Between State-of-Art Results and those by the Image Classification Proposal for Cresci-2017. In the paper introducing TwitBot-20 [13], the authors consider two other datasets, Cresci-2017, already introduced by us above, and PAN-19[5]. To all 3 datasets, the authors of TwitBot-20 apply state-of-the-art detection methods, to evaluate the difference in classification performances. The best result obtained on the Cresci-2017 dataset by [13] is reported in Table 1, first row. The same table, in the second row, shows the performance results obtained by applying our method based on image classification, with ResNet50, where the loss decreases to 0.114. From that table, we can note how our perfomance results are equal to state-of-the-art results, with a slight improvement in MCC.

Table 1. Performances' comparisons on Cresci-2017: state-of-art *vs* image classification.

Metric	Cresci 2017			
	Accuracy	Recall	F1	MCC
Feng et al.	0.98	–	0.98	0.96
Image classification	0.98	0.98	0.98	**0.98**

Figure 3 shows the training and validation losses: the two losses have similar trends, and they decrease until they stabilize, after a number of epochs. Since the model behaves similarly in both the validation and training set, there is no overfitting [46].

Comparison Between State-of-Art Results and Those by the Image Classification Proposal for Cresci-Stock. In this case, the results in [2] will be taken as a

---

[5] https://pan.webis.de/clef19/pan19-web/author-profiling.html.

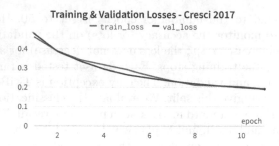

Fig. 3. Training and validation losses for Cresci-2017.

Table 2. Performances' comparisons on Cresci-Stock: state-of-art *vs* image classification. Results of *Antenore et al.* are taken from the Table 4 of [2].

Metric	Cresci stock 2018			
	Accuracy	Recall	F1 score	MCC
Antenore et al.	0.77	0.96	0.82	–
Image classification	**0.89**	0.88	**0.89**	**0.78**

reference. The comparison between the results are reported in Table 2, and, as it is possible to see, the images approach improved the *accuracy*, the *F1 score* and the *MCC*.

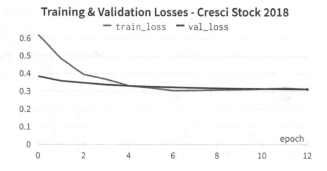

Fig. 4. Training and validation loss for Cresci-Stock.

In Fig. 4, the two losses have a similar macroscopic behavior; in the training phase the loss stabilizes in fewer epochs, which is due to the larger number of samples used (60% in training, 30% in validation). This trend rules out overfitting [46].

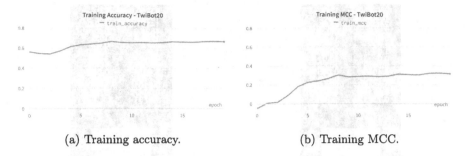

(a) Training accuracy.   (b) Training MCC.

Fig. 5. Training accuracy and MCC considering the images originated from the Twit-Bot20 account timelines

Table 3. Performances' comparison on TwitBot20: state-of-art *vs* image classification *vs* image classification where the image is enriched with SuperTML features [37].

Metric	TwiBot20			
	Accuracy	Recall	F1 score	MCC
Feng et al.	0.81	–	0.85	0.67
Images approach	0.67	0.66	0.61	0.34
Images approach with SuperTML	0.81	0.80	0.80	0.67

Image Classification for the TwiBot20 Accounts. After applying the Algorithm 1 to the timelines of the TwiBot20 accounts and after the CNN training phase, the classification performances are disappointing due to the limited set of tweets per user. It can be seen in Fig. 5b and Fig. 5a in which the accuracy does not improve so much compared to the initial phase, and MCC stabilizes between 0.3 and 0.4. The exact numerical results are in Table 3, second row.

Thus, we decide to use more account features and attach them to the user timeline. Interestingly enough, the article in [37], by Sun et al., proposes an algorithm to represent tabular data as images and then proceeds with image classification. Classification achieves state-of-art results on both large and small datasets, like the Iris dataset[6] and the Higgs Boson Machine Learning(See footnote 6).

Given the effective approach of [37], we enlarge our feature set: the digital DNA plus all the features listed in Table 4. Table 3 shows the resulting images, for a genuine and a bot account. The third row in Table 3 shows the performance results of the classification, when the image is formed with the enlarged feature set (Fig. 6).

Finally, Fig. 7 shows the trends of the training and validation losses, which stabilize around 0.6. Even in this case, the losses have very similar behavior, thus, the model is not overfitting.

---

[6] Iris dataset homepage, Higgs Boson Machine Learning challenge on Kaggle..

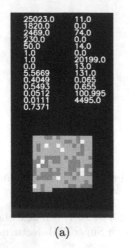

(a)   (b)

Fig. 6. (a): image format for a genuine account; (b): image format for a bot account. Both images have been created following the SuperTML algorithm, proposed in [37]. The upper part of the image is the list of features also reported in Table 4. The lower part of the image is the representation of the Digital DNA.

Table 4. List of features used.

Features
statuses_count, followers_count, friends_count, listed_count, default_profile, favourites_count, profile_use_background_image, verified, followers_growth_rate, friends_growth_rate, favourites_growth_rate, listed_growth_rate, followers_friends_rate, screen_name_length, screen_name_digits_count, description_length, description_digits_count, name_length, name_digits_count, total_tweets_chars_count, total_urls_in_tweets, total_mentions_in_tweets, urls_tweets_rate, mentions_tweets_rate, chars_tweets_rate, account_age

Fig. 7. Training and validation losses for TwitBot20 Image Classification + SuperTML

7   Conclusions

Research in bot classification is still open, mainly due to the continuous evolution of these kind of accounts. This work proposed a novel method for the task, based on image classification. The proposed approach has been proven aligned with state-of-art results. Since it is tough to acquire fully representative benchmark datasets (e.g., social platforms often block scraping, the APIs have a limited number of calls), the natural way to follow is to achieve full advantage of the available data. In the case of TwitBot20, for example, the original dataset offers numerous other pieces of information that were not exploited in the present work to obtain the images, such as, e.g., the network of interactions between accounts. As future work, we might consider exploiting this extra information to better evaluate the proposed approach.

References

1. Ali Alhosseini, S., Bin Tareaf, R., Najafi, P., Meinel, C.: Detect me if you can: spam bot detection using inductive representation learning. In: 2019 World Wide Web Conference, Companion, pp. 148–153. WWW 2019, ACM (2019). https://doi.org/10.1145/3308560.3316504
2. Antenore, M., Rodriguez, J.M.C., Panizzi, E.: A comparative study of bot detection techniques with an application in Twitter COVID-19 discourse. Soc. Sci. Comput. Rev. (2022). https://doi.org/10.1177/08944393211073733
3. Biewald, L.: Experiment tracking with weights and biases (2020). https://www.wandb.com/
4. Cao, Q., Yang, X., Yu, J., Palow, C.: Uncovering large groups of active malicious accounts in online social networks. In: ACM SIGSAC Conference on Computer and Communications Security, pp. 477–488. ACM (2014)
5. Ceron, A.: Internet, news, and political trust: the difference between social media and online media outlets. J. Comput.-Mediat. Commun. **20**(5), 487–503 (2015)
6. Cresci, S.: A decade of social bot detection. Commun. ACM **63**(10), 72–83 (2020)
7. Cresci, S., Di Pietro, R., Petrocchi, M., Spognardi, A., Tesconi, M.: DNA-inspired online behavioral modeling and its application to spambot detection. IEEE Intell. Syst. **31**(5), 58–64 (2016)
8. Cresci, S., Di Pietro, R., Petrocchi, M., Spognardi, A., Tesconi, M.: The paradigm-shift of social spambots: evidence, theories, and tools for the arms race. In: 26th International Conference on World Wide Web Companion, pp. 963–972. ACM (2017). https://doi.org/10.1145/3041021.3055135
9. Cresci, S., Di Pietro, R., Petrocchi, M., Spognardi, A., Tesconi, M.: Social finger-printing: detection of spambot groups through DNA-inspired behavioral modeling. IEEE Trans. Dependable Secur. Comput. **15**(4), 561–576 (2018)
10. Cresci, S., Lillo, F., Regoli, D., Tardelli, S., Tesconi, M.: $FAKE: evidence of spam and bot activity in stock microblogs on Twitter. In: ICWSM (2018). https://doi.org/10.1609/icwsm.v12i1.15073
11. Cresci, S., Pietro, R.D., Petrocchi, M., Spognardi, A., Tesconi, M.: DNA-inspired online behavioral modeling and its application to spambot detection. IEEE Intell. Syst. **31**(5), 58–64 (2016). https://doi.org/10.1109/MIS.2016.29

12. Efthimion, P.G., Payne, S., Proferes, N.: Supervised machine learning bot detection techniques to identify social twitter bots. SMU Data Sci. Rev. **1**(2), 5 (2018)
13. Feng, S., Wan, H., Wang, N., Li, J., Luo, M.: TwiBot-20: a comprehensive Twitter bot detection benchmark. In: CIKM 2021, pp. 4485–4494. ACM (2021). https://doi.org/10.1145/3459637.3482019
14. Ferrara, E., Varol, O., Davis, C., Menczer, F., Flammini, A.: The rise of social bots. Commun. ACM **59**(7), 96–104 (2016)
15. Gangware, C., Nemr, W.: Weapons of mass distraction: foreign state-sponsored disinformation in the digital age. Park Advisors (2019)
16. Gilmary, R., et al.: DNA-influenced automated behavior detection on Twitter through relative entropy. Sci. Rep. **12**, 8022 (2022)
17. Gu, J., et al.: Recent advances in convolutional neural networks. Pattern Recogn. **77**, 354–377 (2018). https://doi.org/10.1016/j.patcog.2017.10.013
18. Guo, B., Ding, Y., Yao, L., Liang, Y., Yu, Z.: The future of misinformation detection: new perspectives and trends. CoRR abs/1909.03654 (2019). http://arxiv.org/abs/1909.03654
19. Hayawi, K., et al.: DeeProBot: a hybrid deep neural network model for social bot detection based on user profile data. Soc. Netw. Anal. Min. **12**(1), 43 (2022). https://doi.org/10.1007/s13278-022-00869-w
20. He, K., Zhang, X., Ren, S., Sun, J.: Deep residual learning for image recognition. CoRR abs/1512.03385 (2015). http://arxiv.org/abs/1512.03385
21. Jeffrey, H.: Chaos game representation of gene structure. Nucleic Acids Res. **18**(8), 2163–2170 (1990). https://doi.org/10.1093/nar/18.8.2163
22. Krizhevsky, A., Sutskever, I., Hinton, G.E.: Imagenet classification with deep convolutional neural networks. Commun. ACM **60**(6), 84–90 (2017). https://doi.org/10.1145/3065386
23. Kudugunta, S., Ferrara, E.: Deep neural networks for bot detection. Inf. Sci. **467**, 312–322 (2018). https://doi.org/10.1016/j.ins.2018.08.019
24. La, S., et al.: DNA sequence recognition using image representation. Res. Comput. Sci. **148**, 105–114 (2019). https://doi.org/10.13053/rcs-148-3-9
25. Liu, Z., Mao, H., Wu, C.Y., Feichtenhofer, C., Darrell, T., Xie, S.: A ConvNet for the 2020s. In: Computer Vision and Pattern Recognition, pp. 11966–11976 (2022). https://doi.org/10.1109/CVPR52688.2022.01167
26. Mazza, M., Avvenuti, M., Cresci, S., Tesconi, M.: Investigating the difference between trolls, social bots, and humans on Twitter. Comput. Commun. **196**, 23–36 (2022)
27. Meel, P., Vishwakarma, D.K.: Fake news, rumor, information pollution in social media and web: a contemporary survey of state-of-the-arts, challenges and opportunities. Expert Syst. Appli. **153** (2020). https://doi.org/10.1016/j.eswa.2019.112986
28. Mustafaraj, E., Metaxas, P.T.: From obscurity to prominence in minutes: political speech and real-time search. In: Web Science: Extending the Frontiers of Society On-Line (2010)
29. Najari S., Salehi M., F.R.: GANBOT: a GAN-based framework for social bot detection. Soc. Netw. Anal. Min. **12**(4) (2022). https://doi.org/10.1007/s13278-021-00800-9
30. Olteanu, A., Castillo, C., Diaz, F., Kıcıman, E.: Social data: biases, methodological pitfalls, and ethical boundaries. Front. Big Data **2**, 13 (2019)
31. Rauchfleisch, A., Kaiser, J.: The false positive problem of automatic bot detection in social science research. PLoS One **15**(10), 1–20 (2020)

32. Sayyadiharikandeh, M., et al.: Detection of novel social bots by ensembles of specialized classifiers. In: CIKM '20: The 29th ACM International Conference on Information and Knowledge Management, pp. 2725–2732. ACM (2020)
33. Shao, C., et al.: Anatomy of an online misinformation network. PLoS ONE **13**(4), e0196087 (2018)
34. Simonyan, K., Zisserman, A.: Very deep convolutional networks for large-scale image recognition. In: Bengio, Y., LeCun, Y. (eds.) Learning Representations (2015)
35. Suarez-Lledo, V., Alvarez-Galvez, J.: Prevalence of health misinformation on social media: systematic review. J. Med. Internet Res. **23**(1), e17187 (2021). https://doi. org/10.2196/17187, http://www.jmir.org/2021/1/e17187/
36. Sultana, F., Sufian, A., Dutta, P.: Advancements in image classification using convolutional neural network. In: 2018 Fourth International Conference on Research in Computational Intelligence and Communication Networks (ICRCICN), pp. 122–129. IEEE (2018). http://arxiv.org/abs/1905.03288
37. Sun, B., et al.: SuperTML: two-dimensional word embedding and transfer learning using imagenet pretrained CNN models for the classifications on tabular data. CoRR abs/1903.06246 (2019). http://arxiv.org/abs/1903.06246
38. Tan, Z., et al.: BotPercent: estimating Twitter bot populations from groups to crowds. arXiv:2302.00381 (2023)
39. Valkenburg, P.M., Peter, J.: Comm research-views from Europe—five challenges for the future of media-effects research. Int. J. Commun. **7**, 19 (2013)
40. Wei, F., Nguyen, U.T.: Twitter bot detection using bidirectional long short-term memory neural networks and word embeddings. In: Trust, Privacy and Security in Intelligent Systems and Applications, pp. 101–109 (2019). https://doi.org/10. 1109/TPS-ISA48467.2019.00021
41. Wu, Y., Fang, Y., Shang, S., Jin, J., Wei, L., Wang, H.: A novel framework for detecting social bots with deep neural networks and active learning. Knowl.-Based Syst. **211** (2021). https://doi.org/10.1016/j.knosys.2020.106525
42. Yang, C., Harkreader, R., Gu, G.: Empirical evaluation and new design for fighting evolving twitter spammers. IEEE Trans. Inf. Forensics Secur. **8**(8), 1280–1293 (2013). https://doi.org/10.1109/TIFS.2013.2267732
43. Yang, K., Varol, O., Davis, C.A., Ferrara, E., Flammini, A., Menczer, F.: Arming the public with AI to counter social bots. CoRR abs/1901.00912 (2019). http:// arxiv.org/abs/1901.00912
44. Yang, Y., et al.: RoSGAS: adaptive social bot detection with reinforced self-supervised GNN architecture search. Trans. Web (2022). https://doi.org/10.1145/ 3572403
45. Yardi, S., Romero, D., Schoenebeck, G., et al.: Detecting spam in a Twitter network. First Monday (2010). https://doi.org/10.5210/fm.v15i1.2793
46. Ying, X.: An overview of overfitting and its solutions. In: Journal of physics: Conference Series, vol. 1168, p. 022022. IOP Publishing (2019)
47. Yu, R., He, X., Liu, Y.: GLAD: group anomaly detection in social media analysis. ACM Trans. Knowl. Discov. Data (TKDD) **10**(2), 1–22 (2015)

Linking Scholarly Datasets—The EOSC Perspective

Marcin Wolski[1], Antoni Klorek[1], Cezary Mazurek[1],
and Anna Kobusińska[2]([⊠])

[1] Poznan Supercomputing and Networking Center, Poznań, Poland
[2] Poznań University of Technology, Poznań, Poland
Anna.Kobusinska@cs.put.poznan.pl

Abstract. A plethora of publicly available, open scholarly data has paved the way for many applications and advanced analytics on science. However, a single dataset often contains incomplete or inconsistent records, significantly hindering its use in real-world scenarios. To address this problem, we propose a framework that allows linking scientific datasets. The resulting connections can increase the credibility of information about a given entity and serve as a link between different scholarly graphs. The outcome of this work will be used in the European Open Science Cloud (EOSC) as a base for introducing new recommendation features.

Keywords: Big scholarly datasets · Entity linking · EOSC · Microsoft Academic Graph · OpenAIRE Graph

1 Introduction

In recent years, with the development of many novel research fields, and the rapid growth of digital publishing, the term Big Scholarly Data (BSD) has been coined and become increasingly popular [23]. Big Scholarly Data was introduced to reflect the size, diversity and complexity of the vast amounts of data associated with scholarly undertakings, such as journal papers, conference proceedings, degree theses, books, patents, presentation slides, and experimental data from research projects [8,23]. These data collections have millions of co-authors, papers, citations, figures and tables, and massive scale-related data produced by scholarly networks and digital libraries. The use of the BSD has gained immense importance lately, particularly with the advent of multi-disciplinary research projects, which use BSD to discover research collaboration, expert finder and recommender systems [9].

One of the ongoing efforts to deliver a virtual, distributed research data repository and related services is the EOSC - European Open Science Cloud. The EOSC resources comprise outcomes of research efforts, such as published papers.

Supported by the *EOSC Future* project, co-funded by the EU Horizon 2020 Programme INFRAEOSC-03-2020/101017536.

As a primary dataset of scientific resources for the EOSC [1], the OpenAIRE Research Graph (OARG) dataset was employed. However, they also contain software and e-infrastructure services, such as computational power, storage, and network to support scientific experiments [7], which sets EOSC apart from other environments.

In this paper, we propose the OARGLink framework that enhances the OARG dataset with the content provided by open scholarly datasets to improve the accessibility and composability of the EOSC resources. AMiner, Microsoft Academic Graph (MAG) [6], and DBLP are examples of the datasets which are widely used for various research purposes [23]. Unfortunately, linking the OARG and open records is a demanding task. Despite these datasets offering information about the corresponding scientific resources, such information usually varies from one dataset to another. For instance, one dataset can provide only the basic information about scholarly resources, such as title, authors, published year etc. At the same time, the other dataset can store such information as the number of citations or the content of the abstract. As a result, the enrichment of the EOSC resources will further enhance its features, such as the intelligent discovery of the EOSC resources and smart recommendations [20] and, in the future, introduce new capabilities such as suggesting research collaborations. By combining resources of EOSC and open datasets, we will make the corresponding resources more credible and enable the usage of scientific networks in EOSC services.

The structure of this paper is as follows. In Sect. 2, the EOSC and the chosen open scholarly datasets are introduced. Section 3 presents the related work. Next, in Sect. 4, we outline a problem of entity linking, and discuss the general idea of the proposed OARGLink framework for connecting the OARG and MAG datasets. Furthermore, Sect. 5 presents the data flow in the proposed solution, and Sect. 6 describes how the data processing algorithms were customized based on a few trials with sample databases. Section 7, presents results that describe the efficiency of the implemented solution, known limitations and the method of results verification. Next, Sect. 8 provides the discussion on the proposed solution. Finally, Sect. 9 presents conclusions and future work.

2   The EOSC and Open Scholarly Datasets

2.1   The EOSC and the EOSC Future Project

The EOSC is an ongoing effort to connect the existing European e-infrastructures, integrate cloud solutions and provide a coherent point of access to various public and commercial services in the field of academic research [4]. The EOSC also is a key acronym for various European R&D projects related to Open Science on the national, regional and European levels. These projects aim at engaging researchers to utilize a web of scientific resources that are open and Findable, Accessible, Interoperable, and Reusable (FAIR). The regional EOSC initiatives and aggregators contribute to the data collection on the users of available resources (e.g. the EOSC-Nordic). Along with the variety of stakeholders of the EOSC ecosystem, several different roles address the needs of these stakeholders, such as research infrastructures, technology providers, service providers, data managers, researchers, policymakers (including funders), and everyday users [3].

The vision of the EOSC Future project is to deliver an operational EOSC Platform with an integrated environment consisting of data, professionally provided services, available research products and infrastructure that will be accessed and used by the European research community. At the heart of the EOSC Platform are the users who provide and exploit EOSC resources. Users include researchers, resource providers, research and technology enablers, trainers and policymakers. The EOSC Portal is considered a universal open hub for EOSC users. The EOSC portal offers public and commercial e-infrastructure services, including distributed and cloud computing resources and the EOSC Research Products. The number of currently registered services in the EOSC Portal is almost 400[1]. The OARG is now the provider of Research Products for the EOSC Platform, delivering around 150 million publications, datasets, research-supporting software, configurations and other products[2]. The estimated size of the target population of the EOSC is roughly 2 million, including 1.7m researchers covering all major fields of science and levels of seniority [2].

2.2 Scholarly Datasets

In general, scholarly datasets are published by organisations which own and develop the platforms for scholarly data management. Such platforms are usually exposed to the scientific community as academic search engines or digital libraries [23]. Their primary function is to crawl documents from the Web, extract useful information from them, and then store and index them in a coherent repository [12,22]. Typically, there are various scholarly applications implemented on top of these repositories.

Table 1. Basic features of a few of the selected open scholarly datasets

Dataset	Entities
OpenAIRE Research Graph (OARG)	140m publications, 16m research data, 286k research software items, 175k organizations
Open Academic Graph (OAG)	MAG: 240m publications, 243m authors & AMiner: 184m publications, 113m authors
AminerNetwork	2m publications (8m citations), 1.7m authors, 4.3m collaboration relationships
DBLP	6.22m publications
OpenAlex	249m publications, 103m authors, 226k venues, 108k institutions

---

[1] https://marketplace.eosc-portal.eu/services.
[2] https://graph.openaire.eu/.

Table 1 presents the basic statistics of the selected datasets. In addition to search engines and digital libraries, many datasets with scientific and social networks have been published so far. Scientific and academic social networks, such as Mendeley, Academia, LinkedIn or ResearchGate, are being utilized by the users to enhance their knowledge about other users in the networks and find new collaborators for their current or future projects [11].

As of the end of 2021, Microsoft stopped developing MAG. The end of this vast and prospering data source raised concerns, so the nonprofit organisation OurResearch developed the OpenAlex (OA) [14], a fully open catalogue of the global research system. The primary OpenAlex source of data is MAG, but the developers make use of other sources such as Crossref or Pubmed. The OA data is constantly updated using available repositories, databases, and internet scrapers. Despite storing information, the authors created a website and API that researchers may easily use to obtain desired records. To the best of our knowledge, the OA is the most extensive open–source platform currently available.

Every dataset has a different graph structure of its entities. OpenAIRE shares entities about publications, datasets, software and other research products. Information about authors is kept inside these tables. OAG, made from Aminer and MAG, contains, despite a publication entity with essential authors' data, a separate table with authors' details, a table with affiliations data and a table with venues. MAG also has different tables with publications and authors. Moreover, this dataset contains many other entities such as venues, journals, affiliations, conferences, etc. AminerNetwork shares publication data, and there is additional information about authors in a separate entity. What is more, there is an entity that contains connections between coauthors. DBLP contains entities about publications, authors, journals and conferences. In turn, MAG provides publications, authors, venues, institutions and concepts. These entities have connections between them all.

3 Related Work

In the existing solutions, a single dataset is often linked with another one, creating a new dataset that combines entities from these sources. To link two large scholarly datasets, MAG and AMiner, the scientists developed a framework to create the Open Academic Graph (OAG) [24]. The resulting dataset contains data from sources and links between publications, authors and venues. This work aimed to build a large, open-knowledge, linked entity graph. Another example is a dataset containing publications from various scientific disciplines built on the base of the arXiv.org resources [16]. The primary purpose of this effort was to link the publications to the MAG to enrich the metadata information. As a result, a freely available dataset with annotated and extracted citations was proposed to be used by researchers and practitioners. Furthermore, a dedicated database model, based on the ResearchGate (RG) data source, was prepared for implementing collaborators finding system [15]. The model comprises two parts: one for designing a consistent, collaborator-finding system and the other that contains different relations between the pair of users. RG dataset has been collected from Jan. 2019 to April 2019 and includes raw data of 3980 RG users.

In all mentioned systems and approaches to linking scholarly data, the difficulty lies in dealing with entity matching, which aims to identify records that belong to the same entity. Entity matching (record linkage) is very important for data integration and cleaning, e.g. removing duplicates. The difficulty arises mainly from heterogeneous and poor-quality data. The existing algorithms for linking entities can be divided into two categories: classification-based and rule-based [10]. The first one tries to determine if two records are the same by classifying them into the same entity or assigning them the same label. Here, Machine Learning and Deep Learning methods are often used. The models learn patterns from training data and then apply them to solve the given problem of entity matching. The challenge in this method is the preparation of high-quality training examples. The ruled-based category is the deterministic approach to link entities. By setting the number of rules, the records are compared against them and if the rules are fulfilled, the records are classified as the same entity. The difficulties are in setting rules not too strict and not too loose to link, preferably all the same records and not omitting the correct ones. Also, many rule-based approaches cannot cope with missing values and require numerous preprocessing steps.

4   The OARGLink Framework—Problem Statement and General Idea

In this paper, *the OARGLink framework* dedicated to linking OARG records (publications) with the records from open scholarly datasets was proposed. The framework aims to create a coherent database of the pair (identical) publications[3].

There are some common problems that usually have to be faced during the integration of heterogeneous databases [19]. Among them, based on analyzing the scholarly datasets' structure and content, we identified problems that can also impact the process of connecting the OARG and open scholarly data. Firstly, there is high dimensional scholarly data that, among other issues, impose computational challenges. Moreover, some of the scholarly data is incomplete, inaccurate or unreasonable. Finally, a problem with data integrity occurs. Unique identifiers between records of two scholarly datasets often do not exist. Moreover, different scholarly datasets have other data structures and fields, meaning they are encoded differently. Moreover, the problem of linking scholarly data usually deals with differences in spelling, formatting and proper fulfilment of the metadata of a given record. The existing datasets often contain incomplete or noisy records (resulting from, e.g. encoding issues in the source PDFs). There does not also exist standard schema used for storing key attributes among the datasets (such as the author's name can be saved as a full name or as abbreviations). Furthermore, connecting scholarly datasets consisting of millions of

---

[3] https://gitlab.pcss.pl/eosc-extra/scholarlydata.

records always poses computational challenges. As a result, the process of linking scholarly datasets is a non-trivial and challenging task [13].

Taking the above arguments into account, we focused our efforts on the identification of the same publications among different datasets and thereby omitting other existing entities such as venues, journals, etc. A journal article can be identified by the journal name, volume, issue number, and starting page number. However, for a large fraction of open-access scholarly papers crawled from the Web, such information is usually not available. Empirically, a paper entity can be uniquely identified by four header fields: title, authors, year, and venue, in which the venue is a conference or a journal name [17].

In the paper, the Record Linkage Method for combining data was proposed to find the connections between OARG publications with the records from the MAG scholarly dataset. The proposed solution can also be considered a relaxed Deterministic (Exact) Matching Method [18]. The publication's year and the number of authors must be exact; one title has to be a part of the second title, and at least one author has to be matched precisely.

For the entity linking among OARG and MAG, we took advantage of a dedicated dump of OARG with a pre-selected set of resources for EOSC (OARG–EOSC). This dump contains 1.7 m papers, 2.63 m datasets, 223k software and 19.6k other research products[4]. The DOI in the OARG-EOSC dataset is present in 1558130 publications, which is 91.037% of records. As a result, taking this field as one to connect entities from different datasets would result in omitting many records already at the start. However, other fields of interest, i.e. publication year, title and authors, are present in every record so that they can be used to identify the same entity among datasets.

5   Data Processing Flow in the OARGLink

The proposed algorithm processes the scholarly data in three phases depicted in Fig. 1. In the algorithm the following general approach for connecting entities in heterogeneous datasets is applied:

1. The fields that might be useful during connection are identified. By leaving only essential fields and eliminating the not needed ones, it is ensured that the next steps would not suffer from high-dimensional data.
2. Data cleaning process is applied—the records with essential fields missing are deleted; any data inconsistencies are identified.
3. Data are modified and unified, e.g. it is checked whether in both datasets a date is stored in the same type and order and changes are applied if needed.

Currently, the framework is implemented and evaluated for connecting the records of OARG with AminerNetwork and OARG with MAG.

---

[4] https://sandbox.zenodo.org/record/1094615#.Y_8k6tLMKRQ.

PHASE 0: Data Extraction and Cleaning

The structure of the publications' metadata differs depending on the dataset, and the number of fields varies from publication to publication. However, some fields are usually shared among them and almost always present. The aim of Phase 0 is to leave only essential fields. These are the paper id, title, authors and publication date. If needed, adequate processing is performed to convert the above fields to be in the proper format:

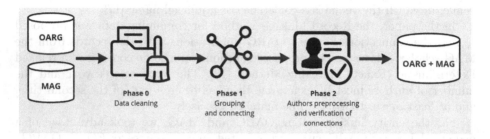

Fig. 1. Data processing in the OARGLink framework

- if the date is written fully, only the year is left,
- if the name and surname of an author are separated, e.g. by a semicolon, they are split into separate words,
- publications with no authors or year are erased

This cleaned data is saved in JSON format.

PHASE 1: Linking Datasets by Publications' Title

In the first phase of the linking procedure, the following steps are performed:

1. The cleaned data is loaded from both datasets—OARG and the one to connect with it.
2. Dictionaries are created for each dataset—the publications in the dictionaries are grouped by the number of authors and the publication year. This grouping minimizes the number of needed comparisons between publications' titles. When the year or the number of authors of publications do not match, there is no point in comparing the titles of these records. Simultaneously, during the same iteration, titles are changed to lowercase.
3. Publications are linked by checking if the title from the external dataset is contained in the OARG publication's title (checking performed only between the same number of authors and year groups, as described in the second point).
4. Ids of connected pairs are stored and saved as JSON files.

PHASE 2: Authors' Preprocessing and Verification of Connected Pairs by Authors' Comparison

The pair of records obtained in Phase 1 links the publications with the same titles, number of authors, and year of publications. These pairs might not always be exact, as authors may differ. Consequently, the second phase verifies and performs additional processing of the authors by performing the following steps:

1. The linked pairs of publications are loaded.
2. Authors from these publications are extracted.
3. Authors are preprocessed in the following way: first, the authors are concatenated into a single string, and the punctuation characters, double spaces (if present), and space from the end of a string (if present) are deleted. Next, all characters are lowercase, and a string is split into separate words. Finally, single–letter words are deleted, and words are sorted alphabetically.
4. Pairs of related publications are selected in which at least one first or last name matches. The proposed solution does not verify whether all authors match because, even with complex preprocessing, many typos or special foreign characters influence the obtained results. After the analysis, this method allows for the maximisation of the number of connected pairs and, at the same time, minimizes wrongly connected ones.
5. Obtained pairs of ids are saved in the JSON file.

The Python programming language was the primary technology to implement *the OARGLink*. Built-in libraries made it possible to implement compelling data reading, cleaning and manipulation methods. We used dictionaries as the primary data structures to store publications' information.

6   The OARGLink Customization

Before connecting the complete datasets of the OARG and the MAG, the data processing algorithms of *the OARGLink* were customized by making a few trials with data samples.

Firstly, an attempt to connect the OpenAIRE with the AminerNetwork was made. Next, the same approach was used to connect the OpenAIRE with the MAG dataset taken as a part of the OAG. After linking each pair of dataset samples, the results were evaluated against the accuracy (i.e. checked if the connected results contain identical publications) and the processing time. Then, the basic algorithm was modified, and the experiments were re-run to check the results.

6.1   Linking Samples of the OARG and AminerNetwork

The existing OpenAIRE dump file containing 189000 publications was used during the linking process. All publications were taken as a sample to conduct the trial run. The full Aminer dump contains over 2 million publications—a sample

of half a million was used to perform the linking. AminerNetwork publication
data was downloaded from the Aminer website[5].

After the completion of two phases, the proposed algorithm found 60 connections. Every connected record was checked manually, and a single mistake was
identified. An incorrectly connected pair is the following:

- Publication from OARG:
  Title: *"Introduction"*, Authors: *Salim Yusuf, Michael Gent, Genell Knatterud,
  Michael Terrin,* Year: *2002.*
- Publication from AminerNetwork:
  Title: *"Introduction"*, Authors: *Michael Bauer, Gene Hoffnagle, Howard John-
  son, Gabriel Silberman,* Year: *2002.*

The mistake originated from the fact that despite having the same titles, number of authors and published year, there are authors with the same names in
both entities. Although the found problem rarely occurs, the proposed algorithm
classified these publications as the same. Summing up this trial, the algorithm
achieved 98.3% precision. However, some publications might have been omitted,
e.g. titles had typos, the same publications were stored under slightly different
titles, etc.

The linking procedure ran with an Intel Core i5 CPU with 8 GB RAM and
lasted less than 25 s.

6.2   Linking Samples of the OpenAIRE and MAG

The EOSC OpenAIRE dump contains 1.7 m publications. The whole dataset
was included in the experiment as it will give us correct judgment for future
work.

The MAG was chosen because it contains far more records than the Aminer
part of the OAG, so looking from a higher perspective, it should result in more
linkings and better results and may be used as a separate data source for the
EOSC portal in the future. Ten batches of 100k publications were sampled from
the MAG (about 240m publications in total) for experimental purposes and
connected one by one with the OARG.

Five hundred and seven connections were found, and only one error occurred,
so the algorithm achieved 99.8% precision. The error is as follows:

- Publication from OARG:
  Title: *"Approximate Solution Of Some Mixed Boundary Value Problems
  Of The Generalized Theory Of Couple-Stress Thermo-Elasticity"*, Authors:
  Chumburidze, Manana, Lekveishvili, David, Year: *2014.*
- Publication from MAG:
  Title: *"Mixed Boundary Value Problems"*, Authors: *Fioralba Cakoni, David
  Colton,* Year: *2014.*

---

[5] https://www.aminer.org/aminernetwork.

The error occurred because the second title is included in the first one, the number of authors is the same, the published year is identical, and there are authors with the same name in both examples.

The OARG connection time, with a single 100k batch and the Intel Core i5 CPU configuration with 8 GB RAM was less than 90 s.

7 The Overall Results and Their Verification

After setting up the optimal configuration for the proposed *the OARGLink* framework, we ran it on sample data of OARG-EOSC and MAG. This intermediate step determined the estimated time needed to connect the complete datasets. Also, the additional experiments did ensure that the algorithm could cope with the more significant amount of data and possible differences in the metadata. The results from the experiment and estimations are presented in Table 2.

Table 2 presents the estimated number of connections and the time needed to accomplish the whole processing. These estimations were calculated based on experimental results and with the assumption of linear growth of these two factors. However, the sample of data taken to the experiment may slightly differ from the whole dataset. As a result, the final number of obtained connections, as well as the time needed, may vary. It's worth mentioning that times were measured on a device with Intel Core i5 CPU and 8 GB RAM and with fixed sizes of batches with publications. So, by presenting the estimations, we also assumed that the final operations would be performed on the same or similar machine and using batches of the same sizes. Eventually, we proceeded to the last step to make connections between EOSC OARG (1.7 m publications) and full MAG (240 m publications). As we calculated from the estimations, it should take about 60 h and around 121k connected pairs should be obtained.

Table 2. Experiments results and estimations

	Results		Estimations
	OARG (1.7 m) with MAG (100k)	OARG (1.7 m) with 10x MAG (100k)	OARG (1.7 m) with whole MAG (240 m)
# of final connections	50	507	121k
time assuming batches 1.7 m and 100k	90 s	15 min	60 h

The linking process started with downloading the whole MAG dataset to the local computer. Compressed files had almost 150 GB; this number increased over three times after uncompressing. Next, all data had to be split into smaller files. We divided each file, which contained over 5 m publications, into ones that had 500k of them. Then, the cleaning was done, and the connecting procedure started. The results are presented in Table 3.

Table 3. Full connection results

	OARG	MAG	OARG+MAG
# of records	1.7 m	240 m	496168

As a result, over 29% of OARG's publications were connected with their MAG's equivalents (around four times more than we had estimated). The procedure took 4 days and ran on two machines with 4-core processors and 8 GB RAM. The processing time was 50% longer than we had estimated. In some batches, there were far more publications to compare. It was caused by the vast number of the same publications' year or the same number of authors among the documents.

Table 4. The presence of DOI in the connected records

# of connected publications (OARG+MAG)	496168
# of OARG's publications with DOI	458438
# of MAG's publications with DOI	433208

We took advantage of DOI (Digital Object Identifier) to verify if the obtained results were correct and whether the connected pairs of the papers from both datasets referred to the same article. The basic statistics about related publications and the presence of DOI are presented in Table 4.

The proposed verification procedure comprised the following steps:

1. The pairs of connected records in which both publications possess DOIs were found
2. DOIs in these pairs were compared and it was calculated how many of them are exact and how many differ.
3. A subset of connected pairs where DOI is different was taken and the related publications were manually analyzed

The results show that 92.4% of connected publications on the OARG's side have DOI. Moreover, in 87.3% of connected publications on the MAG's side, DOI is present.

Table 5 presents the found records (OARG and MAG), which possess DOI in both (corresponding) publications. The connected records were thoroughly analyzed, where DOI was different, although the proposed framework had merged the corresponding publications together. We found out, among others, that both publications can be the same in the title, authors, published date and content, but they were published at different conferences or in other journals. As a result, the document's formatting or structure may differ, or some additional footnotes may be present. So, the articles are the same but published under different DOIs. Nevertheless, the obtained results reveal that 82.5% of connected pairs simultaneously have DOI in both publications. Also, in 93% of these pairs, DOIs are exact, and in 7% of these pairs, DOIs differ.

Table 5. The presence of DOI in BOTH of the connected publications

# of pairs with DOI in both publications	409575
# of pairs where DOIs are exact	380933
# of pairs with different DOIs	28642

8   Discussion

The quality of the entity matching task is a trade-off between precision and coverage. In the case of EOSC, the target was to achieve high precision (we accept the risk of omitting some theoretically existing connections between records). This is because when the target datasets are significant, a small fraction of false positives in the sample data may eventually lead to many false matching. In turn, the wrongly connected records would lead to degrading user experience (measured in EOSC as, e.g. a mean consumer feedback satisfaction score for the presented publication records).

The most apparent method to connect entities with publications in different datasets is to use a universal identifier. Currently, the DOI is recognized as Persistent Identifier (PID) for publications [5]. DOI is a unique and never-changing string assigned to online (journal) articles, books and other works, and it is usually present in the datasets. But, since DOI is not always present in every record of analysed datasets, we argue that such an identifier cannot be used unambiguously for entity linking in scholarly systems.

Our chosen method for connecting common records from different scholarly datasets follows rule-based patterns for integrating heterogeneous databases. Among the considered approaches for implementing the entity linking system, a solution proposed by [24] was the most promising to be re-used. The authors implemented an algorithm based on locality-sensitive hashing (LSH) and Convolution Neural Networks (CNN) to connect publications. However, considering the structure and the format of the publications' metadata in OARG (i.e. lack of some fields in OARG, e.g. venue in the publication's metadata), the considered solution would not work out of the box. The algorithm would not be precise, and the results wouldn't be satisfied without the significant changes in the existing source codes and additional efforts to construct the artificial training set.

Another problem indicated while analyzing the considered approach was related to the time consumption and the extensive need for computational resources of this solution. A robust GPU-based hardware infrastructure would be needed to run the proposed algorithm efficiently.

Furthermore, it is still a challenge to evaluate the entity linking algorithms formally. In the paper [24], the authors manually label venue training data (around 1000 records) and construct artificial complex training data for papers and venues. A similar approach was used by [21], who matched CiteSeerX against DBLP and obtained 236 matching pairs. They also noticed the need to apply data cleaning tasks to filter out highly unhealthy data on top of a supervised approach.

Taking the above into account, we argue that the most effective method for linking Big Scholarly Datasets for EOSC is to combine two different approaches: (1) a rule-based approach based on the precise data matching (which includes preprocessing phase with data cleaning) and (2) the DOI pairing to identify the same publications. As a result, a more significant number of connections are meant to be identified. What is essential intersecting these two approaches should minimize potential errors. The algorithm used in the proposed framework found more connections than the method of connecting publications by only existing DOIs. Incorrect DOIs or their absence doesn't determine the accuracy of our solution.

The results of our experimental work with item-based recommendations based on connected records (MAG-OARG and OpenAlex-OARG) clearly show its usefulness in supporting EOSC-related scenarios. In particular, the OpenAlex dataset has multiple fields present in the metadata of a publication, so adding them to OARG records and using them in the EOSC portal brings additional benefits. Data about citations, references and related works are especially interesting here. This information could enhance the network of interconnections between publications and positively influence the quality of recommendations. Other data, e.g. venues, publishers, and different identifiers, could enrich the presented information at the portal and allow for a deeper understanding of a given record. As a result, EOSC users will see more information about the record and be able to use the portal as the primary source of their research.

The previously mentioned poor quality of data in source datasets may raise concerns if missing fields, such as some authors or wrongly stated year of publication, will result in omitting some matching which should be matched as equal entities. The methods for dealing with some data inconsistencies were stated in the framework's description, but unfortunately, it is impossible to consider all possible errors in input data. The OARGLink mainly focuses on correctness, not returning false positive examples. As a result, precision should be high. Conversely, omitted examples may be classified as false negatives so that the recall measure may be slightly lower.

9   Summary and Future Work

This paper studied the vital problem of linking large-scale scholarly datasets. Connecting heterogeneous scholarly datasets is not a trivial task. Firstly, they consist of various entities in academic graphs, such as author, paper, or organisation entities. Moreover, the structure of familiar entities such as papers or authors differs. Secondly, observing ambiguous values of the same attributes is widespread, such as authors' names or publications' titles. Finally, the scale of datasets is large, usually with millions of entities.

As a part of our work, we analyzed a few scholarly datasets to obtain more information about their content and structure. On top of the OARG, which is the primary data source for EOSC, we selected a few others: the OAG, which is one of the most significant sources of publications and other scholarly information currently available on the Internet; AminerNetwork as its structure is well defined and potentially ideal for the experimental purposes, and eventually DBLP, which constitutes a popular among scientists open scholarly dataset. We narrowed down the work to papers and authors. Firstly, we evaluated the implemented solution with samples of data. Experimental results show that our solution OARGLink achieved at least 98.3% precision for connecting OARG with Aminer and 99.8% accuracy for connecting OARG with MAG. The coverage hasn't been evaluated as a part of this work.

In the final work of connecting OARG for EOSC (1.7 million publications) with MAG (240 million publications), we achieved almost half a million connected pairs, over 29% of the used OARG dataset. The results indicate that the database with linked records can be used as a valuable source of information for further usage in EOSC.

The method proposed in this paper can be reused for linking other scholarly repositories. For example, social network databases can be linked and explored to resolve more accurate bindings between users by taking advantage of specific academic relationships. Furthermore, we have also recognized the need for applying a similar approach in digital libraries. Digital libraries often combine various sources with digital books from different providers into coherent records. In such a case, multiple editions of the same book, sometimes having other titles, must be merged into one entry and presented to the reader.

The primary goal for future work is to use the connected pair of records as a base for implementing new recommendation scenarios for EOSC. On top of that, there are many other potential directions for further development on *the OARGLink framework*. For example, the framework can be adapted to a large-scale computational environment (i.e. adapted to using scalable technologies such as Apache Spark and Hadoop) and run to connect the full OARG (140 million) and MAG datasets. In the future, additional experiments will be conducted to perform an extended evaluation of the implemented method, such as experiments with noisy data to determine the actual coverage.

References

1. Almeida, A.V.D., Borges, M.M., Roque, L.: The European open science cloud: a new challenge for Europe. In: Proceedings of the 5th International Conference on Technological Ecosystems for Enhancing Multiculturality. TEEM 2017, CM (2017)
2. Anca Hienola (ICOS), John Shepherdson (CESSDA ERIC), B.W.C.: D5.2a eosc front-office requirements analysis. Technical report (2022)
3. Barker, M., et al.: Digital skills for fair and open science: report from the EOSC executive board skills and training working group (2021)
4. Budroni, P., Claude-Burgelman, J., Schouppe, M.: Architectures of knowledge: the European open science cloud. ABI-Technik **39**(2), 130–141 (2019)
5. Cousijn, H., et al.: Connected research: the potential of the PID graph. Patterns **2**(1), 100180 (2021)
6. Färber, M., Ao, L.: The microsoft academic knowledge graph enhanced: author name disambiguation, publication classification, and embeddings. Quant. Sci. Stud. **3**(1), 51–98 (2022)
7. Ferrari, T., Scardaci, D., Andreozzi, S.: The open science commons for the European research area. Earth Obs. Open Sci. Innov. ISSI Sci. Rep. Ser. **15**, 43–68 (2018)
8. Giles, C.L.: Scholarly big data: information extraction and data mining. In: Proceedings of the 22nd ACM International Conference on Information & Knowledge Management, pp. 1–2 (2013)
9. Khan, S., Liu, X., Shakil, K., Alam, M.: A survey on scholarly data: from big data perspective. Inf. Process. Manag. **53**, 923–944 (2017)
10. Kong, C., Gao, M., Xu, C., Qian, W., Zhou, A.: Entity matching across multiple heterogeneous data sources. In: Navathe, S.B., Wu, W., Shekhar, S., Du, X., Wang, X.S., Xiong, H. (eds.) DASFAA 2016. LNCS, vol. 9642, pp. 133–146. Springer, Cham (2016). https://doi.org/10.1007/978-3-319-32025-0_9
11. Kong, X., Shi, Y., Yu, S., Liu, J., Xia, F.: Academic social networks: modeling, analysis, mining and applications. J. Netw. Comput. Appl. **132**, 86–103 (2019)
12. Manghi, P., et al.: The OpenAIRE research graph data model. Zenodo (2019)
13. Nasar, Z., Jaffry, S.W., Malik, M.K.: Information extraction from scientific articles: a survey. Scientometrics **117**(3), 1931–1990 (2018). https://doi.org/10.1007/s11192-018-2921-5
14. Priem, J., Piwowar, H., Orr, R.: Openalex: a fully-open index of scholarly works, authors, venues, institutions, and concepts (2022)
15. Roozbahani, Z., Rezaeenour, J., Shahrooei, R., et al.: Presenting a dataset for collaborator recommending systems in academic social network. J. Data, Inf. Manag. **3**, 29–40 (2021). https://doi.org/10.1007/s42488-021-00041-7
16. Saier, T., Färber, M.: unarXive: a large scholarly data set with publications' full-text, annotated in-text citations, and links to metadata. Scientometrics **125**, 3085–3108 (2020). https://doi.org/10.1007/s11192-020-03382-z
17. Sefid, A., et al.: Cleaning noisy and heterogeneous metadata for record linking across scholarly big datasets. In: Proceedings of the AAAI Conference on Artificial Intelligence, vol. 33, pp. 9601–9606 (2019)
18. Shlomo, N.: Overview of data linkage methods for policy design and evaluation. In: Crato, N., Paruolo, P. (eds.) Data-Driven Policy Impact Evaluation, pp. 47–65. Springer, Cham (2019). https://doi.org/10.1007/978-3-319-78461-8_4
19. Wang, L.: Heterogeneous data and big data analytics. Autom. Control Inf. Sci. **3**(1), 8–15 (2017)

20. Wolski, M., Martyn, K., Walter, B.: A recommender system for EOSC. Challenges and possible solutions. In: Guizzardi, R., Ralyte, J., Franch, X. (eds.) Research Challenges in Information Science. RCIS 2022. LNBIP, vol. 446, pp. 70–87. Springer, Cham (2022). https://doi.org/10.1007/978-3-031-05760-1_5

21. Wu, J., Sefid, A., Ge, A.C., Giles, C.L.: A supervised learning approach to entity matching between scholarly big datasets. In: Proceedings of the Knowledge Capture Conference, pp. 1–4 (2017)

22. Wu, Z., et al.: Towards building a scholarly big data platform: challenges, lessons and opportunities. In: IEEE/ACM Joint Conference on Digital Libraries, pp. 117–126 (2014)

23. Xia, F., Wang, W., Bekele, T.M., Liu, H.: Big scholarly data: a survey. IEEE Trans. Big Data 3(1), 18–35 (2017)

24. Zhang, F., et al.: OAG: toward linking large-scale heterogeneous entity graphs. In: Proceedings of the 25th ACM SIGKDD International Conference on Knowledge Discovery and Data Mining, pp. 2585–2595 (2019)

Memory-Efficient All-Pair Suffix-Prefix Overlaps on GPU

Sayan Goswami[✉][iD]

Ahmedabad University, Ahmedabad, India
sayan.goswami@ahduni.edu.in

Abstract. Obtaining overlaps between all pairs from billions of strings is a fundamental computational challenge in de novo whole genome assembly. This paper presents a memory-efficient, massively parallel algorithm that uses GPUs to find overlaps from large sequencing datasets with a very low probability of false positives. Here we use a Rabin-fingerprint-based indexing method which stores the strings with their fingerprints and uses them to generate the fingerprints of suffixes and prefixes. We then sort these fingerprints in hybrid CPU-GPU memory and stream them in GPU to find matches. Experiments show that our implementation can detect a trillion highly probable overlaps within 1.5 billion DNA fragments in just over two hours. Compared to the existing CPU-based approach, it routinely achieves speedups of 5–15x while having a collision rate of 1 in 20 million.

Keywords: Big data · Parallel processing · Memory management · GPU · Genome assembly

1 Introduction

The knowledge of the structure of DNA can be utilized in a vast range of applications such as personalized medicine, epidemiology, evolution, food safety and many more [6]. Deciphering the order of nucleotides in the DNA of a novel organism consists of two steps. First, a biochemical process called *sequencing* clones the DNA, fragments them randomly, and parses fragments into short (100–300 characters long) strings called *reads* or *short-reads*. Next, a computational process called *de novo assembly* is employed to find suffix-prefix overlaps between all pairs of reads and merge overlapping reads to recreate the original sequence. Sequencing projects routinely produce datasets with billions of reads. Assembling such datasets, and more specifically, finding all-pair suffix-prefix overlaps (*APSPO*) between billions of strings, is computationally expensive. On a large dataset, this step could take nearly a day, even on high-end servers.

In recent times, Graphics Processing Units (GPUs) have seen widespread adoption in general-purpose data processing tasks owing to a higher performance-per-dollar [4] and performance-per-watt [3] compared to CPUs. In

© The Author(s), under exclusive license to Springer Nature Switzerland AG 2023
J. Mikyška et al. (Eds.): ICCS 2023, LNCS 14073, pp. 624–638, 2023.
https://doi.org/10.1007/978-3-031-35995-8_44

bioinformatics, several programs that have been ported to GPUs report significant performance improvements [13–15]. GPU-based implementations of hash tables [12] and suffix arrays [1] have also been used in bioinformatics applications. However, for de novo assembly, fewer GPU-based implementations have been reported in the literature, and most of them have only been evaluated on small datasets owing to their limited memory capacities.

In this paper, we present *Empanada*[1], a faster and more memory-efficient GPU-based approach to finding APSPO in large sequencing datasets using an inverted index of $\mathcal{O}(n)$ bits. Our main contributions in this work are as follows. (1) We reduce the index size using a previously unexplored approach of obtaining prefix hashes and avoiding storing them explicitly. (2) We implement this method on GPUs using NVIDIA Thrust, which also enables it to also run in parallel on CPUs. (3) Our implementation can handle datasets much larger than what can be stored in GPU memory and is amenable to out-of-core execution using secondary storage. (4) We evaluate our implementation using several real datasets and observe speedups of two orders of magnitude compared to others. Using our approach, we found overlaps in datasets with about 3 billion strings in a little over 3 h on a single node where other tools ran out of time/space. (5) Finally, our implementations for sorting and searching in hybrid (host-device) memory may be helpful in other applications where the data fits in the host but not the device.

The rest of the paper is organized as follows. In Sect. 2, we give some background on de novo assembly and provide a brief exposition of existing works. Next, we explain our methodology in Sect. 3, followed by a description of our implementation details in Sect. 4. In Sect. 5, we evaluate our implementation using real-world datasets of various sizes and compare the execution times with previous works and finally present our conclusions in Sect. 6.

2   Background and Related Work

A *base* is an element of the set $\Sigma = \{A, C, G, T\}$. A *read* is a string over the alphabet Σ. The *Watson-Crick (WC) Complement*, also called *reverse complement*, of a read p of length l is a read p' of the same length such that $p'[i] = p[l - 1 - i]'$, where $A' = T$, $T' = A$, $C' = G$ and $G' = C$. An *overlap graph* is a directed weighted graph $\mathcal{G}$ obtained using the APSPO in a set of short-reads R and their WC complements R' such that a vertex corresponds to $s_i \in R \cup R'$ and an edge (u, v, w) exists from u to v iff the w-length suffix of the read corresponding to u is the same as the w-length prefix of the read at v. Naturally, for every edge $e = (u, v, w)$ in $\mathcal{G}$, there must also exist a complementary edge $e' = (v', u', w)$.

Given three vertices p, q, and r and edges $e_i = (p, q, w_i)$, $e_j = (q, r, w_j)$, and $e_k = (p, r, w_k)$ in $\mathcal{G}$, e_k is a *transitive edge* if $w_k = w_i - (|q| - w_j)$. Transitive edges can be removed from an overlap graph without any loss of information. A *string graph* is an overlap graph with all transitive edges removed. In de novo assembly, the string graph is refined and chains in the graph (i.e. vertices with

---
[1] The code is available at https://github.com/sayangoswami/empanada.

degree ≤ 1) are collapsed. The reads corresponding to the collapsed vertices are merged using their edge weights, and the resulting strings are called *contigs*.

The problem of finding APSPO has been studied extensively, and a time-optimal solution [8] using suffix trees has been known since 1992. The high memory overhead of suffix trees has led to approaches using enhanced suffix arrays which are faster and more memory efficient. Edena [10], the first available implementation of a string graph-based de novo assembler, uses a suffix array.

Although suffix arrays are more efficient than suffix trees, their space usage is still prohibitively large. This necessitated the development of compressed datastructures such as the compressed suffix array and other compressed full-text indices like the FM-index. SGA [16] is another string graph assembler that uses the FM-index and is capable of handling reasonably large datasets.

Dinh et al. have proposed a memory-efficient prefix tree to represent exact-match overlap graphs and use it to build the LEAP assembler [2]. Compact prefix trees have also been used by Rachid et al., who use an enhanced B-Tree to find strings with a specified prefix [9]. One of the most popular string graph assemblers, Readjoiner [5], avoids building the entire overlap graph by partitioning the suffixes and prefixes and processing each partition independently, resulting in significant improvements in time and memory usage.

GAMS [11] is the first GPU-based string graph assembler and can run on multiple nodes but is only evaluated on small datasets. LaSAGNA [7] was the first reported GPU-based string-graph builder that could handle large datasets using an inverted index from the fingerprints of suffixes and prefixes. It builds the index out-of-core and stores it on disk, thus reducing the memory usage on GPU. However, this index takes $\mathcal{O}(n \log n)$ bits of space on disk for $2n$ bits of data. Therefore this approach spends most of the execution time transferring the index from disk to GPU memory, resulting in low GPU utilization.

This work presents a memory-efficient, parallel, scalable solution that improves upon the fingerprint-based indexing of LaSAGNA by reducing the size of the index to $\mathcal{O}(n)$ bits. This version of the index stores the fingerprints of entire reads from which the fingerprints of suffixes/prefixes are obtained in $\mathcal{O}(n)$ time.

3   Methodology

In this section we present our approach for a space-efficient inverted index. We assume that reads in a dataset are of the same length l, which is the case for Next Generation Sequencing (NGS) machines such as Illumina. We use Rabin-fingerprint to create hashes of all reads in $\mathcal{R} \cup \mathcal{R}'$ as follows. We choose a large prime q and define an encoding $e : \Sigma \rightarrow \mathbb{N}$ which maps every base to $[0, \sigma)$. The fingerprint of a base b is given by $e(b)$. Given the fingerprint $f_{i,m}$ of some m-length substring starting at index i of some read s, we have the following:

$$f_{i,m+1} = (f_{i,m}\sigma + e(s[i+m])) \bmod q \tag{1}$$

$$f_{i+1,m-1} = (f_{i,m} + q - e(s[i])\sigma^{m-1} \bmod q) \bmod q \tag{2}$$

Typically, the rolling hash window expands towards its right and shrinks from left. Using Eq. 1, if we start with the fingerprint $f_{0,1}$ of the leftmost base of a read and successively expand the fingerprint-window, we get the fingerprints of all prefixes and eventually that of the entire read $f_{0,l}$. Similarly, using Eq. 2, if we start from the fingerprint of the entire read and successively shrink the window, we will get the fingerprints of all suffixes.

To store the fingerprints of all prefixes and suffixes as is in LaSAGNA [7], one needs $\mathcal{O}(n \log q) = \mathcal{O}(n \log n)$ bits (q is typically in the order of $\mathcal{O}(r^2)$, where $r = |\mathcal{R} \cup \mathcal{R}'| = n/l$). Instead, our approach only uses $\mathcal{O}(n)$ bits by storing the fingerprints of entire reads and regenerating the fingerprints of prefixes and suffixes from them. Regenerating the suffix fingerprints is straightforward (from Eq. 2). Our contribution is the procedure for regenerating prefix fingerprints from reads and their fingerprints which is as follows.

Given a fingerprint $f_{i,m+1}$ of a read s, our goal is to find $f_{i,m}$. Note that from Eq. 1, we have $\sigma f_{i,m} \equiv (f_{i,m+1} - e(s[i+m])) \pmod{q}$. Expressed as a linear diophantine equation,

$$\sigma f_{i,m} - qk = f_{i,m+1} - c(s[i+m]) \text{ for some } k \in \mathbb{N} \qquad (3)$$

Since σ and q are co-primes, the solutions for $f_{i,m}$ in Eq. 3 belong to exactly one congruence class modulo q, which is

$$f_{i,m} \equiv \sigma^{-1}(f_{i,m+1} - e(s[i+m])) \pmod{q}$$

where σ^{-1} is the multiplicative inverse of σ in the field $\mathbb{Z}/q$. As q is a prime and σ is non-zero, σ^{-1} exists and is equal to $\sigma^{q-2} \pmod q$ which can be calculated using binary exponentiation in $\log q$ steps. Moreover, by definition $f_{i,m} < q$ which implies

$$f_{i,m} = (\sigma^{-1}(f_{i,m+1} - e(s[i+m]))) \bmod q \qquad (4)$$

We can now use the reads and their fingerprints to find overlaps as follows. We iterate through every read $s_i \in \mathcal{R} \cup \mathcal{R}'$ and create a tuple $(f_{0,l}^{(i)}, i)$, where $f_{0,l}^{(i)}$ is the fingerprint of s_i. Each fingerprint $f_{0,l}^{(i)}$ is appended to an array $\mathcal{S}$ and each tuple $(f_{0,l}^{(i)}, i)$ is appended to an array $\mathcal{P}$. We also store all reads in an array such that any base $b_{i,j}$ at position j of a read s_i can be obtained in constant time.

Next, we find suffix-prefix overlaps in descending order of overlap lengths like so. We replace each item $f_{0,l}^{(i)} \in \mathcal{S}$ with $f_{1,l-1}^{(i)}$ using Eq. 2. We also replace each tuple $(f_{0,l}^{(i)}, i) \in \mathcal{P}$ with $(f_{0,l-1}^{(i)}, i)$ using Eq. 4. Finally, we sort the tuples in $\mathcal{P}$ by their fingerprints and perform a binary search of all elements of $\mathcal{S}$ in $\mathcal{P}$. If some $f_{1,l-1}^{(i)} \in \mathcal{S}$ is equal to $f_{0,l}^{(j)} \in \mathcal{P}$ we report with high confidence that read s_i has an $(l-1)$-length overlap with s_j. This process is repeated successively for lengths $l-2, l-3, \cdots, l_{min}$, where l_{min} is the minimum overlap length.

Calculating the fingerprints of reads take $\mathcal{O}(n)$ time, where n is the number of bases. At each overlap-detection round, obtaining the fingerprints for suffixes and prefixes for r reads takes $\mathcal{O}(r)$ time and sorting and searching take $\mathcal{O}(r \log r)$ time. Since there are $\mathcal{O}(l)$ rounds, the total time required to find occurrences is $\mathcal{O}(n \log r)$. Checking whether the occurrences are true positives and adding the edges in the overlap graph take $\mathcal{O}(el)$ time, where e is the total number of edges. Thus, the total time required is $\mathcal{O}(n \log r + el)$. A greedy approach of converting an overlap graph to a string graph is to only retain the longest overlaps for each read. Here, in each round we could filter out those reads from $\mathcal{S}$ and $\mathcal{P}$ which already have an edge. In that case, $e = \mathcal{O}(r)$ and the total time is $\mathcal{O}(n \log r)$. Storing $\mathcal{S}$ and $\mathcal{P}$ takes $(2r \log q + r \log r)$ bits of space which is $\mathcal{O}(n)$ since $\log r$ and $\log q$ are typically less than l. Storing the original reads requires an additional $2n$ bits, since $|\Sigma| = 4$.

4   GPU Implementation

In this section, we discuss how we adapt our methodology for the PRAM model of computation and implement it on GPU. The entire algorithm is implemented using the NVIDIA Thrust library (included with the CUDA toolkit), which can execute the program on NVIDIA GPUs or CPUs parallelized using OpenMP or Intel's Thread Building Blocks (TBB) with minimal change.

4.1   Overview

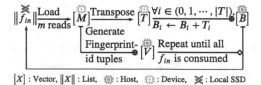

$[X]$: Vector, $\|X\|$: List, ▥ : Host, ▦ : Device, ✂ : Local SSD

Fig. 1. Generating fingerprint-id tuples

Figure 1 depicts the first stage of building the index - generating fingerprints of reads. The input file f_{in} is processed in chunks of m reads such that the intermediate data fits in device memory. Next, for each read in chunk $\mathcal{M}$, we generate fingerprints, create fingerprint-id tuples, and append them to host vector $\mathcal{V}$. We also generate the fingerprints of reverse complement reads and append the corresponding tuples to $\mathcal{V}$ (not shown in the figure for brevity). The original read s_j is assigned the id $2j$ and its reverse complement $2j + 1$.

Note that getting the fingerprints of suffixes and prefixes of length $k-1$ from those of length k only requires the bases at two positions - $k-1$ for prefixes and $l-k$ for suffixes. Thus, if $\mathcal{R}$ is stored in the host in a column-major order, only the bases used to generate fingerprints can be bulk-transferred to device and removed after the fingerprints are generated. Therefore, the reads are transposed and appended to host vector $\mathcal{B}$ so that $\mathcal{B}_{ij}$ is the ith base of read j. Fingerprints are generated using Thrust's `scan_by_key` function, and the transpose and reverse complements are generated using Thrust's `scatter` function.

Figure 2 illustrates the second stage - regenerating suffix/prefix fingerprints and using them to find overlaps. It takes the vector $\mathcal{V}$ of fingerprint-id pairs generated in the previous stage, sorts the pairs, and removes those with identical fingerprints. The sorting by read-ids ensures that for any read that's removed, its reverse complement is removed as well. The main challenge here is that $\mathcal{V}$ does

not fit in GPU memory, so we partition it in main memory and stream partitions to device memory, where they can be processed independently (discussed in
Algorithms 1 and 2). The partitioning involves only an $\mathcal{O}(\log r)$ number of random accesses in the host memory. Most random access operations are offloaded
to GPU.

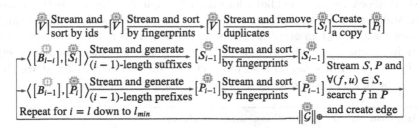

Fig. 2. Finding overlaps

One consequence of partitioning is that both $\mathcal{P}$ and $\mathcal{S}$ must now store tuples
of fingerprints and read-ids, instead of the method described in Sect. 3, where $\mathcal{S}$
only consisted of fingerprints. The fingerprint generation can be done in bulk if
both $\mathcal{P}$ and $\mathcal{S}$ are sorted by read ids since the bases in $\mathcal{B}$ are already arranged
according to their read ids. After the fingerprints are regenerated as per Eq. 2 and
Eq. 4, both the vectors of suffix- and prefix-pairs are sorted by their fingerprints.

The sorted vectors of suffix and prefix pairs are partitioned again, and the
partitions are streamed to GPU to search for overlaps. We use Thrust's vector-
ized binary search functions to search for each suffix fingerprint in the list of
prefix fingerprints. Specifically for each suffix fingerprint f_i, we first search for
the lower bound and then for the upper bound of f_i in the array of prefix fingerprints. The difference between the upper and lower bounds gives the number
of times f_i occurs in the array of prefix fingerprints. We then filter out the read-
ids whose corresponding fingerprints have non-zero occurrences in both vectors,
convert them to an edge-list format and write to the output file.

4.2 Sorting and Searching in Hybrid-Memory

The sorting of vectors and searching for keys are performed in hybrid-memory,
i.e., both the host and device memories. As stated before, most work is done on
the GPU, and the CPU is responsible for partitioning the vectors so that the
GPU can process each partition independently. Furthermore, since the vectors
can be enormous, we want to perform these operations using a constant amount
of auxiliary space in the host memory.

To achieve this, we use a non-contiguous host vector where data is stored in
blocks. This lets us allocate and free chunks of data whenever necessary so that
the peak memory usage is kept in check. The host-vector consists of a queue of
smaller arrays and supports appending new data at the end and reading (and
popping) data from the front. When writing to vectors, the data is appended to

the last block in the queue. When the last block is full, a new one is allocated and pushed into the queue. Data is read from the first block in the queue, and the number of items read is tracked. Once a block is consumed, it's popped from the front of the queue, and the memory is freed. In the actual implementation, we use a memory pool of blocks where freed blocks are stored and reused so that the repeated allocations and deallocations do not negatively impact performance.

In addition to sequential reads and writes, these vectors also support random accesses with the [] operator. This is useful in partitioning sorted vectors for streaming them into device memory, particularly for the `lower_bound` function. The lower bound of a key in a sorted vector is the smallest index at which the key can be inserted without invalidating the sorted order.

---

1 **function** *get_merge_partition_sizes* (A, B, m)
2      $p \leftarrow$ **list<pair<int, int>>**, $o_A \leftarrow 0$, $o_B \leftarrow 0$
3   **while** $o_A < A.n$ **and** $o_B < B.n$
4          $n_A \leftarrow min(A.n - o_A, m/2)$, $n_B \leftarrow min(B.n - o_B, m/2)$
5   **if** $n_A < m/2$ **and** $n_B < m/2$ **then**:
6              $p.push\_back((n_A, n_B))$
7   **break**;
8   **else**:
9              $k \leftarrow min(A[o_A + n_A - 1], B[o_B + n_B - 1])$
10             $n_A \leftarrow A.lower\_bound(o_A, o_A + n_A, k) - o_A$
11             $n_B \leftarrow B.lower\_bound(o_B, o_B + n_B, k) - o_B$
12             $p.push\_back((n_A, n_B))$
13             $o_A \leftarrow o_A + n_A$, $o_B \leftarrow o_B + n_B$
14   **if** $o_A = A.n$ **or** $o_B = B.n$ **then**:
15   **if** $o_A = A.n$ **and** $o_B < B.n$ **then** $p.push\_back((0, nB - o_B))$
16   **else** $p.push\_back((nA - o_A, 0))$
17   **return** p

Algorithm 1. Partitioning a non-contiguous host vector

The pseudocode for partitioning a pair of sorted non-contiguous host vectors is shown in Algorithm 1. The partitioning function accepts an argument m which specifies the maximum size of each partition. If then slides an $(m/2)$-sized window over each vector and resizes the windows based on the minimum of the last elements in both (lines 9–11). The partitions resulting from this algorithm splits the key-ranges of both the vectors in a way such that if a key k is present in partition j of one of the vectors A, then it will not be present in any other partition in A and if k is present in vector B, then it will only be present in partition j of B.

The vectors are sorted in device memories as shown in Algorithm 2. The vectors are loaded on the GPU in chunks of m and sorted using Thrust's `sort_by_key` function. These sorted chunks are then iteratively merged to obtain a single sorted vector. The merge algorithm uses the partition function in Algorithm 1 to split the vector of keys into independent partitions, loads these partitions into GPU and merges them by keys (fingerprints). In practice, the merging is also performed using radix sort. During the sorting and merging phases, the data blocks popped from the vectors are freed, thereby keeping the auxiliary memory usage limited to $\mathcal{O}(b)$, where b is the block size of the vector.

```
1  function sort_by_key (K, V)
2      s ← queue<pair<vector, vector>>
3      dK and dV are m-length arrays of keys and vals on the device
4      repeat until K is empty
5          s.emplace_back()
6          load upto m items from K and V into dK and dV
7          gpu_sort_by_key (dK, dV)
8          s.back().first.push_back (dK, length(dK))
9          s.back().second.push_back (dV, length(dV))
10     while length (s) > 1
11         Ka, Va ← s.pop_front()
12         Kb, Vb ← s.pop_front()
13         Kc, Vc ← merge_by_key (Ka, Va, Kb, Vb)
14         s.push_back ((Kc, Vc))
15     K ← s.back().first, V ← s.back().second
16
17 function merge_by_key (Ka, Va, Kb, Vb)
18     Kc ← vector, Vc ← vector
19     dK and dV are m-length arrays of keys and vals on the device
20     P ← get_merge_partition_sizes (Ka, Kb)
21     ∀p ∈ P:
22         nA ← p.first, nB ← p.second
23         if nA and nB
24             kA→pop_front (dK, nA)
25             kB→pop_front (dK + nA, nB)
26             vA→pop_front (dV, nA)
27             vB→pop_front (dV + nA, nB)
28             gpu_sort_by_key (dK, dV)
29             kC→push_back (dK, nA + nB)
30             vC→push_back (dV, nA + nB)
31         else if nA
32             pop nA items from Ka and Va and push into Kc and Vc
33         else
34             pop nA items from Ka and Va and push into Kc and Vc
35     return Kc, Vc
```

Algorithm 2. Sorting in hybrid memory

5   Results

5.1   Datasets and Testbed

We evaluate our implementation (Empanada) on several real-world datasets whose Sequence Read Experiment (SRX) IDs are given in Table 1. All datasets were generated using Illumina genome sequencing machines and were obtained from the NCBI Sequence Read Archive[2]. They are chosen so that they have similar (or the same) read lengths, and the number of bases roughly doubles in size from one dataset to the next.

Table 1. Datasets used.

SRX ID	Name	l	$|R|$	n
10829778	Fruitfly	152	83.6M	12.7 GB
14756245	Salamander	151	165M	24.9 GB
10572708	Butterfly	151	343.7M	51.9 GB
10301361	Starling	151	723.8M	109.3 GB
1382207{4,6}	Pig	151	1355.4M	204.7 GB

l: Read-length, $|R|$: Number of reads, not including reverse complements, $n = l \times |R|$: number of bases

All experiments were performed on the Expanse cluster in San Diego Supercomputing Centre[3]. Each GPU node has 384 GB RAM, two 20-core 2.5 GHz processors, four NVIDIA V100, each with 12 GB device memory, and a 1.6TB NVME SSD. Empanada uses a single CPU thread and a single GPU in a shared-node setting. The maximum memory available for Empanada was limited to 90 GB for the first 4 datasets (Fruitfly, Salamander, Building, Starling) and 180 GB for the Pig dataset. Each Compute node has two 64-core 2.5 GHz processors, 256 GB RAM and a 1 TB NVME SSD. Compute nodes were used to run SGA and Readjoiner in exclusive mode, with access to all cores and the full available memory. Both types of nodes have access to 12PB Lustre filesystem, but we used SSDs for I/O.

5.2 Execution Times

The total execution times of Empanada are reported in Table 2. We use a minimum overlap length of 55% of the read length, which is at the lower end of what is used in most string-graph-based assemblers. As previously mentioned, creating a string graph requires the sparsification of an overlap graph which can be done by removing transitive edges or retaining only the longest overlaps per read. We report the execution times with and without the greedy longest-edge retention approach. The time elapsed for fingerprint generation (t_f) includes that of reading input data. $t_o^{(g)}$ and $t_o^{(f)}$ denote the time required to obtain overlaps with and without the greedy edge retention, respectively. They include the time taken to remove duplicate reads and write the edges on the disk.

For the largest dataset with 2.3 billion unique reads (r_u), each 151 bases long, Empanada finishes in under 4.5 h $(t^{(f)})$. When it retains only the longest overlaps for each read, Empanada takes under 2.5 h $(t^{(g)})$. As expected, when only the heaviest edges are retained, the number of edges $E^{(g)}$ grows linearly with r_u, whereas when no edges are discarded, the number of edges $E^{(f)}$ explodes.

We define the collision rate for strings of length j as (1 - the number of unique fingerprints generated ÷ the total number of unique suffixes/prefixes of length j). The total collision rate $(n_c^{(g)})$ is defined as the sum of collision rates

[2] https://www.ncbi.nlm.nih.gov/sra/.

[3] https://www.sdsc.edu/support/user_guides/expanse.html.

Table 2. Reads, Edges and Execution times

Data	r	t_f	r_u	$t_o^{(g)}$	$t^{(g)}$	$E^{(g)}$	$t_o^{(f)}$	$t^{(f)}$	$E^{(f)}$
FFY	164M	33 s	107M	2 m 26 s	2 m 59 s	83M	5 m 7 s	5 m 40 s	104.75B
SMR	329M	1 m 39 s	261M	10 m 41 s	12 m 20 s	73M	13 m 15 s	14 m 54 s	4.55B
BFY	687M	3 m 51 s	504M	5 m 34 s	9 m 25 s	417M	28 m 9 s	32 m 0 s	26.72B
SLG	1.45B	9 m 14 s	1.37B	1 h 7 m 59 s	1 h 17 m 13 s	618M	1 h 54 m 50 s	2 h 4 m 4 s	1.67T
PIG	2.71B	17 m 37 s	2.34B	1 h 56 m 42 s	2 h 14 m 19 s	1.4B	3 h 55 m 58 s	4 h 13 m 35 s	127.79B

r: number of error-free reads and reverse complements, t_f: time taken to generate fingerprints, r_u: number of unique reads (and reverse complements), $t_o^{(g)}$: time taken to find overlaps with greedy edges, $t^{(g)}$: total time with greedy edges, $E^{(g)}$: number of greedy edges created, $t_o^{(f)}$: time taken to find overlaps without any edge-removal, $t^{(f)}$: total time without any edge-removal, $E^{(f)}$: number of edges created. M = Million, B = Billion, T = Trillion

of all suffixes/prefixes from l_{min} to l, when edges are selected greedily. In all our experiments, the fingerprints were generated using two 29-bit pseudo-Mersenne primes and combined into one 58-bit integer (stored as a 64-bit integer) to avoid integer overflows during multiplications. In this setup, $n_c^{(g)}$ was found to be 1×10^{-8} for Butterfly, 2×10^{-8} for Starling, and 5×10^{-8} for the others. Note that the other tools do not have the issue of false positives, but we claim that the false positive rate and be made arbitrarily close to 0 by randomly matching characters of the *supposedly* overlapping reads.

5.3 Scalability

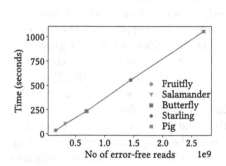

Fig. 3. Fingerprint generation time vs number of reads

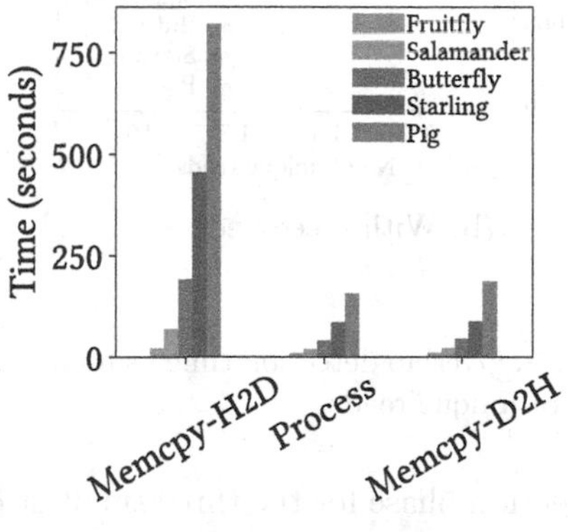

Fig. 4. Time spent of each stage during fingerprint generation

Figure 3 shows the time taken to generate fingerprints w.r.t. the number of reads. It is evident from the graph that this phase is highly-scalable across a wide range of data sizes. This is because this phase consists of finding reverse complements and transposing bases, both of which are based on the scatter pattern, which is embarrassingly parallel. The fingerprint generation is based on parallel prefix-scan, which has a linearithmic time complexity.

Figure 4 presents the time taken by the different parts of the fingerprint generation phase. The 'process' stage in the figure includes all the computations mentioned before. Most of the execution time is spent reading data from disk to device and copying fingerprint-read tuples and transposed bases from device to host. Both of these stages depend only on the data size and hence scale almost perfectly. This figure demonstrates that the major bottleneck in this phase is I/O. The performance of the data-copying could be improved with GPU-Direct which bypasses host memory when transferring data from disk to GPU. We plan to explore this in future work.

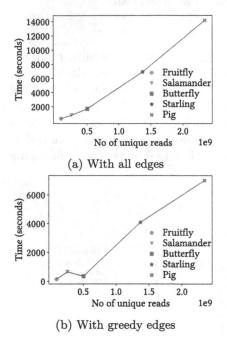

(a) With all edges

(b) With greedy edges

Fig. 5. Overlap detection time vs number of unique reads

Figure 5 plots the time taken to remove duplicates and find overlaps of up to 55% of the read lengths vs the number of unique reads. Figure 5 (a) considers the execution times when all edges are retained. In this scenario, the time spent on overlap detection grows almost linearly with the number of unique reads, with a slight upward trend starting from the Butterfly dataset. This is because, for smaller datasets, the fingerprint-id pairs can be loaded into the GPU and sorted at once, whereas for Starling and Pig, sorting the lists requires multiple passes over the data.

In Fig. 5 (b), the execution times are reported for the cases when only the heaviest edges per vertex were retained. In this case, we observe that the overlap detection time does not always grow predictably with an increase in the number of reads. For instance, the Butterfly dataset finishes quicker than Salamander - a dataset half its size.

To investigate the reason behind this, we study the progression of the overlap detection phase for the three smallest datasets in Fig. 6. For the three subfigures in Fig. 6, the x-axis depicts the overlap round i, during which the program obtains matching suffix-prefix pairs of length $(i + 1)$. Figure 6 (a) shows the number of edges created at each overlap round. We can observe that for the butterfly dataset, about 100 million edges are created during the first overlap round, which is much larger than the two smaller datasets. The vertices from which these edges emerge are removed from the list before the start of the next overlap round. This effect can be seen in Fig. 6 (b), which plots the number of active

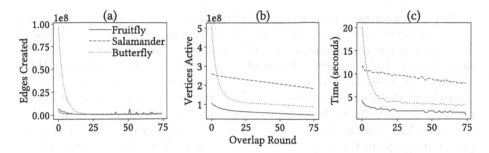

Fig. 6. Edges created vs overlap detection time

vertices per round. Due to the large number of edges discovered in the first few rounds, the number of active vertices decreases sharply, reducing the search space considerably. Consequently, the search time also decreases within the first few rounds. Eventually, searching in the Butterfly dataset finishes faster than in the smaller Salamander dataset.

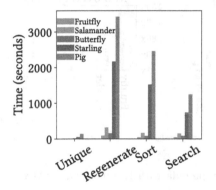

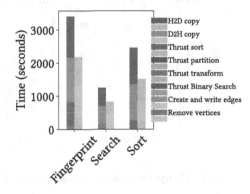

Fig. 7. Time spent of each stage during overlap detection

Fig. 8. Time spent for computation and data-copy for Starling and Pig

In Fig. 7, we examine the individual stages of the edge-building phase across all overlap lengths and study their contributions towards the execution time. The plot shows that most of the time is spent on regenerating fingerprints and sorting them. These two phases display a quasilinear trend consistent with their respective algorithms having multiple merge passes. Note that regenerating fingerprints also involves sorting, which takes up a considerable portion of its time (discussed later). The search phase also significantly contributes to the execution time but grows more slowly (almost linearly) with increasing data sizes. Again, this is consistent with the algorithm, which comprises creating vertex pairs and writing them to disk. The stage 'unique' is a one-time removal of duplicates before the start of overlap detection and takes very little time.

In Fig. 8, we inspect the components of the three most time-consuming stages of overlap detection - fingerprint regeneration, sorting, and searching in light of their share of execution times for Pig (deeply shaded bars) and Starling (lightly shared bars). To begin with, it is readily apparent that host-to-device (H2D) and device-to-host (D2H) data transfers take up most of the time across all three stages for both datasets. As indicated before, regenerating fingerprints in a batch requires sorting them by their read-ids, which for large datasets requires multiple passes. Indeed, sorting and the associated data transfers are the most time-consuming steps in this stage, as seen in the figure. The actual computation of fingerprints which involves Thrust's partition and transform methods, is almost insignificant in comparison. A similar trend can be observed in the sorting and search stages, where memory transfers dominate the actual computation. We are exploring avenues for alleviating this bottleneck from an algorithmic perspective.

5.4 Comparison of Execution Times with Other Tools

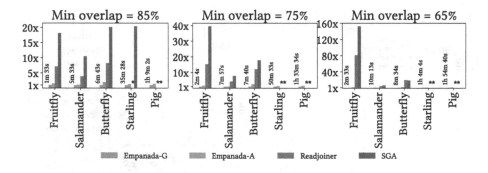

Fig. 9. Comparison of execution times of Empanada with Readjoiner and SGA

Figure 9 compares the performance of Empanada with Readjoiner and SGA, the only string graph assemblers that can handle large short-read datasets. We only use the results of preprocess, index, filter and overlap stages of SGA and the pre-filter and overlap phases of Readjoiner. For each dataset, we report the results of 3 experiments where we set the minimum overlaps to 85%, 75%, and 65% of the read lengths. For Empanada, we study the performance both with and without transitive edge removal. Note that Readjoiner could not process the Starling dataset due to some unknown error and the Pig dataset due to insufficient disk space. SGA could not process Pig due to some unknown error and could process Starling within the maximum allowed time (72 h) only when the minimum overlap was set to 85%. It can be seen that even when Readjoiner is run on 128 cores, Empanada with greedy edge retention (Empanada-G) is at least 4 times faster (in case of Salamander, 85%) and has a median speedup of about 8x. Compared to SGA, Empanada-G has a speedup of 7x at minimum with a median of

18.5x. All of these results are not only significantly better than SGA, but also outperforms our previous tool (LaSAGNA) by a considerable margin. Although we couldn't test LaSAGNA on the Expanse cluster on the large datasets due to a lack of local disk space, for the Fruitfly and Salamander datasets, Empanada finished about thrice as fast as the single-node implementation of LaSAGNA.

6 Conclusions

In this work, we introduce a new memory-efficient, parallel and highly scalable suffix-prefix indexing method using Rabin fingerprints that can find a trillion overlaps in a billion strings in a couple of hours on a single node using a single GPU. Our approach uses a non-contiguous vector to perform an in-place parallel merge sort of data in a two-level memory model, thereby allowing it to process data much larger than GPU memory. Experimental results demonstrate that our implementation outperforms existing approaches by a significant margin and exhibits almost linear scalability on a wide range of data sizes. In the future, we plan to implement a multi-gpu version, and also study the possibility of a heterogeneous cpu-gpu implementation. As an extension to this work, we will also perform more detailed analyses on the assembly accuracy.

Acknowledgements. This work used Expanse CPU and GPU nodes and Expanse Projects Storage at San Diego Supercomputing Centre through allocation CIS220052 from the Advanced Cyberinfrastructure Coordination Ecosystem: Services & Support (ACCESS) program, which is supported by National Science Foundation grants #2138259, #2138286, #2138307, #2137603, and #2138296.

References

1. Büren, F., Jünger, D., Kobus, R., Hundt, C., Schmidt, B.: Suffix array construction on Multi-GPU systems. In: Proceedings of the 28th International Symposium on High-Performance Parallel and Distributed Computing, pp. 183–194 (2019)
2. Dinh, H., Rajasekaran, S.: A memory-efficient data structure representing exact-match overlap graphs with application for next-generation DNA assembly. Bioinformatics **27**(14), 1901–1907 (2011)
3. Dong, T., Dobrev, V., Kolev, T., Rieben, R., Tomov, S., Dongarra, J.: A step towards energy efficient computing: redesigning a hydrodynamic application on CPU-GPU. In: 2014 IEEE 28th International Parallel and Distributed Processing Symposium, pp. 972–981. IEEE (2014)
4. Fan, Z., Qiu, F., Kaufman, A., Yoakum-Stover, S.: GPU cluster for high performance computing. In: SC'04: Proceedings of the 2004 ACM/IEEE Conference on Supercomputing, pp. 47–47. IEEE (2004)
5. Gonnella, G., Kurtz, S.: Readjoiner: a fast and memory efficient string graph-based sequence assembler. BMC Bioinform. **13**(1), 1–19 (2012)
6. Goswami, S., Lee, K., Park, S.J.: Distributed de novo assembler for large-scale long-read datasets. In: 2020 IEEE International Conference on Big Data (Big Data), pp. 1166–1175. IEEE (2020)

7. Goswami, S., Lee, K., Shams, S., Park, S.J.: GPU-accelerated large-scale genome assembly. In: 2018 IEEE International Parallel and Distributed Processing Symposium (IPDPS), pp. 814–824. IEEE (2018)
8. Gusfield, D., Landau, G.M., Schieber, B.: An efficient algorithm for the all pairs suffix-prefix problem. Inf. Process. Lett. **41**(4), 181–185 (1992)
9. Haj Rachid, M., Malluhi, Q.: A practical and scalable tool to find overlaps between sequences. BioMed. Res. Int. **2015** (2015)
10. Hernandez, D., François, P., Farinelli, L., Østerås, M., Schrenzel, J.: De novo bacterial genome sequencing: millions of very short reads assembled on a desktop computer. Genome Res. **18**(5), 802–809 (2008)
11. Jain, G., Rathore, L., Paul, K.: GAMS: genome assembly on Multi-GPU using string graph. In: 2016 IEEE 18th International Conference on High Performance Computing and Communications; IEEE 14th International Conference on Smart City; IEEE 2nd International Conference on Data Science and Systems (HPCC/SmartCity/DSS), pp. 348–355. IEEE (2016)
12. Jünger, D., et al.: General-purpose GPU hashing data structures and their application in accelerated genomics. J. Parallel Distrib. Comput. **163**, 256–268 (2022)
13. Klus, P., et al.: Barracuda-a fast short read sequence aligner using graphics processing units. BMC Res. Notes **5**(1), 1–7 (2012)
14. Liu, C.M., et al.: SOAP3: ultra-fast GPU-based parallel alignment tool for short reads. Bioinformatics **28**(6), 878–879 (2012)
15. Liu, Y., Schmidt, B., Maskell, D.L.: CUSHAW: a CUDA compatible short read aligner to large genomes based on the burrows-wheeler transform. Bioinformatics **28**(14), 1830–1837 (2012)
16. Simpson, J.T., Durbin, R.: Efficient construction of an assembly string graph using the FM-index. Bioinformatics **26**(12), i367–i373 (2010)

Ensemble Based Learning for Automated Safety Labeling of Prescribed Fires

Li Tan[1]📷, Raymond A. de Callafon[1](✉)📷, Mai H. Nguyen[2]📷, and Ilkay Altıntaş[2]📷

[1] Department of Mechanical and Aerospace Engineering, University of California San Diego, La Jolla, CA, USA
{ltan,callafon}@eng.ucsd.edu
[2] San Diego Supercomputer Center, University of California San Diego, La Jolla, CA, USA
{mhnguyen,ialtintas}@ucsd.edu

Abstract. Prescribed fires are controlled burns of vegetation that follow a burn plan to reduce fuel build-up and mitigate unanticipated wildfire impacts. To understand the risks associated to a prescribed burn, modern fire simulation tools can be used to simulate the progression of a prescribed fire as a function of burn conditions that include ignition patterns, wind conditions, fuel moisture and terrain information. Although fire simulation tools help characterize fire behavior, the unknown nonlinear interactions between burn conditions requires the need to run multiple fire simulations (ensembles) to formulate an allowable range on burn conditions for a burn plan. Processing the ensembles is often a labor intensive process run by user-domain experts that interpret the simulation results and carefully label the safety of the prescribed fire. The contribution of this paper is an algorithm of ensemble based learning that automates the safety labeling of ensembles created by a modern fire simulation tool. The automated safety labeling in this algorithm is done by first extracting important prescribed fire performance metrics from the ensembles and learn the allowable range of these metrics from a subset of manually labeled ensembles via a gradient free optimization. Subsequently, remaining ensembles can be labeled automatically based on the learned threshold values. The process of learning and automatic safety labeling is illustrated on 900 ensembles created by QUIC-Fire of a prescribed fire in the Yosemite, CA region. The results show a performance of over 80% matching of learned automated safety labels in comparison to manually generated safety labels created by fire domain experts.

Keywords: Prescribed Fire · QUIC-Fire · Safety · Automated Labeling · Gradient-free Optimization

---

Work is supported by WIFIRE Commons and funded by NSF 2040676 and NSF 2134904 under the Convergence Accelerator program.

J. Mikyška et al. (Eds.): ICCS 2023, LNCS 14073, pp. 639–652, 2023.
https://doi.org/10.1007/978-3-031-35995-8_45

1   Introduction

As the extent of landscapes burned by wildfires continuously grow, it is important to take advantage of prescribed fires to manage the risk of uncontrollable wildfires. A prescribed fire is a controlled burn of vegetation and ignited intentionally to meet fuel and vegetation management objectives, such as reducing hazardous fuels, sustain the natural landscapes, and avoid extreme wildfires. Compared to a wildfire that is unplanned, prescribed fire can be controlled by reducing the risk of a fire escape.

There are many positive effects of a prescribed fire on soil, vegetation, or even some cultural artifacts, and periodic fire plays an important role in the balance of many ecosystems [1–3, 12, 14]. Therefore, prescribed fire can be used as a tool to manage the forest area in various ecological aspects, such as preventing invasive vegetation and facilitating the recovery of specific species [6]. However, people are averse to the risk of a prescribed fire due to the lack of scientific knowledge about the benefit of a prescribed fire in an ecosystem management [13].

Environmental or burn conditions that include the landscape, terrain, fuel moisture, wind speed, wind direction and ignition pattern are important factors for the progression of the fire. Extensive modeling efforts have been documented that help with the prediction of the fire spread as a function of the burn conditions [4, 5, 8, 9, 11, 16]. With the advance of the science and technology, various software tools have been developed to numerically simulate the progression of a prescribed fire. QUIC-Fire [10] is a three-dimensional fire simulation tool that provides dynamic fuel consumption over time.

Although the progression of the consumed fuel can be simulated by QUIC-Fire, the trade-off between controlled fuel consumption and the safety of the prescribed burn must be taken into account when deciding on the allowable burn conditions. In practice, the unknown non-linear interactions between burn conditions requires the need to run multiple QUIC-Fire simulations (ensembles). The ensembles can be labeled as safe, marginal and unsafe by fire domain experts manually to formulate an allowable range on burn conditions. The manual labeling process is labor intensive and time consuming, and a fast and accurate automatic labeling algorithm that incorporates and learns the expertise of a fire domain expert is desirable.

The contribution of this paper is an algorithm of ensemble based learning that automates the safety labeling of ensembles created by a modern fire simulation tool. The automated safety labeling in this algorithm is done by first extracting important prescribed fire performance metrics from the ensembles based on a desired burn boundary within a burn plan. Any fire escapes outside the desired burn boundary is characterized as a *slop-over* and performance metrics identify the size, spacing and the number of slop-overs. Subsequently, manually labeled ensembles are used to learn the allowable range of the slop-over metrics to distinguish between safe, marginal and unsafe fire conditions. With some integer-valued metrics, the learning is formulated via an gradient-free optimization based on a genetic algorithm [7] that has the capability to deal with integer-valued functions.

The optimized (learned) allowable range of the slop-over metrics and the environmental conditions such as wind speed and fuel moisture are configured as parameters in the automatic labeling. The numerical values of these parameters are used in the automatic labeling algorithm. The optimization ensures an optimized prediction accuracy of the automatic safety labeling of the ensembles. In order to authenticate the performance of the automatic labeling, the use cases of 900 ensembles of a prescribed fire in the Yosemite, CA area are utilized. Learning and matching 100% of the manually assigned safety labels of a subset with 48 out of the 900 ensembles, the automatic labeling is used to provide safety labels for the remaining ensembles. With a success rate above 80%, the proposed automatic labeling algorithm works efficiently and accurately, and can be used as a tool to design the burn plan of the prescribed fire.

2 QUIC-Fire Output Data

With the information of the surface moisture, fuel type, wind conditions, and ignition pattern, QUIC-Fire [10] can simulate the spread of the prescribed fire. The typical output produced by QUIC-Fire at each simulation step is the fuel consumption as depicted in Fig. 1.

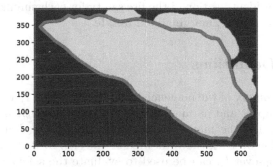

Fig. 1. Output (fuel consumption) of QUIC-Fire. The green line represents the desired boundary and the red line represents the allowable boundary. The yellow area is the burn area (fuel consumed) with $y = 1$ and the dark blue area is the unburned area (fuel not consumed) with $y = 0$. (Color figure online)

Similar to [15], the burn area is represented by the yellow area with $y = 1$, and the unburned area is represented by dark blue area with $y = 0$. The value of each pixel can be expressed as

$$y_{i,j} = f([i,j], b) = \begin{cases} 0, & \text{if } b < 0.001 \\ 1, & \text{if } b \geq 0.001 \end{cases} \tag{1}$$

where $[i, j]$ describes the position of the target pixel in the image, b is the absolute difference value between the fuel densities before the prescribed fire starts and after the prescribed fire ends, and y is the value of pixel at $[i, j]$ and used to distinguish between the burn area and unburned area.

In Fig. 1, the provided desired boundary is drawn by a green line, which defines the area inside the desired boundary that is expected to burn, and the allowable boundary is drawn by red line, which separates buffer area and non-allowable burn area where fire is definitely considered to be unsafe. The allowable boundary is determined by the size of the fuel domain used for the simulation in QUIC-Fire. Without loss of generality, whether a fire escapes outside the allowable boundary can be distinguished by checking whether there is a pixel with $y = 1$ outside the closed polygon representing the allowable boundary. On account of the fact that the shape of the fire is arbitrary, deciding the fire safety by only depending on predetermined boundaries is not enough.

The yellow area outside the desired boundary in Fig. 1 is regarded as the slop-over, and the number of the disconnected yellow area outside the desired boundary is regarded as the number of slop-overs. Hence, the simulation shown as Fig. 1 includes three slop-overs. With the definitions of desired boundary and allowable boundary, slop-over plays an important role in evaluating the safety of the prescribed fire. If the slop-over has the potential to spread outside the allowable boundary, and is hard to control, the corresponding prescribed fire can be unsafe. For identification of the fire safety for each simulation (ensemble), three levels are used: safe, marginal and unsafe.

3   Feature Definitions

Following the summary of the nomenclature given in Table 1, a short explanation is given for the inputs and parameters used in the automatic labeling of ensembles. After collecting the manual labels provided by fire domain experts, the number of the slop-over k_s, the total area of the slop-over A_s, and the distance between each slop-over l_s, can be used to evaluate the safety of the prescribed fire. The total area of the slop-over directly reflects the result of a simulated prescribed fire. Hence, it is an important factor in measuring the fire safety. Limited by the number of firefighters, large number of slop-over or large distance between slop-overs can both result in an uncontrollable prescribed fire.

To quantitatively measure these three terms, some parameters are created in the automatic labeling algorithm. A_{max} and A_{mar} represent the maximum and marginally allowable total area of slop-over, k_{max} denotes the maximum allowable number of slop-over, and l_{max} and l_{mar} indicate the maximum and marginally allowable distance between each slop-over. In addition to simply exploiting the information of slop-over, the complex environmental conditions are also taken into consideration.

For additional flexibility, the parameters α and β are utilized as the constant amplification coefficients to enlarge the potential risk of the slop-over. As a prescribed fire can be more dangerous when the wind speed is higher, the surface

Table 1. Nomenclature of inputs and parameters for automatic labeling

Inputs:	
A_s	total area of slop-overs
k_s	number of slop-overs
l_s	distance between each slop-over
w_s	wind speed
s_s	surface moisture
Parameters:	
A_{max}	maximum allowed total area of slop-overs
A_{mar}	marginally allowed total area of slop-overs
k_{max}	maximum allowed number of slop-overs
α	expansion coefficient
β	expansion coefficient
l_{max}	maximum allowed distance between each slop-over
l_{mar}	marginally allowed distance between each slop-over
w_t	threshold value of wind speed
s_t	threshold value of surface moisture
k_{t_1}	the first threshold value of number of slop-over
k_{t_2}	the second threshold value of number of slop-over

moisture is lower, and the number of the slop-over is larger, four threshold values are created to better distinguish the effect of the total area of slop-over and the distance between each slop-over in different situations. The parameters k_{t_1} and k_{t_2} are established as the number of the slop-over when the risk level of prescribed fire varies significantly, while s_t and w_t are the threshold values for the surface moisture and wind speed respectively. If wind speed is larger than w_t, or surface moisture is smaller than s_t, more caution is required to decide the risk level of the prescribed fire. With these parameters, an automatic safety labeling algorithm can be formulated.

4   Automatic Labeling Algorithm

4.1   Postprocessing of QUIC-Fire Output

To calculate the previously mentioned A_s, k_s and l_s for each ensemble of prescribed burn, the slop-overs should be characterized by removing the burn area inside the desired boundary as shown in Fig. 2. Following the definition of y in (1), all slop-overs have $y = 1$ as shown in Fig. 2(a).

To further distinguish the slop-overs, different non-zero values for y are assigned to different slop-overs. For the numerical implementation, `label` function in the package of scikit-image [17] in Python is a good tool to achieve this goal. It first detects the slop-overs according to the connectivity, and then assigns different values of y to different slop-overs. From Fig. 2(b) it can be observed that three slop-overs are plotted by different colors, where yellow, magenta, and cyan correspond to $y = 1$, $y = 2$, and $y = 3$ respectively.

Afterwards, the number of the slop-over can be determined by the number of different non-zero values of y, and the area of each slop-over can be calculated by summing up the number of pixels with corresponding y. Finally, the distances between the centers of the smallest vertically oriented rectangles that separately contain each slop-over serves as the distances between each slop-over.

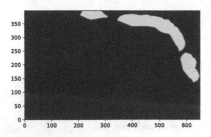   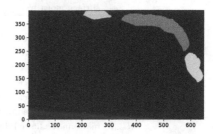

(a) Plot of slop-overs with same y.   (b) Plot of slop-overs with different y.

Fig. 2. Extracted slop-overs.

4.2 Process of Automatic Labeling

Since a distance between slop-overs exists only when there are more than one slop-over, the number and the total area of the slop-over are more important and are utilized first for labeling the fire safety. Due to the limited resource of the fire fighting, it is impossible to control the slop-overs of one prescribed fire simultaneously if multiple slop-overs are far away from each other. Therefore, the distance between slop-overs should also be measured.

Additionally, wind speed and surface moisture around the prescribed fire also affect the fire spread. Even a small slop-over can grow out of control in a short time when the wind speed is large and the surface moisture is low. To account for these situations, two expansion coefficients α and β are applied on the total area of the slop-overs to reflect the emphasis on the effect of the extreme environment. For each ensemble, with computed metrics A_s, k_s and l_s, and provided data of w_s and s_s, the automatic process can be described as follows.

At first, the prescribed fire ensemble is assumed to be safe. Any prescribed fire ensemble with the total area $A_s > A_{max}$, or number of slop-overs $k_s > k_{max}$ is labeled to be unsafe. When the wind speed $w_s > w_t$, the surface moisture $s_s < s_t$, and the number of the slop-over $k_{t_1} < k_s \leq k_{max}$, the prescribed fire is more likely to be unsafe. For that purpose, αA_s is compared to A_{max}. If $\alpha A_s > A_{max}$, the prescribed fire ensemble is regarded as an unsafe fire.

To evaluate the safety of a prescribed fire ensemble by the distance between slop-overs, a prescribed fire ensemble with the number of the slop-over $k_{t_2} < k_s \leq$

k_{max}, and the maximum distance between each slop-over $max(l_s) > l_{max}$ is classified as unsafe. If a prescribed fire is not unsafe, then it will be checked whether it is marginal. The process of judging whether a prescribed fire is marginal is similar, and another expansion coefficient β is set up to put more cautions in the judgement when the environment is more suitable for the spread of the prescribed fire. The automatic labeling algorithm is summarized in Algorithm 1.

Algorithm 1. Automatic Labeling Algorithm

Inputs: A_s, k_s, l_s, w_s, s_s
Parameters: A_{max}, A_{mar}, k_{max}, α, β, l_{max}, l_{mar}, w_t, s_t, k_{t_1}, k_{t_2}
Output: Label of the simulated prescribed fire
1: Assume the prescribed fire is safe at the beginning.
2: **if** the prescribed fire move outside the allowable boundary **then**
3:    the prescribed fire is unsafe
4: **else if** $A_s > A_{max}$ **then**
5:    the prescribed fire is unsafe
6: **else if** $k_s > k_{max}$ **then**
7:    the prescribed fire is unsafe
8: **else if** $k_{t_1} < k_s \le k_{max}$ and $w_s > w_t$ and $s_s < s_t$ and $\alpha A_s > A_{max}$ **then**
9:    the prescribed fire is unsafe
10: **else if** $k_{t_2} < k_s \le k_{max}$ and $max(l_s) > l_{max}$ **then**
11:    the prescribed fire is unsafe
12: **else if** $A_s > A_{mar}$ **then**
13:    the prescribed fire is marginal
14: **else if** $k_{t_1} < k_s \le k_{max}$ and $w_s > w_t$ and $s_s < s_t$ and $\beta A_s > A_{mar}$ **then**
15:    the prescribed fire is marginal
16: **else if** $k_{t_2} < k_s \le k_{max}$ and $max(l_s) > l_{mar}$ **then**
17:    the prescribed fire is marginal
18: **end if**

---

5 Optimization

It is clear that the accuracy of the automatic labeling is dependent on the numerical values of the parameters listed in Table 1. The numerical values of the parameters can be optimized by using safety labels created by fire domain experts. The formal problem of learning the numerical parameters on the basis of manually labeled fire safety data can be stated as the optimization

$$\min_{\mathbf{u}} \sum_{i=1}^{N} d_i \frac{u_i}{q_i} - c(\mathbf{u}),$$

$$\text{subject to: } p_i \le u_i \le q_i \text{ for } i = 1, 2, \dots, N \tag{2}$$

$$A_{mar} \le A_{max}, \quad l_{mar} \le l_{max}, \quad k_{t_1}, k_{t_2} \le k_{max}$$

$$c(\mathbf{u}), A_{mar}, A_{max}, k_{t_1}, k_{t_2}, k_{max}, w_t \in \mathbb{Z}$$

where $\mathbf{u} = [\alpha, \beta, A_{mar}, A_{max}, k_{max}, k_{t_1}, k_{t_2}, l_{mar}, l_{max}, s_t, w_t]$, N is the number of parameter, and d_i is the weighting coefficient with $\sum_{i=1}^{N} d_i = 1$. u_i represents the i_{th} parameter in $\mathbf{u}$, and p_i and q_i are the lower bound and upper bound of the i_{th} parameter respectively. The value of p_i and q_i can be obtained from the burn plan that includes the information of the fuel domain, the wind conditions and the surface moisture for the simulated prescribed fire.

In (2), $c(\mathbf{u})$ denotes the number of safety match between the automatic labels created by Algorithm 1 and the manual labels created by a fire domain expert, where $\mathbf{u}$ represents the parameters. With $u_i \leq q_i$ from Eq. 2 and $\sum_{i=1}^{N} d_i = 1$, it can be verified that

$$\sum_{i=1}^{N} d_i \frac{u_i}{q_i} \leq \sum_{i=1}^{N} d_i = 1 \tag{3}$$

since $c(\mathbf{u})$ is the number of matches between the automatic labeling and manual labeling, any change in $c(\mathbf{u})$ when $\mathbf{u}$ varies is greater than or equal to one.

With (3), an inequality can be derived for the change in $c(\mathbf{u})$, denoted by $\Delta c(\mathbf{u})$, when varying $\mathbf{u}$. The value of $\Delta c(\mathbf{u})$ is bounded by

$$\sum_{i=1}^{N} d_i \frac{u_i}{q_i} \leq 1 \leq \Delta c(\mathbf{u}) \tag{4}$$

and therefore the optimization will first focus on increasing the number of matches between the automatic labels and manual labels, and then decrease the numerical value of the parameters. As a result, the parameters obtained by the optimization will achieve the goal of gaining the maximum match number with the necessary minimum values of the parameters, representing the allowable range on the slop-over metrics.

Because $c(\mathbf{u})$ is an integer-valued function, and there is no analytic expression of $c(\mathbf{u})$, a gradient-free optimization method that can also deal with the integer-valued function should be applied. The genetic algorithm is an explicit and effective solution to this problem. The genetic algorithm will repeatedly modify the population of individual solution. Three steps are included in the genetic algorithm. At first, a random initial population is created. Then, a sequence of new populations are created iteratively based on the previous populations by scoring the fitness of each member of the population, selecting pairs of the members relied on the fitness, and generating the new population by applying crossover and mutation. The last step is to stop the algorithm when the change in value of the fitness function for the best member is less than a tolerance value, or after a predetermined maximum number of iteration. The procedure of the genetic algorithm is summarized in Algorithm 2.

6   Numerical Results

6.1   Ensemble Based Learning

For illustration of the ensemble based learning for automated safety labeling, two fire domain experts work together to manually label the fire safety of 900

Algorithm 2. Genetic Algorithm

Input: Population size m, maximum number of iterations t_{max}, and stopping criterion ϵ

Output: Global optimal solution, $\mathbf{u}_{opt}$

1: Create the initial m members $\mathbf{u}_j (j = 1, 2, \ldots, m)$ of population, and let $t = 0$.
2: Scoring the fitness value of each member, and find the member with best fitness value $f(t)$.
3: select pairs of members from previous population based on fitness value.
4: Apply crossover and mutation to generate the new population.
5: $t = t + 1$.
6: Stop when $t = t_{max}$ or $f(t) - f(t-1) < \epsilon$; Otherwise, go back to step 2 to repeatedly modify the population.

---

ensembles of a prescribed fire in the Yosemite, CA region. QUIC-Fire simulations for the 900 ensembles are created by varying ignition patterns, wind speed, wind direction and fuel moisture for each of the ensembles.

To ensure the validity of the manual labels used for learning, 48 out of the 900 ensembles are labeled by two fire domain experts separately and carefully. Some typical cases in these 48 ensembles with same manual labels are shown as Fig. 3. In Figs. 3(a), 3(b), 3(c) and 3(d), the total area of the slop-overs are small enough and there is a certain distance between the slop-overs and the allowable boundary. Hence, they are labeled as safe prescribed fires. It can be noticed that the sizes of the slop-overs in Figs. 3(e) and 3(f) are relatively large, and the top parts of the slop-overs are fairly close to the allowable boundary. Therefore, the safeties of these two prescribed fire are labeled to be marginal. For Fig. 3(g), the prescribed fire crosses the allowable boundary, and for Fig. 3(h), the total area of the slop-overs is larger than A_{max} despite that the fire does not escape outside the allowable boundary. As a result, both of them are considered to be unsafe. At last, the slop-overs in Fig. 3(i) are large and cross the allowable boundary. Therefore, these slop-overs obviously constitute unsafe prescribed fire conditions.

The ensemble based learning only uses the 48 out of the 900 ensembles, whereas the remaining 852 labels will be labeled automatically based on the optimized parameter values of Table 1 obtained by the learning. For cross validation of the accuracy of learning and labeling, both the 48 ensembles used for learning and the 852 ensembles not used for learning but used for labeling only are compared.

From this cross validation, a 100% accuracy has been achieved for both sets (48 ensembles) of the manual labels created by two fire domain experts respectively. As expected, the optimized parameters for each fire domain expert are slightly different, as each fire domain expert will interpret and label the ensembles slightly differently. As such, the rules used by the two fire domain experts are not completely the same, but either of them can be captured by the learning algorithm. The parameters, $\mathbf{u}^1$ and $\mathbf{u}^2$, optimized by the two sets of manual labels are summarized in Table 2.

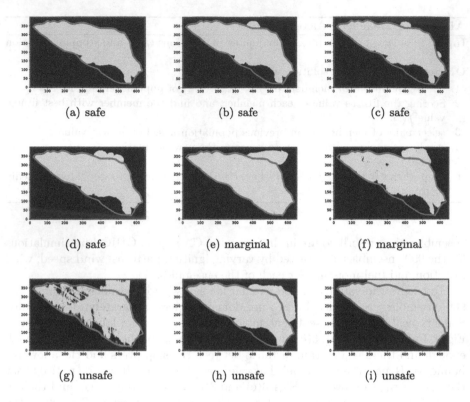

Fig. 3. Typical cases in the training data set.

By inspecting the numerical values of the parameters in Table 2, it can be observed that fire domain expert, from whom $\mathbf{u}^1$ is learned, is more cautious than the fire domain expert, from whom $\mathbf{u}^2$ is learned, because $\mathbf{u}^1$ has smaller values for A_{mar} and A_{max}. Furthermore, α in $\mathbf{u}^1$ is larger than one, and β in $\mathbf{u}^1$ is close to one. This means $\mathbf{u}^1$ is more focused on identifying unsafe prescribed fire conditions. In addition, $\mathbf{u}^1$ takes advantage of the distance between each slop-over to judge the risk level of the prescribed fire. In comparison, $\mathbf{u}^2$ has a higher tolerance for the threat of a prescribed fire, and $\mathbf{u}^2$ has higher probability to assess the risk level of prescribed fire as marginal instead of unsafe with α close to one and β larger than one. Since both l_{mar} and l_{max} in $\mathbf{u}^2$ are close to zero, it can be assumed that the distance between each slop-over is not utilized, which is also supported by $k_{t_2} = k_{max}$.

6.2   Automated Safety Labeling

Due to the large workload for a single fire domain expert to label the remaining 852 ensembles, the 852 manual labels are created together by the two fire domain

Table 2. Parameters, $\mathbf{u}^1$ and $\mathbf{u}^2$, optimized by the two sets of 48 manual labels respectively.

	A_{mar}	A_{max}	α	β	w_t	s_t	l_{mar}	l_{max}	k_{t_1}	k_{t_2}	k_{max}
$\mathbf{u}^1$	8280	10655	1.31	1.01	0	0.2001	133.31	263.72	1	2	3
$\mathbf{u}^2$	9723	17719	1.04	1.37	3	0.3501	1.06	1.96	0	3	3

experts, effectively mixing their fire safety labeling expertise in the remaining data set of ensembles. To cross validate the performance of the automatic labeling, manual labels of the 852 ensembles, not used for learning, are compared to the labels created by Algorithm 1 with the parameters values $\mathbf{u}^1$ and $\mathbf{u}^2$ listed in Table 2 separately.

The match accuracy between the manual labels and the automatic labels created using $\mathbf{u}^1$ is 76.76%; the match accuracy between the manual labels and the automatic labels created using $\mathbf{u}^2$ is 76.88%; the match accuracy between the manual labels and the automatic labels created using either $\mathbf{u}^1$ or $\mathbf{u}^2$ is 80.52%. As a consequence, more than 80% manual labels can be captured by the automatic labeling using either $\mathbf{u}^1$ or $\mathbf{u}^2$.

6.3 Re-evaluation of Manual Labeling

To investigate the inconsistency between manual and automatic labeling, 12 ensembles (Fig. 4) are chosen as canonical cases from the 852 ensembles, in which the manual labels are different from the automatic labels created using either $\mathbf{u}^1$ or $\mathbf{u}^2$. Without loss of generality and with the purpose of reducing the workload, only one fire expert, by whose manual labels $\mathbf{u}^1$ was optimized, relabeled these 12 ensembles and 10 revised manual labels were the same as the automatic labels created by Algorithm 1 with $\mathbf{u}^1$.

In Figs. 4(a) and 4(b), the total area of the slop-overs is small enough. Therefore, both of them should be regarded as safe prescribed fires. In addition, Fig. 4(b) is similar to Fig. 4(c), which further confirms that the fire shown in Fig. 4(b) is safe. For Figs. 4(c) and 4(d), since the slop-overs are larger and hard to control, they should be unsafe. For Figs. 4(e) to 4(i), the prescribed fires cross the allowable boundary in different locations and are unsafe. It is worthwhile to note that the re-evaluation helped to further improve the number of match between manually and automatically created labels by correcting the previous manually applied safety labels.

Since there is no ensemble with four or more slop-overs included in the 48 ensembles used for learning and more slop-overs will lead to more dangerous fire conditions, k_{max} is optimized as 3, and all the prescribed fires with four or more slop-overs will be considered as unsafe fires in the automatic labeling. For Figs. 4(j) and 4(k), both of them have four slop-overs and the difference between them is that all four slop-overs in Fig. 4(j) stay together while one slop-over is far away from the other three slop-overs in Fig. 4(k), which further increase the

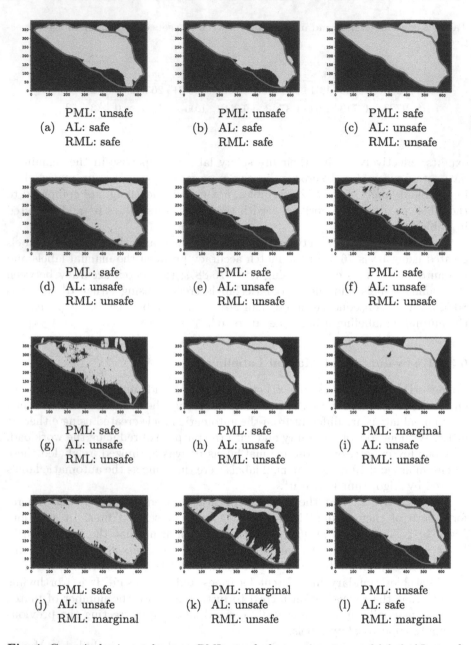

PML: unsafe
(a) AL: safe
RML: safe

PML: unsafe
(b) AL: safe
RML: safe

PML: safe
(c) AL: unsafe
RML: unsafe

PML: safe
(d) AL: unsafe
RML: unsafe

PML: safe
(e) AL: unsafe
RML: unsafe

PML: safe
(f) AL: unsafe
RML: unsafe

PML: safe
(g) AL: unsafe
RML: unsafe

PML: safe
(h) AL: unsafe
RML: unsafe

PML: marginal
(i) AL: unsafe
RML: unsafe

PML: safe
(j) AL: unsafe
RML: marginal

PML: marginal
(k) AL: unsafe
RML: unsafe

PML: unsafe
(l) AL: safe
RML: marginal

Fig. 4. Canonical mismatch cases. PML stands for previous manual label, AL stands for automatic label by applying $\mathbf{u}^1$, and RML stands for revised manual label.

difficulty in controlling the prescribed fire. Hence, the prescribed fires shown by Fig. 4(j) and Fig. 4(k) are considered to be marginal and unsafe respectively by the fire expert.

6.4   Further Improvements

To further improve the automatic labeling, a user-defined marginally allowed number of slop-overs k_{mar} and a user-defined maximum allowed number of slop-overs k_{max} can be imported into Algorithm 1. Expanding the training data to include more scenarios can also improve the performance of automatic labeling at the price of having to provide more manually labeled ensembles.

To make sure the user-defined k_{mar} and k_{max} will not change during the optimization process, two more linear equality constraints on k_{mar} and k_{max} can be added to (2). At last, for Fig. 4(l), even the fire domain expert cannot give an exact answer based on the current data. It means more information, like topography, vegetation, and contingency resources are needed.

In summary, the automatic labeling, Algorithm 1, has a good ability to create the label for the safety of the prescribed fire. Since the label is created by measuring the number of slop-overs, the total area of the slop-over, and the distance between each slop-over, the automatic labeling can not only create the label but also give a feedback about which rule is used to create the label so that people can get access to the interpretation of the automatic labeling.

7   Conclusions

This paper introduces an automatic labeling algorithm to establish the safety label for each ensemble of a simulated prescribed fire. The automatic labeling is based on prescribed fire safety metrics that include the number of slop-overs, the total surface area of slop-overs, and the distance between slop-overs. In addition to the safety label, the automatic labeling algorithm can provide an explanation why a prescribed fire is considered to be safe, marginal, or unsafe. Necessary parameters are optimized in the automatic labeling algorithm via a genetic algorithm to assist in determining the label of each ensemble of the simulated prescribed fire. A numerical validation based on 900 ensembles with manually generated safety labels of a prescribed fire in the Yosemite, CA area showed a 100% match of safety labels for the training data (48 out of 900 ensembles) and a larger than 80% match on the cross validation of safety labels not used in the training data (852 out of 900 ensembles).

Acknowledgement. We thank Matthew Snider and J. Kevin Hiers for providing the manual safety labels of the fire simulations.

References

1. Abrams, M.D.: Fire and the development of oak forests. Bioscience **42**(5), 346–353 (1992)
2. Agee, J.K.: Fire Ecology of Pacific Northwest Forests. Island press, Washington (1996)
3. Alcañiz, M., Outeiro, L., Francos, M., Úbeda, X.: Effects of prescribed fires on soil properties: a review. Sci. Total Environ. **613**, 944–957 (2018)

4. Banerjee, T., Heilman, W., Goodrick, S., Hiers, J.K., Linn, R.: Effects of canopy midstory management and fuel moisture on wildfire behavior. Sci. Rep. **10**(1), 1–14 (2020)
5. Cheney, N., Gould, J., Catchpole, W.: The influence of fuel, weather and fire shape variables on fire-spread in grasslands. Int. J. Wildland Fire **3**(1), 31–44 (1993)
6. Francos, M., Úbeda, X.: Prescribed fire management. Current Opin. Environ. Sci. Health **21**, 100250 (2021)
7. Katoch, S., Chauhan, S.S., Kumar, V.: A review on genetic algorithm: past, present, and future. Multimedia Tools Appl. **80**(5), 8091–8126 (2021)
8. Kerby, J.D., Fuhlendorf, S.D., Engle, D.M.: Landscape heterogeneity and fire behavior: scale-dependent feedback between fire and grazing processes. Landscape Ecol. **22**(4), 507–516 (2007)
9. Linn, R.R., Cunningham, P.: Numerical simulations of grass fires using a coupled atmosphere-fire model: basic fire behavior and dependence on wind speed. J. Geophys. Res. Atmos. 110(D13) (2005)
10. Linn, R.R., et al.: QUIC-fire: a fast-running simulation tool for prescribed fire planning. Environ. Model. Softw. **125**, 104616 (2020)
11. Moinuddin, K., Khan, N., Sutherland, D.: Numerical study on effect of relative humidity (and fuel moisture) on modes of grassfire propagation. Fire Saf. J. **125**, 103422 (2021)
12. Pausas, J.G., Keeley, J.E.: A burning story: the role of fire in the history of life. Bioscience **59**(7), 593–601 (2009)
13. Ryan, K.C., Knapp, E.E., Varner, J.M.: Prescribed fire in north American forests and woodlands: history, current practice, and challenges. Front. Ecol. Environ. **11**(s1), e15–e24 (2013)
14. Scharenbroch, B., Nix, B., Jacobs, K., Bowles, M.: Two decades of low-severity prescribed fire increases soil nutrient availability in a midwestern, USA oak (Quercus) forest. Geoderma **183**, 80–91 (2012)
15. Tan, L., de Callafon, R.A., Altıntaş, I.: Characterizing wildfire perimeter polygons from quic-fire. In: Groen, D., de Mulatier, C., Paszynski, M., Krzhizhanovskaya, V.V., Dongarra, J.J., Sloot, P.M.A. (eds.) Computational Science–ICCS 2022, vol. 13350, pp. 611–622. Springer, Cham (2022). https://doi.org/10.1007/978-3-031-08751-6_44
16. Tan, L., de Callafon, R.A., Block, J., Crawl, D., Çağlar, T., Altıntaş, I.: Estimation of wildfire wind conditions via perimeter and surface area optimization. J. Comput. Sci. **61**, 101633 (2022)
17. Van der Walt, S., et al.: scikit-image: image processing in python. PeerJ **2**, e453 (2014)

SLAM Methods for Augmented Reality Systems for Flight Simulators

Onyeka J. Nwobodo(✉) ⓘ, Kamil Wereszczyński ⓘ, and Krzysztof Cyran ⓘ

Department of Computer Graphics, Vision and Digital Systems, Silesian University of Technology, Akademicka 2A, 44-100 Gliwice, Poland
{onyeka.nwobodo,kamil.wereszczynski,krzysztof.cyran}@polsl.pl

Abstract. In this paper, we present the review and practical evaluation of the flight simulators of Simultaneous Localization and Mapping methods. We present a review of recent research and development in the SLAM application in a wide range of domains, like autonomous driving, robotics and augmented reality (AR). Then we focus on the methods selected from the perspective of their usefulness in the AR systems for training and servicing the flight simulators. The localization and mapping in such an environment are much more complex than in others since the flight simulator is relatively small and close area. Our previous experiments showed that the built-in SLAM system in HoloLens is insufficient for such areas and has to be enhanced with additional elements, like QR codes. Therefore, the presented study on other methods can improve the localization and mapping of AR systems in flight simulators.

Keywords: SLAM · flight simulators · Deep Learning · LiDAR · Augmented Reality

1 Introduction

Simultaneous localization and mapping(SLAM) have attracted much interest recently, particularly in intelligent systems and robotics. There are problems with implementing SLAM due to its complexity. This problem has existed for over 30 years, especially when finding a solution [1]. However, various approximations have come close to resolving this challenging algorithmic problem following decades of mathematical and computational effort [2]. Solving the SLAM problem will enable a wide range of potential applications for autonomous robots [3]. A robot that can navigate its environment without human intervention is said to be autonomous. For a robot to successfully navigate, it must have a

The authors would like to acknowledge that this paper has been written based on the results achieved within the WrightBroS project. This project has received funding from the European Union's Horizon 2020 research and innovation program under the Marie Skłodowska Curie grant agreement No 822483. Disclaimer: The paper reflects only the author's view, and the Research Executive Agency (REA) is not responsible for any use that may be made of the information it contains.

J. Mikyška et al. (Eds.): ICCS 2023, LNCS 14073, pp. 653–667, 2023.
https://doi.org/10.1007/978-3-031-35995-8_46

thorough understanding of its surroundings and be able to track its position in that environment consistently and accurately. Scientists employ various methods to enhance the autonomy and self-discovery of robot navigation, leading to the development of augmented reality(AR) systems. The state-of-the-art SLAM technology uses multiple sensors, such as LIDAR, cameras, and IMU, to create highly accurate and detailed maps.

Simultaneous localization and mapping(SLAM) is a method robotics and computer vision professionals use to locate a robot while simultaneously creating a map of its surroundings. SLAM aims to estimate the agent's location and the environment's structure using sensor data such as lidar, cameras, GNSS receivers and antennas, an Inertial Measurement Unit(IMU), and odometry. It is known as "visual SLAM" when the SLAM algorithm is based on camera sensors, and "LiDAR SLAM" is based on laser scanners [4].

Microsoft HoloLens is a head-mounted device that projects augmented reality(AR) into the users' field of view. It is a direct-based feature with built-in visualization and no camera trajectory module. The HoloLens is an optical see-through gadget with numerous sensors, including an accelerometer, magnetometer, and gyroscope. Numerous good SLAM techniques have greatly aided the development of SLAM technology, including MonoSLAM [5], DTAM [6], LSD SLAM [7], RATSLAM [8], KinetFusion [9], RGB-D SLAM [10]. DeepVo [11]. Various SLAM methods target different objectives; Microsoft HoloLens uses a variant of the SLAM technology to enable its augmented reality capabilities. Therefore, in this work, we discuss different SLAM methods for Microsoft HoloLens and flight simulators, furthermore compare these methods and the deep learning SLAM methodology in a flight simulator environment.

The SLAM problem has numerous current solutions, which can be categorized as filter-based and global optimization approaches. The filter-based approach is the classical approach that recursively performs prediction and update steps. They are typically thought of as a maximum posterior(MAP) method where the robot's previous distribution is estimated using data from sensors like the IMU. The likelihood distribution is constructed by combining the IMU data with those made by a camera or LiDAR [12]. The IMU sensor measurements utilize the prediction step of a standard SLAM filtering technique to forecast the vehicle's motion(odometry) [13]. At the same time, the estimated camera posture and measured image attributes are employed as a likelihood distribution to update the predictions in the update step [12]. This includes all the Kalman filter families (EKF, UKF and SEIF) and particle filters. The global optimization method relies on keeping a few keyframes in the environment and estimates the motion through bundle adjustment(BA). Kummer et al. [14] estimate the robot trajectory like the SLAM graph by processing all sensor measurements. They typically use least-squares adjustment methods and optimization, considered more accurate than the filtering process [15]. This is currently a popular approach for vision-based SLAM such as ORB-SLAM. Following this, other researchers have attempted to increase the effectiveness of SLAM systems by utilizing new feature extraction and matching algorithms, including orientated FAST and rotated BRIEF(ORB) [16] and Speeded-Up Robust Features [17].

SLAM has various applications in decision-making processes, including autonomous driving (on land, air, sea, and underwater), robotics, and augmented reality.

SLAM-based Augmented Reality(AR) is a technology that mixes real-world imagery and computer-generated graphics to improve user experience [18]. SLAM-based AR uses cameras and sensors to map the surroundings in real-time and precisely put virtual items in the real world. This enhances the stability and precision of virtual objects in the real environment and makes for a more immersive AR experience. To provide the appearance that the virtual and real worlds are completely integrated, displaying AR items in the proper place and adhering to the user's perspective demands a solution to numerous static and dynamic registration issues [19]. Recently, scientists have employed SLAM's accuracy and real-time performance for virtual registration in AR [20]. Regarding direct approaches [7], a camera-based method, LSD-SLAM, enables the construction of large-scale, semi-dense maps that do not require adjusting bundles of features. ORB SLAM is another approach to visual SLAM proposed by [21] that uses a feature-based approach to simultaneously estimate the camera pose and build a map of the environment. To partially resolve the scale ambiguity and provide motion cues without the use of visual features, Visual Inertial SLAM (VISLAM) can combine VSLAM with Inertial Measurement Unit(IMU) sensors(accelerometer, gyroscope, and magnetometer) [22]. The HoloLens device's ability to localize has been demonstrated by a recent study on the technology's position tracking [23]. According to [24], with the help of Microsoft's newly released spatial mapping capability, the HoloLens can map scenes in its immediate surroundings. Environmental objects such as walls, floors, and obstacles can be found using spatial mapping.

SLAM on Microsoft HoloLens is used to build 3D maps of an environment and track the device's position within it in real-time. It uses inbuilt cameras and sensors to detect and track distinctive environmental features and estimate the device's pose. The HoloLens device combines RGB cameras, depth sensors, and IMUs(Inertial Measurement Units) that can be used for SLAM. The device can use visual SLAM, visual-inertial SLAM, or RGB-D SLAM to estimate the camera pose and build an environment map. Using SLAM, the HoloLens device can provide a more accurate and stable AR experience, as it can track the user's movements and the position of objects in the environment. More recent work at Microsoft has concentrated on large-scale scene reconstruction utilizing voxel hashing [25] and RGB-D camera re-localization [26] for recovering from tracking failures. In the area of SLAM methods in flight simulators, Skurowski et al. [27] uses the QR codes for marking the crucial points in the simulator. We made this study to avoid including the additional and non-natural elements of the flight simulator cockpit.

The Following is this Paper's Main Contributions

1. The analysis is completed by running/discussing five selected states of the SLAM methods, which have been chosen to represent the diversity of the existing SLAM methods on Microsoft HoloLens and flight simulator,
2. The accuracy of the existing SLAM methods on Microsoft HoloLens is compared with the presented SLAM and deep learning methods in a flight simulator environment.

In numerous studies, SLAM techniques are reviewed and contrasted. Depending on the robot's goals, they provide information on the dependability, real-time estimation, and accurate depiction of the surroundings, as well as the location and orientation of the robot. As reviewed and compared by the following authors [28–32]. It is discovered, nonetheless, that none of the articles compares SLAM techniques using Microsoft HoloLens in a flight simulator environment. The primary **motivation** supporting this research is to evaluate the Microsoft HoloLens SLAM techniques and compare them to the chosen SLAM and deep learning techniques in flight simulator scenarios. The outcome will make selecting the SLAM methods most appropriate to HoloLens in a flight simulator environment easier for accurate, reliable, and long-term motion estimation to render smooth and stable holograms.

2   Materials and Methods

RGB-D SLAM: It is a system that creates a real-time 3D model of the surroundings by fusing RGB and D(depth) information. The additional depth information aids in overcoming some of the difficulties associated with using standard RGB-only SLAM techniques, such as coping with repeated or texture-less situations. The system uses an RGB-D camera, which records information about each pixel's color and depth, and algorithms to predict the camera's position and create a 3D environment model as the camera moves [33]. Modern RGB-D SLAM systems align point features whose spatial coordinates are determined by corresponding sensor depth data using the Iterative Closest Point (ICP) technique. However, because visual features typically lie at the edges of actual objects, noise frequently contaminates the depth measurements of features. Utilizing the benefits of the RGB-D camera's depth image is one technique to deal with this problem. For instance, planes can be inferred from the depth information, and the locations on these planes have less noise than the corners. Instead of using point characteristics as primitives, researchers have proposed using planes. Towards that end, Gao and Zhang [34] provide a technique that extracts and chooses reliable depth values or planer point features. Dai et al. [35] use Delaunay triangulation, in which changes in the triangle edges in adjacent frames are compared to assess the correlation of feature points and differentiate between dynamic from static map points. To create a realistic and immersive experience in a flight simulator environment, the RGB-D SLAM approach employing depth data from an RGB-D camera for HoloLens can be a solid choice.

Advantages

1. Improved spatial awareness: The HoloLens cameras' RGB and depth data can create a more realistic depiction of the surroundings, giving users enhanced depth perception and spatial awareness.
2. Real-time mapping: The HoloLens' ability to update its map of the surroundings in real-time thanks to RGB-D SLAM makes it perfect for usage in dynamic simulation scenarios like flight simulators.
3. Increased interaction: The HoloLens can offer customers a more interactive and immersive experience in the flight simulator environment with an accurate representation of the environment.

Limitations

1. High computational requirements: High computing power is needed to produce an accurate map in real-time using the RGB-D SLAM algorithm. This presents a challenge for processing-constrained devices like the HoloLens.
2. Sensitivity to lighting conditions: RGB-D SLAM methods depend on precise depth data, which is susceptible to variations in lighting. The maps that the HoloLens produces may need to be more accurate.
3. Occlusions: It can be challenging for the HoloLens to effectively acquire depth information when environmental objects overlap in certain circumstances. The maps produced by the RGB-D SLAM algorithm may need to be revised.

ORB-SLAM(oriented Fast Rotation Brief SLAM): One of the latest monocular vision-based SLAM techniques with an open-source implementation is ORB-SLAM [36] using a real-time SLAM library called ORB-SLAM, monocular, stereo, and RGB-D cameras can calculate their camera trajectories and sparse 3D reconstructions. It uses ORB Rotated BRIEF(Binary Robust Independent Elementary Features) and Oriented FAST(Acccleratcd Segment Test) feature detectors developed by [16]. According to Mur-Artal et al. [21], this technique estimates position and maps from an image sequence in real-time. The primary ORB-SLAM process creates an environmental map comprising keyframes and map points. Each keyframe stores a list of 2D features and their location in ORB-SLAM coordinates. Tracking, local mapping, and loop closure are the three parallel threads that comprise the ORB-SLAM process.1)Using motion-only BA to reduce re-projection error and camera localization with each frame by finding matching features on the local map. 2)The local mapping to maintain control over, enhance local mapping and perform local BA. 3)Loop closure uses position graph optimization to find large loops and correct accumulated drift. After optimizing the position graph, this thread triggers a fourth thread to perform a full BA to determine the best solution for structure and motion. To address the issue of the low number of feature point extraction and the simple keyframe loss, Cai et al. [37] proposed an enhanced visual SLAM based on affine transformation for ORB feature extraction. However, the environments used in these methods are frequently static, which is insufficient to meet the demands of the complex task in a dynamic environment. Various approaches address the dynamic problem [38, 39].

Advantages

1. Real-time performance: In a flight simulator scenario where quick judgments are necessary, ORB-SLAM can give real-time mapping and localization of the environment, which is essential.
2. Robustness: ORB-SLAM excels in a flight simulator environment because it is highly resilient to changes in illumination conditions and can endure abrupt motions and shocks.
3. Open-source: Open-source platform ORB SLAM is easily customized and adaptable to fulfill unique requirements.

Limitations

1. Dependence on keypoints: ORB-SLAM relies on locating and following the keypoints in the image, which things like occlusions and changes in lighting can impact. The effectiveness of ORB-SLAM may be hampered if keypoints are not precisely identified or tracked.
2. Limited map-building capabilities: Real-time tracking is the purpose of ORB-SLAM. In comparison to other SLAM algorithms that place a higher priority on creating a comprehensive map. As a result, its map-building capabilities may need to be revised.
3. High computational requirements: In real-time applications, especially in a resource-restricted setting like a HoloLens, ORB SLAM takes a lot of processing resources, which can be difficult.

LSD-SLAM(Large scale direct monocular SLAM): It is a direct monocular SLAM method that tracks and maps objects directly using picture intensities rather than key points. Direct image alignment is used to follow the camera, and semi-dense depth maps created by filtering through several pixelwise stereo comparisons are used to estimate geometry [7]. Then, using a Sim (3) pose-graph of keyframes creates large-scale maps with scale-drift corrections, including loop closures. LSD-SLAM is a real-time executable on CPUs and even on contemporary smartphones. Traditional monocular SLAM techniques have limitations in accuracy, scalability, and performance in vast and dynamic situations. LSD-SLAM was created to address these issues. There are two primary phases to this method: 1)Keyframe selection: Keyframes are chosen based on the visualization information of the incoming image changes, and 2)Direct Image Alignment: The best transformation(i.e. rotation and translation) that aligns the two images is found by directly comparing the current image to the keyframes in the map. Forster et al. [40] suggested SVO(semi-direct visual odometry), a visual odometry(VO) without loop closure detection and re-localization, as a method for measuring motion. SVO follows the features from accelerated segment test(FAST) feature points and surrounding pixels by minimizing the photometric error to determine the camera's motion. Bergmann et al. [41] proposed an online photometric calibration that dynamically estimates the photometric parameters by solving the least squares equation of the feature tracker and modifies the exposure situation of the input sequence to improve the performance of direct visual odometry. It marks a significant advancement in direct formulation placement and mapping precision. Peixin Liu et al. [42] improved the visual slam technique based on the sparse direct method, in which input sequences' silhouette and response functions were optimized based on the camera's photometric configuration. Optimization equations were developed by tracking the input sequence's Shi-Tomasi corners with sparse direct visual odometry (VO) pixel tracking. Additionally, the joint optimization equation was solved using the Levenberg-Marquardt(L-M) method, and the photometric calibration parameters in the VO were updated to realize real-time dynamic compensation of the input sequences' exposure. This reduced the impact of light variations on SLAM's accuracy and robustness. A Shi-Tomasi corner-filtered strategy further reduced the computational complexity of the suggested technique, and the loop closure detection was realized using the orientated FAST and rotated BRIEF(ORB) features.

Advantages

1. Real-time performance: LSD-SLAM is appropriate for usage in flight simulators because it is made to work in real-time.
2. High accuracy: Regarding motion tracking, LSD-LAM is a dependable source because it can precisely follow the simulator's movement.
3. Low computational power requirement: LSD-SLAM is appropriate for low-end computer systems because it does not require much computational resources.

Limitations

1. Difficulty handling dynamic environments: LSD-SLAM can struggle with dynamic environments, where objects rapidly move or change in real-time.
2. Sensitivity to lighting conditions: LSD-SLAM relies heavily on visual information, making it susceptible to lighting conditions and camera noise.
3. Need for good initialization: LSD-SLAM requires an accurate initial pose of the simulator, which can be challenging to achieve in specific flight scenarios.
4. Limited robustness: LSD-SLAM can be affected by insufficient or inaccurate data, leading to incorrect results or lost track.

VINS(Virtual Inertial SLAM): Virtual inertial SLAM is a hybrid technique that employs data from visual and inertial sensors, such as gyroscopes and accelerometers, to map the surroundings and estimate the camera's pose. The technique tracks the camera's movements and calculates its position in the surroundings using the visual elements seen in photos. The inertial measurements help increase the posture estimate's precision by supplying information about recent motion. The map is updated when new features are added, and old features are removed as the camera moves through the surroundings. Several benefits over conventional monocular or stereo-visual SLAM techniques are provided by utilizing visual and inertial information in VINS algorithms. For instance, VINS algorithms can accurately estimate the camera's motion even without visual input, which is frequently the case in chaotic or obscured surroundings. Additionally, VINS algorithms are suited for use in applications that call for long-term mapping or tracking because they may offer drift-free estimations of the camera's position and orientation over time. According to their method [43], inertial measurements should be tightly integrated into a keyframe-based visual SLAM structure; in the visual SLAM system, both the re-projection error terms and the cost function IMU are optimized. The technique to integrate visual information and IMU can be classified into loosely coupled [44] and tightly coupled [43] techniques. This can be achieved by evaluating whether the state vector includes the visual information. The vision-based SLAM and the inertia-based modules are often executed separately as part of the loosely connected techniques. Additionally, the measures are estimated using the combination of the results. The loosely connected approach still suffers from the monocular SLAM drift issue. In a tightly coupled technique, the depth information from the monocular SLAM can be calculated, and the IMU deviation can be fixed by adding visual information to the state vector. This approach is more reliable and accurate than the traditional single-vision-sensor-based SLAM. Yin et al. [45] present a stereo visual-inertial SLAM system, using a technique to detect dynamic features that loosely coupled the stereo scene flow with an inertial measurement unit(IMU) and tightly coupled the dynamic and static characteristics with the IMU measurements for nonlinear optimization.

The Microsoft HoloLens SLAM algorithm uses inertial measurements from the HoloLens' IMU and high-quality visual information from the virtual environment to produce more precise and stable estimates of the device's position and orientation in a flight simulator environment than using just one of these sources alone.

Advantages

1. High precision: Visual and inertial sensors fused can accurately estimate location and orientation.
2. Robustness: The system can operate effectively in low-light conditions or without lighting. It can still deliver accurate position estimates when a temporary loss of visual data occurs.
3. Performance in real-time: The system's ability to perform real-time estimation, allowing for smooth and responsive navigation in the virtual environment, suits it for robotics, augmented reality, and autonomous vehicle applications.
4. Autonomy: The device is equipped with Visual Inertial SLAM, which permits independent operation without external tracking systems.

Limitations

1. Sensitivity to starting conditions: The system's accuracy is strongly influenced by the initial conditions, which might impact the system's operation if not properly calibrated.
2. Cost of computation: The system is challenging to implement on low-power devices due to the high computational resource requirements.
3. Drift over time: The system's mistakes may build up, causing drift in the estimated position and orientation.

Deep Learning SLAM Methods

Deep learning has shown increasingly clear advantages in image processing, especially in data fitting, feature extraction, and spatial transformation, which has produced exciting results. The traditional manual algorithm compared with studies on recognition [46], image segmentation [47], object detection [48] and image classification [49] all of which showed the algorithm to be significantly better. Slam's deep learning implementation can circumvent the limitations imposed on visual odometry and scene recognition by manually creating features. Additionally, it helps build high-level semantics and develop the agent's knowledge base, improving the agent's perceptual and cognitive abilities. The model has undergone data-driven training to make it more consistent with people's interactions and settings. The evolution of visual SLAM from geometry-based methods to deep learning methods occurs. Recently, visual SLAM issues, including visual odometry [50] and loop closure [51], have been addressed using both supervised deep learning and unsupervised approaches. Due to these recent developments, deep learning techniques have a significant potential to address the complex problems of visual SLAM by incorporating adaptive and learning capabilities. Using an unsupervised end-to-end learning framework, Geng et al. [52] propose an ORB-SLAM method that better estimates camera motion and monocular depth. Li et al. [53] employed SegNet, a well-known semantic segmentation network, to segment the images and further separate dynamic objects. Zhang et al. [54] use YOLO [55] to recognize objects in the environment, and to increase the system's accuracy, they used a semantic map to filter dynamic feature points. The scientists' current work suggests fusing SLAM and deep learning to speed up the various visual SLAM system elements. Tateno et al. [56] describes a technique for accurate and dense monocular reconstruction where CNN-predicted dense depth maps from a deep neural network combine depth measurements with direct monocular SLAM. Li et al. [57], using UnDeepVO perform unsupervised training on stereo images, then utilize monocular images to estimate pose and create maps. Vijayanarasimhan et al. [58] present SfM-Net, a neural network that trains the image generated from the geometry to extract 3D structure, ego-motion, segmentation, object rotations, and translations in videos.

Advantages

1. Fusion of SLAM with Deep Learning: One method integrates deep learning-based techniques with conventional SLAM algorithms. For instance, a deep neural network that enhances the accuracy of a traditional SLAM algorithm's output or a deep neural network that refines the output of a typical SLAM method.
2. Increased accuracy: Deep learning can potentially enhance the accuracy of maps produced by conventional SLAM techniques. To create more precise 3D maps, a deep neural network can be trained to estimate the depth of objects in a given environment.
3. Increasing Robustness: Deep learning can also improve the robustness of conventional SLAM systems. For instance, a deep neural network can be taught

to recognize and rectify faults in the SLAM output, improving its ability to handle difficult situations like low light levels or rapid motions.

Limitations

1. Data collection and annotation: Deep learning algorithms need a lot of labeled data to be trained, which might be challenging to gather and label in a flight simulator environment.
2. High computational requirements: Deep learning techniques demand a lot of computer power, which can be difficult for a device like the Hololens, which has a relatively low processing capacity.
3. Model size: Deep learning models can get rather large, making it difficult to store and deploy them on a system like Hololens.
4. Deep learning algorithms can be expensive to train, making them less desirable for commercial flight simulator applications.

3   Results and Discussion

We used HoloLens 2 to run SLAM techniques in the flight simulator, and we got access to real-time live-streaming video of the simulator cockpits from every camera. Utilizing the research mode (RM) depicted in Fig. 1, data is collected from the depth sensors, IMU, and gray-scale cameras. We used the Python library to connect to the HoloLens 2 server, send commands and configuration data, receive and decode data, and analyze the data for usage with other libraries.

The main goal of utilizing HoloLens to run the SLAM methods on the cockpit was to determine how well each method could be customized to find patterns and structures in the cockpit or map the flight simulator cockpit for precise and reliable pose estimation. Our test results are displayed below. In our result, as compared in Table 1, the SLAM methods discussed in our methods section, we inferred with our experimental result, as shown in Fig. 1 the performance of SLAM methods in a flight simulator environment. RGB-D SLAM is considered moderate compared with other SLAM techniques because flight simulators often have a controlled and structured environment, so there are fewer variables and obstacles to consider when detecting features in the environment. ORB and LSD SLAM are visual-based techniques that use features extracted from the images to perform SLAM. ORB SLAM is considered poor because it is not robust to lighting changes or occlusion; it makes it difficult for the SLAM algorithm to distinguish between different objects in the environment, while LSD SLAM is poor in terms of accuracy in a large-scale environment. VINS fuses information from visual and inertial sensors; therefore, it is considered good because it provides accurate results without additional sensors. Deep Learning (DL) is considered good because it can learn to extract useful features from the sensor data and adapt to a different environment.

Table 1. Comparative table of SLAM methods and flight simulator environment

Methods	Map quality	Localization Accuracy	Computational Cost	Robustness	Efficiency	In flight simulator
RGB-D	Moderate	Good	High	Poor	Fair	Moderate
ORB	Low	Good	Moderate	Good	Good	Poor
LSD	High	Excellent	Low	Fair	Fair	Poor
VINS	High	Excellent	High	Excellent	Good	Good
DL	Good	Excellent	Very high	Good	Low	Good

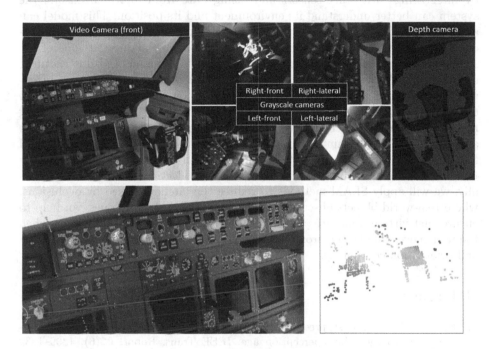

Fig. 1. Top picture shows the simultaneous capture from all the Hololens 2 sensors. The left picture shows the feature detection, while the right one - the 3D Map of the cockpit.

4   Conclusion

This paper discussed and analyzed different SLAM methods in a flight simulator environment. We obtained a high-resolution live stream video of a flight simulator cockpit with a high dynamic range using HoloLens 2. We perform feature detection on the video stream, and the SLAM algorithm then extracts features from the cockpit structures. These futures can be used to estimate the robot's position and orientations relative to the environment's features, track the robot's movement, and build a map of the environment in real-time. Comparing different SLAM techniques is challenging because their performance depends on the complexity of the surrounding environment and the type of sensors being employed.

Nevertheless, this work still suggests selecting a suitable SLAM technique for HoloLens-based flight simulator scenarios. Future research should concentrate on enhancing the effectiveness of Visual Inertial SLAM and upgrading it based on the findings of this publication. We will demonstrate how the proposed techniques may be utilized to develop more accurate and reliable algorithms for feature detection, tracking, and aligning virtual objects in the real world while using augmented reality goggles. This will help to reduce the incidence of tracking failures. We will present the development of better sensor fusion techniques by integrating with other sensors and fusing data from multiple sensors; the system can better understand its environment and its position. This model can aid in improving the SLAM system's accuracy and help it better forecast aircraft behavior under various flight conditions. In augmented reality, one of the significant challenges is the ability to operate in a dynamic and unstructured environment. We will create an algorithm capable of a wider range of environments in handling challenging conditions, such as low light, moving objects and dynamic scenes. We will also investigate how to make algorithms better by using machine learning to track the location and orientation of an AR device in the actual world. The program can recognize features in real-world surroundings and follow their movement to estimate the position and orientation of the device by training machine learning on an extensive data collection of images. Additionally, we will apply SLAM methods to reduce the latency caused by occlusions, where real-world objects obscure virtual objects. SLAM techniques can help to ensure that virtual items are perfectly aligned with the actual world, even when the user moves and the environment changes, by updating the device's position and orientation continually.

References

1. Cadena, C., et al.: Past, present, and future of simultaneous localization and mapping: toward the robust-perception age. IEEE Trans. Robot. **32**(6), 1309–1332 (2016)
2. Chen, Y.: Algorithms for simultaneous localization and mapping, vol. 3, pp. 1–15, February 2013
3. Bresson, G., Alsayed, Z., Yu, L., Glaser, S.: Simultaneous localization and mapping: a survey of current trends in autonomous driving. IEEE Trans. Intell. Veh. **2**(3), 194–220 (2017)
4. Nava, Y.: Visual-LiDAR SLAM with loop closure. PhD thesis, KTH Royal Institute of Technology (2018)
5. Sun, T., Liu, Y., Wang, Y., Xiao, Z.: An improved monocular visual-inertial navigation system. IEEE Sens. J. **21**(10), 11728–11739 (2020)
6. Newcombe, R.A., Lovegrove, S.J., Davison, A.J.: DTAM: dense tracking and mapping in real-time. In: 2011 International Conference on Computer Vision, pp. 2320–2327. IEEE (2011)
7. Engel, J., Schöps, T., Cremers, D.: LSD-SLAM: large-scale direct monocular SLAM. In: Fleet, D., Pajdla, T., Schiele, B., Tuytelaars, T. (eds.) ECCV 2014. LNCS, vol. 8690, pp. 834–849. Springer, Cham (2014). https://doi.org/10.1007/978-3-319-10605-2_54

8. Berkvens, R., Vandermeulen, D., Vercauteren, C., Peremans, H., Weyn, M.: Feasibility of geomagnetic localization and geomagnetic RatSLAM. Int. J. Adv. Syst. Meas. **7**(1–2), 44–56 (2014)

9. Newcombe, R.A., et al.: Kinectfusion: real-time dense surface mapping and tracking. In: 2011 10th IEEE International Symposium on Mixed and Augmented Reality, pp. 127–136. IEEE (2011)

10. Meng, X., Gao, W., Hu, Z.: Dense RGB-D SLAM with multiple cameras. Sensors **18**(7), 2118 (2018)

11. Wang, S., Clark, R., Wen, H., Trigoni, N.: DeepVO: towards end-to-end visual odometry with deep recurrent convolutional neural networks. In: 2017 IEEE International Conference on Robotics and Automation (ICRA), pp. 2043–2050. IEEE (2017)

12. Mohamed, S.A., Haghbayan, M.-H., Westerlund, T., Heikkonen, J., Tenhunen, H., Plosila, J.: A survey on odometry for autonomous navigation systems. IEEE Access **7**, 97466–97486 (2019)

13. Karam, S., Lehtola, V., Vosselman, G.: Integrating a low-cost mems imu into a laser-based slam for indoor mobile mapping. Int. Arch. Photogramm. Remote Sens. Spat. Inf. Sci. **42**, 149–156 (2019)

14. R. Kümmerle, R., Grisetti, G., Strasdat, H., Konolige, K., Burgard, W.: G^2o: a general framework for graph optimization. In: 2011 IEEE International Conference on Robotics and Automation, pp. 3607–3613. IEEE (2011)

15. Deschaud, J.-E.: IMLS-SLAM: scan-to-model matching based on 3D data. In: 2018 IEEE International Conference on Robotics and Automation (ICRA), pp. 2480–2485. IEEE (2018)

16. Rublee, E., Rabaud, V., Konolige, K., Bradski, G.: ORB: an efficient alternative to sift or surf. In: 2011 International Conference on Computer Vision, pp. 2564–2571. IEEE (2011)

17. Bay, H., Ess, A., Tuytelaars, T., Van Gool, L.: Speeded-up robust features (surf). Comput. Vis. Image Underst. **110**(3), 346–359 (2008)

18. Chi, H.C., Tsai, T.H., Chen, S.Y.: Slam-based augmented reality system in interactive exhibition. In: 2020 IEEE Eurasia Conference on IOT, Communication and Engineering (ECICE), pp. 258–262. IEEE (2020)

19. Azuma, R.T.: The most important challenge facing augmented reality. Presence **25**(3), 234–238 (2016)

20. Zhang, Z., Shu, M., Wang, Z., Wang, H., Wang, X.: A registration method for augmented reality system based on visual slam. In: 2019 International Conference on Electronic Engineering and Informatics (EEI), pp. 408–411. IEEE (2019)

21. Mur-Artal, R., Montiel, J.M.M., Tardos, J.D.: ORB-SLAM: a versatile and accurate monocular slam system. IEEE Trans. Robot. **31**(5), 1147–1163 (2015)

22. Liu, H., Chen, M., Zhang, G., Bao, H., Bao, Y.: ICE-BA: incremental, consistent and efficient bundle adjustment for visual-inertial slam. In: Proceedings of the IEEE Conference on Computer Vision and Pattern Recognition, pp. 1974–1982 (2018)

23. Cyrus, J., Krcmarik, D., Moezzi, R., Koci, J., Petru, M.: Hololens used for precise position tracking of the third party devices-autonomous vehicles. Commun.-Sci. Lett. Univ. Zilina **21**(2), 18–23 (2019)

24. Hoffman, M.A.: Microsoft hololens development edition. Science **353**(6302), 876–876 (2016)

25. Nießner, M., Zollhöfer, M., Izadi, S., Stamminger, M.: Real-time 3D reconstruction at scale using voxel hashing. ACM Trans. Graph. (ToG) **32**(6), 1–11 (2013)

26. Glocker, B., Shotton, J., Criminisi, A., Izadi, S.: Real-time RGB-D camera relocalization via randomized ferns for keyframe encoding. IEEE Trans. Vis. Comput. Graph. **21**(5), 571–583 (2014)

27. Skurowski, P., Nurzyńska, K., Pawlyta, M., Cyran, K.A.: Performance of QR code detectors near Nyquist limits. Sensors **22**, 7230 (2022)

28. Cheng, J., Zhang, L., Chen, Q., Hu, X., Cai, J.: A review of visual slam methods for autonomous driving vehicles. Eng. Appl. Artif. Intell. **114**, 104992 (2022)

29. Juneja, A., Bhandari, L., Mohammadbagherpoor, H., Singh, A., Grant, E.: A comparative study of slam algorithms for indoor navigation of autonomous wheelchairs. In: 2019 IEEE International Conference on Cyborg and Bionic Systems (CBS), pp. 261–266. IEEE (2019)

30. Zou, Q., Sun, Q., Chen, L., Nie, B., Li, Q.: A comparative analysis of lidar slam-based indoor navigation for autonomous vehicles. IEEE Trans. Intell. Transp. Syst. **23**(7), 6907–6921 (2021)

31. Khan, M.U., Zaidi, S.A.A., Ishtiaq, A., Bukhari, S.U.R., Samer, S., Farman, A.: A comparative survey of lidar-slam and lidar based sensor technologies. In: 2021 Mohammad Ali Jinnah University International Conference on Computing (MAJICC), pp. 1–8. IEEE (2021)

32. Zhou, X., Huang, R.: A state-of-the-art review on SLAM. In: Intelligent Robotics and Applications. ICIRA 2022. LNCS, vol. 13457, pp. 240–251. Springer, Cham (2022). https://doi.org/10.1007/978-3-031-13835-5_22

33. Klose, S., Heise, P., Knoll, A.: Efficient compositional approaches for real-time robust direct visual odometry from RGB-D data. In: 2013 IEEE/RSJ International Conference on Intelligent Robots and Systems, pp. 1100–1106. IEEE (2013)

34. Gao, X., Wang, R., Demmel, N., Cremers, D.: LDSO: direct sparse odometry with loop closure. In: 2018 IEEE/RSJ International Conference on Intelligent Robots and Systems (IROS), pp. 2198–2204. IEEE (2018)

35. Dai, W., Zhang, Y., Li, P., Fang, Z., Scherer, S.: RGB-D SLAM in dynamic environments using point correlations. IEEE Trans. Pattern Anal. Mach. Intell. **44**(1), 373–389 (2020)

36. Kiss-Illés, D., Barrado, C., Salamí, E.: GPS-SLAM: an augmentation of the ORB-SLAM algorithm. Sensors **19**(22), 4973 (2019)

37. Cai, L., Ye, Y., Gao, X., Li, Z., Zhang, C.: An improved visual slam based on affine transformation for orb feature extraction. Optik **227**, 165421 (2021)

38. Bescos, B., Fácil, J.M., Civera, J., Neira, J.: DynaSLAM: tracking, mapping, and inpainting in dynamic scenes. IEEE Robot. Autom. Lett. **3**(4), 4076–4083 (2018)

39. Cheng, J., Sun, Y., Meng, M.Q.-H.: Improving monocular visual slam in dynamic environments: an optical-flow-based approach. Adv. Robot. **33**(12), 576–589 (2019)

40. Forster, C., Pizzoli, M., Scaramuzza, D.: SVO: fast semi-direct monocular visual odometry. In: 2014 IEEE International Conference on Robotics and Automation (ICRA), pp. 15–22. IEEE (2014)

41. Bergmann, P., Wang, R., Cremers, D.: Online photometric calibration of auto exposure video for realtime visual odometry and slam. IEEE Robot. Autom. Lett. **3**(2), 627–634 (2017)

42. Liu, P., Yuan, X., Zhang, C., Song, Y., Liu, C., Li, Z.: Real-time photometric calibrated monocular direct visual slam. Sensors **19**(16), 3604 (2019)

43. Qin, T., Li, P., Shen, S.: VINS-Mono: a robust and versatile monocular visual-inertial state estimator. IEEE Trans. Robot. **34**(4), 1004–1020 (2018)

44. Weiss, S., Achtelik, M.W., Lynen, S., Chli, M., Siegwart, R.: Real-time onboard visual-inertial state estimation and self-calibration of MAVs in unknown environments. In: 2012 IEEE International Conference on Robotics and Automation, pp. 957–964. IEEE (2012)

45. Yin, H., Li, S., Tao, Y., Guo, J., Huang, B.: Dynam-SLAM: an accurate, robust stereo visual-inertial SLAM method in dynamic environments. IEEE Trans. Robot. (2022)

46. Cheng, Q., Zhang, S., Bo, S., Chen, D., Zhang, H.: Augmented reality dynamic image recognition technology based on deep learning algorithm. IEEE Access **8**, 137370–137384 (2020)

47. Chen, L.-C., Papandreou, G., Kokkinos, I., Murphy, K., Yuille, A.L.: Semantic image segmentation with deep convolutional nets and fully connected crfs. arXiv preprint arXiv:1412.7062 (2014)

48. Redmon, J., Divvala, S., Girshick, R., Farhadi, A.: You only look once: unified, real-time object detection. In: Proceedings of the IEEE Conference on Computer Vision and Pattern Recognition, pp. 779–788 (2016)

49. Huang, G., Liu, Z., Van Der Maaten, L., Weinberger, K.Q.: Densely connected convolutional networks. In: Proceedings of the IEEE Conference on Computer Vision and Pattern Recognition, pp. 4700–4708 (2017)

50. Zhou, T., Brown, M., Snavely, N., Lowe, D.G.: Unsupervised learning of depth and ego-motion from video. In: Proceedings of the IEEE Conference on Computer Vision and Pattern Recognition, pp. 1851–1858 (2017)

51. Gao, X., Zhang, T.: Unsupervised learning to detect loops using deep neural networks for visual SLAM system. Auton. Robot. **41**, 1–18 (2017)

52. Geng, M., Shang, S., Ding, B., Wang, H., Zhang, P.: Unsupervised learning-based depth estimation-aided visual slam approach. Circuits Syst. Signal Process. **39**, 543–570 (2020)

53. Li, F., et al.: A mobile robot visual slam system with enhanced semantics segmentation. IEEE Access **8**, 25442–25458 (2020)

54. Zhang, L., Wei, L., Shen, P., Wei, W., Zhu, G., Song, J.: Semantic SLAM based on object detection and improved octomap. IEEE Access **6**, 75545–75559 (2018)

55. Redmon, J., Farhadi, A.: Yolov3: an incremental improvement, arXiv preprint arXiv:1804.02767 (2018)

56. Tateno, K., Tombari, F., Laina, I., Navab, N.: CNN-SLAM: real-time dense monocular slam with learned depth prediction. In: Proceedings of the IEEE Conference on Computer Vision and Pattern Recognition, pp. 6243–6252 (2017)

57. Li, R., Wang, S., Long, Z., Gu, D.: UnDeepVO: monocular visual odometry through unsupervised deep learning. In: 2018 IEEE International Conference on Robotics and Automation (ICRA), pp. 7286–7291. IEEE (2018)

58. Vijayanarasimhan, S., Ricco, S., Schmid, C., Sukthankar, R., Fragkiadaki, K.: SFM-NET: learning of structure and motion from video, arXiv preprint arXiv:1704.07804 (2017)

Improving the Performance of Task-Based Linear Algebra Software with Autotuning Techniques on Heterogeneous Architectures

Jesús Cámara[1]([✉])(iD), Javier Cuenca[2](iD), and Murilo Boratto[3](iD)

[1] Department of Informatics, University of Valladolid, Valladolid, Spain
jesus.camara@infor.uva.es
[2] Department of Engineering and Technology of Computers, University of Murcia, Murcia, Spain
jcuenca@um.es
[3] Department of Sciences, University of Bahia, Salvador, Brazil
muriloboratto@uneb.br

Abstract. This work presents several self-optimization strategies to improve the performance of task-based linear algebra software on heterogeneous systems. The study focuses on Chameleon, a task-based dense linear algebra software whose routines are computed using a tile-based algorithmic scheme and executed in the available computing resources of the system using a scheduler which dynamically handles data dependencies among the basic computational kernels of each linear algebra routine. The proposed strategies are applied to select the best values for the parameters that affect the performance of the routines, such as the tile size or the scheduling policy, among others. Also, parallel optimized implementations provided by existing linear algebra libraries, such as Intel MKL (on multicore CPU) or cuBLAS (on GPU) are used to execute each of the computational kernels of the routines. Results obtained on a heterogeneous system composed of several multicore and multiGPU are satisfactory, with performances close to the experimental optimum.

Keywords: autotuning · linear algebra · heterogeneous computing · task-based scheduling

1 Introduction

In recent years, the increasing heterogeneity of computational systems has made the efficient execution of scientific applications a major challenge. Tipically, these systems consist of a set of compute nodes made up of several parallel devices, such as multicore processors, graphics accelerators, or programmable devices, with a different number of processing elements and memory hierarchy levels each. Thus, efficiently exploiting the computational capacity of the whole system is not an easy task, and even more so in parallel. In many cases, traditional numerical algorithms need to be redesigned to perform the computations

© The Author(s), under exclusive license to Springer Nature Switzerland AG 2023
J. Mikyška et al. (Eds.): ICCS 2023, LNCS 14073, pp. 668–682, 2023.
https://doi.org/10.1007/978-3-031-35995-8_47

parallel and efficiently. The advances in the development of highly optimized libraries have made it possible to tackle this problem, improving the overall performance of the routines involved. In numerical dense linear algebra, several libraries have been developed for specific parallel devices. The most commonly used are PLASMA [7] or Intel MKL [10] for multicore processors, cuBLAS [13] for NVIDIA GPUs or MAGMA [15] for multicore+GPU systems. Depending on the problem to be solved, these libraries can be used in an isolated way (for a specific device) or combined together when the workload is distributed among the different parallel computing devices. Also, some of them are designed to better exploit the different levels of parallelism of the architecture by using vectorization or multi-threading features (such as Intel MKL) and others, such as PLASMA, focus on using fine-grained parallelism to execute operations in parallel, representing them as a sequence of small tasks that can be dynamically scheduled. Regardless of the purpose of the library, they all provide reasonably good performance without taking into account the values for a set of adjustable parameters related to the hardware platform or the library routines themselves. However, in order to exploit the whole system efficiently, it is also necessary to search for the best values for these parameters.

In the literature, two general researching lines can be found that aim at this objective. On the one hand, there are techniques based on an empirical search supported heuristically by a certain knowledge of the platform architecture. This autotuning approach was introduced by ATLAS [17] and its predecessor PHiPAC [6]. On the other hand, code that is competitive with implementations generated via empirical search could also be produced by analytical proposals [11], either modelling the behaviour of the hardware when the software is executed [12] or, at higher level, modelling the performance of the software when it is running on this hardware [8]. In addition, the emergence of heterogeneous systems integrating accelerator devices with a wide variety of parallel architectures, such as GPUs, has led to the development of proposals focused on how to optimize linear algebra kernels for such platforms [4]. Also, hybrid methods can also be found, such as applying a hierarchical autotuning method at both hardware and software levels [9]. Similarly, wizard frameworks for developers have emerged, such as BOAST [16], which aim to minimise the effort of migrating high-performance computing kernels between platforms of different nature by providing an embedded domain-specific language to describe the kernels and their possible optimization.

The working scenario differs when the software includes a set of components that work together to execute the routines. This is the case for task-based libraries, which are of interest in the field of high-performance computing. This work focuses on Chameleon [1], a task-based dense linear algebra software for heterogeneous architectures. This library is based on the Sequential Task Flow (STF) paradigm and runs on top of a runtime system, which dynamically distributes and executes the tasks (basic computational kernels) among the available computing resources (multicore or GPUs) using a Directed Acyclic Graph (DAG) of tasks. Since the execution of tasks is handled by the runtime system, the self-optimization strategies proposed to select the best values for the adjustable parameters should be extended to consider, in addition, the configurable scheduling parameters.

In previous approaches used to tune the Chameleon library [2], the values of several adjustable parameters (block size and scheduling policy) were selected to get the best asymptotic performance for the whole set of problem sizes. In a preliminary comparative study carried out with the Cholesky routine on multicore CPU and CPU+multiGPU systems, the best overall values selected for these adjustable parameters have been $\{nb = 512, \, sched = \text{"eager"}\}$ and $\{nb = 576, \, sched = \text{"dmdas"}\}$, respectively. However, the performance of the routine improves when the values are properly selected for each problem size, as shown in Fig. 1. By default, Chameleon considers a fixed value for these adjustable parameters regardless of the problem size, which leads to a loss in performance as the problem size increases. With the asymptotic values, instead, there is a loss in performance for small problem sizes. Therefore, these results demonstrate the importance of having a self-optimization process that allows to overcome the performance shortcomings offered by both approaches.

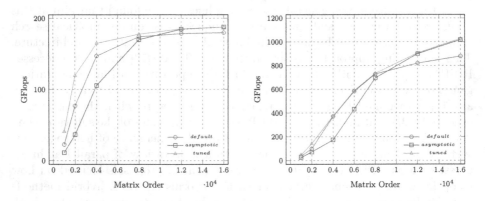

Fig. 1. Performance obtained on multicore CPU (left) and CPU+multiGPU (right) systems with the Cholesky routine of Chameleon when using the default values for the adjustable parameters (`default`), the overall values (`asymptotic`) and the best values selected (`tuned`) with the self-optimization methodology.

These first pieces of evidence together with other preliminary results [3] have led to undertake this work of developing a collaborative framework that jointly addresses two self-optimization methodologies: on the one hand, at installation time by using a training engine and, on the other hand, at execution time through the use of the dynamic task-based scheduling provided by Chameleon. The training engine can use different search strategies depending on the number and type of the adjustable parameters: routine parameters (block size and inner block size), system parameters (number and type of computing resources) and scheduling parameters (policies used by the runtime system). With this approach, at installation time, a performance map of the routine is obtained and stored in a database. After that, at execution time, this information is used to quickly make better decisions for any specific problem to be solved.

The rest of the paper is organized as follows. Section 2 shows the general scheme of execution of a linear algebra routine in Chameleon. In Sect. 3, the self-optimization methodology for selecting the best values of a set of adjustable parameters when working with task-based libraries is presented. Section 4 shows the experimental study carried out with different routines of Chameleon when applying the proposed methodology. Finally, Sect. 5 concludes the paper and outlines future research lines.

2   The Chameleon Library

As mentioned, this paper focuses on Chameleon, a task-based dense linear algebra software that internally uses StarPU [5], a runtime system to dynamically manage the execution of the different computational kernels on the existing hybrid computing resources.

Figure 2 shows the steps to execute a linear algebra routine of Chameleon using the StarPU runtime system. As Chameleon is derived from PLASMA, each routine is computed by following a tile-based algorithm. First, the tasks to compute each of the data tiles in the matrix layout are created based on the STF paradigm and the DAG is generated with dependencies between them. These tasks correspond to the basic algebraic operations (called kernels) involved in the computation of the routine and act on different blocks of data, whose size depends on the value specified for the tile size (nb in the Figure). Then, the tasks are scheduled using one of the scheduling policies provided by StarPU and executed on the available computing resources using efficient implementations of the basic kernels, such as those provided by the Intel MKL and cuBLAS libraries.

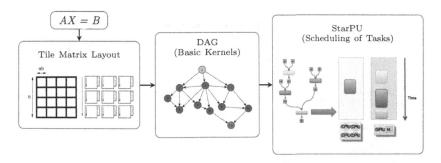

Fig. 2. Execution of a linear algebra operation in Chameleon using StarPU.

3   Self-optimization Methodology

This section describes the proposed self-optimization methodology for selecting the best values for the adjustable parameters that affect the performance of the routines in task-based libraries. Figure 3 shows the general operation scheme. It consists of three main phases: selection of the search strategy, training of the routine with the selected strategy for different problem sizes and values for the set of parameters, and evaluation of the performance obtained by the routine using a different set of problem sizes.

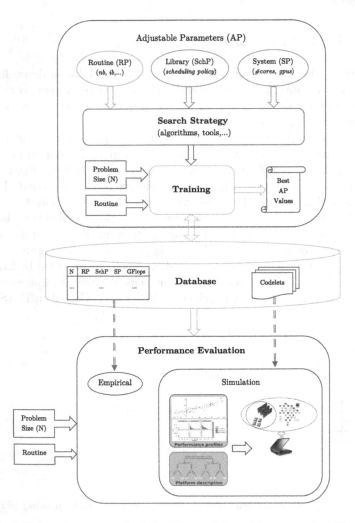

Fig. 3. Self-optimization methodology for task-based linear algebra libraries.

3.1   Selecting the Search Strategy

In task-based libraries, the performance of the routines depends mainly on the values selected for three subsets of adjustable parameters, AP:

- Routines Parameters, RP: block size and inner block size.
- System Parameters, SP: number and type of computing resources.
- Scheduling Parameters, $SchP$: scheduling policy used by the runtime system.

In order to find the best values for these AP, it is necessary to make use of search strategies. Nevertheless, the strategy chosen will be determined by the number of AP considered and the way it is conducted (exhaustive or guided). The proposed self-optimization methodology includes several search strategies to select the best values for RP:

- **Exhaustive_NB**: the routine is executed by varying the block size, nb, for each problem size, n. As a result, the execution time and the performance obtained for each (n, nb) is stored.
- **Exhaustive_NB+IB**: the routine is executed by varying the block size, nb, for each problem size, n, by increasing the value of the inner block size, ib, until it reaches the value of the current block size. As a result, the execution time and the performance obtained for each (n, nb, ib) is stored.
- **Guided_NB**: the routine is executed starting with the first problem size and the first block size considered, nb_1, whose value is increased until the performance decreases. The process continues with the next problem size (in ascending order), taking as starting point the value of nb for which the best performance was obtained for the previous problem size. Then, a bidirectional search is performed on the nb value (with increasing and decreasing values) until the performance obtained worsens the one obtained so far. The process then continues with the remaining problem sizes. As a result, for each problem size, the execution time and the performance obtained with the best value for nb is stored.
- **Guided_NB+IB_2D**: the routine is executed starting with the first problem size and the first block size considered, (n_1, nb_1), and the inner block size, ib, is increased until the value of nb_1 is reached or the performance obtained worsens the one obtained so far. To prevent the process from falling into a local minimum, a percentage value is used as tolerance to decide if the search continues to the next value of ib. The process then continues with the next value of nb for the same problem size, taking as starting point the ib value with which the best performance was obtained for the previous block size. Then, a bidirectional search is performed on the ib value (with increasing and decreasing values) using the percentage value considered until the performance decreases or the value of nb is reached. Once the search has finished for the last value of nb, the execution time and the performance obtained with the best pair (nb, ib) for the current problem size is stored. Next, the following problem size is selected (in ascending order) and the search process is repeated for ib, starting with the first block size, nb_1, and the best ib obtained for the previous problem size.

- **Guided_NB_1D+IB_2D**: the routine is executed in a similar way to the previous strategy, but when the process moves to the next problem size, it uses as starting point the best (nb, ib) values obtained for the previous problem size. The percentage value is also used to prevent the process from stalling at a local minimum, but only when varying the inner block size, ib.
- **Guided_NB_2D+IB_2D**: the routine is executed in a similar way to the previous strategy, but when the process moves to the next problem size, the bi-directional search is applied both in the block size, nb, and the inner block size, ib, taking as starting point the best (nb, ib) values obtained for the previous problem size. Again, the percentage value is used, but in this case for both the bidirectional search of nb and ib.

In addition, the system parameters (SP) can be considered. By default, the Chameleon library only uses the CPU cores available on the system. Thus, a non-expert user might consider using all the computing resources in the system. However, this does not always offer the best performance. Therefore, a search strategy should be used to select the appropriate number of GPUs and CPU cores to use for each problem size. The proposed strategy consists of adding computing resources in increasing order of computational capacity by applying at each step a guided strategy to select the best values of the routine parameters (RP) for the first problem size considered. Then, the process continues with the next problem size, using as a starting point the best values obtained for the RP for the previous problem size and keeps adding new computing resources as long as the performance does not worsen the one obtained so far.

The same idea can be applied to select the best values for the scheduling parameters $(SchP)$. In this case, the parameter to consider in the Chameleon library is the scheduling policy used by the runtime system. Therefore, the search strategy will consist of selecting the best one among those offered by StarPU. This strategy can be used in conjunction with the previous one if the best values for all AP have to be obtained.

3.2 Training the Routines

In the training phase, the routines are executed in the system by varying the values of the AP for each input problem size according to the selected search strategy. This phase is run only once for each routine and set of problem sizes. As a result, the selected values for the AP together with the problem size, n, and the performance obtained (in million floating point operations per second) are stored in a database for further use. As will be shown in the experiments, guided strategies will considerably reduce the training time of the routines compared to the use of exhaustive ones.

3.3 Validating the Methodology

Once the routines have been trained with a set of problem sizes and varying the values for the AP, the proposed methodology can then be validated. To do so, a validation set of problem sizes is used. As shown in Fig. 3, both an empirical and a simulated approach can be considered:

- Empirical: the routine is executed for each problem size, n, of the validation set using the best AP values stored in the database for the problem size closest to n.
- Simulated: for each problem size, n, both the AP values and the *codelets* of the computational kernels involved in the routine are obtained from the database to be used by SimGrid [14] to predict the performance. The *codelets* are automatically created for each basic kernel during the training phase and store performance data related with the execution of each basic kernel for the values used for the problem size and the routine parameters.

Finally, to check the goodness of the decisions taken (AP values), the performances obtained are compared with those that would have been obtained if an exhaustive search for the best AP values would have been carried out directly for each problem size in the validation set.

4   Experimental Study

The experimental study focuses mainly on two representative Chameleon routines: the LU and QR factorizations, although some results will also be shown for the Cholesky routine. We consider the LU routine without pivoting because the one with pivoting is not available in Chameleon as the cost of finding the pivot and executing the swap cannot be efficiently handled. Since both the LU and QR routines are implemented by following a tile-based algorithmic scheme, the impact of the block size on performance will be analyzed. In addition, the inner block size will be also considered, since LU and QR routines uses it to perform the matrix factorization.

The experiments are conducted with the search strategies described in Sect. 3.1 to select the best values of the different adjustable parameters (RP, SP, $SchP$). First, the RP values are selected. Next, the SP values and finally, the $SchP$ values. Training times are also analyzed and an example of validation for the proposed methodology is shown.

The heterogeneous platform used consists of five hybrid computing nodes, but the experiments have been performed on **jupiter**, the most heterogeneous one, which is composed of 2 Intel Xeon hexa-core CPUs (12 cores) and 6 NVIDIA GPUs (4 GeForce GTX590 and 2 T C2075). Also, the Intel MKL and the cuBLAS library are used to run the basic computational kernels scheduled by the runtime system on multicore and GPUs, respectively.

In the experiments, the following set of problem sizes {2000, 4000, 8000, 12000, 16000, 20000, 24000, 28000} and block sizes {208, 256, 288, 320, 384, 448, 512, 576} are used. The values chosen for the block sizes are a representative subset of those that allow to improve the performance of the **gemm** kernel for the problem sizes considered. This kernel covers most of the computing time of the routines under study, therefore, the impact of the block sizes selected for this kernel will have an impact on the overall performance of the routines. Nevertheless, the study could have been carried out with any other set of values.

4.1 Selecting the Routine Parameters

As shown in Fig. 3, the selection of the best values for the RP is carried out during the training phase and then stored in the database. Depending on the search strategy used, the values selected for the RP may differ, but also the time needed to obtain these values. The study will focus on selecting the best values for the block size and the inner block size in the LU and QR factorization routines by applying the search strategies described in Sect. 3.1.

First, the study is carried out with the LU factorization routine. Figure 4 (top left) shows the training time (in seconds) for each problem size when using the Exhaustive_NB+IB strategy, with a total of 6 h and 10 min. Figure 4 (top right), on the other hand, shows the pair (nb, ib) with which the best performance is obtained for each problem size. It is noted that for problem sizes above 8000, the best value for nb always corresponds to the largest value considered in the set of block sizes. In contrast, the values selected for ib are small and vary between 16 and 56 depending on the selected block size.

As shown, the exhaustive search strategy provides good performances, but the training time is high. To reduce it, the guided search strategies (Guided_NB+IB_2D and Guided_NB_2D+IB_2D) are used to select the best values for nb and ib. Figure 4 (middle left) shows the training time (in seconds) spent for each problem size, and Fig. 4 (middle right) shows the performance obtained with the (nb, ib) values selected by the Guided_NB+IB_2D search strategy. In this case, the total time spent is 32 min. Similarly, Fig. 4 (bottom) shows, respectively, the training time (in seconds) and the performance obtained with the Guided_NB_2D+IB_2D strategy for each problem size. Now, the training time spent is only 6 min, less than the required by the Guided_NB+IB_2D strategy and much less than the 6 h required by the Exhaustive_NB+IB.

Next, the same study is carried out with the QR factorization routine for the same set of problem and block sizes. Figure 5 (top left) shows the training time (in seconds) for each problem size when using the Exhaustive_NB+IB strategy. The time is about 29 h and 57 min, which is longer than in the LU routine because the LU factorization does not perform pivoting (as indicated before). Figure 5 (top right) shows the performance obtained for each problem size when using the best (nb, ib) values. Again, for problem sizes above 8000, the best value for the block size, nb, always corresponds to the largest value considered in the set of block sizes. However, the values selected for ib are sligthly higher than the ones selected for the LU routine, varying from 72 to 160. Next, the Guided_NB+IB_2D and Guided_NB_1D+IB_2D search strategies are applied to reduce the training time. Figure 5 (middle left) shows the time spent (in seconds) for each problem size, and Fig. 5 (middle right) shows the results obtained when using the Guided_NB+IB_2D strategy. In this case, the time spent is 8 h and 24 min, 64% less than the exhaustive one. Similarly, Fig. 5 (bottom left) shows the training time (in seconds) for each problem size, and Fig. 5

(bottom right) shows the performance and the (nb, ib) values obtained for each problem size when using the Guided_NB_1D+IB_2D strategy. Now, the time spent is 87 min, 86% and 97% less than using the Guided_NB+IB_2D and Exhaustive_NB+IB_2D strategies, respectively.

Finally, Fig. 6 shows a comparison of the performance results obtained for the LU and QR routines when using the exhaustive and guided search strategies. In Fig. 6 (left) it is observed that the performance obtained by the LU routine with the guided search strategies overlaps with that obtained using the exhaustive search. A similar behaviour is observed in Fig. 6 (right) for the QR routine, especially when the Guided_NB_1D+IB_2D strategy is used and the problem size increases. Therefore, the use of guided strategies will allow to obtain satisfactory performance results with the selected RP values and with reasonable training times.

4.2   Selecting the System and Scheduling Parameters

Besides the routine parameters, there are a set of AP whose values may be selected during the training phase to further improve the performance. In task-based libraries, such as Chameleon, the number and type of computing resources to use and the scheduling policy used by the dynamic task scheduler can be specified when executing a routine. Since the experiments are performed on a heterogeneous node, both the number of CPU cores and GPUs will be considered. In addition, as the execution of the computational kernels of the routines is handled at run-time by StarPU, which decides how to manage the execution of the kernels on the different computing resources of the system, different scheduling policies are considered. Some of them (such as dm, $dmda$ or $dmdas$) use the information from the $codelets$ generated during the training phase. Other policies, however, are only based on priorities (such as $eager$ and $random$), the load of task queues (such as ws and lws) or the availability of the computing resources. Therefore, the proposed methodology could be also applied to select the best values for the system parameters as well as the best scheduling policy among all those offered by StarPU.

Table 1 shows the values obtained for the Cholesky routine for a set of problem sizes. The GPU IDs are displayed to show the order in which they are chosen by the search strategy described in Sect. 3.1, as the scheduler assigns tasks to computing resources following a scheme based on priorities and dependencies between the data required by the different computational kernels, but it does not take into account the computational capacity of these resources. The results show how an appropriate selection of all the adjustable parameters (block size, number of CPU cores and number of GPUs, and scheduling policy) allows to obtain a performance improvement between 10% and 50% with respect to that obtained by the Chameleon library with the default running configuration. Also, the selected values differ for most problem sizes, hence the importance of using a self-optimization methodology that allows to select the best AP values for each problem size to improve the performance of the routines.

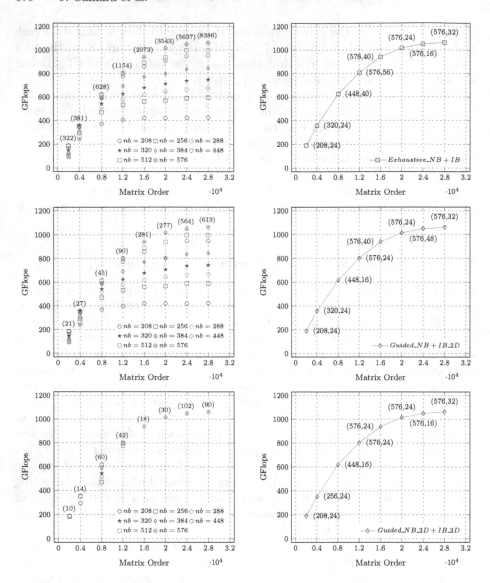

Fig. 4. LU factorization. Training time (left, in seconds) for each problem size when using the exhaustive (top left) and guided (middle and bottom) search strategies, and performance obtained (right) with each strategy when using the best (*nb, ib*) values for each problem size.

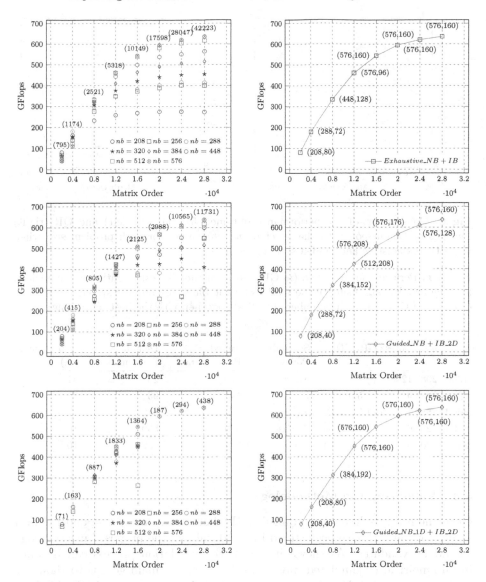

Fig. 5. QR factorization. Training time (left, in seconds) for each problem size when using the exhaustive (top left) and guided (middle and bottom) search strategies, and performance obtained (right) with each strategy when using the best (nb, ib) values for each problem size.

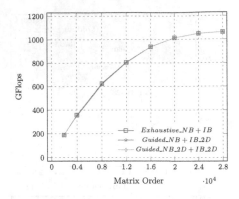
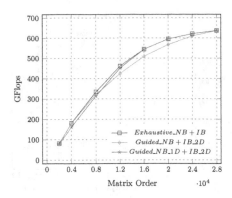

Fig. 6. Comparison of the performance obtained for the LU (left) and QR (right) routines of Chameleon with the selected (*nb*, *ib*) values for each problem size when using the exhaustive and guided search strategies.

Table 1. Performance obtained (in GFlops) with the Cholesky routine of Chameleon for each problem size using the best values for the block size and the system parameters (number of CPU cores and number of GPUs) together with the best scheduling policy.

n	nb	Cores	GPU-IDs	Scheduling Policy	Tuned Performance	Default Performance	Improvement (%)
1000	208	12	{}	eager	47	23	51
2000	208	9	{1,5,0}	lws	143	93	35
4000	288	7	{1,5,0,2,3}	ws	374	312	17
6000	320	6	{1,5,0,2,3,4}	dmdas	590	530	10
8000	320	6	{1,5,0,2,3,4}	dmdas	734	662	10

4.3   Validating the Methodology

Once the routines have been trained in the heterogeneous system and the best values for the *AP* have been stored in the database, the proposed methodology should be validated. As shown in Fig. 3, the validation process can be done in an empirical or simulated way. In both cases, the main goal is to show the performance obtained by the routine when the information stored in the database is used to select the values of the *AP* for a given problem size. The selection of the *AP* values is carried out based on the performance information stored for the closest problem size to the current one. Next, the routine is executed (or simulated) with the selected values.

Table 2 shows the results obtained with the QR routine of Chameleon using the empirical approach. A different set of problem sizes from the one used for the previously trained routine is considered to analyze how far the selected values for the *AP* and the performance obtained are from the experimental optima. It is noted that only two selected values slightly differ from the optimum ones, but its impact on performance is negligible. Therefore, it proves that the methodology works as expected and could be applied to other linear algebra routines.

Table 2. Performance comparison for the QR routine of Chameleon when using the values for nb and ib selected by the methodology for each problem size vs. using the optimal ones after training the routine.

	Autotuned			Optimal			Variation
n	nb	ib	GFlops	nb	ib	GFlops	%
10000	512	112	399	512	112	399	0
14000	576	128	507	512	128	508	0
18000	576	160	573	576	160	573	0
22000	576	160	610	576	160	610	0
26000	576	160	626	576	192	626	0

5 Conclusions

This work presents a methodology to self-optimize task-based libraries, such as Chameleon. A set of strategies to search the best values for different adjustable parameters have been applied during the training phase to several linear algebra routines, such as the Cholesky, LU and QR factorizations.

On the one hand, the impact on performance of certain parameters of the routines, such as the block size and the inner block size, is analysed. The proposed search strategies allow to obtain good performance, reducing even the training time of the routines by guiding the search for the optimal values of these parameters. Also, additional adjustable parameters have been considered, such as the number of computing resources of the heterogeneous platform and the scheduling policy of the runtime system, showing the importance of properly selecting the value of each of them to further reduce the execution time of the routines. Our intention is to extend this work by using the SimGrid simulator in the training and validation phases of the proposed methodology to analyze the potential benefits of its use, as well as to integrate the hierarchical autotuning approach proposed in [9] within the Chameleon library.

Acknowledgements. Grant RTI2018-098156-B-C53 funded by MCIN/AEI and by "ERDF A way of making Europe".

References

1. Agullo, E., Augonnet, C., Dongarra, J., Ltaief, H., Namyst, R., Thibault, S.: Faster, cheaper, better - a hybridization methodology to develop linear algebra software for GPUs. In: GPU Computing Gems, vol. 2 (2010)
2. Agullo, E., et al.: Achieving high performance on supercomputers with a sequential task-based programming model. IEEE Trans. Parallel Distrib. Syst. (2017)
3. Agullo, E., Cámara, J., Cuenca, J., Giménez, D.: On the autotuning of task-based numerical libraries for heterogeneous architectures. In: Advances in Parallel Computing, vol. 36, pp. 157–166 (2020)

4. Anzt, H., Haugen, B., Kurzak, J., Luszczek, P., Dongarra, J.: Experiences in auto-tuning matrix multiplication for energy minimization on GPUs. Concurr. Comput. Pract. Exp. **27**, 5096–5113 (2015)
5. Augonnet, C., Thibault, S., Namyst, R., Wacrenier, P.A.: STARPU: a unified platform for task scheduling on heterogeneous multicore architectures. Concurr. Comput. Pract. Exp. **23**(2), 187–198 (2011)
6. Bilmes, J., Asanovic, K., Chin, C.W., Demmel, J.: Optimizing matrix multiply using PHiPAC: a portable, high-performance, ANSI C coding methodology. In: Proceedings of the 11th International Conference on Supercomputing, pp. 340–347 (1997)
7. Buttari, A., Langou, J., Kurzak, J., Dongarra, J.: A class of parallel tiled linear algebra algorithms for multicore architectures. Parallel Comput. **35**, 38–53 (2009)
8. Cámara, J., Cuenca, J., García, L.P., Giménez, D.: Empirical modelling of linear algebra shared-memory routines. Procedia Comput. Sci. **18**, 110–119 (2013)
9. Cámara, J., Cuenca, J., Giménez, D.: Integrating software and hardware hierarchies in an autotuning method for parallel routines in heterogeneous clusters. J. Supercomput. **76**(12), 9922–9941 (2020). https://doi.org/10.1007/s11227-020-03235-9
10. Intel MKL. http://software.intel.com/en-us/intel-mkl/
11. Kelefouras, V., Kritikakou, A., Goutis, C.: A matrix-matrix multiplication methodology for single/multi-core architectures using SIMD. J. Supercomput. **68**(3), 1418–1440 (2014)
12. Low, T.M., Igual, F.D., Smith, T.M., Quintana-Orti, E.S.: Analytical modeling is enough for high-performance BLIS. ACM Trans. Math. Softw. **43**(2), 1–18 (2016)
13. NVIDIA cuBLAS. https://docs.nvidia.com/cuda/cublas/index.html
14. Stanisic, L., Thibault, S., Legrand, A.: Faithful performance prediction of a dynamic task-based runtime system for heterogeneous multi-core architectures. Concurr. Comput. Pract. Exp. **27**, 4075–4090 (2015)
15. Tomov, S., Dongarra, J.: Towards dense linear algebra for hybrid GPU accelerated manycore systems. Parallel Comput. **36**, 232–240 (2010)
16. Videau, B., et al.: BOAST: a metaprogramming framework to produce portable and efficient computing kernels for HPC applications. Int. J. High Perform. Comput. Appl. **32**, 28–44 (2018)
17. Whaley, C., Petitet, A., Dongarra, J.: Automated empirical optimizations of software and the ATLAS project. Parallel Comput. **27**, 3–35 (2001)

Tempo and Time Signature Detection of a Musical Piece

Daniel Kostrzewa[✉][iD] and Marek Zabialowicz

Department of Applied Informatics, Silesian University of Technology,
Gliwice, Poland
daniel.kostrzewa@polsl.pl

Abstract. Tempo and time signature detection are essential tasks in the field of Music Information Retrieval. These features often affect the perception of a piece of music. Their automatic estimation unlocks many possibilities for further audio processing, as well as supporting music recommendation systems and automatic song tagging. In this article, the main focus is on building a two-phase system for extracting both features. The influence of many parameters of known methods was investigated. The results were also compared with the well-known and scientifically recognized Librosa library for music processing.

Keywords: Tempo · Time Signature · Music Information Retrieval · Audio Features · Sound Analysis · GTZAN

1 Introduction

Music information retrieval (MIR) is a field that has become extremely popular among researchers since the beginning of the 21st century, becoming one of the most intensively developed in recent years. An increased interest due to more accessible access to music, primarily due to portable players and smartphones [18], as well as music streaming services and the competition of those for customers can be observed. In order to find similar songs, a service must acquire information about each song individually, and it must do so in a very short time.

Tempo and time signature detection are fundamental to digital audio processing and music information retrieval. It has applications for the previously mentioned streaming services and music producers who find marking bar lines helpful when editing a recording. Knowledge of tempo and meter is also needed by DJs creating live music, who need to align parallel tracks and samples, i.e., fragments of other songs, broadcasts, or recordings. Knowing a song's tempo and time signature can also help when trying to catalog it automatically. By having information about the artist, title, and genre, and additionally detecting tempo, time signature, and, for example, mood, it is possible to create a very informative database of songs. Such a database would be able to respond to a query based on the listener's tastes, current mood, or playlist destination.

This work was supported by the Statutory Research funds of Department of Applied Informatics, Silesian University of Technology, Gliwice, Poland.

The remainder of the paper is as follows. Section 1 describes the undertaken task, provides related works, and outlines the paper's contribution. Section 2 describes the tempo and time signature detection method in detail. The performed numerical experiments are presented in Sect. 3. Conclusions with the future work comprise Sect. 4.

1.1 Related Work

There are many papers and studies on tempo detection, but the algorithms described are very similar. Metre detection is an issue less frequently described in literature and articles [1].

The most commonly used method for tempo detection is to create a filterbank using the discrete Fourier transform (DFT). A filterbank can be created with the following, among others: a rectangular filter [4,16], with a finite impulse response (FIR) filter [8,10] or a filter with an infinite impulse response (IIR) [18]. Instead of a filter ensemble, the signal envelope [18] or the spectrogram flow function [14] can be calculated. The function calculated in this way changes more slowly than the original signal and is more suitable for processing. Compared to the filter ensemble, these methods allow for better results in music without a rhythm section.

One of the ways to detect tempo is to create a series of comb filters in which the pulsation frequency tells the beat frequency of the [4,16] tempo, and then find the maximum sum of the products of the signal spectra from the filter ensemble and the filter [4,16]. The filter pulse frequency for which the maximum sum of energy was calculated is most likely the beat frequency of the pulses in the song, i.e., the tempo of the song. Another option for detecting tempo based on the comb filter is to use the autocorrelation function (ACF) [2,8,18]. ACF can also be used to detect pulses in a signal to calculate their probability of occurrence over time [10,14].

Foote and Cooper [6] and Foote [5] in their paper presented an interesting way to represent a piece of music graphically. Using short-time Fourier transform (STFT) and spectrogram or Mel-frequency cepstral coefficients (MFCC), they create a two-dimensional matrix that represents the similarity between two excerpts of a song. This idea was used by Pikrakis et al. [15], who used the self-similarity matrix (SSM) created using MFCC to detect time signature and tempo. Gainza [7,9] also used the SSM. Calculating a song's spectrogram creates an audio signal similarity matrix (ASM) and then a beat similarity matrix (BSM). Then, by processing the average values on the diagonals, it calculates how many rhythmic values are in a single bar, finally calculating what value is the base of the meter.

Böck et al. [3] approached the problem in a complex way, not only wanting to detect tempo, but also the position of beats over time. Improving the accuracy of detecting one value directly flows into improving the accuracy of detecting the other. To do this, they use machine learning and a temporal convolutional network (TCN). In order to extract features for learning by the algorithm, they process the spectrogram of the piece. In 2019, Davies and Böck [11] presented a state-of-the-art way to tag bars in music based on TCN, improving on the previous [17] algorithm.

However, despite the significant development of artificial intelligence and deep learning, these methods still need improvement. Therefore, this paper focuses on improving the ability to detect tempo and time signature based on the beat similarity matrix.

1.2   Contribution

The contribution of the paper is two-fold. Primarily, we used algorithms known in the literature for both tempo and time signature detection to develop a two-phase detection system which is not known in the literature. Secondly, the influence of many different parameters was examined on the detection accuracy. Moreover, the tempo detection accuracy was compared to the well-known and recognized Librosa library.

2   Tempo and Time Signature Detection Method

Periodic local maxima can be observed in the graphical representation of a musical piece. These maxima are correlated with the pulse of the song and, therefore, with its tempo. Tempo detection on such a sample can be inaccurate due to overlapping sounds of multiple instruments. To compensate this effect, the signal of a piece of music is transferred to the frequency domain using a fast Fourier transform (FFT). The signal is then divided into subbands by band-pass filters. The divided signal forms a filter ensemble. Each of the spectra from the filter ensemble is transferred back to the time domain using the inverse fast Fourier transform (IFFT). However, the spectra cannot be directly transferred because the signal from which the original spectrum was created was not infinite. In addition, the second spectra of the filter set are discontinuous because they were extracted with a rectangular filter. For this reason, each spectrum must first be smoothed with a Hann time window and then transferred to the time domain. This representation contains some local maxima, but they are stretched out in time. In order to apply the comb filter, the signal has to be transformed into individual peaks. To do this, the derivative of the signal is calculated, where the value of each sample is the difference between this sample with the previous one. Figure 1 demonstrates the signal derivative of one of the filters. It clearly shows the regular beats of the song's pulse.

A comb filter signal is created for each tempo. The distance of the *step* filter pulses is calculated from the formula 1. The length of the comb filter signal is constant for all BPM values and is equal to the length of the song fragment from which the filter ensemble was created.

$$step = \frac{60}{bpm} * f_s \tag{1}$$

where: bpm - the rate expressed in beats per minute, f_s - the final sampling frequency of the signal.

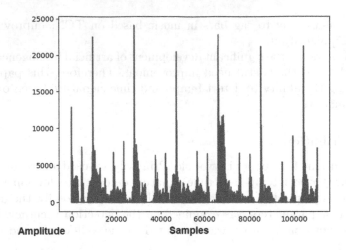

Fig. 1. Derivative of the smoothed signals from the second filter.

The filter signal for a given tempo is then transferred to the frequency domain. The calculated spectrum is successively multiplied by the spectra of the derived signals from the filter ensemble, and the products are summed. In this way, the sum of the energies of the products of the signal and filter derivative spectra is calculated. During the loop run, the maximum sum of energies is sought. The tempo corresponding to the maximum sum of energies is considered the tempo of the song.

In order to detect the time signature of a musical piece, it is necessary to know the tempo to extract the individual metrical values, divide the piece into bars and then calculate their similarity. The repetition of accents and similar bars within a song will enable effective time signature detection. The algorithm is based on the BSM method [8].

The first step is to calculate the spectrogram of a piece of music. To compute the spectrogram, the function *spectrogram* from the library *scipy.signal*[1] is used. The result is a spectrogram whose length of each segment equals a quarter note, the basic metric unit in a time signature with a base of 4. The spectrogram represents the intensity of the sound for frequencies between 0 and half the sampling frequency of the signal, which is usually 22 kHz or 11 kHz. Instruments belonging to the rhythmic group play at lower frequencies (around 6 kHz), so the frequencies of the spectrogram are truncated to this value.

The sounds in a piece of music can be divided into harmonic and rhythmic. The former are sounds from the group of melodic instruments (e.g., guitar, piano, human voice) that create the melody of a song. In contrast, rhythmic sounds are sounds from a group of rhythmic instruments that build the rhythm, providing the foundation and support for the melodic section. In a song, they deftly

---

[1] https://docs.scipy.org/doc/scipy/reference/generated/scipy.signal.spectrogram.
html.

intertwine to form a coherent whole, but two characteristics distinguish them: the frequency and the duration of the sound played. Nobutaka Dno et al. [13] proposed an algorithm to extract the components using the song's spectrogram and a median filter. The method is quite simple in its operation. It resembles the median filter image smoothing known from image processing methods, except that it does not use a square window but vertical and horizontal ones. Percussive components are obtained using a median filter with a vertical smoothing window. This is because the sounds of rhythmic instruments are short and can cover all frequencies. The final step of extracting the harmonic and rhythmic components is to apply binary masking and split the original spectrogram into two parts.

The next step is the calculation of the BSM bar similarity matrix. For the calculation of the distance between each frame, three distance metrics were used and tested [9]: Euclidean, cosine, and Kullback-Leibler. Using the aforementioned metrics, a bar similarity matrix is created. The distance between each frame of the spectrogram is calculated, the width of which is equal to a quarter note at the given rate. These distances are recorded into a two-dimensional BSM matrix.

To investigate the existence of similar metrical structures in a song, the diagonals of the previously obtained BSM matrix are examined. Each diagonal represents the similarity between sounds that are a different number of metric values away. This similarity can be measured by calculating the mean component of each diagonal of the BSM matrix. The function thus obtained is then inverted to build a function that shows the peaks on the diagonals where the components show maximum similarity.

Many candidates are considered in the presented approach to detect the time signatures. This includes simple and regular signatures, as well as complex and irregular ones. The resulting function gives the greatest weight to closely spaced quarter notes. The waveform of the function is presented in Fig. 2. It clearly shows the maximum value for signature 5, so the time signature of the song is 5/4.

3   Quantitative Experiments

3.1   Dataset

Most musical works are protected by copyright, so making them available in the public domain is challenging, even for research purposes. For this reason, most collections do not contain audio files but only meta-data about the contained works. A dataset called GTZAN [19] was used to test the effectiveness of tempo detection. It comes from Tzanetakis and Cook's paper [20]. It contains a collection of one thousand 30-s excerpts from songs. The collection also has a file where the tempo of each song is given in BPM. All songs are 16-bit audio files, sampled 22050 Hz on one channel, saved in .wav format. This collection does not include title or artist information. Time signature information was also not included, so using our knowledge of music theory, 86 songs were drawn for which the time signature was empirically classified. Almost all the pieces are 4/4, and one is 3/4.

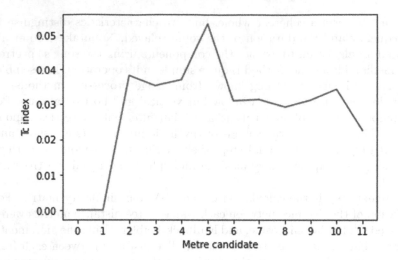

Fig. 2. An example of the time signature detection function.

The pieces included in GTZAN have little variation in time signature, so it was decided to expand the collection with an additional 14 pieces from the in-house collection. The proposal for the first ten items comes from [7]. The final collection of additional songs is presented in Table 1. 100 songs were collected to evaluate the time signature detection algorithm, with a significant prevalence of 4/4.

It has to be noted that we are aware that the skewness (imbalanced classes) of the dataset may lead to doubtful results. However, to the best of our knowledge, there is no well-known and well-prepared dataset for time signature detection. In fact, the author of this paper is working on preparing such a dataset.

3.2 Effectiveness Evaluation Criteria

Two criteria for evaluating tempo detection have been adopted in the research community – Accuracy1 and Accuracy2 [3,8]. Accuracy1 is more stringent, and with this criterion, only the case where the detected BPM value is equal to the original tempo of the song $\pm 2\%$ is considered a correct tempo detection. Accuracy2 allows for a margin of error. When the detected value is a multiple of the original tempo, such a result is also considered correct. This is because if a song is played at 70 BPM and a value of 140 BPM is detected, the pulse of the metronome set to 140 BPM will coincide with the pulse of the song, except that it will pound out eighth notes rather than quarter notes. An analogous situation occurs when a tempo of 70 is detected for a track played at 140 BPM.

Similar assumptions were made when evaluating the performance of time signature detection. With Accuracy1, only the case where an accurate measure of time signature is detected is considered correct. With Accuracy2, the arrangement of accents in the bar is ignored, and the basis of the bar measure

Table 1. A collection of additional songs with different time signatures.

Title	Performer	Time signature	Tempo
Eleven	Primus	11/8	230
Windows To The Soul	Steve Vai	11/8	243
Watermelon In Easter Hay	Frank Zappa	9/4	55
ScatterBrain	Jeff Beck	9/8	250
Take It To The Limit	The Eagles	3/4	90
Doing It All For My Baby	Huey Lewis & The News	12/8	275
Forces...Darling	Koop	8/8	200
Sliabh	Danu	6/8	190
Money	Pink Floyd	7/8	120
Whirl	The Jesus Lizard	5/4	150
Take Five	Dave Brubeck	5/4	174
The Sky Is Red	Leprous	11/4	140
Hey Ya!	OutKast	11/4	158
Ageispolis	Aphex Twin	4/4	101

is simplified to quarter notes. Thus, the values considered the same time signature are: 4/4, 2/4, 2/2, 8/8, 8/4, etc. For both tempo and time signature, pieces detected correctly for Accuracy2 are included in the results for Accuracy1.

In fact, the authors of this paper will focus mainly on the quantitative results of Accuracy2 in both cases (tempo and time signature detection) since the difference between the two can be difficult to discern, even for a musician or a skillful listener.

3.3 Tempo Detection

In testing the effectiveness of tempo detection, the effect of the length of the fragment of the analyzed song and the number of comb filter pulses on the results was examined. The results were also compared with those obtained using the popular Librosa library. Librosa is an open-source[2], free[3] library for audio and music analysis. It provides functions that can be used to develop a MIR system [12]. The library includes functions from digital signal processing, graphical representation of the signal, pulse, tempo analysis of the song, and extraction of various features.

An investigation of the effectiveness of tempo detection of a song began by looking at how the number of pulses of the comb filter affects the results obtained. A 25-s excerpt from each song was used to investigate this relationship, and a number of pulses ranging from 1 to 25 were used to create the signal. The results are presented in Fig. 3.

---

[2] https://github.com/librosa/librosa.

[3] https://librosa.org/doc/latest/index.html.

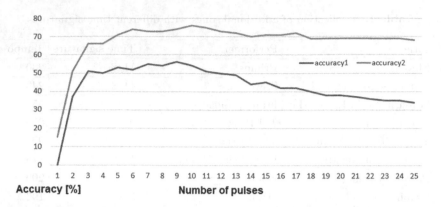

Fig. 3. The influence of the number of comb filter pulses on accuracy.

The graph (Fig. 3) shows that the maximum Accuracy2 performance was obtained for ten pulses. It is surprising to obtain an Accuracy2 of 50% already with only two pulses. With three pulses, the correct Accuracy2 result is obtained with two tracks out of three. The effectiveness increases with the number of pulses, up to a maximum Accuracy1 of 56% for nine pulses and Accuracy2 of 76% for ten pulses. Above ten pulses, the effectiveness decreases. This may be due to the fact that for comb filters with a higher number of pulses, after a while, the beats of the song begin to mismatch with the pulses of the comb filter. Accuracy2 is less susceptible to this relationship and shows a reasonably constant effectiveness of 69%. In contrast, Accuracy1, which for three and more pulses achieves a value of 50% already from 13 pulses, shows worse effectiveness than with three pulses, to finally achieve a result of only 30% with 25 pulses of the comb filter.

The next part of the study examined how the length of the analyzed song affected performance. It was observed that its length significantly influences the calculation time. In order to test this relationship, 100 test pieces were used with a fixed number of ten pulses.

The minimum tempo dictates the minimum duration of the excerpt the algorithm can detect. This value is 60 BPM, so each filter pulse is one second apart. This means that a comb filter signal with ten pulses at 60 BPM is needed to create a signal with a duration of at least 10 s. The results obtained are presented in Fig. 4.

It can be seen from the graph (Fig. 4) that, with a fixed number of ten comb filter pulses, the best effectiveness is provided by the use of a 26-s excerpt from a song. The accuracy increases steadily with a longer song fragment, but the increase is stopped at the 26-s value.

The research provided information on the optimal parameters of the proposed algorithm. In order to confront our method with other works, the results obtained were compared with the accuracy provided by the Librosa library, recognized in the scientific community. A 25-s excerpt and a comb filter with ten pulses were used to detect the tempo of each song.

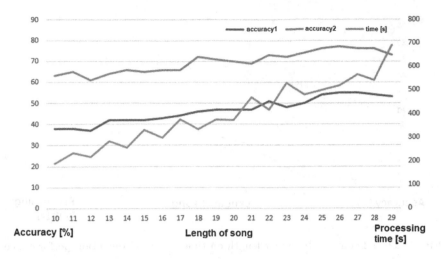

Fig. 4. The influence of the song's length on tempo detection effectiveness.

Table 2. Results of tempo detection with comb filter compared to Librosa library.

Comb filter		Librosa	
Accuracy1 [%]	Accuracy2 [%]	Accuracy1 [%]	Accuracy2 [%]
42,8	74,7	42,6	60

Comparing the results obtained with those obtained by using the Librosa library (Table 2), it can be seen that both methods provide the same performance for the Accuracy1 criterion. However, a significantly better result was obtained for the Accuracy2 criterion (by almost 15% points). Given its popularity in the scientific community, it is surprising to see such a low detection accuracy rate from the Librosa library.

3.4   Time Signature Detection

When investigating the effectiveness of time signature detection, it was examined how the length of the excerpt from the piece being analyzed, the frequency range of the spectrogram, and the metrics of determining distance affect the results.

The research began by examining how the cut fragment's length affects a musical piece's meter detection performance. As with tempo detection, the processing time is significantly influenced. The songs from the GTZAN collection are only 30 s long, so they cannot be used to test how a longer excerpt impacts the effectiveness of the song's time signature analysis. The results were conducted for all 100 songs and fragment lengths ranging from 5 to 30 s (Fig. 5).

It can be seen from the graph (Fig. 5) that the best results were achieved for a 25-s fragment of the song. Therefore, it was decided that this length would be chosen in the final algorithm proposal.

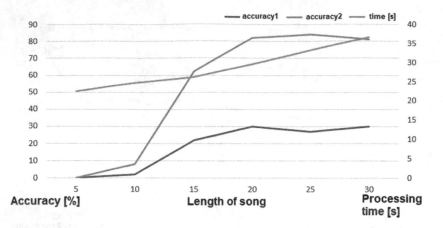

Fig. 5. The influence of fragment length on time signature detection performance.

The maximum performance for the 25-s fragment is encouraging in terms of execution time. The time required to perform the calculations shows linear complexity. Analysis of longer fragments would have taken even longer, and the efficiency of such a solution would have been negligible.

It was also examined how changing the frequency range of the spectrogram affected the results obtained and the time taken to analyze the entire collection. For this purpose, all 100 tracks were used, and, working on their 25-s excerpts, the efficiency and execution time of the computations were measured for a maximum frequency in the range from 3 to 12 kHz. The results are shown in Fig. 6.

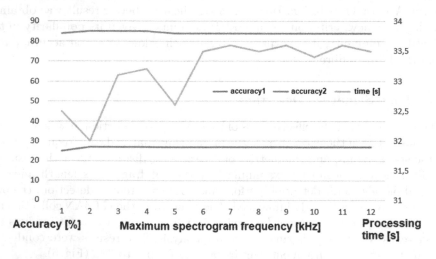

Fig. 6. The impact of reducing the frequency range of the spectrogram on the accuracy.

Studies have shown that reducing the frequency range of the spectrogram has a negligible effect on the efficiency of time signature detection and the time required for calculation. A deterioration in efficiency of one percentage point (in terms of Accuracy2) from a frequency of 5 kHz can be observed, so in the final algorithm, the spectrogram frequency range is reduced to 4 kHz. Reducing the frequency range of the spectrogram has no significant impact on the computation time required.

Three distance measures were used in [10] when creating the BSM matrix: the Euclidean distance (EDM), the cosine distance (CD), and the Kullback-Leibler distance (K-L). It was tested what efficiency is provided by implementing each of these measures on a set of all tracks. The results are presented in Fig. 7.

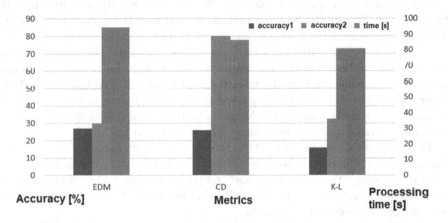

Fig. 7. The influence of the distance measure on the accuracy of time signature detection.

The results show that the Euclidean distance is the most accurate method for determining distances. Not only does the EDM show the best accuracy, but it is also supported by the shortest calculation time. As in the case of the analysis of the effect of fragment length on the accuracy, the results show that the most efficient method is also the fastest, so the proposed algorithm is correct in its performance and optimal in terms of calculation time.

The experiments delivered the most optimal and effective tempo and time signature detection algorithm. However, tests of the effectiveness of the time signature detection were carried out using the tempo information of the song taken from the test set. To test the effectiveness of the overall system and its two-stage operation, a 100-element set of songs and previously tested parameters were used to provide the maximum tempo and time signature detection performance. The results are presented in Table 3.

The presented system provides an efficiency of approximately 70% for both tempo and time signature detection in terms of Accuracy2. During a review of the available literature, no work was found that proposed a system that not only detects tempo but also time signature.

Table 3. Results obtained in detecting the tempo and time signature of a song.

Tempo		Time signature	
Accuracy1 [%]	Accuracy2 [%]	Accuracy1 [%]	Accuracy2 [%]
46	70	21	74

4   Conclusions and Future Work

This work presents an algorithm that effectively detects a musical piece's tempo and time signature. A test set was completed to develop and test the effectiveness. The next step was to implement the adopted algorithm and study the dependencies. The study made it possible to optimize the parameters in terms of the efficiency achieved and the time needed to process the piece.

Overall, the gathered results for both tempo and time signature detection are satisfactory. Surprisingly, the proposed and optimized method for tempo detection achieved better results than the well-known and widely used Librosa library. Moreover, it has to be mentioned that time signature detection is one of the most challenging tasks in the Music Information Retrieval domain. That is the reason why only a very few papers were written on that topic.

Further work needs to start with completing a comprehensive dataset, which has not been created to date. The GTZAN collection does not have information on the song's time signature, and the song excerpts included are only 30 s long. However, the research conducted in this work has shown that even shorter excerpts provide better performance.

It would be ideal to empirically classify the time signature of all the pieces included in GTZAN or any other dataset. Such a procedure would take much time, but this work seems necessary to create the best possible system. It would be interesting to see if different genres of music require different parameters to detect tempo and time signature more accurately. Creating an ensemble of classifiers, where each genre had its parameters, could achieve a better result. Earlier detection of a song's genre based on, e.g., MFCC features and a neural network would provide the needed information, and such knowledge can be further used in a more comprehensive MIR system.

The proposed tempo detection method using a comb filter is acceptable but needs further research. An efficiency of 70% is a result that is too small and takes too long to process. It is possible to consider what results will be obtained from using other types of filters (e.g., triangular ones) to create an ensemble of filters.

The results obtained are satisfactory and motivate further work on this problem. The time signature detection method using the BSM matrix still has a very high potential. The methods and results presented in this work can be incorporated into a broader MIR system, which will be able to detect the genre of a song based on knowledge of tempo and time signature, as well as automatically tag and recommend songs for users of music services.

References

1. Abimbola, J., Kostrzewa, D., Kasprowski, P.: Time signature detection: a survey. Sensors **21**(19), 6494 (2021)
2. Alonso, M., David, B., Richard, G.: Tempo and beat estimation of musical signals. In: ENST-GET, Département TSI (2004)
3. Böck, S., Davies, M., Knees, P.: Multi-task learning of tempo and beat: learning one to improve the other. In: ISMIR (2019)
4. Cheng, K., Nazer, B., Uppuluri, J., Verret, R.: Beat this - a beat synchronization project. https://www.clear.rice.edu/elec301/Projects01/beat_sync/beatalgo.html
5. Foote, J.: Visualizing music and audio using self-similarity, pp. 77–80. MULTI-MEDIA 1999, Association for Computing Machinery, New York, NY, USA (1999). https://doi.org/10.1145/319463.319472
6. Foote, J., Cooper, M.: Visualizing musical structure and rhythm via self-similarity (2001)
7. Gainza, M., Coyle, E.: Time signature detection by using a multi resolution audio similarity matrix. J. Audio Eng. Soc. (2007)
8. Gainza, M., Coyle, E.: Tempo detection using a hybrid multiband approach. IEEE Trans. Audio Speech Lang. Process. **19**(1), 57–68 (2011)
9. Gainza, M.: Automatic musical meter detection. In: 2009 IEEE International Conference on Acoustics, Speech and Signal Processing, pp. 329–332 (2009)
10. Lu, L., Liu, D., Zhang, H.J.: Automatic mood detection and tracking of music audio signals. IEEE Trans. Audio Speech Lang. Process. **14**(1), 5–18 (2006)
11. MatthewDavies, E.P., Böck, S.: Temporal convolutional networks for musical audio beat tracking. In: 2019 27th European Signal Processing Conference (EUSIPCO), pp. 1–5 (2019). https://doi.org/10.23919/EUSIPCO.2019.8902578
12. McFee, B., et al.: librosa/librosa: 0.8.1rc2 (2021). https://doi.org/10.5281/zenodo.4792298
13. Ono, N., Miyamoto, K., Le Roux, J., Kameoka, H., Sagayama, S.: Separation of a monaural audio signal into harmonic/percussive components by complementary diffusion on spectrogram. In: 2008 16th European Signal Processing Conference, pp. 1–4 (2008)
14. Peeters, G.: Time variable tempo detection and beat marking. In: IRCAM - Analysis/Synthesis Team (2005)
15. Pikrakis, A., Antonopoulos, I., Theodoridis, S.: Music meter and tempo tracking from raw polyphonic audio (2004)
16. Scheirer, E.D.: Tempo and beat analysis of acoustic musical signals. J. Acoust. Soc. Am. **103**(1), 588–601 (1998)
17. Schlüter, J., Böck, S.: Improved musical onset detection with convolutional neural networks. In: 2014 IEEE International Conference on Acoustics, Speech and Signal Processing (ICASSP), pp. 6979–6983 (2014). https://doi.org/10.1109/ICASSP.2014.6854953
18. Schuller, B., Eyben, F., Rigoll, G.: Tango or waltz?: putting ballroom dance style into tempo detection. EURASIP J. Audio Speech Music Process. **8**(1), 1–12 (2008)
19. Tzanetakis, G.: Marsyas (music analysis, retrieval and synthesis for audio signals). http://marsyas.info/index.html
20. Tzanetakis, G., Cook, P.: Musical genre classification of audio signals. IEEE Trans. Speech Audio Process. **10**(5), 293–302 (2002). https://doi.org/10.1109/TSA.2002.800560

Correction to: Alternative Platforms and Privacy Paradox: A System Dynamics Analysis

Ektor Arzoglou and Yki Kortesniemi

Correction to:
Chapter "Alternative Platforms and Privacy Paradox:
A System Dynamics Analysis" in J. Mikyška et al. (Eds.):
Computational Science – ICCS 2023, **LNCS 14073,**
https://doi.org/10.1007/978-3-031-35995-8_12

The original version of the book was inadvertently published with the below-mentioned typesetting errors in the Appendix in Chapter 12:

1. Equation 2 -> "Pw(0) = 1000" belongs to Equation 1, not Equation 2. Therefore, the line should be moved below the row "Pw(0) = 1000", rather than above it.

2. There is no line between the row "Privacy concerns erosion time" and "Reference pragmatism".

3. Equations 11a, 11b, and 11c -> The comment "Step input function" should start from right below the "=" signs of the equations, rather than after the equations.

This has been corrected in chapter 12 accordingly.

The updated original version of this chapter can be found at
https://doi.org/10.1007/978-3-031-35995-8_12

J. Mikyška et al. (Eds.): ICCS 2023, LNCS 14073, p. C1, 2023.
https://doi.org/10.1007/978-3-031-35995-8_49

Author Index

© The Editor(s) (if applicable) and The Author(s), under exclusive license
to Springer Nature Switzerland AG 2023
J. Mikyška et al. (Eds.): ICCS 2023, LNCS 14073, pp. 697–699, 2023.
https://doi.org/10.1007/978-3-031-35995-8

Printed in the United States
by Baker & Taylor Publisher Services